MECHANICS OF MATERIALS

MECHANICS OF MATERIALS

Madhukar Vable
Michigan Technological University

New York   Oxford
OXFORD UNIVERSITY PRESS
2002

Oxford University Press

Oxford   New York
Auckland Bangkok Buenos Aires Cape Town Chennai
Dar es Salaam Delhi Hong Kong Istanbul Karachi Kolkata
Kuala Lumpur Madrid Melbourne Mexico City Mumbai Nairobi
São Paulo Shanghai Singapore Taipei Tokyo Toronto

and an associated company in Berlin

Published by Oxford University Press, Inc.
198 Madison Avenue, New York, New York, 10016
http://www.oup-usa.org

Oxford is a registered trademark of Oxford University Press

Library of Congress Cataloging-in-Publication Data

Vable, Madhukar.
    Mechanics of materials / Madhukar Vable.
        p. cm.
    ISBN 0-19-513337-4
    1. Materials—Mechanical properties. 2. Mechanics, Applied. I. Title: Mechanics of
materials. II. Title.

    TA405 .V28 2002
    620.1'1292—dc21

                                                        2001032144

Printing number: 9 8 7 6 5 4 3 2 1

Printed in the United States of America
on acid-free paper

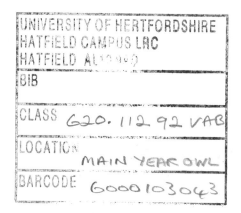

DEDICATED TO MY FATHER
Krishna Rao Vable

AND MY MOTHER
Saudamini Gautam Vable

CONTENTS

*This section contains advanced material and can be omitted without loss of the basic content of the course.

CHAPTER THREE
Mechanical Properties of Materials

CHAPTER FOUR
Axial Members

CHAPTER FIVE
Torsion of Shafts

CHAPTER SIX
Symmetric Bending of Beams

CHAPTER SEVEN
Deflection of Symmetric Beams

CHAPTER EIGHT
Stress Transformation

CHAPTER NINE
Strain Transformation

CHAPTER TEN
Design and Failure

CHAPTER ELEVEN
Stability of Columns

Appendixes

APPENDIX A STATICS REVIEW

APPENDIX B   ALGORITHMS FOR NUMERICAL METHODS

APPENDIX C   CHARTS OF STRESS CONCENTRATION FACTORS

APPENDIX D   PROPERTIES OF SELECTED MATERIALS

APPENDIX E   GEOMETRIC PROPERTIES OF STRUCTURAL STEEL MEMBERS

PREFACE

Mechanics is the body of knowledge that deals with the relationships between forces and the motion of points through space, including the material space. Material science is the body of knowledge that deals with the properties of materials, including their mechanical properties. Mechanics is very deductive—having defined some variables and given some basic premises, one can logically deduce relationships between the variables. Material science is very empirical—having defined some variables one establishes the relationships between the variables experimentally. Mechanics of materials synthesizes the empirical relationships of materials into the logical framework of mechanics, to produce formulas for use in the design of structures and other solid bodies.

In the past twenty-five years there has been a tremendous growth in mechanics, material science, and in new applications of mechanics of materials. Twenty-five years ago, techniques such as the finite-element method and Moiré interferometry were research topics in mechanics, but today these techniques are used routinely in engineering design and analysis. Twenty-five years ago, wood and metal were the preferred materials in engineering design, but today machine components and structures may be made of plastics, ceramics, polymer composites, and metal-matrix composites. Twenty-five years ago, mechanics of materials was primarily used for structural analysis in aerospace, civil, and mechanical engineering, but today mechanics of materials is used in electronic packaging, medical implants, the explanation of geological movements, and the manufacturing of wood products to meet specific strength requirements. Though the principles in mechanics of materials have not changed in the past twenty-five years, the presentation of these principles must evolve to provide the students with a foundation that will permit them to readily incorporate the growing body of knowledge as an extension of the fundamental principles and not as something added on, and vaguely connected to what they already know. This has been my primary motivation for writing this book.

Often one hears arguments that seem to suggest that intuitive development comes at the cost of mathematical logic and rigor, or the generalization of a mathematical approach comes at the expense of intuitive understanding. Yet the icons in the field of mechanics of materials, such as Cauchy, Euler, and Saint-Venant, were individuals who successfully gave physical meaning to the mathematics they used. Accounting of shear stress in the bending of beams is a beautiful demonstration of how the combination of intuition and experimental observations can point the way when self-consistent logic does not. Intuitive understanding is a must—not only for creative engineering design but also for choosing the marching path of a mathematical development. By the same token, it is not the heuristic-based arguments of the older books, but the logical development of arguments and ideas that provides students with the skills and principles necessary to organize the deluge of information in modern engineering. Building a complementary connection between intuition, experimental observations, and mathematical generalization is central to the design of this book.

Learning the course content is not an end in itself, but a part of an educational process. Some of the serendipitous development of theories in mechanics of materials, the mistakes made and the controversies that arose from these mistakes, are all part of the human drama that has many educational values, including learning from others' mistakes, the struggle in understanding difficult concepts, and the fruits of perseverance. The connection of ideas and concepts discussed in a chapter to advanced modern techniques also has educational value, including continuity and integration of subject material, a starting reference point in a literature search, an alternative perspective, and an application of the subject material. Incorporating these educational values without distracting the student from the central ideas and concepts of mechanics of materials is an important complementary objective of this book.

The achievement of these educational objectives is intricately tied to the degree to which the book satisfies the pedagogical needs of the students. The Note to Students describes some of the features that address their pedagogical needs. The Note to the Instructor outlines the design and format of the book to meet the described objectives.

I welcome any comments, suggestions, concerns, or corrections you may have that will help me improve the book. You could relay your input either to the publisher or to me. My e-mail address is mavable@mtu.edu.

ACKNOWLEDGMENTS

A book is shaped by the numerous ideas, events, and people that have influenced an author. In my case, the greatest influence has been that of the late Professor D. L. Sikarskie. He was my Ph.D. advisor, my colleague, and a friend of my family through life's many travails. He was not only a great teacher of ideas, but he brought the same enthusiastic optimism in teaching a novice like me the joys of white-water canoeing. Among the many things I learned from Professor Sikarskie was the modular character of logic that underlies the theories in mechanics of materials, which set me on the path of writing this book. I wish I had finished this book while he was alive.

I am greatly indebted to Professor C. R. Vilmann. The many pedagogical discussions while fishing in the pristine and serene environments of the Boundary Waters, the thorough review of each chapter, and the constructive criticisms that only a friend can give have all improved the quality of this textbook tremendously.

To Professor I. Miskioglu I owe many discussions on exams, proposals, and papers, which occur when two faculty members academically grow together as friends. His ready willingness to construct an experimental setup so that I could take some of the photographs included in this book is deeply appreciated.

Thanks to Professor J. B. Ligon, who not only gave me the various specimens I photographed and included in this book, but who—equally significantly—helped me understand why two-dimensional pictures of shafts in torsion are often confused by students as problems in bending.

Thanks also to Professor G. Jayaraman for his infectious enthusiasm in using my notes that are now this book and for his input on biomechanics problems.

Thanks to Professors L. B. Sandberg and W. M. Bulleit for their help with sections on Reliability and LRFD.

Thanks to my students Paul Miller, Steve Andrasko, Korey Kiepert, Emily Preston, Mary Gieliske, and Tiew Yoon for their help in constructing solutions to the problems and checking problems and examples for accuracy.

Thanks to Peter Gordon, editor of this book, for his tremendous support, particularly in the early phases of the review.

Thanks to Ms. Karen Shapiro, Managing Editor, for her patience and her ability to send manuscripts to remote areas of national parks.

To my wife Pushpa, my daughter Anusha, and my son Adhiraj, what can I say, except IT IS DONE!

A NOTE TO STUDENTS

Some of the features that should help you meet the learning objectives of this book are summarized here briefly.

- A course in statics is a prerequisite for this course. Appendix A reviews the concepts of statics from the perspective of this course. If you had statics a few terms ago, then you may need to review your statics textbook before the brevity of presentation in Appendix A serves you adequately. If you feel comfortable with your knowledge of statics, then you can assess for yourself what you need to review by using the *Statics Review Exams* given in Appendix A.

- All internal forces and moments are printed in **bold italics**. This is to emphasize that the internal forces and moments must be determined by making an imaginary cut, drawing a free-body diagram, and using equilibrium equations or by using methods that are derived from this approach.

- Every chapter starts with an *Overview,* which describes the motivation for studying the chapter and states the major learning objective(s).

- Every chapter ends with *Points and Formulas to Remember,* a one-page synopsis of nonoptional topics. This brings greater focus to the material that must be learned.

- Every *Example* problem starts with a *Plan* and ends with *Comments,* both of which are specially set off to emphasize the importance of these two features. Developing a plan before solving a problem is essential for the development of analysis skills. Comments are observations deduced from the example, highlighting concepts discussed in the text preceding the example, or observations that suggest the direction of development of concepts in the text following the example.

- *Quick Tests* with solutions are designed to help you diagnose your understanding of the text material. To get the maximum benefit from these tests, take them only after you feel comfortable with your understanding of the text material.

- After a major topic you will see a box called *Consolidate Your Knowledge.* It will suggest that you either write a synopsis or derive a formula. "Consolidate Your Knowledge" is a learning device that is based on the observation that it is easy to follow someone else's reasoning but significantly more difficult to develop one's own reasoning. By deriving a formula with the book closed or by writing a synopsis of the text, you force yourself to think of details you would not otherwise. When you know your material well, writing will be easy and will not take much time.

- Every chapter has a section called *General Information,* where connections of the chapter material to historical development and advanced topics are made. History shows that concepts are not an outcome of linear logical thinking, but rather a struggle in the dark in which mistakes were often made but the perseverance of pioneers has left us with a rich inheritance. Connection to advanced topics is an extrapolation of the concepts studied. Other reference material that may be helpful in the future can be found in problems labeled "Stretch yourself."

- Every chapter ends with *Closure,* which serves as a connecting link to the topics in subsequent chapters. Of particular importance are closure sections in Chapters 3 and 7, as these are the two links connecting together the three major parts of the book.

- On the inside back cover is a *Formula Sheet* for easy reference. Only equations of nonoptional topics are listed. There are no explanations of the variables or the equations in order to give your instructor the option of permitting the use of the formula sheet in an exam.

I hope you will enjoy reading this book as much as I did writing it.

A NOTE TO THE INSTRUCTOR

The best way I can show you how the presentation of this book meets the objectives stated in the Preface is by drawing your attention to certain specific features. Described hereafter are the underlying design and motivation of presentation in the context of the development of theories of one-dimensional structural elements and the concept of stress. The same design philosophy and motivation permeate the rest of the book.

Figure 3.15 depicts the logic relating displacements—strains—stresses—internal forces and moments—external forces and moments. The logic is intrinsically very modular—equations relating the fundamental variables are independent of each other. Hence, complexity can be added at any point without affecting the other equations. This is brought to the attention of the reader in Example 3.5, where the stated problem is to determine the force exerted on a car carrier by a stretch cord holding a canoe in place. The problem is first solved as a straightforward application of the logic shown in Figure 3.15. Then, in comments following the example, it is shown how different complexities (in this case nonlinearities) can be added to improve the accuracy of the analysis. Associated with each complexity are posttext problems (numbers written in parentheses) under the headings "Stretch yourself" or "Computer problems," which are well within the scope of students willing to stretch themselves. Thus the central focus in Example 3.5 is on learning the logic of Figure 3.15, which is fundamental to mechanics of materials. But the student can appreciate how complexities can be added to simplified analysis, even if no "Stretch yourself" problems are solved.

This philosophy, used in Example 3.5, is also used in developing the simplified theories of axial members, torsion of shafts, and bending of beams. The development of the theory for structural elements is done rigorously, with assumptions identified at each step. A footnote associated with an assumption directs the reader to examples, optional sections, and "Stretch yourself" problems, where the specific assumption is violated. Thus in Section 5.2 on the theory of the torsion of shafts, Assumption 5 of linearly elastic material has a footnote directing the reader to see "Stretch yourself" problems 5.42 and 5.80 for nonlinear material behavior; Assumption 7 of material homogeneity across a cross section has a footnote directing the reader to see the optional Section 5.4 on composite shafts; and Assumption 9 of untapered shafts is followed by statements directing the reader to Example 5.9 on tapered shafts. Table 7.2 gives a synopsis of all three theories (axial, torsion, and bending) on a single page to show the underlying pattern in all theories in mechanics of materials that the students have seen three times. The central focus in all three cases remains the simplified basic theory, but the presentation in this book should help the students develop an appreciation of how different complexities can be added to the theory, even if no "Stretch yourself" problems are solved or optional topics covered in class.

Compact organization of information seems to some engineering students like an abstract reason for learning theory. Some students have difficulty visualizing a continuum as an assembly of infinitesimal elements whose behavior can be approximated or deduced. There are two features in the book that address these difficulties. I have included sections called *Prelude to Theory* in 'Torsion of Circular Shafts' and 'Symmetric Bending of Beams.' Here numerical problems are presented in which discrete bars welded to rigid plates are considered. The rigid plates are subjected to displacements that simulate the kinematic behavior of cross sections in torsion or bending. Using the logic of Figure 3.15, the problems are solved—effectively developing the theory in a very intuitive manner. Then the section on theory consists essentially of formalizing the observations of the numerical problems in the prelude to theory. The second feature are actual photographs showing nondeformed and deformed grids due to axial, torsion, and bending loads. Seeing is believing is better than accepting on faith that a drawn deformed geometry represents an actual situation. In this manner the complementary connection between intuition, observations, and mathematical generalization is achieved in the context of one-dimensional structural elements.

Double subscripts are used with all stresses and strains. The use of double subscripts has three distinct benefits. (i) It provides students with a procedural way to compute the direction of a stress component which they calculate from a stress formula. The procedure of using subscripts is explained in Section 1.2.1 and elaborated in Example 1.11. This procedural determination of the direction of a stress component on a surface can help many students overcome any shortcomings in intuitive ability. (ii) Computer programs, such as the finite-element method or those that reduce full-field experimental data, produce stress and strain values in a specific coordinate system that must be properly interpreted, which is possible if students know how to use subscripts in determining the direction of stress on a surface. (iii) It is consistent with what the student will see in more advanced courses such as those on composites, where the material behavior can challenge many intuitive expectations.

But it must be emphasized that the use of subscripts is to complement not substitute an intuitive determination of stress direction. Procedures for determining the direction of a stress component by inspection and by subscripts are briefly described at the end of each theory section of structural elements. Examples such as 4.2 on axial members, 5.6 and 5.9 on torsional shear stress, and 6.8 on bending normal stress emphasize both approaches. Similarly there are sets of problems in which the stress direction must be determined by inspection as there are no numbers given—problems such as 5.14 through 5.17 on the direction of torsional shear stress; 6.29 through 6.31 on the tensile and compressive nature of bending normal stress; and 8.1 through 8.9 on the direction of normal and shear stresses on an inclined plane.

I developed a similar section for axial members in Chapter 4, but found that most of the ideas had been covered in Section 3.2. So in axial members the very first example is on the logic of Figure 3.15, and the first four posttext problems are also based on it. An instructor could choose to cover this material before developing the theory for axial members, as is done in torsion and bending.

Many authors use double subscripts with shear stress but not for normal stress. Hence they do not adequately elaborate the use of these subscripts when determining the direction of stress on a surface from the sign of the stress components.

If subscripts are to be used successfully in determining the direction of a stress component obtained from a formula, then the sign conventions for drawing internal forces and moments on free-body diagrams must be followed. Hence there are examples (such as 6.6) and problems (such as 6.26 and 6.27) in which the signs of internal quantities are to be determined by sign conventions. Thus, once more, the complementary connection between intuition and mathematical generalization is enhanced by using double subscripts for stresses and strains.

Other features that you may find useful are described briefly.

All optional topics are marked by an asterisk (*) to account for instructor interest and pace. Skipping these topics can at most affect the student's ability to solve some posttext problems in subsequent chapters, and these problems are easily identifiable.

General Information is an optional section in all chapters. In some examples and posttext problems, reference is made to a topic that is described under general information. The only purpose of this reference is to draw attention to the topic, but knowledge about the topic is not needed for solving the problem.

The topics of stress and strain transformation can be moved before the discussion of structural elements (Chapter 4), as shown in the two syllabi in the instructor's manual. I strived to eliminate confusion regarding maximum normal and shear stress at a point with the maximum values of stress components calculated from the formulas developed for structural elements.

The posttext problems are categorized for ease of selection for discussion and assignments. Generally speaking, the starting problems in each problem set are single-concept problems. This is particularly true in the later chapters, where problems are designed to be solved by inspection to encourage the development of intuitive ability. Design problems involve the sizing of members, selection of materials (later chapters) to minimize weight, determination of maximum allowable load to fulfill one or more limitations on stress or deformation, and construction and use of failure envelopes in optimum design (Chapter 10)—and are in color. "Stretch yourself" problems are optional problems for motivating and challenging students who have spent time and effort understanding the theory. These problems often involve an extension of the theory to include added complexities. "Computer" problems are also optional problems and require a knowledge of spreadsheets, or of simple numerical methods such as numerical integration, roots of a nonlinear equation in some design variable, or use of the least-squares method. Additional categories such as "Stress concentration factor," "Stress intensity factor," "Fatigue," and "Transmission of power" problems are chapter-specific *optional* problems associated with optional text sections.

The book can be viewed in three major parts. The first part consists of the first three chapters, where in addition to introducing the basic concepts of stress, strain, and Hooke's law, an important underlying theme is to build the links of the logic of relating displacements to external forces (as shown in Figure 3.15). The second part consists of Chapters 4 through 7, where the theories and formulas of one-dimensional structural elements are developed and used in the analysis and design of these structural elements. The third part consists of Chapters 8 through 10. Here stress and strain transformation are preludes to Chapter 10 on the design and failure of structures made from one-dimensional elements. The closure sections in Chapters 4 and 7 are important connecting links between the three parts.

Many textbooks on mechanics of materials have such topics as asymmetric bending, energy methods, and principal stresses and strains in three dimensions. To do full justice to these important topics I am writing an intermediate mechanics of materials textbook. But it is possible that there are some instructors who, for a variety of pedagogical considerations, would like to introduce one or more of these topics in their introductory courses. Problems on these topics can be found in the "Stretch yourself" segments of the problem sets. These problems can be used by individual instructors to address some of their needs. It is my hope that you will find this an acceptable compromise between your needs, the need to maintain a focus on the central concepts, and the need to keep the textbook to a reasonable length.

I have tried to achieve a mix of practical problems, as shown by photographs, and problems that are designed for understanding the fundamental principles. If you can suggest other types of problems that will be useful in teaching this course material, please let me know. I will gratefully acknowledge your contribution.

CHAPTER ONE
STRESS

1.0 OVERVIEW

The ship S.S. Schenectady, shown in Figure 1.1, broke in two because the forces acting on the ship were greater than the ship's strength.[1] But what exactly is the strength of a structure? To answer this question, we will introduce a variable called *stress,* which is a measure of strength. By defining a variable that is a measure of strength, we take the first step toward developing formulas that can be used in strength analysis and the design of structural members.

   To develop formulas for stress in structural members, we will use a logic that will be fully developed in Section 3.2. Two of the links of that logic are shown in Figure 1.2. What motivates the construction of these two links is something you learned in statics—analysis is simpler if, before writing equilibrium equations, any distributed forces that may be on the free-body diagram are replaced by equivalent forces and moments.[2] Formulas developed in mechanics of materials relate stresses to internal forces and moments. Free-body diagrams can then be used to relate internal forces and moments to external forces and moments.

Figure 1.1 Failure of S.S. Schenectady.

Figure 1.2 Two-step process of relating stresses to external forces and moments.

---

[1]There are many mechanisms of strength failure. In the case of the S.S. Schenectady the failure occurred because of crack growth, which started at a hatchway on deck.
[2]See Appendix A.6 for refreshing the concept of static equivalency.

1

The two learning objectives of this chapter are:

1. Understanding the concept of stress.
2. Understanding the two-step analysis of relating stresses to external forces and moments.

1.1 STRESS ON A SURFACE

In this section it will be shown that the stress on a surface is an internally distributed force system that can be resolved into two components: normal (perpendicular) to the imaginary cut surface, called *normal* stress, and tangent (parallel) to the imaginary cut surface, called *shear* stress.

1.1.1 Normal Stress

In Figure 1.3 the cable of the chandelier and the columns supporting the building must be strong enough to support the weight of the chandelier and the weight of the building, respectively. If we make an imaginary cut as shown and draw the free-body diagrams, we see that forces normal to the imaginary cut are needed to balance the weight in each free-body diagram. If we divide this *internal* normal force N by the area of the cross section exposed by the imaginary cut A, then we obtain the intensity of an internal normal force distribution. The magnitude of this distributed normal

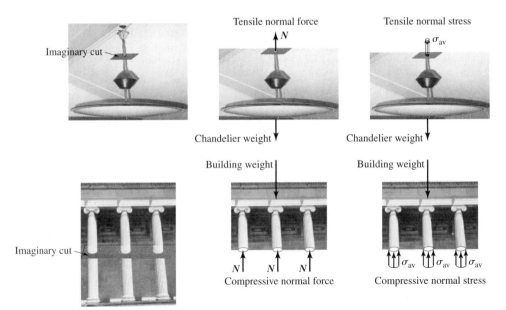

Figure 1.3 Examples of normal stress distribution.

force is called *average* normal stress and is given by the equation

$$\boxed{\sigma_{av} = N/A}$$ (1.1)

where σ is the Greek letter sigma used to designate normal stress and the subscript av emphasizes that the normal stress is an average value. We may view σ_{av} as a uniformly distributed normal force, as shown in Figure 1.3, which can be replaced by a statically equivalent internal normal force. We will develop this viewpoint further in Section 1.1.3. Notice that N is in boldface italics, as are all internal forces (and moments) in this book.

Equation (1.1) is consistent with our intuitive understanding of strength. Consider the following observations. (i) We know that if we keep increasing the force on a body, then the body will eventually break. Thus we expect the quantifier for strength (stress) to increase in value with the increase of force until it reaches a critical value. In other words, we expect stress to be directly proportional to force, as in Equation (1.1). (ii) If we compare two bodies that are identical in all respects except that one is thicker than the other, then we know that the thicker body is stronger. This implies that as the body gets thicker (larger cross-sectional area), we move away from the critical breaking value, and the value of the quantifier of strength should decrease. In other words, stress should vary inversely with the cross-sectional area, as in Equation (1.1).

Equation (1.1) shows that the unit of stress is force per unit area. Table 1.1 lists the various units of stress used in this book. It should be noted that 1 psi is equal to 6.895 kPa, or approximately 7 kPa. Alternatively, 1 kPa is equal to 0.145 psi, or approximately 0.15 psi.

We note that the normal force in the cable of the chandelier in Figure 1.3 is tensile, that is, the force pulls the imaginary cut surface away from the rest of the cable. The normal forces in the columns are compressive, that is, a force pushes the imaginary cut surface into the rest of the column. This leads to the following definition:

Definition 1 Normal stress that pulls the surface away from the body is called *tensile* stress. Normal stress that pushes the surface into the body is called *compressive* stress.

In other words, tensile stress acts in the direction of the outward normal to the surface whereas compressive stress is opposite to the direction of the outward normal to the surface. Normal stress is usually reported as tensile or compressive and not as positive or negative. Thus $\sigma = 100$ MPa (T) or $\sigma = 10$ ksi (C) are the correct ways of reporting tensile or compressive normal stresses.

TABLE 1.1   Units of stress

Abbreviation	Units	Basic Units
psi	Pounds per square inch	lb/in^2
ksi	Kilopounds per square inch	$10^3 \ lb/in^2$
Pa	Pascal	N/m^2
kPa	Kilopascal	$10^3 \ N/m^2$
MPa	Megapascal	$10^6 \ N/m^2$
GPa	Gigapascal	$10^9 \ N/m^2$

Figure 1.4 Girl in a swing, Example 1.2.

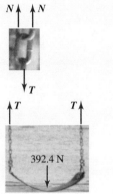

Figure 1.5 Free-body diagram of swing.

EXAMPLE 1.1

A girl whose mass is 40 kg is using a swing set. The diameter of the wire used for constructing the links of the chain is 5 mm. Determine the maximum average normal stress in the links, assuming that the inertial forces can be neglected.

PLAN

At any point, except at the bottom of the swing, only a component of the girl's weight is supported by the chains. Thus the maximum tension in the chains will be at the bottom of the swing. We can make an imaginary cut through the chains, draw a free-body diagram, and find the tension T in each chain. We note that the imaginary cut in each chain cuts the link at two surfaces. On each surface there is an internal normal force N, which is equal to $T/2$. Once N is known, we can divide it by the cross-sectional area of the wire to get the average normal stress.

Solution   The wire forming the links is of circular cross section and the cross-sectional area can be found as follows:

$$A = \frac{\pi d^2}{4} = \frac{\pi 5^2}{4} = 19.6 \text{ mm}^2 = 19.6 \times 10^{-6} \text{ m}^2 \qquad \text{(E1)}$$

The weight of the girl is $W = 40g = 392.4$ N. We can make an imaginary cut through the chains and calculate the tension, which we can relate to the normal force at each surface where the link material intersects the imaginary cut (Figure 1.5) using the following equations:

$$T = 2N \qquad \text{(E2)}$$

$$2T = 392.4 \text{ N} \qquad \text{(E3)}$$

or

$$4N = 392.4 \text{ N} \qquad \text{or} \qquad N = 98.1 \text{ N}$$

Then

$$\sigma_{av} = \frac{N}{A} = \frac{98.1}{19.6 \times 10^{-6}} = 4.996 \times 10^6 \text{ N/m}^2$$

ANS. $\sigma_{av} = 5.0$ MPa (T)

COMMENT

An alternative view is to think that the total material area of the link in each chain is $2A = 39.2 \times 10^{-6} \text{ m}^2$. The internal normal force in each chain is $T = 196.2$ N thus the average normal stress is $\sigma_{av} = T/2A = 196.2/39.2 \times 10^{-6} = 5 \times 10^6 \text{ N/m}^2$, as before.

TABLE 1.2   Fracture stress magnitudes

Material	ksi	MPa	Relative to Wood
Metals	$90 \pm 90\%$	$630 \pm 90\%$	7.0
Granite	$30 \pm 60\%$	$210 \pm 60\%$	2.5
Wood	$12 \pm 25\%$	$84 \pm 25\%$	1.0
Glass	$9 \pm 90\%$	$63 \pm 90\%$	0.89
Nylon	$8 \pm 10\%$	$56 \pm 10\%$	0.67
Rubber	$2.7 \pm 20\%$	$19 \pm 20\%$	0.18
Bones	$2 \pm 25\%$	$14 \pm 25\%$	0.16
Concrete	$6 \pm 90\%$	$42 \pm 90\%$	0.03
Adhesives	$0.3 \pm 60\%$	$2.1 \pm 60\%$	0.02

Tensile and compressive are among several adjectives used to qualify the word stress. Two other adjectives introduced in this chapter are defined next.

Definition 2 The normal stress acting in the direction of the axis of a slender member (rod, cable, bar, column) is called *axial* stress.

Definition 3 The compressive normal stress that is produced when one surface presses against another is called *bearing* stress.

The normal stresses in the cable and the columns in Figure 1.3 are axial stresses. The compressive stress in a column would not be labeled bearing stress, but the compressive stress on the base of the column or on the floor is called bearing stress. The difference is that at the base of the column there are two physical surfaces in contact.

An important consideration in all analyses is to know whether the calculated values of the variables are reasonable. A simple mistake, such as forgetting to convert feet to inches or millimeters to meters, can result in values of stress that are incorrect by orders of magnitude. Less dramatic errors can also be caught if one has a sense of the limiting stress values for a material. Table 1.2 shows fracture stress values for a few common materials. Fracture stress is an experimentally measured value at which a material breaks. The numbers are approximate and $\pm$ indicates variations of the stress values in each class of material. The order of magnitude and the relative strength with respect to wood are shown to help you in acquiring a feel for the numbers.

1.1.2   Shear Stress

In Figure 1.6*a* the double-sided tape used for sticking a hook on the wall must have sufficient bonding strength to support the weight of the clothes hung from the hook. The tape could debond either at the wall surface or at the hook surface. The free-body diagram shown is created by making an imaginary cut at the wall surface. In Figure 1.6*b* the paper in the ring binder will tear out if the pull of the hand overcomes the strength of the paper. The free-body diagram shown is created by making an imaginary cut along the path of the rings as the paper would be torn out. In both free-body diagrams the internal force necessary for equilibrium is parallel (tangent) to the imaginary cut surface. If we divide this internal shear force V by the cross sectional area A exposed by the imaginary cut, then we obtain the intensity of an internal shear force distribution. The magnitude

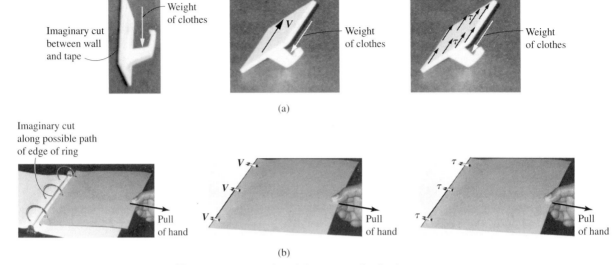

(a)

(b)

Figure 1.6 Examples of shear stress distribution.

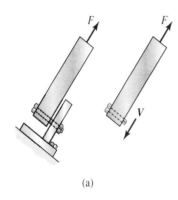

(a)

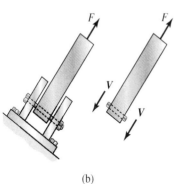

(b)

Figure 1.7 Pins in (*a*) single and (*b*) double shear.

of this distributed shear force is called *average* shear stress and is given by the equation

$$\tau_{av} = V/A$$ (1.2)

where τ is the Greek letter tau used to designate shear stress and the subscript av emphasizes that the shear stress is an average value. We may view τ_{av} as a uniformly distributed shear force, which can be replaced by a statically equivalent internal normal force V. We will develop this viewpoint further in Section 1.1.3.

Pins are one of the most common example of a structural member in which shear stress is assumed uniform. Pins are used for transferring forces between two or more solid members. In pins the shear stress is the dominant stress component on the imaginary surface perpendicular to the pin axis. Bolts, screws, nails, and rivets are often approximated as pins if the primary function of these mechanical fasteners is the transfer of forces from one member to another. However, if the primary function of these mechanical fasteners is to press two solid bodies into each other (seals), then the normal stress must be accounted for and these fasteners cannot be approximated as pins. Bolts used in bridge trusses can be approximated as pins, but bolts used for holding the lids on gas tanks and boilers cannot. Sometimes pins are also used to protect critical components in the same way a fuse protects an electric circuit. In this protective function, the pins are designed to break when the force being transferred exceeds a level that would damage a critical component. The attachment of blades to the transmission shaft in a lawn mower is an example of this protective function. The pin is supposed to break if the blades hit a large rock that may bend the transmission shaft. In pins the normal stress is neglected and the shear stress is assumed constant on a surface perpendicular to the axis of the pin.

Figure 1.7 shows two types of connections at a support. In Figure 1.7*a* a single cut between the support and the member will break the connection. In this mode the pins are

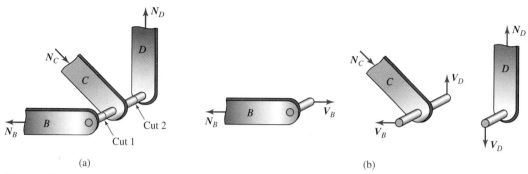

Figure 1.8 Multiple forces on a pin.

said to be in *single* shear. In Figure 1.7b two cuts are needed to break the connection. In this mode the pins are said to be in *double* shear. For the same reaction force, the pin in double shear has a smaller shear stress.

When more than two members (forces) are acting on a pin, it is important to define the imaginary surface on which the shear stress is to be calculated. Figure 1.8a shows a magnified view of a pin connection between three members. The shear stress on the imaginary cut surface 1 will be different from that on the imaginary cut surface 2, as shown by the free-body diagrams in Figure 1.8b.

EXAMPLE 1.2

Two possible configurations for the assembly of a joint in a bridge are to be evaluated. The two configurations with the forces in the bridge members are shown in Figure 1.9. The radius of the pin is $\frac{1}{2}$ in. Determine which joint assembly is preferred by calculating the maximum shear stress in the pin for each case.

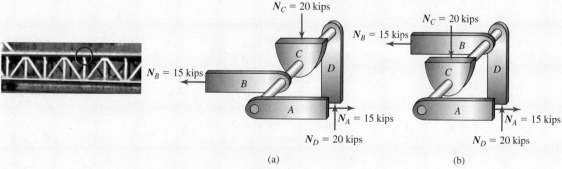

Figure 1.9 Forces on a joint and different joining configurations.

PLAN

We can make imaginary cuts between individual members for the two configurations and draw the corresponding free-body diagrams. Using force balance we can determine the shear force at each cut, from which we can determine the shear stresses. We can then compare the shear stresses and determine the maximum shear stress in each configuration.

Solution The area of the pin is $A = \pi(0.5)^2 = 0.7854 \text{ in}^2$. Making imaginary cuts between members we can draw the free-body diagrams and calculate the internal shear force at the imaginary cut, as shown in Figure 1.10.

Imaginary cut between members A and B

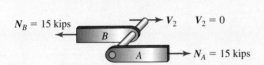

$V_1 = 15$ kips

$N_A = 15$ kips

Imaginary cut between members A and C

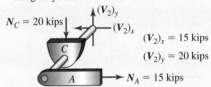

$V_1 = 15$ kips

$N_A = 15$ kips

Imaginary cut between members B and C

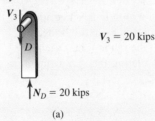

$N_B = 15$ kips

$V_2 = 0$

$N_A = 15$ kips

Imaginary cut between members C and B

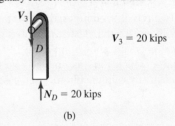

$N_C = 20$ kips

$(V_2)_y$

$(V_2)_x$

$(V_2)_x = 15$ kips

$(V_2)_y = 20$ kips

$N_A = 15$ kips

Imaginary cut between members C and D

V_3

$V_3 = 20$ kips

$N_D = 20$ kips

Imaginary cut between members B and D

V_3

$V_3 = 20$ kips

$N_D = 20$ kips

(a)

(b)

Figure 1.10 Free-body diagrams.

From the free-body diagrams in Figure 1.10*a* we see that the maximum shear force exists between members C and D. Thus the maximum shear stress in the pin in this configuration is $\tau_{max} = V_3/A = 25.46$ ksi. From the free-body diagrams in Figure 1.10*b* we note that the shear force at the section between C and B has two components, which results in a resultant shear force value of $V_2 = \sqrt{15^2 + 20^2} = 25$. Thus the maximum shear stress will exist on a section between C and B and is $\tau_{max} = V_2/A = 31.8$ ksi.

The configuration of Figure 1.10*a* is preferred, as it will result in smaller shear stresses.

COMMENTS

1. The problem emphasizes the importance of visualizing the imaginary cut surface in the calculation of stresses.

2. A simple change in assembly sequence can cause a joint to fail. This observation is true any time more than two members are joined together.

EXAMPLE 1.3

All members of the truss shown in Figure 1.11 have a cross-sectional area of 500 mm² and all pins have a diameter of 20 mm. Determine:

(a) The axial stresses in members *BC* and *DE*,

(b) The shear stress in the pin at *A*, assuming the pin is in double shear.

PLAN

(a) We can find the internal force in member *DC* by drawing the free-body diagram of joint *D*. The internal force in member *BC* can be found by drawing a free-body diagram after making an imaginary cut through *CB*, *CF*, and *EF*.

(b) Noting that *AB* is a two-force member, we can draw the free-body diagram of the entire truss and find the support reaction at *A*, from which we can calculate the shear stress in the pin at *A*.

Solution   The cross-sectional areas of all pins is $A_p = \pi (0.02)^2/4 = 314.2 \times 10^{-6}\,\text{m}^2$ and the cross-sectional areas of all members are $A_m = 500 \times 10^{-6}\,\text{m}^2$.

(a) We draw the free-body diagram of joint *D* (Figure 1.12) and calculate the internal force N_{DE}, from which we can find the axial stress in *DE*, using the following equations:

$$\sum F_y \qquad N_{DC}\sin 45 - 21 = 0 \qquad \text{or} \qquad N_{DC} = 29.7\ \text{kN}$$

$$\sum F_x \qquad -N_{DE} - N_{DC}\cos 45 = 0 \qquad \text{or} \qquad N_{DE} = -21\ \text{kN}$$

Then

$$\sigma_{DE} = \frac{N_{DE}}{A_m} = -42 \times 10^6\ \text{N/m}^2$$

$$\text{ANS.} \qquad \sigma_{DE} = 42\ \text{MPa (C)}$$

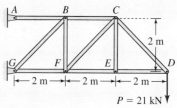

Figure 1.11   Truss.

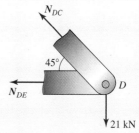

Figure 1.12   Free-body diagram of joint *D*.

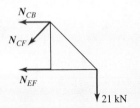

Figure 1.13 Free-body diagram.

We can make an imaginary cut through members *CB*, *CF*, and *EF* and draw the free-body diagram (Figure 1.13). By taking the moment about point *F* we can find the internal force in member *CB* by using the following equations:

$$\sum M_F \qquad N_{CB} \times 2 - 21 \times 4 = 0 = 42 \text{ kN}$$

Then

$$\sigma_{CB} = \frac{N_{CB}}{A_m} = 42 \times 10^6 \text{ N/m}^2$$

ANS. $\sigma_{CD} = 84 \text{ MPa (T)}$

Intuitive check: The force *P* is pulling the structure downward. Thus line *DEFG* will become shorter and line *ABC* will become longer. Therefore members on the top, including *BC*, will be in tension and members at the bottom, which includes *DE*, will be in compression, which is consistent with our answer.

(b) We can draw the free-body diagram of the entire truss and take the moment at point *G* to find the reaction force at *A*. Because the pin at *A* is in double shear, we will need two cuts to free the member *AB* from the support. Each of the two cuts will have half the force of A_x. The calculation for the shear stress in pin *A* is shown in Figure 1.14 and given by the equations.

$$\sum M_G \qquad A_x \times 2 - 21 \times 6 = 0 \qquad \text{or} \qquad A_x = 63 \text{ kN}$$

Then

$$\tau_A = \frac{A_x/2}{A_p} = 100 \times 10^6 \text{ N/m}^2$$

ANS. $\tau_A = 100 \text{ MPa}$

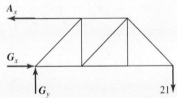

Figure 1.14 Free-body diagram for calculating shear stress in pin *A*.

COMMENT

In part (a) we could have solved for the force in *BC* by noting that *EC* is a zero force member and by drawing the free-body diagram of joint *C*.

1.1.3 Internally Distributed Force Systems

In Sections 1.1.1 and 1.1.2 the normal stress and the shear stress were introduced as the average intensity of an internal normal and shear force distribution, respectively. But what if there are internal moments at a cross section? Would there be normal and shear stresses at such sections? How are the normal and shear stresses related to internal moments? To answer these questions and to get a better understanding of the character of normal stress and shear stress, we now consider an alternative view, which is more fundamental.

The forces of attraction and repulsion between two particles (atoms or molecules) in a body are assumed to act along the line that joins the two particles.[3] The forces vary inversely as an exponent of the radial distance separating the two particles. Thus every particle exerts a force on every other particle, as shown symbolically in Figure 1.15a on an imaginary surface of a body. These forces between particles hold the body together and are referred to as internal forces. When we apply external forces on a body, the shape of the body changes. The change in shape implies that the distance between particles must change, which implies further that the forces between particles (internal forces) must change. When the change in the internal forces exceeds some characteristic material value, the body will break. Thus the strength of the material can be characterized by the measure of *change in the intensity* of internal forces. This measure of change in the intensity of internal forces is what we call stress.

In Figure 1.15b we replace all forces that are exerted on any single particle by the resultants of these forces on that particle. The magnitude and direction of these resultant forces will vary with the location of the particle (point), as shown in Figure 1.15b.[4] In other words, we have an internally distributed force system that is generated in the material when external forces are applied. It is the intensity of this internally distributed force system that we are trying to measure as stress. Furthermore, we note that force is a vector. Hence the internally distributed forces (stress on a surface) can be resolved into normally (perpendicular to the surface) and tangentially (parallel to the surface) distributed forces, as shown in Figure 1.15b. We summarize the description in this paragraph with the following definition.

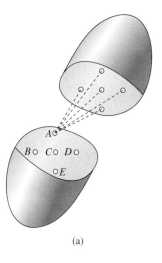
(a)

Definition 4 The intensity of internally distributed forces on an imaginary cut surface of a body is called the *stress* on the surface. The intensity of an internally distributed force that is normal to the surface of an imaginary cut is called the *normal stress* on the surface. The intensity of an internally distributed force that is parallel to the surface of an imaginary cut surface is called the *shear stress* on the surface.

Normal stress on a surface may be viewed as the internal forces that develop due to the material resistance to the pulling apart or pushing together of two adjoining planes of an imaginary cut. Like pressure, normal stress is always perpendicular to the surface of the imaginary cut. But unlike pressure, which can only be compressive, normal stress can be tensile.

Shear stress on a surface may be viewed as the internal forces that develop due to the material resistance to the sliding of two adjoining planes along the imaginary cut. Like friction, shear stresses act tangent to the plane in the direction opposite to the impending motion of the surface. But unlike friction, shear stress is not related to the normal forces (stresses).

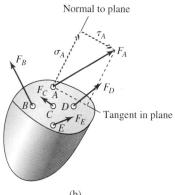
(b)

Figure 1.15 Internal forces between particles on two sides of an imaginary cut. (*a*) Forces between particles in a body, shown on particle A. (*b*) Resultant force on each particle.

---

[3]Forces that act along the line joining two particles are called *central* forces. The concept of central forces started with Newton's universal gravitation law, which states: "the force between two particles is inversely proportional to the square of the radial distance between two particles and acts along the line joining the two particles." At atomic levels the central forces do not vary with the square of the radial distance but with an exponent, which is a power of 8 or 10.
[4]Should there be a resultant moment also? See Section 1.3 for a description of couple stress.

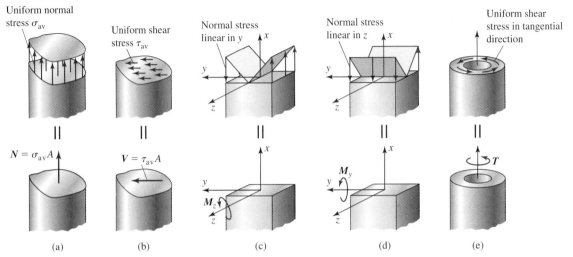

Figure 1.16 Static equivalency.

Now that we have established that the stress on a surface is an internally distributed force system, we are in a position to answer the questions raised at the beginning of the section. If the normal and shear stresses are constant in magnitude and direction across the cross section, as shown in Figure 1.16a and b, then these can be replaced by statically equivalent normal and shear forces. [We obtain the equivalent forms of Equations (1.1) and (1.2).] But if either the magnitude or the direction of the normal and shear stresses changes across the cross section, then internal bending moments M_y, M_z and the internal torque T may be necessary for static equivalency, as shown in Figure 1.16c, d, and e.

Figure 1.16 shows some of the stress distributions we will see in this book. But how do we deduce the variation of stress on a surface when stress is an internal quantity that cannot be measured directly? The theories in this book that answer this question were developed over a long period of time using experimental observations, intuition, and logical deduction in an iterative manner. Assumptions have to be made regarding loading, geometry, and material properties of the structural member in the development of the theory. If the theoretical predictions do not conform to experimental observations, then assumptions have to be modified to include added complexities until the theoretical predictions are consistent with experimental observations. In Chapters 4 through 6 we shall see how complexities can be added to the simplified theories for structural members that we will develop.

Section 4.7 discusses the clues one can use to deduce stress behavior across a cross section. But a more general approach is to approximate the behavior of displacements[5] of points on a cross section and use the logic described in Section 3.2 to develop a theory. This process will be shown in Chapters 4 through 6. Two of the links in the logic, shown in Figure 1.2, are based on an idea you learned in statics, namely,

[5]In 1853 Saint-Venant was the first to observe that if displacements as a function of position are known, then a mechanics of materials problem is completely solved. See Section 10.5 on the various contributions of Saint-Venant.

analysis is simpler if the distributed forces are first replaced by equivalent forces and moments before the equilibrium equations are applied.

- Examples 1.4 through 1.6 emphasize the two-step simplification in the context of stress–internal force–external force.

- Example 1.7 emphasizes the two-step simplification in the context of stress–internal moment–external force.

- Examples 1.8 and 1.9 demonstrate how formulas can be developed once a stress behavior is known or assumed.

- Example 1.10 emphasizes the importance of the orientation of the imaginary cut surface and the direction of the stress on the surface that is formalized as stress at a point in Section 1.2.

EXAMPLE 1.4

A T cross section has a uniform normal stress distribution of 10 ksi (T), as shown in Figure 1.17. Determine the equivalent internal normal force and the location at which it should be placed.

PLAN

A distributed force can be replaced by a single force acting at the centroid of the distribution. For uniform loads the centroid of the distribution coincides with the centroid of the area on which the distributed force is acting. We can consider the T section as two rectangles and in each rectangle replace the distribution by a single force acting at the centroid of the rectangle. We can then replace the two forces by a single force.

Solution   We replace the distributed forces in each cross section by equivalent forces, as shown in Figure 1.18.

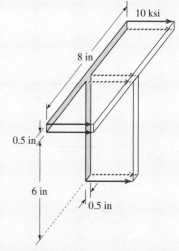

Figure 1.17   Normal stress distribution in Example 1.4.

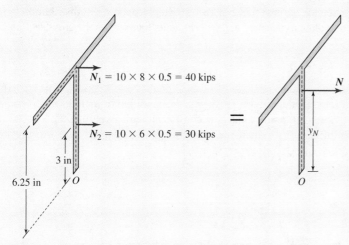

$N_1 = 10 \times 8 \times 0.5 = 40$ kips

$N_2 = 10 \times 6 \times 0.5 = 30$ kips

Figure 1.18   Statically equivalent diagrams.

If these two diagrams are to be statically equivalent, then the total force on the right must equal the total force on the left, and the moments from the forces around any line must be the same. We obtain

$$\sum F_x \qquad N = N_1 + N_2 = 70$$

ANS. $N = 70$ kips

$$\sum M_O \qquad N \times y_N = N_1 \times 6.25 + N_2 \times 3 = 340 \qquad \text{or} \qquad y_N = \frac{340}{70}$$

ANS. $y_N = 4.875$ in

COMMENT

An alternative calculation is to find the centroid of the cross section directly using Equation (A.4) of Appendix A and place the equivalent internal force at the centroid.

EXAMPLE 1.5

A wooden rod reinforced with steel has a normal stress distribution on a cross section as shown in Figure 1.19. Replace the stress distribution by an equivalent axial force and determine its location on the cross section.

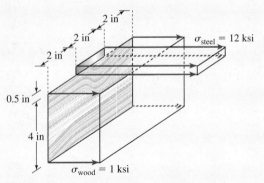

Figure 1.19   Normal stress distribution in Example 1.5.

PLAN

We can replace the normal stress in steel and wood by the normal forces N_{steel} and N_{wood} at the centroid of the areas in steel and wood, respectively, and then replace the two forces by a single equivalent force.

Solution We replace the stresses in each material by the equivalent internal force, as shown in Figure 1.20. If the left figure is to be statically equivalent to the right figure, then the total force on the right must equal the total force on the left, and the moments from the forces around any line must be the same. We obtain

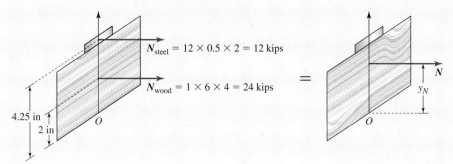

Figure 1.20 Statically equivalent diagrams.

$$\sum F_x \qquad N = N_{steel} + N_{wood} = 36$$

ANS. $N = 36$ kips

$$\sum M_O \qquad N \times y_N = N_{steel} \times 4.25 + N_{wood} \times 2 = 99 \quad \text{or} \quad y_N = \frac{99}{36}$$

ANS. $y_N = 2.75$ in

COMMENT

The centroid of the cross section is at 3.08, which is different from the location at which we have to place our equivalent force if we do not want to include a moment to represent our internal force distribution. The effect of material properties on the stress distribution and, hence, on the location and magnitude of equivalent internal forces will be developed further in Chapter 3.

EXAMPLE 1.6

A bar is axially loaded such that there is no rotation of the cross section, as shown in Figure 1.21. Determine the external forces F_1 and F_4, assuming that the stress distribution and the cross section AA are those given in:

(a) Example 1.4.

(b) Example 1.5.

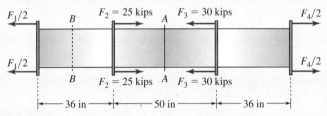

Figure 1.21   Axially loaded bar.

PLAN

We can draw the free-body diagrams (as in Figure 1.22) after we make an imaginary cut at section AA and relate the internal axial forces to the external forces. Using the results for the internal forces in the previous examples, we can calculate the external forces.

Solution

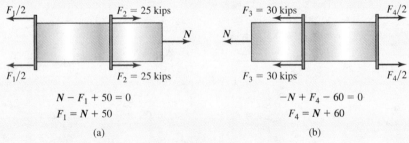

$$N - F_1 + 50 = 0 \qquad\qquad -N + F_4 - 60 = 0$$
$$F_1 = N + 50 \qquad\qquad\qquad F_4 = N + 60$$

(a)   (b)

Figure 1.22   Free-body diagrams.

(a) In Example 1.4, $N = 70$ kips, which implies

ANS. $F_1 = 120$ kips $F_4 = 130$ kips

(b) In Example 1.5, $N = 36$ kips, which implies

ANS. $F_1 = 86$ kips $F_4 = 96$ kips

COMMENTS

1. We could have calculated F_1 as shown in Figure 1.23. We would then have to draw a similar free-body diagram for each cases, leading to a very cumbersome process. By replacing stresses with internal forces and then relating internal forces to external forces, we broke the complexity down into a sequence of simpler steps. In other words, the relationship of external forces (and moments) to internal forces and the relationship of internal forces to stress distributions are two distinct ideas, as depicted in Figure 1.2.

2. Suppose we wished to find the normal stress at section *BB* in Figure 1.21. In case of the homogeneous cross section of Example 1.4, we will have to find the internal force at *BB* and divide it by the cross-sectional area. But how can we get the two-part stress distribution for the two dissimilar material cross sections? Thus if we know the stress behavior across the cross section we can always relate it to external forces, but the reverse may not be possible when material models become complex.

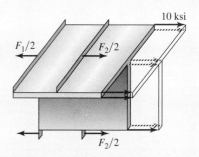

Figure 1.23 Elaboration of comments.

EXAMPLE 1.7

A simply supported beam is uniformly loaded, as shown in Figure 1.24. The cross section *AA* was found to have a normal stress distribution. It also has an average shear stress τ_{av}.

(a) Replace the normal stress distribution by a statically equivalent internal moment M_z.

(b) Determine the intensity *w* of the external load and the average shear stress τ_{av}.

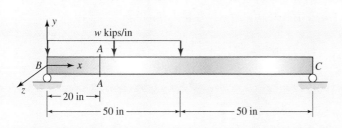

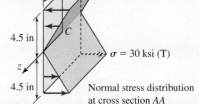

Figure 1.24 Simply supported beam.

PLAN

(a) We can replace each triangular distributed force by a single force at two-thirds the distance from the apex. We can then replace the two forces by an equivalent moment.

(b) By drawing a free-body diagram of the entire beam we can find the reactions at the support in terms of w. By making an imaginary cut at AA and taking the left side of the cut for a free-body diagram, we can relate the internal moment and the internal shear force to the reaction force and find both w and the shear stress.

Solution   Each of the triangular prisms can be replaced by an equivalent force N representing the volume of the prism, as shown in Figure 1.25. The two internal forces N form a couple and can be replaced by an equivalent moment, as given by the equation

$$M_z = 2 \times N \times 3 = 2430 \qquad \text{(E1)}$$

$$\text{ANS.} \qquad M_z = 2430 \text{ in} \cdot \text{kips}$$

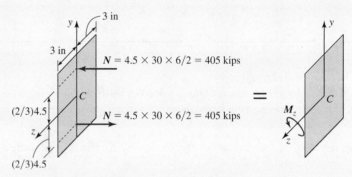

Figure 1.25   Statically equivalent internal moment.

We can replace the distributed force by an equivalent force and draw the free-body diagram of the entire beam (Figure 1.26). By balancing the moment about support C we can obtain the reaction force at B,

$$R_B \times 100 - 50w \times 75 = 0 \qquad \text{or} \qquad R_B = 37.5w$$

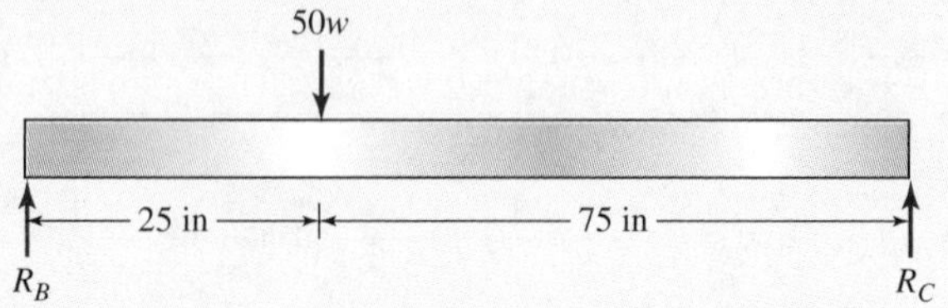

Figure 1.26   Free-body diagram of entire beam.

The cross-sectional area of the beam is $A = (9)(6) = 54 \text{ in}^2$. Thus the average shear force acting at section AA is $V_A = 54\tau_{av}$.

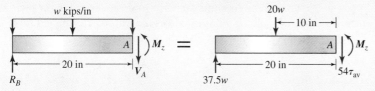

Figure 1.27 Free-body diagrams.

Taking the moment about point A in Figure 1.27, we obtain

$$37.5w \times 20 - 20w \times 10 - M_z = 0 \quad \text{or} \quad 550w = M_z \quad \text{(E2)}$$

Substituting Equation (E1) into Equation (E2) we obtain

$$w = \frac{2430}{550} \quad \text{or} \quad \text{ANS.} \quad w = 4.418 \text{ kips/in}$$

Balancing forces in the y direction, we obtain

$$37.5w - 20w - 54\tau_{av} = 0 \quad \text{or} \quad \tau_{av} = 0.324w \quad \text{(E3)}$$

Substituting the value of w, we obtain

$$\tau_{av} = 0.324 \times 4.418 \quad \text{or} \quad \text{ANS.} \quad \tau_{av} = 1.43 \text{ ksi}$$

COMMENTS

1. In Figure 1.25 we note that the total internal force in the x direction is zero, that is, the total tensile force balances the total compressive force. We will see this type of stress distribution in the bending of beams.

2. Equation (E1) is a static equivalency statement, whereas Equations (E2) and (E3) are equilibrium equations. If the stress distribution were different, it would change Equation (E1), but not Equations (E2) or (E3). Later on we will develop formulas for static equivalency that relate the bending normal stress to the internal bending moment, and this formula will be independent of external loads.

3. The problem once more emphasizes the distinction between the relationship of stresses to internal forces and moments and the relationship of external loads to internal forces and moments. The alternative analysis, shown in Figure 1.28, would be very tedious.

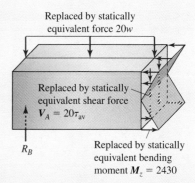

Replaced by statically equivalent force 20w

Replaced by statically equivalent shear force $V_A = 20\tau_{av}$

R_B

Replaced by statically equivalent bending moment $M_z = 2430$

Figure 1.28 Elaboration of comment.

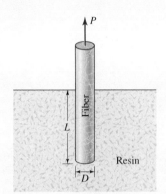

Figure 1.29 Fiber pull-out.

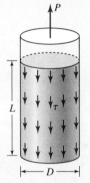

Figure 1.30 Calculation of shear stress in fiber–resin interface.

EXAMPLE 1.8

A fiber pull-out test is to be conducted to determine the shear strength of the interface between the fiber and the resin matrix in a composite material (Figure 1.29).[6] Assuming a uniform shear stress τ at the interface, derive a formula for the shear stress in terms of the applied force P, the length of fiber L, and the fiber diameter D.

PLAN

The shear stress is acting on the cylindrical surface area of the embedded fiber. The shear stress is uniform and hence can be replaced by an equivalent shear force V, which we can equate to P.

Solution We can draw the cylindrical surface of the fiber and show the uniform stress on the surface (Figure 1.30). The surface area A is equal to the circumference multiplied by the fiber length L. The shear force is the shear stress multiplied by the surface area, and by equilibrium it is also equal to P, as shown by the equations

$$A = \pi DL$$
$$V = \tau A = (\pi DL)\tau = P$$

Then

ANS. $\qquad \tau = \dfrac{P}{\pi DL}$ $\qquad$ (E1)

COMMENTS

1. In this problem we viewed the shear stress in the interface like a frictional force.

2. In the preceding test it is implicitly assumed that the strength of the fiber is greater than the interface strength. Otherwise the fiber would break before it gets pulled out.

3. In a test the force P is increased slowly until the fiber is pulled out. The pull-out force is recorded, and the shear strength can be calculated.

4. Suppose we have determined the shear strength from Equation (E1) for specific dimensions D and L of the fiber. Now we should be able to predict the force P that a fiber with different dimensions would support. If on conducting the test the experimental value of P is significantly different from the value predicted, then our assumption of uniform shear stress in the interface is most likely incorrect.

---

[6]A composite material is any material made from more than one constituent. We shall discuss the growing field of reinforced plastic (composite) in Section 3.12.3 in greater detail.

EXAMPLE 1.9

In a test to determine the shear strength of an adhesive, a torque (a moment along the axis) is applied to two thin cylinders joined together with the adhesive (Figure 1.31). Assuming a uniform shear stress τ in the adhesive, develop a formula for the adhesive shear stress τ in terms of the applied torque T, the cylinder radius R, and the cylinder thickness t.

PLAN

We can make an imaginary cut through the adhesive layer and first calculate the internal shear force on a differential area. We can then calculate the moment from the internal shear force on the differential area. By integrating we can find the total internal moment acting on the adhesive, which we can equate to the applied external moment T.

Solution We make an imaginary cut through the adhesive and draw the shear stress in the tangential direction, as shown in Figure 1.32. The differential area is the differential arc length ds multiplied by the thickness t. The differential tangential shear force dV is the shear stress multiplied by the differential area. The differential internal torque (moment) dT is the moment arm R multiplied by dF. By integrating over the entire circumference we obtain the total internal torque, which we equate to the external torque,

$$T = \int R\,dV = \int R(\tau t)R\,d\theta = \int_0^{2\pi} \tau t R^2\,d\theta = \tau t R^2 (2\pi)$$

Then

ANS. $\qquad \tau = \dfrac{T}{2\pi R^2 t}$

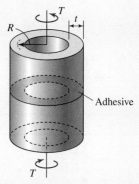

Figure 1.31 Torque on two adhesively joined cylinders.

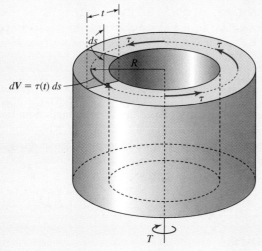

Figure 1.32 Free-body diagram.

COMMENTS

1. By recording the value of the torque at which the top half of the cylinder separates from the bottom half, we can calculate the shear strength of the adhesive.

2. The assumption of uniform shear stress can only be justified for thin cylinders, as we shall show in Section 5.6.

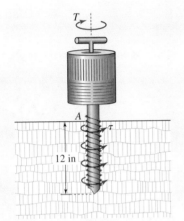

Figure 1.33 Torque on a drill.

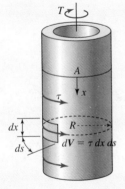

Figure 1.34 Torque calculations.

EXAMPLE 1.10

A 12-in-deep hole is to be made for placing explosive charges in a granite rock (Figure 1.33). The shear strength of the granite is 5 ksi. Determine the minimum torque T that must be applied to a 2-in-diameter drill assuming a uniform shear stress along the length of the drill. Neglect the taper at the end.

PLAN

The imaginary cut surface is the surface of the hole in the granite. The shear stress on the surface of the hole would act like a distributed frictional force on the cylindrical surface of the drill. This observation can help us to calculate the force due to the distributed frictional force. We find the moment from this frictional force and relate it to the applied torque.

Solution   The shear stress acts tangential to the cylindrical surface of the drill, as shown in Figure 1.34. By multiplying the shear stress by the differential surface area $ds\, dx$ we obtain the differential tangential shear force dV. By multiplying dV by the moment arm R, we obtain the internal torque dT, which is due to the shear stress over the differential surface area. By integrating over the circumference $[ds = R\, d\theta = 1\, d\theta]$ and the length of the drill, we obtain the total internal torque, which we can equate to the external torque T. By performing integration, we obtain the value of the external torque,

$$T = \int R\, dV = \int_0^{12}\int_0^{2\pi} R(\tau)R\, d\theta\, dx = \int_0^{12}\int_0^{2\pi} 5\, d\theta\, dx$$

$$= \int_0^{12} 5 \times 2\pi\, dx = 10\pi \times 12 = 120\pi$$

ANS. $T = 377$ in·kips

COMMENTS

1. In this example and in Example 1.8 shear stress acted on the outside cylindrical surface. But due to the direction of the shear stresses, in Example 1.8 we replaced the shear stresses by just an internal shear

force, whereas in this example we replaced the shear stresses by an internal torque.

2. In Example 1.9 and in this example the surfaces on which the shear stresses are acting are different. Yet in both examples we replaced the shear stresses by the equivalent internal torque.

3. The two preceding comments emphasize that before we can define which internal force or which internal moment is statically equivalent to the internal stress distribution, we must specify the direction of stress and the orientation of the surface on which the stress is acting. We shall develop this concept further in Section 1.2.

Consolidate your knowledge   Write in one page all you understand about stress on a surface.

QUICK TEST 1.1   Time: 15 minutes/Total: 20 points

Answer true or false and justify each answer in one sentence. Grade yourself with the answers given in Appendix G. Give yourself one point for each correct answer (true or false) and one point for every correct explanation.

1.   You can measure stress directly with an instrument the way you measure temperature with a thermometer.

2.   There can be only one normal stress component acting on the surface of an imaginary cut.

3.   If a shear stress component on the left surface of an imaginary cut is upward, then on the right surface it will be downward.

4.   If a normal stress component puts the left surface of an imaginary cut in tension, then the right surface will be in compression.

5.   The correct way of reporting shear stress is $\tau = 70$ kips.

6.   The correct way of reporting positive axial stress is $\sigma = +15$ MPa.

7.   1 GPa equals 10^6 Pa.

8.   1 psi is approximately equal to 7 Pa.

9.   A common failure stress value for metals is 10,000 Pa.

10.   Stress on a surface is the same as pressure on a surface as both quantities have the same units.

PROBLEM SET 1.1

Figure P1.1

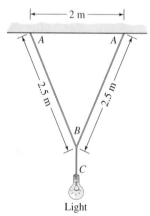

Figure P1.5

Tensile stress

1.1   In a tug of war, each person shown in Figure P1.1 exerts a force of 200 lb. If the diameter of the rope is $\frac{1}{2}$ in, determine the axial stress in the rope.

1.2   A weight is being raised using a cable and a pulley, as shown in Figure P1.2. If the weight $W = 200$ lb, determine the axial stress assuming: (a) the cable diameter is $\frac{1}{8}$ in (b) the cable diameter is $\frac{1}{4}$ in.

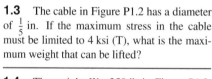
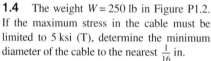

Figure P1.2

1.3   The cable in Figure P1.2 has a diameter of $\frac{1}{5}$ in. If the maximum stress in the cable must be limited to 4 ksi (T), what is the maximum weight that can be lifted?

1.4   The weight $W = 250$ lb in Figure P1.2. If the maximum stress in the cable must be limited to 5 ksi (T), determine the minimum diameter of the cable to the nearest $\frac{1}{16}$ in.

1.5   A 6-kg light shown in Figure P1.5 is hanging from the ceiling by wires of 0.75-mm diameter. Determine the tensile stress in wires AB and BC.

1.6   A 8-kg light shown in Figure P1.5 is hanging from the ceiling by wires. If the tensile stress in the wires cannot exceed 50 MPa, determine the minimum diameter of the wire, to the nearest tenth of a millimeter.

1.7   Wires of 0.5-mm diameter are to be used for hanging lights such as the one shown in Figure P1.5. If the tensile stress in the wires cannot exceed 80 MPa, determine the maximum mass of the light that can be hung using these wires.

1.8   A 3-lb picture is hung using a wire of $\frac{1}{8}$-in diameter, as shown in Figure P1.8. What is the average normal stress in the wires.

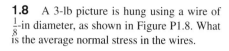
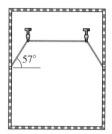

Figure P1.8

1.9   A 5-lb picture is hung using a wire, as shown in Figure P1.8. If the tensile stress in the wires cannot exceed 800 psi, determine the minimum diameter of the wire to the nearest $\frac{1}{16}$ in.

1.10   Wires of 16-mil diameter are used for hanging a picture, as shown in Figure P1.8. If the tensile stress in the wire cannot exceed 750 psi, determine the maximum weight of the picture that can be hung using these wires. 1 mil $= \frac{1}{1000}$ in.

1.11   A board is raised to lean against the left wall using a cable and pulley, as shown in Figure P1.11. Determine the axial stress in the cable in terms of the length L of the board, the specific weight γ per unit length of the board, the cable diameter d, and the angles θ and α, shown in Figure P1.11.

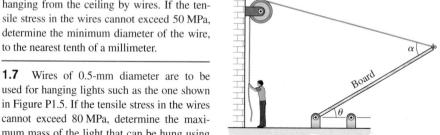

Figure P1.11

Compressive and bearing stresses

A hollow circular column supporting a building is attached to a metal plate and bolted into the concrete foundation, as shown in Figure P1.12. The column outside diameter is 4 in and an inside diameter is 3.5 in. Use this figure to solve Problems 1.12 and 1.13.

1.12 The metal plate dimensions are 8 in × 8 in × 0.5 in. The load P is estimated at 150 kips. Determine: (a) the compressive stress in the column; (b) the average bearing stress between the metal plate and the concrete.

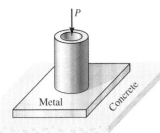

Figure P1.12

1.13 The metal plate dimensions are 10 in × 10 in × 0.75 in. If the allowable average compressive stress in the column is 30 ksi and the allowable average bearing stress in concrete is 2 ksi, determine the maximum load P that can be applied to the column.

1.14 A hollow square column supporting a building is attached to a metal plate and bolted into the concrete foundation, as shown in Figure P1.14. The column has outside dimensions of 120 mm × 120 mm and a thickness of 10 mm. The load P is estimated at 600 kN. The metal plate dimensions are 250 mm × 250 mm × 15 mm. Determine: (a) the compressive stress in the column; (b) the average bearing stress between the metal plate and the concrete.

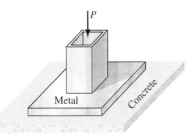

Figure P1.14

1.15 A column with the cross section shown in Figure P1.15 supports a building. The column is attached to a metal plate and bolted into the concrete foundation. The load P is estimated at 750 kN. The metal plate dimensions are 300 mm × 300 mm × 20 mm. Determine: (a) the compressive stress in the column; (b) the average bearing stress between the metal plate and the concrete.

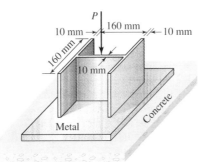

Figure P1.15

1.16 A 70-kg person is standing on a bathroom scale that has dimensions of 150 mm × 100 mm × 40 mm (Figure P1.16). Determine the bearing stress between the scale and the floor.

1.17 A 30-ft-tall brick chimney has an outside diameter of 3 ft and a wall thickness of 4 in (Figure P1.17). If the specific weight of the bricks is 80 lb/ft³, determine the average bearing stress at the base of the chimney.

1.18 Determine the average bearing stress at the bottom of the block shown in Figure P1.18 in terms of the specific weight γ and the length dimensions a and h.

1.19 The Washington Monument is an obelisk with a hollow rectangular cross section that tapers along its length. An approximation of the monument geometry is shown in Figure P1.19. The thickness at the base is 4.5 m and at top it is 2.5 m. The monument is constructed from marble and granite. Using a specific weight of 28 kN/m³ for these materials, determine the average bearing stress at the base of the monument.

Figure P1.16

Figure P1.17

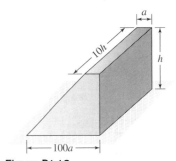

Figure P1.18

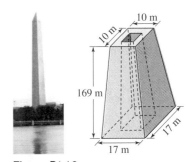

Figure P1.19

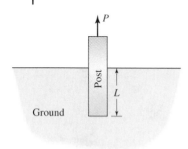

Figure P1.20

Figure P1.23

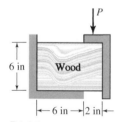

Figure P1.24

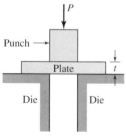

Figure P1.26

Shear stress

1.20 The post shown in Figure P1.20 has a rectangular cross section of 2 in × 4 in. The length L of the post buried in the ground is 12 in and the average shear strength of the soil is 2 psi. Determine the force P needed to pull the post out of the ground.

1.21 The post shown in Figure P1.20 has a circular cross section of 100-mm diameter. The length L of the post buried in the ground is 400 mm. It took a force of 1250 N to pull the post out of the ground. What was the average shear strength of the soil?

1.22 The cross section of the post shown in Figure P1.20 is an equilateral triangle with each side of dimension a. If the average shear strength of the soil is τ, determine the force P needed to pull the post out of the ground in terms of τ, L, and a.

1.23 A force $P = 10$ lb is applied to the handle of a hammer in an effort to pull a nail out of the wood, as shown in Figure P1.23. The nail has a diameter of $\frac{1}{8}$ in and is buried in wood to a depth of 2 in. Determine the average shear stress acting on the nail.

1.24 The device shown in Figure P1.24 is used for determining the shear strength of the wood. The dimensions of the wood block are 6 in × 8 in × 1.5 in. If the force required to break the wood block is 12 kips, determine the average shear strength of the wood.

1.25 The dimensions of the wood block in Figure P1.24 are 6 in × 8 in × 2 in. Estimate the force P that should be applied to break the block if the average shear strength of the wood is 1.1 ksi.

The punch and die arrangement shown schematically in Figure P1.26 is used to punch out thin plate objects of different shapes.

1.26 The cross section of the punch and die shown in Figure P1.26 is a circle of 1-in diameter. A force $P = 6$ kips is applied to the punch. If the plate thickness $t = \frac{1}{8}$ in, determine the average shear stress in the plate along the path of the punch.

1.27 The cross section of the punch and die shown in Figure P1.26 is a square of 10 mm × 10 mm. The plate shown has a thickness $t = 3$ mm and an average shear strength of 200 MPa. Determine the average force P needed to drive the punch through the plate.

1.28 The schematic of a punch and die for punching washers is shown in Figure P1.28. Determine the force P needed to punch out washers, in terms of the plate thickness t, the average plate shear strength τ, and the inner and outer diameters of the washers d_i and d_o.

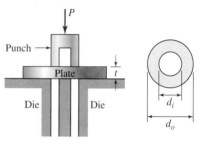

Figure P1.28

1.29 Two cast-iron pipes are adhesively bonded together over a length of 200 mm (Figure P1.29). The outer diameters of the two pipes are 50 mm and 70 mm, and the wall thickness of each pipe is 10 mm. The two pipes separated while transmitting a force of 100 kN. What was the average shear stress in the adhesive just before the two pipes separated?

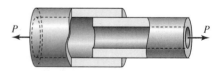

Figure P1.29

1.30 Two cast-iron pipes are adhesively bonded together over a length of 200 mm (Figure P1.30). The outer diameters of the two pipes are 50 mm and 70 mm, and the wall thickness of each pipe is 10 mm. The two pipes separated while transmitting a torque of 2 kN · m. What was the average shear stress in the adhesive just before the two pipes separated?

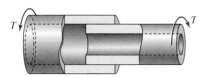

Figure P1.30

1.31 Two cast-iron pipes are held together by a bolt, as shown in Figure P1.31. The outer diameters of the two pipes are 50 mm and 70 mm, and the wall thickness of each pipe is 10 mm. The diameter of the bolt is 15 mm. The bolt broke while transmitting a torque of 2 kN·m. On what surface(s) did the bolt break? What was the average shear stress in the bolt on the surface where it broke?

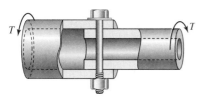

Figure P1.31

1.32 The forces acting on a pin in a truss joint are shown in Figure P1.32. The diameter of the pin is 20 mm. Determine the maximum transverse shear stress in the pin.

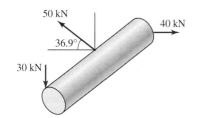

Figure P1.32

Normal and shear stresses

1.33 A weight $W = 200$ lb. is being raised using a cable and a pulley, as shown in Figure P1.33. The cable diameter is $\frac{1}{4}$ in and the pin in the pulley has a diameter of $\frac{3}{8}$ in. Determine the axial stress in the cable and the shear stress in the pin, assuming the pin is in double shear.

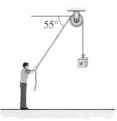

Figure P1.33

1.34 The cable in Figure P1.33 has a diameter of $\frac{1}{5}$ in and the pin in the pulley has a diameter of $\frac{3}{8}$ in. If the maximum normal stress in the cable must be limited to 4 ksi (T) and the maximum shear stress in the pin is to be limited to 2 ksi, determine the maximum weight that can be lifted to the nearest lb. The pin is in double shear.

1.35 The manufacturer of the plastic carrier for drywall panels shown in Figure P1.35 prescribes a maximum load P of 200 lb. If the cross-sectional areas at sections AA and BB are 1.3 in² and 0.3 in², respectively, determine the average shear stress at section AA and the average normal stress at section BB at the maximum load P.

1.36 A bolt passing through a piece of wood is shown in Figure P1.36. Determine: (a) the axial stress in the bolt; (b) the average shear stress in the bolt head; (c) the average bearing stress between the bolt head and the wood; (d) the average shear stress in the wood.

1.37 The axial force $P = 12$ kips acts on a rectangular member, as shown in Figure P1.37. Determine the average normal and shear stresses on the inclined plane AA.

1.38 Two rectangular bars of 10-mm thickness are loaded as shown in Figure P1.38. If the normal stress on plane AA is 150 MPa (C), determine the force F_1 and the normal and shear stresses on plane BB.

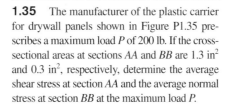

Figure P1.38

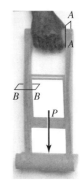

Figure P1.35

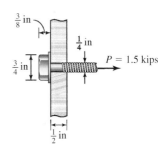

Figure P1.36

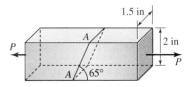

Figure P1.37

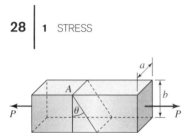

Figure P1.39

Figure P1.40

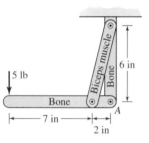

Figure P1.41

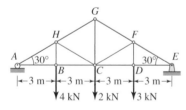

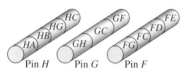

Figure P1.42

1.39   (a) In terms of P, a, b, and θ determine the average normal and shear stresses on the inclined plane AA shown in Figure P1.39. (b) Plot the normal and shear stresses as a function of θ and determine the maximum values of the normal and shear stresses. (c) At what angles of the inclined plane do the maximum normal and maximum shear stresses occurs.

1.40   An axial load is applied to a 1-in-diameter circular rod (Figure P1.40). The shear stress on section AA was found to be 20 ksi. The section AA is at 45° to the axis of the rod. Determine the applied force P and the average normal stress acting on section AA.

1.41   A simplified model of a child's arm lifting a weight is shown in Figure P1.41. The cross-sectional area of the biceps muscle is estimated as 2 in². Determine the average normal stress in the muscle and the average shear force at the elbow joint A.

Figure P1.42 shows a truss and the sequence of assembly of members at pins H, G, and F. All members of the truss have cross-sectional areas of 250 mm² and all pins have diameters of 15 mm.

1.42   Determine the axial stresses in members HA, HB, HG, and HC of the truss shown in Figure P1.42.

1.43   Determine the maximum shear stress in pin H.

1.44   Determine the axial stresses in members FG, FC, FD, and FE of the truss shown in Figure P1.42.

1.45   Determine the maximum shear stress in pin F.

1.46   Determine the axial stresses in members GH, GC, and GF of the truss shown in Figure P1.42.

1.47   Determine the maximum shear stress in pin G.

1.48   The pin at C in Figure P1.48 is has a diameter of $\frac{1}{2}$ in and is in double shear. The cross-sectional areas of members AB and BC are 2 in² and 2.5 in². Determine the axial stress in member AB and the shear stress in pin C.

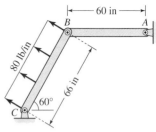

Figure P1.48

1.49   All pins shown in Figure P1.49 are in single shear and have diameters of 40 mm. All members have square cross sections. Determine the maximum shear stresses in the pins and the axial stress in member BD.

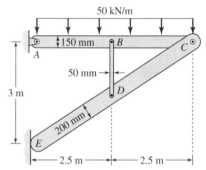

Figure P1.49

Internally distributed force system

1.50   The C cross section in Figure P1.50 has a uniform normal stress distribution of 10 ksi (T). Determine the equivalent internal normal force N and the location y_N from the bottom at which it should be placed so as to require no internal moment.

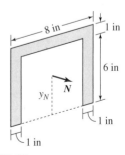

Figure P1.50

1.51 A steel strip is attached to a wooden bar, as shown in Figure P1.51. The axial stresses in steel and wood were found to be at uniform values of 15 ksi (T) and 1 ksi (T), respectively. Determine the equivalent axial force N and its location y_N from the bottom so as to require no internal moment.

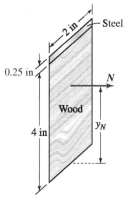

Figure P1.51

1.52 Determine the values of F_2 and F_3 in Figure P1.52 assuming section AA is of Problem 1.50.

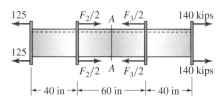

Figure P1.52

1.53 Determine the values of F_2 and F_3 in Figure P1.52 assuming section AA is of Problem 1.51.

1.54 Determine the axial forces F_1 and F_4 in the composite bar shown in Figure P1.54. The normal stresses at section AA, in steel and wood are 180 MPa (C) and 12 MPa (C), respectively.

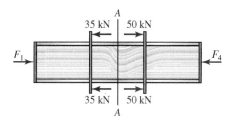

Figure P1.54

1.55 A cantilever beam with a square cross section is loaded with a linearly varying distributed load of maximum intensity w N/m, as shown in Figure P1.55a. The section AA has an average shear stress τ and a normal stress distribution, as shown in Figure P1.55b. (a) Replace the normal stress distribution by an equivalent internal force N and internal moment M_z acting at the centroid C; (b) Determine the intensity w and the average shear stress τ on section AA.

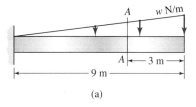

(a)

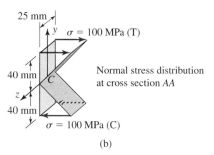

(b)

Figure P1.55

1.56 A simply supported beam has a moment couple M_{ext} applied at one end, as shown in Figure P1.56a. The section AA has an average shear stress τ and a normal stress distribution as given in Figure P1.56b. (a) Replace the normal stress distribution by an equivalent internal force N and internal moment M_z acting at the centroid C; (b) Determine the external moment M_{ext}, the external force P, and the average shear stress τ on section AA.

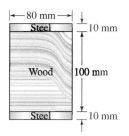

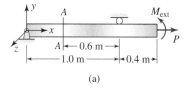

(a)

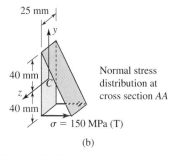

(b)

Figure P1.56

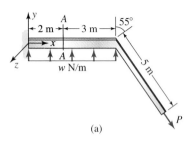

(a)

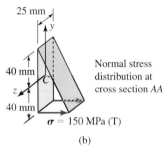

Normal stress distribution at cross section *AA*

$\sigma = 150$ MPa (T)

(b)

Figure P1.57

1.57 The cross section *AA* on the structure shown in Figure P1.57a has an average shear stress τ and a normal stress as shown in Figure P1.57b. (a) Replace the normal stress distribution by an equivalent internal normal force N and an internal moment M_z acting at the centroid C. (b) Determine the value of the average shear stress τ, the intensity of the externally distributed force w, and the force P.

1.58 A simply supported beam with a C cross section has a moment couple M_{ext} applied at one end, as shown in Figure P1.58a. The section *AA* has an average shear stress τ and a normal stress distribution as given in Figure P1.58b. (a) Replace the normal stress distribution by an equivalent internal moment M_z acting at the centroid C. (b) Determine the external moment M and the average shear stress τ on section *AA*.

1.59 The cross section *AA* on the structure shown in Figure P1.59a has an average shear stress τ and a normal stress as shown in Figure P1.59b. (a) Replace the normal stress distribution by an equivalent internal normal force N and an internal moment M_z acting at the centroid C of the T cross section. (b) Determine the value of the average shear stress τ, the intensity of the externally distributed force w, and the force P.

Design problems

1.60 The bottom screw in the hook shown in Figure P1.60 supports 60% of the load P while the remaining 40% of P is carried by the top screw. The shear strength of the screws is 50 MPa. Develop a table for the maximum load P that the hook can support for screw diameters that vary from 1 mm to 5 mm in steps of 1 mm.

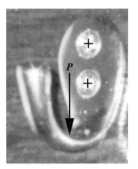

Figure P1.60

1.61 A tire swing is suspended using three chains, as shown in Figure P1.61. Each chain makes an angle of 12° with the vertical. The chain is made from links as shown. For design purposes assume that more than one person may use the swing, and hence the swing is to

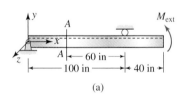

(a)

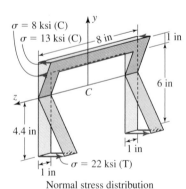

Normal stress distribution at cross section *AA*

(b)

Figure P1.58

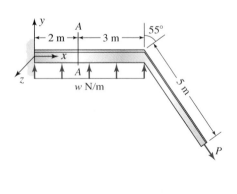

(a)

Figure P1.59

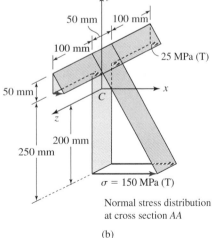

Normal stress distribution at cross section *AA*

(b)

be designed to carry a weight of 500 lb. If the maximum average normal stress in the links is not to exceed 10 ksi, determine to the nearest $\frac{1}{16}$ in the diameter of the wire that should be used for constructing the links.

Figure P1.61

1.62 Two cast-iron pipes are held together by a bolt, as shown in Figure P1.62. The outer diameters of the two pipes are 50 mm and 70 mm and the wall thickness of each pipe is 10 mm. The diameter of the bolt is 15 mm. What is the maximum force P this assembly can transmit if the maximum permissible stresses in the bolt and the cast iron are 200 MPa in shear and 150 MPa in tension, respectively.

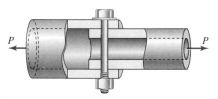

Figure P1.62

1.63 A normal stress of 20 ksi is to be transferred from one plate to another by riveting a plate on top, as shown in Figure P1.63.

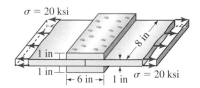

Figure P1.63

The shear strength of the $\frac{1}{2}$ in rivets used is 40 ksi. Assuming all rivets carry equal shear stress, determine the minimum even number of rivets that must be used.

1.64 Two possible joining configurations are to be evaluated. The forces on joint A in a truss were calculated and are shown in Figure P1.64. The pin diameter is 20 mm. Determine which joint assembly is better by calculating the maximum shear stress in the pin for each case.

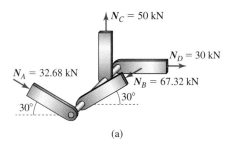

(a)

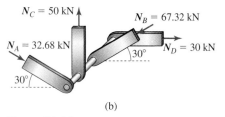

(b)

Figure P1.64

1.65 Truss analysis showed the forces at joint A given in Figure P1.65. Determine the sequence in which the three members at joint A should be assembled so that the shear stress in the pin is minimum.

1.66 An 8 in × 8 in reinforced concrete bar needs to be designed to carry a compressive axial force of 235 kips. The reinforcement is done using $\frac{1}{2}$-in round steel bars. Assuming the normal stress in concrete to be a uniform maximum value of 3 ksi and in steel bars to be a uniform value of 20 ksi, determine the minimum number of iron bars that are needed.

1.67 A wooden axial member has a cross section of 2 in × 4 in. The member was glued along line AA, as shown in Figure P1.67. Determine the maximum force P that can be applied to the repaired axial member if the maximum normal stress in the glue cannot

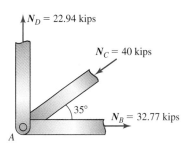

Figure P1.65

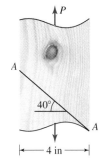

Figure P1.67

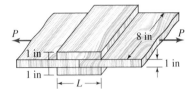

Figure P1.68

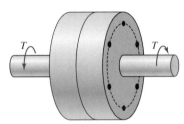

Figure P1.70

Figure P1.71

exceed 800 psi and the maximum shear stress in the glue cannot exceed 350 psi.

1.68 An adhesively bonded joint in wood is fabricated as shown in Figure P1.68. The length of the bonded region $L = 5$ in. Determine the maximum force P the joint can support if the shear strength of the adhesive is 300 psi and the wood strength is 6 ksi in tension.

1.69 The joint in Figure P1.68 is to support a force $P = 25$ kips. What should be the length L of the bonded region if the adhesive strength in shear is 300 psi?

1.70 It is proposed to use $\frac{1}{2}$-in bolts in a 10-in-diameter coupling for transferring a torque of 100 in · kips from one 4-in-diameter shaft onto another (Figure P1.70). The maximum average shear stress in the bolts is to be limited to 20 ksi. How many bolts are needed, and at what radius should the bolts be placed on the coupling?

1.71 A human hand can comfortably apply a torsional moment of 15 in · lb (Figure P1.71).

(a) What should be the breaking shear strength of a seal between the lid and the bottle, assuming the lid has a diameter of 1.5 in and a height of $\frac{1}{2}$ in?

(b) If the same sealing strength as in part (a) is used on a lid that is 1 in. in diameter and $\frac{1}{2}$ in. in height, what would be the torque needed to open the bottle?

The normal stress in the members of the truss shown in Figure P1.72 is to be limited to 160 MPa in tension or compression. All members have circular cross sections. The shear stress in the pins is to be limited to 250 MPa. Use this information in solving Problems 1.72 through 1.75.

1.72 Determine the minimum diameters to the nearest millimeter of members *ED*, *EG*, and *EF* of the truss shown in Figure P1.72.

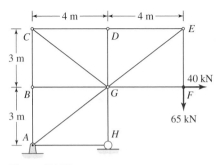

Figure P1.72

1.73 Determine the minimum diameter of pin *E* to the nearest millimeter and the sequence of assembly of members *ED*, *EG*, and *EF*.

1.74 Determine the minimum diameters to the nearest millimeter of members *CG*, *CD*, and *CB* of the truss shown in Figure P1.72.

1.75 Determine the minimum diameter of pin *C* to the nearest millimeter and the sequence of assembly of members *CG*, *CD*, and *CB*.

Stretch yourself

1.76 Truss analysis showed the forces at joint *A* given in Figure P1.76. Determine the sequence in which the four members at joint *A* should be assembled to minimize the shear stress in the pin.

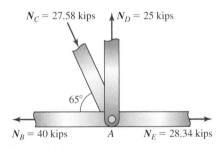

Figure P1.76

1.2 STRESS AT A POINT

The breaking of a structure starts at the point where the internal force intensity, that is, stress, exceeds some material characteristic value. This implies that we need to refine our definition of stress on a surface to that of 'Stress at a Point.' But an infinite number of planes (surfaces) can pass through a point. Then, which imaginary surface do we shrink to zero? When we shrink the surface area to zero, which equation should we use, (1.1) or (1.2)? Both difficulties can be addressed by assigning directions to the orientation of the imaginary surface and to the internal force on this surface and carrying the description of the directions as subscripts of the stress components in the same way we carried x, y, and z as subscripts to describe the components of vectors.

Figure 1.35 shows a body cut by an imaginary plane that has an outward normal in the i direction. On this surface we have a differential area ΔA_i on which a resultant force[7] acts. $\Delta \mathbf{F}_j$ is the component of the force in the j direction. A component of average stress is $\Delta \mathbf{F}_j / \Delta A_i$. If we shrink ΔA_i to zero we get the definition of a stress component at a point, as shown by the equation

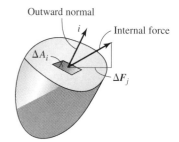

Outward normal

Figure 1.35 Stress at a point.

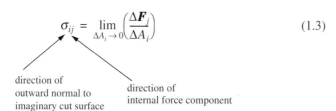

$$\sigma_{ij} = \lim_{\Delta A_i \to 0} \left(\frac{\Delta \mathbf{F}_j}{\Delta A_i} \right) \qquad (1.3)$$

direction of outward normal to imaginary cut surface

direction of internal force component

Now when we look at a stress component, the first subscript tells us the orientation of the imaginary surface and the second the direction of the internal force.

In three dimensions each subscript i and j can refer to an x, y, or z direction. In other words, there are nine possible combinations of the two subscripts. This is shown in the stress matrix in Figure 1.36.

$$\begin{bmatrix} \sigma_{xx} & \tau_{xy} & \tau_{xz} \\ \tau_{yx} & \sigma_{yy} & \tau_{yz} \\ \tau_{zx} & \tau_{zy} & \sigma_{zz} \end{bmatrix}$$

Figure 1.36 Stress matrix in three dimensions.

The diagonal elements in the stress matrix are normal stresses and all off-diagonal elements represent shear stresses.

Note that the stress at a point needs a magnitude and two directions to be specified. Thus stress is unlike any quantity we have seen so far. In three dimensions we need nine components of stress, and in two dimensions we need four components of stress to completely specify the stress at a point. Table 1.3 shows the number of components needed to specify a scalar, a vector, and stress. Now force, moment, velocity, and acceleration are all different quantities, but they all are called vectors. In a similar manner, stress belongs to a category called *tensors*. More specifically, stress is a second-order tensor,[8] where the second order refers to the exponent in the last row. In this terminology, a vector is a tensor of order 1, and a scalar is a tensor of order 0.

Sign convention for stress

ΔA_i will be considered positive if the outward normal to the surface is in the positive i direction. If the outward normal is in the negative i direction, then ΔA_i will be considered

---

[7]Should there be a resultant moment also? See Section 1.3 for a description of couple stress.
[8]To be labeled as tensor, a quantity must also satisfy certain coordinate transformation properties, which will be discussed briefly in Chapter 8.

TABLE 1.3   Comparison of number of components

Quantity	One Dimension	Two Dimensions	Three Dimensions
Scalar	$1 = 1^0$	$1 = 2^0$	$1 = 3^0$
Vector	$1 = 1^1$	$2 = 2^1$	$3 = 3^1$
Stress	$1 = 1^2$	$4 = 2^2$	$9 = 3^2$

negative. With this convention in mind, we can immediately deduce the sign for stress. A stress component can be positive in two ways. Both the numerator and the denominator are positive in Equation (1.3), or both the numerator and the denominator are negative in Equation (1.3). This leads to the following.

Definition 5 A positive stress component multiplied by a surface area that has an outward normal in the positive direction produces an internal force in the positive direction.

or

A positive stress component multiplied by a surface area that has an outward normal in the negative direction produces an internal force in the negative direction.

We conclude this section with the following points to remember. Stress is an internal quantity that has units of force per unit area. A stress component at a *point* is specified by magnitude and two directions (i.e., stress is a second-order tensor). Stress on a *surface* needs a magnitude and only one direction to be specified (i.e., stress on a surface is a vector). The first subscript on stress gives the direction of the outward normal of the imaginary cut surface. The second subscript gives the direction of the internal force. The sign of a stress component is determined from the direction of the internal force and the direction of the outward normal to the imaginary cut surface.

1.2.1 Stress Elements

The discussion in the previous section shows that stress at a point is an abstract quantity, and developing an intuitive feel for it will be difficult. Stress on a surface, however, is easier to visualize as the intensity of a distributed force on a surface.

Definition 6 A stress element is an imaginary object that helps us visualize stress at a point by constructing surfaces that have outward normals in the coordinate directions.

In Cartesian coordinates the stress element is a cube; in cylindrical or a spherical coordinates the stress element is a fragment of a cylinder or a sphere, respectively. We start our discussion with the construction of a stress cube to emphasize the basic principles in the construction of stress elements. We can use a similar process to draw stress elements in cylindrical and spherical coordinate systems.[9]

Construction of a stress cube

Consider the point at which we want to describe stress. Around this point imagine an object that has sides parallel to the coordinate system. The cube has six surfaces with

---

[9]See Example 1.12 for stress elements in spherical coordinates.

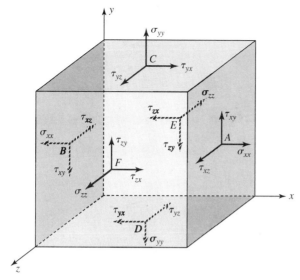

Figure 1.37 Stress cube.

outward normals that are either in the positive or in the negative coordinate direction, as shown in Figure 1.37. In other words, we have now accounted for the first subscript in our stress definition. We know that force is in the positive or negative direction of the second subscript. We use our sign convention to show the stress in the direction of the force on each of the six surfaces.

To demonstrate this construction, we will assume that all nine stress components in the stress matrix

$$
\begin{bmatrix}
\sigma_{xx} & \tau_{xy} & \tau_{xz} \\
\tau_{yx} & \sigma_{yy} & \tau_{yz} \\
\tau_{zx} & \tau_{zy} & \sigma_{zz}
\end{bmatrix}
$$

are positive. Let us consider the first row. The first subscript gives us the direction of the outward normal, which is the x direction. Surfaces A and B in Figure 1.37 have outward normals in the x direction, and it is on these surfaces that the stress component of the first row will be shown.

The direction of the outward normal on surface A is in the positive x direction [the denominator is positive in Equation (1.3)]. For the stress component to be positive on surface A, the force must be in the positive direction [the numerator must be positive in Equation (1.3)], as shown in Figure 1.37.

The direction of the outward normal on surface B is in the negative x direction [the denominator is negative in Equation (1.3)]. For the stress component to be positive on surface B, the force must be in the negative direction (the numerator must be negative), as shown in Figure 1.37.

Let us now consider row 2 in the stress matrix. From the first subscript we know that the normal to the surface is in the y direction. Surface C has an outward normal in the positive y direction, therefore all forces on surface C are in the positive direction of the second subscript, as shown in Figure 1.37. Surface D has an outward normal in the negative y direction, therefore all forces on surface D are in the negative direction of the second subscript, as shown in Figure 1.37.

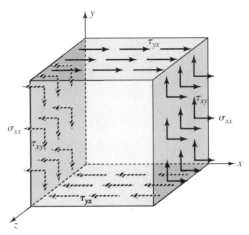

Figure 1.38 Stress components are distributed forces on a surface.

Using the same logic, the components of row 3 in the stress matrix are shown on surfaces E and F in Figure 1.37.

It should be emphasized that the single arrow used to show the stress component does not imply that the stress component is a force. The alternative illustration in Figure 1.38, showing the stress components as distributed forces on surfaces A and B, is visually more accurate but very tedious to draw every time we need to visualize stress. We will show stress components using single arrows as in Figure 1.37, but visualize them as shown in Figure 1.38.

The positive normal stress components (e.g., σ_{xx}) are pulling the cube in opposite directions, that is, the cube is in tension due to a positive normal stress component. As mentioned earlier, normal stresses are reported as tension or compression and not as positive or negative.

1.2.2 Plane Stress

Plane stress is one of the two types of two-dimensional simplifications in mechanics of materials. In Chapter 2 we will study the other type of two-dimensional simplification, plane strain. In Chapter 3 we will study the difference between the two types of simplifications. By two dimensional we imply that one of the coordinates does not play a role in the description of the problem. If we choose z to be the coordinate, we set all stresses with subscript z to zero:

$$\begin{bmatrix} \sigma_{xx} & \tau_{xy} & 0 \\ \tau_{yx} & \sigma_{yy} & 0 \\ 0 & 0 & 0 \end{bmatrix}$$

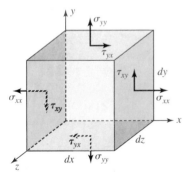

Figure 1.39 Stress cube in plane stress.

Figure 1.39 shows the stress cube for a point in plane stress. We note that the plane with outward normal in the z direction is stress-free. Stress-free surfaces are also called *free* surfaces, and these surfaces play an important role in stress analysis, which we shall discuss in Section 4.7.

If a body is in equilibrium, then all points on the body are in equilibrium. A question immediately arises: is the stress cube that represents a point on the body in equilibrium? To answer this question we need to convert the stresses into forces by multiplying by the surface area. We take a simple problem of plane stress and assume that the cube in Figure 1.39 has lengths of dx, dy, and dz in the coordinate directions. We draw a two-dimensional picture of the stress cube after multiplying each stress component by the surface area and get the force diagram of Figure 1.40.[10]

In Figure 1.40 we note that the equations of force equilibrium are satisfied by the assumed state of stress at a point. We consider the moment about point O and obtain

$$(\tau_{xy}\,dy\,dz)dx = (\tau_{yx}\,dx\,dz)dy$$

or

$$\boxed{\tau_{xy} = \tau_{yx}} \qquad (1.4a)$$

In a similar manner we can show

$$\boxed{\tau_{yz} = \tau_{zy}} \qquad (1.4b)$$

$$\boxed{\tau_{zx} = \tau_{xz}} \qquad (1.4c)$$

Equations (1.4a) through (1.4c) emphasize that shear stress is symmetric. The symmetry of shear stress implies that in three dimensions there are only six independent stress components out of the nine components necessary to specify stress at a point, and in two dimensions there are only three independent stress components out of the four components necessary to specify stress at a point. In Figure 1.39 notice that the shear stress components τ_{xy} and τ_{yx} point either toward the corners or away from the corners. This observation can be used in drawing the symmetric pair of shear stresses after drawing the shear stress on one of the surfaces of the stress cube.

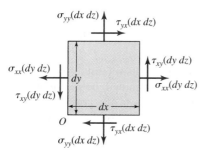

Figure 1.40 Force diagram for plane stress.

---

[10]Figure 1.40 is only valid if we assume that the stresses are varying very slowly with the x and y coordinates. If this were not true, we would have to account for the increase in stresses over a differential element. But a more rigorous analysis will also reveal that shear stresses are symmetric, see Problem 1.98.

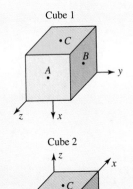

Cube 1

Cube 2

Figure 1.41 Cubes in different coordinate systems.

EXAMPLE 1.11

Show the following stress components on surfaces *A*, *B*, and *C* of the two cubes shown in different coordinate systems in Figure 1.41.

$$\begin{bmatrix} \sigma_{xx} = 80 \text{ MPa (T)} & \tau_{xy} = 30 \text{ MPa} & \tau_{xz} = -70 \text{ MPa} \\ \tau_{yx} = 30 \text{ MPa} & & \\ \tau_{zx} = -70 \text{ MPa} & & \sigma_{zz} = 40 \text{ MPa (C)} \end{bmatrix}$$

PLAN

We can identify the surface with the outward normal in the direction of the first subscript. If the outward normal is in the positive coordinate direction, then the denominator in Equation (1.3) is positive; otherwise it is negative. Knowing the signs of the denominator and of the stress component, we can determine the sign of the internal force in Equation (1.3). We can show the internal force on the identified surface in the positive coordinate direction of the second subscript if the sign of the internal force is positive and in the negative coordinate direction if the sign of the internal force is negative.

Solution **Cube 1** The first subscript of σ_{xx}, τ_{xy}, and τ_{xz} shows that the outward normal is in the *x* direction; hence these components will be shown on surface *C*. The outward normal on surface *C* is in the negative *x* direction; hence the denominator in Equation (1.3) is negative. Therefore:

- The internal force has to be in the negative *x* direction to produce a positive (tensile) σ_{xx}.
- The internal force has to be in the negative *y* direction to produce a positive τ_{xy}.
- The internal force has to be in the positive *z* direction to produce a negative τ_{xz}.

The first subscript of τ_{yx} shows that the outward normal is in the *y* direction; hence this component will be shown on surface *B*. The outward normal on surface *B* is in the positive *y* direction; hence the denominator in Equation (1.3) is positive. Therefore:

- The internal force has to be in the positive *x* direction to produce a positive τ_{yx}.

The first subscript of τ_{zx}, σ_{zz} shows that the outward normal is in the *z* direction; hence these components will be shown on surface *A*. The outward normal on surface *A* is in the positive *z* direction; hence the denominator in Equation (1.3) is positive. Therefore:

- The internal force has to be in the negative x direction to produce a negative τ_{zx}.
- The internal force has to be in the negative z direction to produce a negative (compressive) σ_{zz}.

The solution of this part of the example is depicted in Figure 1.42*a*.

Cube 2 The first subscript of σ_{xx}, τ_{xy}, and τ_{xz} shows that the outward normal is in the x direction; hence these components will be shown on surface A. The outward normal on surface A is in the negative x direction; hence the denominator in Equation (1.3) is negative. Therefore:

- The internal force has to be in the negative x direction to produce a positive (tensile) σ_{xx}.
- The internal force has to be in the negative y direction to produce a positive τ_{xy}.
- The internal force has to be in the positive z direction to produce a negative τ_{xz}.

The first subscript of τ_{yx} shows that the outward normal is in the y direction; hence this component will be shown on surface B. The outward normal on surface B is in the negative y direction; hence the denominator in Equation (1.3) is negative. Therefore:

- The internal force has to be in the negative y direction to produce a positive τ_{yx}.

The first subscript of τ_{zx}, σ_{zz} shows that the outward normal is in the z direction; hence these components will be shown on surface C. The outward normal on surface C is in the positive z direction; hence the denominator in Equation (1.3) is negative. Therefore:

- The internal force has to be in the negative x direction to produce a negative τ_{zx}.
- The internal force has to be in the negative z direction to produce a negative (compressive) σ_{zz}.

The solution of this part of the example is depicted in Figure 1.42*b*.

COMMENT

In drawing the normal stresses we could have made use of the fact that σ_{xx} is tensile, hence pulls the surface outward, and σ_{zz} is compressive, hence pushes the surface inward, which is a quicker way of getting the directions of these stress components than the arguments based on signs and subscripts.

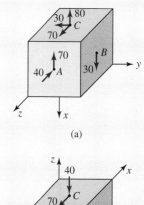

(a)

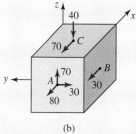

(b)

Figure 1.42 Solution of Example 1.11.

EXAMPLE 1.12

Show the following positive stress components:

$$\begin{bmatrix} \sigma_{rr} & \tau_{r\theta} & \tau_{r\phi} \\ \tau_{\theta r} & \sigma_{\theta\theta} & \tau_{\theta\phi} \\ \tau_{\phi r} & \tau_{\phi\theta} & \sigma_{\phi\phi} \end{bmatrix}$$

on a stress element drawn in the spherical coordinate system in Figure 1.43.

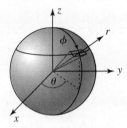

Figure 1.43 Stresses in spherical coordinates.

PLAN

We can construct a stress element with surfaces that have outward normals in the r, θ, and ϕ directions. The first subscript will identify the surface on which the row of stress components is to be shown. The second subscript then will show the direction of the stress component on the surface.

Solution We draw a stress element with lines in the directions of r, θ, and ϕ, as shown in Figure 1.44.

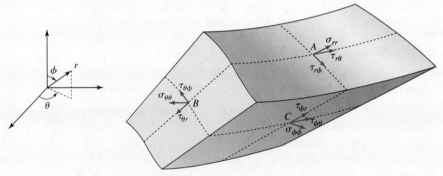

Figure 1.44 Stress element in spherical coordinates.

- The stresses σ_{rr}, $\tau_{r\theta}$, and $\tau_{r\phi}$ will be on surface A in Figure 1.44.
- The outward normal on surface A is in the positive r direction. Thus the forces have to be in the positive r, θ, and ϕ directions to result in positive σ_{rr}, $\tau_{r\theta}$, and $\tau_{r\phi}$.

- The stresses $\tau_{\theta r}$, $\sigma_{\theta\theta}$, and $\tau_{\theta\phi}$ will be on surface B in Figure 1.44.
- The outward normal on surface B is in the negative θ direction. Thus the forces have to be in the negative r, θ, and ϕ directions to result in positive $\tau_{\theta r}$, $\sigma_{\theta\theta}$, and $\tau_{\theta\phi}$.
- The stresses $\tau_{\phi r}$, $\tau_{\phi\theta}$, and $\sigma_{\phi\phi}$ will be on surface C in Figure 1.44.
- The outward normal on surface C is in the positive ϕ direction. Thus the forces have to be in the positive r, θ, and ϕ directions to result in positive $\tau_{\phi r}$, $\tau_{\phi\theta}$, and $\sigma_{\phi\phi}$.

Consolidate your knowledge   Write in one page all you understand about stress at a point.

QUICK TEST 1.2   Time: 15 minutes/Total: 20 points

Answer true or false and justify each answer in one sentence. Grade yourself with the answers given in Appendix G. Give yourself one point for every correct answer (true or false) and one point for every correct explanation.

1. Stress at a point is a vector like stress on a surface.
2. In three dimensions stress has nine components.
3. In three dimensions stress has six independent components.
4. At a point in plane stress there are three independent stress components.
5. At a point in plane stress there are always six zero stress components.
6. If the shear stress on the left surface of an imaginary cut is upward and defined as positive, then on the right surface of the imaginary cut it is downward and negative.
7. A stress element can be drawn to any scale.
8. A stress element can be drawn at any orientation.
9. Stress components are opposite in direction on the two surfaces of an imaginary cut.
10. Stress components have opposite signs on the two surfaces of an imaginary cut.

PROBLEM SET 1.2

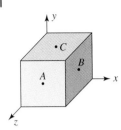

Figure P1.77

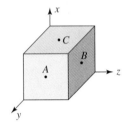

Figure P1.78

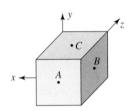

Figure P1.79

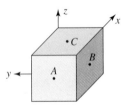

Figure P1.80

1.77 Show the nonzero stress components on the A, B, and C faces of the cube in Figure P1.77.

$$\begin{bmatrix} \sigma_{xx} = 100 \text{ MPa (T)} & \tau_{xy} = 200 \text{ MPa} & \tau_{xz} = -125 \text{ MPa} \\ \tau_{yx} = 200 \text{ MPa} & \sigma_{yy} = 175 \text{ MPa (C)} & \tau_{yz} = 225 \text{ MPa} \\ \tau_{zx} = -125 \text{ MPa} & \tau_{zy} = 225 \text{ MPa} & \sigma_{zz} = 150 \text{ MPa (C)} \end{bmatrix}$$

1.78 Show the nonzero stress components on the A, B, and C faces of the cube in Figure P1.78.

$$\begin{bmatrix} \sigma_{xx} = 90 \text{ MPa (C)} & \tau_{xy} = -200 \text{ MPa} & \tau_{xz} = 0 \\ \tau_{yx} = -200 \text{ MPa} & \sigma_{yy} = 175 \text{ MPa (C)} & \tau_{yz} = 225 \text{ MPa} \\ \tau_{zx} = 0 & \tau_{zy} = 225 \text{ MPa} & \sigma_{zz} = 150 \text{ MPa (C)} \end{bmatrix}$$

1.79 Show the nonzero stress components on the A, B, and C faces of the cube in Figure P1.79.

$$\begin{bmatrix} \sigma_{xx} = 0 & \tau_{xy} = -15 \text{ksi} & \tau_{xz} = 0 \\ \tau_{yx} = -15 \text{ ksi} & \sigma_{yy} = 10 \text{ ksi (C)} & \tau_{yz} = 25 \text{ ksi} \\ \tau_{zx} = 0 & \tau_{zy} = 25 \text{ ksi} & \sigma_{zz} = 20 \text{ ksi (T)} \end{bmatrix}$$

1.80 Show the nonzero stress components on the A, B, and C faces of the cube in Figure P1.80.

$$\begin{bmatrix} \sigma_{xx} = 0 & \tau_{xy} = -15 \text{ ksi} & \tau_{xz} = 0 \\ \tau_{yx} = -15 \text{ ksi} & \sigma_{yy} = 10 \text{ ksi (C)} & \tau_{yz} = 25 \text{ ksi} \\ \tau_{zx} = 0 & \tau_{zy} = 25 \text{ ksi} & \sigma_{zz} = 20 \text{ ksi (T)} \end{bmatrix}$$

1.81 Show the nonzero stress components on the A, B, and C faces of the cube in Figure P1.81.

$$\begin{bmatrix} \sigma_{xx} = 70 \text{ MPa (T)} & \tau_{xy} = -40 \text{ MPa} & \tau_{xz} = 0 \\ \tau_{yx} = -40 \text{ MPa} & \sigma_{yy} = 85 \text{ MPa (C)} & \tau_{yz} = 0 \\ \tau_{zx} = 0 & \tau_{zy} = 0 & \sigma_{zz} = 0 \end{bmatrix}$$

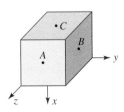

Figure P1.81

1.82 Show the nonzero stress components on the A, B, and C faces of the cube in Figure P1.82.

$$\begin{bmatrix} \sigma_{xx} = 70 \text{ MPa (T)} & \tau_{xy} = -40 \text{ MPa} & \tau_{xz} = 0 \\ \tau_{yx} = -40 \text{ MPa} & \sigma_{yy} = 85 \text{ MPa (C)} & \tau_{yz} = 0 \\ \tau_{zx} = 0 & \tau_{zy} = 0 & \sigma_{zz} = 0 \end{bmatrix}$$

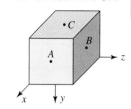

Figure P1.82

1.83 Show the stress components of a point in plane stress on the square in Figure P1.83.

$$\begin{bmatrix} \sigma_{xx} = 100 \text{ MPa (T)} & \tau_{xy} = -75 \text{ MPa} \\ \tau_{yx} = -75 \text{ MPa} & \sigma_{yy} = 85 \text{ MPa (T)} \end{bmatrix}$$

Figure P1.83

1.84 Show the stress components of a point in plane stress on the square in Figure P1.84.

$$\begin{bmatrix} \sigma_{xx} = 100 \text{ MPa (T)} & \tau_{xy} = -75 \text{ MPa} \\ \tau_{yx} = -75 \text{ MPa} & \sigma_{yy} = 85 \text{ MPa (T)} \end{bmatrix}$$

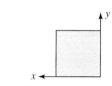

Figure P1.84

1.85 Show the stress components of a point in plane stress on the square in Figure P1.85.

$$\begin{bmatrix} \sigma_{xx} = 27 \text{ ksi (C)} & \tau_{xy} = 18 \text{ ksi} \\ \tau_{yx} = 18 \text{ ksi} & \sigma_{yy} = 85 \text{ ksi (T)} \end{bmatrix}$$

Figure P1.85

1.86 Show the stress components of a point in plane stress on the square in Figure P1.86.

$$\begin{bmatrix} \sigma_{xx} = 27 \text{ ksi (C)} & \tau_{xy} = 18 \text{ ksi} \\ \tau_{yx} = 18 \text{ ksi} & \sigma_{yy} = 85 \text{ ksi (T)} \end{bmatrix}$$

Figure P1.86

Show the nonzero stress components in the r, θ, and x cylindrical coordinate system on the A, B, and C faces of the stress elements shown in Figures P1.87 and P1.88.

1.87 (Figure P1.87).

$$\begin{bmatrix} \sigma_{rr} = 145 \text{ MPa (C)} & \tau_{r\theta} = 100 \text{ MPa} & \tau_{rx} = -125 \text{ MPa} \\ \tau_{\theta r} = 100 \text{ MPa} & \sigma_{\theta\theta} = 160 \text{ MPa (T)} & \tau_{\theta x} = 165 \text{ MPa} \\ \tau_{xr} = -125 \text{ MPa} & \tau_{x\theta} = 165 \text{ MPa} & \sigma_{xx} = 150 \text{ MPa (T)} \end{bmatrix}$$

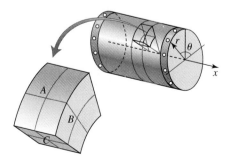

Figure P1.87

1.88 (Figure P1.88).

$$\begin{bmatrix} \sigma_{rr} = 10 \text{ ksi (C)} & \tau_{r\theta} = 22 \text{ ksi} & \tau_{rx} = 32 \text{ ksi} \\ \tau_{\theta r} = 22 \text{ ksi} & \sigma_{\theta\theta} = 0 & \tau_{\theta x} = 25 \text{ ksi} \\ \tau_{xr} = 32 \text{ ksi} & \tau_{x\theta} = 25 \text{ ksi} & \sigma_{xx} = 20 \text{ ksi (T)} \end{bmatrix}$$

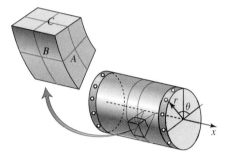

Figure P1.88

1.89 Show the stress components of a point in plane stress on the stress element in polar coordinates in Figure P1.89.

$$\begin{bmatrix} \sigma_{rr} = 125 \text{ MPa (T)} & \tau_{r\theta} = -65 \text{ MPa} \\ \tau_{\theta r} = -65 \text{ MPa} & \sigma_{\theta\theta} = 90 \text{ MPa (C)} \end{bmatrix}$$

Figure P1.89

1.90 Show the stress components of a point in plane stress on the stress element in polar coordinates in Figure P1.90.

$$\begin{bmatrix} \sigma_{rr} = 125 \text{ MPa (T)} & \tau_{r\theta} = -65 \text{ MPa} \\ \tau_{\theta r} = -65 \text{ MPa} & \sigma_{\theta\theta} = 90 \text{ MPa (C)} \end{bmatrix}$$

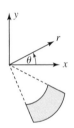

Figure P1.90

1.91 Show the stress components of a point in plane stress on the stress element in polar coordinates in Figure P1.91.

$$\begin{bmatrix} \sigma_{rr} = 25 \text{ ksi (C)} & \tau_{r\theta} = 12 \text{ ksi} \\ \tau_{\theta r} = 12 \text{ ksi} & \sigma_{\theta\theta} = 18 \text{ ksi (T)} \end{bmatrix}$$

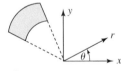

Figure P1.91

1.92 Show the stress components of a point in plane stress on the stress element in polar coordinates in Figure P1.92.

$$\begin{bmatrix} \sigma_{rr} = 25 \text{ ksi (C)} & \tau_{r\theta} = 12 \text{ ksi} \\ \tau_{\theta r} = 12 \text{ ksi} & \sigma_{\theta\theta} = 18 \text{ ksi (T)} \end{bmatrix}$$

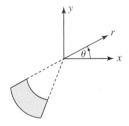

Figure P1.92

Show the nonzero stress components in the r, θ, and φ spherical coordinate system on the A, B, and C faces of the stress elements shown in Figure P1.93 and P1.94.

1.93 (Figure P1.93).

$$\begin{bmatrix} \sigma_{rr} = 135 \text{ MPa (C)} & \tau_{r\theta} = 100 \text{ MPa} & \tau_{r\phi} = -125 \text{ MPa} \\ \tau_{\theta r} = 100 \text{ MPa} & \sigma_{\theta\theta} = 160 \text{ MPa (C)} & \tau_{\theta\phi} = 175 \text{ MPa} \\ \tau_{\phi r} = -125 \text{ MPa} & \tau_{\phi\theta} = 175 \text{ MPa} & \sigma_{\phi\phi} = 150 \text{ MPa (T)} \end{bmatrix}$$

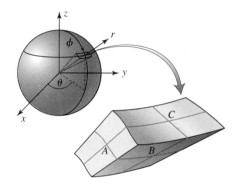

Figure P1.93

1.94 (Figure P1.94).

$$\begin{bmatrix} \sigma_{rr} = 0 & \tau_{r\theta} = -18 \text{ ksi} & \tau_{r\phi} = 0 \\ \tau_{\theta r} = -18 \text{ ksi} & \sigma_{\theta\theta} = 10 \text{ ksi (C)} & \tau_{\theta\phi} = 25 \text{ ksi} \\ \tau_{\phi r} = 0 & \tau_{\phi\theta} = 25 \text{ ksi} & \sigma_{\phi\phi} = 20 \text{ ksi (T)} \end{bmatrix}$$

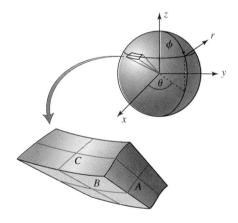

Figure P1.94

Stretch yourself

1.95 Show that the normal stress σ_{xx} on a surface can be replaced by the equivalent internal normal force N and internal bending moments M_y and M_z as shown in Figure P1.95 and given by the equations

$$N = \int_A \sigma_{xx} \, dA \qquad (1.5a)$$

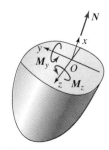

Figure P1.95

$$M_y = -\int_A z\sigma_{xx} \, dA \tag{1.5b}$$

$$M_z = -\int_A y\sigma_{xx} \, dA \tag{1.5c}$$

1.96 The normal stress on a cross section is given by $\sigma_{xx} = a + by$, where y is measured from the centroid of the cross section. If A is the cross-sectional area, I_{zz} is the area moment of inertia about the z axis, and N and M_z are the internal axial force and the internal bending moment given by Equations (1.5a) and (1.5c), respectively, show the following:

$$\sigma_{xx} = \frac{N}{A} - \left(\frac{M_z}{I_{zz}}\right)y \tag{1.6}$$

We will encounter Equation (1.6) in combined axial and symmetric bending problems in later Chapter 10.

1.97 The normal stress on a cross section is given by $\sigma_{xx} = a + by + cz$, where y and z are measured from the centroid of the cross section using Equations (1.5a), (1.5b), and (1.5c) show that

$$\sigma_{xx} = \frac{N}{A} - \left(\frac{M_z I_{yy} - M_y I_{yz}}{I_{yy} I_{zz} - I_{yz}^2}\right)y - \left(\frac{M_y I_{zz} - M_z I_{yz}}{I_{yy} I_{zz} - I_{yz}^2}\right)z \tag{1.7}$$

where I_{yy}, I_{zz}, and I_{yz} are the area moment of inertias.

Equation (1.7) is used in the unsymmetrical bending of beams. Note that if either y or z is an axis of symmetry, then $I_{yz} = 0$. In such a case Equation (1.7) simplifies considerably.

1.98 An infinitesimal element in plane stress is shown in Figure P1.98. F_x and F_y are the body forces acting at the point and have the dimensions of force per unit volume. By converting stresses into forces and writing equilibrium equations show that

$$\frac{\partial \sigma_{xx}}{\partial x} + \frac{\partial \tau_{yx}}{\partial y} + F_x = 0 \tag{1.8a}$$

$$\frac{\partial \tau_{xy}}{\partial x} + \frac{\partial \sigma_{yy}}{\partial y} + F_y = 0 \tag{1.8b}$$

$$\tau_{xy} = \tau_{yx} \tag{1.8c}$$

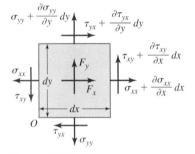

Figure P1.98

*1.3 GENERAL INFORMATION

This section describes briefly the 500 years of struggle in formulating the concept of stress. In hindsight, the long evolution of the concept of stress is not surprising because stress is not a single idea but a package of ideas that may be repackaged in many ways, depending on the needs of the analysis. Our chapter dealt with only one such package, called *Cauchy's stress,* which is used most in engineering design and analysis.

The concept of stress as a quantifier of the strength of a material evolved over a long period of time. The first formal treatment on the subject of strength is seen in the notes of the famous inventor and artist Leonardo da Vinci (1452–1519). Leonardo da Vinci conducted several experiments on the strength of structural materials. In his notes on "testing the strength of iron wires of various lengths" he shows a sketch of how to measure the strength of wire experimentally. The dependence of the strength of a material on length as now recognized is due to variations in manufacturing defects along the length.

The first indication of the concept of stress is found in Galileo (1564–1642). Galileo was born in Pisa and became a professor of mathematics at the age of twenty-five. For his belief in the Copernican theory on the motion of planets, which contradicted the interpretation of scriptures at that time, Galileo was put under house arrest for the last eight years of his life. During that period he wrote the book *Two New Sciences,* in which he discusses his various contributions to the field of mechanics. One such discussion is on the strength of a cantilever beam bending under the action of its own weight. Galileo viewed strength as the absolute resistance to fracture and concluded that the strength of a bar depends on its cross-sectional area but is independent of its length. We thus see that the first rudiments of the concept of stress are visible in Galileo's work. We will discuss Galileo's work on beam bending in Section 6.9.

Coulomb (1736–1806) was the first person to differentiate between normal stress and shear stress. Charles-Augustin Coulomb was born in Angoulême. In 1781 he received the French Academy of Sciences award for his memoir *Theorie des machines simples,* in which he discussed friction between bodies. The theory of dry friction is named after him. Considering the similarities between shear stress and friction it seems natural that Coulomb would be the first to understand and differentiate between normal stress and shear stress. We will see other works of Coulomb in the chapters on failure theory and on the torsion of circular shafts.

Navier (1785–1857) started with Newton's concept of central force between two particles and initiated the mathematical development of the concept of stress, which Cauchy brought to its final form. Results obtained using Navier's approach and those using other methods led to a controversy that took eighty years to be resolved, as we shall see in Section 3.12.

Cauchy (1789–1857) has been credited with the concept of stress as we studied it in this chapter. Augustin Cauchy (Figure 1.45) was born in Paris but had to leave during the French Revolution. He lived in the village of Arcueil where other famous mathematicians and scientists of the period were taking refuge. At the age of twenty-one Cauchy was doing engineering work at the port of Cherbourg, which certainly must have enhanced his understanding of the hydrodynamic concept of pressure. Pressure acts always normal to the surface. But Cauchy assumed that on an internal surface it acts at an angle and hence can be resolved into components, which are normal stress and shear stress. Combining this idea with his natural mathematical abilities, Cauchy developed a concept of stress called Cauchy's stress. We shall see Cauchy's genius in chapters on strain, material properties, and stress and strain transformation.

Figure 1.45   Augustin Cauchy.

We see that unlike force, which is indivisible into more elementary ideas, the concept of stress is a package of ideas. This package can be assembled with related but different elementary ideas. If instead of using the cross-sectional area of an undeformed body, we use the cross-sectional area of a deformed body, then we get *true stress.* If we use the cross-sectional area of a deformed body and take the image (component) of this area in the undeformed configuration, then we get *Kirchhoff's stress.* There are other stress measures that can be defined and are used in nonlinear analysis.

The fact that the symmetry of shear stress as given by Equations (1.4a) through (1.4c) is a consequence of there being no body moments was recognized by Maxwell (1831–1879). If a body moment is present, such as in electromagnetic fields, then shear stresses will not be symmetric.

In Figure 1.15 we replaced the internal forces on a particle by a resultant force but no moment because we started with the argument that the force between two particles was along the line joining the two particles (central force). Voigt (1850–1919) did extensive work with crystals and is credited with introducing the concept of stress tensor. Voigt recognized that in some cases a moment (couple) vector should be included when representing the interaction between particles by equivalent internal loads, as shown in Figure 1.15b. If stress analysis is being conducted at a very small scale,[11] then the moment transmitted by bonds between molecules may need to be included in the concept of stress. *Couple stress* is the term sometimes used to indicate that the concept of stress includes the presence of a moment (couple) vector.

The discussion in this section points out that there are many definitions of stress. We choose the definition depending on the problem and the information we are seeking from our analysis. Most engineering analysis is linear and deals with large bodies for which Cauchy's stress gives very good results. Cauchy's stress is sometimes referred to as engineering stress. Unless stated otherwise, the term "stress" always implies Cauchy's stress in the field of mechanics of materials and in this book.

1.4 CLOSURE

In this chapter we have established the linkage between stresses, internal forces and moments, and external forces and moments. We have seen that to replace stresses by internal forces and internal moments requires knowledge of how the stress varies at each point on the surface. Although we can deduce simple stress behavior on a cross section, we would like to have other alternatives, in particular ones in which the danger of assuming physically impossible deformations is eliminated. This can be achieved if we can establish a relationship between stresses and deformations. Before we can discuss this relationship we need to understand the measure of deformation, which is the subject of Chapter 2. We will then relate stresses and strains in Chapter 3 and finally complete the logical sequence in Section 3.2, which discusses the logic in mechanics of materials.

All analyses in mechanics are conducted in a coordinate system, which is chosen for simplification (whenever possible). Thus the stresses we obtain are in a given coordinate system. Now our motivation for learning about stress is to define a measure of strength. Thus we can conclude that material will fail when the stress at a point reaches some critical maximum value. There is no reason to expect that the stresses will be maximum in the arbitrarily chosen coordinate system. To determine the maximum stress at a point thus implies that we establish a relationship between stresses in different coordinate systems, as we shall do in Chapter 8.

We have seen that the concept of stress is a difficult one. If this concept is to be internalized so that an intuitive understanding is developed, then it is imperative that a discipline be developed to visualize the imaginary surface on which the stress is being considered.

---

[11]Nanostructures are a new research frontier where material strength and other properties are being manipulated by manipulating individual atoms.

POINTS AND FORMULAS TO REMEMBER

- Stress is an internal quantity.
- The internally distributed force on an imaginary cut surface of a body is called stress on a surface.
- Stress has units of force per unit area.
- 1 psi is equal to 6.95 kPa, or approximately 7 kPa. 1 kPa is equal to 0.145 psi, or approximately 0.15 psi.
- The internally distributed force that is normal (perpendicular) to the surface of an imaginary cut is called normal stress on a surface.
- Normal stress that pulls the surface away from the body is called tensile stress.
- Normal stress that pushes the surface into the body is called compressive stress.
- Normal stress is reported as tensile or compressive and not as positive or negative.
- The normal stress acting in the direction of the axis of a slender member is called axial stress.
- The compressive normal stress that is produced when one surface presses against another is called bearing stress.
- The internally distributed force that is parallel (tangent) to the surface of an imaginary cut surface is called shear stress on the surface.
- Stress on a surface:

$$\sigma_{av} = N/A \quad (1.1) \qquad \tau_{av} = V/A \quad (1.2)$$

  where σ_{av} is the average normal stress, τ_{av} is the average shear stress, N is the internal normal force, V is the internal shear force, and A is the cross-sectional area of the imaginary cut on which N and V act.
- The relationship of external forces (and moments) to internal forces and the relationship of internal forces to stress distributions are two distinct ideas.
- Stress at a point:

$$\sigma_{ij} = \lim_{\Delta A_i \to 0}\left(\frac{\Delta F_j}{\Delta A_i}\right) \quad (1.3)$$

  where i is the direction of the outward normal to the imaginary cut surface, and j is the outward normal to the direction of the internal force.
- Stress at a point needs a magnitude and two directions to specify it, i.e., stress at a point is a second-order tensor.
- The first subscript on stress denotes the direction of the outward normal of the imaginary cut surface. The second subscript denotes the direction of the internal force.
- The sign of a stress component is determined from the direction of the internal force and the direction of the outward normal to the imaginary cut surface.

- Stress element is an imaginary object that helps us visualize stress at a point by constructing surfaces that have outward normals in the coordinate directions.

$$\tau_{xy} = \tau_{yx} \quad (1.4a) \qquad \tau_{yz} = \tau_{zy} \quad (1.4b) \qquad \tau_{zx} = \tau_{xz} \quad (1.4c)$$

- Shear stress is symmetric.
- In three dimensions there are nine stress components, but only six are independent.
- In two dimensions there are four stress components, but only three are independent.
- The pair of symmetric shear stress components point either toward the corner or away from the corner on a stress element.
- In plane stress all stress components on a plane are zero, i.e., all stress components with subscript z are zero.
- A point on a free surface is said to be in plane stress.

CHAPTER TWO
STRAIN

2.0 OVERVIEW

The explosion of the space shuttle Challenger (Figure 2.1a) on January 28, 1986, was attributed to a combustible gas leak in the O-ring joint near the bottom of the right solid rocket booster. The solid rocket boosters shown on the shuttle Atlantis, like the Challenger, are assembled using the O-ring joints illustrated in Figure 2.1c. Due to deformation, the gap in the joint was significantly in excess of the allowable design value, causing the combustible gas to escape. Strain is a measure of the intensity of deformation, which is an important variable in the development of formulas used in the design against deformation failures.

The change in the shape of a structure can be described by the displacements of points on the structure. The relationship of strain to displacement is a problem in geometry (kinematics). As depicted in Figure 2.2, this problem represents a link in the logical chain of relating displacements to external forces, which is discussed in greater

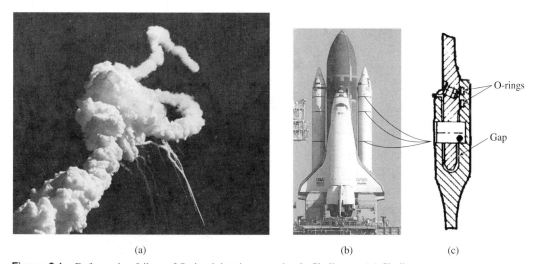

(a)   (b)   (c)

Figure 2.1   Deformation failure of O-ring joints in space shuttle Challenger. (*a*) Challenger explosion during flight. (*b*) Shuttle Atlantis. (*c*) O-ring joint.

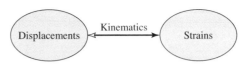

Figure 2.2 Strains and displacements.

detail in Section 3.2. Drawing an approximate deformed body shape is the primary analysis tool used to obtain relationships between displacements and strains. The drawing of an approximate deformed shape is analogous to drawing a free-body diagram to obtain forces. Drawing an approximate deformed shape is also important for the development of an intuitive understanding of displacements and strains.

The two major learning objective are:

1. Understanding the concept of strain.
2. Understanding the use of approximate deformed shapes for calculating strains from displacements.

2.1 DISPLACEMENT AND DEFORMATION

The motion of points on a body due to applied forces are of two types: (i) the body as a whole moves without changing shape, that is, motion of a rigid body, and (ii) motion due to a change of shape of the body, that is, motion due to deformation. The question we confront is: as observers, how do we decide if a moving body is undergoing deformation?

A rigid body by definition is a body in which the distance between any two points does not change. If we observe the trajectory of any two points on the body and the distance between the trajectories does not change, then we will have two parallel trajectories. In other words, the motion is that of the translation of a rigid body. If an observer notices that the distance between the trajectories of two points is changing, then the body is deforming.

In addition to translation, a rigid body can also rotate. In the rotation of rigid bodies all lines on the body rotate by equal amounts. If the motion of two lines, however, reveals that the angle between the lines is changing, then the body is deforming.

Whether it is the distance between two points or the angle between two lines that is changing, deformation can only be described in terms of the relative movements of points on the body. This leads to the following distinction between displacement and deformation.

Definition 1 The total movement of a point with respect to fixed reference coordinates is called *displacement*. The relative movement of a point with respect to another point on the body is called *deformation*.

Several examples and problems in this chapter will emphasize the distinction between deformation and displacement.

2.2 LAGRANGIAN AND EULERIAN STRAIN

A handbook cost $L_0 = \$100$ a year ago. Today it costs $L_f = \$125$. What is the percentage change in the price of the handbook?

Either of the two answers is correct. (i) The book costs 25% *more* than what it cost a year ago. (ii) The book cost 20% *less* a year ago than what it costs today.

The first answer was computed using the original value as a reference value as follows: change $= [(L_f - L_0)/L_0] \times 100$. The second answer was computed using the final value as the reference value as follows: change $= [(L_0 - L_f)/L_f] \times 100$. The two arguments emphasize that in describing change it is necessary to specify the reference value from which change is to be calculated. In the contexts of deformation and strain, this leads to the following definition:

Definition 2 *Lagrangian strain* is computed from deformation by using the original undeformed geometry as the reference geometry. *Eulerian strain* is computed from deformation by using the final deformed geometry as the reference geometry.

The Lagrangian description is usually used in solid mechanics. The Eulerian description is usually used in fluid mechanics. When a material undergoes very large deformations, such as in soft rubber or projectile penetration of metals, then either description may be used, depending on the need of the analysis. We will use Lagrangian strain in this book, except in a few "stretch yourself" problems.

2.3 AVERAGE STRAIN

In Section 2.1 we saw that to differentiate the motion of a point due to translation from deformation we need to measure changes in length. To differentiate the motion of a point due to rotation from deformation we need to measure changes in angle. In this section we discuss the two variables *normal* strain and *shear* strain, which are measures of changes in length and angle, respectively.

2.3.1 Normal Strain

Figure 2.3 shows a line on the surface of a balloon that grows from its original length L_0 to its final length L_f as the balloon expands. The change in length $L_f - L_0$ represents the deformation of the line, which after being divided by its original length gives the intensity of deformation. This intensity of change in length is called the *average* normal strain and is described by the equation

$$\varepsilon_{\text{av}} = \frac{L_f - L_0}{L_0} \tag{2.1}$$

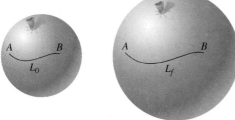

Figure 2.3 Normal strain and change in length.

where ε is the Greek symbol epsilon used to designate normal strain and the subscript av emphasizes that the normal strain is an average value. The following sign convention follows from Equation (2.1).

Definition 3 Elongations ($L_f > L_0$) result in positive normal strains. Contractions ($L_f < L_0$) result in negative normal strains.

An alternative form of Equation (2.1) is

$$\varepsilon_{av} = \frac{\delta}{L_0} \qquad (2.2)$$

where the Greek letter delta (δ) designates deformation of the line and is equal to $L_f - L_0$.

We now consider a special case in which the displacements are in the direction of a straight line. Consider two points A and B on a line in the x direction, as shown in Figure 2.4. Points A and B move to A_1 and B_1, respectively. The coordinates of the point change from x_A and x_B to $x_A + u_A$ and $x_B + u_B$, respectively. From Equation (2.1) and Figure 2.4 we obtain

$$\varepsilon_{av} = \frac{u_B - u_A}{x_B - x_A} \qquad (2.3)$$

where u_A and u_B are the displacements of points A and B, respectively. Hence $u_B - u_A$ is the relative displacement, that is, it is the deformation of the line.

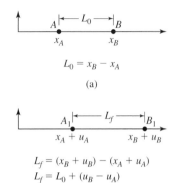

$L_0 = x_B - x_A$

(a)

$L_f = (x_B + u_B) - (x_A + u_A)$
$L_f = L_0 + (u_B - u_A)$

(b)

Figure 2.4 Normal strain and displacement.

2.3.2 Shear Strain

Figure 2.5 shows an elastic band with a grid attached to two wooden bars using masking tape. The top wooden bar is slid to the right causing the grid to deform. As can be seen, the angle between lines ABC changes. The measure of this change of angle is defined by shear strain, usually designated by the Greek symbol gamma (γ). The average Lagrangian shear strain is defined as the change of angle from a right angle, as shown:

$$\gamma_{av} = \frac{\pi}{2} - \alpha \qquad (2.4)$$

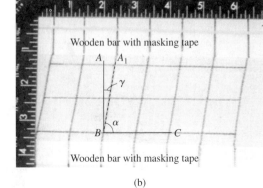

(a)   (b)

Figure 2.5 Shear strain and angle changes. (*a*) Undeformed grid. (*b*) Deformed grid.

where the Greek symbol alpha (α) designates the final angle measured in radians (rad), and the Greek symbol pi (π) equals 3.14159 rad.

Definition 4 Decreases in angle ($\alpha < \pi/2$) result in positive shear strains. Increases in angle ($\alpha > \pi/2$) result in negative shear strains.

2.3.3 Units of Average Strain

Equation (2.1) shows that normal strain is dimensionless, hence should have no units. However, to differentiate average strain and strain at a point (discussed in Section 2.5), units of length per unit length are used for average normal strains. Thus average normal strains are reported with units such as in/in, cm/cm, or m/m. Radians are used as units in reporting average shear strains.

In reporting experimental results and to describe very large deformations, a percentage change is used for strains, that is, the right side in Equation (2.1) or (2.4) is multiplied by 100 before reporting the results. Thus a normal strain of 0.5% is equal to a strain of 0.005. In reporting small strains the Greek symbol mu (μ) is used to represent the prefix of micro ($\mu = 10^{-6}$). Thus a strain of 1000 μin/in is the same as a normal strain of 0.001 in/in.

EXAMPLE 2.1

The displacements of the rigid plates in the x direction due to a set of axial forces were observed as follows:

Determine the axial strains in the rods in sections AB, BC, and CD (Figure 2.6).

$$u_A = -0.0100 \text{ in}$$
$$u_B = 0.0080 \text{ in}$$
$$u_C = -0.0045 \text{ in}$$
$$u_D = 0.0075 \text{ in}$$

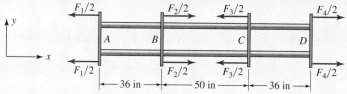

Figure 2.6 Axial displacements in Example 2.1.

PLAN
We calculate the relative movement of the points in each section, and then we can calculate the normal strains from Equation (2.3).

Solution
$$\varepsilon_{AB} = \frac{u_B - u_A}{x_B - x_A} = \frac{0.018}{36}$$

ANS. $\varepsilon_{AB} = 500 \ \mu\text{in/in}$

$$\varepsilon_{BC} = \frac{u_C - u_B}{x_C - x_B} = \frac{-0.0125}{50}$$

ANS. $\varepsilon_{BC} = -250 \ \mu\text{in/in}$

$$\varepsilon_{CD} = \frac{u_D - u_C}{x_D - x_C} = \frac{0.012}{36}$$

ANS. $\varepsilon_{CD} = 333.3 \ \mu\text{in/in}$

COMMENT
The problem brings out the difference between the displacements, which were given, and the deformations, which we calculated before finding the strains.

EXAMPLE 2.2
A bar of hard rubber is attached to a rigid bar, which is moved to the right as shown in Figure 2.7. Determine the average shear strain at point A.

PLAN
The rectangle will become a parallelogram as the rigid bar moves. We can draw an approximate deformed shape and calculate the change of angle to determine the shear strain.

Solution   Point B moves to point B_1, as shown in Figure 2.8. The right angle that is changed is the angle between AB and the horizontal line.

$$\tan \gamma = \frac{BB_1}{AB} = \frac{0.5}{100}$$

ANS. $\gamma = 5000 \ \mu\text{rad}$

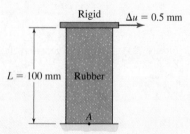

Figure 2.7   Geometry in Example 2.2.

COMMENTS

1. In drawing the deformed shape we assumed that line *AB* remained straight during the deformation. If this assumption were not valid, then the shear strain would vary in the vertical direction. To determine the varying shear strain we would need additional information. Thus our assumption of line *AB* remaining straight is the simplest assumption that accounts for the given information.

2. The value of γ and tan γ is the same, which occurs whenever the argument of the tangent function is small. Thus for small shear strains the tangent function can be approximated by its argument.

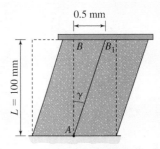

Figure 2.8 Exaggerated deformed shape.

EXAMPLE 2.3

A thin ruler, 12 in long, is deformed into a circular arc with a radius of 30 in that subtends an angle of 23° at the center. Determine the average normal strain in the ruler.

PLAN

We know the original length. The final length is the length of a circular arc, which we can find. We then calculate the normal strain from Equation (2.1).

Solution   The original length $L_0 = 12$ in. The angle subtended by the circular arc shown in Figure 2.9 can be found in terms of radians as $\Delta\theta = [(23°)\pi]/180° = 0.4014$ rad. The length of the arc can be found, from which the normal strain can be calculated as follows:

$$L_f = R\Delta\theta = 12.04277 \text{ in}$$
$$\varepsilon = \frac{L_f - L_0}{L_0} = \frac{0.04277}{12}$$

ANS. $\varepsilon = 3564 \ \mu\text{in/in}$

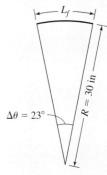

Figure 2.9 Deformed geometry in Example 2.3.

COMMENTS

1. In Example 2.1 the normal strain was generated by the displacements in the axial direction. In this example the normal strain is being generated by bending.

2. In Chapter 6 on the symmetric bending of beams we shall consider a beam made up of lines that will bend like the ruler and calculate the normal strain due to bending as we calculated it in this example.

EXAMPLE 2.4

A belt and a pulley system in a VCR has the dimensions shown in Figure 2.10. To ensure adequate but not excessive tension in the belts, the average normal strain in the belt must be a minimum of 0.019 mm/mm and a maximum of 0.034 mm/mm. What should be the minimum and maximum undeformed lengths of the belt to the nearest millimeter?

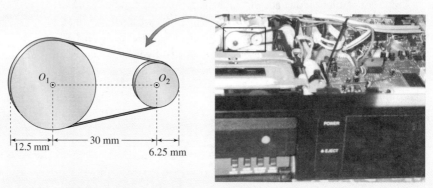

Figure 2.10 Belt and pulley in a VCR.

PLAN

The belt must be tangent at the point where it comes in contact with the pulley. The deformed length of the belt is the length of belt between the tangent points on the pulleys, plus the length of belt wrapped around the pulleys. To find the length of belt wrapped around a pulley we need to find the angle subtended by the tangent points at the center and multiply it by the radius of each pulley. Once we know the deformed length of the belt, we can find the original length by using Equation (2.1) and the given limits on normal strain.

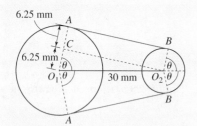

Figure 2.11 Analysis of geometry.

Solution We can draw the radial lines from the center to the tangent points A and B, as shown in Figure 2.11. The radial lines O_1A and O_2B must be perpendicular to the belt AB, hence both lines are parallel and at the same angle θ with the horizontal. We can draw a line parallel to AB through point O_2 to get line CO_2. Noting that CA is equal to O_2B, we can obtain CO_1 as the difference between the two radii. Triangle O_1CO_2 is a right-angle triangle, and we can find side CO_2 and the angle θ as follows:

$$AB = CO_2 = \sqrt{30^2 - 6.25^2} = 29.342 \text{ mm} \qquad (E1)$$

$$\cos \theta = \frac{CO_1}{O_1O_2} = \frac{6.25}{30}$$

$$\theta = \text{arc cos } 0.2083 = 1.3609 \text{ rad} \qquad (E2)$$

The deformed length L_f of the belt is the sum of arcs AA and BB and twice the length AB. The arc lengths are the angle $(2\pi - 2\theta)$ times the radius of each pulley:

$$AA = 12.5(2\pi - 2\theta) = 44.517 \text{ mm}$$

$$BB = 6.25(2\pi - 2\theta) = 22.258 \text{ mm}$$

$$L_f = 2(AB) + AA + BB = 125.46 \text{ mm}$$

We are given that $0.019 \leq \varepsilon \leq 0.034$. From Equation (2.1) we obtain the limits on the original length:

$$\varepsilon = \frac{L_f - L_0}{L_0} \leq 0.034, \qquad L_0 \geq \frac{125.46}{1 + 0.034}$$

$$L_0 \geq 121.33 \qquad\qquad\qquad \text{(E3)}$$

$$\varepsilon = \frac{L_f - L_0}{L_0} \geq 0.019, \qquad L_0 \leq \frac{125.46}{1 + 0.019}$$

$$L_0 \leq 123.1 \qquad\qquad\qquad \text{(E4)}$$

To satisfy Equations (E3) and (E4) to the nearest millimeter we obtain the following limits on the original length L_0:

$$\text{ANS.} \qquad 122 \text{ mm} \leq L_0 \leq 123 \text{ mm}$$

COMMENTS

1. We rounded upward in Equation (E3) because rounding downward would violate the limit in Equation (E4).

2. Tolerances on dimensions have to be specified for manufacturing. Thus in our problem we have a tolerance range of 1 mm.

3. Though the problem seems more difficult than the previous three examples, it should be noted that the difficulty is in the analysis of the geometry rather than in the concept of strain. This example once more emphasizes that the analysis of deformation and strain is a problem in geometry, hence drawing the approximate deformed shape is essential in the analysis of strain.

2.4 SMALL-STRAIN APPROXIMATION

In many problems of engineering interest a body undergoes only small deformations. A significant simplification can be achieved by making the approximation of small strains, as demonstrated by the simple example shown in Figure 2.12. Due to a force acting on the bar, point P moves by an amount D at an angle θ to the direction of the

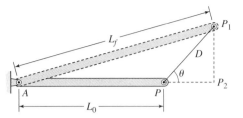

Figure 2.12 Small normal-strain calculations.

bar. Using the cosine rule in triangle APP_1 the length L_f can be found in terms of L_0, D, and θ as

$$L_f = \sqrt{L_0^2 + D^2 + 2L_0 D \cos\theta} = L_0 \sqrt{1 + \left(\frac{D}{L_0}\right)^2 + 2\left(\frac{D}{L_0}\right)\cos\theta}$$

From Equation (2.1) we obtain the average normal strain in bar AP as

$$\varepsilon = \frac{L_f - L_0}{L_0} = \sqrt{1 + \left(\frac{D}{L_0}\right)^2 + 2\left(\frac{D}{L_0}\right)\cos\theta} - 1 \qquad (2.5)$$

Equation (2.5) is valid regardless of the magnitude of the deformation. Now consider the situation where D/L_0 is small. In such a case we can neglect the $(D/L_0)^2$ term and expand the radical by binomial[1] expansion as

$$\varepsilon \approx \left(1 + \frac{D}{L_0}\cos\theta + \cdots + \cdots\right) - 1$$

Neglecting the higher order terms, we obtain an approximation for small strain,

$$\varepsilon_{small} = \frac{D\cos\theta}{L_0} \qquad (2.6)$$

In Equation (2.6) deformation and strain are linearly related, whereas in Equation (2.5) strain and deformation are nonlinearly related. This implies that small-strain calculations will result in a linear system of analysis, which is a significant simplification.

    Another interesting observation from Equation (2.6) is that small-strain calculation requires the component of deformation in the direction of the original line element. This is a very useful observation, and we shall use it significantly. Another way of looking at small-strain approximation is to say that the deformed length AP_1 is being approximated by the length AP_2.

    We now consider the question of what we mean by small strain. We compare strains from Equation (2.6) to those from Equation (2.5) in the following manner. For different values of small strain and for $\theta = 45°$ the ratio of D/L is found from Equation (2.6), and the strain from Equation (2.5) is calculated as shown in Table 2.1. Equation (2.6) is an approximation of Equation (2.5), and the error in the approximation is shown in the third column of Table 2.1. It is seen from Table 2.1 that when the strain is less than 0.01, then the error is less than 1%, which is acceptable for most engineering analyses.

---

[1]For small d the binomial expansion can be written as $(1 + d)^{1/2} = 1 + d/2 +$ terms d^2 and higher order.

TABLE 2.1   Small-strain approximation

2.4   SMALL-STRAIN
APPROXIMATION

61

ε_{small}, Equation (2.6)	ε, Equation (2.5)	% Error, $\left(\dfrac{\varepsilon - \varepsilon_{small}}{\varepsilon}\right) \times 100$
1.0	1.23607	19.1
0.5	0.58114	14.0
0.1	0.10454	4.3
0.05	0.005119	2.32
0.01	0.01005	0.49
0.005	0.00501	0.25
0.001	0.00100	0.05

We conclude this section with summary of our observations.

1. Small-strain approximation may be used for strains less than 0.01.

2. Small-strain calculations result in linear deformation analysis.

3. Small normal strains are calculated by using the deformation component in the original direction of the line element, regardless of the orientation of the deformed line element.

4. In small shear strain (γ) calculations the following approximations may be used for the trigonometric functions: $\tan \gamma \approx \gamma$, $\sin \gamma \approx \gamma$, and $\cos \gamma \approx 1$.

*2.4.1   Vector Approach to Small-Strain Approximation

To calculate strains from known displacements of the pins in truss problems is algebraically difficult using the small-strain approximation given by Equation (2.6). Similar algebraic difficulties are encountered in three-dimensional problems. To address these difficulties we develop a vector approach, which is procedural but less intuitive than the scalar approach of Equation (2.6).

The deformation of the bar in Equation (2.6) is given by $\delta = D \cos \theta$ and can be written in vector form using the dot product as follows:

$$\delta = \overline{\mathbf{D}}_{AP} \cdot \overline{\mathbf{i}}_{AP} \tag{2.7}$$

where $\overline{\mathbf{D}}_{AP}$ is the deformation vector of the bar AP and $\overline{\mathbf{i}}_{AP}$ is the unit vector in the original direction of bar AP. In Figure 2.12 the vector $\overline{\mathbf{D}}_{AP}$ is also the displacement vector of point P as point A is fixed. If point A is also being displaced, then the deformation vector is obtained by taking the difference between the displacement vectors of point P and point A. If points A and P are two points in space with coordinates (x_A, y_A, z_A) and (x_P, y_P, z_P), respectively, which are displaced by amounts (u_A, v_A, w_A) and (u_P, v_P, w_P) in the x, y, and z directions, respectively, then the deformation vector $\overline{\mathbf{D}}_{AP}$ and the unit vector $\overline{\mathbf{i}}_{AP}$ can be written as

$$\begin{aligned} \overline{\mathbf{D}}_{AP} &= (u_P - u_A)\overline{\mathbf{i}} + (v_P - v_A)\overline{\mathbf{j}} + (w_P - w_A)\overline{\mathbf{k}} \\ \overline{\mathbf{i}}_{AP} &= (x_P - x_A)\overline{\mathbf{i}} + (y_P - y_A)\overline{\mathbf{j}} + (z_P - z_A)\overline{\mathbf{k}} \end{aligned} \tag{2.8}$$

where $\overline{\mathbf{i}}$, $\overline{\mathbf{j}}$, and $\overline{\mathbf{k}}$ are the unit vectors in the x, y, and z directions, respectively. The important point to remember about the calculation of $\overline{\mathbf{D}}_{AP}$ and $\overline{\mathbf{i}}_{AP}$ is the following: the same reference point must be used in the calculation of the deformation vector and the unit vector.

- Example 2.5 presents strain calculations without small-strain approximation and with small-strain approximation using Equations (2.6) and (2.7).
- Example 2.6 demonstrates the use of small-strain approximation in a simple structure.
- Example 2.7 demonstrates small-strain approximation for shear strain.
- Example 2.8 demonstrates the procedure of using Equation (2.7) for small-strain calculations in a truss problem.

EXAMPLE 2.5

Two bars are connected to a roller that slides in a slot, as shown in Figure 2.13. Determine the strains in bar AP by:

(a) Finding the deformed length of AP without small-strain approximation.
(b) Using Equation (2.6).
(c) Using Equation (2.7).

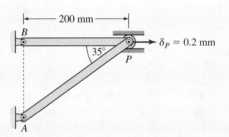

Figure 2.13   Small-strain calculations.

PLAN

(a) We can draw an exaggerated deformed shape of the two bars and find the deformed length of bar AP using the cosine rule.
(b) We can draw the exaggerated deformed shape and drop a perpendicular from the final position of point P onto the original direction of bar AP and, using geometry, find the deformation of bar AP.
(c) We can find the unit vector in the direction of AP and take the dot product with the displacement vector of point P, which is the same as the deformation vector of bar AP.

The length AP used in all three methods can be found as $AP = 200/\cos 35 = 244.155$ mm.

Solution

(a) Let point P move to point P_1, as shown in Figure 2.14. The angle APP_1 is $145°$. From the triangle APP_1 we can find the length AP_1

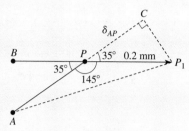

Figure 2.14 Exaggerated deformed shape.

using the cosine formula and find the strain using Equation (2.1),

$$AP_1 = \sqrt{AP^2 + PP_1^2 - 2(AP)(PP_1)\cos 145} = 244.3188$$

ANS. $\qquad \varepsilon_{AP} = \dfrac{AP_1 - AP}{AP} = \dfrac{0.1639}{244.15} = 671.2 \ \mu\text{mm/mm}$

(b) We want the component of PP_1 in the direction of AP. We drop a perpendicular from P_1 onto the line in direction AP and find the deformation of AP as

$$\delta_{AP} = 0.2\cos 35 = 0.1638$$

ANS. $\qquad \varepsilon_{AP} = \dfrac{0.1638}{244.155} = 671.01\mu\text{mm/mm}$

(c) Let the unit vectors in the x and y directions be given by $\bar{\mathbf{i}}$ and $\bar{\mathbf{j}}$. We can find the unit vector in direction AP and the deformation vector $\bar{\mathbf{D}}$ and calculate the strain using Equation (2.7),

$$\bar{\mathbf{i}}_{AP} = \cos 35\,\bar{\mathbf{i}} + \sin 35\,\bar{\mathbf{j}}, \qquad \bar{\mathbf{D}} = 0.2\bar{\mathbf{i}},$$

$$\delta_{AP} = \bar{\mathbf{D}} \cdot \bar{\mathbf{i}}_{AP} = 0.2\cos 35 = 0.1638$$

ANS. $\qquad \varepsilon_{AP} = \dfrac{0.1638}{244.155} = 671.01 \ \mu\text{mm/mm}$

COMMENTS

1. The calculations for parts (b) and (c) are identical since there is no difference in the approximation between the two approaches. The strain value for part (a) differs from that in parts (b) and (c) by 0.028% which, for most engineering calculations, is insignificant.

2. The implication of a small-strain approximation is that the final length AP_1 is being approximated by length AC.

EXAMPLE 2.6

A gap of 0.18 mm exists between the rigid plate and bar B before the load P is applied on the system shown in Figure 2.15. After load P is applied, the axial strain in rod B is $-2500\ \mu m/m$. Determine the axial strain in rods A.

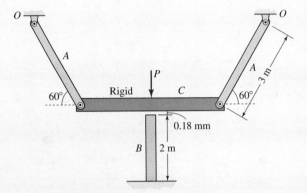

Figure 2.15   Undeformed geometry in Example 2.6.

PLAN

From the strain in bar B the deformation of bar B can be found and related to the displacement of the rigid plate by drawing an approximate deformed shape. We can then relate the displacement of the rigid plate to the deformation of bar A using small-strain approximation.

Solution   From the given strain of bar B we can find the deformation of bar B as $\delta_B = \varepsilon_B L_B = 0.005$ m. Let points D and E be points on the rigid plate. Let the position of these points be D_1 and E_1 after the load P has been applied, as shown in Figure 2.16. From geometry we can relate

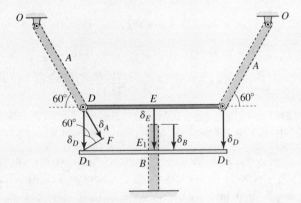

Figure 2.16   Deformed position.

the movement of point E to the deformation of bar B, as given by Equation (E1) and find the displacement of point E. The displacements of points D and E are the same as the rigid plate moves horizontally downward without rotation and can be calculated as shown by Equation (E2). We can drop a perpendicular from D_1 (final position) to the line in the original direction OD and relate the deformation of bar A to the displacement of point D, as shown by Equation (E3):

$$\delta_E = \delta_B + 0.00018 = 0.00518 \text{ m} \qquad \text{(E1)}$$

$$\delta_D = \delta_E = 0.00518 \text{ m} \qquad \text{(E2)}$$

$$\delta_A = \delta_D \sin 60 \qquad \text{(E3)}$$

We can then find the strain in A using Equation (2.2),

$$\delta_A = 0.00518 \sin 60 = 0.004486 \text{ m}$$

$$\varepsilon_A = \frac{\delta_A}{L_A} = \frac{0.004330}{3}$$

ANS.　　$\varepsilon_A = 1495 \ \mu\text{m/m}$

COMMENTS

1. Equation (E2) is the relationship of points on the rigid bar, whereas Equations (E1) and (E3) are the relationship between the movement of points on the rigid bar and the deformation of the bar. This two-step process simplifies deformation analysis as it reduces the possibility of mistakes in the calculations.

2. We dropped the perpendicular from D_1 to OD and not from D to OD_1 because OD is the original direction, and not OD_1.

EXAMPLE 2.7

Two bars of hard rubber are attached to a rigid disk of radius 20 mm (Figure 2.17). The rotation of the rigid disk by an angle $\Delta\phi$ causes a shear strain at point A of 2000 μrad. Determine the rotation $\Delta\phi$ and the shear strain at point C.

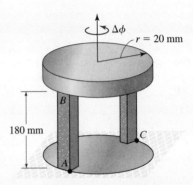

Figure 2.17 Geometry in Example 2.7.

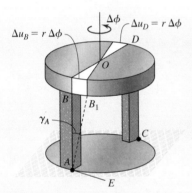

Figure 2.18 Deformed geometry.

PLAN

We can relate the displacement of point B to shear strain at point A as we did in Example 2.2. The rotation of the rigid disk will cause all radial lines on the disk to rotate by equal amounts, that is, by $\Delta\phi$. If we use this information and assume small strains and deformation, we can relate the displacement of point B to $\Delta\phi$ and hence find $\Delta\phi$. We repeat the calculation for the bar at C to find the strain at C.

Solution The shear strain at A is $\gamma_A = 2000\ \mu = 0.002$. We draw the approximated deformed shape of the two bars as shown in Figure 2.18. The displacement of point B is approximately equal to the arc length BB_1, which is related to the rotation of the disk, as shown in Figure 2.18 and given by Equation (E1). The displacement of point B can also be related to the shear strain at A, and we can find $\Delta\phi$ as shown by Equations (E2) and (E3).

$$\Delta u_B = 20(\Delta\phi) \tag{E1}$$

$$\tan\gamma_A \approx \gamma_A = \frac{BB_1}{AB} = 20\frac{\Delta\phi}{180} \tag{E2}$$

$$\Delta\phi = 9\gamma_A = 9 \times 0.002 \tag{E3}$$

ANS. $\Delta\phi = 0.018$ rad

Knowing $\Delta\phi$, we can find the displacement of point D and divide it by the length of CD to find the shear strain at C,

$$\gamma_C = \frac{\Delta u_C}{CD} = \frac{20 \times 0.018}{180}$$

ANS. $\gamma_C = 2000\ \mu$rad

COMMENTS

1. We approximated the arc BB_1 by a straight line, which is only valid if the deformations are small.

2. The change in angle that we used in determining the shear strain is the angle formed by the tangent line AE and the axial line AB.

3. In Chapter 5, on the torsion of circular shafts, we will consider a shaft made up of bars and calculate shear strain due to torsion as we calculated it in this example.

---

*EXAMPLE 2.8

The displacements of pins u and v in the x and y directions, respectively, were computed by the finite-element method[2] and are given as

$$u_B = 2.700 \text{ mm} \qquad v_B = -9.025 \text{ mm}$$

$$u_C = 5.400 \text{ mm} \qquad v_C = -14.000 \text{ mm}$$

$$u_G = 8.000 \text{ mm} \qquad v_G = -14.000 \text{ mm}$$

$$u_H = 9.200 \text{ mm} \qquad v_H = -9.025 \text{ mm}$$

Determine the axial strains in members BC, HB, HC, and HG (Figure 2.19).

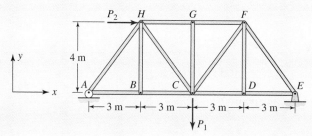

Figure 2.19 Truss in Example 2.8.

---

[2]The finite-element method is a computer method in which displacements inside an element (such as axial members) are described by polynomials, and coefficients of the polynomials are displacements of points (such as pins) which are determined numerically. The method is described briefly in Section 4.9.

PLAN

We can find the deformation vectors for each of the bars from the given displacements. The unit vectors for bars HG and BC are $\overline{\mathbf{i}}$, the unit vector in the x direction. The unit vector for rod BH is $\overline{\mathbf{j}}$, the unit vector in the y direction. We can find the unit vector for bar HC by first finding the position vector and dividing it by the magnitude of the position vector. The deformation of each bar can be found using Equation (2.7). We then find the strains for each bar using Equation (2.2).

Solution   Let the unit vectors in the x and y directions be given by $\overline{\mathbf{i}}$ and $\overline{\mathbf{j}}$, respectively. The deformation vectors for each bar can be found for the given displacement as

$$\overline{\mathbf{D}}_{BC} = (u_C - u_B)\overline{\mathbf{i}} + (v_C - v_B)\overline{\mathbf{j}} = 2.7\,\overline{\mathbf{i}} - 4.975\,\overline{\mathbf{j}}\,\text{mm}$$

$$\overline{\mathbf{D}}_{HB} = (u_B - u_H)\overline{\mathbf{i}} + (v_B - v_H)\overline{\mathbf{j}} = -6.5\,\overline{\mathbf{i}}\,\text{mm}$$

$$\overline{\mathbf{D}}_{HC} = (u_C - u_H)\overline{\mathbf{i}} + (v_C - v_H)\overline{\mathbf{j}} = -3.8\,\overline{\mathbf{i}} - 4.975\,\overline{\mathbf{j}}\,\text{mm}$$

$$\overline{\mathbf{D}}_{HG} = (u_G - u_H)\overline{\mathbf{i}} + (v_G - v_H)\overline{\mathbf{j}} = -1.2\,\overline{\mathbf{i}} - 4.975\,\overline{\mathbf{j}}\,\text{mm}$$

The unit vectors in the directions of bars BC, HB, and HG can be written by inspection as these bars are horizontal or vertical. The position vector from point H to C is $\overline{\mathbf{HC}} = 3\overline{\mathbf{i}} - 4\overline{\mathbf{j}}$. By dividing the position vector by its magnitude, we can obtain the unit vector in the direction of bar HC,

$$\overline{\mathbf{i}}_{BC} = \overline{\mathbf{i}} \qquad \overline{\mathbf{i}}_{HB} = -\overline{\mathbf{j}} \qquad \overline{\mathbf{i}}_{HG} = \overline{\mathbf{i}}$$

$$\overline{\mathbf{i}}_{HC} = \frac{HC}{|\overline{\mathbf{HC}}|} = \frac{3\overline{\mathbf{i}} - 4\overline{\mathbf{j}}}{\sqrt{3^2 + 4^2}} = 0.6\,\overline{\mathbf{i}} - 0.8\,\overline{\mathbf{j}}$$

We can find the deformation of each bar from Equation (2.7),

$$\delta_{HC} = \overline{\mathbf{D}}_{BC} \cdot \overline{\mathbf{i}}_{BC} = 2.7\ \text{mm}$$

$$\delta_{HG} = \overline{\mathbf{D}}_{HG} \cdot \overline{\mathbf{i}}_{HG} = -1.2\ \text{mm}$$

$$\delta_{HB} = \overline{\mathbf{D}}_{HB} \cdot \overline{\mathbf{i}}_{HB} = 0$$

$$\delta_{HC} = \overline{\mathbf{D}}_{HC} \cdot \overline{\mathbf{i}}_{HC} = 0.6 \times -3.8 + -4.975 \times -0.8 = 1.7\ \text{mm}$$

The strains in each bar can be found from Equation (2.2),

$$\varepsilon_{BC} = \frac{\delta_{BC}}{L_{BC}} = \frac{2.7}{3 \times 10^3} = 0.9 \times 10^{-3}$$

ANS. $\varepsilon_{BC} = 900 \ \mu\text{mm/mm}$

$$\varepsilon_{HG} = \frac{\delta_{HG}}{L_{HG}} = \frac{-1.2}{3 \times 10^3} = -0.4 \times 10^{-3}$$

ANS. $\varepsilon_{HG} = -400 \ \mu\text{mm/mm}$

$$\varepsilon_{HB} = \frac{\delta_{HB}}{L_{HB}} = 0$$

ANS. $\varepsilon_{HB} = 0$

$$\varepsilon_{HC} = \frac{\delta_{HC}}{L_{HC}} = \frac{1.7}{5 \times 10^3} = 0.340 \times 10^{-3}$$

ANS. $\varepsilon_{HC} = 340 \ \mu\text{mm/mm}$

COMMENTS

1. The zero strain in *HB* is not surprising. By looking at joint *B* we can see that *HB* is a zero-force member. Though we have yet to establish the relationship between internal forces and deformation, we know intuitively that if a body deforms, then internal forces will develop.

2. We took a very procedural approach in solving the problem and, as a consequence, did several additional computations. For horizontal bars *BC* and *HG* we could have found the deformation by simply subtracting the *u* components, and for the vertical bar *HB* we can find the deformation by subtracting the *v* component. But care must be exercised in determining whether the bar is in extension or in contraction, otherwise an error in sign can occur.

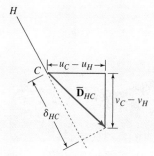

Figure 2.20 Visualization of the deformation vector for bar *HC*.

3. In Figure 2.20 point *H* is held fixed (reference point), and an exaggerated relative movement of point *C* is shown by the vector $\overline{\mathbf{D}}_{HC}$. The calculation of the deformation of bar *HC* is shown graphically.

4. Suppose that instead of finding the relative movement of point *C* with respect to *H*, we had used point *C* as our reference point and found the relative movement of point *H*. The deformation vector would be $\overline{\mathbf{D}}_{CH}$, which is equal to $-\overline{\mathbf{D}}_{HC}$. But the unit vector direction would also reverse, that is, we would use $\overline{\mathbf{i}}_{CH}$, which is equal to $-\overline{\mathbf{i}}_{HC}$. Thus the dot product to find the deformation would yield the same number and the same sign. The result being independent of the point being used as reference is true only for small strains, which we have implicitly assumed in this analysis.

PROBLEM SET 2.1

Average normal strains

2.1 An 80-cm stretch cord is used to tie the rear of a canoe to the car hook, as shown in Figure P2.1. In the stretched position the cord forms the side AB of the triangle shown. Determine the average normal strain in the stretch cord.

2.2 The diameter of a spherical balloon changes from 250 mm to 252 mm (Figure P2.2). Determine the change in the average circumferential normal strain.

2.3 Two rubber bands are used for packing an air mattress for camping (Figure P2.3). The undeformed length of a rubber band is 7 in. Determine the average normal strain in the rubber bands if the diameter of the mattress is 4.1 in at the section where the rubber bands are on the mattress.

2.4 A canoe on top of a car is tied down using rubber stretch cords, as shown in Figure P2.4a. The undeformed length of the stretch cord is 40 in. Determine the average normal strain in the stretch cord assuming that the path of the stretch cord over the canoe can be approximated as shown in Figure P2.4b.

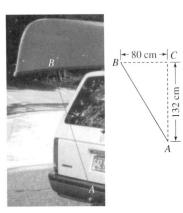

Figure P2.1

Figure P2.2

Figure P2.3

(a)

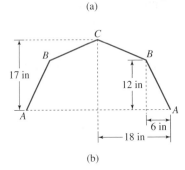

(b)

Figure P2.4

2.5 Due to the application of the forces in Figure P2.5, the displacements of the rigid plates in the x direction were observed as $u_B = -1.8$ mm, $u_C = 0.7$ mm, and $u_D = 3.7$ mm. Determine the axial strains in the rods in sections AB, BC, and CD.

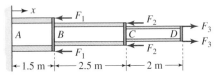

Figure P2.5

2.6 Due to the application of the forces in Figure P2.5, the average normal strains in the bars were found to be $\varepsilon_{AB} = -800\ \mu$, $\varepsilon_{BC} = 600\ \mu$, and $\varepsilon_{CD} = 1100\ \mu$. Determine the movement of point D with respect to the left wall.

2.7 Due to the application of the forces, the rigid plate in Figure P2.7 is observed to move 0.0236 in to the right. Determine the average normal strains in bars A and B.

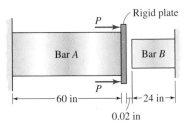

Figure P2.7

2.8 Due to the application of the forces, the average normal strain in bar A in Figure P2.7 was found to be 2000 μin/in. Determine the normal strain in bar B.

2.9 Due to the application of force P, point B in Figure P2.9 was observed to move upward by 0.06 in. If the length of bar A is 24 in, determine the average normal strain in bar A.

2.10 Due to the application of force P, the average normal strain in bar A in Figure P2.9 was found to be $-6000\ \mu$. If the length of bar A is 36 in, determine the movement of point B.

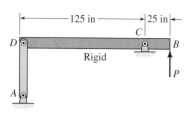

Figure P2.9

2.11 Due to the application of force P, point B in Figure P2.11 was observed to move upward by 0.06 in. If the length of bar A is 24 in, determine the average normal strain in bar A.

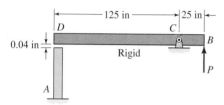

Figure P2.11

2.12 Due to the application of force P, the average normal strain in bar A in Figure P2.11 was found to be $-6000\ \mu$. If the length of bar A is 36 in, determine the movement of point B.

2.13 Due to the application of force P, point B in Figure P2.13 was observed to move upward by 0.06 in. If the lengths of bars A and F are 24 in, determine the average normal strain in bars A and F.

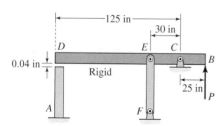

Figure P2.13

2.14 Due to the application of force P, the average normal strain in bar A in Figure P2.13 was found to be $-6000\ \mu$. If the lengths of bars A and F are 36 in, determine the movement of point B and the average normal strain in bar F.

2.15 Due to the application of force P, point B in Figure P2.15 was observed to move left by 0.75 mm. If the length of bar A is 1.2 m, determine the average normal strain in bar A.

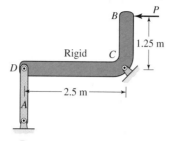

Figure P2.15

2.16 Due to the application of force P, the average normal strain in bar A in Figure P2.15 was found to be $-2000\ \mu$. If the length of bar A is 2 m, determine the movement of point B.

2.17 Due to the application of force P, point B in Figure P2.17 was observed to move left by 0.75 mm. If the length of bar A is 1.2 m, determine the average normal strain in bar A.

2.18 Due to the application of force P, the average normal strain in bar A in Figure P2.17 was found to be $-2000\ \mu$. If the length of bar A is 2 m, determine the movement of point B.

2.19 Due to the application of force P, point B in Figure P2.19 was observed to move left by 0.75 mm. If the lengths of bars A and F are 1.2 m, determine the average normal strains in bars A and F.

2.20 Due to the application of force P, the average normal strain in bar A in Figure P2.19 was found to be $-2000\ \mu$. Bars A and F are 2 m long. Determine the movement of point B and the average normal strain in bar F.

Average shear strains

A rectangular plastic plate deforms into a shaded shape, as shown. In Problems 2.21 through 2.26 determine the average shear strains at point A.

2.21 Determine the average shear strain at point A in Figure P2.21.

2.22 Determine the average shear strain at point A in Figure P2.22.

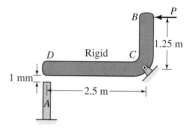

Figure P2.17

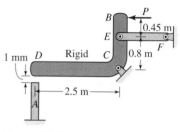

Figure P2.19

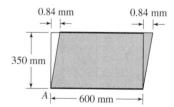

Figure P2.21

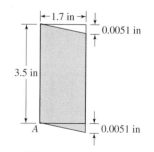

Figure P2.22

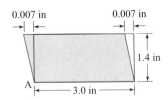

Figure P2.23

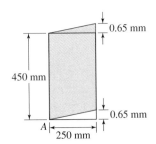

Figure P2.24

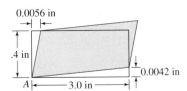

Figure P2.25

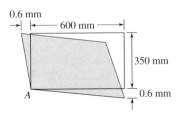

Figure P2.26

2.23 Determine the average shear strain at point A in Figure P2.23.

2.24 Determine the average shear strain at point A in Figure P2.24.

2.25 Determine the average shear strain at point A in Figure P2.25.

2.26 Determine the average shear strain at point A in Figure P2.26.

A thin triangular plate ABC forms a right angle at point A, as shown. During deformation, point A moves vertically down by δ_A, which is given in each problem. Determine the average shear strains at point A in Problems 2.27 through 2.29.

2.27 $\delta_A = 0.005$ in

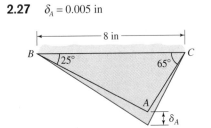

Figure P2.27

2.28 $\delta_A = 0.008$ in

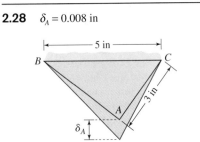

Figure P2.28

2.29 $\delta_A = 0.75$ mm

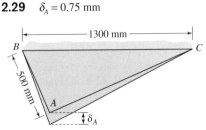

Figure P2.29

A thin triangular plate ABC forms a right angle at point A. During deformation, point A moves

horizontally by δ_A, which is given in each problem. Determine the average shear strains at point A in Problems 2.30 through 2.32.

2.30 $\delta_A = 0.005$ in

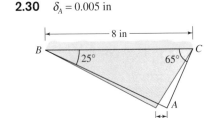

Figure P2.30

2.31 $\delta_A = 0.008$ in

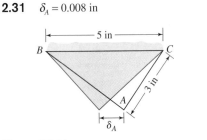

Figure P2.31

2.32 $\delta_A = 0.75$ mm

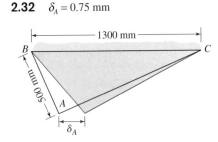

Figure P2.32

The diagonals of two squares form a right angle at point A in Figure P2.33. The two rectangles are pulled horizontally to a deformed shape, shown by colored lines. Determine the average shear strain at point A if the displacements of points A and B are as given in Problems 2.33 and 2.34.

2.33 $\delta_A = 0.4$ mm and $\delta_B = 0.8$ mm in Figure P2.33.

2.34 $\delta_A = 0.3$ mm and $\delta_B = 0.9$ mm in Figure P2.33.

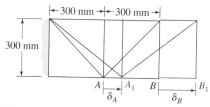

Figure P2.33

Small-strain approximations

In Problems 2.35 and 2.36 the roller at P can slide in the slot only by the given amount. Determine the strains in bar AP by (a) finding the deformed length of AP without small-strain approximation, (b) using Equation (2.6), and (c) using Equation (2.7).

2.35

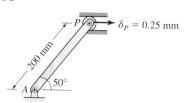

Figure P2.35

2.36

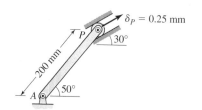

Figure P2.36

In Problems 2.37 through 2.42 a roller at P slides in a slot as shown in each problem. Determine the deformation in bars AP and BP in each problem by using small-strain approximation.

2.37 Determine the deformation in bars *AB* and *BP* shown in Figure P2.37.

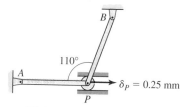

Figure P2.37

2.38 Determine the deformation in bars *AB* and *BP* shown in Figure P2.38.

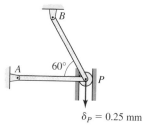

Figure P2.38

2.39 Determine the deformation in bars *AB* and *BP* shown in Figure P2.39.

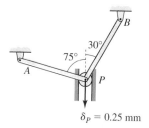

Figure P2.39

2.40 Determine the deformation in bars *AB* and *BP* shown in Figure P2.40.

2.41 Determine the deformation in bars *AB* and *BP* shown in Figure P2.41.

2.42 Determine the deformation in bars *AB* and *BP* shown in Figure P2.42.

2.43 A gap of 0.004 in exists between the rigid bar and bar *A* before the load *P* is applied (Figure P2.43). The rigid bar is hinged at point *C*. Due to force *P*, the strain in bar *A* was found to be $-500\ \mu$. Determine the strain in bar *B*. The lengths of bars *A* and *B* are 30 and 50 in, respectively.

Vector approach to small-strain approximation

For the truss shown in Figure P2.44 the pin displacements in the x and y directions given by u and v, respectively, were computed by the finite-element method and are as follows:

$u_B = 12.6$ mm	$v_B = -24.48$ mm
$u_C = 21.0$ mm	$v_C = -69.97$ mm
$u_D = -16.8$ mm	$v_D = -119.65$ mm
$u_E = -12.6$ mm	$v_E = -69.97$ mm
$u_F = -8.4$ mm	$v_F = -28.68$ mm

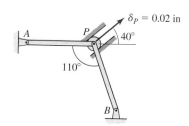

Figure P2.40

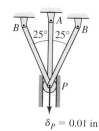

Figure P2.41

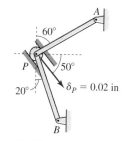

Figure P2.42

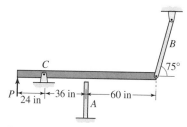

Figure P2.43

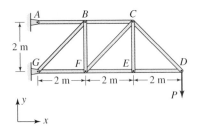

Figure P2.44

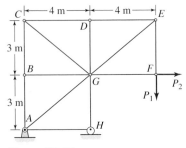

Figure P2.47

Use these values to solve Problems 2.44, 2.45, and 2.46.

2.44 Determine the axial strains in members *AB*, *BF*, *FG*, and *GB*.

2.45 Determine the axial strains in members *BC*, *CF*, and *FE*.

2.46 Determine the axial strains in members *ED*, *DC*, and *CE*.

For the truss shown in Figure P2.47 the pin displacements in the x and y directions given by u and v, respectively, were computed by the finite-element method and are as follows:

$u_B = 7.00$ mm	$v_B = 1.500$ mm
$u_C = 17.55$ mm	$v_C = 3.000$ mm
$u_D = 20.22$ mm	$v_D = -4.125$ mm
$u_E = 22.88$ mm	$v_E = -32.250$ mm
$u_F = 9.00$ mm	$v_F = -33.750$ mm
$u_G = 7.00$ mm	$v_G = -4.125$ mm
$u_H = 0$	$v_H = 0$

Use these values to solve Problems 2.47, 2.48, and 2.49.

2.47 Determine the axial strains in members *AB*, *BG*, *GA*, and *AH*.

2.48 Determine the axial strains in members *BC*, *CG*, *GB*, and *CD*.

2.49 Determine the axial strains in members *GF*, *FE*, *EG*, and *DE*.

2.50 Three poles are pin connected to a ring at *P* and to the supports on the ground. The ring slides on a vertical rigid pole by 2 in, as shown in Figure P2.50. The coordinates of the four points are as given. Determine the normal strain in each bar due to the movement of the ring.

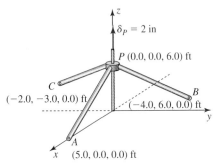

Figure P2.50

2.5 STRAIN COMPONENTS

Let *u*, *v*, and *w* be the displacements in the *x*, *y*, and *z* directions, respectively. Figure 2.21 and Equations (2.9a) through (2.9i) define average engineering strain components.

$$\varepsilon_{xx} = \frac{\Delta u}{\Delta x} \qquad (2.9a)$$

$$\varepsilon_{yy} = \frac{\Delta v}{\Delta y} \qquad (2.9b)$$

$$\varepsilon_{zz} = \frac{\Delta w}{\Delta z} \qquad (2.9c)$$

$$\gamma_{xy} = \frac{\Delta u}{\Delta y} + \frac{\Delta v}{\Delta x} \qquad (2.9d)$$

$$\gamma_{yx} = \frac{\Delta v}{\Delta x} + \frac{\Delta u}{\Delta y} = \gamma_{xy} \qquad (2.9e)$$

$$\gamma_{yz} = \frac{\Delta v}{\Delta z} + \frac{\Delta w}{\Delta y} \qquad (2.9f)$$

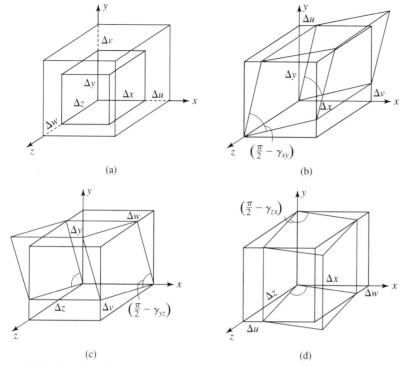

Figure 2.21 (*a*) Normal strains. (*b*) Shear strain γ_{xy}. (*c*) Shear strain γ_{yz}. (*d*) Shear strain γ_{zx}.

$$\gamma_{zy} = \frac{\Delta w}{\Delta y} + \frac{\Delta v}{\Delta z} = \gamma_{yz} \tag{2.9g}$$

$$\gamma_{zx} = \frac{\Delta w}{\Delta x} + \frac{\Delta u}{\Delta z} \tag{2.9h}$$

$$\gamma_{xz} = \frac{\Delta u}{\Delta z} + \frac{\Delta w}{\Delta x} = \gamma_{zx} \tag{2.9i}$$

$$\begin{bmatrix} \varepsilon_{xx} & \gamma_{xy} & \gamma_{xz} \\ \gamma_{yx} & \varepsilon_{yy} & \gamma_{yz} \\ \gamma_{zx} & \gamma_{zy} & \varepsilon_{zz} \end{bmatrix}$$

Figure 2.22 Engineering strain matrix.

Equations (2.9a) through (2.9i) show that strain has nine components in three dimensions, but only six components are independent because of the symmetry of shear strain. The symmetry of shear strain makes intuitive sense. The change of angle between the x and y directions is obviously the same as between the y and x directions. In the Equations (2.9a) through (2.9i) the first subscript is the direction of displacement and the second the direction of the line element. But because of the symmetry of shear strain, the order of the subscripts is immaterial. The engineering strain components can be shown as an engineering strain matrix (Figure 2.22). The matrix is symmetric because of the symmetry of shear strain.

2.5.1 Plane Strain

Plane strain is one of the two types of two-dimensional idealizations in mechanics of materials. In Chapter 1 we saw the other type of two-dimensional idealization, plane stress. We will see the difference between the two types of two-dimensional idealizations

$$\begin{bmatrix} \varepsilon_{xx} & \gamma_{xy} & 0 \\ \gamma_{yx} & \varepsilon_{yy} & 0 \\ 0 & 0 & 0 \end{bmatrix}$$

Figure 2.23 Plane-strain matrix.

in Chapter 3. By two-dimensional we imply that one of the coordinates does not play a role in the solution of the problem. Choosing z to be that coordinate, we set all strains with subscript z to be zero, as shown in the strain matrix in Figure 2.23. Notice that in plane strain, four components of strain are needed though only three are independent because of the symmetry of shear strain.

The assumption of plane strain is often made in analyzing very thick bodies, such as points around tunnels or mine shafts in earth, or a point in the middle of a thick cylinder, such as a submarine hull. In thick bodies it is argued that a point has to push a lot of material in the thickness direction to move. Hence the strains can be expected to be small in the thickness direction. Thus the strain in the thickness direction is not zero, but it is small enough to be neglected. Plane strain is a mathematical approximation made to simplify analysis.

EXAMPLE 2.9

Displacements u and v in the x and y directions, respectively, were measured at many points on a body by the geometric Moiré method.[3] The displacements of four points on the body of Figure 2.24 are as follows:

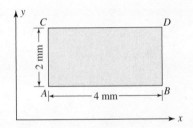

Figure 2.24 Undeformed geometry in Example 2.9.

$$u_A = -0.0100 \text{ mm} \qquad v_A = 0.0100 \text{ mm}$$
$$u_B = -0.0050 \text{ mm} \qquad v_B = -0.0112 \text{ mm}$$
$$u_C = 0.0050 \text{ mm} \qquad v_C = 0.0068 \text{ mm}$$
$$u_D = 0.0100 \text{ mm} \qquad v_D = 0.0080 \text{ mm}$$

Determine strains ε_{xx}, ε_{yy}, and γ_{xy} at point A.

PLAN

The calculation of strains requires the calculation of deformation, that is, the relative movement of points on the body. We can use point A as our reference point and calculate the relative movement of points B and C and then find the strains from Equations (2.9a), (2.9b), and (2.9c).

Solution The relative movements of points A, B, and C with respect to A are:

$$u_B - u_A = 0.0050 \qquad v_B - v_A = -0.0212$$
$$u_C - u_A = 0.0150 \qquad v_C - v_A = -0.0032$$

The normal strains in the directions AB and AC can be calculated using the components of deformation in the AB and AC directions, as per

---

[3]See Section 2.7 for a brief description of the Moiré methods.

Equations (2.9a) and (2.9b),

$$\varepsilon_{xx} = \frac{u_B - u_A}{x_B - x_A} = \frac{0.0050}{4} = 0.00125$$

ANS. $\varepsilon_{xx} = 1250 \, \mu$

$$\varepsilon_{yy} = \frac{v_C - v_A}{y_C - y_A} = \frac{-0.0032}{2} = -0.0016$$

ANS. $\varepsilon_{yy} = -1600 \, \mu$

From Equation (2.9c) the shear strain can be found,

$$\gamma_{xy} = \frac{v_B - v_A}{x_B - x_A} + \frac{u_C - u_A}{y_C - y_A} = \frac{-0.0212}{4} + \frac{0.0150}{2} = 0.0022$$

ANS. $\gamma_{xy} = 2200 \, \mu$

COMMENT
Figure 2.25 shows an exaggerated deformed shape of the rectangle. Point A moves to point A_1; similarly the other points move to B_1, C_1, and D_1. By drawing the undeformed rectangle from point A we can show the relative movements of the three points. We could have calculated the length of A_1B from the Pythagorean theorem as $A_1B_1 = \sqrt{(4 - 0.005)^2 + (-0.0212)^2} = 3.995056$, which would yield the following strain value:

$$\varepsilon_{xx} = \frac{A_1B_1 - AB}{AB} = 1236 \, \mu.$$

The difference between the two calculations is 1.1%. We will have to perform similar tedious calculations to find the other two strains if we want to gain an additional accuracy of 1% or less. But notice the simplicity of the calculations that come from small-strain approximation.

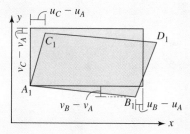

Figure 2.25 Elaboration of comment.

2.6 STRAIN AT A POINT

In Section 2.5 the lengths Δx, Δy, and Δz were finite. If we shrink these lengths to zero in Equations (2.9a) through (2.9i), we obtain the definition of strain at a point,

$$\varepsilon_{xx} = \lim_{\Delta x \to 0}\left(\frac{\Delta u}{\Delta x}\right) = \frac{\partial u}{\partial x} \qquad (2.10a)$$

$$\varepsilon_{yy} = \lim_{\Delta y \to 0}\left(\frac{\Delta v}{\Delta y}\right) = \frac{\partial v}{\partial y} \qquad (2.10b)$$

$$\varepsilon_{zz} = \lim_{\Delta z \to 0}\left(\frac{\Delta w}{\Delta z}\right) = \frac{\partial w}{\partial z} \tag{2.10c}$$

$$\gamma_{xy} = \gamma_{yx} = \lim_{\substack{\Delta x \to 0 \\ \Delta y \to 0}}\left(\frac{\Delta u}{\Delta y} + \frac{\Delta v}{\Delta x}\right) = \frac{\partial u}{\partial y} + \frac{\partial v}{\partial x} \tag{2.10d}$$

$$\gamma_{yz} = \gamma_{zy} = \lim_{\substack{\Delta y \to 0 \\ \Delta z \to 0}}\left(\frac{\Delta v}{\Delta z} + \frac{\Delta w}{\Delta y}\right) = \frac{\partial v}{\partial z} + \frac{\partial w}{\partial y} \tag{2.10e}$$

$$\gamma_{zx} = \gamma_{xz} = \lim_{\substack{\Delta x \to 0 \\ \Delta z \to 0}}\left(\frac{\Delta w}{\Delta x} + \frac{\Delta u}{\Delta z}\right) = \frac{\partial w}{\partial x} + \frac{\partial u}{\partial z} \tag{2.10f}$$

Because the limiting operation is in a given direction, we obtain partial derivatives and not the ordinary derivatives.

From the preceding equations we see that engineering strain has two subscripts to indicate both the direction of deformation and the direction of the line element that is being deformed. Thus it would seem that engineering strain is also a second-order tensor. However, unlike stress, engineering strain does not satisfy certain coordinate transformation laws, which we will study in Chapter 8. Hence it is not a second-order tensor but is related to it as follows:

$$\text{tensor normal strains} = \text{engineering normal strains}$$

$$\text{tensor shear strains} = \frac{\text{engineering shear strains}}{2}$$

In Chapter 8 we shall see that the factor $1/2$, which changes engineering shear strain to tensor shear strain, plays an important role in strain transformation.

2.6.1 Strain at a Point on a Line

In axial members we shall see that the displacement u is only a function of x. Hence the partial derivative in Equation (2.10a) becomes an ordinary derivative, and we obtain

$$\boxed{\varepsilon_{xx} = \frac{du(x)}{dx}} \tag{2.11}$$

If the displacement is given as a function of x, then we can obtain the strain as a function of x by differentiating it.

If strain is given as a function of x, then by integrating we can obtain the deformation between two points, that is, the relative displacement of two points. If we know the displacement of one of the points then we can find the displacement of the other point. Alternatively stated, the integration of Equation (2.11) generates an integration constant. To determine the integration constant, we need to know the displacement at a point on the line.

EXAMPLE 2.10

Calculations using the finite-element method[4] show that the displacement in a quadratic axial element is given by

$$u(x) = 125.0(x^2 - 3x + 8)10^{-6} \text{ cm}, \qquad 0 \le x \le 2 \text{ cm}$$

Determine the normal strain ε_{xx} at $x = 1$ cm.

PLAN

We can find the strain by using Equation (2.11) at any x and obtain the final result by substituting the value of $x = 1$.

Solution   Differentiating the given displacement, we obtain

$$\varepsilon_{xx} = \frac{du}{dx} = 125.0(2x - 3)10^{-6}$$

Substituting $x = 1$ in the preceding strain expression, we obtain the strain.

ANS. $\varepsilon_{xx} = -125 \, \mu$

---

[4]The finite-element method is a computer-based method in which displacements inside an element (axial members) are described by polynomials, and coefficients of the polynomials are displacements of points which are determined numerically. The method is described briefly in Section 4.9.

EXAMPLE 2.11

A bar that is hanging by its own weight has axial strain $\varepsilon_{xx} = K(L - x)$, where K is a constant for a given material (Figure 2.26). Find the total extension of the bar.

PLAN

The elongation of the bar corresponds to the displacement of point B. We know that the displacement at point A is zero. We start with Equation (2.11) and integrate.

Solution $\varepsilon_{xx} = \dfrac{du}{dx} = K(L - x)$ (E1)

$$\int_{u_A}^{u_B} du = \int_{x_A=0}^{x_B=L} K(L - x)\, dx$$

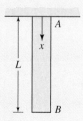

Figure 2.26   Bar in Example 2.11.

Then

$$u_B - u_A^{\;0} = K\left(Lx - \frac{x^2}{2}\right)\Big|_0^L \qquad \text{or} \qquad \text{ANS.} \qquad u_B = \frac{KL^2}{2}$$

COMMENTS

1. From strains we obtain (deformation) relative displacement $u_B - u_A$ and not absolute displacement. To get absolute displacement we choose a point on the body that did not move.

2. We could integrate Equation (E1) to obtain $u(x) = K(Lx - x^2/2) + C_1$. We can then find the integration constant C_1. Using the condition that the displacement u at $x = 0$ is zero, we obtain $C_1 = 0$. We could then substitute $x = L$ to obtain the displacement of point B. The integration constant C_1 represents rigid-body translation, which we eliminate by fixing the bar to the wall.

Consolidate your knowledge   Write all you know about deformation and strain.

QUICK TEST 2.1	Time: 15 minutes/Total: 20 points

Grade yourself using the answers in Appendix G. Each problem is worth 2 points.

1. What is the difference between displacement and deformation?
2. What is the difference between Lagrangian and Eulerian strains?
3. In decimal form, what is the value of normal strain that is equal to 0.3%?
4. In decimal form, what is the value of normal strain that is equal to 2000 μ?
5. Does the right angle increase or decrease with positive shear strains?
6. If the left end of a rod moves more than the right end in the negative x direction, will the normal strain be negative or positive? Justify your answer.
7. Can a 5% change in length be considered to be small normal strain? Justify your answer.
8. How many nonzero strain components are there in three dimensions?
9. How many nonzero strain components are there in plane strain?
10. How many independent strain components are there in plane strain?

Strain components

A rectangle deforms into the colored shapes shown in Problems 2.51 through 2.53. Determine the average values of strain components ε_{xx}, ε_{yy}, and γ_{xy} at point A.

2.51 Determine ε_{xx}, ε_{yy}, and γ_{xy} at point A in Figure P2.51.

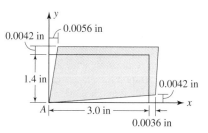

Figure P2.51

2.52 Determine ε_{xx}, ε_{yy}, and γ_{xy} at point A in Figure P2.52.

2.53 Determine ε_{xx}, ε_{yy}, and γ_{xy} at point A in Figure P2.53.

Displacements u and v in the x and y directions, respectively, were measured by the Moiré interferometry method at many points on a body. The displacements of four points on a body are given in Problems 2.54 through 2.57. Determine the average values of the strain components ε_{xx}, ε_{yy}, and γ_{xy} at point A in Figure P2.54.

2.54

$u_A = 0$	$v_A = 0$
$u_B = 0.625\ \mu mm$	$v_B = -0.3125\ \mu mm$
$u_C = -0.500\ \mu mm$	$v_C = -0.5625\ \mu mm$
$u_D = 0.250\ \mu mm$	$v_D = -1.125\ \mu mm$

2.55

$u_A = 0.625\ \mu mm$	$v_A = -0.3125\ \mu mm$
$u_B = 1.500\ \mu mm$	$v_B = -0.5000\ \mu mm$
$u_C = 0.250\ \mu mm$	$v_C = -1.125\ \mu mm$
$u_D = 1.250\ \mu mm$	$v_D = -1.5625\ \mu mm$

2.56

$u_A = -0.500\ \mu mm$	$v_A = -0.5625\ \mu mm$
$u_B = 0.250\ \mu mm$	$v_B = -1.125\ \mu mm$
$u_C = -1.250\ \mu mm$	$v_C = -1.250\ \mu mm$
$u_D = -0.375\ \mu mm$	$v_D = -2.0625\ \mu mm$

2.57

$u_A = 0.250\ \mu mm$	$v_A = -1.125\ \mu mm$
$u_B = 1.250\ \mu mm$	$v_B = -1.5625\ \mu mm$
$u_C = -0.375\ \mu mm$	$v_C = -2.0625\ \mu mm$
$u_D = 0.750\ \mu mm$	$v_D = -2.7500\ \mu mm$

Strain at a point

2.58 In a tapered circular bar that is hanging vertically, the axial displacement due to its weight was found as

$u(x) =$

$$\left(-19.44 + 1.44x - 0.01x^2 - \frac{933.12}{72 - x}\right)10^{-3} \text{ in}$$

Determine the axial strain ε_{xx} at $x = 24$ in.

2.59 In a tapered rectangular bar that is hanging vertically, the axial displacement due to its weight was found as

$u(x) =$
$$[-50x + 20x^2 + 2.5\ \ln(1 - 0.8x)]10^{-6} \text{ mm}$$

Determine the axial strain ε_{xx} at $x = 150$ mm.

2.60 The axial displacement in the quadratic one-dimensional finite element shown in Figure P2.60 is

$$u(x) = \frac{u_1}{2a^2}(x - a)(x - 2a) - \frac{u_2}{a^2}x(x - 2a)$$
$$+ \frac{u_3}{2a^2}x(x - a)$$

Determine the strain at node 2.

2.61 Due to the applied load, the strain in the tapered bar in Figure P2.61 was found to be $\varepsilon_{xx} = 0.2/(40 - x)^2$. Determine the extension of the bar.

2.62 The axial strain in a bar of length L was found to be $\varepsilon_{xx} = KL/(3L - 2x)$, $0 \le x \le L$, where K is a constant for a given material, loading, and cross-sectional dimension. Determine the total extension in terms of K and L.

2.63 The axial strain in a bar of length L due to its own weight was found to be $\varepsilon_{xx} = K[4L - 2x - 8L^3/(4L - 2x)^2]$, $0 \le x \le L$, where K is a constant for a given material and cross-sectional dimension. Determine the total extension in terms of K and L.

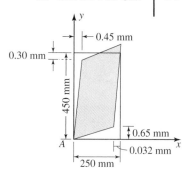

Figure P2.52

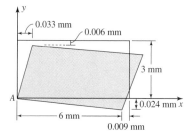

Figure P2.53

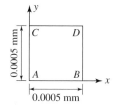

Figure P2.54

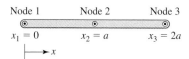

Figure P2.60

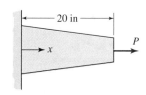

Figure P2.61

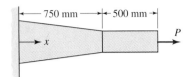

Figure P2.64

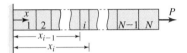

Figure P2.65

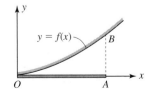

Figure P2.68

2.64 A bar has a tapered and a uniform section securely fastened, as shown in Figure P2.64. Determine the total extension of the bar if the axial strain in each section is

$$\varepsilon = \frac{1500 \times 10^3}{1875 - x} \mu, \qquad 0 \le x \le 750 \text{ mm}$$

$$\varepsilon = 1500 \mu, \qquad 750 \text{ mm} \le x \le 1250 \text{ mm}$$

Stretch yourself

2.65 N axial bars are securely fastened together. Determine the total extension of the composite bar shown in Figure P2.65 if the strain in the ith section is

$$\varepsilon_i = a_i, \qquad x_{i-1} \le x \le x_i$$

2.66 True strain ε_T is calculated from $d\varepsilon_T = du / (L_0 + u)$, where u is the deformation at any given instant and L_0 is the original undeformed length. Thus the increment in true strain is the ratio of change in length at any instant to the length at that given instant. If ε represents engineering strain, show that at any instant the relationship between true strain and engineering strain is given by the following equation:

$$\varepsilon_T = \ln(1 + \varepsilon) \qquad (2.12)$$

2.67 The displacements in a body are given by

$$u = [0.5(x^2 - y^2) + 0.5xy]10^{-3} \text{ mm}$$

$$v = [0.25(x^2 - y^2) - xy]10^{-3} \text{ mm}$$

Determine strains ε_{xx}, ε_{yy}, and γ_{xy} at $x = 5$ mm and $y = 7$ mm.

A metal strip is to be pulled and bent to conform to a rigid surface such that the length of the strip OA fits the arc OB of the surface (Figure P2.68). The equation of the surface $y = f(x)$ and the length OA are given for Problems 2.68 through 2.70. Determine the average normal strain in the metal strip.

2.68 $f(x) = 0.04x^{3/2}$ in; length $OA = 9$ in.

2.69 $f(x) = 625x^{3/2}$ μmm; length $OA = 200$ mm.

Computer problems

2.70 $f(x) = 0.04x^{3/2} - 0.005x$ in; length $OA = 9$ in. Use numerical integration.

2.71 Measurements made along the path of the stretch cord that is stretched over the canoe in Problem 2.4 (Figure P2.71) are shown in Table P2.71. The y coordinate was measured to the closest $\frac{1}{32}$ in. Between points A and B the cord path can be approximated by a straight line. Determine the average strain in the stretch cord if its original length it is 40 in. Use a spread sheet and approximate each 2-in x interval by a straight line.

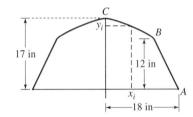

Figure P2.71

TABLE P2.71

x_i	y_i
0	17
2	$16\frac{30}{32}$
4	$16\frac{29}{32}$
6	$16\frac{19}{32}$
8	$16\frac{3}{32}$
10	$15\frac{16}{32}$
12	$14\frac{24}{32}$
14	$13\frac{28}{32}$
$x_B = 16$	$y_B = 12$
$x_A = 18$	$y_A = 0$

*2.7 GENERAL INFORMATION

Normal strain, as a ratio of deformation over length, is seen to appear in experiments conducted as far back as the thirteenth century. Thomas Young (1773–1829) who, as we shall see in the next chapter, has a material constant named after him, is credited with having been the first to consider shear as an elastic strain. He called the shear elastic strain *detrusion*. Cauchy (1789–1857), who introduced the concept of stress we use in this book, also introduced the mathematical definition of engineering strain given by Equations (2.9a) through (2.9i). The nonlinear Lagrangian strain written in tensor form was introduced by Green (1793–1841) and is called Green's strain tensor. The nonlinear Eulerian strain written in tensor form was introduced by Almansi and is called Almansi's strain tensor. Green's and Almansi's strain tensors are often referred to as strain tensors in Lagrangian and Eulerian coordinates, respectively.

Displacements at different points on a solid body can be measured or analyzed by a variety of methods. We will discuss briefly the principles behind two modern techniques for finding displacements in a solid body.

2.7.1 Moiré Fringe Method

The Moiré fringe method is an experimental technique of measuring displacements at a point that uses the phenomenon of light interference produced by two equally spaced gratings. Figure 2.27 shows equally spaced parallel bars in two gratings. Suppose we start with the position when the bars of the right gratings are positioned to overlap the spacings of the left grating. An observer on the right will be in a dark region as no light ray can pass through both gratings. Let us suppose the left grating moves less than the spacing between the bars. Now we will have space between each pair of bars and we will have regions of dark and light. The lines of light and dark are called *fringes*. When the left grating has moved through one pitch (spacing between bars), the observer will once more be in the dark. Counting the number of times the regions of light and dark (i.e., the number of fringes) pass through a point, the displacement can be found by multiplying the pitch and the number of fringes. It should be noted that any motion of the left grating parallel to the direction of the bars will not result in change in light intensity. Hence displacements calculated from Moiré fringes are always perpendicular to the lines in the grating. We will need a grid of perpendicular lines to find the two components of

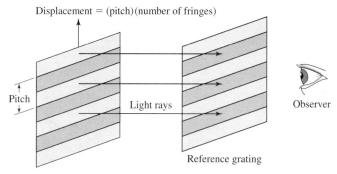

Figure 2.27 Destructive light interference by two equally spaced gratings.

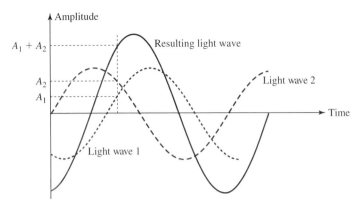

Figure 2.28 Superposition of two light waves.

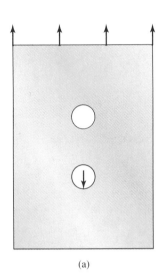

(a)

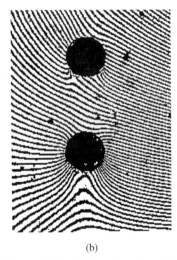

(b)

Figure 2.29 Deformation of a grid obtained from optical Moiré interferometry.

displacements in a two-dimensional problem. The left grating is cemented, etched, printed, photographed, stamped, or scribed onto a specimen. The right grating is referred to as the reference grating. Clearly, the order of displacement that can be measured depends on the number of lines in the grating.

Figure 2.27 explains the interference of light by mechanical means. The corresponding method is called geometric Moiré method. The geometric Moiré method is used for displacement measurements in the range of 1 mm to as small as 10 μm, which corresponds to a grid of 1 to 100 lines per millimeter. In U.S. customary units the range is of 0.1 in to as small as 0.001 in, corresponding to grids having from 10 to 1000 lines per inch.

Light interference can also be produced optically. Techniques based on optical light interference are generally termed optical interferometry. Consider two light rays of the same frequency arriving at a point, as shown in Figure 2.28. The amplitude of the resulting light wave is the sum of the two amplitudes. If the crest of one light wave falls on the trough of another light wave, then the resulting amplitude will be zero and we will have darkness at that point. If the crests of two waves arrive at the same time, then we will have light brighter than either of the two waves. This addition and subtraction is called *constructive* and *destructive* interference and is used in interferometry for measurements in a variety of ways. In Moiré interferometry a reference grid may be created by reflecting the light of the grid fixed to the specimen by using two identical light sources. As the grid on the specimen moves, the reflective light and the incident light interfere constructively and destructively to produce Moiré fringes. Displacements as small as 10^{-5} in, which corresponds to 100,000 lines per inch, can be measured. In the metric system, the order of displacements is 25×10^{-5} mm, which corresponds to 4000 lines per millimeter.

In an experiment to study mechanically fastened composites, load was applied on one end of the joint and equilibrated by applying a load on the lower hole, as shown in Figure 2.29a. Moiré fringes parallel to the applied load on the top plate are shown in Figure 2.29b.

2.8 CLOSURE

In this chapter we saw that the relation between displacement and strains is derived by studying the geometry of the deformed body. Drawing an approximate deformed body is to make assumptions regarding the displacements of points on the body. The simplest assumption for a displacement component is either to assume it to be a constant in a coordinate direction, or to assume it to be a linear function of the coordinate. From the displacements we can obtain the strains. It should be emphasized that the strain–displacement relation is independent of material properties. In the next chapter we shall introduce material properties and the relationship between stresses and strains. Thus from displacements we deduce the strains, from which we will deduce stress variations, from which we shall then find internal forces and relate the internal forces to external forces, as we did in Chapter 1. We shall see this complete logic in Chapter 3.

We will study strains again in Chapter 9 on strain transformation. Strain transformation is relating strains in different coordinate systems. This is important as experimental measurements of strains as well as analyses are usually performed in a coordinate system chosen to simplify calculations. Developing a discipline of drawing deformed shapes has the same importance as drawing a free-body diagram for calculating forces. Drawing deformed shapes is important for developing an intuitive understanding of deformation and strain as well as for reducing mistakes in strain and deformation calculations.

POINTS AND FORMULAS TO REMEMBER

- The total movement of a point with respect to a fixed reference coordinate is called displacement.
- The relative movement of a point with respect to another point on the body is called deformation.
- The displacement of a point is the sum of rigid body motion and motion due to deformation.
- Lagrangian strain is computed from deformation by using the original undeformed geometry as the reference geometry.
- Eulerian strain is computed from deformation by using the final deformed geometry as the reference geometry.

$$\varepsilon = \frac{L_f - L_0}{L_0} \quad (2.1) \qquad \varepsilon = \frac{\delta}{L_0} \quad (2.2) \qquad \varepsilon = \frac{u_B - u_A}{x_B - x_A} \quad (2.3)$$

where ε is the average normal strain, L_0 is the original length of a line, L_f is the final length of a line, δ is the deformation of the line, and u_A and u_B are displacements of points x_A and x_B, respectively.

- Elongations result in positive normal strains. Contractions result in negative normal strains.

$$\gamma = \pi/2 - \alpha \quad (2.4)$$

where α is the final angle measured in radians and $\pi/2$ is the original right angle.

- Decreases in angle result in positive shear strain. Increases in angle result in negative shear strain.
- Small-strain approximation may be used for strains less than 0.01.
- Small-strain calculations result in linear deformation analysis.
- Small normal strains are calculated by using the deformation component in the original direction of the line element, regardless of the orientation of the deformed line element.

- In small shear strain (γ) calculations the following approximation may be used for the trigonometric functions:

$$\tan \gamma \approx \gamma \qquad \sin \gamma \approx \gamma \qquad \cos \gamma \approx 1$$

- In small strain,

$$\delta = \bar{\mathbf{D}}_{AP} \cdot \bar{\mathbf{i}}_{AP} \quad (2.7)$$

where $\bar{\mathbf{D}}_{AP}$ is the deformation vector of the bar AP and $\bar{\mathbf{i}}_{AP}$ is the unit vector in the original direction of the bar AP.

- The same reference point must be used in the calculations of the deformation vector and the unit vector.

Average strain

$$\varepsilon_{xx} = \frac{\Delta u}{\Delta x} \qquad \gamma_{xy} = \gamma_{yx} = \frac{\Delta u}{\Delta y} + \frac{\Delta v}{\Delta x}$$

$$\varepsilon_{yy} = \frac{\Delta v}{\Delta y} \qquad \gamma_{yz} = \gamma_{zy} = \frac{\Delta v}{\Delta z} + \frac{\Delta w}{\Delta y} \qquad \begin{matrix}(2.9a)\\ \text{through}\\ (2.9i)\end{matrix}$$

$$\varepsilon_{zz} = \frac{\Delta w}{\Delta z} \qquad \gamma_{zx} = \gamma_{xz} = \frac{\Delta w}{\Delta x} + \frac{\Delta u}{\Delta z}$$

Strain at a point

$$\varepsilon_{xx} = \frac{\partial u}{\partial x} \qquad \gamma_{xy} = \gamma_{yx} = \frac{\partial u}{\partial y} + \frac{\partial v}{\partial x}$$

$$\varepsilon_{yy} = \frac{\partial v}{\partial y} \qquad \gamma_{yz} = \gamma_{zy} = \frac{\partial v}{\partial z} + \frac{\partial w}{\partial y} \qquad \begin{matrix}(2.10a)\\ \text{through}\\ (2.10f)\end{matrix}$$

$$\varepsilon_{zz} = \frac{\partial w}{\partial z} \qquad \gamma_{zx} = \gamma_{xz} = \frac{\partial w}{\partial x} + \frac{\partial u}{\partial z}$$

where u, v, and w are the displacements of a point in the x, y, and z directions, respectively.
- Shear strain is symmetric.
- In three dimensions there are nine strain components but only six are independent.
- In two dimensions there are four strain components but only three are independent.

If u is only a function of x,

$$\varepsilon_{xx} = \frac{du(x)}{dx} \quad (2.11)$$

CHAPTER THREE
MECHANICAL PROPERTIES
OF MATERIALS

3.0 OVERVIEW

A metal wire and a rubber stretch cord that have the same undeformed lengths and are subjected to the same loads deform by significantly different amounts, as shown in Figure 3.1. That rubber deforms significantly more than metal is not a surprising result, but it emphasizes that material properties must play an important role in developing formulas relating deformation to applied forces. How do we describe the mechanical properties of materials? This chapter discusses the qualitative as well as the quantitative characterization of a material's mechanical properties.

Qualitative descriptions of a material by adjectives such as elastic, ductile, or tough have very specific meanings that must be understood, for these adjectives form the engineering language of material description. Quantitative material descriptions are the equations of the curves that pass through the experimental observations of stress and strain values. A material model is a specific equation with a certain number of parameters that is chosen to describe the relationship between stresses and strains, as shown in Figure 3.2. The parameters in the equation are usually determined by fitting a curve through experimental observations in a least-squares manner.[1]

Figure 3.2 shows the last link that we need to construct the chain of logic relating deformation to forces. As mentioned in earlier chapters, we will use this logic to develop the theories for relating stresses and deformation to applied loads for axial members, shafts, and beams.

Figure 3.1 Material impact on deformation.

Figure 3.2 Relationship of stresses and strains.

---

[1]See Appendix B.3 for details on fitting a curve using least squares.

The two major learning objectives of this chapter are:

1. Understand the qualitative and quantitative descriptions of mechanical properties of materials.
2. Learn the logic of relating deformation to external forces.

3.1 MATERIAL CHARACTERIZATION

The American Society for Testing and Materials (ASTM) describes test procedures for determining the various properties of a material. These descriptions are guidelines used by experimentalists to obtain reproducible results for material properties. Many parameters relating stresses and strains are determined by the tension test discussed next.

3.1.1 Tension Test

Two standard tension test specimens are shown in the Figure 3.3. The flat white plastic specimen with the rectangular cross section is shown in the tension test machine. The specimen with the circular cross section is referred to in the description of the tension test that follows.

The ends of the specimen are gripped and pulled in the axial direction. The tightness of the grip, the symmetry of the grip, friction, and other local effects are assumed and are observed to die out rapidly with the increase in distance[2] from the ends. This dissipation of local effects is further facilitated by the gradual change in the width or diameter of the

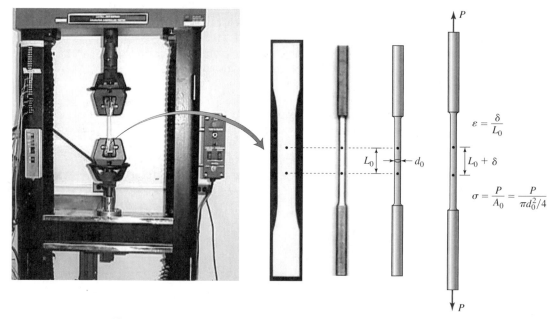

Figure 3.3 Tension test machine and specimen. (Courtesy Professor I. Miskioglu.)

---

[2]This phenomenon is discussed in Section 3.8 on Saint-Venant's principle.

specimen. The specimen is designed so that its central region is in a uniform state of axial stress. The initial diameter d_0 is measured, and the specimen's initial area of cross section A_0 can be found. For a given applied force P, the corresponding normal stress can be calculated by dividing the force by the cross-sectional area A_0. Two marks are made in the central region, and the length between the marks is called the gage length. It is symbolized by L_0. The movement of the two marks is recorded for each value of force P. The normal strain is calculated from the deformation δ by recording the movement of the two marks in the central region. For metals ASTM recommends a gage length $L_0 = 2$ in and an initial diameter $d_0 = 0.5$ in. The tension test may be conducted by controlling the force P and measuring the corresponding deformation or by controlling the deformation (movement of the grips) and measuring the corresponding force P. The stress σ and the strain ε are then plotted to obtain the stress–strain curve, as shown in Figure 3.4.

The stress–strain curve in Figure 3.4 may be obtained for a metal such as aluminum or steel. As the force P is applied to the specimen, the straight line OA in the plot is obtained initially. The end of this linear region is called the proportional limit. Following the proportional limit, there may be a slight decrease in stress for some metals, as shown in region AB. The stress then starts increasing once again, as shown in region BD. In a force-controlled experiment, the specimen will suddenly break as the applied load is increased when the stress reaches point D. Although we will be able to calculate the strain at rupture point E, we will not know the corresponding stress. In a displacement-controlled experiment we will see the decrease in stress and obtain the curve DE. The following definitions are used to describe the observations reported in this paragraph.

Definition 1 The point up to which stress and strain are related linearly is called *proportional limit*.

Definition 2 The largest stress in the stress–strain curve is called *ultimate stress*.

Definition 3 The stress at the point of rupture is called *fracture* or *rupture stress*.

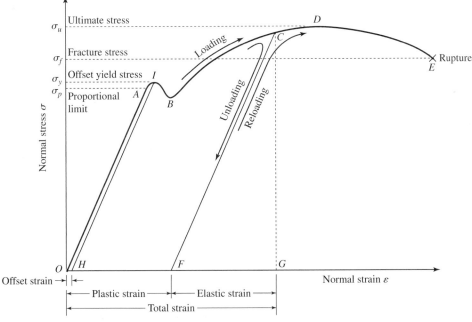

Figure 3.4 Stress–strain curve.

Elastic and plastic regions

If we load the specimen up to any point along line *OA* and then start unloading, we will observe that we retrace the stress–strain curve and return to point *O*. In other words, the material regains its original shape when the applied force is removed. If we started unloading after reaching point *C*, then we will come down the straight line *FC*, which will be parallel to line *OA*. We note that at point *F*, though the stress is zero, the strain is nonzero, reflecting the observation that the material is deformed permanently. The point at which permanent deformation starts lies somewhere in the region *AB* and is not clearly defined. The following definitions are associated with the observations in this paragraph.

Definition 4 The region of the stress–strain curve in which the material returns to the undeformed state when applied forces are removed is called *elastic region*.

Definition 5 The region in which the material deforms permanently is called *plastic region*.

Definition 6 The point demarcating the elastic from the plastic region is called *yield point*. The stress at yield point is called *yield stress*.

Definition 7 The permanent strain when stresses are zero is called *plastic strain*.

The strain at point *C*, which is in the plastic region, is the sum of the plastic strain given by *OF* and an elastic strain given by *FG*. Thus the total strain at a point is the sum of elastic strain and plastic strain.

For many materials the yield point is not clearly defined. It is close to the proportional limit but may be different. For such materials the following procedure is followed to determine the yield point. A prescribed value of offset strain is marked on the horizontal axis to get point *H* in Figure 3.4. A line parallel to the linear part of the stress–strain curve is drawn. Point *I* represents the intersection of the straight line and the stress–strain curve. The stress at point *I* is the offset yield strength. ASTM recommends the value of offset strain to be used for a particular material. Usually the offset strain is given as a percentage. A strain of 0.2% equals $\varepsilon = 0.002$, as described in Chapter 2. Another way of looking at offset yield is as follows. Suppose we were at a stress level at point *I*; then the offset strain would correspond to the plastic strain at point *I*.

Definition 8 The *offset* yield stress is a stress that would produce a plastic strain corresponding to the specified offset strain.

For most metals the elastic region and the linear region are nearly the same because of the proximity of the yield point to the proportional limit. But it should be emphasized that elastic and linear are two distinct material descriptions. Figure 3.5*a*

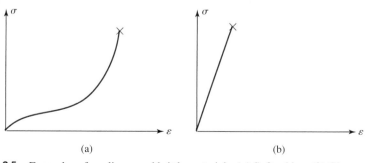

(a) (b)

Figure 3.5 Examples of nonlinear and brittle materials. (*a*) Soft rubber. (*b*) Glass.

shows the stress–strain curve for a soft rubber that can stretch several times its original length and yet return to its original geometry. Thus soft rubber is an elastic but a non-linear material.

Ductile and brittle materials

Soft rubber can undergo large deformations but is not called a ductile material. Aluminum and copper, which undergo large plastic deformations, are called ductile materials. Glass, which has a stress–strain curve of the type shown in Figure 3.5b, exhibits nearly no plastic deformation and is called a brittle material. Thus the key distinction between a ductile and a brittle material is the degree of plastic deformation before rupture. The degree of ductility of a material is usually described as percent elongation before rupture. Thus 17% elongation for aluminum and 35% elongation for copper refer to the large plastic strains before rupture, even though there is a small amount of elastic deformation in the total deformation before rupture. We record the following definitions:

Definition 9 A material that can undergo a large plastic deformation before fracture is called *ductile* material.

Definition 10 A material that exhibits little or no plastic deformation at failure is called *brittle* material.

One of the uses of the classification of materials as ductile and brittle is for the purpose of design. When we compute stresses at a point, we need to know whether maximum shear stress or maximum normal stress should be used for characterizing failure.[3] The following have been observed to be true:

- A ductile material usually yields when the maximum shear stress exceeds the yield shear stress of the material.
- A brittle material usually ruptures when the maximum tensile normal stress exceeds the ultimate tensile stress of the material.

Hard and soft materials

The difference between hardness and softness is not the strength (ultimate stress) of the material, but rather the resistance a material offers to scratches and indentation. The most common hardness[4] test consist of using a hard indenter of standard shape that is pressed into the material using a specified load. The depth of indentation is measured and assigned a numerical scale for comparing different materials.

A soft material can be made into a harder material using the phenomenon of strain hardening. In Figure 3.4, if the load is applied up to the stress level at point C and unloaded, then the material has a permanent deformation given by the strain at F. If now the material is reloaded, then the material behaves elastically until point C, after which additional plastic strain is recorded. In other words, during the reloading phase, point C is the new yield point, which is higher than the yield point of the original material. This phenomenon of increasing the yield point each time the stress value exceeds the yield stress is called strain hardening.

---

[3]The calculation of maximum normal and shear stresses is discussed in Chapter 8 and failure theories are discussed in Section 10.4.

[4]Rockwell and Brinell are the two most common methods of measuring hardness.

Figure 3.6 Specimen showing necking. (Courtesy Professor J. B. Ligon.)

Strain hardening is used very effectively in the manufacturing of such items as pots and pans. Aluminum is a relatively a soft metal and thus can be deformed with smaller forces than other metals. In deep drawing, the technique used in manufacturing pots and pans, the aluminum undergoes a large plastic deformation. Due to strain hardening, the large plastic deformation hardens the material, which makes the pots and pans more durable. It should be noted that as the yield point increases, the remaining plastic deformation before fracture decreases, and hence, by definition, it becomes more brittle. We record the following definitions.

Definition 11 *Hardness* is the resistance to indentation.

Definition 12 Raising the yield point with increasing strain is called *strain hardening*.

True stress and true strain

In Figure 3.4 the stress–strain curve between the ultimate stress and rupture (region *DE*) shows a decrease in stress with increasing strain. This decrease in stress with increasing strain is seen only if we plot Cauchy's stress versus engineering strain. An alternative is to plot true stress versus true strain.[5] Unlike Cauchy's stress, which is calculated by dividing the load *P* by the original undeformed cross-sectional area, true stress is calculated by dividing the load *P* by the area of the deformed cross section at the load value. Similarly, true strain is calculated on the basis of actual length rather than on the basis of the original undeformed length used in engineering strain. If we plot true stress versus true strain, then the stress will continue to increase with increasing strain and region *DE* will be a continuation of region *BD* in Figure 3.4.

In region *DE* of Figure 3.4, the specimen undergoes significant reduction in the cross-sectional area. Figure 3.6 shows a broken specimen from a tension test. Notice that the decrease in cross-sectional area makes the specimen look as if it had an elongated neck. This phenomenon is called necking and is defined next.

Definition 13 The sudden decrease in cross-sectional area after ultimate stress is called *necking*.

3.1.2 Material Constants

The relationship between normal stress and normal strain in a tension test for the linear region can be written as

$$\sigma = E\varepsilon \tag{3.1}$$

---

[5]True stress and true strain were briefly discussed in Section 1.3 and Problem 2.66, respectively.

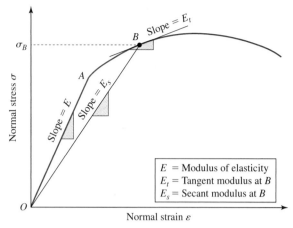

Figure 3.7 Different material moduli.

Equation (3.1) is known as *Hooke's law*.[6] The material constant E represents the slope of the straight line in a stress–strain curve, as shown in Figure 3.7, and is called *Young's modulus*[7] or *modulus of elasticity*.

In the nonlinear regions a number of ways are used to represent the stress–strain curve. The choice depends on the need of the analysis being performed. In Section 3.11 we will consider several approximations of nonlinear stress–strain curves and see how these approximations affect the analysis. The following definitions are often used in nonlinear analysis and are shown in Figure 3.7.

Definition 14 The slope of the tangent drawn to the stress–strain curve at a given stress value is called *tangent modulus*.

Definition 15 The slope of the line that joins the origin to the point on the stress–strain curve at a given stress value is called *secant modulus*.

Figure 3.8 shows that the elongation of a cylindrical specimen in the longitudinal direction causes contraction in the lateral (perpendicular) direction and vice versa. The ratio of the normal strains in the two directions is a material constant designated by the Greek symbol v (Nu) and is called *Poisson's*[8] *ratio*,

$$v = -\left(\frac{\varepsilon_{\text{lateral}}}{\varepsilon_{\text{longitudinal}}}\right) \tag{3.2}$$

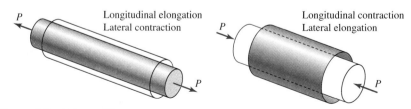

Figure 3.8 Poisson effect.

---

[6]Named after Robert Hooke. See Section 3.12 for additional details.
[7]Named after Thomas Young. See Section 3.12 for additional details.
[8]Named after Denis Poisson. See Section 3.12 for additional details.

Poisson's ratio is a dimensionless quantity that has a value between 0 and $\frac{1}{2}$ for most materials. However, some composite materials can have negative values for v, but not less than -1, that is, $-1 \leq v \leq \frac{1}{2}$.

To establish a relationship between shear stress and shear strain, a torsion test has to be conducted using a machine of the type shown in Figure 3.9. On plotting shear stress τ versus shear strain γ we would obtain a curve similar to that shown in Figure 3.4. In the linear region we would obtain

$$\tau = G\gamma \tag{3.3}$$

where the material constant G is called *shear modulus of elasticity* or *modulus of rigidity*. Table 3.1 shows the moduli of elasticity of some typical engineering materials. Wood, being the most familiar material to many of us, is used as a basis of comparison.

Figure 3.9   Torsion testing machine. (Courtesy Professor I. Miskioglu.)

TABLE 3.1   Comparison of moduli of elasticity for typical materials

Material	Modulus of Elasticity (10^3 ksi)	Modulus Relative to Wood
Rubber	0.12	0.06
Nylon	0.6	0.30
Adhesives	1.1	0.55
Soil	1.0	0.50
Bones	1.86	0.93
Wood	2.0	1.00
Concrete	4.6	2.30
Granite	8.7	4.40
Glass	10.0	5.00
Aluminum	10.0	5.00
Steel	30.0	15.00

3.1.3 Compression Test

Analysis is greatly simplified by assuming material behavior in tension and compression to be the same. This assumption of similar tension and compression properties works well for the values of material constants (such as E and v). Hence stress and deformation formulas developed in this book can be applied to members in tension and in compression. But compression tests show that for many brittle materials material strength in compression can be very different from that in tension, and for ductile materials the stress reversal from tension to compression in the plastic region can cause failure that must be avoided in strength-based design.

Figure 3.10a shows the stress–strain diagrams of two brittle materials. The compressive strength of cast iron is four times its tensile strength. Concrete has negligible tensile strength but can carry compressive stresses up to 5 ksi. Reinforcing concrete with steel bars helps alleviate the tensile stress in concrete by making the steel bars carry most of the tensile stresses. Notice the slopes of the lines, that is, the moduli of elasticity in tension and in compression are the same.

Figure 3.10b shows the stress–strain diagrams for a ductile material such as mild steel. If the compression test is conducted without unloading, then the tensile material behavior and the compressive material behavior are nearly identical, that is, modulus of elasticity, yield stress, and ultimate stress are nearly the same in tension and in compression. However, if material is loaded past the yield stress (point A) up to point B and unloaded, then as one crosses point C into the compressive region, one observes that the stress–strain diagram starts curving. After reaching point D, which is at least $2\sigma_{\text{yield}}$ below point B, suppose we once more reverse direction to reach point F, at which there is no applied load acting on the material. Notice that the plastic strain at point F is smaller than that at point C, and it is conceivable that the loading–unloading cycle is such that the material is returned to point O with no plastic strain. Does that mean we have the same material as the one we started with? No. The internal structure of the material has been altered significantly. The repetition of the loading–unloading cycle can cause the material to break at stress levels smaller than the ultimate stress. This phenomenon is called the *Bauschinger effect*.[9] Design usually precludes cyclic loading into the plastic region. Even in the elastic region cyclic loading can cause failure, as discussed in Section 3.10 on fatigue.

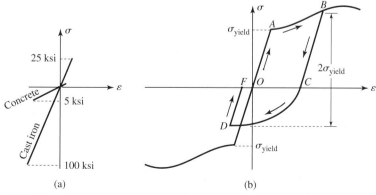

(a)   (b)

Figure 3.10 Differences in tension and compression. (*a*) Brittle material. (*b*) Ductile material.

[9]Named after Johann Bauschinger (1833–1893), who first studied this phenomenon.

EXAMPLE 3.1

A tension test was conducted on a circular specimen of titanium alloy. The gage length of the specimen was 2 in and the diameter in the test region before loading was 0.5 in. Some of the data from the tension test are given in Table 3.2, where P is the applied load and δ is the corresponding deformation. Calculate the following quantities:

(a) Stress at proportional limit.

(b) Ultimate stress.

(c) Yield stress at offset strain of 0.4%.

(d) Modulus of elasticity.

(e) Tangent and secant moduli of elasticity at a stress of 136 ksi.

(f) Plastic strain at a stress of 136 ksi.

TABLE 3.2 Tension test data in Example 3.1

#	P (kips)	δ (10^{-3} in)
1	0.0	0.0
2	5.0	3.2
3	15.0	9.5
4	20.0	12.7
5	24.0	15.3
6	24.5	15.6
7	25.0	15.9
8	25.2	16.9
9	25.4	19.7
10	26.0	28.5
11	26.5	36.9
12	27.0	46.5
13	27.5	58.3
14	28.0	75.2
15	28.2	87.1
16	28.3	100.0
17	28.2	112.9
18	28.0	124.8

PLAN

We can divide the column of load P by the cross-sectional area to get the values of stress. We can divide the column of deformation δ by the gage length of 2 in to get strain. We can plot the values to obtain the stress–strain curve and calculate the quantities, as described in Section 3.1.

Solution We divide the load column by the cross-sectional area $A = \pi 0.5^2/4 = 0.1964$ in^2 to obtain stress σ, and the deformation column by the gage length of 2 in to obtain strain ε, as shown in Table 3.3, which is obtained using a spread sheet. Figure 3.11 shows the corresponding stress–strain curve.

TABLE 3.3 Stress and strain in Example 3.1

#	σ (ksi)	ε (10^{-3})
1	0.0	0.0
2	25.5	1.6
3	76.4	4.8
4	101.9	6.4
5	122.2	7.6
6	124.8	7.8
7	127.3	8.0
8	128.3	8.5
9	129.9	10.5
10	132.4	14.3
11	135.0	18.4
12	137.5	23.3
13	140.1	29.1
14	142.6	37.6
15	143.6	43.5
16	144.0	50.0
17	143.6	56.5
18	142.6	62.4

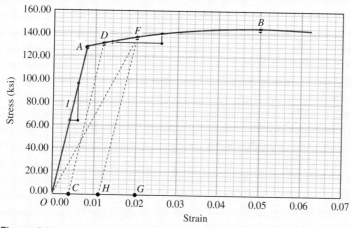

Figure 3.11 Stress–strain curve.

(a) Point A in Figure 3.11 represents the end of the linear region. Hence, by Definition 1, it is the proportional limit. The stress at point A is $\sigma_{prop} = 128$ ksi.   ANS.

(b) The stress at point B in Figure 3.11 is the largest stress on the stress–strain curve. Hence, by Definition 2, it is the ultimate stress, $\sigma_{ult} = 144$ ksi.   ANS.

(c) The offset strain of 0.004 (or 0.4%) corresponds to point C. We can draw a line parallel to OA from point C, which intersects the stress–strain curve at point D. Per Definition 8, the stress at point D is the offset yield stress that has a value of $\sigma_{yield} = 132$ ksi.   ANS.

(d) Per Equation (3.1), the modulus of elasticity E is the slope of line OA. Using the triangle at point I we can find E,

$$E = \frac{96 - 64}{0.006 - 0.004}$$

ANS. $E = 16,000$ ksi

(e) At point F the stress is 136 ksi. Per Definition 14, we can find the tangent modulus by finding the slope of the tangent at F,

$$E_t = \frac{140 - 132}{0.026 - 0.014}$$

ANS. $E_t = 666.7$ ksi

Per Definition 15, the slope of line OF is the secant modulus of elasticity at 136 ksi. We can use triangle OFG to calculate the slope of OF,

$$E_s = \frac{136 - 0}{0.02 - 0}$$

ANS. $E_s = 6800$ ksi

(f) To find the plastic strain at 136 ksi, we draw a line parallel to OA through point F. Following the description in Figure 3.4, OH represents the plastic strain. We know that the value of plastic strain will be between 0.01 and 0.012. We can do a more accurate calculation by noting that the plastic strain OH is the total strain OG minus the elastic strain HG. We find the elastic strain by dividing the stress at F (136 ksi) by the modulus of elasticity E,

$$\varepsilon_{plastic} = \varepsilon_{total} - \varepsilon_{elastic} = 0.02 - \frac{136}{16,000} = 0.0115$$

ANS. $\varepsilon_{plastic} = 11,500 \, \mu$

Consolidate your knowledge   Write all you know about the tension test.

*3.1.4   Strain Energy

In spring and damper designs, the energy stored or dissipated is as significant as are stress and deformation. Similarly in impact, such as in a crash, the dissipation of kinetic energy through plastic deformation is important. Some of the failure theories are based on energy rather than on maximum stress or strain. Minimum-energy principles are an alternative to force balance as a statement of equilibrium. Several material properties and descriptions are associated with the energy stored in the material during deformation.

Definition 16 The energy stored in a body due to deformation is called *strain energy*.

Definition 17 The strain energy per unit volume is called *strain energy density* and is the area underneath the stress–strain curve up to the point of deformation.

The relationship between strain energy and strain energy density is

$$U = \int_V U_0 \, dV \tag{3.4}$$

where U is the strain energy, U_0 is the strain energy density, and V is the volume of the body. Noting that the strain energy density is the area under the curve shown in Figure 3.12, we obtain the following equation:

$$U_0 = \int_0^\varepsilon \sigma \, d\varepsilon \tag{3.5}$$

Equation (3.5) shows that strain energy density has the same dimensions as stress because strain is dimensionless. But the units of strain energy density are units of energy per unit volume, which are different from those of stress. The units for strain energy density are $N \cdot m/m^3$, J/m^3, $in \cdot lb/in^3$, or $ft \cdot lb/ft^3$.

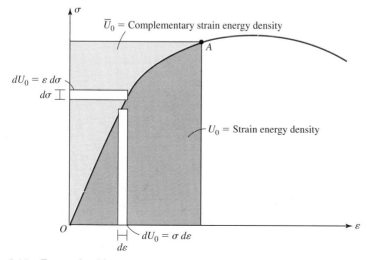

Figure 3.12   Energy densities.

Another related concept is the complementary strain energy density $\bar{U}_0$, also shown in Figure 3.12 and defined as

$$\bar{U}_0 = \int_0^\sigma \varepsilon \, d\sigma \qquad (3.6)$$

Two material constants are now defined and shown in Figure 3.13.

Definition 18 The strain energy density at the yield point is called *modulus of resilience.*

Definition 19 The strain energy density at rupture is called *modulus of toughness.*

Modulus of resilience is a measure of recoverable (elastic) energy per unit volume that can be stored in a material. Since a spring is designed to operate in the elastic range, the higher the modulus of resilience, the more energy it can store.

Modulus of toughness is a measure of the energy per unit volume that can be absorbed by a material without breaking. The modulus of toughness is important in resistance to cracks and crack propagation. The plot in Figure 3.13c shows the important difference between a strong material (high ultimate stress) and a tough material (large area under the stress–strain curve).

It should be noted that strain energy density, complementary strain energy density, modulus of resilience, and modulus of toughness are all quantities with units of energy per unit volume.

3.1.5 Linear Strain Energy Density

Most engineering structures are designed to function without permanent deformation. Thus most of the problems we will work with involve linear–elastic material. In the linear region stress and strain are related by Hooke's law. Substituting $\sigma = E\varepsilon$ in Equation (3.5) and integrating, we obtain $U_0 = \int_0^\varepsilon E\varepsilon \, d\varepsilon = E\varepsilon^2/2$, which, using Hooke's law, can be rewritten as

$$U_0 = \frac{1}{2}\sigma\varepsilon \qquad (3.7)$$

Equation (3.7) reflects that the strain energy density is equal to the area of the triangle underneath the stress–strain curve in the linear region. If, instead of a normal stress–strain

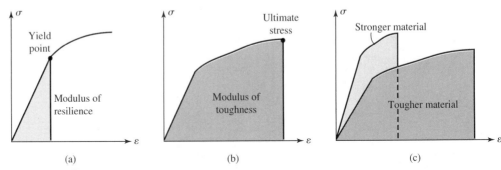

Figure 3.13 Energy-related moduli.

curve, we had a shear stress–strain curve, we could have written a similar expression for strain energy density in terms of shear stress and shear strain,

$$U_0 = \frac{1}{2}\tau\gamma \qquad (3.8)$$

Because strain energy, and hence strain energy density, is a scalar quantity, we can add the strain energy density due to the individual stress and strain components to obtain the following equation:

$$U_0 = \frac{1}{2}[\sigma_{xx}\varepsilon_{xx} + \sigma_{yy}\varepsilon_{yy} + \sigma_{zz}\varepsilon_{zz} + \tau_{xy}\gamma_{xy} + \tau_{yz}\gamma_{yz} + \tau_{zx}\gamma_{zx}] \qquad (3.9)$$

EXAMPLE 3.2

For the titanium alloy in Example 3.1, determine:

(a) Modulus of resilience. Use proportional limit as an approximation for yield point.

(b) Strain energy density at a stress level of 136 ksi.

(c) Complementary strain energy density at a stress level of 136 ksi.

(d) Modulus of toughness.

PLAN

(a) We can identify the proportional limit (yield point) on the curve and find the area of the triangle to obtain the modulus of resilience per Definition 18.

(b) On the stress–strain curve we locate the stress value of 136 ksi. The area underneath the curve gives us the strain energy density.

(c) We can subtract the area of part (b) from the rectangle formed by the axis and lines parallel to the axis that pass through the 136-ksi stress value to obtain the complementary strain energy density.

(d) We can identify the ultimate stress point on the stress–strain curve and find the area underneath it to obtain the modulus of toughness per Definition 19.

Solution

(a) The yield point corresponds to point A in Figure 3.14. The area of the triangle OAA_1 can be calculated as shown by the equation

$$AOA_1 = \frac{128 \times 0.008}{2} = 0.512 \qquad (E1)$$

ANS.   Per Definition 18, the modulus of resilience is 0.512 in · kips/in^3.

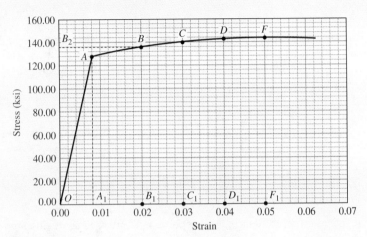

Figure 3.14 Area under curve in Example 3.2.

(b) Point B in Figure 3.14 is at 136 ksi. The strain energy density at point B is the area AOA_1 plus the area AA_1BB_1. The area AA_1BB_1 is the area of a trapezoid and can be found as shown by Equation (3.2),

$$AA_1BB_1 = \frac{(128 + 136)\,0.012}{2} = 1.584 \qquad \text{(E2)}$$

$$BB_1CC_1 = \frac{(136 + 140)\,0.010}{2} = 1.38 \qquad \text{(E3)}$$

$$CC_1DD_1 = \frac{(140 + 142)\,0.010}{2} = 1.41 \qquad \text{(E4)}$$

$$DD_1FF_1 = \frac{(142 + 144)\,0.010}{2} = 1.43 \qquad \text{(E5)}$$

Thus the strain energy density at B is $U_B = 0.512 + 1.584$

ANS. $\qquad U_B = 2.1 \text{ in} \cdot \text{kips/in}^3$.

(c) The complementary strain energy density at B can be found by subtracting U_B from the area of the rectangle OB_2BB_1. Thus, $\overline{U}_B = 136 \times 0.02 - 2.1$

ANS. $\qquad \overline{U}_B = 0.62 \text{ in} \cdot \text{kips/in}^3$.

(d) The ultimate stress corresponds to point F on the graph. The area underneath the curve can be calculated by approximating the curve by a series of straight lines AB, BC, CD, and DF, as shown in Figure 3.14. The total area is the sum of the areas given by

Equations (E1) through (E5). The total sum of the areas is 6.316.

ANS.   Per Definition 19, the Modulus of toughness is 6.32 in ·
kips/in³.

COMMENTS

Approximation of the curve by a straight line for the purpose of finding areas is the same as using the trapezoidal rule of integration. We
can obtain more accurate results if we use the data in Table 3.3 and
approximate by a straight line between two consecutive data points.
This would become tedious unless we use a spread sheet or a computer
program, as discussed in Appendix B.1.

3.2 LOGIC IN MECHANICS OF MATERIALS

In Chapter 1 we studied the two steps of relating stresses to internal forces and
relating internal forces to external forces. In Chapter 2 we studied the relationship
of strains and displacements. In Section 3.1 we studied the relationship of stresses
and strains. In this section we integrate all these concepts and study the mechanics
of relating displacements to external forces.

The logic that relates the displacement of points on a body to the external
forces is shown symbolically in Figure 3.15. It is possible to start at any point in
the logic and move either in the clockwise direction shown by the filled
arrows ➡ or in the counterclockwise direction shown by the hollow arrows
▷. It is not possible to relate displacement directly to external forces without
imposing limitations and making assumptions regarding the geometry of the body,
material behavior, and external loading. This is emphasized in Figure 3.15 by the
absence of arrows directly linking displacement and external forces. In other
words, displacements cannot be related to external forces directly but require
assumptions on geometry, material behavior, and external forces.

The starting point in the logical progression depends on the information we
have, or information we can deduce about a particular variable. If the material
model is simple, then it is possible to deduce the behavior of stresses, as we did in
Chapter 1. But as the complexity in material models grows, so does the complexity
of stress distributions, and deducing stress distribution becomes increasingly difficult. Unlike stresses, displacements can be measured directly or observed or
deduced from geometric considerations. The theories for axial rods, torsion of
shafts, and bending of beams described in later chapters will be developed by
approximating displacements and relating these displacements to external forces
and moments using the logic shown in Figure 3.15.

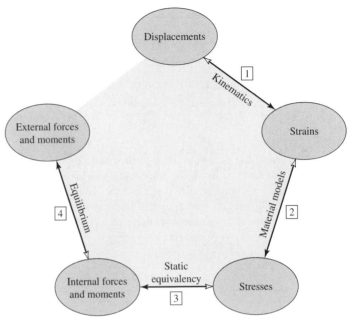

Figure 3.15 Logic in structural analysis.

Examples 3.3 and 3.4 demonstrate the flow of logic shown in Figure 3.15. The modular character of the logic permits the addition of complexities without changing the logical progression of derivation or solution of the problem, as demonstrated by Example 3.5.

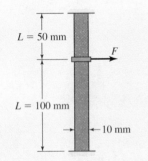

Figure 3.16 Geometry in Example 3.3.

EXAMPLE 3.3

A rigid plate is attached to two 10 mm × 10 mm square bars (Figure 3.16). The bars are made of hard rubber with a shear modulus $G = 1.0$ MPa. The rigid plate is constrained to move horizontally due to action of the force F. If the horizontal movement of the plate is 0.5 mm, determine the force F assuming uniform shear strain in each bar.

PLAN

We can draw an approximate deformed shape and calculate the shear strain in each bar. Using Hooke's law we can find the shear stress in each bar. By multiplying the shear stress by the area we can find the equivalent internal shear force. By drawing the free-body diagram of the rigid plate we can relate the internal shear force to the external force F and determine F.

Solution

1. *Strain calculation:* Assuming small strain we can find the shear strain in each bar by drawing an approximate deformed shape, as shown in Figure 3.17,

$$\tan \gamma_{AB} \approx \gamma_{AB} = \frac{0.5}{100} = 5000 \ \mu\text{rad}$$

$$\tan \gamma_{CD} \approx \gamma_{CD} = \frac{0.5}{50} = 10{,}000 \ \mu\text{rad}$$

2. *Stress calculation:* From Hooke's law $\tau = G\gamma$ we can find the shear stress in each bar,

$$\tau_{AB} = 10^6 \times 5000 \times 10^{-6} = 5000 \ \text{N/m}^2$$

$$\tau_{CD} = 10^6 \times 10{,}000 \times 10^{-6} = 10{,}000 \ \text{N/m}^2$$

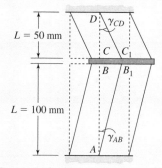

Figure 3.17 Deformed geometry.

3. *Internal force calculation:* The cross-sectional area of the bar is $A = 100 \ \text{mm}^2$. Assuming uniform shear stress, we can find the shear force in each bar,

$$V_{AB} = \tau_{AB} A = 5000 \times 100 \times 10^{-6} = 0.5 \ \text{N}$$

$$V_{CD} = \tau_{CD} A = 10{,}000 \times 100 \times 10^{-6} = 1.0 \ \text{N}$$

4. *External force calculation:* We can make imaginary cuts on either side of the rigid plate and draw the free-body diagram (Figure 3.18). From equilibrium of the rigid plate we can obtain the external force F,

$$F = V_{AB} + V_{CD} = 1.5 \ \text{N}$$

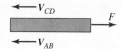

Figure 3.18 Free-body diagram.

EXAMPLE 3.4

The steel bars ($E = 200$ GPa) in the truss shown in Figure 3.19 have cross-sectional area of 100 mm². The displacements u and v of the pins in the x and y directions, respectively, were found to be

$$u_B = -0.500 \ \text{mm} \qquad v_B = -2.714 \ \text{mm}$$

$$u_C = -1.000 \ \text{mm} \qquad v_C = -6.428 \ \text{mm}$$

$$u_D = 1.300 \ \text{mm} \qquad v_D = -2.714 \ \text{mm}$$

Determine the forces F_1 and F_2.

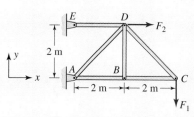

Figure 3.19 Pin displacements in Example 3.4.

PLAN

We can find strains using small-strain approximation as in Example 2.8.
Following the logic in Figure 3.15 we can find stresses and then the
internal force in each member. We can then draw free-body diagrams of
joints C and D to find the forces F_1 and F_2.

Solution

1. *Strain calculations:* The strains in the horizontal and vertical members can be found directly from the displacements,

$$\varepsilon_{AB} = \frac{u_B - u_A}{L_{AB}} = -0.250 \times 10^{-3} \qquad \varepsilon_{BC} = \frac{u_C - u_B}{L_{BC}} = -0.250 \times 10^{-3}$$

$$\varepsilon_{ED} = \frac{u_D - u_E}{L_{ED}} = 0.650 \times 10^{-3} \qquad \varepsilon_{BD} = \frac{v_D - v_B}{L_{BD}} = 0$$

For the inclined member AD we can find the deformation using the
dot product, as was done in Example 2.8. We first find the relative
displacement vector $\overline{\mathbf{D}}_{AD}$ and then take a dot product with the unit
vector $\overline{\mathbf{i}}_{AD}$,

$$\overline{\mathbf{D}}_{AD} = (u_D\overline{\mathbf{i}} + v_D\overline{\mathbf{j}}) - (u_A\overline{\mathbf{i}} + v_A\overline{\mathbf{j}}) = (1.3\overline{\mathbf{i}} - 2.714\overline{\mathbf{j}}) \text{ mm}$$

$$\overline{\mathbf{i}}_{AD} = \cos 45\ \overline{\mathbf{i}} + \sin 45\ \overline{\mathbf{j}} = 0.707\overline{\mathbf{i}} + 0.707\overline{\mathbf{j}}$$

$$\delta_{AD} = \overline{\mathbf{D}}_{AD} \cdot \overline{\mathbf{i}}_{AD} = 1.3 \times 0.707 + -2.714 \times 0.707 = -1.000 \text{ mm}$$

$$L_{AD} = 2.828 \text{ m}$$

Thus

$$\varepsilon_{AD} = \frac{\delta_{AD}}{L_{AD}} = -0.3535 \times 10^{-3}$$

Similarly for member CD we obtain

$$\overline{\mathbf{D}}_{CD} = (u_D\overline{\mathbf{i}} + v_D\overline{\mathbf{j}}) - (u_C\overline{\mathbf{i}} + v_C\overline{\mathbf{j}}) = 2.3\overline{\mathbf{i}} + 3.714\overline{\mathbf{j}} \text{ mm}$$

$$\overline{\mathbf{i}}_{CD} = -\cos 45\ \overline{\mathbf{i}} + \sin 45\ \overline{\mathbf{j}} = -0.707\overline{\mathbf{i}} + 0.707\overline{\mathbf{j}}$$

$$\delta_{CD} = \overline{\mathbf{D}}_{CD} \cdot \overline{\mathbf{i}}_{CD} = 2.3(-0.707) + 3.714 \times 0.707 = 1.000 \text{ mm}$$

$$L_{CD} = 2.828 \text{ m}$$

Thus

$$\varepsilon_{CD} = \frac{\delta_{CD}}{L_{CD}} = 0.3535 \times 10^{-3}$$

2. *Stress calculations:* From Hooke's law $\sigma = E\varepsilon$, we can find stresses in each member as follows:

$$\sigma_{AB} = 200 \times 10^9(-0.250 \times 10^{-3}) \text{ N/m}^2 = 50 \text{ MPa (C)}$$

$$\sigma_{BC} = 200 \times 10^9(-0.250 \times 10^{-3}) \text{ N/m}^2 = 50 \text{ MPa (C)}$$

$$\sigma_{ED} = 200 \times 10^9(0.650 \times 10^{-3}) \text{ N/m}^2 = 130 \text{ MPa (T)}$$

$$\sigma_{BD} = 200 \times 10^9(0.000 \times 10^{-3}) \text{ N/m}^2 = 0$$

$$\sigma_{AD} = 200 \times 10^9(-0.3535 \times 10^{-3}) \text{ N/m}^2 = 70.7 \text{ MPa (C)}$$

$$\sigma_{CD} = 200 \times 10^9(0.3535 \times 10^{-3}) \text{ N/m}^2 = 70.7 \text{ MPa (T)}$$

3. *Internal force calculations:* The internal normal force can be found from $N = \sigma A$, where the cross-sectional area is $A = 100 \times 10^{-6} \text{ m}^2$. This yields the following internal forces:

$$N_{AB} = 5 \text{ kN (C)} \qquad N_{BC} = 5 \text{ kN (C)}$$
$$N_{ED} = 13.0 \text{ kN (T)} \qquad N_{BD} = 0$$
$$N_{AD} = 7.07 \text{ kN (C)} \qquad N_{CD} = 7.07 \text{ kN (T)}$$

4. *External forces:* We draw free-body diagrams of pins C and D and calculate the external forces (Figures 3.20 and 3.21).

$$\sum F_y \qquad N_{CD} \sin 45 - F_1 = 0$$

ANS. $F_1 = 5$ kN

Figure 3.20 Free-body diagram of joint C.

Check
The forces in the x direction must also be in equilibrium. Thus $N_{BC} = N_{CD} \cos 45 = 7.07 \cos 45 = 5$, which checks with the value we calculated.

$$\sum F_x \qquad F_2 + N_{CD} \sin 45 + N_{AD} \sin 45 - N_{ED} = 0$$

ANS. $F_2 = 3$ kN

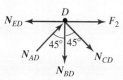

Figure 3.21 Free-body diagram of joint D.

Check
The forces in the y direction must also be in equilibrium. With N_{BD} equal to zero the force balance in the y direction implies that N_{AD} should be equal to N_{CD}, which checks with the values calculated.

COMMENT
Notice the direction of the internal forces. Forces that are pointed into the joint are compressive and the forces pointed away from the joint are tensile.

(a)

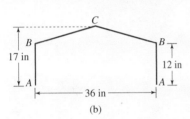

(b)

Figure 3.22 Approximation of stretch cord path on top of canoe in Example 3.5.

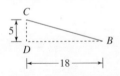

EXAMPLE 3.5

A canoe on top of a car is tied down using rubber stretch cords, as shown in Figure 3.22a. The undeformed length of the stretch cord is 40 in. The initial diameter of the cord is $d = 0.5$ in and the modulus of elasticity of the cord is $E = 510$ psi. Assume that the path of the stretch cord over the canoe can be approximated as shown in Figure 3.22b. Determine the approximate force exerted by the cord on the carrier of the car.

PLAN

We can find the stretched length L_f of the cord from geometry. Knowing L_f and $L_0 = 40$ in, we can find the average normal strain in the cord from Equation (2.1). Using the modulus of elasticity we can find the average normal stress in the cord from Hooke's law, given by Equation (3.1). Knowing the diameter of the cord, we can find the cross-sectional area of the cord and multiply it with the normal stress to obtain the tension in the cord. If we make an imaginary cut in the cord just above A, we see that the tension in the cord is the force exerted on the carrier.

Solution

1. *Strain calculations:* We can find the length BC from the Pythagorean theorem and, noting the symmetry, find the total length L_f of the stretched cord and the average normal strain,

$$BC = \sqrt{5^2 + 18^2} = 18.68$$

$$L_f = 2(AB + BC) = 61.36 \text{ in} \tag{E1}$$

$$\varepsilon = \frac{L_f - L_0}{L_0} = \frac{61.36 - 40}{40} = 0.5341 \tag{E2}$$

2. *Stress calculation:* From Hooke's law we can find the stress as

$$\sigma = E\varepsilon = 510 \times 0.5341 = 272.38 \text{ psi} \tag{E3}$$

3. *Internal force calculations:* We can find the cross-sectional area from the given diameter $d = 0.5$ in and multiply it with the stress to obtain the internal tension,

$$A = \frac{\pi d^2}{4} = \frac{\pi 0.5^2}{4} = 0.1963 \text{ in}^2 \tag{E4}$$

$$T = \sigma A = 0.1963 \times 272.38 = 53.5 \text{ lb} \tag{E5}$$

4. *Reaction force calculation:* We can make a cut just above A and draw the free-body diagram to calculate the force R exerted on the carrier,

$$\text{ANS.} \quad R = T = 53.5 \text{ lb} \tag{E6}$$

COMMENTS

1. Unlike the previous two examples, where relatively accurate solutions would be obtained, in this example we have large strains and several other approximations as elaborated in the next comment. The only thing we can say with some confidence is that the answer has the right order of magnitude.

2. The following approximations were made in this example:

   (a) The path of the cord should have been an inclined straight line between the carrier rail and the point of contact on the canoe, and then the path should have been the contour of the canoe.

   (b) The strain along the cord is nonuniform, which we approximated by a uniform average strain.

   (c) The stress–strain curve of the rubber cord is nonlinear. Thus as the strain changes along the length, so does the modulus of elasticity E, and we need to account for this variation of E in the calculation of stress.

   (d) The cross-sectional area for rubber will change significantly with strain and must be accounted for in the calculation of the internal tension.

3. Depending on the need of our accuracy we can include additional complexities to address the error from the preceding approximations.

   (a) Suppose we did a better approximation of the path as described in part (2a) but made no other changes. In such a case the only change would be in the calculation of L_f in Equation (E1) (see Problem 2.71), but the rest of the equations would remain the same.

   (b) Suppose we make marks on the cord every 2 in before we stretch it over the canoe. We can then measure the distance between two consecutive marks when the cord is stretched. Now we have L_f for each segment and can repeat the calculation for each segment (see Problem 3.59).

   (c) Suppose, in addition to the above two, we have the stress–strain curve of the stretch cord material. Now we can use the tangent modulus in Hooke's law for each segment, and hence we can get more accurate stresses in each segment. We can then calculate the internal force as before (see Problem 3.60).

   (d) Rubber has a Poisson's ratio of 0.5. Knowing the longitudinal strain from Equation (E2) for each segment, we can compute the transverse strain in each segment and find the diameter of the cord in the stretched position in each segment. This will give us a more accurate area of cross section, and hence a more accurate value of internal tension in the cord (see Problem 3.61).

4. The preceding comments demonstrate how complexities can be added one at a time to improve the accuracy of a solution. In a similar manner, complexities can be added to the theories which we shall derive for axial members, shafts, and beams in Chapters 4 through 6. "Stretch yourself" problems in these chapters are problems based on adding complexities to the derived theory. Which complexity to include depends on the individual case and our need for accuracy.

QUICK TEST 3.1   Time: 15 minutes/Total: 20 points

Grade yourself using the answers given in Appendix G. Each question is worth two points.

1. What are the typical units of modulus of elasticity and Poisson's ratio in the metric system?
2. Define offset yield stress.
3. What is strain hardening?
4. What is necking?
5. What is the difference between proportional limit and yield point?
6. What is the difference between a brittle material and a ductile material?
7. What is the difference between linear material behavior and elastic material behavior?
8. What is the difference between strain energy and strain energy density?
9. What is the difference between modulus of resilience and modulus of toughness?
10. What is the difference between a strong material and a tough material?

Stress-strain curves

A tensile test specimen having a diameter of 10 mm and a gage length of 50 mm was tested to fracture. The stress–strain curve from the tension test is shown in Figure P3.1. The lower plot is the expanded region OAB and associated with the strain values given on the lower scale. Solve Problems 3.1 through 3.8 using this graph.

3.1 Determine the axial force acting on the specimen when it is extended by (a) 0.2 mm; (b) 4.0 mm.

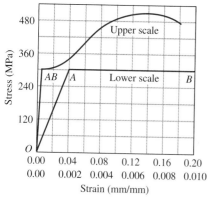

Figure P3.1

3.2 Determine the extension of the specimen when the axial force on the specimen is 33 kN.

3.3 Determine: (a) the ultimate stress; (b) the fracture stress.

3.4 Determine: (a) the modulus of elasticity; (b) the proportional limit.

3.5 Determine the total strain, the elastic strain, and the plastic strain when the axial force on the specimen is 33 kN.

3.6 After the axial load was removed, the specimen was observed to have a length of 54 mm. What was the maximum axial load applied to the specimen?

3.7 Determine the offset yield stress at 0.2%.

3.8 At the stress level of 420 MPa, determine: (a) the tangent modulus; (b) the secant modulus.

A tensile test specimen having a diameter of $\frac{5}{8}$ in and a gage length of 2 in was tested to fracture. The stress–strain curve from the tension test is shown in Figure P3.9. The lower plot is the expanded region OAB and associated with the strain values given on the lower scale. Solve Problems 3.9 through 3.16 using this graph.

3.9 Determine the axial force acting on the specimen when it is extended by (a) 0.006 in; (b) 0.120 in.

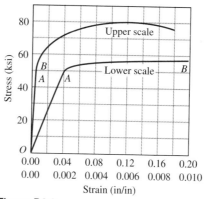

Figure P3.9

3.10 Determine the extension of the specimen when the axial force on the specimen is 20 kips.

3.11 Determine: (a) the ultimate stress; (b) the fracture stress.

3.12 Determine: (a) the modulus of elasticity; (b) the proportional limit.

3.13 Determine the total strain, the elastic strain, and the plastic strain when the axial force on the specimen is 20 kips.

3.14 After the axial load was removed, the specimen was observed to have a length of 2.12 in. What was the maximum axial load applied to the specimen?

3.15 Determine the offset yield stress at 0.1%.

3.16 At the stress level of 72 kips, determine: (a) the tangent modulus; (b) the secant modulus.

3.17 A 12 mm × 12 mm square metal alloy having a gage length of 50 mm was tested in tension. The results are given in Table P3.17. Draw the stress–strain curve and calculate the following quantities. (Use of spread sheet is recommended.)

(a) Modulus of elasticity.
(b) Proportional limit.
(c) Yield stress at 0.2% offset.
(d) Tangent modulus at a stress level of 1400 MPa.
(e) Secant modulus at a stress level of 1400 MPa.
(f) Plastic strain at a stress level of 1400 MPa.

TABLE P3.17

Load (kN)	Change in Length (mm)
0.00	0.00
17.32	0.02
60.62	0.07
112.58	0.13
147.22	0.17
161.18	0.53
168.27	1.10
176.03	1.96
182.80	2.79
190.75	4.00
193.29	4.71
200.01	5.80
204.65	7.15
209.99	8.88
212.06	9.99
212.17	11.01
208.64	11.63
204.99	12.03
199.34	12.31
192.15	12.47
185.46	12.63
Break	

3.18 A mild steel specimen of 0.5 in diameter and a gage length of 2 in was tested in tension. The test results are reported Table P3.18. Draw the stress–strain curve and calculate the following quantities. (Use of spread sheet is recommended.)

(a) Modulus of elasticity.
(b) Proportional limit.
(c) Yield stress at 0.05% offset.
(d) Tangent modulus at a stress level of 50 ksi.
(e) Secant modulus at a stress level of 50 ksi.
(f) Plastic strain at a stress level of 50 ksi.

TABLE P3.18

Load (10^3 lb)	Change in Length (10^{-3} in)
0.00	0.00
3.11	1.2
7.24	3.4
7.50	22.2
7.70	8.76
7.90	19.05
8.16	28.70
8.46	37.73
8.82	47.18
9.32	59.06
9.86	70.85
10.40	84.23
10.82	97.85
11.18	112.10
11.72	140.40
11.99	161.21
12.27	192.65
12.41	214.22
12.55	245.93
12.70	283.47
12.77	316.36
12.84	363.10
12.04	385.34
11.44	396.03
10.71	406.42
9.96	414.72
Break	

Material constants

3.19 A rectangular bar has a cross-sectional area of 2 in² and an undeformed length of 5 in, as shown in Figure 3.19. When a load $P = 50,000$ lb is applied, the bar deforms to a position shown by the colored shape. Determine the modulus of elasticity and the Poisson's ratio of the material.

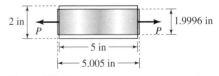

Figure P3.19

0 A force $P = 20$ kips is applied to a
plate that is attached to a square bar, as
wn in Figure P3.20. If the plate moves a
ance of 0.005 in, determine the modulus
lasticity.

1 A force $P = 20$ kips is applied to a
plate that is attached to a square bar, as
wn in Figure P3.21. If the plate moves a
ance of 0.0125 in, determine the shear
lulus of elasticity.

2 A circular bar of 200-mm length and
nm diameter is subjected to a tension test.
to an axial force of 77 kN, the bar is seen
longate by 4.5 mm and the diameter is
to reduce by 0.162 mm. Determine the
ulus of elasticity and the shear modulus
asticity. Assume line AB remains straight.

3 A circular bar of 6-in length and 1-in
neter is made from a material with a mod-
of elasticity $E = 30,000$ ksi and a Poisson's
$v = \frac{1}{3}$. Determine the change in length
diameter of the bar when a force of
ips is applied to the bar.

4 A circular bar of 15-in length and
-in diameter is made from a material with
odulus of elasticity $E = 28,000$ ksi and a
son's ratio $v = 0.32$. Due to a force the
is seen to elongate by 0.04 in. Determine
change in diameter and the applied force.

5 A 1 in × 1 in square bar is 20 in long
is made from a material that has a Pois-
is ratio of $\frac{1}{3}$. In a tension test, the bar is
to elongate by 0.03 in. Determine the
entage change in volume of the bar.

6 A circular bar of 50-in length and
diameter is made from a material with a
ulus of elasticity $E = 28,000$ ksi and a
son's ratio $v = 0.32$. Determine the per-
age change in volume of the bar when an
force of 20 kips is applied.

7 An aluminum rectangular bar has a
section of 25 mm × 50 mm and a length of
mm. The modulus of elasticity $E = 10,000$
nd the Poisson's ratio $v = 0.25$. Determine
percentage change in the volume of the bar
an axial force of 300 kN is applied.

3.28 A circular bar of length L and diameter
d is made from a material with a modulus of
elasticity E and a Poisson's ratio v. Assuming
small strain, show that the percentage change in
the volume of the bar when an axial force P is
applied and given as $400P(1 - 2v)/(E\pi d^2)$. Note
the percentage change is zero when $v = 0.5$.

3.29 A rectangular bar has a cross-sectional
dimensions of $a \times b$ and a length L. The bar
material has a modulus of elasticity E and a
Poisson's ratio v. Assuming small strain,
show that the percentage change in the volume of the bar when an axial force P is
applied given by $100P(1 - 2v)/(Eab)$. Note
the percentage change is zero when $v = 0.5$.

Strain energy

3.30 What is the strain energy in the bar of
Problem 3.19?

3.31 What is the strain energy in the bar of
Problem 3.20?

3.32 What is the strain energy in the bar of
Problem 3.21?

3.33 A circular bar of length L and diameter of d is made from a material with a modulus of elasticity E and a Poisson's ratio v. In
terms of the given variables, what is the linear
strain energy in the bar when axial load P is
applied to the bar?

3.34 A rectangular bar has a cross-
sectional dimensions of $a \times b$ and a length L.
The bar material has a modulus of elasticity
E and a Poisson's ratio v. In terms of the
given variables, what is the linear strain energy
in the bar when axial load P is applied to
the bar?

3.35 For the material having the stress–
strain curve shown in Figure P3.1, determine:
(a) Modulus of resilience. Use proportional
    limit as an approximation for yield point.
(b) Strain energy density at a stress level of
    420 MPa.
(c) Complementary strain energy density at a
    stress level of 420 MPa.
(d) Modulus of toughness.

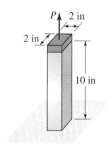

Figure P3.20

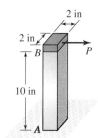

Figure P3.21

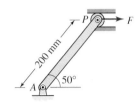

Figure P3.39

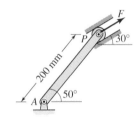

Figure P3.40

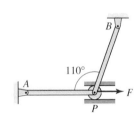

Figure P3.41

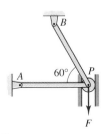

Figure P3.42

3.36 For the material having the stress–strain curve shown in Figure P3.9, determine:
(a) Modulus of resilience. Use proportional limit as an approximation for yield point.
(b) Strain energy density at a stress level of 72 ksi.
(c) Complementary strain energy density at a stress level of 72 ksi.
(d) Modulus of toughness.

3.37 For the metal alloy given in Problem 3.17, determine:
(a) Modulus of resilience. Use proportional limit as an approximation for yield point.
(b) Strain energy density at a stress level of 1400 MPa.
(c) Complementary strain energy density at a stress level of 1400 MPa.
(d) Modulus of toughness.

3.38 For the mild steel given in Problem 3.18, determine:
(a) Modulus of resilience. Use proportional limit as an approximation for yield point.
(b) Strain energy density at a stress level of 50 ksi.
(c) Complementary strain energy density at a stress level of 50 ksi.
(d) Modulus of toughness.

Logic in mechanics

In Problems 3.39 and 3.40 the roller at P slides in the slot by an amount $\delta_P = 0.25$ mm due to the force F. Member AP has a cross-sectional area A = 100 mm² and a modulus of elasticity E = 200 GPa. If the roller moves by the amount given, determine the force F.

3.39 Determine the applied force F in Figure P3.39.

3.40 Determine the applied force F in Figure P3.40.

In Problems 3.41 through 3.43 a roller slides in a slot by the amount $\delta_P = 0.25$ mm in the direction of the force F. Each bar has a cross-sectional area A = 100 mm² and a modulus of elasticity E = 200 GPa. Bars AP and BP have lengths $L_{AP} = 200$ mm and $L_{BP} = 250$ mm, respectively. Determine the applied force F.

3.41 Determine the applied force F in Figure P3.41.

3.42 Determine the applied force F in Figure P3.42.

3.43 Determine the applied force F in Figure P3.43.

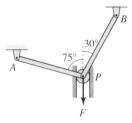

Figure P3.43

3.44 A little boy is shooting paper darts at his friends using a rubber band that has an unstretched length of 7 in. The piece of rubber band between points A and B is pulled to form the two sides AC and CB of a triangle, as shown in Figure P3.44. Assume the same normal strain in AC and CB, and the rubber band around the thumb and forefinger is a total of 1 in. The cross-sectional area of the band is $\frac{1}{128}$ in² and the rubber has a modulus of elasticity E = 150 psi. Determine the approximate force F and the angle θ at which the paper dart leaves the boy's hand.

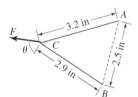

Figure P3.44

3.45 Three poles are pin connected to a ring at P and to the supports on the ground. The coordinates of the four points are given in Figure P3.45. All poles have cross-sectional areas A = 1 in² and a modulus of elasticity E = 10,000 ksi. If under the action of force F the ring at P moves vertically by the distance $\delta_P = 2$ in, determine the force F.

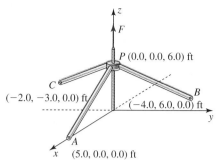

Figure P3.45

3.46 A gap of 0.004 in exists between a rigid bar and bar A before a force F is applied (Figure P3.46). The rigid bar is hinged at point C. Due to force F the strain in bar A was found to be -500 μin/in. The lengths of bars A and B are 30 and 50 in, respectively. Both bars have cross-sectional areas $A = 1$ in^2 and a modulus of elasticity $E = 30{,}000$ ksi. Determine the applied force F.

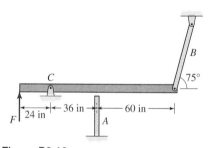

Figure P3.46

The pins in the truss shown in Figure P3.47 are displaced by u and v in the x and y directions, respectively, as given:

$u_A = -4.6765$ mm	$v_A = 0$
$u_B = -3.3775$ mm	$v_B = -8.8793$ mm
$u_C = -2.0785$ mm	$v_C = -9.7657$ mm
$u_D = -1.0392$ mm	$v_D = -8.4118$ mm
$u_E = 0.0000$ mm	$v_E = 0.0000$ mm
$u_F = -3.2600$ mm	$v_F = -8.4118$ mm
$u_G = -2.5382$ mm	$v_G = -9.2461$ mm
$u_H = -1.5500$ mm	$v_H = -8.8793$ mm

All rods in the truss have cross-sectional areas $A = 100$ mm^2 and a modulus of elasticity $E = 200$ GPa. Solve Problems 3.47 through 3.49.

3.47 Determine the external forces P_1 and P_2 in the truss.

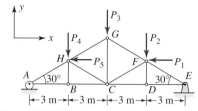

Figure P3.47

3.48 Determine the external force P_3 in the truss.

3.49 Determine the external forces P_4 and P_5 in the truss.

Stretch yourself

A circular rod of 15-mm diameter is acted upon by a distributed force p(x) that has the units of kN/m, as shown in Figure P3.50. The modulus of elasticity of the rod is 70 GPa. In Problems 3.50 and 3.51 determine the distributed force p(x) if the displacement u(x) in the x direction is as given. (x is measured in meters.)

3.50 $u(x) = 30(x - x^2)10^{-6}$ m.

3.51 $u(x) = 50(x^2 - 2x^3)10^{-6}$ m.

3.52 Consider the beam shown in Figure P3.52. Due to the action of the forces, the displacement in the x direction was found to be $u = [(60x + 80xy - x^2y)/180]10^{-3}$ in. The modulus of elasticity of the beam is 30,000 ksi. Determine: The statically equivalent internal normal force N and the internal bending moment M_z acting at point O at a section at $x = 20$ in. Assume an unknown shear stress is acting on the cross-section.

Computer problems

3.53 Using numerical integration and the data in Problem 3.17, determine the strain energy density when the specimen shows an average strain of 0.18 mm/mm.

3.54 Using numerical integration, determine the modulus of toughness for the material described in Problem 3.17.

Figure P3.50

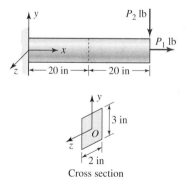

Cross section

Figure P3.52

3.55   Assume that the stress–strain curve
after yield stress in Problem 3.17 is described
by the quadratic equation $\sigma = a + b\varepsilon + c\varepsilon^2$.
(a) Determine the coefficients a, b, and c by
   the least-squares method.
(b) Find the strain energy density when the
   specimen shows an average strain of
   0.18 mm/mm.
(c) Find the tangent modulus of elasticity at a
   stress level of 1400 MPa.

3.56   Using numerical integration and the
data in Problem 3.18, determine the strain
energy density when the specimen shows an
average strain of 0.15 in/in.

3.57   Using numerical integration, determine the modulus of toughness for the material described in Problem 3.18.

3.58   Assume that the stress–strain curve
after yield stress in Problem 3.18 is described
by the quadratic equation $\sigma = a + b\varepsilon + c\varepsilon^2$.
(a) Determine the coefficients a, b, and c by
   the least-squares method.
(b) Find the strain energy density when
   the specimen shows an average strain of
   0.15 in/in.
(c) Find the tangent modulus of elasticity at a
   stress level of 50 ksi.

*Marks were made on the cord used for tying the
canoe on top of the car in Example 3.5. These
marks were made every 2 in to produce a total
of 20 segments. The stretch cord is symmetric
with respect to the top of the canoe. The starting
point of the first segment is on the carrier rail of
the car and the end point of the tenth segment is*

TABLE P3.59

Segment Number	Deformed Length (inches)
1	3.4
2	3.4
3	3.4
4	3.4
5	3.4
6	3.4
7	3.1
8	2.7
9	2.3
10	2.2

*on the top of the canoe. The measured length of
each segment is as shown in Table 3.59. Using
this information and the data given in Problems 3.59 through 3.61, determine: (a) the tension in the cord of each segment; (b) the force
exerted by the cord on the carrier of the car.*

3.59   Use the modulus of elasticity $E =$
510 psi and the diameter of the stretch cord
as 0.5 in.

3.60   Use the diameter of the stretch cord as
0.5 in and the following equation for the
stress–strain curve:

$$\sigma = \begin{cases} 1020\varepsilon - 1020\varepsilon^2 \text{ psi} & \varepsilon < 0.5 \\ 255 \text{ psi} & \varepsilon \ge 0.5 \end{cases}$$

3.61   Use the Poisson's ratio $\nu = \frac{1}{2}$ and the
initial diameter of 0.5 in and calculate the
diameter in the deformed position for each
segment. Use the stress–strain relationship
given in Problem 3.60.

3.3   ISOTROPY AND HOMOGENEITY

The description of a material as isotropic or homogeneous has connotations that are
acquiring greater significance with the growth of new materials, in particular composites,[10] in which two or more materials are combined together to produce a stronger or stiffer material than if only a single material were used. As will be seen in
this section, both material descriptions are approximations that are influenced by
several factors.

---

[10]See Section 3.12.3 for a brief description of this rapidly growing area.

The number of material constants that need to be measured depends on the material model we want to incorporate into our analysis. A material model is the relationship[11] between stresses and strains. The simplest model is a linear relationship between stresses and strains. With no additional assumptions, the linear relationship of the six strain components to six stress components can be written as follows:

$$
\begin{aligned}
\varepsilon_{xx} &= C_{11}\sigma_{xx} + C_{12}\sigma_{yy} + C_{13}\sigma_{zz} + C_{14}\tau_{yz} + C_{15}\tau_{zx} + C_{16}\tau_{xy} \\
\varepsilon_{yy} &= C_{21}\sigma_{xx} + C_{22}\sigma_{yy} + C_{23}\sigma_{zz} + C_{24}\tau_{yz} + C_{25}\tau_{zx} + C_{26}\tau_{xy} \\
\varepsilon_{zz} &= C_{31}\sigma_{xx} + C_{32}\sigma_{yy} + C_{33}\sigma_{zz} + C_{34}\tau_{yz} + C_{35}\tau_{zx} + C_{36}\tau_{xy} \\
\gamma_{yz} &= C_{41}\sigma_{xx} + C_{42}\sigma_{yy} + C_{43}\sigma_{zz} + C_{44}\tau_{yz} + C_{45}\tau_{zx} + C_{46}\tau_{xy} \\
\gamma_{zx} &= C_{51}\sigma_{xx} + C_{52}\sigma_{yy} + C_{53}\sigma_{zz} + C_{54}\tau_{yz} + C_{55}\tau_{zx} + C_{56}\tau_{xy} \\
\gamma_{xy} &= C_{61}\sigma_{xx} + C_{62}\sigma_{yy} + C_{63}\sigma_{zz} + C_{64}\tau_{yz} + C_{65}\tau_{zx} + C_{66}\tau_{xy}
\end{aligned}
\tag{3.10}
$$

Equation (3.10) implies that we need 36 material constants to describe the most general linear relationship between stress and strain. However, it can be shown that the matrix formed by the constants C_{ij} is a symmetric matrix (i.e., $C_{ij} = C_{ji}$, where i and j can be any number from 1 to 6). This symmetry can be proven by using the requirement that the strain energy always be positive,[12] but the proof of symmetry is beyond the scope of this book. The symmetry requirement reduces the number of independent constants to 21 for the most general linear relationship between stress and strain.

Equation (3.10) presupposes that the relation between stress and strain in the x direction is different from the relation between stress and strain in the y or z direction. Alternatively stated, Equation (3.10) implies that if we apply a force (stress) in the x direction and observe the deformation (strain), then this deformation will be different from the deformation that will be produced if we apply the same force in the y direction. This phenomenon is not observable by the naked eye for most metals, but if we were to look at the metals at the crystal-size level, then the number of constants needed to describe the stress–strain relationship depends on the crystal structure. Thus at what level are we conducting the analysis—eye level or crystal-size level? If we average the impact of the crystal structure at the eye level, then we have defined the simplest material description:

Definition 20   An *isotropic* material has a stress–strain relationships that are independent of the orientation of the coordinate system at a point.

An *anisotropic* material is a material that is not isotropic. The most general anisotropic material requires 21 independent material constants to describe a linear stress–strain relationships. An isotropic body requires only two[13] independent material constants to describe a linear stress–strain relationships. In between the isotropic material and the most general anisotropic material there are several types of materials, which are discussed briefly in Section 3.11.2.

---

[11]Some call this constitutive equations.

[12]See Section 3.12.1 regarding a controversy over the number of independent constants required in a linear stress–strain relationship.

[13]The general proof is beyond the scope of this book. Example 9.8 and Problem 9.81 show it for isotropic materials starting from simpler equations than Equation (3.10).

The degree of difference in material properties with orientation, the scale at which the analysis is being conducted, and the kind of information that is desired from the analysis are some of the factors that influence whether we treat a material as isotropic or anisotropic.

The three constants[14] that we shall encounter most in this book are the modulus of elasticity E, the shear modulus of elasticity G, and the Poisson's ratio v. In Example 9.8 we shall show that the following relationship exists for isotropic materials:

$$G = \frac{E}{2(1 + v)} \tag{3.11}$$

Homogeneity is another approximation that is often used to describe a material behavior.

Definition 21 A material is said to be *homogeneous* if the material properties are the same at all points in the body. Alternatively, if the material constants C_{ij} are functions of the coordinates x, y, or z, then the material is called nonhomogeneous.

Most materials at the atomic level, the crystalline level, or the grain-size level are nonhomogeneous. The treatment of a material as homogeneous or nonhomogeneous depends once more on the type of information that is to be obtained from the analysis. Homogenization of material properties is a process of averaging[15] different material properties by an overall material property. Any body can be treated as a homogeneous body if the scale at which the analysis is conducted is made sufficiently large.

Isotropic–homogeneous, anisotropic–homogeneous, isotropic–nonhomogeneous, and anisotropic–nonhomogeneous are all possible descriptions of material behavior.

3.4 GENERALIZED HOOKE'S LAW FOR ISOTROPIC MATERIALS

Definition 22 The relationship between stresses and strains in three dimensions is called *generalized Hooke's law*.

We can develop the relationship from the definitions of the three material constants E, v, and G and the assumption of isotropy. The generalized Hooke's law is a stress–strain relationship at a point; hence no assumption of homogeneity needs to be made. In Figure 3.23 normal stresses are applied one at a time. From the definition of the modulus of elasticity we can obtain the strain in the direction of the applied stress, which then is used to get the strains in the perpendicular direction by using the definition of Poisson's ratio.

---

[14]There are other constants used to describe material properties (see Problems 3.81 and 3.93), but for isotropic materials only two are independent constants, that is, all other constants can be found if any two constants are known.

[15]See Problem 4.52 on the rule of mixture, which is used for obtaining an effective modulus of elasticity in composites.

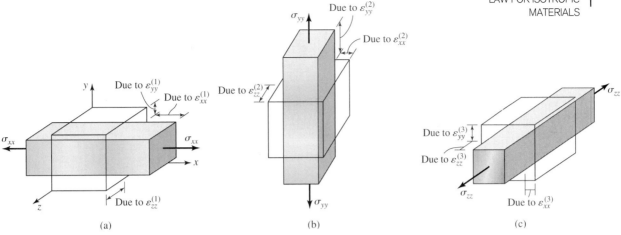

Figure 3.23 Derivation of generalized Hooke's law.

From Figure (3.23*a*) we obtain

$$\varepsilon_{xx}^{(1)} = \frac{\sigma_{xx}}{E}$$

$$\varepsilon_{yy}^{(1)} = -\nu\varepsilon_{xx}^{(1)} = -\nu\left(\frac{\sigma_{xx}}{E}\right)$$

$$\varepsilon_{zz}^{(1)} = -\nu\varepsilon_{xx}^{(1)} = -\nu\left(\frac{\sigma_{xx}}{E}\right)$$

From Figure (3.23*b*) we obtain

$$\varepsilon_{xx}^{(2)} = -\nu\varepsilon_{yy}^{(2)} = -\nu\left(\frac{\sigma_{yy}}{E}\right)$$

$$\varepsilon_{yy}^{(2)} = \frac{\sigma_{yy}}{E}$$

$$\varepsilon_{zz}^{(2)} = -\nu\varepsilon_{xx}^{(2)} = -\nu\left(\frac{\sigma_{yy}}{E}\right)$$

From Figure (3.23*c*) we obtain

$$\varepsilon_{xx}^{(3)} = -\nu\varepsilon_{zz}^{(3)} = -\nu\left(\frac{\sigma_{zz}}{E}\right)$$

$$\varepsilon_{yy}^{(3)} = -\nu\varepsilon_{zz}^{(3)} = -\nu\left(\frac{\sigma_{zz}}{E}\right)$$

$$\varepsilon_{zz}^{(3)} = \frac{\sigma_{zz}}{E}$$

Use of the same E and ν to relate stresses and strains in different directions implicitly assumes isotropy. Notice that no change occurs in the right angles from the application of normal stresses. Thus no shear strain is produced due to normal stresses in a fixed coordinate system for an isotropic material.

Assuming the material is linearly elastic, we can use the principle of superposition to obtain the total strain when all three normal stresses are present simultaneously (i.e., $\varepsilon_{ii} = \varepsilon_{ii}^{(1)} + \varepsilon_{ii}^{(2)} + \varepsilon_{ii}^{(3)}$), as shown in Equations (3.12a) through (3.12c). From the definition of shear modulus given in Equation (3.3), we obtain

Equations (3.12d) through (3.12f).

Generalized Hooke's law:

$$\varepsilon_{xx} = \frac{\sigma_{xx} - v(\sigma_{yy} + \sigma_{zz})}{E} \tag{3.12a}$$

$$\varepsilon_{yy} = \frac{\sigma_{yy} - v(\sigma_{zz} + \sigma_{xx})}{E} \tag{3.12b}$$

$$\varepsilon_{zz} = \frac{\sigma_{zz} - v(\sigma_{xx} + \sigma_{yy})}{E} \tag{3.12c}$$

$$\gamma_{xy} = \frac{\tau_{xy}}{G} \tag{3.12d}$$

$$\gamma_{yz} = \frac{\tau_{yz}}{G} \tag{3.12e}$$

$$\gamma_{zx} = \frac{\tau_{zx}}{G} \tag{3.12f}$$

Stresses and strains are defined at a point, and hence these equations are defined for a point in the material that is isotropic and linearly elastic. The equations are valid for nonhomogeneous material. The nonhomogeneity will make the material constants E, v, and G functions of the spatial coordinates. The use of Poisson's ratio to relate strains in perpendicular directions is valid not only for Cartesian coordinates but for any orthogonal coordinate system. Thus the generalized Hooke's law may be written for any orthogonal coordinate system, such as spherical and polar coordinate systems.

An alternative form[16] for Equations (3.12a) through (3.12c), which may be easier to remember, is the matrix form

$$\left\{ \begin{array}{c} \varepsilon_{xx} \\ \varepsilon_{yy} \\ \varepsilon_{zz} \end{array} \right\} = \frac{1}{E} \left[\begin{array}{ccc} 1 & -v & -v \\ -v & 1 & -v \\ -v & -v & 1 \end{array} \right] \left\{ \begin{array}{c} \sigma_{xx} \\ \sigma_{yy} \\ \sigma_{zz} \end{array} \right\} \tag{3.13}$$

3.5 PLANE STRESS AND PLANE STRAIN

In Chapter 1 we discussed a two-dimensional problem that we called plane stress problem. In Chapter 2 we discussed a two-dimensional problem called plane strain problem. If we take those two definitions and apply them to Equations (3.12a) through (3.12f), we obtain the matrices shown in Figure 3.24. The difference between the two two-dimensional idealizations of material behavior is in the zero and nonzero values of the normal strain and normal stress in the z direction. In plane

---

[16]Another alternative is $\varepsilon_{ii} = [(1 + v)\sigma_{ii} - vI_1]/E$, where $I_1 = \sigma_{xx} + \sigma_{yy} + \sigma_{zz}$.

Plane stress ⟶ $\begin{bmatrix} \sigma_{xx} & \tau_{xy} & 0 \\ \tau_{yx} & \sigma_{yy} & 0 \\ 0 & 0 & 0 \end{bmatrix}$ $\xrightarrow{\text{Generalized Hooke's law}}$ $\begin{bmatrix} \varepsilon_{xx} & \gamma_{xy} & 0 \\ \gamma_{yx} & \varepsilon_{yy} & 0 \\ 0 & 0 & \varepsilon_{zz} = -\dfrac{v}{E}(\sigma_{xx} + \sigma_{yy}) \end{bmatrix}$

Plane strain ⟶ $\begin{bmatrix} \varepsilon_{xx} & \gamma_{xy} & 0 \\ \gamma_{yx} & \varepsilon_{yy} & 0 \\ 0 & 0 & 0 \end{bmatrix}$ $\xrightarrow{\text{Generalized Hooke's law}}$ $\begin{bmatrix} \sigma_{xx} & \tau_{xy} & 0 \\ \tau_{yx} & \sigma_{yy} & 0 \\ 0 & 0 & \sigma_{zz} = v(\sigma_{xx} + \sigma_{yy}) \end{bmatrix}$

Figure 3.24 Stress and strain matrices in plane stress and plane strain.

stress $\sigma_{zz} = 0$, which from Equation (3.12c) implies that the normal strain in the z direction is $\varepsilon_{zz} = -v(\sigma_{xx} + \sigma_{yy})/E$. In plane strain $\varepsilon_{zz} = 0$, which from Equation (3.12c) implies that the normal stress in the z direction is $\sigma_{zz} = v(\sigma_{xx} + \sigma_{yy})$.

To better appreciate the difference between plane stress and plane strain, consider the following example of two plates, shown in Figure 3.25, on which only compressive normal stresses in the x and y directions are applied.

The top and bottom surfaces on the plate in Figure 3.25a are free surfaces (plane stress), but because the plate is free to expand, the deformation (strain) in the z direction is not zero. The plate in Figure 3.25b is constrained from expanding in the z direction by the rigid surfaces. As the material pushes on the plate, a reaction force develops, and this reaction force results in a nonzero value of normal stress in the z direction. Though the example in Figure 3.25 helps explain the difference, it should be emphasized that plane stress or plane strain are often approximations to simplify analysis.

It should be recognized that in plane strain and plane stress conditions there are only three independent quantities, even though the nonzero quantities number more than three. For example, if we know σ_{xx}, σ_{yy}, and τ_{xy}, then we can calculate ε_{xx}, ε_{yy}, γ_{xy}, ε_{zz}, and σ_{zz} for plane stress and plane strain. Similarly, if we know ε_{xx}, ε_{yy}, and γ_{xy}, then we can calculate σ_{xx}, σ_{yy}, τ_{xy}, σ_{zz}, and ε_{zz} for plane stress and plane strain. Thus in both plane stress and plane strain the number of independent stress or strain components is three, although the number of nonzero components is greater than three. Examples (3.6) and (3.7) elaborate on the difference between plane stress and plane strain conditions and the difference between nonzero and independent quantities.

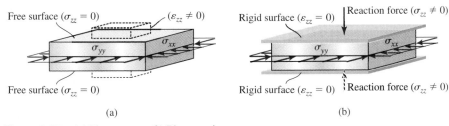

Figure 3.25 (a) Plane stress. (b) Plane strain.

EXAMPLE 3.6

The stresses at a point on steel were found to be $\sigma_{xx} = 15$ ksi (T), $\sigma_{yy} = 30$ ksi (C), and $\tau_{xy} = 25$ ksi. Using $E = 30{,}000$ ksi and $G = 12{,}000$ ksi, determine the strains ε_{xx}, ε_{yy}, γ_{xy}, ε_{zz} and the stress σ_{zz} assuming:

(a) The point is in a state of plane stress.

(b) The point is in a state of plane strain.

PLAN

In both cases the shear strain is the same and can be calculated using Equation (3.12d).

(a) For plane stress $\sigma_{zz} = 0$ and the strains ε_{xx}, ε_{yy}, and ε_{zz} can be found from Equations (3.12a), (3.12b), and (3.12c), respectively.

(b) For plane strain $\varepsilon_{zz} = 0$ and Equation (3.12c) can be used to find σ_{zz}. The stresses σ_{xx}, σ_{yy}, and σ_{zz} can be substituted into Equations (3.12a) and (3.12b) to calculate the normal strains ε_{xx} and ε_{yy}.

Solution   From Equation (3.12d),

$$\gamma_{xy} = \frac{\tau_{xy}}{G} = \frac{25}{12 \times 10^3}$$

ANS. $\gamma_{xy} = 2083 \; \mu$

The Poisson's ratio can be found from Equation (3.11), $G = E / 2(1 + v)$,

$$12{,}000 = \frac{30{,}000}{2(1 + v)} \qquad \text{or} \qquad v = 0.25$$

(a) *Plane stress*: The normal strains in the x, y, and z directions are found from Equations (3.12a), (3.12b), and (3.12c), respectively.

$$\varepsilon_{xx} = \frac{\sigma_{xx} - v(\sigma_{yy} + \sigma_{zz})}{E} = \frac{15 - 0.25(-30)}{30{,}000}$$

ANS. $\varepsilon_{xx} = 750 \; \mu$

$$\varepsilon_{yy} = \frac{\sigma_{yy} - v(\sigma_{zz} + \sigma_{xx})}{E} = \frac{-30 - 0.25 \times 15}{30{,}000}$$

ANS. $\varepsilon_{yy} = -1125 \; \mu$

$$\varepsilon_{zz} = \frac{\sigma_{zz} - v(\sigma_{xx} + \sigma_{yy})}{E} = \frac{0 - 0.25(15 - 30)}{30{,}000}$$

ANS. $\varepsilon_{zz} = 125 \; \mu$

(b) *Plane strain:* From Equation (3.12c), we have $\varepsilon_{zz} = [\sigma_{zz} - \nu(\sigma_{xx} + \sigma_{yy})]/E = 0$ or

$$\sigma_{zz} = \nu(\sigma_{xx} + \sigma_{yy}) = 0.25(15 - 30)$$

$$\text{ANS.} \qquad \sigma_{zz} = 3.75 \text{ ksi (C)}$$

The normal strains in the x and y directions are found from Equations (3.12a) and (3.12b),

$$\varepsilon_{xx} = \frac{\sigma_{xx} - \nu(\sigma_{yy} + \sigma_{zz})}{E} = \frac{15 - 0.25(-30 - 3.75)}{30,000}$$

$$\text{ANS.} \qquad \varepsilon_{xx} = 781.2 \ \mu$$

$$\varepsilon_{yy} = \frac{\sigma_{yy} - \nu(\sigma_{zz} + \sigma_{xx})}{E} = \frac{-30 - 0.25(15 - 3.75)}{30,000}$$

$$\text{ANS.} \qquad \varepsilon_{yy} = 124 - 1094 \ \mu$$

COMMENTS

1. The three independent quantities in this problem were σ_{xx}, σ_{yy}, and τ_{xy}. Knowing these we were able to find all the strains in plane stress and plane strain.

2. The difference in the values of the strains came from the zero value of σ_{zz} in plane stress and a value of 3.75 ksi (C) in plane strain.

EXAMPLE 3.7

The strains at a point on aluminum ($E = 70$ GPa, $G = 28$ GPa, and $\nu = 0.25$) were found to be $\varepsilon_{xx} = 650 \ \mu$, $\varepsilon_{yy} = 300 \ \mu$, and $\gamma_{xy} = 750 \ \mu$. Determine the stresses σ_{xx}, σ_{yy}, and τ_{xy} and the strain ε_{zz} assuming the point is in plane stress.

PLAN

The shear strain can be calculated using Equation (3.12d). If we note that $\sigma_{zz} = 0$ and the strains ε_{xx} and ε_{yy} are given, the stresses σ_{xx} and σ_{yy} can be found by solving Equations (3.12a) and (3.12b) simultaneously. The strain ε_{zz} can then be found from Equations (3.12c).

Solution From Equations (3.12d),

$$\tau_{xy} = G\gamma_{xy} = 28 \times 10^9 \times 750 \times 10^{-6}$$

ANS. $\tau_{xy} = 21$ MPa

Equations (3.12a) and (3.12b) can be rewritten with $\sigma_{zz} = 0$,

$$\sigma_{xx} - v\sigma_{yy} = E\varepsilon_{xx} = 70 \times 10^9 \times 650 \times 10^{-6} \text{ N/m}^2 \qquad \text{or}$$

$$\sigma_{xx} - v\sigma_{yy} = 45.5 \text{ MPa} \qquad\qquad\qquad (E1)$$

$$\sigma_{yy} - v\sigma_{xx} = E\varepsilon_{yy} = 70 \times 10^9 \times 300 \times 10^{-6} \text{ N/m}^2 \qquad \text{or}$$

$$\sigma_{yy} - 0.25\sigma_{xx} = 21 \text{ MPa} \qquad\qquad\qquad (E2)$$

Solving Equations (E1) and (E2) we obtain

ANS. $\sigma_{xx} = 54.1$ MPa (T) $\qquad \sigma_{yy} = 34.5$ MPa (T)

From Equation (3.12c) we obtain ε_{zz},

$$\varepsilon_{zz} = \frac{\sigma_{zz} - v(\sigma_{xx} + \sigma_{yy})}{E} = \frac{0 - 0.25(54.13 + 34.53)10^6}{70 \times 10^9}$$

ANS. $\varepsilon_{zz} = -317 \, \mu$

COMMENTS

1. Equations (E1) and (E2) have a very distinct structure. If we multiply either equation by v and add the product to the other equation, the result will be to eliminate one of the unknowns. Equation (3.15) in Problem 3.88 is developed in this manner and can be used for solving this problem. But this would imply remembering one more formula. We can avoid this by remembering the defined structure of Hooke's law, which is applicable to all types of problems and not just plane stress.

2. Equation (3.16) in Problem 3.89 gives $\varepsilon_{zz} = -[v/(1 - v)](\varepsilon_{xx} + \varepsilon_{yy})$. Substituting $v = 0.25$ and $\varepsilon_{xx} = 650 \, \mu$, $\varepsilon_{yy} = 300 \, \mu$, we obtain $\varepsilon_{zz} = -(0.25/0.75)(650 + 300) = 316.7 \, \mu$, as before. This formula is useful if we do not need to calculate stresses, and we will use it in Chapter 9.

QUICK TEST 3.2   Time: 15 minutes/Total: 20 points

Grade yourself using the answers given in Appendix G. Each question is worth two points.

1.   What is the difference between an isotropic and a homogeneous material?

2.   What is the number of independent material constants needed in a linear stress–strain relationship for an isotropic material?

3.   What is the number of independent material constants needed in a linear stress–strain relationship for the most general anisotropic materials?

4.   What is the number of independent *stress* components in plane stress problems?

5.   What is the number of independent *strain* components in plane stress problems?

6.   How many nonzero *strain* components are there in plane stress problems?

7.   What is the number of independent *strain* components in plane strain problems?

8.   What is the number of independent *stress* components in plane strain problems?

9.   How many nonzero *stress* components are there in plane strain problems?

10. Is the value of E always greater than G, less than G, or does it depend on the material? Justify your answer.

PROBLEM SET 3.2

3.62   Write the generalized Hooke's law for isotropic material in cylindrical coordinates (r, θ, z).

3.63   Write the generalized Hooke's law for isotropic material in spherical coordinates (r, θ, ϕ).

The stresses and two material constants are given. In Problems 3.64 through 3.69 calculate ε_{xx}, ε_{yy}, γ_{xy}, ε_{zz}, and σ_{zz}: (a) assuming plane stress; (b) assuming plane strain.

3.64
$\sigma_{xx} = 100$ MPa (T)
$\sigma_{yy} = 150$ MPa (T)
$\tau_{xy} = -125$ MPa

$E = 200$ GPa
$v = 0.32$

3.65
$\sigma_{xx} = 225$ MPa (C)
$\sigma_{yy} = 125$ MPa (T)
$\tau_{xy} = 150$ MPa
$E = 70$ GPa
$G = 28$ GPa

3.66
$\sigma_{xx} = 22$ ksi (C)
$\sigma_{yy} = 25$ ksi (C)
$\tau_{xy} = -15$ ksi
$E = 30{,}000$ ksi
$v = 0.3$

Figure P3.76

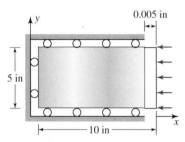

Figure P3.77

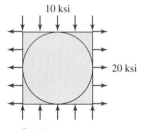

Figure P3.78

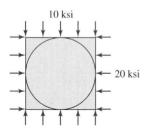

Figure P3.79

3.67

$\sigma_{xx} = 15$ ksi (T)
$\sigma_{yy} = 12$ ksi (C)
$\tau_{xy} = -10$ ksi
$E = 10,000$ ksi
$G = 3900$ ksi

3.68

$\sigma_{xx} = 300$ MPa (C)
$\sigma_{yy} = 300$ MPa (T)
$\tau_{xy} = 150$ MPa
$G = 15$ GPa
$v = 0.2$

3.69

$\sigma_{xx} = 100$ psi (T)
$\sigma_{yy} = 125$ psi (C)
$\tau_{xy} = -50$ psi
$E = 2000$ psi
$G = 800$ psi

*The strains and two material constants are
given. In Problems 3.70 through 3.75 calcu-
late σ_{xx}, σ_{yy}, τ_{xy}, σ_{zz}, and ε_{zz} assuming the
point is in plane stress.*

3.70

$\varepsilon_{xx} = 500 \mu$
$\varepsilon_{yy} = 400 \mu$
$\gamma_{xy} = -300 \mu$
$E = 200$ GPa
$v = 0.32$

3.71

$\varepsilon_{xx} = -3000 \mu$
$\varepsilon_{yy} = 1500 \mu$
$\gamma_{xy} = 2000 \mu$
$E = 70$ GPa
$G = 28$ GPa

3.72

$\varepsilon_{xx} = -800 \mu$
$\varepsilon_{yy} = -1000 \mu$
$\gamma_{xy} = -500 \mu$
$E = 30,000$ ksi
$v = 0.3$

3.73

$\varepsilon_{xx} = 1500 \mu$
$\varepsilon_{yy} = -1200 \mu$
$\gamma_{xy} = -1000 \mu$
$E = 10,000$ ksi
$G = 3900$ ksi

3.74

$\varepsilon_{xx} = -2000 \mu$
$\varepsilon_{yy} = 2000 \mu$
$\gamma_{xy} = 1200 \mu$
$G = 15$ GPa
$v = 0.2$

3.75

$\varepsilon_{xx} = 50 \mu$
$\varepsilon_{yy} = 75 \mu$
$\gamma_{xy} = -25 \mu$
$E = 2000$ psi
$G = 800$ psi

3.76 The cross section of the wooden
piece that is visible in Figure P3.76 is
40 mm × 25 mm. The clamped length of the
wooden piece in the vice is 125 mm. The
modulus of elasticity of wood is $E = 14$ GPa
and the Poisson's ratio $v = 0.3$. The jaws of
the vice exert a uniform pressure of 3.2 MPa
on the wood. Determine the average change
of length of the wood.

3.77 A thin plate ($E = 30,000$ ksi, $v =
0.25$) under the action of uniform forces
deforms to the shaded position, as shown in
Figure P3.77. Assuming plane stress, determine the average normal stresses in the x and
y directions.

3.78 A 2 in × 2 in square with a circle
inscribed is stressed as shown Figure P3.78.
The plate material has a modulus of elasticity
$E = 10,000$ ksi and a Poisson's ratio $v = 0.25$.
Assuming plane stress, determine the major
and minor axes of the ellipse formed due to
deformation.

3.79 A 2 in × 2 in square with a circle
inscribed is stressed as shown Figure P3.79.
The plate material has a modulus of elasticity
$E = 10,000$ ksi and a Poisson's ratio $v = 0.25$.
Assuming plane stress, determine the major
and minor axes of the ellipse formed due to
deformation.

3.80 A 50 mm × 50 mm square with a circle inscribed is stressed as shown Figure P3.80.
The plate material has a modulus of elasticity
$E = 70$ GPa and a Poisson's ratio $v = 0.25$.
Assuming plane stress, determine the major
and minor axes of the ellipse formed due to
deformation.

3.81 Derive the following relations of normal stresses in terms of normal strain from the generalized Hooke's law:

$$\sigma_{xx} =$$
$$[(1-v)\varepsilon_{xx} + v\varepsilon_{yy} + v\varepsilon_{zz}]\frac{E}{(1-2v)(1+v)}$$

$$\sigma_{yy} =$$
$$[(1-v)\varepsilon_{yy} + v\varepsilon_{zz} + v\varepsilon_{xx}]\frac{E}{(1-2v)(1+v)}$$

$$\sigma_{zz} =$$
$$[(1-v)\varepsilon_{zz} + v\varepsilon_{xx} + v\varepsilon_{yy}]\frac{E}{(1-2v)(1+v)}$$
$$(3.14)$$

An alternative form that is easier to remember is $\sigma_{ii} = 2G\varepsilon_{ii} + \lambda(I_1)$, where i can be x, y, or z; $I_1 = \varepsilon_{xx} + \varepsilon_{yy} + \varepsilon_{zz}$; G is the shear modulus; and $\lambda = 2Gv/(1-2v)$ is called Lame's constant.[17]

3.82 For a point in plane stress show that

$$\sigma_{xx} = [\varepsilon_{xx} + v\varepsilon_{yy}]\frac{E}{1-v^2}$$
$$\sigma_{yy} = [\varepsilon_{yy} + v\varepsilon_{xx}]\frac{E}{1-v^2}$$
$$(3.15)$$

3.83 For a point in plane stress show that

$$\varepsilon_{zz} = -\left(\frac{v}{1-v}\right)(\varepsilon_{xx} + \varepsilon_{yy}) \qquad (3.16)$$

3.84 Using Equations (3.15) and (3.16) solve for σ_{xx}, σ_{yy}, and ε_{zz} in Problem 3.70.

3.85 Using Equations (3.15) and (3.16) solve for σ_{xx}, σ_{yy}, and ε_{zz} in Problem 3.71.

3.86 Using Equations (3.15) and (3.16) solve for σ_{xx}, σ_{yy}, and ε_{zz} in Problem 3.72.

3.87 Using Equations (3.15) and (3.16) solve for σ_{xx}, σ_{yy}, and ε_{zz} in Problem 3.73.

3.88 Using Equations (3.15) and (3.16) solve for σ_{xx}, σ_{yy}, and ε_{zz} in Problem 3.74.

3.89 Using Equations (3.15) and (3.16) solve for σ_{xx}, σ_{yy}, and ε_{zz} in Problem 3.75.

---

[17]Named after G. Lame (1795–1870).

3.90 For a point in plane strain show that

$$\varepsilon_{xx} = [(1-v)\sigma_{xx} - v\sigma_{yy}]\frac{1+v}{E}$$
$$\varepsilon_{yy} = [(1-v)\sigma_{yy} - v\sigma_{xx}]\frac{1+v}{E}$$
$$(3.17)$$

3.91 For a point in plane strain show that

$$\sigma_{xx} = [(1-v)\varepsilon_{xx} + v\varepsilon_{yy}]\frac{E}{(1-2v)(1+v)}$$
$$\sigma_{yy} = [(1-v)\varepsilon_{yy} + v\varepsilon_{xx}]\frac{E}{(1-2v)(1+v)}$$
$$(3.18)$$

3.92 A differential element subjected to only normal strains is shown in Figure P3.92. The ratio of change in a volume ΔV to the original volume V is called the volumetric strain ε_V, or *dilation*.

For small strain prove

$$\varepsilon_V = \frac{\Delta V}{V} = \varepsilon_{xx} + \varepsilon_{yy} + \varepsilon_{zz} \quad (3.19)$$

3.93 Prove

$$p = -K\varepsilon_V \qquad p = -\left(\frac{\sigma_{xx} + \sigma_{yy} + \sigma_{zz}}{3}\right)$$
$$K = \frac{E}{3(1-2v)} \qquad (3.20)$$

where K is the *bulk modulus* and p is the *hydrostatic pressure* because at a point in a fluid the normal stresses in all directions are equal to $-p$. Note that at $v = \frac{1}{2}$ there is no change in volume, regardless of the value of the stresses. Such materials are called *incompressible materials*.

Stretch yourself

An orthotropic material[18] has the following stress–strain relationship at a point in plane stress:

$$\varepsilon_{xx} = \frac{\sigma_{xx}}{E_x} - \frac{v_{yx}}{E_y}\sigma_{yy} \qquad \varepsilon_{yy} = \frac{\sigma_{yy}}{E_y} - \frac{v_{xy}}{E_x}\sigma_{xx}$$
$$\gamma_{xy} = \frac{\tau_{xy}}{G_{xy}} \qquad \frac{v_{yx}}{E_y} = \frac{v_{xy}}{E_x}$$
$$(3.21)$$

Use Equations (3.21) to solve Problems 3.94 through 3.101.

---

[18]See Section 3.12.3 for an additional description of orthotropic materials.

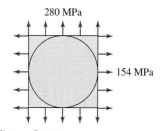

280 MPa

154 MPa

Figure P3.80

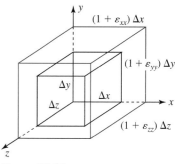

$(1 + \varepsilon_{xx})\Delta x$

$(1 + \varepsilon_{yy})\Delta y$

$(1 + \varepsilon_{zz})\Delta z$

Δy

Δx

Δz

Figure P3.92

The stresses at a point on a free surface of an orthotropic material are given in Problems 3.94 through 3.97. Also given are the material constants. Using Equations (3.21) solve for the strains ε_{xx}, ε_{yy}, and γ_{xy}.

Problem	σ_{xx}	σ_{yy}	τ_{xy}	E_x	E_y	ν_{xy}	G_{xy}
3.94	4 ksi (T)	10 ksi (C)	4 ksi	7500 ksi	2500 ksi	0.3	1250 ksi
3.95	25 ksi (C)	5 ksi (C)	−8 ksi	25,000 ksi	2000 ksi	0.32	1500 ksi
3.96	200 MPa (C)	80 MPa (C)	−54 MPa	53 GPa	18 GPa	0.25	9 GPa
3.97	300 MPa (T)	50 MPa (T)	60 MPa	180 GPa	15 GPa	0.28	11 GPa

The strains at a point on a free surface of an orthotropic material are given in Problems 3.98 through 3.101. Also given are the material constants. Using Equations (3.21) solve for the stresses σ_{xx}, σ_{yy}, and τ_{xy}.

Problem	ε_{xx}	ε_{yy}	γ_{xy}	E_x	E_y	ν_{xy}	G_{xy}
3.98	−1000 μ	500 μ	−250 μ	7500 ksi	2500 ksi	0.3	1250 ksi
3.99	−750 μ	−250 μ	400 μ	25,000 ksi	2000 ksi	0.32	1500 ksi
3.100	2000 μ	−800 μ	300 μ	53 GPa	18 GPa	0.25	9 GPa
3.101	1500 μ	−750 μ	−450 μ	180 GPa	15 GPa	0.28	11 GPa

3.102 Using Equations (3.21), show that on a free surface of an orthotropic material, the following relationships can be written:

$$\sigma_{xx} = \frac{E_x(\varepsilon_{xx} + \nu_{yx}\varepsilon_{yy})}{1 - \nu_{yx}\nu_{xy}} \qquad \sigma_{yy} = \frac{E_y(\varepsilon_{yy} + \nu_{xy}\varepsilon_{xx})}{1 - \nu_{yx}\nu_{xy}} \tag{3.22}$$

3.6 FAILURE AND FACTOR OF SAFETY

There are many types of failures. The breaking of the ship S.S. Schenectady shown in Figure 1.1 was a failure of strength, whereas the failure of the O-ring joints in the shuttle Challenger, shown in Figure 2.1, was due to excessive deformation. A general definition of failure is given as follows:

Definition 23 Failure implies that a component or a structure does not perform the function for which it was designed.

Attention must be paid to the stiffness design of a component or structure to prevent failure due to excessive or insufficient displacement. A machine component, because of excessive deformation, may cause interference with other moving parts; a chair may feel rickety because of poor joint design; a gasket seal leaks because of insufficient deformation of the gasket at some points; lock washers, which are used to keep bolted joints from becoming loose, may not provide sufficient spring force to perform their function unless it deflects sufficiently; a building undergoing excessive deformation may become aesthetically displeasing. These are some examples of failure caused by too little or too much deformation.

The stiffness of a component (structural element) depends on the modulus of elasticity of the material as well as on the geometric properties of the member, such as cross-sectional area, area moments of inertia, polar moments of inertia, and length of the components. The design of joints has a significant effect on the overall stiffness of a structure. The use of adhesives in place of, or in conjunction with, mechanical fasteners is one way of increasing the stiffness of joints. The use of carpenter's glue in the joints of a chair to prevent the rickety feeling is a simple example.

Attention to prevent breaking of a component is an obvious design objective based on strength. But equally important at times is the attention that must be paid to ensure that a component is not too strong. The adhesive bond between the lid and a sauce bottle must break so that the bottle may be opened by hand; shear pins must break before critical components get damaged; the steering column of an automobile must collapse rather than impale the passenger in a crash. These are some examples where too much strength is detrimental. Ultimate stress is used for assessing failure due to breaking or rupture. Ultimate normal stress is also used for assessing failure of a brittle material.

Permanent deformation rather than rupture is another stress-based failure. Dents or stress lines in the body of an automobile; locking up of bolts and screws because of permanent deformation of threads; slackening of tension wires holding a structure in place—these are some examples in which plastic deformation is the cause of failure. Yield stress or proportional limit are used as strength indices for assessing failure due to plastic deformation. Shear stress at yield is also used for assessing failure of a ductile material.

A support in a bridge may fail but the bridge can still carry traffic. In other words, the failure of a component does not imply failure of the entire structure. Thus the strength of a structure, or the deflection of the entire structure, may depend on a large number of variables. In such cases loads on the structure are used to characterize failure. Failure loads may be based on the stiffness, the strength, or both.

A margin of safety must be built into any design to account for uncertainties or a lack of knowledge, lack of control over the environment, and the simplifying assumptions made to obtain results. The measure of this margin of safety is the factor of safety K_{safety}, which is defined as follows:

$$K_{safety} = \frac{\text{failure-producing value}}{\text{computed (allowable) value}} \qquad (3.23)$$

Equation (3.23) implies that the factor of safety must always be greater than 1. The numerator of Equation (3.23) could be the value of failure deflection, failure stress, or failure load and is assumed known. In analysis, the denominator is determined, and from it the factor of safety is found. In design, the factor of safety is specified and the variables affecting the denominator are determined such that the denominator value is not exceeded. Thus in design the denominator is often referred to as the allowable value.

There are several issues that must be considered in determining the appropriate factor of safety in design, as will be discussed. No single issue dictates the choice of factor of safety. The value chosen for the factor of safety is a compromise of the various issues, which is arrived at from experience.

Cost considerations are the primary reason for using a low factor of safety. Large fixed cost could be due to the use of an expensive material, or to using a large quantity of material specified in design to meet a given factor of safety requirement. Weight resulting in higher fuel consumption is an example of higher running costs. In aerospace industries the running costs supersede material costs. Material costs dominate the furniture industry. The automobile industry seeks a compromise between fixed and running costs.

Liability cost considerations push for a greater factor of safety. Though liability is a consideration in all design, the building industry is most conscious of it in determining the factor of safety.

Lack of control or lack of knowledge concerning the operating environment will push for higher factors of safety. Uncertainties in predicting earthquakes, cyclones, or tornadoes will require higher safety factors for the design of buildings located in regions prone to these natural calamities. A large scatter in material properties as usually seen in newer materials is an uncertainty that will require the use of a larger factor of safety.

Human safety consideration not only push the factor of safety higher but often result in government regulations of the factors of safety such as reflected in building codes.

This list of issues affecting the factor of safety is by no means complete, but is an indication of the subjectivity that goes into the choice of the factor of safety. The factors of safety that may be recommended for most applications range from 1.1 to 6.

EXAMPLE 3.8

In the leaf spring design in Figure 3.26 the formulas for the maximum stress σ and deflection δ are derived from bending theory[19] of beams,

$$\sigma = \frac{3PL}{nbt^2}$$

$$\delta = \frac{3PL^3}{4Enbt^3}$$

where P is the load supported by the spring, L is the length of the spring, n is the number of leaves, b is the width of each leaf, t is the

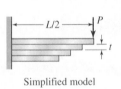

$L/2$ — P — t

Simplified model

Leaf spring

Figure 3.26   Leaf spring in Example 3.8.

---

[19]The formulas are given by Equation (7.3) and derived in Example 7.4.

thickness of each leaf, and E is the modulus of elasticity. A spring has the following data: $L = 20$ in, $b = 2$ in, $t = 0.25$ in, and $E = 30,000$ ksi. The failure stress is 120 ksi and the failure deflection is 0.5 in.

(a) If the spring is estimated to carry a maximum force $P = 250$ lb and is to have a factor of safety of 4, determine the minimum number of leaves.

(b) For the answer in part (a) what is the real factor of safety?

PLAN

(a) We can compute the allowable stress and allowable deflection using Equation (3.23) and the given factor of safety of 4, the failure stress, and the failure deflection. We can find two values of n from the two formulas given, which ensure that the allowable values of normal stress and deflection are not exceeded. The higher of the two values of n is the minimum number of leaves in the spring design.

(b) Substituting the value of n in these equations, we can compute the maximum stress and deflection. Substituting these computed values in Equation (3.23), we can find the two factors of safety corresponding to failure stress and failure deflection and choose the lower factor of safety.

Solution

(a) Substituting the given values of the variables in the stress formula we obtain

$$\sigma = \frac{3PL}{nbt^2} = \frac{3 \times 250 \times 20}{n \times 2 \times 0.25^2} = \frac{120 \times 10^3}{n} \tag{E1}$$

From Equation (3.23) we find that the allowable stress is $\sigma_{allow} = 120 / 4 = 30$ ksi. Thus from Equation (E1) we obtain

$$\frac{120 \times 10^3}{n} \leq 30 \times 10^3 \qquad \text{or} \qquad n \geq 4 \tag{E2}$$

Substituting the given values of the variables in the deflection formula we obtain

$$\delta = \frac{3PL^3}{4Enbt^3} = \frac{3 \times 250 \times 20^3}{4 \times 30 \times 10^6 \times n \times 2 \times 0.25^3} = \frac{1.6}{n} \tag{E3}$$

From Equation (3.23) we can find that the allowable deflection is $\delta_{allow} = 0.5 / 4 = 0.125$ in. Thus from Equation (E3) we obtain

$$\frac{1.6}{n} \leq 0.125 \qquad \text{or} \qquad n \geq 12.8 \tag{E4}$$

Thus the minimum number of leaves that will satisfy Equations (E2)
and (E4) is $n = 13$.   ANS.

(b) Substituting $n = 13$ in Equations (E1) and (E3) we can find
the computed values of stress and deflection. Using these along
with the failure values we can find the factors of safety from Equation (3.23),

$$\sigma_{comp} = 9.23 \times 10^3 \text{ psi} \qquad K_\sigma = \frac{120 \times 10^3}{9.23 \times 10^3} = 13 \qquad \text{(E5)}$$

$$\delta_{comp} = 0.1232 \text{ in} \qquad K_\delta = \frac{0.5}{0.1232} = 4.06 \qquad \text{(E6)}$$

The factor of safety for the system is governed by the lowest factor
of safety, which in our case is given by Equation (E6). Thus the
factor of safety is $K_\delta = 4.06$.   ANS.

COMMENTS

1. This problem demonstrates the difference between allowable values, which are used in design decisions based on a specified factor
of safety, and computed values, which are used in analysis for finding the factor of safety.

2. For purposes of design, formulas are initially obtained based on
simplified models, such as shown in Figure 3.26. Once the preliminary relationship between variables has been established, then complexities are often incorporated by using factors that are determined
empirically (experimentally). Thus the deflection of the spring,
accounting for curvature, end support, variation of thickness, and so
on, is given by $\delta = K(3PL^3/4Enbt^3)$, where the factor K is determined experimentally as function of the complexities not accounted
for in the simplified model. This comment highlights that mechanics of materials theories provide a guide to developing formulas for
complex realities.

*3.7 STRESS CONCENTRATION

We define stress concentration as follows:

Definition 24 Large stress gradients in a small region are called *stress concentration*.

These large gradients could be due to sudden changes in geometry, material properties, or loading, as has been mentioned earlier. We know from Saint-Venant's principle,

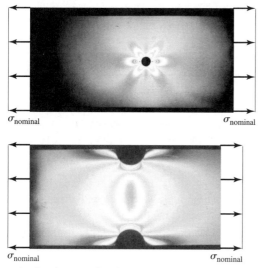

$\sigma_{nominal}$ $\sigma_{nominal}$

$\sigma_{nominal}$ $\sigma_{nominal}$

Figure 3.27 Photoelastic pictures showing stress concentration. (Courtesy Professor I. Miskioglu.)

which will be discussed in the next section, that we can use our theoretical models to calculate stress away from the regions of large stress concentration. These stress values predicted by the theoretical models away from regions of stress concentration are called nominal stresses. Figure 3.27 shows photoelastic pictures[20] of two structural members under uniaxial tension. Large stress gradients near the circular cutout boundaries cause fringes to be formed. Each color boundary represents a fringe order that can be used in the calculation of the stresses.

Definition 25 The stress predicted by theoretical models away from the regions of stress concentration is called *nominal* stress.

Stress concentration factor is an engineering concept that permits us to extrapolate the results of our elementary theory into the region of large stress concentration where the assumptions on which the theory is based are violated. The stress concentration factor K_{conc} is defined as

$$K_{conc} = \frac{\text{maximum stress}}{\text{nominal stress}} \qquad (3.24)$$

The stress concentration factor K_{conc} is found from charts, tables, or formulas that have been determined experimentally, numerically, analytically, or from a combination of the three. Appendix C shows several graphs that can be used in the calculation of stress concentration factors for problems in this book. Additional graphs can be found in handbooks describing different situations. Knowing the nominal stress and the stress concentration factor, the maximum stress can be estimated and used in

---

[20]See Section 8.4.1 for a brief description of photoelasticity.

design or to estimate the factor of safety. Example 3.9 demonstrates the use of the stress concentration factor.

*3.8 SAINT-VENANT'S PRINCIPLE

Theories in mechanics of materials are constructed by making assumptions regarding load, geometry, and material variations. These assumptions are usually not valid near concentrated forces or moments such as those that may be present near supports, near corners or/and holes, near interfaces of two materials, and in flaws such as cracks. Disturbance in the stress and displacement fields, however, dissipates rapidly as one moves away from the regions where the assumptions of the theory are violated. The statement about the dissipation of a disturbance with distance is stated in the following manner by Saint-Venant's principle:

> Two statically equivalent load systems produce nearly the same stress in regions at a distance that is at least equal to the largest dimension in the loaded region.

Consider the two statically equivalent load systems shown in Figure 3.28. In Figure 3.28*a* we have a concentrated force *P*. If we draw a stress cube under this force, we expect that the stresses will be tending toward infinity because of the infinitesimal area of the stress cube. However, at a cross section that is at a distance *W* below the applied load, the stress distribution would be closer to a uniform distribution, with a value equal to *P* divided by the cross-sectional area. In Figure 3.28*b* we have a distributed force that produces a total force of *P*. Though the distributed force is shown uniform, it could be of any distribution that does not produce a moment. Thus this distributed force is statically equivalent to the one in Figure 3.28*a*. By Saint-Venant's principle the stress at a distance *W* below would be nearly uniform. In the region at a distance less than *W* the stress distribution will be different, and it is possible that there are also shear stress components present. In a similar manner changes in geometry and materials have local effects that can be ignored at distances. We have considered the effect of changes in geometry and an engineering solution to the problem in Section 3.7 on stress concentration.

The importance of Saint-Venant's principle is that we can develop our theories with reasonable confidence away from the regions of stress concentration. These theories provide us with formulas for the calculation of nominal stress. We can then use the stress concentration factor to obtain maximum stress in regions of stress concentration where our theories are not valid.

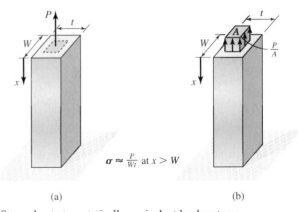

$$\sigma \approx \frac{P}{Wt} \text{ at } x > W$$

(a)   (b)

Figure 3.28 Stress due to two statically equivalent load systems.

EXAMPLE 3.9

Finite-element[21] analysis shows that a long structural component carries a uniform axial stress of 35 MPa (T) (Figure 3.29). A hole in the center needs to be drilled for passing cables through the structural component. The yield stress of the material is 200 MPa. If failure due to yielding is to be avoided, determine the maximum diameter of the hole that can be drilled using a factor of safety of 1.6.

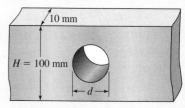

Figure 3.29 Component geometry in Example 3.9.

PLAN

From the given factor of safety of 1.6, the failure stress of 200 MPa, and Equation (3.23), we can compute the allowable stress. This allowable stress is the maximum stress in Equation (3.24). Knowing the maximum stress and the given gross nominal stress of 35 MPa, we can obtain the permissible stress concentration factor from Equation (3.24). From the plot of K_{gross} in Figure C.1 of Appendix C we can estimate the ratio of d/H. Knowing that $H = 100$ mm, we can find the maximum diameter d of the hole.

Solution The failure stress of 200 MPa is divided by the factor of safety of 1.6, as per Equation (3.23), to obtain the allowable stress,

$$\sigma_{allow} = \frac{200}{1.6} = 125 \text{ MPa} \qquad \text{(E1)}$$

The permissible stress concentration factor can be calculated by dividing σ_{allow} by the nominal stress of 35 MPa, as per Equation (3.24), to obtain

$$K_{conc} \leq \frac{125}{35} \leq 3.57 \qquad \text{(E2)}$$

From Figure C.1 of Appendix C we estimate the ratio of d/H as 0.367. Substituting $H = 100$ mm we obtain

$$\frac{d}{100} \leq 0.367 \quad \text{or} \quad d \leq 36.7 \text{ mm} \qquad \text{(E3)}$$

Thus the maximum permissible diameter to the nearest millimeter is $d_{max} = 36$ mm.   ANS.

COMMENTS

1. The value of $d/H = 0.367$ was found from linear interpolation between the value of $d/H = 0.34$, where the stress concentration factor is 0.35, and the value of $d/H = 0.4$, where the stress concentration

---

[21]See Section 4.9 for a brief description of the finite-element method.

factor is 0.375. These points were used as they are easily read from the graph. Because we are rounding downward in Equation (E3), any value between 0.36 and 0.37 is acceptable. In other words, the third place of the decimal value is immaterial.

2. As we used the maximum diameter of 36 mm instead of 36.7 mm, the effective factor of safety will be slightly higher than the specified value of 1.6, which makes this design a conservative design.

3. Creating the hole will change the stress around its. As per Saint-Venant's principle the stress for field from the hole will not be significantly affected. This justifies the use of nominal stress without the hole in our calculation.

*3.9 EFFECT OF TEMPERATURE

A material expands with an increase in temperature and contracts with a decrease in temperature. If the change in temperature is uniform, and if the material is isotropic and homogeneous, then all lines on the material will change dimensions by equal amounts. This will result in a normal strain, but there will be no change in the angles between any two lines, and hence there will be no shear strain produced. Experimental observations confirm this deduction. Experiments also show that the change in temperature ΔT is related to the thermal normal strain ε_T,

$$\varepsilon_T = \alpha \Delta T \tag{3.25}$$

where the Greek letter alpha α is the linear coefficient of thermal expansion. The linear relationship given by Equation (3.25) is valid for metals at temperatures well below the melting point. In this linear region the strains for most metals are small and the usual units for α are $\mu/°F$ or $\mu/°C$, where $\mu = 10^{-6}$. Throughout the discussion in this section it is assumed that the material is in the linear region.

The tension test described in Section 3.1 will be conducted at some ambient temperature. We expect the stress–strain curve to have the same character at two different ambient temperatures. If we raise the temperature before we apply the force P on the specimen, then the specimen will expand, resulting in a thermal strain, as given by Equation (3.25). But as there is no external force, there will be no resulting internal forces and hence no stresses. Thus the increase in temperature before the application of the force causes the stress–strain curve starting point to move from point O to point O_1, as shown in Figure 3.30. The total strain at any point is the sum of mechanical strain and thermal strains, which can be written as follows:

$$\varepsilon = \frac{\sigma}{E} + \alpha \Delta T \tag{3.26}$$

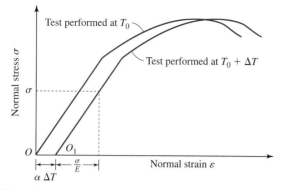

Figure 3.30 Effect of temperature on stress–strain curve.

If there are no internal forces generated in a body due to a change in temperature, then no stresses will be produced. Material nonhomogeneity, material anisotropy, nonuniform temperature distribution, or reaction forces from body constraints are the reasons for the generation of stresses from temperature changes. Alternatively, no thermal stresses are produced in a homogeneous, isotropic, unconstrained body due to uniform temperature changes.

The generalized Hooke's law relates mechanical strains to stresses. The total normal strain, as seen from Equation (3.26), is the sum of mechanical and thermal strains. For isotropic materials undergoing changes in temperature, the generalized Hooke's law is written as

$$\varepsilon_{xx} = \frac{\sigma_{xx} - \nu(\sigma_{yy} + \sigma_{zz})}{E} + \alpha \Delta T \tag{3.27a}$$

$$\varepsilon_{yy} = \frac{\sigma_{yy} - \nu(\sigma_{zz} + \sigma_{xx})}{E} + \alpha \Delta T \tag{3.27b}$$

$$\varepsilon_{zz} = \frac{\sigma_{zz} - \nu(\sigma_{xx} + \sigma_{yy})}{E} + \alpha \Delta T \tag{3.27c}$$

$$\gamma_{xy} = \frac{\tau_{xy}}{G} \tag{3.27d}$$

$$\gamma_{yz} = \frac{\tau_{yz}}{G} \tag{3.27e}$$

$$\gamma_{zx} = \frac{\tau_{zx}}{G} \tag{3.27f}$$

(Mechanical strain) (Thermal strain)

We do not have to comment regarding to the homogeneity of the material or the uniformity of the temperature change because Hooke's law is written for a point and not for the whole body. The relationship of shear stress and shear strain is as before because no shear strains are produced for isotropic bodies due to temperature changes.

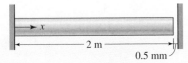

Figure 3.31 Bar in Example 3.10.

EXAMPLE 3.10

A circular bar ($E = 200$ GPa, $v = 0.32$, and $\alpha = 11.7\ \mu/°C$) has a diameter of 100 mm. The bar is built into a rigid wall on the left, and a gap of 0.5 mm exists between the right wall and the bar prior to an increase in temperature, as shown in Figure 3.31. The temperature of the bar is increased uniformly by 80°C. Determine the average axial stress and the change in the diameter of the bar.

Method 1

PLAN

A reaction force in the axial direction will be generated to prevent an expansion greater than the gap. This would generate σ_{xx}. As there are no forces in the y or z direction, the other normal stresses σ_{yy} and σ_{zz} can be approximated to zero in Equation (3.27a). The total deformation is the gap, from which the total average axial strain for the bar can be found. The thermal strain can be calculated from the change in the given temperature. Thus in Equation (3.27a) the only unknown is σ_{xx}. Once σ_{xx} has been calculated, the strain ε_{yy} can be found from Equation (3.27b). From ε_{yy} the change in the diameter can be calculated.

Solution   The total axial strain is the total deformation (gap) divided by the length of the bar,

$$\varepsilon_{xx} = \frac{0.5 \times 10^{-3}}{2} = 250 \times 10^{-6}$$

$$\alpha \Delta T = 11.7 \times 10^{-6} \times 80 = 936 \times 10^{-6}$$

Because σ_{yy} and σ_{zz} are zero, Equation (3.27a) can be written as $\varepsilon_{xx} = \sigma_{xx}/E + \alpha \Delta T$, from which we can obtain σ_{xx},

$$\sigma_{xx} = E(\varepsilon_{xx} - \alpha \Delta T) = 200 \times 10^{9}(250 - 936)10^{-6}$$

$$= -137.2 \times 10^{6}\ N/m^{2}$$

ANS. $\sigma_{xx} = 137.2$ MPa (C)

From Equation (3.27b) we can obtain ε_{yy} and calculate the change in diameter,

$$\varepsilon_{yy} = -v\frac{\sigma_{xx}}{E} + \alpha \Delta T = -0.25\left(\frac{-137.2 \times 10^{6}}{200 \times 10^{9}}\right) + 936 \times 10^{-6}$$

$$= 1.107 \times 10^{-3}$$

$$\Delta D = \varepsilon_{yy}D = 1.107 \times 10^{-3} \times 100\ mm$$

ANS. $\Delta D = 0.1107$ mm increase

COMMENTS

1. If $\alpha \Delta T$ were less than ε_{xx}, then σ_{xx} would come out as tension and our assumption that the gap closes would be invalid. In such a case there would be no stress σ_{xx} generated.

2. The increase in diameter is due partly to Poisson's effect and partly to thermal strain in the y direction.

Method 2

PLAN

We can think of the problem in two steps.

1. Ignore the restraining effect of the right wall and think of the bar as free to expand. The thermal expansion δ_T can be found.

2. Apply the force P to bring the bar back to the restraint position due to the right wall and compute the corresponding stress.

Solution We draw an approximate deformed shape of the bar, assuming there is no right wall to restrain the deformation (Figure 3.32). The thermal expansion δ_T is the thermal strain multiplied by the length of the bar,

$$\delta_T = (\alpha \Delta T)L = 11.7 \times 10^{-6} \times 80 \times 2 = 1.872 \times 10^{-3} \quad \text{(E1)}$$

From Figure 3.32 we can see that by subtracting the gap from the thermal expansion, we will obtain the contraction δ_P we need to satisfy the restraint imposed by the right wall. We can then find the mechanical strain and compute the corresponding stress,

$$\delta_P = \delta_T - 0.5 \times 10^{-3} = 1.372 \times 10^{-3}$$

$$\varepsilon_P = \frac{\delta_P}{L} = \frac{1.372 \times 10^{-3}}{2} = 0.686 \times 10^{-3}$$

$$\sigma_P = E\varepsilon_P = 200 \times 10^9 \times 0.686 \times 10^{-3}$$

ANS. $\sigma_P = 137.2$ MPa (C)

The change in diameter can be found as in Method 1.

COMMENT

In Method 1 we ignored the intermediate steps and conducted the analysis at equilibrium. We implicitly recognized that for a linear system the process of reaching equilibrium is immaterial. In Method 2 we conducted the thermal and mechanical strain calculations separately. Method 1 is more procedural areas Method 2 is more intuitive.

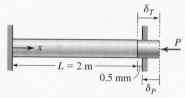

Figure 3.32   Approximate deformed shape of bar.

EXAMPLE 3.11

Solve Example 3.7 with a temperature increase of 20°C. Use $\alpha = 23 \; \mu/°C$.

PLAN

Shear stress is unaffected by temperature change and its value is the same as in Example 3.7. Hence $\tau_{xy} = 21$ MPa. In Equations (3.27a) and (3.27b) $\sigma_{zz} = 0$, $\varepsilon_{xx} = 650 \; \mu$, and $\varepsilon_{yy} = 300 \; \mu$ are known and $\alpha\Delta T$ can be found and substituted to generate two equations in the two unknown stresses σ_{xx} and σ_{yy}, which are found by solving the equations simultaneously. Then from Equation (3.27c), the normal strain ε_{zz} can be found.

Solution We can find the thermal strain as $\Delta T = 20$ and $\alpha\Delta T = 460 \times 10^{-6}$. Equations (3.27a) and (3.27b) and can be rewritten with $\sigma_{zz} = 0$,

$$\sigma_{xx} - \nu\sigma_{yy} = E(\varepsilon_{xx} - \alpha\Delta T) = 70 \times 10^9 (650 - 460)10^{-6} \; \text{N/m}^2$$

$$\sigma_{xx} - 0.25\sigma_{yy} = 13.3 \; \text{MPa} \qquad\qquad\qquad (E1)$$

$$\sigma_{yy} - \nu\sigma_{xx} = E(\varepsilon_{yy} - \alpha\Delta T) = 70 \times 10^9 (300 - 460)10^{-6} \; \text{N/m}^2$$

$$\sigma_{yy} - 0.25\sigma_{xx} = -11.2 \; \text{MPa} \qquad\qquad\qquad (E2)$$

By solving Equations (E1) and (E2) we obtain

> ANS. $\sigma_{xx} = 11.2$ MPa (T) and $\sigma_{yy} = 8.4$ MPa (C).

From Equation (3.27c) with $\sigma_{zz} = 0$ we obtain

$$\varepsilon_{zz} = \frac{-\nu(\sigma_{xx} + \sigma_{yy})}{E} + \alpha\Delta T = \frac{-0.25(11.2 - 8.4)10^6}{70 \times 10^9} + 460 \times 10^{-6}$$

> ANS. $\varepsilon_{zz} = 450 \; \mu$

COMMENT

Equations (E1) and (E2) once more have the same structure as in Example 3.7. The only difference is that in Example 3.7 we were given the mechanical strain and in this example we obtained the mechanical strain by subtracting the thermal strain from the total strain.

*3.10 FATIGUE

To appreciate the difference between static ultimate strength and fatigue strength, try to break a piece of wire (such as a paper clip) by pulling on it by hand. You will not be able to break it as you need to exceed the ultimate static stress of the material. Next take the same piece of wire and bend it one way and then the other a few times, and you find it breaks easily. The explanation of this phenomenon is as follows.

All materials are *assumed* to have microcracks. Because the lengths of the cracks are very small, these cracks are not critical,[22] and hence in static problems the bulk strength of the material corresponds to the ultimate stress determined by a tension test. However, if the material is subjected to cyclic loading, these microcracks can grow until a crack reaches some critical length, at which time the remaining material breaks. The stress value at rupture in a cyclic loading is significantly lower than the ultimate stress of the material.

Definition 26 Failure due to cyclic loading at stress levels significantly lower than the static ultimate stress is called *fatigue*.

Failure due to fatigue is like a brittle failure, irrespective of whether the material is brittle or ductile. We see that there are two phases of failure. In the first phase the microcracks grow. These regions of crack growth can be identified by striation marks, also called beach marks, as shown in Figure 3.33. On examination of a fractured surface, this region of microcrack growth shows only small deformation. In phase 2, which is after the critical crack length has been reached, the failure surface of the region shows significant deformation.

The following strategy is used in design to account for fatigue failure. Experiments are conducted at different magnitude levels of cyclic stress, and the number of cycles at which the material fails is recorded. A plot is made of stress versus the number of cycles to failure (*S–N*). There is always significant scatter in the data. A curve through the data is constructed. This curve is called the *S–N* curve. At low level of stress the failure may occur in millions and, at times, billions of cycles. To accommodate this large scale, a log scale is used for the number of cycles. Figure 3.34 shows some typical *S–N* curves. Notice that the curve approaches a stress level asymtotically, implying that if stresses are kept below this level, then the material would not fail under cyclic loading. This asymtotic level is called endurance limit or fatigue strength.

Figure 3.33 Failure of lead solder due to fatigue. (Courtesy Professor I. Miskioglu.)

Striation marks

Definition 27 The highest stress level for which the material would not fail under cyclic loading is called *endurance limit* or *fatigue strength*.

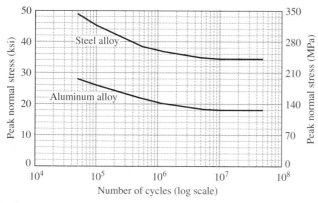

Figure 3.34 *S–N* curves.

[22]See Section 10.3 for additional details regarding growth of cracks and critical crack length.

It should be emphasized that a particular *S–N* curve for a material depends on many factors, such as manufacturing process, machining process, surface preparation, and operating environment. Thus two specimens made from the same steel alloys, but with a different history, will result in different *S–N* curves. Care must be taken to use an *S–N* curve that corresponds as closely as possible to the actual situation.

In a typical preliminary design, static stress analysis would be conducted using the peak load of the cyclic loading. In other words, we would find what would be the highest stress level in a cyclic loading. Using an appropriate *S–N* curve, the number of cycles to failure for the computed stress value is then calculated. This number of cycles to failure is the predicted life of the structural component. If the predicted life is unacceptable, then the component will be redesigned to lower the stress level and hence increase the number of cycles to failure.

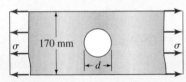

Figure 3.35 Uniaxially loaded plate with a hole in Example 12.

EXAMPLE 3.12

The steel plate shown in Figure 3.35 has the *S–N* curve given in Figure 3.34.

(a) Determine the maximum diameter of the hole to the nearest millimeter if the predicted life of one-half million cycles is desired for a uniform far-field stress $\sigma = 75$ MPa.

(b) For the hole radius in part (a), what percentage reduction in far-field stress must occur if the predicted life is to be increased to 1 million cycles?

PLAN

(a) From Figure 3.34 we can find the maximum stress that the material can carry for one-half million cycles. By dividing this maximum stress by the gross nominal stress of 75 MPa we obtain the gross stress concentration factor K_{gross} as per Equation (3.24). From the plot of K_{gross} in Figure C.1 of Appendix C we can estimate the ratio d/H. Knowing that the value of *H* is 170 mm, we can find the diameter *d* of the hole.

(b) The percentage reduction in the gross nominal stress σ is the same as that in the maximum stress values in Figure 3.34 from one million cycles to one-half million cycles.

Solution

(a) From Figure 3.34 the maximum allowable stress for one-half million cycles is estimated as 273 MPa. From Equation (3.24) the gross stress concentration factor is $K_{gross} = 273/75 = 3.64$. From Figure C.1 of Appendix C the value of the ratio d/H corresponding to $K_{gross} = 3.64$ is 0.374. Knowing that $H = 170$ mm, we obtain that *d* should be less than 63.58 mm. Thus the maximum permissible diameter to the nearest millimeter is $d = 63$ mm.   ANS.

(b) From Figure 3.34 the maximum allowable stress for one million cycles is estimated as 259 MPa. Thus the percentage reduction in maximum allowable stress is $[(273 - 259)/273]100 = 5.13\%$. As the geometry is the same as in part (a), the percentage reduction in far-field stress should be the same as in the maximum allowable stress. **The percentage reduction required is 5.13%.** ANS.

COMMENT

A 5.13% reduction in stress value causes the predicted life cycle to double. There are many factors that can cause small changes in stress values, resulting in a very wide range of predictive life cycles. Our estimates of allowable stress in Figure 3.34, of the ratio d/H from Figure C.1 of Appendix C, of the far-field stress σ, and the tolerances of drilling the hole are some factors that can affect significantly our life prediction of the component. This emphasizes that the data used in predicting life cycles and failure due to fatigue must be of much higher accuracy than generally used in traditional engineering analysis.

*3.11 NONLINEAR MATERIAL MODELS

In previous sections the stress–strain relationship at a point was assumed to be represented by Hooke's law, which implies that the material behavior is linear and elastic. Rubber, plastics, muscles, and other organic tissues exhibit nonlinearity in the stress–strain relationship, even at small strains. Metals also exhibit nonlinearity after yield stress. In this section we consider various nonlinear material models, that is, various forms of equations that are used for representing the stress–strain nonlinear relationship. The constants in the equation relating stress and strain are material constants. Material constants are found by matching the stress–strain equation to the experimental data in a least-squares sense. For the sake of simplicity we shall assume that the material behavior is the same in tension and in compression.

In the next three sections we will consider three material models:

1. Elastic–perfectly plastic model, in which the nonlinearity is approximated by a constant.

2. Linear strain-hardening (also called bilinear) model, in which the nonlinearity is approximated by a linear function.

3. Power law model, in which the nonlinearity is approximated by a one-term nonlinear function.

There are other material models described in the problem set. The choice of material model to use depends not only on the material stress–strain curve, but also on the need for accuracy and the resulting complexity of analysis.

3.11.1 Elastic–Perfectly Plastic Material Model

Figure 3.36 shows the stress–strain curves describing an elastic–perfectly plastic behavior of a material. It is assumed that the material has the same behavior in tension and in compression. Similarly for shear stress–strain, the material behavior is the same for positive and negative stresses and strains.

Before yield stress the stress–strain relationship is given by Hooke's law, and after yield stress the stress is a constant. The elastic–perfectly plastic material behavior is a simplifying[23] approximation used to conduct an elastic–plastic analysis. The approximation is a conservative approximation as it ignores the material capacity to carry higher stresses than the yield stress. The equations describing the stress–strain curve are

$$
\sigma = \begin{cases} \sigma_{\text{yield}}, & \varepsilon \geq \varepsilon_{\text{yield}} \\ E\varepsilon, & -\varepsilon_{\text{yield}} \leq \varepsilon \leq \varepsilon_{\text{yield}} \\ -\sigma_{\text{yield}}, & \varepsilon \leq -\varepsilon_{\text{yield}} \end{cases}
$$

$$
\tau = \begin{cases} \tau_{\text{yield}}, & \gamma \geq \gamma_{\text{yield}} \\ G\gamma, & -\gamma_{\text{yield}} \leq \gamma \leq \gamma_{\text{yield}} \\ -\tau_{\text{yield}}, & \gamma \leq -\gamma_{\text{yield}} \end{cases}
$$

(3.28)

Definition 28 The set of points forming the boundary between the elastic and plastic regions on a body is called *elastic–plastic boundary*.

Determining the location of the elastic–plastic boundary is one of the critical issues in elastic–plastic analysis. As shall be shown by examples, the location of the elastic–plastic boundary is determined using the following observations:

1. On the elastic–plastic boundary the strain must be equal to the yield strain, and stress equal to yield stress.

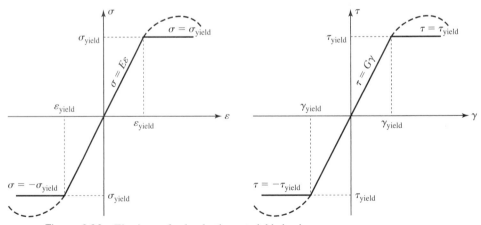

Figure 3.36 Elastic–perfectly plastic material behavior.

---

[23]There is an analysis technique called *limit analysis* that is based on the elastic–perfectly plastic material behavior. Using limit analysis one can predict the maximum load of a complex structure such as a truss.

2. Deformations and strains are continuous at all points, including points at the elastic–plastic boundary.

3.11 NONLINEAR MATERIAL MODELS | **145**

If deformation is not continuous, then it is implied that holes or cracks are being formed in the material. If strains, which are derivative displacements, are not continuous, then corners are being formed during deformation.

3.11.2 Linear Strain-Hardening Material Model

Figure 3.37 shows the stress–strain curve for a linear strain-hardening model, also referred to as *bilinear* material[24] model. It is assumed that the material has the same behavior in tension and in compression. Similarly for shear stress and strain, the material behavior is the same for positive and negative stresses and strains.

This is another conservative, simplifying approximation of material behavior in which we once more ignore the material ability to carry higher stresses than shown by straight lines. The location of the elastic–plastic boundary is once more a critical issue in the analysis, and it is determined as described in the previous section.

The equations describing the stress–strain curve are

$$
\sigma = \begin{cases}
\sigma_{\text{yield}} + E_2(\varepsilon - \varepsilon_{\text{yield}}), & \varepsilon \geq \varepsilon_{\text{yield}} \\
E_1 \varepsilon, & -\varepsilon_{\text{yield}} \leq \varepsilon \leq \varepsilon_{\text{yield}} \\
-\sigma_{\text{yield}} + E_2(\varepsilon + \varepsilon_{\text{yield}}), & \varepsilon \geq \varepsilon_{\text{yield}}
\end{cases}
$$

$$
\tau = \begin{cases}
\tau_{\text{yield}} + G_2(\gamma - \gamma_{\text{yield}}), & \gamma \geq \gamma_{\text{yield}} \\
G_1 \gamma, & -\gamma_{\text{yield}} \leq \gamma \leq \gamma_{\text{yield}} \\
-\tau_{\text{yield}} + G_2(\gamma + \gamma_{\text{yield}}), & \gamma \leq \gamma_{\text{yield}}
\end{cases}
$$

(3.29)

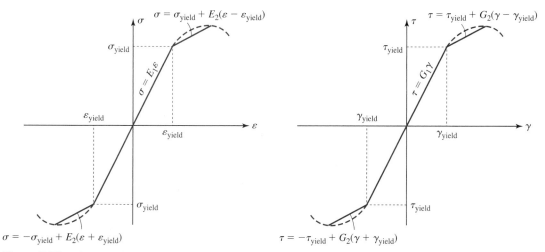

Figure 3.37 Linear strain-hardening model.

---

[24]There is a numerical analysis technique called *incremental plasticity* that uses a similar approximation. The major difference is that the nonlinear stress–strain curve is not approximated by a single straight line, but by a series of straight lines over small intervals.

3.11.3 Power-Law Model

Figure 3.38 shows a power-law representation of a nonlinear stress–strain curve. It is assumed that the material has the same behavior in tension and in compression. Similarly for shear stress and strain; the material behavior is the same for positive and negative stresses and strains. The equations describing the stress–strain curve are

$$\sigma = \begin{cases} E\varepsilon^n, & \varepsilon \geq 0 \\ -E(-\varepsilon)^n, & \varepsilon < 0 \end{cases} \qquad \tau = \begin{cases} G\gamma^n, & \gamma \geq 0 \\ -G(-\gamma)^n, & \gamma < 0 \end{cases} \qquad (3.30)$$

The constants E and n are the strength coefficient and the strain-hardening coefficient, respectively. They are determined to fit the experimental stress–strain curve in a least-squares sense. Materials that strain-harden, such as most metals in the plastic region or most plastics, are represented by the solid curve with a strain-hardening coefficient of less than 1. Materials that show strain softening, such as soft rubber, muscles, and other organic materials, are represented by the dashed line with a strain-hardening coefficient greater than 1.

From Equation (3.30) we note that when strain is negative, the term in parentheses becomes positive, permitting evaluation of the number to fractional powers. Furthermore with negative strain we obtain negative stress, as we should.

In Section 3.11.2 we discussed that the stress–strain relationship could be written using different equations for different stress levels. We could, in a similar manner, combine a linear equation for the linear part and a nonlinear equation for the nonlinear part, or we could combine two nonlinear equations, thus creating additional material models. Several other material models are considered in the problem set with this section.

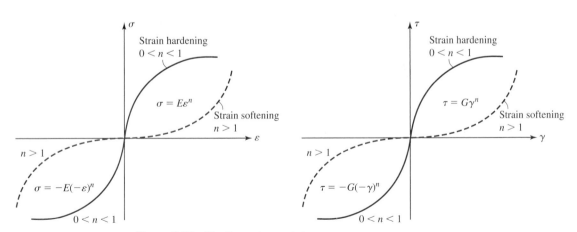

Figure 3.38 Nonlinear stress–strain curves.

EXAMPLE 3.13

Aluminum has a yield stress $\sigma_{yield} = 40$ ksi in tension, a yield strain $\varepsilon_{yield} = 0.004$, an ultimate stress $\sigma_{ult} = 45$ ksi, and the corresponding ultimate strain $\varepsilon_{ult} = 0.17$. Determine the material constants and plot the corresponding stress–strain curves for the following models:

(a) Elastic–perfectly plastic model.

(b) Linear strain-hardening model.

(c) Nonlinear power-law model.

PLAN

We have coordinates of three points on the curve: $P_0 (\sigma_0 = 0.00, \varepsilon_0 = 0.000)$, $P_1 (\sigma_1 = 40.0, \varepsilon_1 = 0.004)$, and $P_2 (\sigma_2 = 45.0, \varepsilon_2 = 0.017)$. Using these data we can find the various constants in the material models.

Solution

(a) The modulus of elasticity E is the slope between points P_0 and P_1. After yield stress, the stress is a constant. The stress–strain behavior can be written as

$$\text{ANS.} \qquad E_1 = \frac{\sigma_1 - \sigma_0}{\varepsilon_1 - \varepsilon_0} = \frac{40}{0.004} = 10{,}000 \text{ ksi}$$

$$\text{ANS.} \qquad \sigma = \begin{cases} 10{,}000\varepsilon \text{ ksi}, & |\varepsilon| \le 0.004 \\ 40 \text{ ksi}, & |\varepsilon| \ge 0.004 \end{cases} \qquad \text{(E1)}$$

(b) In the linear strain-hardening model the slope of the straight line before yield stress is as calculated in part (a). After the yield stress, the slope of the line can be found from the coordinates of points P_1 and P_2. The stress–strain behavior can be written as

$$\text{ANS.} \qquad E_2 = \frac{\sigma_2 - \sigma_1}{\varepsilon_2 - \varepsilon_1} = \frac{5}{0.013} = 384.6 \text{ ksi}$$

$$\qquad \qquad \qquad \qquad \qquad \qquad \qquad \qquad \qquad \qquad \qquad \qquad \text{(E2)}$$

$$\text{ANS.} \qquad \sigma = \begin{cases} 10{,}000\varepsilon \text{ ksi}, & |\varepsilon| \le 0.004 \\ 40 + 384.6(\varepsilon - 0.004) \text{ ksi}, & |\varepsilon| \ge 0.004 \end{cases}$$

(c) The two constants E and n in $\sigma = E\varepsilon^n$ can be found by substituting the coordinates of the two points P_1 and P_2 to generate the following two equations:

$$40 = E(0.004)^n \qquad \text{(E3)}$$

$$45 = E(0.017)^n \qquad \text{(E4)}$$

Dividing Equation (E4) by Equation (E3) and taking the logarithm of both sides, we can solve for n,

$$\ln\left(\frac{0.017}{0.004}\right)^n = \ln\left(\frac{45}{40}\right) \quad \text{or} \quad n\ln(4.25) = \ln(1.125)$$

$$\text{or} \quad \text{ANS.} \quad n = 0.0814 \quad \text{(E5)}$$

Substituting Equation (E5) into Equation (E3), we can obtain the value of E,

$$E = \frac{40}{0.004^{0.0814}} = 62.7 \text{ ksi} \quad \text{(E6)}$$

We can now write the stress–strain equations for the power-law model,

$$\sigma = \begin{cases} 62.7\varepsilon^{0.0814} \text{ ksi}, & \varepsilon \geq 0 \\ -62.7\ (-\varepsilon)^{0.0814} \text{ ksi}, & \varepsilon < 0 \end{cases} \quad \text{(E7)}$$

Stresses at different strains can be found using Equations (E1), (E2), and (E7) and plotted as shown in Figure 3.39.

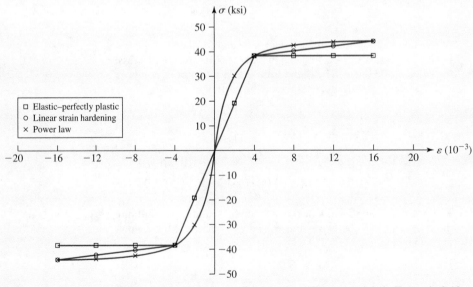

Figure 3.39 Stress–strain curves for different models in Example 3.13.

EXAMPLE 3.14

In a hollow circular shaft the shear strain at a section in polar coordinates was found to be $\gamma_{x\theta} = 3\rho \times 10^{-3}$, where ρ is the radial coordinate measured in inches (Figure 3.40). Write expressions for $\tau_{x\theta}$ as a function of ρ and plot the shear strain $\gamma_{x\theta}$ and shear stress $\tau_{x\theta}$ distributions across the cross section. Assume the shaft is made from elastic–perfectly plastic material that has a yield stress $\tau_{yield} = 24$ ksi and a shear modulus $G = 6000$ ksi.

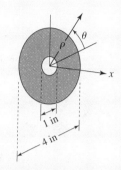

Figure 3.40 Hollow shaft in Example 3.14.

PLAN

We can find the yield strain in shear γ_{yield} from the given yield stress τ_{yield} and the shear modulus G. We can then find the location of the elastic–plastic boundary by finding ρ_y, at which the shear strain reaches the value of γ_{yield}. The shear stress at points before ρ_y can be found from Hooke's law, and after ρ_y it will be the yield stress.

Solution The location of the elastic–plastic boundary can be found as follows:

$$\gamma_{yield} = \frac{\tau_{yield}}{G} = \frac{24 \times 10^3}{6000 \times 10^3} = 0.004 = 0.003\rho_y$$

or

$$\rho_y = \frac{0.004}{0.003} = 1.33 \text{ in}$$

Up to ρ_y stress and strain are related by Hooke's law, and hence

$$\tau_{x\theta} = G\gamma_{x\theta} = 6 \times 10^6 \times 3\rho \times 10^{-3} = 18\rho \times 10^3 \text{ psi}$$

After ρ_y the stress is equal to τ_y, and the shear stress can be written as

$$\text{ANS.} \qquad \tau_{x\theta} = \begin{cases} 18\rho \text{ ksi,} & 0.5 \le \rho \le 1.333 \\ 24 \text{ ksi,} & 1.333 \le \rho \le 2.0 \end{cases} \qquad \text{(E1)}$$

The shear strain and shear stress distributions across the cross section are shown in Figure 3.41.

COMMENT

In this problem we knew the strain distribution and hence could locate the elastic–plastic boundary easily. In most problems we do not know the strains due to a load, and finding the elastic–plastic boundary is significantly more difficult.

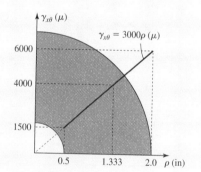

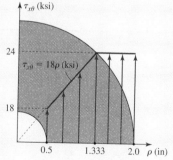

Figure 3.41 Strain and stress distributions.

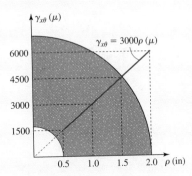

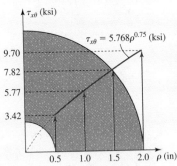

Figure 3.42 Strain and stress distributions in Example 3.15.

EXAMPLE 3.15

Resolve Example 3.14 assuming the shaft material has a stress–strain relationship given by $\tau = 450\gamma^{0.75}$ ksi.

PLAN

Substituting the strain expression into the stress–strain equation we can obtain stress as a function of ρ and plot it.

Solution Substituting the strain distribution into the stress–strain relation we obtain

$$\tau_{x\theta} = 450 \times 0.003^{0.75}\rho^{0.75} \tag{E1}$$

$$\text{ANS.} \quad \tau_{x\theta} = 5.768\rho^{0.75} \text{ ksi}$$

The shear stress can be found at several points and plotted as shown in Figure 3.42.

COMMENT

We see that although the strain distribution is linear across the cross section, the stress distribution is nonlinear due to material nonlinearity. Deducing the stress distribution across the cross section would be difficult, but deducing a linear strain distribution is possible from geometric considerations, as shall be seen in Chapter 5 on the torsion of circular shafts.

EXAMPLE 3.16

At the cross section in a beam the normal strain due to bending about the z axis was found to vary as $\varepsilon_{xx} = -0.0125y$, with y measured in meters (Figure 3.43). Write the expressions for normal stress σ_{xx} as a function of y and plot the σ_{xx} distribution across the cross section. Assume beam is made from elastic–perfectly plastic material that has a yield stress $\sigma_{\text{yield}} = 250$ MPa and a modulus of elasticity $E = 200$ GPa.

PLAN

Points furthest from the origin will be the most strained, and the plastic zone will start from the top and bottom and move inward symmetrically. We can find the yield strain ε_{yield} from the given yield stress σ_{yield} and the modulus of elasticity E. We can then find the location of the elastic–plastic boundary by finding y_y, at which the normal strain reaches the value of ε_{yield}. The normal stress before y_y can be found from Hooke's law, and after y_y it will be the yield stress.

Solution The location of the elastic–plastic boundary can be found as

$$\varepsilon_{yield} = \frac{\sigma_{yield}}{E} = \frac{\pm 250 \times 10^6}{200 \times 10^9} = \pm 1.25 \times 10^{-3} = -0.0125 y_y$$

$$y_y = \frac{\pm 1.25 \times 10^{-3}}{-0.0125} = \mp 0.1 \text{ m}$$

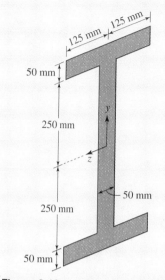

Figure 3.43 Beam cross section in Example 3.16.

Up to the elastic–plastic boundary, i.e., y_y, the material is in the linear range and Hooke's law applies. Thus,

$$\sigma_{xx} = 200 \times 10^9 (-0.0125 y) = -2500 y \text{ MPa} \qquad |y| \leq 0.1 \text{ m}$$

The normal stress as a function of y can be written as

ANS. $\sigma_{xx} = \begin{cases} -250 \text{ MPa}, & 0.1 \leq y \leq 0.3 \\ -2500 y \text{ MPa}, & -0.1 \leq y \leq 0.1 \\ 250 \text{ MPa}, & -0.3 \leq y \leq -0.1 \end{cases}$ (E1)

The normal strain and stress as a function of y can be plotted as shown in Figure 3.44.

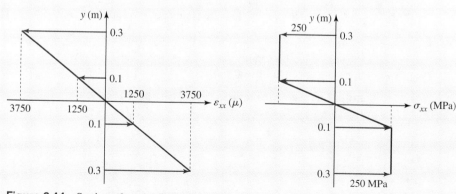

Figure 3.44 Strain and stress distributions in Example 3.16.

COMMENTS

1. To better appreciate the stress distribution we can plot it across the entire cross section, as shown in Figure 3.45.

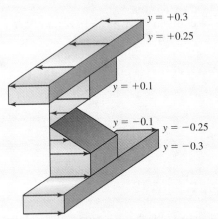

y = +0.3
y = +0.25
y = +0.1
y = −0.1
y = −0.25
y = −0.3

Figure 3.45 Stress distribution across cross section in Example 3.16.

2. Once more we see that the stress distribution across the cross section will be difficult to deduce, but as shall be seen in Chapter 6 on the symmetric bending of beams, we can deduce the approximate strain distribution from geometric considerations.

EXAMPLE 3.17

Resolve Example 3.16 assuming that the stress–strain relationship is given by $\sigma = 9000\varepsilon^{0.6}$ MPa in tension and in compression.

PLAN

We substitute the strain value in Equations (3.30) and obtain the equation for stress in terms of y.

Solution Substituting the strains in the stress–strain relation in Equations (3.30), we obtain

$$\sigma_{xx} = \begin{cases} E\varepsilon_{xx}^{0.6}, & \varepsilon_{xx} \geq 0 \\ -(E(-\varepsilon_{xx})^{0.6}, & \varepsilon_{xx} \leq 0) \end{cases} \qquad \text{ANS. } \sigma_{xx} = \begin{cases} 649.2(-y)^{0.6}, & y \leq 0 \\ -649.2(y)^{0.6}, & y \geq 0 \end{cases}$$

The strains and stresses can be found at different values of y and plotted as shown in Figure 3.46.

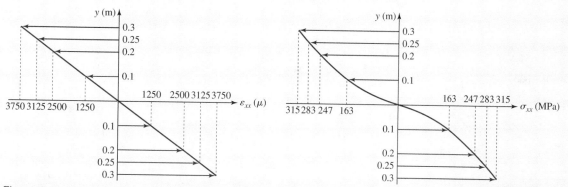

Figure 3.46 Strain and stress distributions in Example 3.17.

COMMENT

To better appreciate the stress distribution we can plot it across the entire cross section, as shown in Figure 3.47.

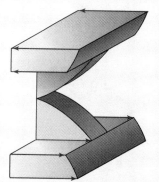

Figure 3.47 Stress distribution across cross section in Example 3.17.

PROBLEM SET 3.3

Factor of safety

3.103 A joint in a wooden structure is to be designed for a factor of safety of 3 (Figure P3.103). If the average failure stress in shear on the surface *BCD* is 1.5 ksi and the average failure bearing stress on the surface *BEF* is 6 ksi, determine the smallest dimensions h and d to the nearest $\frac{1}{16}$ of an inch.

3.104 A 300-lb light is hanging from a ceiling by a chain (Figure P3.104). The links of the chain are loops made from a thick wire. Determine the minimum diameter of

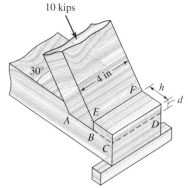

Figure P3.103

Figure P3.104

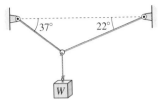

Figure P3.105

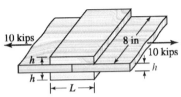

Figure P3.106

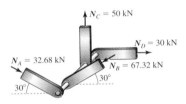

Figure P3.107

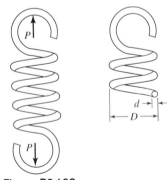

Figure P3.108

the wire to the nearest $\frac{1}{16}$ in for a factor of safety of 4. The normal failure stress for the wire is 28 ksi.

3.105 Determine the maximum weight W that can be suspended using cables, as shown in Figure P3.105, for a factor of safety of 1.2. The cable's fracture stress is 200 MPa and its diameter is 10 mm.

3.106 An adhesively bonded joint in wood is fabricated as shown in Figure P3.106. For a factor of safety of 1.25, determine the minimum overlap length L and dimension h to the nearest $\frac{1}{8}$ in. The shear strength of the adhesive is 400 psi and the wood strength is 6 ksi in tension.

3.107 A joint in a truss has the configuration shown in Figure P3.107. Determine the minimum diameter of the pin to the nearest millimeter for a factor of safety of 2.0. The pin's failure stress in shear is 300 MPa.

3.108 The shear stress on the cross section of the wire of a helical spring (Figure P3.108) is given by $\tau = K(8PC/\pi d^2)$, where P is the force on the spring, d is the diameter of the wire from which the spring is constructed, C is called the spring index, given by the ratio $C = D/d$, D is the diameter of the coiled spring, and K is called the Wahl factor, given by

$$K = \frac{4C-1}{4C-4} + \frac{0.615}{C}$$

The spring is to be designed to resist a maximum force of 1200 N and must have a factor of safety of 1.1 in yield. The shear stress in yield is 350 MPa. Make a table listing admissible values of C and d for 4 mm $\le d \le$ 16 mm in steps of 2 mm.

Stress concentration

3.109 A steel bar is axially loaded, as shown in Figure P3.109. Determine the factor

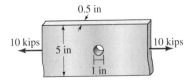

Figure 3.109

of safety for the bar if yielding is to be avoided. The normal yield stress for steel is 30 ksi. Use the stress concentration factor chart in Appendix C.

3.110 The stress concentration factor for a stepped flat tension bar with shoulder fillets (Figure P3.110) was determined as

$$K_{conc} = 1.970 - 0.384\left(\frac{2r}{H}\right) - 1.018\left(\frac{2r}{H}\right)^2$$
$$+ 0.430\left(\frac{2r}{H}\right)^3$$

This equation is valid only if $H/d > 1 + 2r/d$ and $L/H > 5.784 - 1.89r/d$. The nominal stress is P/dt. Make a chart for the stress concentration factor versus H/d for the following values of r/d: 0.2, 0.4, 0.6, 0.8, 1.0. (Use of a spread sheet is recommended.)

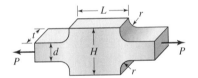

Figure 3.110

3.111 Determine the maximum normal stress in the stepped flat tension bar shown in Figure P3.110 for the following data: $P = 9$ kips, $H = 8$ in, $d = 3$ in, $t = 0.125$ in, and $r = 0.625$ in.

3.112 An aluminum stepped tension bar is to carry a load $P = 56$ kN. The normal yield stress of aluminum is 160 MPa. The bar in Figure P3.110 has $H = 300$ mm, $d = 100$ mm, and $t = 10$ mm. For a factor of safety of 1.6, determine the minimum value r of the fillet radius if yielding is to be avoided.

3.113 The stress concentration factor for a flat tension bar with U-shaped notches (Figure P3.113) was determined as

$$K_{conc} = 3.857 - 5.066\left(\frac{4r}{H}\right) + 2.469\left(\frac{4r}{H}\right)^2$$
$$- 0.258\left(\frac{4r}{H}\right)^3$$

The nominal stress is P/Ht. Make a chart for the stress concentration factor vs. r/d for the following values of H/d: 1.25, 1.50, 1.75, 2.0. (Use of a spread sheet is recommended.)

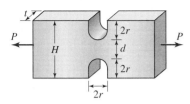

Figure P3.113

3.114 Determine the maximum normal stress in the flat tension bar shown in Figure P3.113 for the following data: $P = 150$ kN, $H = 300$ mm, $r = 15$ mm, and $t = 5$ mm.

3.115 A steel tension bar with U-shaped notches of the type shown in Figure P3.113 is to carry a load $P = 12$ kips. The normal yield stress of steel is 30 ksi. The bar has $H = 9$ in and $t = 0.25$ in. For a factor of safety of 1.4, determine the value of r if yielding is to be avoided.

Temperature effects

3.116 An iron rim ($\alpha = 6.5$ $\mu/°F$) of 35.98-in diameter is to be placed on a wooden cask of 36-in diameter. Determine the minimum temperature increase needed to slip the rim onto the cask.

3.117 The temperature is increased by 60°C in both steel ($E_s = 200$ GPa, $\alpha_s = 12.0$ $\mu/°C$) and aluminum ($E = 72$ GPa, $\alpha = 23.0$ $\mu/°C$). Determine the angle by which the pointer rotates from the vertical position (Figure P3.122).

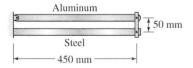

Figure P3.122

3.118 Solve Problem 3.64 if the temperature decrease is 25°C. Use $\alpha = 11.7$ $\mu/°C$.

3.119 Solve Problem 3.65 if the temperature increase is 50°C. Use $\alpha = 23.6$ $\mu/°C$.

3.120 Solve Problem 3.72 if the temperature increase is 40°F. Use $\alpha = 6.5$ $\mu/°F$.

3.121 Solve Problem 3.73 if the temperature decrease is 100°F. Use $\alpha = 12.8$ $\mu/°F$.

3.122 Solve Problem 3.74 if the temperature decrease is 75°C. Use $\alpha = 26.0$ $\mu/°C$.

3.123 A plate ($E = 30,000$ ksi, $v = 0.25$, $\alpha = 6.5 \times 10^{-6}/°F$) cannot expand in the y direction and can expand at most by 0.005 in. in the x direction, as shown in Figure P3.123. Assuming plane stress, determine the average normal stresses in the x and y directions due to a uniform temperature increase of 100°F.

3.124 Derive the following relations of normal stresses in terms of normal strains from Equations (3.27a), (3.27b), and (3.27c):

$$
\sigma_{xx} = [(1-v)\varepsilon_{xx} + v\varepsilon_{yy} + v\varepsilon_{zz}]
$$
$$
\times \frac{E}{(1-2v)(1+v)} - \frac{E\alpha\,\Delta T}{1-2v}
$$
$$
\sigma_{yy} = [(1-v)\varepsilon_{yy} + v\varepsilon_{zz} + v\varepsilon_{xx}]
$$
$$
\times \frac{E}{(1-2v)(1+v)} - \frac{E\alpha\,\Delta T}{1-2v} \qquad (3.31)
$$
$$
\sigma_{zz} = [(1-v)\varepsilon_{zz} + v\varepsilon_{xx} + v\varepsilon_{yy}]
$$
$$
\times \frac{E}{(1-2v)(1+v)} - \frac{E\alpha\,\Delta T}{1-2v}
$$

3.125 For a point in plane stress show

$$
\sigma_{xx} = (\varepsilon_{xx} + v\varepsilon_{yy})\frac{E}{1-v^2} - \frac{E\alpha\,\Delta T}{1-v}
$$
$$
\sigma_{yy} = (\varepsilon_{yy} + v\varepsilon_{xx})\frac{E}{1-v^2} - \frac{E\alpha\,\Delta T}{1-v} \qquad (3.32)
$$

3.126 For a point in plane stress show

$$
\varepsilon_{zz} = -\left(\frac{v}{1-v}\right)(\varepsilon_{xx} + \varepsilon_{yy}) + \left(\frac{1+v}{1-v}\right)\alpha\,\Delta T
$$
$$
(3.33)
$$

Fatigue

3.127 A machine component is made from a steel alloy that has an S–N curve as shown in Figure 3.34. Estimate the service life of the component if the peak stress is reversed at the rates shown.
(a) 40 ksi at 200 cycles per minute.
(b) 36 ksi at 250 cycles per minute.
(c) 32 ksi at 300 cycles per minute.

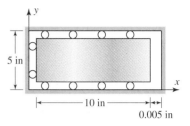

Figure P3.123

3.128 A machine component is made from an aluminum alloy that has an S–N curve as shown in Figure 3.34. What should be the maximum permissible peak stress in MPa for the following situations:
(a) 17 hours of service at 100 cycles per minute.
(b) 40 hours of service at 50 cycles per minute.
(c) 80 hours of service at 20 cycles per minute.

A uniaxial stress acts on an aluminum plate with a hole is shown in Figure P3.129. The aluminum has an S–N curve as shown in Figure 3.34. Use this information to solve Problems 3.129 through 3.131.

3.129 Predict the number of cycles the plate could be used if $d = 3.2$ in and the far-field stress $\sigma = 6$ ksi.

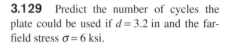

Figure P3.129

3.130 Determine the maximum diameter of the hole (Figure P3.129) to the nearest $\frac{1}{8}$ in, if the predicted service life of one-half million cycles is desired for a uniform far-field stress $\sigma = 6$ ksi.

3.131 Determine the maximum far-field stress σ (Figure P3.129) if the diameter of the hole is 2.4 in and a predicted service life of three-quarters of a million cycles is desired.

Nonlinear material models

3.132 Bronze has a yield stress $\sigma_{yield} = 18$ ksi in tension, a yield strain $\varepsilon_{yield} = 0.0012$, ultimate stress $\sigma_{ult} = 50$ ksi, and the corresponding ultimate strain $\varepsilon_{ult} = 0.50$. Determine the material constants and plot the resulting stress–strain curve for:
(a) Elastic–perfectly plastic model.
(b) Linear strain-hardening model.
(c) Nonlinear power-law model.

3.133 Cast iron has a yield stress $\sigma_{yield} = 220$ MPa in tension, a yield strain $\varepsilon_{yield} = 0.00125$, ultimate stress $\sigma_{ult} = 340$ MPa, and the corresponding ultimate strain $\varepsilon_{ult} = 0.20$. Determine the material constants and plot the resulting stress–strain curve for:
(a) Elastic–perfectly plastic model.
(b) Linear strain-hardening model.
(c) Nonlinear power-law model.

A solid circular shaft of 3-in diameter has a shear strain at a section in polar coordinates
of $\gamma_{x\theta} = 2\rho \times 10^{-3}$, *where ρ is the radial coordinate measured in inches. For the material models in Problems 3.134 through 3.137 write the expressions for $\tau_{x\theta}$ as a function of ρ and plot the shear strain $\gamma_{x\theta}$ and shear stress $\tau_{x\theta}$ distributions across the cross section.*

3.134 The shaft is made from an elastic–perfectly plastic material, which has a yield stress $\tau_{yield} = 18$ ksi and a shear modulus $G = 12,000$ ksi.

3.135 The shaft is made form a bilinear material as shown in Figure 3.37. The material has a yield stress $\tau_{yield} = 18$ ksi and shear moduli $G_1 = 12,000$ ksi and $G_2 = 4800$ ksi.

3.136 The shaft material has a stress–strain relationship given by $\tau = 243\gamma^{0.4}$ ksi.

3.137 The shaft material has a stress–strain relationship given by $\tau = 12,000\gamma - 120,000\gamma^2$ ksi.

A hollow circular shaft has an inner diameter of 50 mm and an outside diameter of 100 mm. The shear strain at a section in polar coordinates was found to be $\gamma_{x\theta} = 0.2\rho$, where ρ is the radial coordinate measured in meters. For the material models in Problems 3.138 through 3.141 write the expressions for $\tau_{x\theta}$ as a function of ρ and plot the shear strain $\gamma_{x\theta}$ and shear stress $\tau_{x\theta}$ distributions across the cross section.

3.138 The shaft is made from an elastic–perfectly plastic material that has a shear yield stress $\tau_{yield} = 175$ MPa and a shear modulus $G = 26$ GPa.

3.139 The shaft is made from a bilinear material as shown in Figure 3.37. The material has a shear yield stress $\tau_{yield} = 175$ MPa and shear moduli $G_1 = 26$ GPa and $G_2 = 14$ GPa.

3.140 The shaft material has a stress–strain relationship given by $\tau = 3435\gamma^{0.6}$ MPa.

3.141 The shaft material has a stress–strain relationship given by $\tau = 26,000\gamma - 208,000\gamma^2$ MPa.

A hollow rectangular beam has the dimensions shown in Figure P3.142. The normal strain due to bending about the z axis was *found to vary as $\varepsilon_{xx} = -0.01y$, with y measured in meters. For the material models in Problems 3.142 through 3.145 write the expressions for normal stress σ_{xx} as a function of y and plot the σ_{xx} distribution across the cross section. Assume similar material behaviour in tension and compression.*

3.142 The beam is made from an elastic–perfectly plastic material that has a yield stress $\sigma_{yield} = 250$ MPa and a modulus of elasticity $E = 200$ GPa.

3.143 The beam is made from a bilinear material as shown in Figure 3.37. The material has a yield stress $\sigma_{yield} = 250$ MPa and moduli of elasticity $E_1 = 200$ GPa and $E_2 = 80$ GPa.

3.144 The beam material has a stress–strain relationship given by $\sigma = 952\varepsilon^{0.2}$ MPa.

3.145 The beam material has a stress–strain relationship given by $\sigma = 200\varepsilon - 2000\varepsilon^2$ MPa.

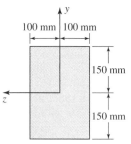

Figure P3.142

*3.12 GENERAL INFORMATION

Material modeling and the determination of constants is rich with controversy. A proper definition of the modulus of elasticity that did not include some geometric parameters of the cross section took a long time to evolve, and several individuals introduced related concepts over a period of 100 years. Who should be given credit for the modulus of elasticity? The correct independent constants that are needed to describe the linear relationship between stress and strain were also very controversial and took an equally long time to resolve. These controversies are described in the section on history.

Material grouping based on the number of constants that are needed to describe a linear stress–strain relation are briefly introduced in Section 3.12.2 as a prelude to the growing area of composites. Composites are lightweight materials that can be very stiff and strong. The material properties of composites can be tailored to fulfill specific design requirements. In the case of polymer composites, complex shapes can be molded with little or no machining. Composites are discussed briefly in Section 3.12.3.

3.12.1 History

Robert Hooke was born in 1625. Even as a child he showed great interest in mechanical toys and drawings. In 1662 he was appointed curator of the experiments of the Royal Society in England for his inventive ability and his willingness to design apparatus for demonstrating his own and other Royal Society Fellows' ideas. Among his many works are experiments on springs and elastic bodies. In 1678 he published the results of his experiments in a paper in which he states: "Ut tensio sic vis" (as the extension so is the force). This proportional linear relationship between force and deformation is the real Hooke's law. As an acknowledgment of his work on elastic bodies, the stress–strain relation of Equation (3.1) is also called Hooke's law.

In 1680, while working in France, Mariotte conducted many experiments with beams and arrived at the linear relationship of force and deformation independent of Robert Hooke. Euler, in his mathematical studies on beam buckling published in 1757, also used Hooke's law and introduced what he called the moment of stiffness. He suggested that this moment of stiffness could be determined experimentally. This moment of stiffness is the bending rigidity, which we shall introduce in the chapter of beam bending. But the idea that there is a material constant that should be determined experimentally was introduced by Giordano Ricardi. In a paper published in 1782 on the first six modes of vibration for chimes of brass and steel he gave values for the modulus of elasticity. The credit for defining and measuring the modulus of elasticity, however, is given to Thomas Young, as E is often referred to as Young's modulus.

Thomas Young was appointed a professor of natural philosophy at the Royal Institute in England. After his resignation from this post in 1803, he published his course material. In this course material he gives a definition for the modulus of elasticity in terms of the pressure produced at the base of a column due to its weight. This definition includes the area of cross section, which is like the axial rigidity we will study in the chapter of axial members. The refinement of the definition of the modulus of elasticity as purely a material property—independent of the geometric features of the cross section—is a later development.

Poisson's ratio is part of a controversy that raged over most of the 19th century. The molecular theory of stress initiated by Navier is based on the central-force concept described in the Section 1.3. Equilibrium equations, written in terms of displacement, which were derived by Navier using the molecular theory of stress, had only one independent material constant for isotropic bodies. In 1839 George Green started from the alternative viewpoint that at equilibrium the potential energy must be minimum, and came to the conclusion that there must be two independent constants for the isotropic stress–strain relationship. From his independent analysis, using Navier's molecular theory of stress, Poisson had concluded that the Poisson's ratio must be $\frac{1}{4}$. With the value of Poisson's ratio of $\frac{1}{4}$, the equilibrium equations of Green and Navier become identical. W. Wertheim's experiments on glass and brass in 1848 did not support the $\frac{1}{4}$ value for Poisson's ratio, but Wertheim continued to believe that only one independent material constant was needed to describe the relationship between stress and strain. Wertheim, however, believed that the value of Poisson's ratio should be $\frac{1}{3}$. The believers in the existence of one independent material constant would dismiss the experimental results that suggested two independent constants on the basis that the material on which the experiment was conducted was not truly isotropic. In the case of anisotropic material, Cauchy and Poisson (using Navier's molecular theory of stress) concluded that there were fifteen independent material constants, whereas Green's analysis showed that there should be twenty-one independent material constants that relate stress to strain. The two viewpoints could be resolved if there were six relationships between the material constants. Voigt's experiments between 1887 and 1889 on single crystals with known anisotropic properties showed that the six relationships between the material constants were untenable. Thus nearly half a century after the deaths of Navier (1785–1836), Poisson (1781–1840), Cauchy (1789–1857), and Green (1793–1841), the experimental results of Voigt resolved the controversy. Today we accept that isotropic materials have two independent constants and anisotropic materials have twenty-one independent constants in the general linear stress–strain relationship.

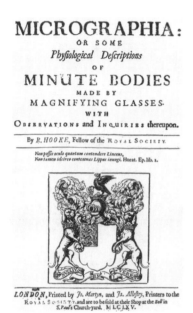

Figure 3.48 The cover of Robert Hooke's *Micrographia,* published in 1665.

Figure 3.49 Thomas Young.

There are thirty-one types of crystals. Body made up from these crystals can be grouped into classes for the purpose of defining the independent material constants in the linear stress–strain relationship. The most general anisotropic material, which requires twenty-one independent constants, is also called *triclinic* material. Three other important material groups besides isotropic are described briefly next.

Monoclinic material requires thirteen independent material constants. The z plane is the plane of symmetry. This implies that the stress–strain relationships are the same in the positive and negative z directions.

Orthotropic material requires nine independent constants. Orthotropic materials have two orthogonal planes of symmetry, that is, if we rotate by 90° about the x or the y axis, we obtain the same stress–strain relationships.

Transversely isotropic material requires five independent material constants. Transversely isotropic material is isotropic in a plane, that is, rotation by an arbitrary angle about the z axis does not change the stress–strain relationship and the material is isotropic in the xy plane.

Isotropic material requires only two independent material constants. Rotation about the x, y, or z axis by any arbitrary angle results in the same stress–strain relationship.

3.12.3 Composite Materials

A body made from more than one material can be called a composite. The ancient Egyptians made composite bricks for building the pyramids by mixing straw and mud. The resulting brick was stronger than the brick made from mud alone. Modern polymer composites rely on the same phenomenological effect in mixing fibers and epoxies to get high strength and high stiffness per unit weight.

Fibers are inherently stiffer and stronger than bulk material. Bulk glass such as in window panes has a breaking strength of a few thousand psi. Glass fibers, however, have a breaking strength on the order of one-half million psi. The increase in strength and stiffness is due to a reduction of defects and the alignment of crystals along the fiber axis. The plastic epoxy holds these high-strength and high-stiffness fibers together.

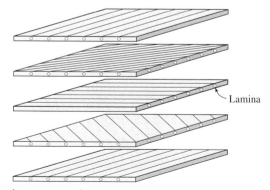
Lamina

Figure 3.50 Laminate construction.

In long-fiber (continuous-fiber) composites, a *lamina* is constructed by laying the fibers in a given direction and pouring epoxy on top. Clearly, each lamina will have different mechanical properties in the direction of the fibers and in the direction perpendicular to the fibers. If the properties of the fibers and the epoxies are averaged (homogenized), then each lamina can be regarded as an orthotropic material. Laminae with different fiber orientations are then put together to create a *laminate*. The overall properties of the laminate can be controlled by the orientation of the fibers and the stacking sequence of the laminae. The designer thus has additional design variables (material properties) that can be tailored to fulfill a given design's requirements. Continuous-fiber composite technology is still very expensive compared to that of metals, but a significant weight reduction justifies its use in the aerospace industry and in specialty sports equipment.

One way of producing short-fiber composites is to spray fibers onto epoxy and cure the mixture. The random orientation of the fibers results in an overall (homogenized) transversely isotropic material whose properties depend on the ratio of the volume of fibers to the volume of epoxy. Chopped fibers are cheaper to produce than the continuous-fiber composites and are finding increasing use in automobile and marine industries for designing secondary structures, such as body panels.

3.13 CLOSURE

In this chapter we studied the many adjectives that can be used in describing material behavior, and we established the empirical relationships between stress and strain. We also saw that the number of material constants that we need depends on the material model we wish to incorporate into our analysis. The simplest material model is the linear, elastic, isotropic material that requires only two material constants.

We also studied how material models can be integrated into a logic by which we can relate displacements to external forces. If we choose a more complex material model, then it changes the stress distribution across the cross section but does not change the relationship (equations) between displacements and strains or the relationship (equations) between stresses and internal forces and moments. Similarly we can add complexity to the relationship between displacements and strains without changing the material model. Thus the modular structure of the logic permits us to add complexities at several points, then carry the complexity forward into the equations that are unchanged from the added complexity.

In Chapters 4 through 7 theories and formulas will be developed, using the logic shown in Figure 3.15, for axial members, torsion of circular shafts, and symmetric bending of beams. The governing motivation will be to develop the simplest possible theories for these structural members. In order to develop these simple theories, we shall impose limitations and make assumptions regarding loading, material behavior, and geometry. The difference between limitations and assumptions is in the degree to which the theory must be modified when a limitation or assumption is not valid. An entire theory must be redeveloped if a limitation is to be overcome. On the other hand, assumptions are points where complexities can be added and the derivation path that was established for the simplified theory can be repeated with the added complexity. Examples, problems, and optional sections will demonstrate the addition of complexities to the simplified theories.

1. The length of the member is significantly greater (approximately 10 times) than the greatest dimension in the cross section. Approximations across the cross section are now possible as the region of approximation is small. We will deduce constant or linear approximations of deformation across the cross section and confirm the validity of an approximation by using photographs of deformed shapes.

2. We are away from regions of stress concentration, where displacements and stresses can be three-dimensional. The results from the simplified theories can be extrapolated into the region of stress concentration, as described in Section 3.7. In a similar manner, results can be extrapolated into regions near cracks using the stress intensity factor described in Section 10.3.

3. The variation of external loads or change in cross-sectional area is gradual, except in regions of stress concentration. The theory of elasticity shows that this limitation is necessary; otherwise the approximations across the cross section would be untenable.

4. The external loads are such that the axial, torsion, and bending problems can be studied individually. This not only requires that the applied loads be in a given direction but, as shall be seen, it requires that the loads pass through a specific point on the cross section.

Often reality is more complex than can be accounted for by even the most sophisticated theory. In such cases, theories provide a relationship between variables that can be modified by empirically determined factors. These empirically modified formulas of mechanics of materials form the basis of most structural and machine design.

POINTS AND FORMULAS TO REMEMBER

- The point up to which stress and strain are linearly related is called proportional limit.
- The largest stress in the stress–strain curve is called ultimate stress.
- The sudden decrease in the cross-sectional area after ultimate stress is called necking.
- The stress at the point of rupture is called fracture or rupture stress.
- The region of the stress–strain curve in which the material returns to the undeformed state when applied forces are removed is called elastic region.
- The region in which the material deforms permanently is called plastic region.
- The point demarcating the elastic from the plastic region is called yield point.
- The stress at yield point is called yield stress.
- The permanent strain when stresses are zero is called plastic strain.
- The offset yield stress is a stress that would produce a plastic strain corresponding to the specified offset strain.

- A material that can undergo large plastic deformation before fracture is called ductile material.
- A material that exhibits little or no plastic deformation at failure is called brittle material.
- Hardness is the resistance to indentation.
- Raising the yield point with increasing strain is called strain hardening.
- A ductile material usually yields when the maximum shear stress exceeds the yield shear stress of the material.
- A brittle material usually ruptures when the maximum tensile normal stress exceeds the ultimate tensile stress of the material.

$$\sigma = E\varepsilon \quad (3.1) \qquad v = -\left(\frac{\varepsilon_{lateral}}{\varepsilon_{longitudinal}}\right) \quad (3.2) \qquad \tau = G\gamma \quad (3.3)$$

where E is the modulus of elasticity, v is Poisson's ratio, and G is the shear modulus of elasticity.

- The slope of the tangent to the stress–strain curve at a given stress value is called tangent modulus.
- The slope of the line that joins the origin to the given stress value is called secant modulus.
- An isotropic material has a stress–strain relation that is independent of the orientation of the coordinate system.
- In a homogeneous material the material constants do not change with the coordinates x, y, or z of a point.
- There are only two independent material constants in a linear stress–strain relationship for an isotropic material, but there can be 21 independent material constants in an anisotropic material.
- Generalized Hooke's law for isotropic materials:

$$\begin{array}{lll}
\varepsilon_{xx} = [\sigma_{xx} - v(\sigma_{yy} + \sigma_{zz})]/E & \gamma_{xy} = \tau_{xy}/G & \\
\varepsilon_{yy} = [\sigma_{yy} - v(\sigma_{zz} + \sigma_{xx})]/E & \gamma_{yz} = \tau_{yz}/G & G = \dfrac{E}{2(1+v)} \quad \text{(3.12a) through (3.12f)} \\
\varepsilon_{zz} = [\sigma_{zz} - v(\sigma_{xx} + \sigma_{yy})]/E & \gamma_{zx} = \tau_{zx}/G &
\end{array}$$

- Failure implies that a component or a structure does not perform the function for which it was designed.
- Failure could be due to too little or too much deformation or strength.
- Factor of safety:

$$K_{safety} = \frac{\text{failure-producing value}}{\text{computed (allowable) value}} \quad (3.23)$$

- The factor of safety must always be greater than 1.
- The failure-producing value could be the value of deformation, yield stress, ultimate stress, or loads on a structure.

CHAPTER FOUR
AXIAL MEMBERS

4.0 OVERVIEW

The simplest structural member is an axial bar, or rod. An axial bar is a long straight body on which the forces are applied along the longitudinal axis. Cables can also be analyzed as axial members if the cables are straight. An axial bar can support tensile or compressive forces whereas a cable cannot support compressive forces. The hydraulic cylinders on the dump truck in Figure 4.1 are examples of axial members in compression, whereas the cables of the suspension bridge are examples of axial members in tension. Connecting rods in an engine, struts in aircraft engine mounts, members of a truss representing a bridge or a building, spokes in bicycle wheels, columns in a building—these are some other examples of structural members that are analyzed as axial members.

The simplest theory for axial members will be rigorously developed following the logic shown in Figure 3.15, but subject to the limitations described in Section 3.13.

(a)

(b)

Figure 4.1 Examples of axial members.

The formulas from the theory will be used in the design and analysis of statically determinate and indeterminate structures. The two most important tools in the analysis of structures are the construction of free-body diagrams and approximate deformed shapes.

The two major learning objectives are:

1. Understand the theory, its limitations, and its applications for the design and analysis of axial members.

2. Develop the discipline to draw free-body diagrams and approximate deformed shapes in the design and analysis of structures.

4.1 THEORY

The theory will be developed subject to the limitations described in Section 3.13, namely; (i) the length of the member is significantly greater than the greatest dimension in the cross section; (ii) we are away from the regions of stress concentration; (iii) the variation of external loads or changes in the cross-sectional areas is gradual, except in regions of stress concentration; (iv) the axial load is applied such that there is no bending.

We will also assume that external forces are not functions of time,[1] that is, we have a static problem. An externally distributed force per unit length $p(x)$ and external forces F_1 and F_2 at each end act on an axial bar, as shown in Figure 4.2. The cross-sectional area $A(x)$ can be of any shape and could be a function of x. We expect that the internal forces will be a function of x as the external forces are a function of x.

The objectives of the theory are:

1. To obtain a formula for the relative displacements $u_2 - u_1$ in terms of the internal force N.

2. To obtain a formula for the axial stress σ_{xx} in terms of the internal force N.

To account for the variation[2] in the distributed load $p(x)$ and cross-sectional area $A(x)$, we will take $\Delta x = x_2 - x_1$ as the infinitesimal distance in which these quantities can be treated as constants. The deformation behavior across the cross section will be approximated. The logic shown in Figure 4.3 (discussed earlier in Section 3.2) will be used to develop the simplest theory for axial members. But assumptions will be identified as we move from one step to the next. These assumptions are the points at which

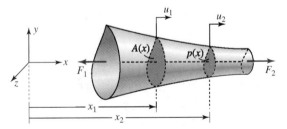

Figure 4.2 Segment of an axial bar.

---

[1]See Problems 4.35 through 4.37 for dynamic problems.
[2]Note that we have imposed the limitation that the variation in $A(x)$ and $p(x)$ is gradual.

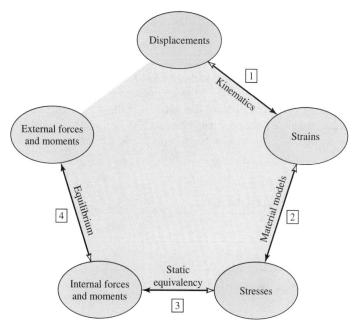

Figure 4.3 Logic in mechanics of materials.

complexities can be added to the theories, as described by footnotes associated with the assumptions.

4.1.1 Kinematics

Figure 4.4 shows a grid on an elastic band that is pulled in the axial direction. The vertical lines remain approximately vertical, but the horizontal distance between the vertical lines changes. Thus all points on a vertical line are displaced by equal amounts. If this surface observation is also true in the interior, then the following assumption is implied:

Assumption 1 Plane sections remain plane and parallel.

The displacement in the x direction is measured as u. Assumption 1 implies that u cannot be a function of y. This does not imply that all cross sections are displaced by equal amounts in the x direction. In fact the strain would be zero if the displacement of the cross section in the x direction were constant at all x. The comments in this

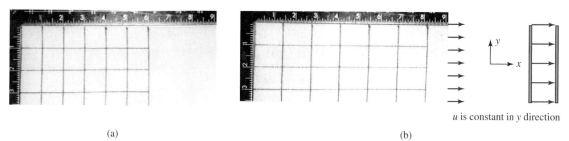

(a)   (b)

Figure 4.4 Axial deformation. (*a*) Original grid. (*b*) Deformed grid.

paragraph may be stated as follows:

$$u = u(x) \tag{4.1}$$

Definition 1 The displacement u is considered positive in the positive x direction.

An alternative perspective is as follows. Because the cross section is significantly smaller than the length, we can approximate a function such as u by a constant (uniform) over a cross section. In Chapter 6, on beam bending, we shall approximate u as a linear function of y.

4.1.2 Strain Distribution

Assumption 2 Strains are small.[3]

If points x_2 and x_1 are close in Figure 4.2, then the strain at any point x can be calculated as

$$\varepsilon_{xx} = \lim_{\Delta x \to 0} \left(\frac{u_2 - u_1}{x_2 - x_1} \right) = \lim_{\Delta x \to 0} \left(\frac{\Delta u}{\Delta x} \right)$$

or

$$\boxed{\varepsilon_{xx} = \frac{du(x)}{dx}} \tag{4.2}$$

Equation (4.2) emphasizes that the axial strain is uniform across the cross section and is only a function of x. In deriving Equation (4.2) we made no statement regarding material behavior. In other words, Equation (4.2) does not depend on the material model if Assumptions 1 and 2 are valid. But clearly if the material or loading is such that Assumptions 1 and 2 are not tenable, then Equation (4.2) will not be valid.

4.1.3 Material Model

Our motivation is to develop a simple theory for axial deformation. Thus we make assumptions regarding material behavior that will permit us to use the simplest material model given by Hooke's law.

Assumption 3 Material is isotropic.

Assumption 4 Material is linearly elastic.[4]

Assumption 5 There are no inelastic strains.[5]

---

[3]See Problem 4.38 for large strains.
[4]See Problems 4.34 and 4.52 for nonlinear material behavior.
[5]Inelastic strains could be due to temperature, humidity, plasticity, viscoelasticity, and so on. We shall consider inelastic strains due to temperature in Section 4.5.

Substituting Equation (4.2) into Hooke's law, that is, $\sigma_{xx} = E\varepsilon_{xx}$, we obtain

$$\sigma_{xx} = E\frac{du}{dx} \qquad (4.3)$$

Though the strain does not depend on y or z, we cannot say the same for the stress in Equation (4.3) since E could change across the cross section, as in laminated or composite bars.

4.1.4 Internal Axial Force

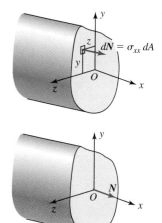

Figure 4.5 Statically equivalent internal axial force.

In order to study only axial problems with no bending, we shall replace the normal stress distribution σ_{xx} given by Equation (4.3) by an equivalent internal axial force and no internal bending moment.[6] To achieve this the equivalent internal axial force N must be placed at a specific point on the cross section, which we will fix as the origin. In Section 4.1.5 it will be seen that this specific point is the centroid for a homogeneous cross section, but for a laminated cross section the variation of E across the cross section will influence the location of the origin.

Consider a cross section of an axial member, as shown in Figure 4.5. The axial force on a differential area is given by $\sigma_{xx}\, dA$. The total contribution can be found by integrating over the entire cross section, obtaining

$$N = \int_A \sigma_{xx}\, dA \qquad (4.4)$$

Definition 2 The internal axial force N is considered positive if it is in the direction of the outward normal to the imaginary cut surface, that is, N is considered positive in tension.

Equation (4.4) is independent of the material model as it represents static equivalency between the normal stress on the entire cross section and the internal axial force. If we were to consider a laminated cross section or nonlinear material behavior, then it would affect the value and distribution of σ_{xx} across the cross section, but Equation (4.4) relating σ_{xx} and N would remain unchanged.

Substituting σ_{xx} from Equation (4.3) and noting that du/dx is a function of x only, whereas the integration is with respect to y and z ($dA = dy\, dz$), we obtain the following:

$$N = \int_A E\frac{du}{dx}\, dA = \frac{du}{dx}\int_A E\, dA \qquad (4.5)$$

4.1.5 Location of Origin

If the normal stress distribution σ_{xx} is to be replaced by only an axial force at the origin, then the internal moments M_y and M_z must be zero at the origin. From Figure 4.5 this yields the following two equations:

$$\int_A y\sigma_{xx}\, dA = 0 \qquad (4.6a)$$

$$\int_A z\sigma_{xx}\, dA = 0 \qquad (4.6b)$$

[6]We shall consider internal moments in the bending of beams in Chapter 6.

Once more it should be emphasized that Equations (4.6a) and (4.6b) are independent of the material models. However, if the material is homogeneous, then we shall see that the axial stress is uniform across the cross section. Equations (4.6a) and (4.6b) can then be simplified to yield

$$\int_A y \, dA = 0 \tag{4.7a}$$

$$\int_A z \, dA = 0 \tag{4.7b}$$

Equations (4.7a) and (4.7b) are satisfied if y and z are measured from the centroid.[7]

If the external forces are collinear and passing through the origin, then no bending moment will be produced in the rod and we will have pure axial deformation. The location of the origin depends on the material model. For a linear elastic homogeneous cross section the stress will be uniform across the cross section, as shall be seen in the next section, and the origin will be the centroid of the cross section, which is a geometric property. For composite bars, discussed in Section 4.2, the stress will be uniform in each material, but will change as the material changes. Hence the location of the origin will depend on the geometric properties and the material properties, as demonstrated in Example 4.11.

4.1.6   Axial Formulas

Consistent with the motivation of developing the simplest possible formulas, we would like to take E outside the integral, that is, E should not change across the cross section. We make the following assumption:

Assumption 6 Material is homogeneous across the cross section.[8]

From Equation (4.5) we obtain

$$N = E \frac{du}{dx} \int_A dA = EA \frac{du}{dx}$$

which can be rewritten as

$$\boxed{\frac{du}{dx} = \frac{N}{EA}} \tag{4.8}$$

The higher the value of EA, the smaller will be the deformation for a given value of the internal force. Thus the rigidity of the bar increases with the increase in EA. This implies that an axial bar can be made more rigid by either choosing a stiffer material (higher value of E) or increasing the cross-sectional area, or both. Example 4.4 brings out the importance of axial rigidity in design.

---

[7]See Appendix A.4.
[8]See Section 4.2 on composite bars, where this assumption is not valid.

Definition 3 The quantity EA is called *axial rigidity*.

If we substitute Equation (4.4) into Equation (4.3) we obtain

$$\boxed{\sigma_{xx} = \frac{N}{A}} \tag{4.9}$$

We have used Equation (4.9) in Chapters 1 and 3, but it must be recognized that this equation is valid only if all the limitations are imposed, and if Assumptions 1 through 6 are valid. But if we start with the viewpoint that the limitations and assumptions are all valid, then we can reverse the logic and obtain Equation (4.8) as follows:

$$\frac{du}{dx} = \varepsilon_{xx} = \frac{\sigma_{xx}}{E} = \frac{N}{EA}$$

N and A do not change across the cross section in Equation (4.9). Hence axial stress does not change across the homogeneous cross section. This implies that the origin must be the centroid of the cross section for no bending to occur. Thus for pure axial deformation all internal and external forces must be collinear and pass through the centroid of the cross section for homogeneous materials. Implicitly this assumes that the centroid must lie on a straight line. This eliminates curved bars but not tapered bars.

We can integrate Equation (4.8) to obtain the deformation between two points,

$$u_2 - u_1 = \int_{u_1}^{u_2} du = \int_{x_1}^{x_2} \frac{N}{EA} \, dx \tag{4.10}$$

where u_1 and u_2 are the displacements of sections at x_1 and x_2, respectively. To obtain a simple formula we would like to take the three quantities N, E, and A outside the integral, which means these quantities should not change with x. To achieve this simplicity, we make the following assumptions:

Assumption 7 The material is homogeneous between x_1 and x_2.

Assumption 8 The bar is not tapered between x_1 and x_2.

Assumption 9 The external (hence internal) axial force does not change with x between x_1 and x_2.

In Example 4.5 the cross-sectional area changes with x and Assumption 8 is violated. In Example 4.6 accounting for the weight makes the internal force N a function of x, and hence Assumption 9 is violated. In both examples we have to use Equation (4.10) or, equivalently, Equation (4.8), to solve the problems and not the simplified equation presented next.

If Assumptions 7 through 9 are valid, then N, E, and A are constant between x_1 and x_2, and from Equation (4.10) we obtain[9]

$$\boxed{u_2 - u_1 = \frac{N(x_2 - x_1)}{EA}} \tag{4.11}$$

---

[9]An alternative derivation is that if N, E, and A are constant, then the slope du/dx is a constant and can be represented by the difference formula as $du/dx = (u_2 - u_1)/(x_2 - x_1) = N/EA$.

If bars are made from two materials joined at some point x_m, or if the cross-sectional area A changes at x_m, or if there is an external force applied at x_m that would cause the internal force to change at x_m, then point x_m should not be between x_1 or x_2. In other words, in Equation (4.11) points x_1 and x_2 must be chosen such that neither N, E, nor A changes between these points.

N is an internal axial force that has to be determined by making an imaginary cut and drawing a free-body diagram. In what direction should N be drawn on the free-body diagram? There are two possibilities, elaborated further in Example 4.2.

1. N is always drawn in tension on the imaginary cut, and the equilibrium equation then gives a positive or a negative value for N. In other words, we are following the sign convention for internal force given by Definition 2. Therefore the sign for relative deformation obtained from Equation (4.11) is positive for extension and negative for contraction. Furthermore, if a displacement value for u is obtained at any section, then it is positive in the positive x direction, as per Definition 1.

2. N is drawn on the imaginary cut in a direction to equilibrate the external forces. Since inspection is being used in determining the direction of N, extension or contraction for the relative deformation in Equation (4.11) must also be determined by inspection.

4.1.7 Axial Stresses and Strains

The surface of an axial rod is usually a free surface. Thus the normal and shear stresses in the y and z directions on the surface are zero. We assume that these stresses are also zero in the interior of the rod because the greatest dimension along the cross section is an order of magnitude less than the length dimension. In the Cartesian coordinate system all stress components except σ_{xx} can be assumed zero.

From the generalized Hooke's law for isotropic materials, given by Equations (3.12a) through (3.12c), we obtain the normal strains for axial members as

$$\varepsilon_{xx} = \frac{\sigma_{xx}}{E} \qquad \varepsilon_{yy} = -\frac{v\sigma_{xx}}{E} = -v\varepsilon_{xx} \qquad \varepsilon_{zz} = -\frac{v\sigma_{xx}}{E} = -v\varepsilon_{xx} \qquad (4.12)$$

where v is the Poisson's ratio. In Equation (4.12), the normal strains in the y and z directions are due to Poisson's effect. Assumption 1 that plane sections remain plane and parallel implies that no right angle would change during deformation, and hence the assumed deformation implies that shear strains in axial members are zero. Alternatively, if shear stresses are zero, then by Hooke's law shear strains are zero.

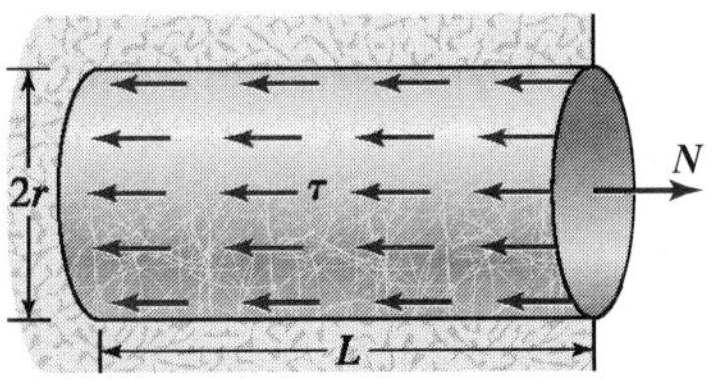

If frictional forces act on an axial member, then clearly the shear stress on the surface is not zero. Can we then still assume zero shear stress in axial members? The answer is, yes. We can analyze such problems with our theory by treating the frictional force as $p(x)$ and neglecting the shear stresses in the cross section. To explain this, we conduct an approximate analysis of a fiber being pulled out of a resin. Let us assume a fiber of radius r has a uniform shear stress τ acting on the surface over a segment L, as shown in Figure 4.6. The normal force[10] at a cross section is equal to the

Figure 4.6 Comparison of normal stress and shear stress acting on the surface.

---

[10]The normal force can also be obtained by integrating Equation (4.14) with $p(x) = -(2\pi r)\tau$.

$$N = (2\pi r L)\tau$$

$$\sigma = \frac{N}{\pi r^2} = 2\left(\frac{L}{r}\right)\tau$$

Noting that the ratio L/r is on the order of 10 or more, we see that the axial stress will be an order of magnitude greater than the shear stress applied on the surface. Thus by treating the interface shear stress as a distributed force $p(x)$ on the fiber (axial member) we will not introduce a significant error in our analysis. We thus conclude the following. (i) Frictional forces on the surface of an axial member can be modeled by $p(x)$. (ii) In the Cartesian coordinate system all stress components except σ_{xx} can be assumed zero, as mentioned earlier.

Consolidate your knowledge

1. Identity five examples of axial members from your daily life.
2. With book closed, derive Equation (4.11), listing all the assumptions as you go along.

EXAMPLE 4.1

Two thin bars are securely attached to a rigid plate, as shown in Figure 4.7. The cross-sectional area of each bar is 20 mm^2. The force F is to be placed such that the rigid plate moves only horizontally by 0.05 mm without rotating. Determine the force F and its location h for the following two cases:

(a) Both bars are made from steel with a modulus of elasticity $E = 200$ GPa.

(b) Bar 1 is made of steel ($E = 200$ GPa) and bar 2 is made of aluminum ($E = 70$ GPa).

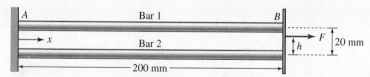

Figure 4.7 Axial bars in Example 4.1.

PLAN

The relative displacement of point B with respect to A is 0.05 mm, from which we can find the axial strain. By multiplying the axial strain by the modulus of elasticity we can obtain the axial stress. By multiplying the axial stress by the cross-sectional area we can obtain the internal axial force in each bar. We can draw the free-body diagram of the rigid plate and by equilibrium obtain the force F and its location h.

Solution

1. *Strain calculations:* The displacement of B is $u_B = 0.05$ mm. Point A is built into the wall, hence it has zero displacement. The normal strain in both rods is the same and can be found as

$$\varepsilon_1 = \varepsilon_2 = \frac{u_B - u_A}{x_B - x_A} = \frac{0.05}{200} = 250 \ \mu\text{mm/mm} \qquad (E1)$$

2. *Stress calculations:* From Hooke's law $\sigma = E\varepsilon$, we can find the normal stress in each bar for the two cases.

   Case (a): As E and ε_1 are the same for both bars, the stress is the same in both bars. We obtain the following:

$$\sigma_1 = E_1\varepsilon_1 = 200 \times 10^9 \times 250 \times 10^{-6}$$

Thus

$$\sigma_1 = 50 \text{ MPa (T)} \qquad (E2)$$

$$\sigma_2 = 50 \text{ MPa (T)} \qquad (E3)$$

   Case (b): Because E is different for the two bars, the stress in the two bars is different and can be found as follows:

$$\sigma_1 = E_1\varepsilon_1 = 200 \times 10^9 \times 250 \times 10^{-6} = 50 \text{ MPa (T)} \qquad (E4)$$

$$\sigma_2 = E_2\varepsilon_2 = 70 \times 10^9 \times 250 \times 10^{-6} = 17.5 \text{ MPa (T)} \qquad (E5)$$

3. *Internal forces:* Assuming that the normal stress is uniform in each bar, we can find the internal normal force as $N = \sigma A$, where $A = 20 \text{ mm}^2 = 20 \times 10^{-6} \text{ m}^2$. The calculations for the two cases are shown next.

   Case (a): Both bars have the same internal force since stress and cross-sectional area are the same,

$$N_1 = \sigma_1 A_1 = 50 \times 10^6 \times 20 \times 10^{-6}$$

Thus,

$$N_1 = 1000 \text{ N (T)} \qquad (E6)$$

$$N_2 = 1000 \text{ N (T)} \qquad (E7)$$

   Case (b): Because stresses are different in each bar, the equivalent internal force is different for each bar, as shown,

$$N_1 = \sigma_1 A_1 = 50 \times 10^6 \times 20 \times 10^{-6} = 1000 \text{ N (T)} \qquad (E8)$$

$$N_2 = \sigma_2 A_2 = 17.5 \times 10^6 \times 20 \times 10^{-6} = 350 \text{ N (T)} \qquad (E9)$$

4. *External force:* We make an imaginary cut through the bars and draw the free-body diagram of the rigid plate shown in Figure 4.8. We note that the internal axial force is tensile in all cases, and this fact is recognized in drawing the free-body diagram. By equilibrium of forces in the x direction we obtain

$$F = N_1 + N_2 \qquad \text{(E10)}$$

By equilibrium of moment about point O we obtain

$$N_1(20 - h) - N_2 h = 0 \qquad \text{(E11)}$$

or

$$h = \frac{20N_1}{N_1 + N_2} \qquad \text{(E12)}$$

We can substitute the value of the internal forces for each case in Equations (E10) and (E12) and obtain F and h as shown.

Case (a): Substituting Equations (E6) and (E7) into Equations (E10) and (E12), we obtain

$$F = 1000 + 1000$$

$$\text{ANS.} \qquad F = 2000 \text{ N}$$

$$h = \frac{20 \times 1000}{1000 + 1000}$$

$$\text{ANS.} \qquad h = 10 \text{ mm}$$

Case (b): Substituting Equations (E8) and (E9) into Equations (E10) and (E12), we obtain

$$F = 1000 + 350$$

$$\text{ANS.} \qquad F = 1350 \text{ N}$$

$$h = \frac{20 \times 1000}{1000 + 350}$$

$$\text{ANS.} \qquad h = 14.81 \text{ mm}$$

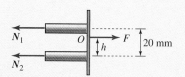

Figure 4.8 Free-body diagram in Example 4.1.

COMMENTS

1. Both bars, irrespective of the material, were subjected to the same axial strain. This is the fundamental kinematic assumption in the development of the theory for axial members, discussed in Section 4.1.

2. The summation on the right in Equation (E10) can be rewritten as $\sum_{i=1}^{n=2} \sigma_i \, \Delta A_i$, where σ_i is the normal stress in the ith bar, ΔA_i is the cross-sectional area of the ith bar, and $n = 2$ reflects that we have two bars in this problem. If we had n bars attached to the rigid plate,

then the total axial force would be given by summation over *n* bars. As we increase the number of bars *n* to infinity, the cross-sectional area ΔA_i will tend to zero (infinitesimal area written as *dA*) as we try to fit an infinite number of bars on the same plate, resulting in a continuous body with the summation replaced by an integral, as given by Equation (4.4).

3. If the external force were located at any point other than that given by the value of *h*, then the plate would rotate. This emphasizes that for pure axial problems with no bending, a point on the cross section must be found that is such that the internal moment from the axial stress distribution is zero. To emphasize this, consider the left side of Equation (E11), which can be rewritten as $\sum_{i=1}^{n} y_i \sigma_i \, \Delta A_i$, where y_i is the coordinate of the *i*th rod's centroid. The summation is an expression of the internal moment that is needed for static equivalency. This internal moment must equal zero if the problem is of pure axial deformation and is represented by Equations (4.6a) and (4.6b).

4. Even though the strains in both bars were the same in both cases, the stresses were different when *E* changed. Case (a) will correspond to a homogeneous cross section whereas case (b) would be analogous to a laminated bar in which the nonhomogeneity affects the stress distribution.

EXAMPLE 4.2

Solid circular bars of brass (E_{br} = 100 GPa, v_{br} = 0.34) and aluminum (E_{al} = 70 GPa, v_{al} = 0.33) having 200-mm diameter are attached to a steel tube (E_{st} = 210 GPa, v_{st} = 0.3) of the same outer diameter, as shown in Figure 4.9. For the loading shown determine:

(a) The movement of the plate at *C* with respect to the plate at *A*.

(b) The change in diameter of the brass cylinder.

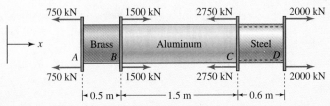

Figure 4.9 Axial member in Example 4.2.

(c) The maximum inner diameter in the steel tube if the factor of safety with respect to failure due to yielding is to be at least 1.2. The yield stress for steel is 250 MPa in tension.

PLAN

(a) We can make imaginary cuts in each segment and determine the internal force in each segment by equilibrium. By using Equation (4.11) we can find relative movements of the cross sections at B with respect to A and at C with respect to B. By adding the two relative displacements we can obtain the relative movement of the cross section at C with respect to the section at A.

(b) We can find the normal stress σ_{xx} in AB from Equation (4.9). From Equation (4.12) we can find ε_{yy}, and multiplying by the diameter we obtain the change in diameter.

(c) We can calculate the allowable axial stress in steel from the given failure values and factor of safety. Knowing the internal force in CD we can find the cross-sectional area from which we can calculate the internal radius.

Solution

(a) We make an imaginary cut in segment AB, that is, the location of the imaginary cut is defined by $0 < x < 0.5$ m. We take the left part of the cut and draw the free-body diagram in Figure 4.10 to find N_{AB},

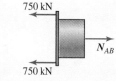

Figure 4.10 Internal force in segment AB in Example 4.2.

$$N_{AB} = 1500 \text{ kN} \tag{E1}$$

$$A_{AB} = \frac{\pi}{4}0.2^2 = 31.41 \times 10^{-3} \text{ m}^2$$

then we use Equation (4.11) to find the relative movement of point B with respect to point A,

$$u_B - u_A = \frac{N_{AB}(x_B - x_A)}{E_{AB}A_{AB}} = \frac{1500 \times 10^3 \times 0.5}{100 \times 10^9 \times 31.41 \times 10^{-3}}$$

$$= 0.2388 \times 10^{-3} \text{ m} \tag{E2}$$

Next we make an imaginary cut in segment BC, that is, the location of the imaginary cut is defined by $0.5 < x < 2$ m. We take the left part of the cut and draw the free-body diagram in Figure 4.11 to find N_{BC},

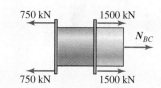

Figure 4.11 Internal force in segment BC in Example 4.2.

$$N_{BC} + 3000 - 1500 = 0$$

$$N_{BC} = -1500 \text{ kN} \tag{E3}$$

$$A_{BC} = A_{AB} = 31.41 \times 10^{-3} \text{ m}^2$$

We then use Equation (4.11) to find the relative movement of point C with respect to point B,

$$u_C - u_B = \frac{N_{BC}(x_C - x_B)}{E_{BC}A_{BC}} = \frac{-1500 \times 10^3 \times 1.5}{70 \times 10^9 \times 31.41 \times 10^{-3}}$$

$$= -1.0233 \times 10^{-3} \text{ m} \qquad \text{(E4)}$$

Adding Equations (E2) and (E4) we obtain

$$u_C - u_A = (u_C - u_B) + (u_B - u_A) = (0.2388 - 1.0233)10^{-3}$$

$$= -0.7845 \times 10^{-3} \text{ m}$$

or

ANS. $\quad u_C - u_A = 0.7845$ mm contraction

(b) We can divide the internal force in AB given by Equation (E1) by the cross-sectional area A_{AB} to find the normal stress σ_{xx} in AB as per Equation (4.9). Substituting σ_{xx}, $E_{br} = 100$ GPa, $\nu_{br} = 0.34$ in Equation (4.12), we can find ε_{yy}. Multiplying ε_{yy} by the diameter of 200 mm we obtain the change in diameter Δd,

$$\sigma_{xx} = \frac{N_{AB}}{A_{AB}} = \frac{1500 \times 10^3}{31.41 \times 10^{-3}} = 47.8 \times 10^6$$

$$\varepsilon_{yy} = -\frac{\nu_{br}\sigma_{xx}}{E_{br}} = -\frac{47.8 \times 10^6 \times 0.34}{100 \times 10^9} = -0.162 \times 10^{-3} = \frac{\Delta d}{200}$$

or

ANS. $\quad \Delta d = -0.032$ mm

(c) We can make an imaginary cut in segment CD, that is, the location of the imaginary cut is defined by $2 < x < 2.6$ m. We take the right part of the cut and draw the free-body diagram in Figure 4.12 to find N_{CD},

$$N_{CD} = 4000 \text{ kN}$$

$$\sigma_{CD} = \frac{N_{CD}}{A_{CD}} = \frac{4000 \times 10^3}{(\pi/4)(0.2^2 - D_i^2)}$$

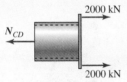

Figure 4.12 Section CD in Example 4.2.

The axial stress in CD can be written in terms of the internal diameter D_i. Using the given factor of safety we determine the value of D_i,

$$k = \frac{\sigma_{yield}}{\sigma_{CD}} = \frac{250 \times 10^6 \times \pi(0.2^2 - D_i^2)}{16,000 \times 10^3} = 49.09(0.2^2 - D_i^2) \geq 1.2$$

Then

$$D_i^2 \leq 0.2^2 - 24.4 \times 10^{-3} \qquad \text{or} \qquad \text{ANS.} \qquad D_i = 124.7 \text{ mm}$$

COMMENTS

1. On a free-body diagram some may prefer to show N in a direction that counterbalances the external forces. In such cases the sign convention in Definition 2 is not being followed, and hence the sign convention for u given in Definition 1 cannot be used. In the calculation of $u_C - u_A$ the addition and subtraction must be done manually to account for contraction and extension of the relative deformation, as shown in Figure 4.13.

2. An alternative perspective of the calculation of $u_C - u_A$ is as follows:

$$u_C - u_A = \int_{x_A}^{x_C} \frac{N}{EA} dx = \underbrace{\int_{x_A}^{x_B} \frac{N_{AB}}{E_{AB}A_{AB}} dx}_{u_B - u_A} + \underbrace{\int_{x_B}^{x_C} \frac{N_{BC}}{E_{BC}A_{BC}} dx}_{u_C - u_B}$$

or, written more compactly,

$$\Delta u = \sum_{i=1}^{n} \frac{N_i \Delta x_i}{E_i A_i} \qquad (4.13)$$

where $\sum$ denotes summation and n is the number of segments on which the summation is performed, which in our case is 2. Equation (4.13) can be used only if the sign convention for the internal force N is followed.

3. Note that $N_{BC} - N_{AB} = -3000$ and the magnitude of the applied external force at the section at B is 3000. Similarly $N_{CD} - N_{BC} = 5500$, which is the magnitude of the applied external force at the section at C. In other words, the internal axial force jumps by the value of the external force as one crosses the external force from left to right. We will make use of this observation in the next section, when we develop a graphical technique for finding the internal axial force.

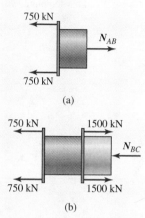

Figure 4.13 Intuitive deformation calculations. (*a*) Extension, $u_B - u_A = 0.2388 \times 10^{-3}$. (*b*) Contraction, $u_C - u_B = 1.0233 \times 10^{-3}$.

4.1.8 Axial Force Diagram

In the previous example we constructed several free-body diagrams to determine the internal axial force at different segments of the axial member. An axial force diagram is a graphical technique for determining internal axial forces, which avoids the repetition of drawing free-body diagrams.

An axial force diagram is a plot of the internal axial force N versus x. To construct an axial force diagram we create a small template to guide us in which direction the internal axial force will jump, as shown in Figure 4.14.

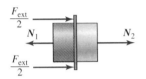

Figure 4.14   Axial bar template.

Definition 4   A template is a free-body diagram of a small segment of an axial bar created by making an imaginary cut just before and just after the section where the external force is applied.

The external force F_{ext} on the template can be drawn either to the left or to the right. The ends represent the imaginary cut just to the left and just to the right of the applied external force. On these cuts the internal axial forces are drawn in tension. An equilibrium equation, that is, the template equation, is written,

$$N_2 = N_1 - F_{ext}$$

If the external force on the axial bar is in the direction of the assumed external force on the template, then the value of N_2 is calculated according to the template equation. If the external force on the axial bar is opposite to the direction shown on the template, then N_2 is calculated by changing the sign of F_{ext} in the template equation. Example 4.3 demonstrates the use of templates in constructing axial force diagrams.

EXAMPLE 4.3

Draw the axial force diagram for the axial member shown in Example 4.2 and calculate the movement of the section at C with respect to the section at A.

PLAN

We can start the process by considering an imaginary extension on the left. In the imaginary extension the internal axial force is zero. Using the template in Figure 4.14 to guide us, we can draw the axial force diagram. Using Equation (4.13), we can find the relative displacement of the section at C with respect to the section at A.

Solution   Let LA be an imaginary extension on the left of the shaft, as shown in Figure 4.15. Clearly the internal axial force in the imaginary segment LA is zero. As one crosses the section at A, the internal force must jump by the applied axial force of 1500 kN. The forces at A are in the opposite direction to the force F_{ext} shown on the template in Figure 4.14;

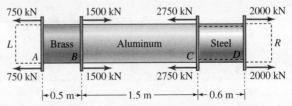

Figure 4.15   Extending the axial bar for an axial force diagram.

thus we must use opposite signs in the template equation. The internal force just after the section at A will be +1500 kN. This is the starting value in the internal axial force diagram.

We approach the section at B with an internal force value of +1500 kN. The force at B is in the same direction as the force shown on the template in Figure 4.14. Hence we subtract 3000 as per the template equation, to obtain a value of −1500 kN, as shown in Figure 4.16.

We now approach the section at C with an internal force value of −1500 kN and note that the forces at C are opposite to those on the template. Hence we add 5500 to obtain +4000 kN.

The force at D is in the same direction as that on the template, and after subtracting we obtain a zero value in the imaginary extended bar DR. The return to zero value must always occur because the bar is in equilibrium.

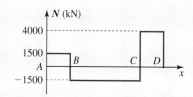

Figure 4.16 Axial force diagram.

From Figure 4.16 the internal axial forces in segments AB and BC are $N_{AB} = 1500$ kN and $N_{BC} = -1500$ kN. Substituting into Equation (4.13) we obtain the relative deformation of the section at C with respect to the section at A,

$$\Delta u = u_C - u_A = \frac{N_{AB}(x_B - x_A)}{E_{AB}A_{AB}} + \frac{N_{BC}(x_C - x_B)}{E_{BC}A_{BC}} \qquad (E1)$$

$$u_C - u_A = \frac{1500 \times 10^3 \times 0.5}{100 \times 10^9 \times 31.41 \times 10^{-3}} + \frac{-1500 \times 10^3 \times 1.5}{70 \times 10^9 \times 31.41 \times 10^{-3}}$$

$$= 0.2388 \times 10^{-3} + (-1.0233 \times 10^{-3})$$
$$= -0.7845 \times 10^{-3} \text{ m} \qquad (E2)$$

or

ANS. $u_C - u_A = 0.7845$ mm contraction

COMMENT

We could have created the template with the external force F_{ext}, as shown in Figure 4.17, and used it to create the axial force diagram.

$$N_2 = N_1 + F_{ext} \qquad (E3)$$

We approach the section at A and note that the +1500 kN is in the same direction as that shown on the template of Figure 4.17. As per the template Equation (E3), we add. Thus our starting value is +1500 kN, as shown in Figure 4.16. As we approach the section at B, the internal force N_1 is +1500 kN, and the applied force of 3000 kN is in the opposite direction to the template of Figure 4.17, so we subtract to obtain N_2 as −1500 kN. We

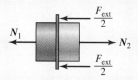

Figure 4.17 Alternative template.

approach the section at C and note that the applied force is in the same direction as the applied force on the template of Figure 4.17. Hence we add 5500 kN to obtain +4000 kN. The force at section D is opposite to that shown on the template of Figure 4.17, so we subtract 4000 to get a zero value in the extended portion DR.

The example shows that the direction of the external force F_{ext} on the template is immaterial.

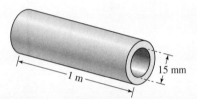

Figure 4.18 Cylindrical rod in Example 4.4.

EXAMPLE 4.4

A 1-m-long hollow rod is to transmit an axial force of 60 kN. The inner diameter of the rod must be 15 mm to fit existing attachments (Figure 4.18). The elongation of the two ends of the rod is limited to 0.1 mm. The shaft can be made of titanium alloy or aluminum. The modulus of elasticity E, the allowable normal stress σ_{allow}, and the density γ for the two materials are given in Table P4.1. Determine the minimum outer diameter to the nearest millimeter of the lightest rod that can be used for transmitting the axial force.

TABLE 4.1   Material properties in Example 4.4

Material	E (GPa)	σ_{allow} (MPa)	γ (mg/m^3)
Titanium alloy	96	400	4.4
Aluminum	70	200	2.8

PLAN

The change in radius affects only the cross-sectional area A and no other quantity in Equation (4.9) and (4.11). For each material we can find the minimum cross-sectional area A needed to satisfy the stiffness and strength requirements. Knowing the minimum A for each material, we can find the minimum outer radius. We can then find the volume and hence the mass of each material and make our decision on the lighter bar.

Solution   We note that for both materials $x_2 - x_1 = 1$ m. From Equations (4.9) and (4.11) we obtain for titanium alloy the following limits on A_{Ti}:

$$(\Delta u)_{Ti} = \frac{60 \times 10^3 \times 1}{96 \times 10^9 A_{Ti}} \leq 2 \times 10^{-3} \quad \text{or} \quad A_{Ti} \geq 0.313 \times 10^{-3} \text{ m}^2 \quad \text{(E1)}$$

$$(\sigma_{max})_{Ti} = \frac{60 \times 10^3}{A_{Ti}} \leq 400 \times 10^6 \quad \text{or} \quad A_{Ti} \geq 0.150 \times 10^{-3} \text{ m}^2 \quad \text{(E2)}$$

Using similar calculations for the aluminum shaft, we obtain the following limits on A_{Al}:

$$(\Delta u)_{Al} = \frac{60 \times 10^3 \times 1}{28 \times 10^9 A_{Al}} \leq 2 \times 10^{-3} \quad \text{or} \quad A_{Al} \geq 1.071 \times 10^{-3} \text{ m}^2 \quad \text{(E3)}$$

$$(\sigma_{max})_{Al} = \frac{60 \times 10^3}{A_{Al}} \leq 200 \times 10^6 \quad \text{or} \quad A_{Al} \geq 0.300 \times 10^{-3} \text{ m}^2 \quad \text{(E4)}$$

Thus if $A_{Ti} \geq 0.313 \times 10^{-3}$ m^2, it will meet both conditions in Equations (E1) and (E2). Similarly if $A_{Al} \geq 1.071 \times 10^{-3}$ m^2, it will meet both conditions in Equations (E3) and (E4). The external diameters D_{Ti} and D_{Al} can be found as

$$A_{Ti} = \frac{\pi}{4}(D_{Ti}^2 - 0.015^2) \geq 0.313 \times 10^{-3} \quad D_{Ti} \leq 24.97 \times 10^{-3} \text{ m} \quad \text{(E5)}$$

$$A_{Al} = \frac{\pi}{4}(D_{Al}^2 - 0.015^2) \geq 1.071 \times 10^{-3} \quad D_{Al} \leq 39.86 \times 10^{-3} \text{ m} \quad \text{(E6)}$$

Rounding upward to the closest millimeter we obtain

$$D_{Ti} = 25 \times 10^{-3} \text{ m} \quad \text{(E7)}$$

$$D_{Al} = 40 \times 10^{-3} \text{ m} \quad \text{(E8)}$$

We can find the mass of each material by taking the product of the material density and the volume of a hollow cylinder,

$$m_{Ti} = 4.4 \times 10^6 \times \frac{\pi}{4}(0.025^2 - 0.015^2)1 = 1382 \text{ g} \quad \text{(E9)}$$

$$m_{Al} = 2.8 \times 10^6 \times \frac{\pi}{4}(0.040^2 - 0.015^2)1 = 3024 \text{ g} \quad \text{(E10)}$$

From Equations (E9) and (E10) we see that the titanium alloy shaft is lighter.

ANS.   A titanium alloy shaft with an outside diameter of 25 mm should be used.

COMMENTS
1. For both materials the stiffness limitation dictated the calculation of the external diameter, as can be seen from Equations (E1) and (E3).
2. Even though the density of aluminum is lower than that of titanium alloy, the mass of titanium is less. Because of the higher modulus of

elasticity of titanium alloy, we can meet the stiffness requirement using less material than with aluminum.

3. The answer may change if cost is a consideration. The cost of titanium per kilogram is significantly higher than that of aluminum. Thus based on material cost we may choose aluminum. However, if the weight affects the running cost, then economic analysis will have to be done to determine whether the material cost or the running cost is higher.

4. If in Equation (E5) we had 24.05×10^{-3} m on the right-hand side, our answer for D_{Ti} would still be 25 mm because we have to round upward to ensure meeting the greater-than sign requirement in Equation (E5).

EXAMPLE 4.5

A rectangular aluminum bar ($E_{al} = 10{,}000$ ksi, $v = 0.25$) of $\frac{3}{4}$-in thickness consists of a uniform and tapered cross section, as shown in Figure 4.19. The depth in the tapered section varies as $h(x) = 2 - 0.02x$. Determine:

(a) The elongation of the bar under the applied loads.

(b) The change in dimension in the y direction in section BC.

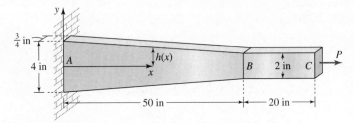

Figure 4.19 Axial member in Example 4.5.

PLAN

(a) The extension of the rod is given by the displacement of the section at C with respect to the wall, that is, we have to find $u_C - u_A$. In segment BC the quantities E, A, and the internal axial force N are all constants, and we can use Equation (4.11) to find the relative displacement of the section at C with respect to the section at B. But in segment AB the depth is changing with x, and hence the cross-sectional area A is changing with x. Thus to obtain the relative displacement of B with respect to A we have to start with

Equation (4.8) and integrate. We combine the two relative displacements to obtain the relative displacement of C with respect to A.

(b) The axial stress in BC can be found. Using Equation (4.12) the normal strain in the y direction can be found. Multiplying by 2 in, the original length in the y direction, the change in depth can be found.

Solution

(a) We can make an imaginary cut in segment BC, that is, the location of the imaginary cut is defined by $50 < x < 70$ in. To avoid the calculation of the wall reaction, we take the right part after the cut and draw the free-body diagram in Figure 4.20 to find N_{BC},

Figure 4.20 Free-body diagram of section BC in Example 4.5.

$$N_{BC} = 10 \text{ kips}$$

$$A_{BC} = \tfrac{3}{4} \times 2 = 1.5 \text{ in}^2 \tag{E1}$$

We can use Equation (4.9) to find the relative movement of point C with respect to point B,

$$u_C - u_B = \frac{N_{BC}(x_C - x_B)}{E_{BC}A_{BC}} = \frac{10 \times 20}{10{,}000 \times 1.5} = 13.33 \times 10^{-3} \text{ in} \tag{E2}$$

We make an imaginary cut in segment AB, that is, the location of the imaginary cut is defined by $0 < x < 50$ in. To avoid the calculation of the wall reaction, we take the right part after the cut and draw the free-body diagram in Figure 4.21 to find N_{AB},

$$N_{AB} = 10 \text{ kips} \tag{E3}$$

Figure 4.21 Free-body diagram of section AB in Example 4.5.

Next we find A_{AB} and, noting that the area is a function of x, we integrate Equation (4.8) to find the relative movement of point B with respect to A,

$$A_{AB} = \tfrac{3}{4} \times 2h = 1.5(2 - 0.02x) \tag{E4}$$

$$\left(\frac{du}{dx}\right)_{AB} = \frac{N_{AB}}{E_{AB}A_{AB}} = \frac{10}{10{,}000 \times 1.5(2 - 0.02x)} \tag{E5}$$

$$\int_{u_A}^{u_B} du = \int_{x_A=0}^{x_B=50} \frac{10^{-3}}{1.5(2 - 0.02x)} \, dx$$

$$u_B - u_A = \frac{10^{-3}}{1.5} \frac{1}{(-0.02)} \ln(2 - 0.02x) \Big|_0^{50}$$

$$= -\frac{10^{-3}}{0.03}[\ln(1) - \ln(2)] = 23.1 \times 10^{-3} \text{ in} \tag{E6}$$

Adding Equations (E2) and (E6) and noting that $u_A = 0$, as point A is built into the wall we obtain

$$u_C - u_A^{\;0} = (23.1 + 13.3)10^{-3} \quad \text{or} \quad \text{ANS. } u_C = 0.036 \text{ in elongation}$$

(b) The axial stress in BC is $\sigma_{AB} = N_{BC}/A_{BC} = 10/1.5 = 6.667$ ksi. From Equation (4.12) the normal strain in the y direction can be found,

$$\varepsilon_{yy} = -\frac{\nu_{AB}\sigma_{AB}}{E_{AB}} = -\frac{0.25 \times 6.667}{10,000} = -0.1667 \times 10^{-3}$$

The change in dimension in the y direction Δv can be found as

$$\Delta v = \varepsilon_{yy}(2) = -0.3333 \times 10^{-3}$$

ANS. $\Delta v = 0.3333 \times 10^{-3}$ in contraction

COMMENT

An alternative approach is to integrate Equation (E5) as follows:

$$u(x) = -\frac{10^{-3}}{0.03}\ln(2 - 0.02x) + c$$

where c is an integration constant. To find the integration constant, we note that at $x = 0$ the displacement $u = 0$ and hence, $c = (10^{-3}/0.03)\ln(2)$. Substituting the value of c, we obtain

$$u(x) = -\frac{10^{-3}}{0.03}\ln\left(\frac{2 - 0.02x}{2}\right)$$

Knowing u at all x, we can obtain the extension by substituting $x = 50$ to get the displacement at C.

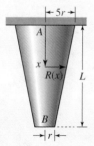

Figure 4.22 Truncated cone in Example 4.6.

EXAMPLE 4.6

The radius of a circular truncated cone varies with x as $R(x) = (r/L)(5L - 4x)$ (Figure 4.22). Determine the extension of the truncated cone due to its own weight in terms of E, L, r, and γ, where E and γ are the modulus of elasticity and the specific weight of the material, respectively.

PLAN

We can make an imaginary cut at location x and take the lower part of the truncated cone as the free-body diagram. In the free-body diagram we can find the volume of the truncated cone as a function of x. Multiplying the volume by the specific weight, we can obtain the weight of the truncated cone and equate it to the internal axial force, thus obtaining the internal force as a function of x. We can then integrate Equation (4.10) to obtain the relative displacement of B with respect to A.

Solution We make an imaginary cut at some location x and take the lower portion to obtain the free-body diagram for calculating the internal axial force, as shown in Figure 4.23. We can find the volume V of the truncated cone by subtracting the volumes of two complete cones[11] between C and D and between B and D. We can find point D by noting that at D the radius $R(x = L + h)$ of the cone is zero. Thus (r/L) $[5L - 4(L + h)] = 0$, or $h = L/4$. Once we know the volume V we can multiply it by the specific weight to get the weight of the truncated cone between B and C and equate it to internal axial force N as shown,

$$V = \frac{1}{3}\pi R^2\left(L - x + \frac{L}{4}\right) - \frac{1}{3}\pi r^2\frac{L}{4} = \frac{\pi}{12}\left[\frac{r^2}{L^2}(5L - 4x)^3 - r^2 L\right] \quad \text{(E1)}$$

$$N = W = \gamma V = \frac{\gamma\pi r^2}{12L^2}[(5L - 4x)^3 - L^3] \quad \text{(E2)}$$

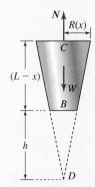

Figure 4.23 Free-body diagram of truncated cone in Example 4.6.

Substituting N from Equation (E2), noting the cross-sectional area at location x (point C) as

$$A = \pi R^2 = \pi\frac{r^2}{L^2}(5L - 4x)^2$$

in Equation (4.8), and integrating from point A to point B, we obtain the relative movement of point B with respect to point A. Since point A is built into the wall, and hence the displacement of point A is zero, we obtain the displacement of point B, which is the extension of the bar,

$$\frac{du}{dx} = \frac{N}{EA} = \frac{\dfrac{\gamma\pi r^2}{12L^2}[(5L - 4x)^3 - L^3]}{E\pi\dfrac{r^2}{L^2}(5L - 4x)^2} \quad \text{(E3)}$$

$$\int_{u_A}^{u_B}du = \int_{x_A=0}^{x_B=L}\frac{\gamma}{12E}\left[(5L - 4x) - \frac{L^3}{(5L - 4x)^2}\right]dx \quad \text{(E4)}$$

$$u_B - u_A = \frac{\gamma}{12E}\left[5Lx - 2x^2 - \frac{L^3}{4(5L - 4x)}\right]\Bigg|_0^L$$

$$= \frac{\gamma L^2}{12E}\left(5 - 2 - \frac{1}{4} + \frac{1}{20}\right) \quad \text{(E5)}$$

$$\text{ANS.} \qquad u_B = \frac{7}{30}\left(\frac{\gamma L^2}{E}\right) \text{ downward}$$

---

[11]The volume of a cone is one-third the cone height multiplied by the cone base area.

Dimension check: The dimensional consistency* of the answer is checked as shown,

$$\gamma \to O\left(\frac{F}{L^3}\right) \qquad L \to O(L) \qquad E \to O\left(\frac{F}{L^2}\right)$$

$$u \to O(L) \qquad \frac{\gamma L^2}{E} \to O\left(\frac{(F/L^3)L^2}{F/L^2}\right) \to O(L) \to \text{checks}$$

COMMENTS

1. An alternative approach to determining the volume of the truncated cone in Figure 4.23 is to find first the volume of an infinitesimal disc (Figure 4.24). We then integrate from point C to point B as shown,

$R(x)$

dx

Figure 4.24 Alternative approach to finding volume of truncated cone.

$$V = \int_x^L dV = \int_x^L \pi R^2 dx = \int_x^L \pi \frac{r^2}{L^2}(5L - 4x)^2 dx$$

$$= -\pi \frac{r^2}{L^2} \frac{(5L - 4x)^3}{3(-4)} \bigg|_x^L$$

On substituting the limits we obtain the volume given by Equation (E1), as before.

2. The advantage of the approach in comment 1 is that it can be used for any complex function representation of $R(x)$, such as given in Problems 4.26 and 4.27, whereas the approach used in solving the example problem is only valid for a linear representation of $R(x)$.

---

*$O(\ )$ represents the dimension of the quantity on the left; F represents the dimension for the force, L represents the dimension for length; thus modulus of elasticity E, which has dimensions of force (F) per unit area (L^2), has dimension shown as $O(F/L^2)$.

*4.1.9   General Approach to Distributed Axial Forces

Distributed axial forces are usually due to inertial forces, gravitational forces, or frictional forces acting on the surface of the axial bar. The internal axial force N becomes a function of x when an axial bar is subjected to a distributed axial force $p(x)$, as seen in Example 4.6. If $p(x)$ is a simple function, then we can find N as a function of x by drawing a free-body diagram, as we did in Example 4.6. However, if the distributed force $p(x)$ is a complex function,[12] it may be easier to use the alternative described in this section.

---

[12]See Problems 4.26 through 4.28, 4.35, 4.41, and 4.42.

Consider an infinitesimal axial element that is created by making two imaginary cuts at a distance dx from each other, as shown in Figure 4.25. By equilibrium of forces in the x direction we obtain

$$(N + dN) + p(x)\,dx - N = 0$$

or

$$\boxed{\frac{dN}{dx} + p(x) = 0} \qquad (4.14)$$

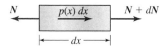

Figure 4.25 Equilibrium of an axial element.

EXAMPLE 4.7
Determine the internal force N in Example 4.6 using the approach outlined in Section 4.1.9.

PLAN
The distributed force $p(x)$ per unit length is the product of the specific weight times the area of cross section. We can integrate Equation (4.14) and use the condition that the value of the internal force at the free end is zero to obtain the internal force as a function of x.

Solution The distributed force $p(x)$ is the weight per unit length and is equal to the specific weight times the area of cross section $A = \pi R^2 = \pi(r^2/L^2)(5L - 4x)^2$ as shown,

$$p(x) = \gamma A = \gamma \pi \frac{r^2}{L^2}(5L - 4x)^2 \qquad (E1)$$

We note that point B ($x = L$) is on a free surface and hence the internal force at B is zero. We integrate Equation (4.14) from L to x after substituting $p(x)$ from Equation (E1) and obtain N as a function of x,

$$\int_{N_B=0}^{N} dN = -\int_{x_B=L}^{x} p(x)\,dx = -\int_{L}^{x} \gamma\left[\pi\frac{r^2}{L^2}(5L - 4x)^2\right]dx$$

$$= -\left(\gamma\pi\frac{r^2}{L^2}\right)\left[\frac{(5L - 4x)^3}{-4 \times 3}\right]\Bigg|_{L}^{x}$$

ANS. $$N = \frac{\gamma\pi r^2}{12L^2}[(5L - 4x)^3 - L^3] \qquad (E2)$$

COMMENT
An alternative approach is to substitute Equation (E1) into Equation (4.14) and integrate to obtain

$$N(x) = \frac{\gamma\pi r^2}{12L^2}(5L - 4x)^3 + c_1 \qquad (E3)$$

To determine the integration constant, we use the boundary condition that at N ($x = L$) = 0, which yields $c_1 = -(\gamma\pi r^2/12L^2)L^3$. Substituting this value into Equation (E3), we obtain N as in Equation (E2).

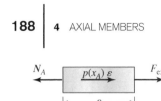

Figure 4.26 Boundary condition on internal axial force.

Equation (4.14) assumes that $p(x)$ is positive in the positive x direction. The sign of N obtained from Equation (4.14) follows the sign convention given in Definition 2. If $p(x)$ is zero in a segment of the axial bar, then the internal force N is a constant in that segment.

Equation (4.14) can be integrated to obtain the internal force N. The integration constant can be found by knowing the value of the internal force N at either end of the bar. To obtain the value of N at the end of the shaft (say, point A), a free-body diagram is constructed after making an imaginary cut at an infinitesimal distance ε from the end (Figure 4.26) and writing the equilibrium equation as

$$\lim_{\varepsilon \to 0} [F_{ext} - N_A - p(x_A)\varepsilon] = 0 \qquad N_A = F_{ext}$$

This equation shows that the distributed axial force does not affect the boundary condition on the internal axial force. The value of the internal axial force N at the end of an axial bar is equal to the concentrated external axial force applied at the end.

Suppose the weight per unit volume, that is, the specific weight of a bar, is γ. By multiplying the specific weight by the cross-sectional area A we would obtain the weight per unit length. Thus $p(x)$ is equal to γA in magnitude. If the x coordinate is chosen in the direction of gravity, then $p(x)$ is positive [$p(x) = +\gamma A$], and if it is opposite to the direction of gravity [$p(x) = -\gamma A$], then $p(x)$ is negative.

QUICK TEST 4.1   Time: 20 minutes/Total: 20 points

Answer true or false and justify each answer in one sentence. Grade yourself with the answers given in Appendix G.

1. Axial *strain* is uniform across a nonhomogeneous cross section.

2. Axial *stress* is uniform across a nonhomogeneous cross section.

3. The formula $\sigma_{xx} = N/A$ can be used for finding the stress on a cross section of a tapered axial member.

4. The formula $u_2 - u_1 = N(x_2 - x_1)/EA$ can be used for finding the deformation of a segment of a tapered axial member.

5. The formula $\sigma_{xx} = N/A$ can be used for finding the stress on a cross section of an axial member subjected to distributed forces.

6. The formula $u_2 - u_1 = N(x_2 - x_1)/EA$ can be used for finding the deformation of a segment of an axial member subjected to distributed forces.

7. The equation $N = \int_A \sigma_{xx} \, dA$ *cannot* be used for nonlinear materials.

8. The equation $N = \int_A \sigma_{xx} \, dA$ *can* be used for a nonhomogeneous cross section.

9. External axial forces must be collinear and pass through the centroid of a homogeneous cross section for no bending to occur.

10. Internal axial forces jump by the value of the concentrated external axial force at a section.

In Problems 4.1 through 4.4 use the logic shown in Figure P4.3 to determine the external forces.

4.1 Aluminum bars $(E = 30{,}000 \text{ ksi})$ are welded to rigid plates, as shown in Figure P4.1. All bars have a cross-sectional area of 0.5 in². Due to the applied forces the rigid plates at A, B, C, and D are displaced in the x direction without rotating by the following amounts: $u_A = -0.0100$ in, $u_B = 0.0080$ in, $u_C = -0.0045$ in, and $u_D = 0.0075$ in. Determine the applied forces F_1, F_2, F_3, and F_4.

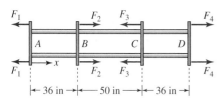

Figure P4.1

4.2 Brass bars between sections A and B, aluminum bars between sections B and C, and steel bars between sections C and D are welded to rigid plates, as shown in Figure P4.2. The rigid plates are displaced in the x direction

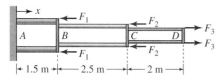

Figure P4.2

without rotating by the following amounts: $u_B = -1.8$ mm, $u_C = 0.7$ mm, and $u_D = 3.7$ mm. Determine the external forces F_1, F_2, and F_3 using the properties given in Table P4.2.

TABLE P4.2

	Brass	Aluminum	Steel
Modulus of elasticity	70 GPa	100 GPa	200 GPa
Diameter	30 mm	25 mm	20 mm

4.3 The ends of four circular steel bars $(E = 200 \text{ GPa})$ are welded to a rigid plate, as

shown in Figure P4.3. The other ends of the bars are built into walls. Due to the action of the external force F the rigid plate moves to the right by 0.1 mm without rotating. If the bars have a diameter of 10 mm, determine the applied force F.

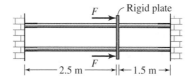

Figure P4.3

4.4 Rigid plates are securely fastened to bars A and B, as shown in Figure P4.4. A gap of 0.02 in exists between the rigid plates before the forces are applied. After application of the forces the normal strain in bar A was found to be 500 μ. The cross-sectional area and the modulus of elasticity for each bar are as follows: $A_A = 1$ in², $E_A = 10{,}000$ ksi, $A_B = 0.5$ in², and $E_B = 30{,}000$ ksi. Determine the applied forces F, assuming that the rigid plates do not rotate.

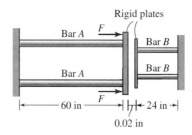

Figure P4.4

4.5 A crane is lifting a mass of 1000-kg, as shown in Figure P4.5. The weight of the iron

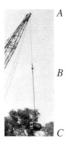

Figure P4.5

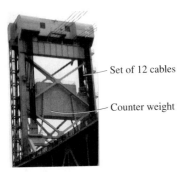

Figure P4.6

Set of 12 cables

Counter weight

ball at *B* is 25 kg. A single cable having a diameter of 25 mm runs between *A* and *B*. Two cables run between *B* and *C*, each having a diameter of 10 mm. Determine the axial stresses in the cables.

4.6 The counterweight in a lift bridge has 12 cables on the left and 12 cables on the right, as shown in Figure P4.6. Each cable has an effective diameter of 0.75 in, a length of 50 ft, a modulus of elasticity of 30,000 ksi, and an ultimate strength of 60 ksi. (a) If the counterweight is 100 kips, determine the factor of safety for the cable. (b) What is the extension of each cable when the bridge is being lifted?

In Problems 4.7 through 4.10 draw the axial force diagrams. Then to check your results find the internal forces in segments AB, BC, and CD by making imaginary cuts and drawing free-body diagrams.

4.7 (Figure P4.7).

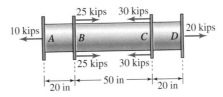

Figure P4.7

4.8 (Figure P4.8).

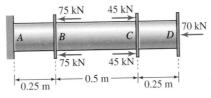

Figure P4.8

4.9 (Figure P4.9).

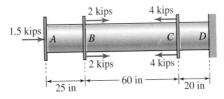

Figure P4.9

4.10 (Figure P4.10).

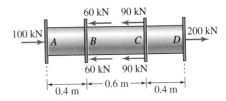

Figure P4.10

4.11 The axial rigidity of the bar in Problem 4.7 is $EA = 8000$ kips. Determine the movement of the section at *D* with respect to the section at *A*.

4.12 The axial rigidity of the bar in Problem 4.8 is $EA = 80,000$ kN. Determine the movement of the section at *C*.

4.13 The axial rigidity of the bar in Problem 4.9 is $EA = 2000$ kips. Determine the movement of the section at *B*.

4.14 The axial rigidity of the bar in Problem 4.10 is $EA = 50,000$ kN. Determine the movement of the section at *D* with respect to the section at *A*.

4.15 Three segments of 4-in × 2-in rectangular wooden bars ($E = 1600$ ksi) are secured together with rigid plates and subjected to axial forces, as shown in Figure P4.15. Determine: (a) the movement of the rigid plate at *D* with respect to the plate at *A*; (b) the maximum axial stress.

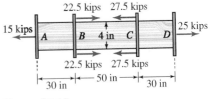

Figure P4.15

4.16 Two circular steel bars ($E_s = 30,000$ ksi, $v_s = 0.3$) of 2-in diameter are securely connected to an aluminum bar ($E_{al} = 10,000$ ksi, $v_{al} = 0.33$) of 1.5-in diameter, as shown in Figure P4.16. Determine: (a) the displacement of the section at *C* with respect to the wall; (b) the maximum change in the diameter of the bars.

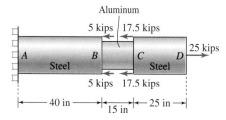

Figure P4.16

4.19 $A = K(4L - 3x)$ in Figure P4.18.

4.20 A tapered and an untapered solid circular steel bar ($E = 30,000$ ksi) are securely fastened to a solid circular aluminum bar ($E = 10,000$ ksi), as shown in Figure P4.20. The untapered steel bar has a diameter of 2 in. The aluminum bar has a diameter of 1.5 in. The diameter of the tapered bars varies from 1.5 in to 2 in. Determine: (a) the displacement of the section at C with respect to the section at A; (b) the maximum axial stress in the bar.

Distributed axial force

The columns shown have length L, modulus of elasticity E, and specific weight γ. In Problems 4.21 through 4.23 determine the contraction of each column in terms of L, E, γ, and the variable describing the cross section.

4.21 The cross section is a circle of radius a (Figure P4.21).

4.22 The cross section is an equilateral triangle of side a (Figure P4.22).

4.23 The cross-sectional area is A (Figure P4.23).

4.24 On the truncated cone of Example 4.6 a force $P = \gamma \pi r^2 L/5$ is also applied, as shown in Figure P4.24. Determine the total elongation of the cone due to its weight and the applied force. (*Hint:* Use superposition.)

4.17 Two cast-iron pipes ($E = 100$ GPa) are adhesively bonded together, as shown in Figure P4.17. The outer diameters of the two pipes are 50 mm and 70 mm and the wall thickness of each pipe is 10 mm. Determine the displacement of end B with respect to end A.

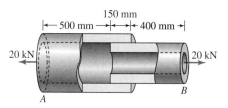

Figure P4.17

Tapered axial members

In Problems 4.18 and 4.19 the tapered bar shown in Figure P4.18 has a cross-sectional area that varies with x as given. Determine the elongation of the bar in terms of P, L, E, and K.

4.18 $A = K(2L = 0.25\ x)^2$ in Figure P4.18.

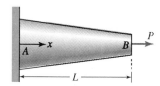

Figure P4.18

Figure P4.21

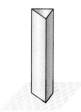

Figure P4.22

Figure P4.23

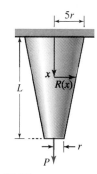

Figure P4.24

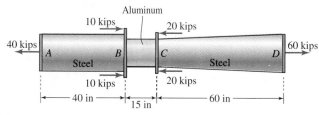

Figure P4.20

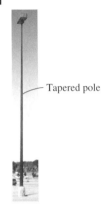

Figure P4.25

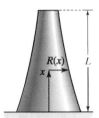

Figure P4.26

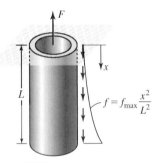

Figure P4.28

Figure P4.29

4.25 A 20-ft-tall thin, hollow tapered tube of a uniform wall thickness of $\frac{1}{8}$ in is used for a light pole in a parking lot, as shown in Figure P4.25. The mean diameter at the bottom is 8 in, and at the top it is 2 in. The weight of the lights on top of the pole is 80 lb. The pole is made of aluminum alloy with a specific weight of 0.1 lb/in³, a modulus of elasticity $E = 11,000$ ksi, and a shear modulus of rigidity $G = 4000$ ksi. Determine: (a) the maximum axial stress; (b) the contraction of the pole. (*Hint*: Approximate the cross-sectional area of the thin-walled tube by the product of circumference and thickness.)

In Problems 4.26 and 4.27 determine the contraction of a column due to its own weight, as shown in Figure P4.26. The specific weight γ, the modulus of elasticity E, the length L, and the radius R are as given in each problem.

4.26 $\gamma = 0.28$ lb/in³, $E = 3600$ ksi, $L = 120$ in, and $R = \sqrt{240 - x}$, where R and x are in inches.

4.27 $\gamma = 24$ kN/m³, $E = 25$ GPa, $L = 10$ m, and $R = 0.5e^{-0.07x}$, where R and x are in meters.

4.28 The frictional force per unit length on a cast-iron pipe being pulled from the ground varies as a quadratic function, as shown in Figure P4.28. Determine the force F needed to pull the pipe out of the ground and the elongation of the pipe before the pipe slips, in terms of the modulus of elasticity E, the cross-sectional area A, the length L, and the maximum value of the frictional force f_{max}.

Design problems

4.29 The spare wheel in an automobile is stored under the vehicle and raised and lowered by a cable, as shown in Figure P4.29. The wheel has a mass of 25 kg. The ultimate strength of the cable is 300 MPa, and it has an effective modulus of elasticity $E = 180$ GPa. At maximum extension the cable length is 36 cm. (a) For a factor of safety of 4, determine to the nearest millimeter the minimum diameter of the cable if failure due to rupture is to be avoided. (b) What is the maximum extension of the cable for the answer in part (a)?

4.30 An adhesively bonded joint in wood ($E = 1800$ ksi) is fabricated as shown in Figure P4.30. If the total elongation of the joint between A and D is to be limited to 0.05 in, determine the maximum axial force F that can be applied.

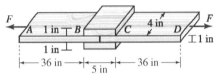

Figure P4.30

4.31 A 5-ft-long hollow rod is to transmit an axial force of 30 kips. The outer diameter of the rod must be 6 in to fit existing attachments. The relative displacement of the two ends of the shaft is limited to 0.024 in. The axial rod can be made of steel or aluminum. The modulus of elasticity E, the allowable axial stress σ_{allow}, and the specific weight γ are given in Table P4.31. Determine the maximum inner diameter in increments of $\frac{1}{8}$ in of the lightest rod that can be used for transmitting the axial force and the corresponding weight.

TABLE P4.31   Material properties

Material	E (ksi)	σ_{allow} (ksi)	γ (lb/in³)
Steel	30,000	24	0.285
Aluminum	10,000	14	0.100

4.32 A hitch for an automobile is to be designed for pulling a maximum load of 3600 lb. A solid square bar fits into a square tube and is held in place by a pin, as shown in Figure P4.32. The allowable axial stress in the

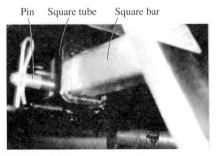

Figure P4.32

bar is 6 ksi, the allowable shear stress in the pin is 10 ksi, and the allowable axial stress in the steel tube is 12 ksi. To the nearest $\frac{1}{16}$ in, determine the minimum cross-sectional dimensions of the pin, the bar, and the tube. (*Hint:* The pin is in double shear.)

Stretch yourself

4.33 An axial rod has a constant axial rigidity EA and is acted upon by a distributed axial force $p(x)$. If at the section at A the internal axial force is zero, show that the relative displacement of the section at B with respect to the displacement of the section at A is given by

$$u_B - u_A = \frac{1}{EA}\left[\int_{x_A}^{x_B}(x - x_B)p(x)\,dx\right] \quad (4.15)$$

4.34 The stress–strain relationship for a nonlinear material is given by the power law $\sigma = E\varepsilon^n$. If all assumptions except Hooke's law are valid, show that

$$u_2 - u_1 = \left(\frac{N}{EA}\right)^{1/n}(x_2 - x_1) \quad (4.16)$$

and the axial stress σ_{xx} is given by Equation (4.9).

4.35 Determine the elongation of a rotating bar in terms of the rotating speed ω, density γ, length L, modulus of elasticity E, and cross-sectional area A (Figure P4.35). (*Hint:* The body force per unit volume is $\rho\omega^2x$.)

Figure P4.35

4.36 Consider the dynamic equilibrium of the differential elements shown in Figure P4.36, where N is the internal force, γ is the density, A is the cross-sectional area, and $\partial^2u/\partial t^2$ is acceleration. By substituting for N from Equation (4.8) into the dynamic equilibrium equation derive the following wave equation:

$$\frac{\partial^2 u}{\partial t^2} = c^2\frac{\partial^2 u}{\partial x^2} \qquad c = \sqrt{\frac{E}{\gamma}} \quad (4.17)$$

The material constant c is the velocity of propagation of sound in the material.

4.37 Show by substitution that the functions $f(x - ct)$ and $g(x + ct)$ satisfy the wave equation, Equation (4.17).

4.38 The strain displacement relationship for large axial strain is given by

$$\varepsilon_{xx} = \frac{du}{dx} + \frac{1}{2}\left(\frac{du}{dx}\right)^2 \quad (4.18)$$

where we recognize that as u is only a function of x, the strain from Equation (4.18) is uniform across the cross section. For a linear, elastic, homogeneous material show that

$$\frac{du}{dx} = \sqrt{1 + \frac{2N}{EA}} - 1 \quad (4.19)$$

The axial stress σ_{xx} is given by Equation (4.9).

Computer problems

4.39 Table P4.39 gives the measured radii at several points along the axis of the solid tapered rod shown in Figure P4.39. The rod is made of aluminum ($E = 100$ GPa) and has a length of 1.5 m. Determine: (a) the elongation of the rod using numerical integration; (b) the maximum axial stress in the rod.

TABLE P4.39

x (m)	R(x) (mm)	x (m)	R(x) (mm)
0.0	100.6	0.8	60.1
0.1	92.7	0.9	60.3
0.2	82.6	1.0	59.1
0.3	79.6	1.1	54.0
0.4	75.9	1.2	54.8
0.5	68.8	1.3	54.1
0.6	68.0	1.4	49.4
0.7	65.9	1.5	50.6

4.40 Let the radius of the tapered rod in Problem 4.39 be represented by the equation $R(x) = a + bx$. Using the data in Table P4.39 determine constants a and b by the least-squares method and then find the elongation of the rod by analytical integration.

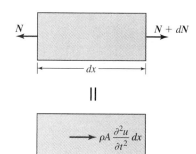

Figure P4.36

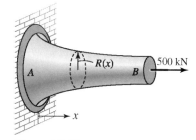

Figure P4.39

Figure P4.41

4.41 Table P4.41 shows the values of the distributed axial force at several points along the axis of the hollow steel rod (E = 30,000 ksi) shown in Figure P4.41. The rod has a length of 36 in, an outside diameter of 1 in, and an inside diameter of 0.875 in. Determine: (a) the displacement of end A using numerical integration; (b) the maximum axial stress in the rod.

4.42 Let the distributed force $p(x)$ in Problem 4.41 be represented by the equation $p(x) = cx^2 + bx + a$. Using the data in Table P4.41 determine constants a, b, and c by the least-squares method and then find the displacement of the section at A by analytical integration.

TABLE P4.41

x (inches)	$p(x)$ (lb/in)	x (inches)	$p(x)$ (lb/in)
0	260	21	−471
3	106	24	−598
6	32	27	−645
9	40	30	−880
12	−142	33	−1035
15	−243	36	−1108
18	−262		

*4.2 COMPOSITE BARS

Bars with cross sections constructed from more than one material are referred to as composite bars. Iron bars inserted in concrete for reinforcement, steel bars attached to wooden bars to increase stiffness, laminated structures, and fibers in resins to form a lamina are some examples. Clearly, Assumption 6 of material homogeneity across the cross section is no longer valid. Hence the formulas that follow after Assumption 6 are no longer valid but, as we shall see, after accounting for the cross-sectional non-homogeneity, the derivation process to get the new formulas is the same as before.

We assume that all materials, such as shown in the laminated cross section in Figure 4.27, are securely bonded to each other so that our kinematic assumption, Assumption 1, which states that plane sections remain plane and parallel, is still valid. We further assume that we still have small strains, that all materials are linear, elastic, isotropic materials, and that there are no inelastic strains. In other words, Assumptions 1 through 5 remain valid. Thus Equation (4.5), which precedes Assumption 6, is still valid and forms our starting point for deriving the axial formulas for composite bars.

Consider the laminated structure shown in Figure 4.27. Each material has a modulus of elasticity E_i that is constant over the material cross-sectional area A_i. Suppose there are n materials in the cross section. We can write the integral over the cross

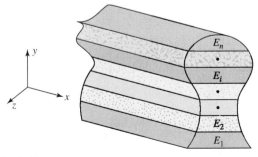

Figure 4.27 Laminated bar.

section in Equation (4.5) as the sum of the integrals over each material,

$$N = \frac{du}{dx}\int_A E \, dA = \frac{du}{dx}\left(\int_{A_1} E_1 \, dA + \int_{A_2} E_2 \, dA + \cdots + \int_{A_n} E_n \, dA\right)$$

Note that E_i is a constant in each integral and can be taken outside the integral. The remaining integral is the area A_i, and we thus obtain

$$N = \frac{du}{dx}(E_1 A_1 + E_2 A_2 + \cdots + E_n A_n)$$

Written more compactly,

$$N = \frac{du}{dx}\left(\sum_{j=1}^{n} E_j A_j\right) \tag{4.20}$$

Equation (4.20) shows that the axial rigidity of the composite bar is the sum of the axial rigidities of all materials. We can write Equation (4.3) for the ith material as $(\sigma_{xx})_i = E_i \, du/dx$, where $(\sigma_{xx})_i$ is the axial stress in the ith material. Substituting Equation (4.20), we obtain

$$\boxed{(\sigma_{xx})_i = \frac{NE_i}{\sum_{j=1}^{n} E_j A_j}} \tag{4.21}$$

We assume that the axial rigidity and the internal axial force do not change with x between x_1 and x_2. In other words, Assumptions 7 through 9 are applicable to composite bars. Then du/dx is constant between x_1 and x_2 and we can write

$$\frac{du}{dx} = \frac{u_2 - u_1}{x_2 - x_1}$$

in Equation (4.20) to obtain

$$\boxed{u_2 - u_1 = \frac{N(x_2 - x_1)}{\sum_{j=1}^{n} E_j A_j}} \tag{4.22}$$

Equations (4.21) and (4.22),[13] which are applicable to composite bars, now replace Equations (4.8) and (4.11) for the homogeneous cross section. The internal axial force N represents the statically equivalent axial force over the entire cross section, as it did for the homogeneous cross section. The analysis techniques for finding the internal axial force N at a cross section remain the same as before, that is, one may use either free-body diagrams or axial force diagrams to find N. However, the location at which the internal axial force acts on the cross section depends on the material distribution across the cross section, as will be emphasized by Example 4.11.

　　Equation (4.20) emphasizes that the normal axial strain ε_{xx} (i.e., du/dx) is uniform across the cross section. But the normal axial stress $(\sigma_{xx})_i$ is uniform only in each material and changes with each material point because the value of E_i changes with each material, as shown in Equation (4.21). Thus the normal axial stress is piecewise uniform across the corss section. This is elaborated further in Example 4.8.

---

[13]If we consider a homogeneous material, then $E_1 = E_2 = \cdots = E_i = \cdots = E_n = E$. Substituting this into Equations (4.21) and (4.22) we obtain Equations (4.8) and (4.11) for homogeneous material, as expected.

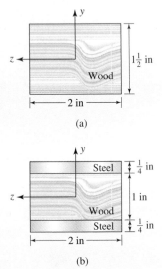

Figure 4.28 Cross sections in Example 4.8. (a) Homogeneous. (b) Laminated.

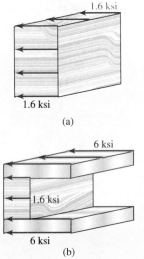

Figure 4.29 Stress distributions in Example 4.8. (a) Homogeneous cross section. (b) Laminated cross section.

EXAMPLE 4.8

A homogeneous wooden cross section and a cross section in which the wood is reinforced with steel are shown in Figure 4.28. The normal strain for both cross sections is uniform, $\varepsilon_{xx} = -200\ \mu$. The moduli of elasticity for steel and wood are $E_{\text{steel}} = 30{,}000$ ksi and $E_{\text{wood}} = 8000$ ksi.

(a) Plot the σ_{xx} distribution for each of the two cross sections shown.

(b) Calculate the equivalent internal axial force N for each cross section using Equation (4.4).

PLAN

(a) Using Hooke's law we can find the stress values in each material. Noting that the stress is uniform in each material, we can plot it across the cross section.

(b) For the homogeneous cross section we can perform the integration in Equation (4.4) directly. For the nonhomogeneous cross section we can write the integral in Equation (4.4) as the sum of the integrals over steel and wood and then perform the integration to find N.

Solution

(a) From Hooke's law we can write

$$(\sigma_{xx})_{\text{wood}} = 8 \times 10^3(-200)10^{-6} = -1.6 \text{ ksi} \qquad (\text{E1})$$

$$(\sigma_{xx})_{\text{steel}} = 30 \times 10^3(-200)10^{-6} = -6 \text{ ksi} \qquad (\text{E2})$$

For the homogeneous cross section the stress distribution is as given in Equation (E1), but for the laminated case it switches from Equation (E1) to (E2), depending on the location of the point where the stress is being evaluated, as shown in Figure 4.29.

(b) *Homogeneous cross section:* Substituting the stress distribution for the homogeneous cross section in Equation (4.4) and integrating, we obtain the equivalent internal axial force,

$$N = \int_A -1.6\, dA = -1.6A = -1.6 \times 2 \times 1.5 = -4.8$$

ANS. $N = 4.8$ kips (C)

Laminated cross section: The stress value changes as we move across the cross section. Let A_{sb} and A_{st} represent the cross-sectional areas of steel at the bottom and the top. Let A_{w} represent the cross-sectional area of wood. We can write the integral in Equation (4.4) as the sum of three integrals, substitute the stress values of Equations (E1) and (E2), and perform the integration,

$$N = \left(\int_{A_{sb}} \sigma_{xx}\, dA + \int_{A_w} \sigma_{xx}\, dA + \int_{A_{st}} \sigma_{xx}\, dA \right)$$

$$= \left(\int_{A_{sb}} -6\, dA + \int_{A_w} -1.6\, dA + \int_{A_{st}} -6\, dA \right) \qquad (E3)$$

or

$$N = [-6A_{sb} + (-1.6)A_w + (-6)A_{st}]$$

$$= \left[-6 \times 2 \times \frac{1}{4} + (-1.6 \times 2 \times 1) + \left(-6 \times 2 \times \frac{1}{4} \right) \right]$$

ANS. $\quad N = 9.2$ kips (C)

COMMENTS

1. Writing the integral in the internal axial force as the sum of integrals over each material, as is done in Equation (E3), is equivalent to calculating the internal force carried by each material and then summing, as shown in Figure 4.30.

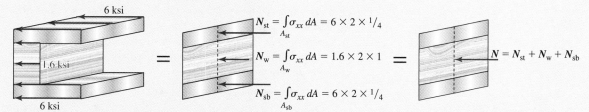

Figure 4.30 Statically equivalent internal force in Example 4.8 for laminated cross section.

2. An alternative calculation for the internal force is as follows. We can find the axial rigidity of the bar as $\sum_{j=1}^{n} E_j A_j = 30{,}000 \times 2 \times \frac{1}{4} \times 2 + 8000 \times 2 \times 1 = 46{,}000$ kips. Substituting the strain and axial rigidity into Equation (4.20), we obtain the internal force $N = 46{,}000(-200)10^{-6} = -9.2$ kips, as before.

3. The cross section is symmetric geometrically as well as materially. Thus we can determine the location of the origin to be on the line of symmetry. Suppose the lower steel strip was not present. Then we will have to determine the vertical distance from the bottom (or the top) where the equivalent force will have to be placed, as given by Equation (4.23).

4. The example demonstrates that although the strain is uniform across the cross section, the stress is not. We considered material non-homogeneity in this example. In a similar manner we can consider other models, such as elastic–perfectly plastic, or material models that have nonlinear stress–strain curves.

EXAMPLE 4.9

A cast-iron pipe ($E_{iron} = 100$ GPa) and a copper pipe ($E_{copper} = 130$ GPa) are adhesively bonded together, as shown in Figure 4.31. The outer diameters of the two pipes are 50 mm and 70 mm and the wall thickness of each pipe is 10 mm. Determine:

(a) The displacement of end D with respect to end A.

(b) The axial stresses in iron and copper in the bonded region.

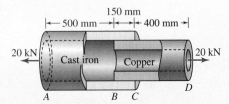

Figure 4.31 Pipes in Example 4.9.

PLAN

(a) Pipe segments AB and CD have homogeneous cross sections and we can use Equation (4.11) to find the relative displacements of the segment ends. However, segment BC is a composite pipe, and we use Equation (4.22) to find the relative displacement of the section at C with respect to the section at B.

(b) We can use Equation (4.21) to find the axial stress in each material.

Solution

(a) The cross-sectional areas of cast iron (A_{iron}) and copper (A_{copper}) can be calculated as follows:

$$A_{iron} = \frac{\pi}{4}(0.07^2 - 0.05^2) = 1.885 \times 10^{-3} \text{ m}^2 \qquad (E1)$$

$$A_{copper} = \frac{\pi}{4}(0.05^2 - 0.03^2) = 1.257 \times 10^{-3} \text{ m}^2 \qquad (E2)$$

The axial rigidities for each segment of pipe can be found as shown,

$$E_{iron}A_{iron} = 1.885 \times 10^{-3} \times 100 \times 10^9 = 188.5 \times 10^6 \qquad (E3)$$

$$E_{copper}A_{copper} = 1.257 \times 10^{-3} \times 130 \times 10^9 = 163.4 \times 10^6 \qquad (E4)$$

$$E_{BC}A_{BC} = \sum E_j A_j = E_{iron}A_{iron} + E_{copper}A_{copper}$$

$$= 351.9 \times 10^6 \qquad (E5)$$

We can see that regardless of where we make our imaginary cut in the pipe, the internal force value is the same—$N = 20$ kN. By substituting Equations (E3), (E4), and (E5) into Equations (4.11) and (4.22) we obtain the relative displacements of the various segments of the pipe,

$$u_B - u_A = \frac{N(x_B - x_A)}{E_{iron}A_{iron}} = \frac{20 \times 10^3 \times 0.5}{188.5 \times 10^6} = 53.05 \times 10^{-6} \quad (E6)$$

$$u_C - u_B = \frac{N(x_C - x_B)}{\sum E_j A_j} = \frac{20 \times 10^3 \times 0.15}{351.9 \times 10^6} = 8.53 \times 10^{-6} \quad (E7)$$

$$u_D - u_C = \frac{N(x_D - x_C)}{E_{copper}A_{copper}} = \frac{20 \times 10^3 \times 0.4}{163.4 \times 10^6} = 48.96 \times 10^{-6} \quad (E8)$$

We can add Equations (E6), (E7), and (E8) to obtain the relative deformation of the section at D with respect to the section at A,

$$u_D - u_A = (53.05 + 8.53 + 48.96)10^{-6} = 110.54 \times 10^{-6} \text{ m}$$

$$\text{ANS.} \quad u_D - u_A = 0.1105 \text{ mm}$$

(b) The stress in each material in the bonded region can be calculated using Equation (4.21),

$$(\sigma_{BC})_{iron} = \frac{NE_{iron}}{\sum E_j A_j} = \frac{20 \times 10^3 \times 100 \times 10^9}{351.9 \times 10^6} = 5.68 \times 10^6$$

$$\text{ANS.} \quad (\sigma_{BC})_{iron} = 5.68 \text{ MPa (T)}$$

$$(\sigma_{BC})_{copper} = \frac{NE_{copper}}{\sum E_j A_j} = \frac{20 \times 10^3 \times 130 \times 10^9}{351.9 \times 10^6} = 7.39 \times 10^6$$

$$\text{ANS.} \quad (\sigma_{BC})_{copper} = 7.39 \text{ MPa (T)}$$

COMMENTS

1. The analysis procedure is the same for homogeneous and nonhomogeneous cross sections. The difference is in the formulas used in the calculation of stress and the relative displacement.

2. We can find the axial force carried by each material in the bonded region by multiplying the respective axial stresses by the cross-sectional area. This yields $N_{iron} = 10.71$ kN and $N_{copper} = 9.23$ kN.

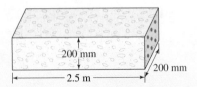

Figure 4.32 Reinforced concrete bar in Example 4.10.

EXAMPLE 4.10

A 200-mm × 200-mm reinforced concrete bar of 2.5-m length needs to be designed to carry a compressive axial force of 1000 kN (Figure 4.32). The allowable contraction of the bar is 3 mm. Determine, to the nearest kilogram, the minimum amount of concrete and cast iron that should be used in constructing the reinforced concrete bar. The material properties are listed in Table 4.2.

TABLE 4.2   Material properties in Example 4.10

	Concrete	Cast Iron
Modulus of elasticity	$E_{conc} = 20$ GPa	$E_{iron} = 170$ GPa
Density	2400 kg/m^3	7350 kg/m^3
Allowable stress	12 MPa	150 MPa

PLAN

From Equation (4.21) we can determine the axial rigidity that is needed to meet the limitation on the allowable stress in concrete and cast iron. From Equation (4.22) we can determine the axial rigidity that is needed to meet the limitation on the allowable deformation. The value of the axial rigidity that satisfies all the limitations on stress and deformation can thus be determined, and from it the cross-sectional area for cast iron and concrete can be found. By multiplying the areas by the given length we can find the volume, which we can multiply by the density to obtain the amount of each constituent in kilograms.

Solution   In the cross section of the reinforced bar, let A_{conc} and A_{iron} represent the cross-sectional areas of concrete and cast iron, respectively. The axial rigidity of the reinforced bar can be written as

$$\sum_{j=1}^{n} E_j A_j = (20A_{conc} + 170A_{iron})10^9 \qquad (E1)$$

Noting that the axial force $N = 1000$ kN, we can write the stress in concrete and cast iron using Equation (4.21). These stresses must be less than the allowable value, and we obtain two conditions on axial rigidity:

$$(\sigma_{xx})_{conc} = \frac{NE_{conc}}{\sum_{j=1}^{n} E_j A_j} = \frac{1000 \times 10^3 \times 20 \times 10^9}{(20A_{conc} + 170A_{iron})10^9}$$

$$= \frac{20 \times 10^6}{20A_{conc} + 170A_{iron}} \leq 12 \times 10^6$$

or

$$20A_{conc} + 170A_{iron} \geq 1.667 \qquad \text{(E2)}$$

$$(\sigma_{xx})_{iron} = \frac{NE_{iron}}{\sum_{j=1}^{n} E_j A_j} = \frac{1000 \times 10^3 \times 170 \times 10^9}{(20A_{conc} + 170A_{iron})10^9}$$

$$= \frac{170 \times 10^6}{20A_{conc} + 170A_{iron}} \leq 150 \times 10^6$$

or

$$20A_{conc} + 170A_{iron} \geq 1.133 \qquad \text{(E3)}$$

The contraction of the bar $u_2 - u_1$ should be less than 3 mm, or 3×10^{-3} m, over a length $x_2 - x_1 = 2.5$ m. From Equation (4.22) we obtain another limitation on axial rigidity,

$$u_2 - u_1 = \frac{N(x_2 - x_1)}{\sum_{j=1}^{n} E_j A_j} = \frac{1000 \times 10^3 \times 2.5}{(20A_{conc} + 170A_{iron})10^9} \leq 3 \times 10^{-3}$$

or

$$20A_{conc} + 170A_{iron} \geq 0.833 \qquad \text{(E4)}$$

If the axial rigidity is equal to 1.667, then we meet the three limitations given by Equations (E2), (E3), and (E4). For the minimum amount we use the equality sign in Equation (E2). Noting that the total cross-sectional area is $A = 200 \times 200$ mm^2, or $A = 0.04$ m^2, we can write

$$20A_{conc} + 170A_{iron} = 1.667 \qquad \text{(E5)}$$

$$A_{conc} + A_{iron} = 0.04 \qquad \text{(E6)}$$

Solving Equations (E5) and (E6), we obtain

$$A_{conc} = 34.2 \times 10^{-3} \, \text{m}^2 \qquad \text{(E7)}$$

$$A_{iron} = 5.8 \times 10^{-3} \, \text{m}^2 \qquad \text{(E8)}$$

Multiplying these areas by the length of 2.5 m, we obtain the volume of each material which, on multiplication, gives us the mass of concrete m_{conc} and the mass of iron m_{iron},

$$m_{conc} = 34.2 \times 10^{-3} \times 2.5 \times 2400 = 205.2 \qquad \text{(E9)}$$

$$\text{ANS.} \qquad m_{conc} = 206 \text{ kg.}$$

$$m_{iron} = 5.8 \times 10^{-3} \times 2.5 \times 7350 = 106.6 \qquad \text{(E10)}$$

$$\text{ANS.} \qquad m_{iron} = 107 \text{ kg.}$$

COMMENTS

1. In Equations (E9) and (E10) we rounded upward. By using more material we increase the axial rigidity, and thus our approximation (rounding) is on the safer side.

2. Suppose we use circular cast-iron rods of 25-mm diameter. Then each rod will have a cross-sectional area of 0.491×10^{-3} m². We can divide Equation (E8) by the area of each rod to obtain a value of 11.8, which we round upward to get 12 iron rods. Now we have an estimate of the number of irons rods and the amount of concrete we need to construct the reinforced concrete bar.

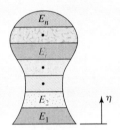

Figure 4.33 Laminated cross section in Example 4.11.

EXAMPLE 4.11

For a laminated, symmetric, linear elastic bar, as shown in Figure 4.33, show that the location of the origin η_c from the bottom can be found from the following formula:

$$\eta_c = \frac{\sum_{i=1}^{n} \eta_i E_i A_i}{\sum_{i=1}^{n} E_i A_i} \tag{4.23}$$

where E_i is the modulus of elasticity of the ith material, A_i is the cross-sectional area of the ith material, η_i is the location of the centroid of the ith material as measured from a common datum line, which in this example is the bottom, and n is the number of materials in the cross section.

PLAN

Equation (4.6a) can be used to find the origin as it is independent of the material model. After substituting Equation (4.3) into Equation (4.6a) we write the integral over the entire cross section as the sum of the integrals over each material. In each material the modulus of elasticity is a constant and can be taken outside the integral and the origin determined.

Solution Substituting Equation (4.3) into Equation (4.6a) and noting that du/dx is a function of x only, while the integration is with respect to y and z ($dA = dy\ dz$), we obtain

$$\int_A yE \frac{du}{dx} dA = \frac{du}{dx} \int_A yE\ dA = 0$$

Since du/dx cannot be zero, we obtain

$$\int_A yE \, dA = 0 \tag{4.24}$$

We can now establish the relation between y and η,

$$y = \eta - \eta_c \tag{E1}$$

See Figure 4.34. Substituting Equation (E1) into Equation (4.24) we obtain

$$\int_A \eta E \, dA - \int_A \eta_c E \, dA = 0$$

Noting that η_c is a constant, we obtain

$$\eta_c = \frac{\int_A \eta E \, dA}{\int_A E \, dA} \tag{E2}$$

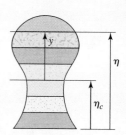

Figure 4.34 Coordinate description in Example 4.11.

Writing the integration over the area as the sum of the integrations over each material and noting that E_i is a constant within each A_i, Equation (E2) can be written as

$$\eta_c = \frac{\sum_{i=1}^{n} \int_{A_i} \eta E_i \, dA}{\sum_{i=1}^{n} \int_{A_i} E_i \, dA} = \frac{\sum_{i=1}^{n} E_i \int_{A_i} \eta \, dA}{\sum_{i=1}^{n} E_i \int_{A_i} dA} \tag{E3}$$

From the definition of centroid for η_i we note that $\int_{A_i} \eta \, dA = \eta_i A_i$ and $\int_{A_i} dA = A_i$. Substituting these two identities into Equation (E3) we obtain Equation (4.23).

COMMENTS

1. If the external axial forces do not pass through the point defined by Equation (4.23), then the laminated rod will bend as well as deform axially.

2. For a homogeneous material the modulus of elasticity E is constant over the cross section. Hence the E in the numerator of Equation (E3) will cancel the E in the denominator and we see that η_c defines the centroid as per Figure A.3 in the Appendix.

3. If $E_1 = E_2 = \cdots = E_i = \cdots = E_n = E$, then the modulus of elasticity in the numerator and the denominator of Equation (4.23) can be taken outside the summation and we obtain

$$\eta_c = \frac{\sum_{i=1}^{n} \eta_i A_i}{\sum_{i=1}^{n} A_i} \tag{4.25}$$

η_c defines here the centroid for a homogeneous body as per the definition in Section A.4.

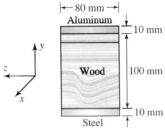

Figure P4.43

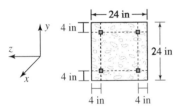

Figure P4.44

PROBLEM SET 4.2

Composite bars

4.43 The strain at a cross section of an axial rod is assumed to have the uniform value $\varepsilon_{xx} = 2000\ \mu$ (Figure P4.43). (a) Plot the stress distribution across the laminated cross section. (b) Determine the equivalent internal axial force N and its location from the bottom of the cross section. Use $E_{alu} = 100$ GPa, $E_{wood} = 10$ GPa, and $E_{steel} = 200$ GPa.

4.44 A reinforced concrete bar is constructed by embedding 2-in × 2-in square iron rods (Figure P4.44). Assuming a uniform strain $\varepsilon_{xx} = -1500\ \mu$ in the cross section: (a) plot the stress distribution across the cross section; (b) determine the equivalent internal axial force N. Use $E_{iron} = 25{,}000$ ksi and $E_{conc} = 3000$ ksi.

4.45 A wooden rod ($E_w = 2000$ ksi) and a steel strip ($E_s = 30{,}000$ ksi) are fastened securely to each other and to the rigid plates, as shown in Figure P4.45. Determine: (a) the location h of the line along which the external forces must act to produce no bending; (b) the maximum axial stress in steel and wood.

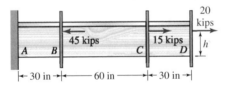

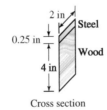

Cross section

Figure P4.45

4.46 A solid-steel circular rod ($E_{steel} = 200$ GPa) of 80-mm diameter is 4 m long. It extends through and is attached to a hollow brass bar ($E_{brass} = 100$ GPa) of 120-mm outside diameter (Figure P4.46). Determine: (a) the displacement of point C with respect to the wall; (b) the maximum axial stress in steel and brass.

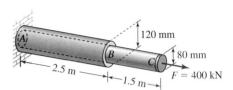

Figure P4.46

4.47 A concrete column is reinforced using nine iron circular bars of 1-in diameter (Figure P4.47). The moduli of elasticity for concrete and iron are $E_{conc} = 4500$ ksi and $E_{iron} = 25{,}000$ ksi. Determine: (a) the maximum axial stress in concrete and iron; (b) the contraction of the column.

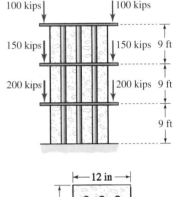

Figure P4.47

4.48 A cast-iron pipe ($E_{iron} = 100$ GPa) and a copper pipe ($E_{copper} = 130$ GPa) are adhesively bonded together, as shown in Figure P4.48. The outer diameters of the two pipes are 30 mm and 50 mm and the wall thickness of each pipe is 10 mm. Determine: (a) the maximum axial stress in the region BC;

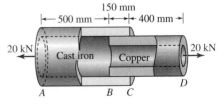

Figure P4.48

(b) the displacement of end D with respect to end A.

4.49 The cross section of a bar is made from two materials, as shown in Figure P4.49. Assume parallel sections remain parallel, that is, $\varepsilon_{xx} = (du/dx)(x)$. In terms of the variables P, E, and h determine: (a) the location y_P of force P on the cross section so that there is only axial deformation and no bending; (b) the axial stress at point A.

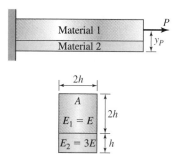

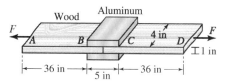

Figure P4.49

4.50 Two 1-in-thick wood panels ($E_W = 1800$ ksi) are joined together using two $\frac{1}{8}$-in-thick aluminum sheets ($E_{Al} = 10,000$ ksi), as shown in Figure P4.50. The allowable stress in wood is 1.5 ksi, the allowable stress in aluminum is 24 ksi, and the total elongation of the joint length AD is to be limited to 0.05 in. Determine the maximum axial force F that can be applied.

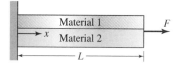

Figure P4.50

Stretch yourself

4.51 The average stress for a composite lamina shown in Figure P4.51 can be obtained by dividing the internal axial force in Equation (4.20) by the total cross-sectional area A. The tension test for a composite lamina would yield Hooke's law as $\sigma_{av} = E_{eff}\varepsilon$, where E_{eff} is the effective modulus of elasticity for the lamina. Show that

$$E_{eff} = \frac{E_{fiber} V_{fiber} + E_{resin} V_{resin}}{V_{fiber} + V_{resin}} \quad (4.26)$$

where E_{fiber} and E_{resin} are the moduli of elasticity of fiber and resin, and. V_{fiber} and V_{resin} are the volumes of fiber and resin in the lamina, respectively. Equation (4.26) is known as the *rule of mixtures*.

Figure P4.51

4.52 An axial rod is made from two materials, as shown in Figure P4.52. The stress–strain relationship in material 1 is given by $\sigma = E_1 \varepsilon^{0.5}$, and in material 2 the relation is $\sigma = E_2 \varepsilon^{0.5}$. Let A_1 and A_2 be the cross-sectional areas of material 1 and material 2, respectively. Determine the elongation of the bar in terms of F, A_1, A_2, E_1, E_2, and L.

Figure P4.52

4.3 STRUCTURAL ANALYSIS

Equation (4.11) assumes that the bar lies in the x direction. Structures are usually made of axial bars in different orientations, and for this reason in structural analysis the form of equation given here is preferred over Equation (4.11),

$$\boxed{\delta = \frac{NL}{EA}} \quad (4.27)$$

where the expression $x_2 - x_1$ in Equation (4.11) is replaced by L, representing the length of the bar, and the expression $u_2 - u_1$ in Equation (4.11) is replaced by δ, representing the deformation of the bar in the undeformed direction. The emphasis here is on highlighting that irrespective of the movement of points on the bar, we will use only the component of deformation that is in the original direction of the bar, according to our approximation of small strain discussed in Chapter 2. It should also be recognized that L, E, and A are positive, hence the sign of δ is the same as that of N. In other words, if N is a tensile force, then δ is elongation; if N is a compressive force, then δ is contraction. Example 4.12 demonstrates the use of Equation (4.27).

4.3.1 Statically Indeterminate Structures[14]

Statically indeterminate structures arise when there are more supports than needed to hold a structure in place. These extra supports are included for considerations of safety or to increase the stiffness of the structures. Each extra support introduces additional unknown reactions, and hence the total number of unknown reactions exceeds the number of static equilibrium equations.

Definition 5 The degree of static redundancy is the number of unknown reactions minus the number of equilibrium equations.

If the degree of static redundancy is zero, then we have a statically determinate structure and all unknowns can be found from equilibrium equations. If the degree of static redundancy is not zero, then we need additional equations to determine the unknown reactions. These additional equations are the relationships between the deformations of bars, derived under the assumption that displacement at any point on the body is a continuous function. If the displacement at a point is discontinuous, then it implies that the structure is broken at that point.

Definition 6 Compatibility equations are geometric relationships between the deformations of bars and are derived from the deformed shapes of the structure.

Drawing the approximate deformed shape of a structure for writing compatibility equations is of the same importance as drawing a free-body diagram for writing equilibrium equations. The deformations shown in the deformed shape of the structure must be consistent with the direction of forces drawn on the free-body diagram. The number of compatibility equations needed is always equal to the degree of static redundancy.

Sometimes we shall choose displacements of points or rotations of a rigid bar to describe the deformed geometry of a structure. The primary motivation for doing this would be to minimize the number of unknowns that must be determined to solve the problem. In such cases the compatibility equations will describe the deformations of the bars in terms of these variables, which is called degrees of freedom.

Definition 7 The variables that describe a deformed geometry are called *degrees of freedom*.

In many structures there are gaps between structural members. These gaps may be by design to permit expansion due to temperature changes, or they may be inadvertent due to improper accounting for manufacturing tolerances. In analyzing structures with gaps, the

---

[14]See Appendix A.2 for an additional discussion of statically indeterminate structures.

following point must be remembered: analysis is being performed at the final equilibrium state, and for a linear system it does not matter how we reach the final equilibrium state. To account for these gaps in the analysis we can take one of the two approaches:

1. We can first calculate the external force that is needed to close the gap and find the corresponding stresses and deformations in all members. The remaining force that needs to be accounted for is the total applied force minus the force necessary to close the gap. We apply this remaining external force and calculate the additional deformations and stresses in the required members. Superposing the results of the two calculations we obtain the total deformations and stresses.

2. We start by assuming that the gap is closed at equilibrium, solve the problem, then check whether the assumption is correct. If the assumption is correct, then we calculate the deformations and stresses. But if the assumption is incorrect, then we set the internal forces equal to zero in those members that do not deform if the gap is not closed and resolve the remaining equations.

Each approach has some advantages. If the gap closes, then approach 2 requires solving the problem only once, whereas approach 1 requires solving the problem twice. If the gap does not close, then approach 1 is simpler and the problem is solved only once. In this book the second approach is preferred, and procedures are described using this approach.

Displacement, strain, stress, and internal force are all related as depicted by the logic shown in Figure 4.3 and incorporated in the formulas developed in Section 4.1. If one of these quantities is found, then the rest could be found for an axial member. Thus theoretically, in structural analysis, any of the four quantities could be treated as an unknown variable. But analysis is traditionally conducted using either forces (internal or reaction) or displacement (deformation or degrees of freedom) as the unknown variables, as described in the two methods that follow.

4.3.2 Force Method, or Flexibility Method

In this method internal forces or reaction forces are treated as the unknowns. If the degree of static redundancy is less than the number of degrees of freedom, then the force method will result in a smaller set of algebraic equations that has to be solved.

In Equation (4.27) the coefficient L/EA, multiplying the internal unknown force, is called the flexibility coefficient. If the unknowns are internal forces (rather than reaction forces), as is usually the case in large structures, then the matrix in the simultaneous equations is called the flexibility matrix. Reaction forces are often preferred in hand calculations because the number of unknown reactions (degree of static redundancy) is either equal to or less than the unknown internal forces.

4.3.3 Displacement Method, or Stiffness Method

In this method the displacements of points are treated as the unknowns. If the number of degrees of freedom is less than the degree of static redundancy, then the displacement method will result in a smaller set of algebraic equations and should be used for solving the problem.

The coefficient multiplying the deformation EA/L is called the stiffness coefficient. Using small-strain approximation, the relationship between the displacement of points and the deformation of the bars is found from the deformed shape and substituted in the compatibility equations. Using Equation (4.27) and equilibrium equations, the displacement and the external forces are related. The matrix multiplying the unknown displacements is called the stiffness matrix.

4.3.4 General Procedure for Indeterminate Structure

The procedure outlined can be used for solving statically indeterminate structure problems by either the force method or by the displacement method.

1. If there is a gap, assume it will close at equilibrium.
2. Draw free-body diagrams, noting the tensile and compressive nature of internal forces. Write equilibrium equations relating internal forces to each other.
   or
   Write equilibrium equations in which the internal forces are written in terms of reaction forces, if the force method is to be used.
3. Draw an exaggerated approximate deformed shape, ensuring that the deformation is consistent with the free body diagrams of step 2. Write compatibility equations relating deformation of the bars to each other.
   or
   Write compatibility equations in terms of unknown displacements of points on the structure, if displacement method is to be used.
4. Write internal forces in terms of deformations using Equation (4.27).
5. Solve the equations of steps 2, 3, and 4 simultaneously for the unknown forces (for force method) or for the unknown displacements (for displacement method).
6. Check whether the assumption of gap closure in step 1 is correct.

In Examples 4.13 and 4.14 both the force method and the displacement method are used to demonstrate the similarities and differences in the two methods.

EXAMPLE 4.12
Three bars made of steel ($E = 30,000$ ksi) have cross-sectional areas of 1 in^2 (Figure 4.35). Determine the displacement of point D with respect to the no-load position.

PLAN
This is a statically determinate problem as we can find the internal forces in all members by static equilibrium. The displacement of point D with respect to point C can be found from Equation (4.27). Similarly the deformation of rod AC or BC can be found from Equation (4.27).

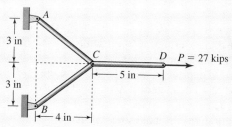

Figure 4.35 Geometry in Example 4.12.

But to relate the displacement of point C to the deformation for rod AC (or BC) we will use small-strain approximation.

Solution Using Equation (4.27), we calculate the displacement of D relative to the displacement at C (Figure 4.36),

$$N_{CD} = 27 \text{ kips} \qquad \text{(E1)}$$

$$\delta_{CD} = u_D - u_C = \frac{N_{CD}L_{CD}}{E_{CD}A_{CD}} = \frac{27 \times 5}{30{,}000 \times 1}$$

$$u_D - u_C = 4.5 \times 10^{-3} \qquad \text{(E2)}$$

Figure 4.36 Bar CD in Example 4.12.

The internal forces in members AC and CB can be found from equilibrium (Figure 4.37),

$$\sum F_y \qquad N_{CA}\sin\theta - N_{CB}\sin\theta = 0 \quad \text{or} \quad N_{CA} = N_{CB} \quad \text{(E3)}$$

$$\sum F_x \qquad -N_{CA}\cos\theta - N_{CB}\cos\theta + 27 = 0 \quad \text{or} \quad 2N_{CA}\left(\frac{4}{5}\right) = 27$$

$$\text{or} \qquad N_{CA} = 16.875 \text{ kips} \qquad \text{(E4)}$$

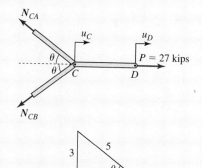

Figure 4.37 Free-body diagram in Example 4.12.

We conduct the deformation analysis using small-strain approximation (Figure 4.38),

$$\delta_{AC} = \frac{N_{CA}L_{CA}}{E_{CA}A_{CA}} = \frac{16.875 \times 5}{30{,}000 \times 1} = 2.8125 \times 10^{-3}$$

$$u_C = \frac{\delta_{AC}}{\cos\theta} = 3.52 \times 10^{-3} \qquad \text{(E5)}$$

Adding Equations (E2) and (E5) we obtain

$$u_D = (4.5 + 3.52)10^{-3}$$

$$\text{ANS.} \qquad u_D = 0.008 \text{ in}$$

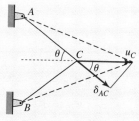

Figure 4.38 Deformed geometry in Example 4.12.

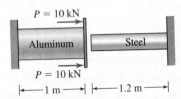

P = 10 kN

Aluminum   Steel

P = 10 kN

|←—— 1 m ——→| |←——— 1.2 m ———→|

Figure 4.39 Geometry in Example 4.13.

EXAMPLE 4.13

An aluminum rod ($E_{al} = 70$ GPa) is securely fastened to a rigid plate that does not rotate during the application of load P, and a steel rod ($E_{st} = 210$ GPa) are shown in Figure 4.39. A gap of 0.5 mm exists between the rigid plate and the steel rod before the load is applied. The aluminum rod has a diameter of 20 mm and the steel rod has a diameter of 10 mm. Determine: (a) the movement of the rigid plate from the unloaded position; (b) the axial stress in steel.

Force Method

PLAN

The movement of the rigid plate corresponds to the deformation of the aluminum rod. We assume that the force P is sufficient to close the gap. After the gap has closed each wall has a reaction force leading to two unknowns. We have one equilibrium equation in the axial direction, hence this problem has 1 degree of redundancy. To solve this problem we will have one equilibrium equation and one equation of compatibility. We follow the procedure outlined in Section 4.3.4.

Solution

Step 1 We assume that the force P is sufficient to close the gap. If this assumption is correct, then steel will be in compression. If the assumption is incorrect, then steel will have to be extended to close the gap, and hence steel will be in tension.

Step 2 The degree of redundancy is 1. Thus we use one unknown reaction to formulate our equilibrium equations. We make imaginary cuts at the equilibrium position and obtain the free-body diagrams in Figure 4.40. By equilibrium of forces we can obtain the internal forces in terms of the wall reactions,

$$N_{al} = R_L \tag{E1}$$

$$N_{st} = 20 \times 10^3 - R_L \tag{E2}$$

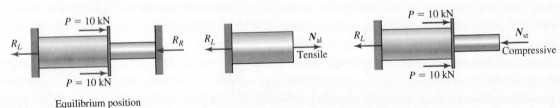

P = 10 kN

R_L R_R R_L N_{al} R_L P = 10 kN N_{st}

Tensile   Compressive

P = 10 kN   P = 10 kN

Equilibrium position

Figure 4.40 Free-body diagrams in Example 4.13.

Step 3 We can also draw the exaggerated deformed shape in Figure 4.41 and show aluminum in extension and steel in contraction, to ensure consistency with the tensile and compressive axial forces shown on the free-body diagrams in Figure 4.40. We can then write the compatibility equation,

$$\delta_{st} = \delta_{al} - 0.0005 \qquad (E3)$$

Step 4 The radius of the aluminum rod is 10 mm or 0.01 m, and the radius of the steel rod is 5 mm or 0.005 m. From Equation (4.27) we can write the deformation of aluminum and steel in terms of the internal forces,

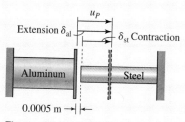

Figure 4.41 Approximate deformed shape in Example 4.13.

$$\delta_{al} = \frac{N_{al}L_{al}}{E_{al}A_{al}} = \frac{N_{al} \times 1}{70 \times 10^9 \times \pi \times 0.01^2} = 0.04547N_{al} \times 10^{-6} \quad (E4)$$

$$\delta_{st} = \frac{N_{st}L_{st}}{E_{st}A_{st}} = \frac{N_{st} \times 1.2}{210 \times 10^9 \times \pi \times 0.005^2} = 0.07277N_{st} \times 10^{-6} \quad (E5)$$

Step 5 Substituting Equations (E1) and (E2) into Equations (E4) and (E5) we obtain deformation in terms of the unknown reactions,

$$\delta_{al} = 0.04547R_L \times 10^{-6} \qquad (E6)$$

$$\delta_{st} = 0.07277(20 \times 10^3 - R_L)10^{-6} = (1455.4 - 0.07277R_L)10^{-6} \qquad (E7)$$

Substituting Equations (E6) and (E7) into Equation (E3) and multiplying by 10^6, we can solve for R_L,

$$1455.4 - 0.07277R_L = 0.04547R_L - 500 \quad \text{or} \quad R_L = 16{,}538 \quad (E8)$$

Substituting Equation (E8) into Equations (E1) and (E2) we obtain the internal forces,

$$N_{al} = 16{,}538 \text{ N} \qquad N_{st} = 3462 \text{ N} \qquad (E9)$$

Step 6 The force in steel resulted in a positive value, and we started with a compressive force on steel. Hence the assumption of the gap being closed is correct.

(a) The movement of the rigid plate u_P is equal to the deformation of aluminum. Substituting Equation (E8) into Equation (E6) we obtain

$$u_P = \delta_{al} = 0.04547 \times 16{,}538 \times 10^{-6} = 0.752 \times 10^{-3} \text{ m}$$

ANS. $\quad u_P = 0.752 \text{ mm}$

(b) The normal stress in steel can be found from Equation (4.9) as

$$\sigma_{st} = \frac{N_{st}}{A_{st}} = \frac{3462}{\pi \times 0.005^2} = 44.1 \times 10^6 \text{ N/m}^2$$

ANS. $\quad \sigma_{st} = 44.1 \text{ MPa (C)}$

COMMENTS

1. The movement of the plate u_P is greater than the gap, which once more confirms that the assumption that the gap is closed is correct.

2. An alternative approach is to use internal forces as the unknowns. We can make a cut on either side of the rigid plate at the equilibrium position with the gap closed and draw a tensile force for aluminum and a compressive force for steel on the free-body diagram, as shown in Figure 4.42. We can then write the equilibrium equation,

$$N_{st} + N_{al} = 20 \times 10^3 \text{ N} \tag{E10}$$

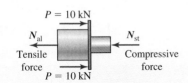

$P = 10$ kN

N_{al} N_{st}

Tensile force   Compressive force

$P = 10$ kN

Figure 4.42 Alternative free-body diagram in Example 4.13.

We could obtain the same equation by eliminating R_L from Equations (E1) and (E2).

Substituting Equations (E4) and (E5) into Equation (E3) and multiplying by 10^6 we obtain

$$0.04547N_{al} - 0.07277N_{st} = 500 \tag{E11}$$

Equations (E10) and (E11) can be written in matrix form as

$$
\begin{matrix} [F] & \{N\} & \{P\} \end{matrix}
$$
$$
\begin{bmatrix} 1 & 1 \\ 0.04547 & -0.07277 \end{bmatrix} \begin{Bmatrix} N_{at} \\ N_{st} \end{Bmatrix} = \begin{Bmatrix} 20 \times 10^3 \\ 500 \end{Bmatrix}
$$

The matrix $[F]$ is called *flexibility matrix.*

3. When we used internal forces as unknowns we had to solve two equations simultaneously, as elaborated in comment 2. But with the reaction force as the unknown we had only one unknown, which is the number of degrees of static redundancy. Thus for hand calculations the reaction forces as unknowns are preferred when using the force method. But in computer programs the process of substitution in step 5 is difficult to implement compared to constructing the equilibrium and compatibility equations in terms of internal forces. Thus in computer methods internal forces are treated as unknowns in force methods.

4. Suppose we had started with the direction of the force in steel as tension (Figure 4.43). Then we would get the following equilibrium equation:

$$-N_{st} + N_{al} = 20 \times 10^3 \text{ N} \qquad (E12)$$

If incorrectly we do not make any changes in Equation (E3) or Equation (E7), that is, we continue to use the deformation in steel as contraction even though the assumed force is tensile, and we then solve Equations (E12) and (E11), we obtain $N_{al} = 34996$ N and $N_{st} = 14996$ N. These answers demonstrate how a simple error in sign produces dramatically different results.

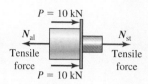

Figure 4.43 Tensile forces in free-body diagram in Example 4.13.

Displacement Method

PLAN
Let the plate move to the right by the amount u_P and assume that the gap is closed. If the result shows that u_P is less than 0.0005, then we will know that our assumption is incorrect. We follow the procedure outlined in Section 4.3.4.

Solution
Step 1   We assume that the gap is closed.

Step 2   We can substitute Equation (E1) into Equation (E2) to eliminate R_L and obtain the equilibrium equation,

$$N_{st} + N_{al} = 20 \times 10^3 \text{ N} \qquad (E13)$$

We could also obtain this equation from the free-body diagram shown in Figure 4.42.

Step 3   We draw the exaggerated deformed shape, as shown in Figure 4.38, and obtain the deformation of the bars in terms of the plate displacement u_P as

$$\delta_{al} = u_P \qquad (E14)$$

$$\delta_{st} = u_P - 0.0005 \qquad (E15)$$

Step 4   Using Equations (4.27) we can write the internal forces in terms of deformation,

$$N_{al} = \delta_{al}\left(\frac{E_{al}A_{al}}{L_{al}}\right) = 21.99 \times 10^6 \times \delta_{al} \qquad (E16)$$

$$N_{st} = \delta_{st}\left(\frac{E_{st}A_{st}}{L_{st}}\right) = 13.74 \times 10^6 \times \delta_{st} \qquad (E17)$$

Step 5   We can substitute Equations (E14) and (E15) into Equations (E16) and (E17) to obtain the internal forces in terms of u_P,

$$N_{al} = 21.99 \times 10^6\, u_P \qquad (E18)$$

$$N_{st} = 13.74\,(u_P - 0.0005)10^6 \qquad (E19)$$

Substituting Equations (E18) and (E19) into Equation (E13) and multiplying by 10^{-6} we obtain

$$21.99u_P + 13.74 \ (u_P - 0.0005) = 20 \times 10^{-3}$$

or

$$u_P = 0.752 \times 10^{-3} \ \text{m}$$

ANS. $u_P = 0.752$ mm

Step 6 As $u_P > 0.0005$, the assumption of gap closing is correct.

Substituting u_P into Equation (E19), we obtain $N_{\text{st}} = 3423.2$ N, which implies that the steel is in compression, as expected. We can now find the axial stress in steel, as before.

COMMENTS

1. In the force method as well as in the displacement method the number of unknowns was 1 as the degree of redundancy and the number of degrees of freedom were 1. This is not always the case. In the next example the number of degrees of freedom is less than the degree of redundancy, and hence the displacement method will be easier to implement.

2. The equilibrium equation, Equation (E13), is in terms of internal forces only. If there are reaction forces, then these must be eliminated by using other equilibrium equations, as we did in step 2 of the displacement method. This will be true in all displacement methods as we do not want to carry reaction forces as other unknowns. By the same logic, if the compatibility equations in the force method contain variables other than deformations of members, then these must be eliminated by using other compatibility equations. For example, we can substitute Equation (E14) into Equation (E15) to obtain Equation (E3).

EXAMPLE 4.14

Three steel bars A, B, and C ($E = 200$ GPa) have lengths $L_A = 4$ m, $L_B = 3$ m, and $L_C = 2$ m, as shown in Figure 4.44. All bars have the same cross-sectional area of 500 mm². Determine: (a) the elongation in bar B; (b) the normal stress in bar C.

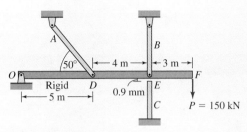

Figure 4.44 Geometry in Example 4.14.

Displacement Method

PLAN

Let the pin at E move down by an amount δ_E and assume that the gap is closed. If the result shows that δ_E is less than 0.9 mm, then we will know that our assumption is incorrect. We follow the procedure outlined in Section 4.3.4 to determine the internal forces in the members.

Solution

(a) The elongation of bar B is the same as δ_E.

(b) By dividing the internal force in bar C by the cross-sectional area we can obtain the normal stress in bar C.

Step 1 We assume that the force P is sufficient to close the gap.

Step 2 We can draw the free-body diagram of the rigid bar in Figure 4.45 with bars A and B in tension and bar C in compression to correspond to the deformation of these bars shown in Figure 4.46. By equilibrium of moment at point O in Figure 4.45 we obtain the equilibrium equation for the internal forces,

$$N_A \sin 50(5) + N_B(9) + N_C(9) - P(12) = 0$$

or

$$3.83 N_A + 9 N_B + 9 N_C = 1800 \times 10^3 \qquad \text{(E1)}$$

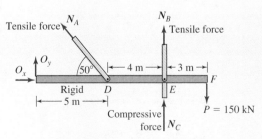

Figure 4.45 Free-body diagram in Example 4.14.

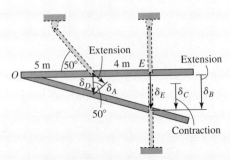

Figure 4.46 Deformed shape in Example 4.14.

Step 3 We can draw an exaggerated deformed shape. Noting that the gap is 0.9 mm = 0.0009 m, we can write the compatibility equations relating the deformations of bars *B* and *C* in terms of the displacement of pin *E*,

$$\delta_B = \delta_E \tag{E2}$$

$$\delta_C = \delta_E - 0.0009 \tag{E3}$$

Using similar triangles we can relate the displacement of point *D* to the displacement of pin *E*,

$$\frac{\delta_D}{5} = \frac{\delta_E}{9} \tag{E4}$$

Using small-strain approximation we can relate the deformation of bar *A* to the displacement of point *D*,

$$\delta_D = \frac{\delta_A}{\sin 50} \tag{E5}$$

Substituting Equation (E5) into Equation (E4), we obtain the deformation of bar *A* in terms of the displacement of pin *E*,

$$\frac{\delta_A / \sin 50}{5} = \frac{\delta_E}{9} \quad \text{or} \quad \delta_A = 0.4256 \delta_E \tag{E6}$$

Step 4 The axial rigidity of all bars is $EA = 200 \times 10^9 \times 500 \times 10^{-6} = 100 \times 10^6$ N. From Equation (4.27),

$$N_A = \frac{100 \times 10^6}{4} \delta_A = 25 \times 10^6 \delta_A \tag{E7}$$

$$N_B = \frac{100 \times 10^6}{3} \delta_B = 33.33 \times 10^6 \delta_B \tag{E8}$$

$$N_C = \frac{100 \times 10^6}{2} \delta_C = 50.00 \times 10^6 \delta_C \tag{E9}$$

Step 5 Substituting Equations (E6), (E2), and (E3) into Equations (E7), (E8), and (E9), we obtain the internal forces in terms of δ_E,

$$N_A = 25 \times 10^6 \times 0.4256\delta_E = 10.64 \times 10^6 \delta_E \tag{E10}$$

$$N_B = 33.33 \times 10^6 \delta_E = 33.33 \times 10^6 \delta_E \tag{E11}$$

$$N_C = 50.00 \times 10^6 (\delta_E - 0.0009) = 50.00 \times 10^6 \delta_E - 45 \times 10^3 \tag{E12}$$

Substituting Equations (E10), (E11), and (E12) in Equation (E1) we obtain the displacement of pin E,

$$3.83(10.64 \times 10^6 \delta_E) + 9(33.33 \times 10^6 \delta_E)$$
$$+ 9(50.00 \times 10^6 \delta_E - 45 \times 10^3) = 1800 \times 10^3$$

or

$$\delta_E = 2.788 \times 10^{-3} \text{ m}$$

ANS. $\delta_E = 2.8$ mm

Step 6   The assumption of gap closure is correct as $\delta_E = 2.8$ mm whereas the gap is only 0.9 mm.

From Equation (E12) we obtain the internal axial force in bar C, from which we obtain the axial stress in bar C,

$$N_C = 50.00 \times 10^6 (2.788 \times 10^{-3}) - 45 \times 10^3 = 94.4 \times 10^3$$

$$\sigma_C = \frac{N_C}{A_C} = \frac{94.4 \times 10^3}{500 \times 10^{-6}} = 188.8 \times 10^6 \text{ N/m}^2$$

ANS. $\sigma_C = 189$ MPa (C)

COMMENTS

1. Equation (E4) is a relationship of points on the rigid bar. Equations (E2), (E3), and (E5) relate the motion of points on the rigid bar to the deformation of the rods. This two-step process helps break the complexity into simpler steps.

2. The degree of freedom for this system is 1. In place of δ_E as an unknown, we could have used the displacement of any point on the rigid bar or the rotation angle of the bar, as all of these quantities are related. The choice of δ_E was based on the fact that we needed to find δ_B, which is the same as δ_E.

Force Method

PLAN

We assume that the force P is sufficient to close the gap. If this assumption is correct, then bar C will be in compression. If the assumption is incorrect, then bar C will have to be pulled up to close the gap and bar C steel will be in tension. We follow the procedure outlined in Section 4.3.4 to determine the internal forces.

Solution

Step 1 Assume that the gap closes.

Step 2 We draw the free-body diagram of the rigid bar, as shown in Figure 4.45, and obtain the equilibrium equation, Equation (E1), rewritten here for convenience,

$$3.83N_A + 9N_B + 9N_C = 1800 \times 10^3 \qquad \text{(E13)}$$

Equation (E13) has three unknowns, hence the degree of redundancy is 2. We will need two compatibility equations.

Step 3 We draw the deformed shape, as shown in Figure 4.46, and obtain relationships between points on the rigid bar and the deformation of the bars. Then by eliminating δ_E from Equations (E3) and (E6) and using Equation (E2) we obtain the compatibility equations,

$$\delta_C = \delta_B - 0.0009 \qquad \text{(E14)}$$

$$\delta_A = 0.4256\,\delta_B \qquad \text{(E15)}$$

Step 4 From Equation (4.27) we obtain the deformations of the bars in terms of the internal forces for each member,

$$\delta_A = \frac{N_A L_A}{E_A A_A} = \frac{N_A(4)}{100 \times 10^6} = 0.04 N_A \times 10^{-6} \qquad \text{(E16)}$$

$$\delta_B = \frac{N_B L_B}{E_B A_B} = \frac{N_B(3)}{100 \times 10^6} = 0.03 N_B \times 10^{-6} \qquad \text{(E17)}$$

$$\delta_C = \frac{N_C L_C}{E_C A_C} = \frac{N_C(2)}{100 \times 10^6} = 0.02 N_C \times 10^{-6} \qquad \text{(E18)}$$

Step 5 Substituting Equations (E16), (E17), and (E18) in Equations (E14) and (E15) and multiplying by 10^6 we obtain two more equations of internal forces,

$$0.02N_C = 0.03N_B - 900 \qquad \text{or} \qquad N_C = 1.5N_B - 45 \times 10^3 \quad \text{(E19)}$$

$$0.04N_A = 0.4256 \times 0.03\ N_B \qquad \text{or} \qquad N_A = 0.3192N_B \qquad \text{(E20)}$$

Solving Equations (E13), (E19), and (E20) we obtain the following values of internal forces:

$$N_A = 29.67 \times 10^3 \qquad N_B = 92.95 \times 10^3 \qquad N_C = 94.43 \times 10^3 \quad \text{(E21)}$$

Step 6 As the internal force N_C is positive, the assumed direction of compression is correct, and the assumption of the gap being closed is valid.

Substituting N_B from Equation (E21) into Equation (E17) we find the deformation of bar B, which is the same as the displacement of pin E,

$$\delta_E = \delta_B = 0.03 \times 92.95 \times 10^3 \times 10^{-6} = 2.788 \times 10^3$$

ANS. $\delta_E = 2.8$ mm

The calculation for the normal stress in bar C is as before.

COMMENTS

Equations (E13), (E19), and (E20) represent three equations in three unknowns. Had we used the reaction forces O_x and O_y in Figure 4.45 as the unknowns, we could have generated two equations in two unknowns, consistent with the fact that this system has a degree of redundancy of 2. But as we saw, the displacement method required only one unknown. Thus our choice of method of solution should be dictated by the number of degrees of freedom and the number of degrees of static redundancy. For fewer degrees of freedom we should use the displacement method; for fewer degrees of static redundancy we should use the force method.

Consolidate your knowledge

1. With book closed, write a procedure for solving a statically indeterminate problem by the force method.
2. With book closed, write a procedure for solving a statically indeterminate problem by the displacement method.

PROBLEM SET 4.3

In Problems 4.53 and 4.54 a rigid bar is hinged at C. The modulus of elasticity of bar A is E = 30,000 ksi, the cross-sectional area is A = 1.25 in², and the length is 24 in. Determine the applied force F if point B moves upward by 0.002 in.

4.53 Determine the applied force F in Figure P4.53.

4.54 Determine the applied force F in Figure P4.54.

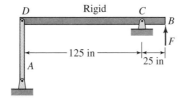

Figure P4.53

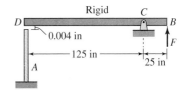

Figure P4.54

In Problems 4.55 and 4.56 a rigid bar is hinged at C. The modulus of elasticity of bar A is E = 100 GPa, the cross-sectional area is

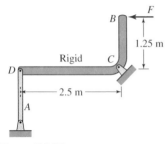

Figure P4.55

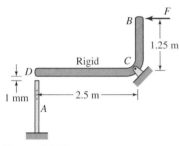

Figure P4.56

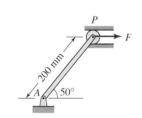

Figure P4.57

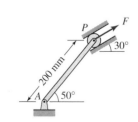

Figure P4.58

$A = 15 \text{ mm}^2$, and the length is 1.2 m. Determine the applied force F if point B moves to the left by 0.75 mm.

4.55 Determine the applied force F in Figure P4.55.

4.56 Determine the applied force F in Figure P4.56.

In Problems 4.57 and 4.58 the roller at P slides in the slot due to the force F = 100 kN. Member AP has a cross-sectional area $A = 100 \text{ mm}^2$ and a modulus of elasticity E = 200 GPa. Determine the displacement of the roller.

4.57 Determine the displacement of the roller in Figure P4.57.

4.58 Determine the displacement of the roller in Figure P4.58.

In Problems 4.59 and 4.60 a rigid bar is hinged at C. The modulus of elasticity of bar A is E = 30,000 ksi, the cross-sectional area is $A = 1.25 \text{ in}^2$, and the length is 24 in. Determine the axial stress in bar A and the displacement of point D on the rigid bar.

4.59 (Figure P4.59).

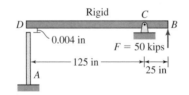

Figure P4.59

4.60 (Figure P4.60).

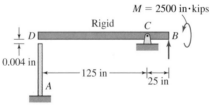

Figure P4.60

In Problems 4.61 through 4.63 a force F = 20 kN is applied to the roller that slides inside a slot. Both bars have a cross-sectional area $A = 100 \text{ mm}^2$ and a modulus of elasticity

E = 200 GPa. Bars AP and BP have lengths $L_{AP} = 200$ mm and $L_{BP} = 250$ mm. Determine the displacement of the roller and the axial stress in bar A.

4.61 (Figure P4.61).

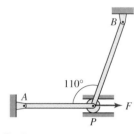

Figure P4.61

4.62 (Figure P4.62).

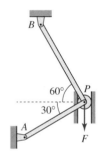

Figure P4.62

4.63 (Figure P4.63).

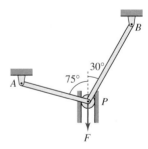

Figure P4.63

An aluminum hollow cylinder ($E_{al} = 10,000$ ksi, $v_{al} = 0.25$) and a steel hollow cylinder ($E_{st} = 30,000$ ksi, $v_{st} = 0.28$) are securely fastened to a rigid plate, as shown in Figure P4.65. Both cylinders are made from $\frac{1}{8}$-in thickness sheet metal. The outer diameters of the aluminum and steel cylinders are 4 in and 3 in, respectively. Use this information to solve Problems 4.64 and 4.65.

4.64 The load $P = 20$ kips in Figure P4.64. Determine: (a) the displacement of the rigid plate; (b) the change in diameter of each cylinder.

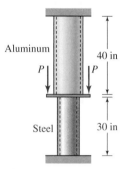

Figure P4.64

4.65 The allowable stresses in aluminum and steel are 10 ksi and 25 ksi, respectively. Determine the maximum force P that can be applied to the assembly shown in Figure P4.64.

A gap of 0.004 inch exists between the rigid bar and bar A before the force F is applied. The rigid bar is hinged at point C. The lengths of bars A and B are 30 and 50 inches respectively. Both bars have an area of cross-section $A = 1$ in² and modulus of elasticity $E = 30,000$ ksi. Use this information to solve Problems 4.66 and 4.67.

4.66 If $P = 100$ kips in Figure P4.66, determine the axial stresses in bars A and B.

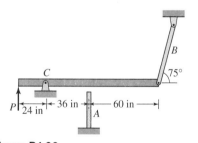

Figure P4.66

4.67 If the allowable normal stress in the bars is 20 ksi in tension or compression, determine the maximum force P that can be applied to the assembly shown in Figure P4.66.

In Figure P4.66 a gap exists between the rigid bar and rod A before force F is applied. The

rigid bar is hinged at point C. The lengths of bars A and B are 1 m and 1.5 m, and the diameters are 50 mm and 30 mm, respectively. The bars are made of steel with a modulus of elasticity $E = 200$ GPa and Poisson's ratio $v = 0.28$. Solve Problems 4.68 and 4.69.

4.68 If $F = 75$ kN in Figure P4.68, determine: (a) the deformation of the two bars; (b) the change in the diameters of the two bars.

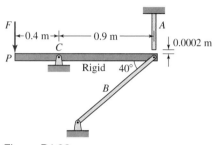

Figure P4.68

4.69 In Figure P4.68 the allowable axial stresses in bars A and B are 110 MPa and 125 MPa, respectively. Determine the maximum force F that can be applied.

A rectangular aluminum bar ($E = 10,000$ ksi), a steel bar ($E = 30,000$ ksi), and a brass bar ($E = 15,000$ ksi) form a structure, as shown in Figure P4.69. All bars have the same thickness of 0.5 in. A gap of 0.02 in exists before the load P is applied to the rigid plate. Assume that the rigid plate does not rotate. Solve Problems 4.70 and 4.71.

4.70 If $P = 15$ kips in Figure P4.70 determine: (a) the axial stress in steel; (b) the displacement of the rigid plate with respect to the right wall.

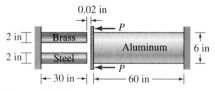

Figure P4.70

4.71 Determine the maximum load P in Figure P4.70, if the allowable axial stresses in brass, steel, and aluminum are 8 ksi, 15 ksi, and 10 ksi, respectively.

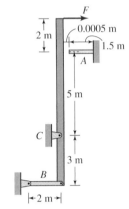

Figure P4.72

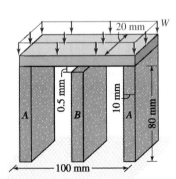

Figure P4.75

Bars A and B have cross-sectional areas of 400 mm² and a modulus of elasticity E = 200 GPa. A gap exists between bar A and the rigid bar before the force F is applied, as shown in Figure P4.72. Use this information to solve Problems 4.72 and 4.73.

4.72 The applied force F = 10 kN in Figure P4.72. Determine: (a) the axial stress in bar B; (b) the deformation of bar A.

4.73 Determine the maximum force F that can be applied if the allowable stress in member B is 120 MPa (C) and the allowable deformation of bar A is 0.25 mm.

4.74 A rectangular steel bar (E = 30,000 ksi, ν = 0.25) of 0.5-in thickness has a gap of 0.01 in between the section at D and a rigid wall before the forces are applied. Assuming that the applied forces are sufficient to close the gap, determine: (a) the movement of rigid plate at C with respect to the left wall; (b) the change in the depth d of segment CD.

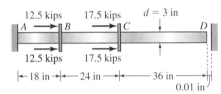

Figure P4.74

Three plastic members of equal cross sections are shown in Figure P4.75. Member B is smaller than members A by 0.5 mm. A distributed force is applied to the rigid plate, which moves downward without rotating. The moduli of elasticity for members A and B are 1.5 GPa and 2.0 GPa, respectively. Use this information to solve Problems 4.75 and 4.76.

4.75 The distributed force W = 20 MPa in Figure P4.75. Determine the axial stress in each member.

4.76 The allowable stresses in members A and B in Figure P4.75 are 50 MPa and 30 MPa. Determine the maximum intensity of the distributed force that can be applied to the rigid plate.

4.77 An aluminum circular bar (E_{al} = 70 GPa) and a steel tapered circular bar (E_{st} = 200 GPa) are securely attached to a rigid plate on which axial forces are applied, as shown in Figure P4.77. Determine: (a) the displacement of the rigid plate; (b) the maximum axial stress in steel.

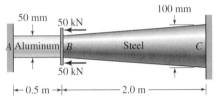

Figure P4.77

Composite bars

4.78 A 2-in × 1-in wooden bar is reinforced with $\frac{1}{4}$-in steel bars to create laminated bars. Two pieces of these laminated bars are securely fastened to a rigid plate loaded as shown in Figure P4.78. The modulus of elasticity of steel and wood are 30,000 ksi and 8000 ksi, respectively. Determine the movement of the rigid plate and the maximum stress in steel and wood.

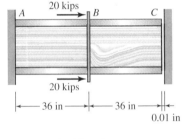

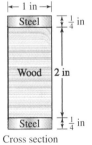

Figure P4.78

4.79 A concrete column is reinforced using nine circular iron bars of 1-in diameter, as shown in Figure P4.79. The moduli of elasticity for concrete and iron are $E_c =$ 4500 ksi and $E_i = 25,000$ ksi. Determine the maximum axial stress in concrete and iron.

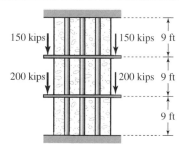

Cross section

Figure P4.79

Design problems

4.80 A rigid bar hinged at point O has a force P applied to it, as shown in Figure P4.80. Bars A and B are made of steel ($E = 30,000$ ksi). The cross-sectional areas of bars A and B are $A_A = 1$ in² and $A_B = 2$ in². If the allowable deflection at point C is 0.01 in and the allowable stress in the bars is 25 ksi, determine the maximum force P that can be applied.

4.81 The structure at the base of a crane is modeled by the pin-connected structure shown in Figure P4.81. The allowable axial

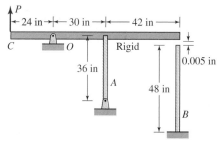

Figure P4.80

stresses in members AC and BC are 15 ksi, and the modulus of elasticity is 30,000 ksi. To ensure adequate stiffness at the base, the displacement of pin C in the vertical direction of the force is to be limited to 0.1 in. Determine the minimum cross-sectional areas for members AC and BC.

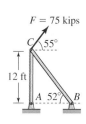

Figure P4.81

4.82 The landing wheel of a plane is modeled as shown in Figure P4.82. The pin at C is in double shear and has an allowable shear stress of 12 ksi. The allowable axial stress for link BC is 30 ksi. Determine the diameter of pin C and the effective cross-sectional area of link BC.

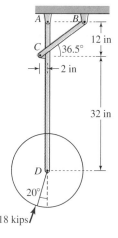

Figure P4.82

(Note: Attachments at *A* and *B* are approximated by pins to simplify analysis. There are two links represented by *BC*, one on either side of the hydraulic cylinder, which we are modeling as a single link with an effective cross-sectional area that is to be determined so that the free-body diagram is two-dimensional.)

Stress concentration

4.83 The allowable shear stress in the stepped axial rod shown in Figure P4.83 is 20 ksi. If *F* = 10 kips, determine the smallest fillet radius that can be used at section *B*. Use the stress concentration graphs given in Appendix C.

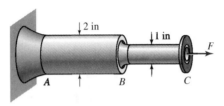

Figure P4.83

4.84 The fillet radius in the stepped circular rod shown in Figure P4.83 is 6 mm. Determine the maximum axial force *F* that can act

on the rigid wheel if the allowable axial stress is 120 MPa and the modulus of elasticity is 70 GPa. Use the stress concentration graphs given in Appendix C.

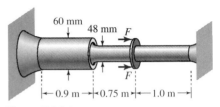

Figure P4.84

Fatigue

4.85 The fillet radius is 0.2 mm in the stepped steel circular rod shown in Figure P4.83. What should be the peak value of the cyclic load *F* to ensure a service life of one-half million cycles? Use the *S–N* curve shown in Figure 3.34.

4.86 The aluminum axial rod in Figure P4.84 is subjected to a cyclic load *F*. Determine the peak value of *F* to ensure a service life of one million cycles. Use the *S–N* curves shown in Figure 3.34 and modulus of elasticity *E* = 70 GPa.

*4.4 INITIAL STRESS OR STRAIN

Members in a statically indeterminate structure may have an initial stress or strain before the loads are applied. These initial stresses or strains may be intentional or unintentional and can be caused by several factors. A good design must account for these factors by calculating the acceptable levels of prestress.

Nuts on a bolt are usually finger-tightened to hold an assembly in place. At this stage the assembly is usually stress free. The nuts are then given additional turns to pretension the bolts. When a nut is tightened by one full rotation, the distance it moves is called the *pitch*. Alternatively pitch is the distance between two adjoining peaks on the threads. One reason for pretensioning is to prevent the nuts from becoming loose and falling off. Another reason is to introduce an initial stress that will be opposite in sign to the stress that will be generated by the loads. For example, a cable in a bridge may be pretensioned by tightening the nut and bolt systems to counter the slackening in the cable that may be caused due to wind or seasonal temperature changes.

If during assembly a member is shorter than required, then it will be forced to stretch, thus putting the entire structure into a prestress. Tolerances for the manufacture of members must be prescribed to ensure that the structure is not excessively prestressed.

In prestressed concrete, metal bars are initially stretched by applying tensile forces, and then concrete is poured over these bars. After the concrete has set, the applied tensile forces are removed. The initial prestress in the bars is redistributed, putting the concrete in compression. Concrete has good compressive strength but poor tensile strength. By prestressing the concrete it can be used in situations where it may be subjected to tensile stresses.

EXAMPLE 4.15

Bars A and B in the mechanism shown in Figure 4.47 are made of steel with a modulus of elasticity $E = 200$ GPa, a cross-sectional area $A = 100$ mm^2, and a length $L = 2.5$ m. Bar A is pulled by 3 mm to fill the gap before the force F is applied.

(a) Determine the initial axial stress in both bars.

(b) If the applied force $F = 10$ kN, determine the total axial stress in both bars.

PLAN

(a) We can use the force method to solve the problem. After the gap has been closed, the two bars will be in tension. The degree of static redundancy for this problem is 1. We can write one compatibility equation and one equilibrium equation of the moment about C and solve the problem.

(b) We can consider calculating the internal forces with just F, assuming the gap has closed and the system is stress free before F is applied. Bar B will be in compression and bar A will be in tension due to the force F. The internal forces in the bar can be found as in part (a). The initial stresses in part (a) can be superposed on the stresses due to solely F, to obtain the total axial stresses.

Solution

(a) We draw the free-body diagram of the rigid bars with bar A in tension and bar B in compression, as shown in Figure 4.48, and take the moment about point C,

$$N_A(5) = N_B(2) \tag{E1}$$

To get the compatibility equation we draw the approximate deformed shape in Figure 4.49. The movements of points E and D on the rigid bar can be related by similar triangles. The movement of point E is equal to the deformation of bar B, and the extension of bar A and the movement of point D are equal to the gap, as shown.

$$\frac{\delta_D}{5} = \frac{\delta_B}{2} \tag{E2}$$

$$\delta_D + \delta_A = 0.003 \tag{E3}$$

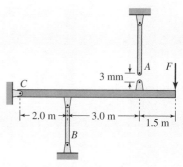

Figure 4.47 Two-bar mechanism in Example 4.15.

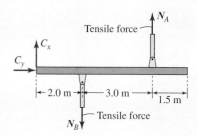

Figure 4.48 Free-body diagram for part (a) in Example 4.15.

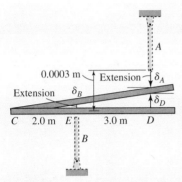

Figure 4.49 Deformed geometry in Example 4.15.

From Equations (E2) and (E3) we obtain

$$2.5\delta_B + \delta_A = 0.003 \tag{E4}$$

From Equation (4.27) we obtain

$$\delta_A = \frac{N_A L_A}{E_A A_A} = \frac{N_A \times 2.5}{20 \times 10^6} = 0.125 N_A \times 10^{-6} \tag{E5}$$

$$\delta_B = \frac{N_B L_B}{E_B A_B} = \frac{N_B \times 2.5}{20 \times 10^6} = 0.125 N_B \times 10^{-6} \tag{E6}$$

Substituting Equations (E5) and (E6) into Equation (E4) we obtain

$$2.5(0.125 N_B \times 10^{-6}) + 0.125 N_A \times 10^{-6} = 0.003$$

or

$$2.5 N_B + N_A = 24{,}000 \tag{E7}$$

Solving Equations (E1) and (E7) we obtain

$$N_A = 3310.3 \text{ N} \qquad N_B = 8275.9 \text{ N}$$

The stresses in A and B can now be found as

$$\sigma_A = \frac{N_A}{A_A} = 33.1 \times 10^6 \tag{E8}$$

$$\text{ANS.} \quad \sigma_A = 33 \text{ MPa (T)}$$

$$\sigma_B = \frac{N_B}{A_B} = 82.7 \times 10^6 \tag{E9}$$

$$\text{ANS.} \quad \sigma_B = 83 \text{ MPa (T)}$$

(b) In the calculations that follow, the purpose of the overbars is to distinguish the variables from those in part (a). We draw the free-body diagram of the rigid bars in Figure 4.50, with bars A in tension and bar B in compression, and we take the moment about point C,

$$F \times 6.5 - \bar{N}_A(5) - \bar{N}_B(2) = 0$$

or

$$5\bar{N}_A + 2\bar{N}_B = 65 \times 10^3 \tag{E10}$$

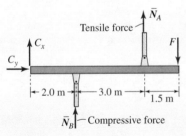

Figure 4.50 Free-body diagram for part (b) in Example 4.15.

To get the compatibility equation we draw the approximate deformed shape in Figure 4.51. For this part of the problem the movements of points D and E are equal to the deformation of the bar. We use it in writing the deformation relationships,

$$\frac{\bar{\delta}_A}{5} = \frac{\bar{\delta}_B}{2} \tag{E11}$$

The relation between deformation and internal forces is as before, as shown in Equations (E5) and (E6). Substituting Equations (E5) and (E6) into Equation (E11) we obtain

$$0.125\bar{N}_B \times 10^{-6} = 0.4(0.125\bar{N}_A \times 10^{-6})$$

or

$$\bar{N}_B = 0.4\bar{N}_A \tag{E12}$$

Solving Equations (E10) and (E12) we obtain

$$\bar{N}_A = 11.20 \times 10^3 \qquad \bar{N}_B = 4.48 \times 10^3$$

The stresses in A and B can now be found,

$$\bar{\sigma}_A = \frac{\bar{N}_A}{A_A} = 112 \text{ MPa (T)} \tag{E13}$$

$$\bar{\sigma}_B = \frac{\bar{N}_B}{A_B} = 44.8 \times 10^6 = 45 \text{ MPa (C)} \tag{E14}$$

The total axial stress can now be obtained by superposing the stresses in Equations (E8) and (E13) for bar A and those in Equations (E9) and (E14) for bar B,

$$(\sigma_A)_{\text{total}} = 145 \text{ MPa (T)} \qquad (\sigma_B)_{\text{total}} = 38 \text{ MPa (T)}$$

COMMENTS

1. We solved the problem twice, to incorporate the initial stress (strain) due to misfit and then to account for the external load. Since the problem is linear, it should not matter how we reach the final equilibrium position. In the next section we will see that it is possible to solve the problem only once, but it would require an understanding of how initial strain is accounted for in the theory.

2. Consider the following problem, shown in Figure 4.52. After the nut is fingertight, it is given an additional quarter turn before the

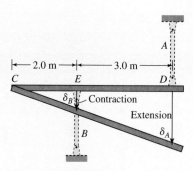

Figure 4.51 Deformed geometry for part (b) in Example 4.15.

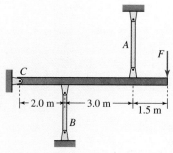

Figure 4.52 Problem similar to Example 4.15.

force F is applied. We are required to find the initial axial stress in both bars and the total axial stress. The pitch of the threads is 12 mm. The mechanisms of introducing the initial strains are different for the problems in Figures 4.47 and 4.52, but the results of the two problems are identical. To appreciate this, we recognize that the nut moves by pitch times the number of turns, that is, 3 mm. In both problems let us initially ignore bar B and force F. Then in both problems the rigid bar at point D moves upward by 3 mm, but at this stage, both bar A and bar B are stress free. Now suppose that by applying a force, we stretch bar B and attach it to the rigid bar at point E. On removal of the force from bar B, the system would seek an equilibrium, which will be dictated by the properties of bars A and B and not by how we initially raised point D. The strain due to the tightening of a nut may be hard to visualize, but the analogous problem of strain due to misfit can be visualized and used as an alternative visualization aid.

*4.5 TEMPERATURE EFFECTS

Length changes due to temperature variations introduce stresses caused by the constraining effects of other members in a statically indeterminate structure. For the purposes of analysis, there are a number of similarities between initial strain and thermal strain. Thus we shall rederive our theory to incorporate initial strain and see how the initial strain affects our analysis. We start with the position that plane sections remain plane and parallel and we have small strain, that is, Assumptions 1 and 2 are valid. Hence the total strain at any cross section is uniform and only a function of x, as in Equation (4.2). Assuming a material is isotropic and linearly elastic (Assumptions 3 and 4 are valid) but due to temperature or any other factor we have an initial strain ε_0 at a point, we obtain the following equation for the stress–strain relationship:

$$\varepsilon_{xx} = \frac{du}{dx} = \frac{\sigma_{xx}}{E} + \varepsilon_0 \tag{4.28}$$

Substituting Equation (4.28) into Equation (4.4) and assuming that the material is homogeneous and the initial strain ε_0 is uniform across the cross section, we obtain

$$N = \int_A \left(E\frac{du}{dx} - E\varepsilon_0 \right) dA = \frac{du}{dx}\int_A E\, dA - \int_A E\varepsilon_0\, dA = \frac{du}{dx}EA - EA\varepsilon_0 \tag{4.29}$$

or

$$\frac{du}{dx} = \frac{N}{EA} + \varepsilon_0 \tag{4.30}$$

Substituting Equation (4.30) into Equation (4.28), we obtain the familiar relationship

$$\sigma_{xx} = \frac{N}{A} \qquad (4.31)$$

If Assumptions 7 through 9 are valid, and if ε_0 does not change with x, then all quantities on the right-hand side of Equation (4.31) are constant between x_1 and x_2, and we obtain by integration

$$u_2 - u_1 = \frac{N(x_2 - x_1)}{EA} + \varepsilon_0(x_2 - x_1)$$

or alternatively,

$$\boxed{\delta = \frac{NL}{EA} + \varepsilon_0 L} \qquad (4.32)$$

Equations (4.31) and (4.32) imply that the initial strain affects the deformation but does not affect the stresses. This seemingly paradoxical result has different explanations for the thermal strains and for strains due to misfits or to pretensioning of the bolts.

First we consider the strain ε_0 due to temperature changes. If a body is homogeneous and unconstrained, then no stresses are generated due to temperature changes, as observed in Chapter 3. This observation is equally true for statically determinate structures. The determinate structure simply expands or adjusts to account for the temperature changes. But in an indeterminate structure, the deformation of various members must satisfy the compatibility equations. The compatibility constraints cause the internal forces to be generated, which in turn affects the stresses.

In thermal analysis $\varepsilon_0 = \alpha \Delta T$. An increase in temperature corresponds to extension, whereas a decrease in temperature corresponds to contraction. Equation (4.32) assumes that N is positive in tension, and hence extensions due to ε_0 are positive and contractions are negative. However, if on the free-body diagram N is shown as a compressive force, then δ is shown as contraction in the deformed shape. Consistency requires that contraction due to ε_0 be treated as positive and extension as negative in Equation (4.32). The sign of $\varepsilon_0 L$ due to temperature changes must be consistent with the force N shown on the free-body diagram.

We now consider the issue of initial strains caused by factors discussed in Section 4.4. If we start our analysis with the undeformed geometry even when there is an initial strain or stress, then the implication is that we have imposed a strain that is opposite in sign to the actual initial strain before imposing external loads. To elaborate this issue of sign, we put $\delta = 0$ in Equation (4.32) to correspond to the undeformed state, and note that N and ε_0 must have opposite signs for the two terms on the right-hand side to combine, yielding a result of zero. But strain and internal forces must have the same sign. For example, if a member is short and has to be pulled to overcome a gap due misfit, then at the undeformed state the bar has been extended and is in tension before external loads are applied. The problem can be corrected only if we think of ε_0 as negative to the actual initial strain. Thus prestrains (stresses) can be analyzed by using ε_0 as negative to the actual initial strain in Equation (4.32).

If in a problem we have external forces in addition to the initial strain, then we can solve the problem in two ways. We can find the stresses and the deformation due to initial strain and due to external forces individually, as we did in Section 4.4, and superpose the solution. The advantage of such an approach is that we have a good intuitive feel for the solution process. The disadvantage is that we have to solve the problem twice. Alternatively we could use Equation (4.32) and solve the problem once, but we need to be careful with our signs, and the approach is less intuitive and more mathematical.

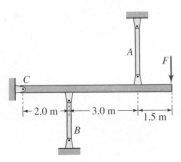

Figure 4.53 Two-bar mechanism in Example 4.16.

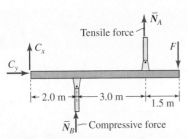

Figure 4.54 Free-body diagram in Example 4.16.

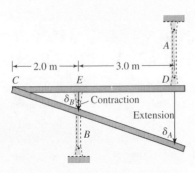

Figure 4.55 Deformed geometry in Example 4.16.

EXAMPLE 4.16

Bars A and B in the mechanism shown in Figure 4.53 are made of steel with a modulus of elasticity $E = 200$ GPa, a coefficient of thermal expansion $\alpha = 12 \ \mu/°C$, a cross-sectional area $A = 100 \ mm^2$, and a length $L = 2.5$ m. If the applied force $F = 10$ kN and the temperature of bar A is decreased by 100°C, find the total axial stress in both bars.

PLAN

We can use the force method to solve this problem. The problem has 1 degree of redundancy. We can write one compatibility equation and, using Equation (4.32), get one equation relating the internal forces. By taking the moment about point C in the free-body diagram of the rigid bar, we can obtain the remaining equation and solve the problem.

Solution   The axial rigidity and the thermal strain can be calculated as shown,

$$EA = 200 \times 10^9 \times 100 \times 10^{-6} = 20 \times 10^6$$

$$\varepsilon_0 = \alpha \Delta T = 12 \times 10^{-6} \times -100 = -1200 \times 10^{-6}$$

We draw the free-body diagram of the rigid bar in Figure 4.54, with bar A in tension and bar B in compression, and take the moment about point C,

$$F(6.5) - N_A(5) - N_B(2) = 0$$

or

$$5N_A + 2N_B = 65 \times 10^3 \tag{E1}$$

To get the compatibility equation we draw the approximate deformed shape in Figure 4.55. In this problem the movements of points D and E are equal to the deformation of the bar. We use this in writing the deformation relationships,

$$\frac{\delta_A}{5} = \frac{\delta_B}{2} \tag{E2}$$

From Equation (4.32) we obtain

$$\delta_A = \frac{N_A L_A}{E_A A_A} + \varepsilon_0 L_A = \frac{N_A(2.5)}{20 \times 10^6} - 1200 \times 2.5 \times 10^{-6}$$

$$= (0.125 N_A - 3000) 10^{-6} \tag{E3}$$

$$\delta_B = \frac{N_B L_B}{E_B A_B} = \frac{N_B(2.5)}{20 \times 10^6} = 0.125 N_B \times 10^{-6} \qquad \text{(E4)}$$

Substituting Equations (E3) and (E4) into Equation (E2) we obtain

$$0.125 N_B \times 10^{-6} = 0.4(0.125 N_A - 3000)10^{-6}$$

or

$$N_B = 0.4 N_A - 9600 \qquad \text{(E5)}$$

Solving Equations (E1) and (E5) we obtain

$$N_A = 14.52 \times 10^3 \qquad N_B = -3.79 \times 10^3 \qquad \text{(E6)}$$

Noting that we assumed that bar B is in compression, the sign of N_B in Equation (E6) implies that it is in tension. The stresses in A and B can now be found by dividing the internal forces by the cross-sectional area to obtain

ANS. $\sigma_A = 145$ MPa (T) $\sigma_B = 38$ MPa (T)   (E7)

COMMENTS

1. In Figure 4.47 the prestrain in member A is $0.0003/2.5 = 1200 \times 10^{-6}$ extension. This means that $\varepsilon_0 = -1200 \times 10^{-6}$. Substituting this value we obtain Equation (E3). Nor will any other equation in this example change for problems represented by Figures 4.47 and 4.52. Thus it is not surprising that the results of this example are identical to those of Example 4.15. But unlike Example 4.15, we solved the problem only once.

2. It would be hard to guess intuitively that bar B will be in tension because the initial strain is greater than the strain caused by the external force F. But this observation is obvious in the two solutions obtained in Example 4.15.

3. To calculate the initial strain using the method in this example, it is recommended that the problem be formulated initially in terms of the force F. Then to calculate initial strain, substitute $F = 0$. This recommendation avoids some of the confusion that will be caused by a change of the sign of ε_0 in the initial strain calculations.

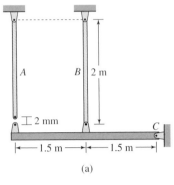

(a)

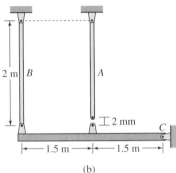

(b)

Figure P4.88

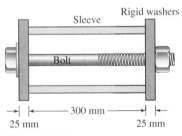

Figure P4.89

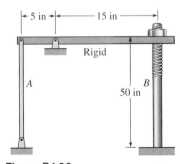

Figure P4.90

PROBLEM SET 4.4

Initial strains

4.87 During assembly of a structure, a misfit between bar A and the attachment of the rigid bar was found, as shown in Figure P4.87. If bar A is pulled and attached, determine the initial stress introduced due to the misfit. The modulus of elasticity of the circular bars A and B is $E = 10,000$ ksi and the diameter is 1 in.

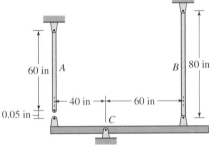

Figure P4.87

4.88 Bar A was manufactured 2 mm shorter than bar B due to an error. The attachment of these bars to the rigid bar would cause a misfit of 2 mm. Calculate the initial stress for each assembly, shown in Figure P4.88. Which of the two assembly configurations would you recommend? Use a modulus of elasticity $E = 70$ GPa and a diameter of 25 mm for the circular bars.

4.89 A steel bolt is passed through an aluminum sleeve as shown in Figure P4.89. After assembling the unit by finger-tightening (no deformation) the nut is given a $\frac{1}{4}$ turn. If the pitch of the threads is 0.3 mm, determine the initial axial stress developed in the sleeve and the bolt. The moduli of elasticity for steel and aluminum are $E_{st} = 200$ GPa and $E_{al} = 70$ GPa and the cross-sectional areas are $A_{st} = 500$ mm² and $A_{al} = 1100$ mm².

4.90 The rigid bar shown in Figure P4.90 is horizontal when the unit is put together by finger-tightening the nut. The pitch of the threads is 0.125 in. The properties of the bars are listed in Table P4.90. Develop a table in steps of quarter turns of the nut that can be used for prescribing the pretension in bar B. The maximum number of quarter turns is limited by the yield stress.

TABLE P4.90 Material properties

	Bar A	Bar B
Modulus of elasticity	10,000 ksi	30,000 ksi
Yield stress	24 ksi	30 ksi
Cross-sectional area	0.5 in²	0.75 in²

Temperature effects

In Problems 4.91 and 4.92 the increase in temperature varies as $\Delta T = T_L x^2 / L^2$. Determine the axial stress and the movement of a point at $x = L/2$ in terms of the length L, the modulus of elasticity E, the cross-sectional area A, the coefficient of thermal expansion α, and the increase in temperature at the end T_L.

4.91 (Figure P4.91).

Figure P4.91

4.92 (Figure P4.92).

Figure P4.92

4.93 The tapered bar shown in Figure P4.93 has a cross-sectional area that varies with x as $A = K(L - 0.5x)^2$. If the increase in temperature of the bar varies as $\Delta T = T_L x^2 / L^2$, determine the axial stress at midpoint in terms of the length L, the modulus of elasticity E, the cross-sectional area A, the parameter K, the coefficient of thermal expansion α, and the increase in temperature at the end T_L.

Figure P4.93

4.94 Three metallic rods are attached to a rigid plate, as shown in Figure P4.94. The temperature of the rods is lowered by 100°F after the forces are applied. Assuming the rigid plate does not rotate, determine the movement of the rigid plate. The material properties are listed in Table P4.94.

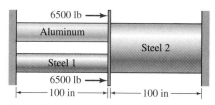

Figure P4.94

4.95 Solve Problem 4.89 assuming that in addition to turning the nut, the temperature of the assembled unit is raised by 40°C. The

TABLE P4.94 Material properties

	Area (in²)	E (ksi)	α (10^{-6}/ F)
Aluminum	4	10,000	12.5
Steel 1	4	30,000	6.6
Steel 2	12	30,000	6.6

coefficients of thermal expansion for steel and aluminum are $\alpha_{st} = 12 \, \mu/°C$ and $\alpha_{al} = 22.5 \, \mu/°C$.

4.96 Determine the axial stress in bar A of Problem 4.90 assuming that the nut is turned 1 full turn and the temperature of bar A is decreased by 80°F. The coefficient of thermal expansion for bar A is $\alpha_{st} = 22.5 \, \mu/°F$.

*4.6 ELASTIC–PERFECTLY PLASTIC AXIAL MEMBERS

Figure 4.56 shows the stress–strain curves describing an elastic–perfectly plastic behavior of a material that was discussed in Section 3.11.1. It is assumed that the material has the same behavior in tension and in compression.

The axial stress is uniform across the cross section. Thus when an axial member in a structure reaches yield stress, then the entire member will become plastic. As can be seen in Figure 4.56, the member that has turned plastic can carry stress equal to the yield stress and can elongate or compress to any amount needed to satisfy the compatibility equations. It is the compatibility equations (continuity of displacement) that are used to determine the elongation or strain in the axial member that has turned plastic.

If we plot the applied load versus the displacement of the point where the load is applied, we obtain the load-deflection curve. In the elastic region the load-deflection curve is a straight line. When one member in a structure goes plastic, the deformation is controlled by the other members, but since the stress is constant in the plastic member, there is a change in the slope of the load-deflection curve. The slope of the load-deflection curve changes each time a member in a structure goes plastic. But if a sufficient number of members go plastic, then we may have uncontrolled deformation and the structure is said to have collapsed. The load at which the structure collapses is called the collapse load. For the simple structures that we will consider, the collapse load would correspond to all members turning plastic. But for more complex structures, a structure can collapse without requiring all members to turn plastic. There exists an analysis technique called limit analysis, which can be used for determining the collapse load, but it is beyond the scope of this book.

We record the following for future use:

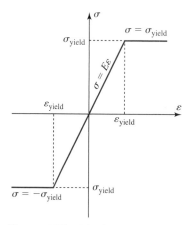

Figure 4.56 Elastic–perfectly plastic material behavior.

Definition 8 The plot of the applied force versus the deflection at that point in the direction of the applied force is called *load-deflection curve*.

Definition 9 The load at which the structure exhibits unbounded deformation is called *collapse load*.

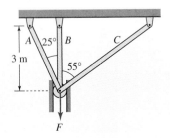

Figure 4.57 Geometry in Example 4.17.

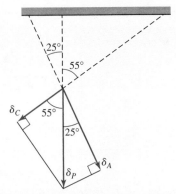

Figure 4.58 Deformation relationship in Example 4.17.

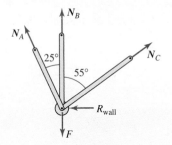

Figure 4.59 Equilibrium of forces in Example 4.17.

EXAMPLE 4.17

The three steel members shown in Figure 4.57 have a cross-sectional area $A = 100$ mm^2, a modulus of elasticity $E = 200$ GPa, and a yield stress $\sigma_{\text{yield}} = 250$ MPa. The three members are attached to a roller that can slide in the slot shown. Plot the load-deflection curve and determine the collapse load.

PLAN

Using the displacement of the roller δ_P, we can write the deformation of each bar in terms δ_P, that is, write the compatibility equations. Using a free-body diagram of the roller we can relate the internal forces in the member to the external forces, namely, write the equilibrium equations. Using Equation (4.27) we can relate the internal forces to the deformation of each bar. By substituting the compatibility equations we can obtain the relationship between internal forces and δ_P. As the yield stresses and the cross-sectional areas for all members are the same, the member that will reach the yield stress first will be the one that has the greatest internal force. At the yield stress of the member the corresponding F and δ_P can be found. The deformation relation and the equilibrium relationship between the internal forces do not change as the load is increased, but the stress in the member that turned plastic is set equal to the yield stress. We repeat the calculations for each member until all members turn plastic.

Solution The length of bar B is given as $L_B = 3$ m, and the lengths of bars A and C can be found from geometry as $L_A = 3/\cos 25 = 3.31$ m and $L_C = 3/\cos 55 = 5.23$ m. The axial rigidity for all three members is $EA = 200 \times 10^9 \times 100 \times 10^{-6} = 20 \times 10^6$ N.

Compatibility equations: We draw an exaggerated deformed shape in Figure 4.58 and relate the displacement of the pin to the deformation of the bars using small-strain approximation,

$$\delta_A = \delta_P \cos 25 = 0.9063 \, \delta_P \tag{E1}$$

$$\delta_B = \delta_P \tag{E2}$$

$$\delta_C = \delta_P \cos 55 = 0.5736 \, \delta_P \tag{E3}$$

Equilibrium equations: We can relate the internal forces in the members to the force F by drawing the free-body diagram of the roller in Figure 4.59 and balancing forces in the y direction. Notice that by balancing forces in the x direction we would include the wall reaction as an additional unknown, and since this is not of interest to us, we will not write the force equilibrium in the x direction.

$$N_A \cos 25 + N_B + N_C \cos 55 - F = 0$$

or

$$0.9063N_A + N_B + 0.5736N_C = F \qquad \text{(E4)}$$

Assuming linear elastic behavior for all members, we can write the internal force in terms of the member deformation using Equation (4.27). Next we substitute Equations (E1), (E2), and (E3) and obtain the internal forces in terms of δ_P,

$$N_A = \frac{E_A A_A}{L_A}\delta_A = \frac{20 \times 10^6}{3.31}0.9063\,\delta_P = 5.476 \times 10^6 \delta_P \qquad \text{(E5)}$$

$$N_B = \frac{E_B A_B}{L_B}\delta_B = \frac{20 \times 10^6}{3}\delta_P = 6.667 \times 10^6 \delta_P \qquad \text{(E6)}$$

$$N_C = \frac{E_C A_C}{L_C}\delta_C = \frac{20 \times 10^6}{5.23}0.5736\,\delta_P = 2.1934 \times 10^6 \delta_P \qquad \text{(E7)}$$

First member turns plastic: From Equations (E5), (E6), and (E7) we see that the internal force in B is maximum. With all members having the same cross-sectional area and yield stress, we know that bar B will be the first to turn plastic. The internal force in B at the yield stress can be obtained as

$$N_B = \sigma_{\text{yield}}A_B = 250 \times 10^6 \times 100 \times 10^{-6} = 25 \times 10^3 \text{ N} \qquad \text{(E8)}$$

At yield stress, Equation (E6) is still valid and we can find the value of δ_{P1} at which member B turns plastic,

$$\delta_{P1} = \frac{25 \times 10^3}{6.667 \times 10^6} = 3.75 \times 10^{-3} \text{ m} \qquad \text{(E9)}$$

Substituting Equation (E9) into Equations (E5) and (E7), we obtain the internal forces in members A and C,

$$N_A = 5.476 \times 10^6 \times 3.75 \times 10^{-3} = 20.53 \times 10^3 \text{ N} \qquad \text{(E10)}$$

$$N_C = 2.1934 \times 10^6 \times 3.75 \times 10^{-3} = 8.225 \times 10^3 \text{ N} \qquad \text{(E11)}$$

Substituting Equations (E8), (E10), and (E11) into Equation (E4), we obtain the force F_1 at which member B turned plastic,

$$0.9063 \times 20.53 \times 10^3 + 25 \times 10^3 + 0.5736 \times 8.225 \times 10^3 = F_1$$

or

$$F_1 = 48.3 \text{ kN} \qquad \text{(E12)}$$

Second member turns plastic: From Equations (E5) and (E7), and the fact that the cross-sectional area and the yield stress are the same, we know that A will turn plastic before C. The internal force in A will be

$$N_A = \sigma_{yield} A_A = 250 \times 10^6 \times 100 \times 10^{-6} = 25 \times 10^3 \text{ N} \quad \text{(E13)}$$

At yield stress, Equation (E5) is still valid and we can find the value of δ_{P2} at which member A turns plastic,

$$\delta_{P2} = \frac{25 \times 10^3}{5.476 \times 10^6} = 4.57 \times 10^{-3} \text{ m} \qquad \text{(E14)}$$

Substituting Equation (E14) into Equations (E5) and (E7), we obtain the internal forces in member C,

$$N_C = 2.1934 \times 10^6 \times 4.57 \times 10^{-3} = 10.01 \times 10^3 \text{ N} \quad \text{(E15)}$$

Substituting Equations (E8), (E13), and (E15) into Equation (E4), we obtain the force F_2 at which member A turned plastic,

$$0.9063 \times 25 \times 10^3 + 25 \times 10^3 + 0.5736 \times 10.01 \times 10^3 = F_2$$

or

$$F_2 = 53.4 \text{ kN} \qquad \text{(E16)}$$

Third member turns plastic: The internal force in C at yield stress can be obtained as

$$N_C = \sigma_{yield} A_C = 250 \times 10^6 \times 100 \times 10^{-6} = 25 \times 10^3 \text{ N} \quad \text{(E17)}$$

At yield stress, Equation (E7) is still valid and we can find the value of δ_{P3} at which member B turns plastic,

$$\delta_{P3} = \frac{25 \times 10^3}{2.1934 \times 10^6} = 11.4 \times 10^{-3} \text{ m} \qquad \text{(E18)}$$

Substituting Equations (E8), (E13), and (E17) into Equation (E4), we obtain the force F_3 at which member C turned plastic,

$$0.9063 \times 25 \times 10^3 + 25 \times 10^3 + 0.5736 \times 25 \times 10^3 = F_3$$

or

$$F_3 = 62 \text{ kN} \tag{E19}$$

We can plot F versus δ_P using Equations (E9), (E12), (E14), (E16), (E18), and (E19) and obtain the load-deflection curve for the structure, as shown in Figure 4.60.

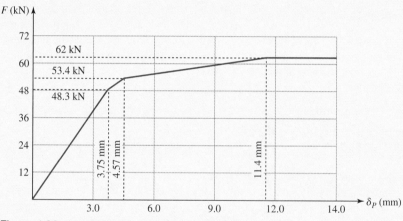

Figure 4.60   Load-deflection curve in Example 4.17.

COMMENTS

1. There is a change in slope in the load-deflection curve at loads where a member turns plastic. This is the nonlinearity in the problem.

2. The calculation for the load-deflection curve can be tedious with an increase in the number of members. However, the calculation of the collapse load is simple as it corresponds to the equilibrium condition when all members are at yield stress.

3. If each member were made from a different material and had a different cross-sectional area, then, from the internal forces in Equations (E5), (E6), and (E7), the stress in each member would have to be found and compared to the yield stress of the member before a decision can be made about which member turns plastic first.

*4.7 STRESS APPROXIMATION

Developing stress formulas by starting with a displacement approximation is a general approach, which permits the incorporation of different complexities, as demonstrated in this chapter. But there are many simple applications that are based on strength design, and deformation formulas or stiffness design are not critical issues. In these simple strength design applications we can obtain stress formulas starting with a stress approximation across the cross section, as was demonstrated in Examples 1.8 and 1.9. But how do we deduce a stress behavior across the cross section? In this section we consider the clues that we can use to deduce approximate stress behavior. In Section 4.8 we will show how to apply the ideas discussed in this section to thin-walled pressure vessels. Section 5.6 on the torsion of thin-walled tubes is another application of ideas discussed in this section.

Think of each stress component as a mathematical function that is to be approximated. The simplest approximation of a function (stress component) is to assume it to be a constant, as was done in Figure 1.16a and b. The next level of sophistication (complexity) is to assume a stress component as a linear function, as was done in Figure 1.16c and d. If we continued this line of thinking, we would next assume a quadratic or higher order of polynomials. The choice of a polynomial for approximating a stress component is dictated by several factors, some of which are discussed in this section.

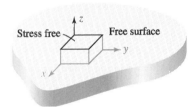

Figure 4.61 Free surface and plane stress.

4.7.1 Free Surface

The more observation points a curve passes through, the more accurate will be the prediction using such a curve. Points where stress components must go to zero are important observation points when approximating a stress component by a curve. The identification of these zero-stress points by inspection is tied intricately to the concept of a free surface.

Definition 10 A surface on which no external forces or moments are acting is called *free surface.*

A segment of a body that has no forces acting on the surface is shown in Figure 4.61. If we consider a point on the surface and draw a stress cube, then the surface with the outward normal in the z direction will have no stresses, and we have a situation of plane stress at that point. Because the points on which no forces are acting can be identified by inspection, these points provide us with a clue to making assumptions regarding stress behavior, as will be demonstrated next.

The drill shown in Figure 4.62 has point A located just outside the material that is being drilled. If we draw a stress cube at point A, then the surface shown is a free surface. Hence all stresses, including the shear stress, must go to zero. Point B is at the tip of the drill, the point at which the material is being sheared off, that is, at point B the shear stress must be equal to the shear strength of the material. Now we have two points of observation. The simplest curve that can be fitted through two points is a straight line. A linear approximation of shear stress, as shown in Figure 4.62, is a better approximation than the

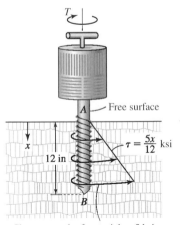

Shear strength of material = 5 ksi

Figure 4.62 Using free surface to guide stress approximation.

uniform behavior we assumed in Example 1.10. It can be confirmed that with linear shear stress behavior, the minimum torque will be 188.5 in·kips, which is half of what we obtained in Example 1.10. Only experiment can confirm whether the stress approximation in Figure 4.62 is correct. If it is not, then the experimental results would suggest other equations to consider.

4.7.2   Thin Bodies

The smaller the region of approximation, the better is the accuracy of the analytical model. If the dimensions of a cross section are small compared to the length of the body, then assuming a constant or a linear stress distribution across the cross section will introduce small errors in the calculation of internal forces and moments, such as in pins discussed in Section 1.1.2. We now consider another small region of approximation, termed thin bodies. A factor of 10 (an order of magnitude) is usually used in engineering to describe attributes such as thin, small, or large in relation to other dimensions of a body.

Figure 4.63 shows a segment of a plate with loads in the x and y directions. The top and bottom surfaces of the plate are free surfaces, that is, plane stress exists on both surfaces. This does not imply that a point in the middle of the two surfaces is also in a state of plane stress, but if the plate is thin compared to its other dimensions then to simplify analysis, it is reasonable to assume that the entire plate is in plane stress. The other stress components are usually assumed uniform or linear in the thickness direction in thin bodies.

The assumption of plane stress is made in thin bodies even when there are forces acting on one of the surfaces in the thickness direction. The assumption is justified if the maximum stresses in the xy plane turn out to be an order of magnitude greater than the applied distributed load. But the validity of the assumption can only be checked posterior, that is, after the stress formula has been developed. Examples of the latter situation are floors and ceilings of buildings and domes, and thin-walled cylindrical and spherical pressure vessels, as will be discussed in Section 4.8.

4.7.3   Axisymmetric Bodies

A necessary requirement to construct simple analytical models is to limit the model's application to a particular geometry and loading. In this chapter we considered structural members that were much longer (>10 times) than the greatest dimension across the cross section and loading that was axial. We now consider another situation in which the geometry is axisymmetric and the external loads are also axisymmetric.

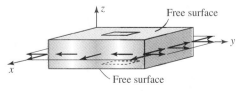

Figure 4.63   Plane stress assumption in thin plates.

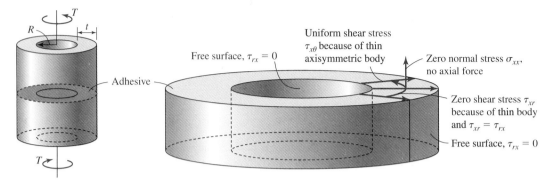

Figure 4.64 Deducing stress behavior in adhesively bonded thin cylinders.

If a body has a cross section that is symmetric about an axis and if the applied external forces or moments are also symmetric about the same axis, then the stresses cannot depend on the angular location of a point. In other words, the stress components must also be symmetric with respect to the axis. By using this argument of axisymmetry in thin bodies, we can get good stress approximation, as will be demonstrated by a simple example and further elaborated in Section 4.8.

We consider all the stress components acting in the adhesive layer between two thin cylinders subjected to a torque,[15] as shown in Figure 4.64. The shear stress in the radial direction τ_{xr} is assumed to be zero because the symmetric counterpart of this shear stress, τ_{rx}, has to be zero on the inside and outside free surfaces of this thin body. Because the problem is axisymmetric, the normal stress σ_{xx} and the tangential shear stress $\tau_{x\theta}$ cannot depend on the angular coordinate. But a uniform axial stress σ_{xx} would produce an internal axial force. Because no external axial force exists, we approximate the axial stress as zero. Because of thinness, the tangential shear stress $\tau_{x\theta}$ is assumed to be constant in the radial direction. In Section 5.6, in a similar manner, we shall deduce the behavior of the shear stress distribution in thin-walled cylindrical bodies of arbitrary cross sections.

4.7.4 Limitations

We make approximations (assumptions) to produce simple analytical models. All analytical models (mathematical representations of nature) have errors in their predictions. Whether or not an approximation is acceptable depends on the accuracy needs of the analysis and the experimental results. If all we are seeking is an order-of-magnitude value for stresses, then assuming a uniform stress behavior in most cases will give us an adequate answer. But constructing sophisticated models based only on stress approximation is difficult, if not impossible. How can we guess the piecewise uniform stress distribution on a cross section for a composite bar, as discussed in Section 4.2? Furthermore, an assumed stress distribution may correspond to a material deformation that is physically impossible. That is, we may be making a stress approximation that would

---

[15]We developed the stress formula in Example 1.9.

require holes and/or corners being formed inside the material.[16] Another difficulty is that we need to approximate six independent stress components, which are difficult to visualize and, being internal, cannot be measured directly to validate the approximation. These difficulties can be overcome by approximating the displacement that can be observed experimentally, and as was shown in this chapter, it permits us to incorporate different complexities in a very logical manner.

We conclude this section with the following observations:

1. On a free surface, a point is in plane stress.
2. Some of the stress components must tend to zero as the point approaches the free surface.
3. A state of plane stress may be assumed for thin bodies.
4. Stress components may be approximated as uniform or linear in the thickness direction for thin bodies.
5. A body that has both a load and a cross section that are symmetric about an axis, must have stresses that are also symmetric about the axis.

*4.8 THIN-WALLED PRESSURE VESSELS

Cylindrical and spherical pressure vessels are used for storage, as shown in Figure 4.65, and for the transportation of fluids and gases. The inherent symmetry and the assumption of thinness make it possible to deduce the behavior of stresses to a first approximation. The argument of symmetry implies that stresses cannot depend on the angular location. By limiting ourselves to thin walls, we can assume uniform stresses in the thickness (radial) direction. The net effect is that all shear stresses in cylindrical or

(a)   (b)

Figure 4.65   Gas storage tanks.

---

[16]A mathematically more rigorous mechanics of materials approach, called theory of elasticity, requires that stresses satisfy the compatibility equations to ensure that no physically impossible deformations are produced.

spherical coordinates are zero, the radial normal stress can be neglected, and the two remaining normal stresses are constant in the radial and circumferential directions. The two unknown stresses can be related to pressure by static equilibrium.

The "thin-wall" limitation implies that the ratio of the inner radius R to the wall thickness t is greater than 10. The higher the ratio of R/t, the better is the prediction of our analysis.

4.8.1 Cylindrical Vessels

We use the cylindrical coordinate system (r, θ, x), as shown in the Figure 4.66. The stress element on the right shows the stress components in the cylindrical coordinate system on four surfaces. The outer surface of the cylinder is stress free. Hence the shear stresses $\tau_{r\theta}$ and τ_{rx} and the normal stress σ_{rr} are all zero on the outer surface (at A). On the inner surface (at B) there is only a radial force due to pressure but there are no tangential forces. Hence on the inner surface the shear stresses $\tau_{r\theta}$ and τ_{rx} are zero. Since the wall is thin, we can assume that the shear stresses $\tau_{r\theta}$ and τ_{rx} are zero across the thickness. The radial normal stress varies from a zero value on the outer surface to a value of the pressure on the inner surface. If we neglect the radial stress and calculate the normal stresses σ_{xx} and $\sigma_{\theta\theta}$, we find that σ_{rr} is an order of magnitude less than the other two normal stresses. Thus we justify the assumption of neglecting σ_{rr} posterior. A nonzero value of $\tau_{\theta x}$ will either have a resultant torque or implies that points will move in the θ direction. Since there is no applied torque, and the movement of a point cannot depend on the angular location because of symmetry, we conclude that the shear stress $\tau_{\theta x}$ will be zero.

Thus all shear stresses are zero, the radial normal stress is neglected, and the axial stress σ_{xx} and the hoop stress $\sigma_{\theta\theta}$ are assumed uniform across the thickness and across the circumference, as shown in Figure 4.67a. We could start with a differential element and find the internal forces by integrating σ_{xx} and $\sigma_{\theta\theta}$ over appropriate areas. But as these two stresses are uniform across the entire circumference, we can reach the same conclusions by considering two free-body diagrams, as shown in Figure 4.67b and c.

By equilibrium of forces on the free-body diagram in Figure 4.67b we obtain $2\sigma_{\theta\theta}(t\,dx) = p(2R)\,dx$, or

$$\sigma_{\theta\theta} = \frac{pR}{t} \tag{4.33}$$

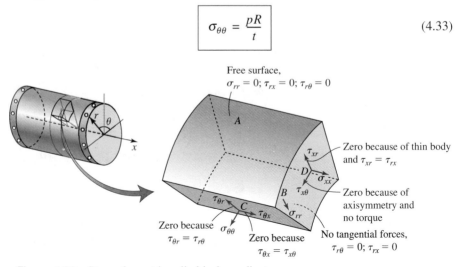

Figure 4.66 Stress element in cylindrical coordinates.

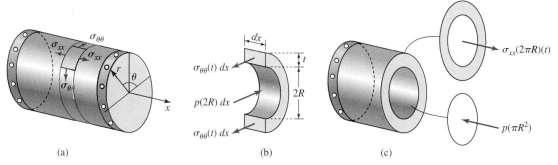

Figure 4.67 Stress analysis in thin cylindrical pressure vessels.

By equilibrium of forces on the free-body diagram in Figure 4.67c we obtain $\sigma_{xx}(2\pi R)(t) = p(\pi R^2)$, or

$$\sigma_{xx} = \frac{pR}{2t} \tag{4.34}$$

With $R/t > 10$ the stresses σ_{xx} and $\sigma_{\theta\theta}$ are greater than the maximum value of radial stress σ_{rr} $(=p)$ by factors of at least 5 and 10, respectively. This justifies our assumption of neglecting the radial stress in our analysis.

The axial stress σ_{xx} and the hoop stress $\sigma_{\theta\theta}$ are always tensile under internal pressure. The formulas may be used for small applied external pressure but with the following caution. External pressure causes compressive normal stresses that can cause the cylinder to fail due to buckling. The buckling phenomenon is discussed in Chapter 11.

Although the normal stresses are assumed not to vary in the circumferential or thickness direction, our analysis does not preclude variations in the axial direction (x direction). But the variations in the x direction must be gradual. If the variations are very rapid, then our assumption that stresses are uniform across the thickness will not be valid, as can be shown by a more rigorous three-dimensional elasticity analysis.

Our analysis will give good results on sections in the middle of the cylinder, but the accuracy will decrease as we approach the ends. The local effect of the lid attachment dominates the behavior of stresses in these regions, as elaborated in detail in Section 3.7 on stress concentration.

4.8.2 Spherical Vessels

We use the spherical coordinate system (r, θ, ϕ) for our analysis, as shown in Figure 4.68a. Proceeding in a manner similar to the analysis of cylindrical vessels, we deduce the following:

1. All shear stresses are zero, that is,

$$\tau_{r\phi} = \tau_{\phi r} = 0 \qquad \tau_{r\theta} = \tau_{\theta r} = 0 \qquad \tau_{\theta\phi} = \tau_{\phi\theta} = 0$$

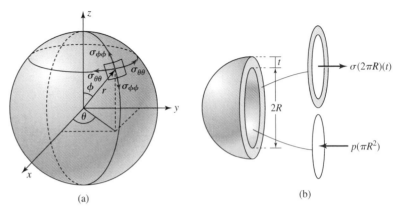

Figure 4.68 Stress analysis in thin spherical coordinates.

2. Normal radial stress σ_{rr} varies from a zero value on the outside to the value of the pressure on the inside. We will once more neglect the radial stress in our analysis and justify it posterior.

3. The normal stresses $\sigma_{\theta\theta}$ and $\sigma_{\phi\phi}$ are equal and are constant over the entire vessel. We set $\sigma_{\theta\theta} = \sigma_{\phi\phi} = \sigma$.

As all imaginary cuts through the center are the same, we consider the free-body diagram shown in Figure 4.68*b*. By equilibrium of forces we obtain $\sigma(2\pi R)(t) = p\pi R^2$, or

$$\sigma = \frac{pR}{2t} \tag{4.35}$$

With $R/t > 10$ the normal stress σ is greater than the maximum value of radial stress $\sigma_{rr}\,(=p)$ by a factor of at least 5. This justifies our assumption of neglecting the radial stress in our analysis. At each and every point the normal stress in any circumferential direction is the same for thin spherical pressure vessels.

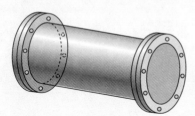

Figure 4.69 Cylindrical tank in Example 4.18.

EXAMPLE 4.18

The lid is bolted to the tank in Figure 4.69 along the flanges using 1-in-diameter bolts. The tank is made from sheet metal that is $\frac{1}{2}$ in thick and can sustain a maximum hoop stress of 24 ksi in tension. The normal stress in the bolts is to be limited to 60 ksi in tension. A manufacturer can make tanks from 2 ft in diameter to 8 ft in diameter in steps of 1 ft. Develop a table that the manufacturer can use to advise customers of the size of tank and the number of bolts per lid needed to hold a desired gas pressure.

PLAN

We can establish a relationship between the pressure p and the radius R of the tank through the limiting value on hoop stress, using Equation (4.33). We can relate the number of bolts needed by noting that the force due to pressure on the lid is carried equally by the bolts (Figure 4.70).

Solution We note that the area of the bolts is $A_{bolt} = \pi(1)^2/4 = \pi/4$. From Equation (4.33) we obtain

$$\sigma_{\theta\theta} = \frac{pR}{1/2} \leq 24{,}000 \text{ psi} \qquad \text{or} \qquad p \leq \frac{12{,}000}{R} \text{ psi}$$

Figure 4.70 shows the free-body diagram of the lid. By equilibrium of forces in the x direction we obtain

$$nN_{bolt} = N_{lid}$$

$$\sigma_{bolt} = \frac{4pR^2}{n} \leq 60{,}000$$

Substituting for p we obtain

$$\frac{4 \times 12{,}000R}{n} \leq 60{,}000 \qquad \text{or} \qquad n \geq 0.8R$$

Writing the inequalities that p and n must satisfy in terms of the diameter D of the tank, we obtain

$$p \leq 24{,}000/D \qquad n \geq 0.4D$$

We consider the values of D from 24 in to 96 in. in steps of 12 in and calculate the values of p and n. We report the values of p by rounding downward to the nearest integer that is a factor of 5, and the values of n are reported by rounding upward to the nearest integer, as given in Table 4.3.[17]

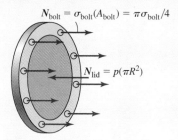

$$N_{bolt} = \sigma_{bolt}(A_{bolt}) = \pi\sigma_{bolt}/4$$

$$N_{lid} = p(\pi R^2)$$

Figure 4.70 Relating forces in bolts and lid in Example 4.18.

TABLE 4.3 Results of Example 4.18

Tank Diameter D (ft)	Maximum Pressure p (psi)	Minimum Number of Bolts n
2	1000	10
3	665	15
4	500	20
5	400	24
6	330	30
7	280	34
8	250	39

---

[17]Why are we rounding downward for the pressure and upward for the number of bolts?

PROBLEM SET 4.5

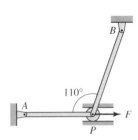

Figure P4.97

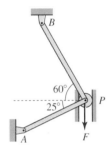

Figure P4.98

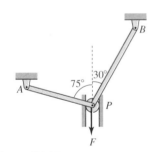

Figure P4.99

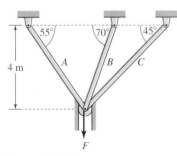

Figure P4.100

Elastic–perfectly plastic response of axial members

In Problems 4.97 through 4.99 a force F is applied to the roller that slides inside a slot. Both bars have a cross-sectional area $A = 100 \text{ mm}^2$, a modulus of elasticity $E = 200$ GPa, and a yield stress of 250 MPa. Bars AP and BP have lengths $L_{AP} = 200$ mm and $L_{BP} = 250$ mm. Draw the load-deflection curve and determine the collapse load.

4.97   Draw the load-deflection curve and determine the collapse load in Figure P4.97.

4.98   Draw the load-deflection curve and determine the collapse load in Figure P4.98.

4.99   Draw the load-deflection curve and determine the collapse load in Figure P4.99.

4.100   Three steel members shown in Figure P4.100 have a cross-sectional area $A = 100 \text{ mm}^2$, a modulus of elasticity $E = 200$ GPa, and a yield stress of 250 MPa. The three members are attached to a roller that can slide in the slot shown. Draw the load-deflection curve and determine the collapse load.

4.101   Three steel bars A, B, and C ($E = 200$ GPa, $\sigma_{yield} = 200$ MPa) have lengths $L_A = 4$ m, $L_B = 3$ m, and $L_C = 2$ m. All bars have the same cross-sectional areas of 500 mm^2. Draw the load-deflection curve for the structure and determine the collapse load.

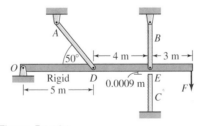

Figure P4.101

4.102   Draw the load-deflection curve and determine the collapse load for the two-bar linkage shown in Figure P4.102. Use the data given in Table P4.102.

4.103   Three poles are pin connected to a ring at D and to supports on the ground. The

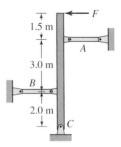

Figure P4.102

TABLE P4.102   Material properties

	Bar *A*	Bar *B*
Modulus of elasticity	200 GPa	170 GPa
Yield stress	250 MPa	220 MPa
Cross-sectional area	100 mm²	75 mm²
Length	1.5 m	1.5 m

ring slides on a vertical rigid pole. The coordinates of the four points are given in Figure P4.103. All poles have cross-sectional areas $A = 1 \text{ in}^2$, a modulus of elasticity $E = 10,000$ ksi, and a yield stress $\sigma_{yield} = 40$ ksi. Draw the load-deflection curve and determine the collapse load.

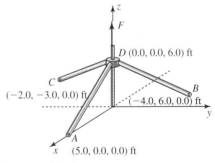

Figure P4.103

Thin-walled pressure vessels

4.104   Fifty rivets of 10-mm diameter are used for attaching caps at each end on a 1000-mm mean diameter cylinder, as shown in Figure P4.104. The wall of the cylinder is 10 mm thick and the gas pressure is 200 kPa. Determine the hoop stress and the axial stress in the cylinder and the shear stress in each rivet.

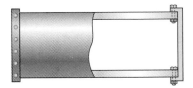

Figure P4.104

4.105 A pressure tank 15 ft long and 40 in in mean diameter is to be fabricated from a $\frac{1}{2}$-in-thick sheet. A 15-ft-long, 8-in-wide, $\frac{1}{2}$-in-thick plate is bonded onto the tank to seal the gap, as shown in Figure P4.105. What is the shear stress in the adhesive when the pressure in the tank is 75 psi? Assume uniform shear stress over the entire inner surface of the attaching plate.

Figure P4.105

Design problems

4.106 A 5-ft mean diameter spherical tank has a wall thickness of $\frac{3}{4}$ in. If the maximum normal stress is not to exceed 10 ksi, determine the maximum permissible pressure.

4.107 In a spherical tank having a 500-mm mean radius and a thickness of 40 mm, a hole of 50-mm diameter is drilled and then plugged using adhesive of 1.2-MPa shear strength to form a safety pressure release mechanism (Figure P4.107). Determine the

maximum allowable pressure and the corresponding hoop stress in the tank material.

Figure P4.107

4.108 A 20-in mean diameter pressure cooker is to be designed for a 15-psi pressure (Figure P4.108). The allowable normal stress in the cylindrical pressure cooker is to be limited to 3 ksi. Determine the minimum wall thickness of the pressure cooker. A $\frac{1}{2}$-lb weight on top of the nozzle is used to control the pressure in the cooker. Determine the diameter d of the nozzle.

4.109 The cylindrical gas tank shown in Figure P4.109 is made from 8-mm-thick sheet metal and must be designed to sustain a maximum normal stress of 100 MPa. Develop a table of maximum permissible gas pressures and the corresponding mean diameters of the tank in steps of 100 mm between diameter values of 400 mm and 900 mm.

4.110 A pressure tank 15 ft long and 40 in in mean diameter is to be fabricated from a $\frac{1}{2}$-in-thick sheet. A 15-ft-long, 8-in-wide, $\frac{1}{2}$-in-thick plate is to be used for sealing the gap by using two rows of 90 rivets each. If the shear strength of the rivets is 36 ksi and the normal stress in the tank is to be limited to 20 ksi, determine the maximum pressure and the minimum diameter of the rivets that can be used.

Figure P4.108

Figure P4.109

Figure P4.110

*4.9 GENERAL INFORMATION

This section describes a powerful and very popular numerical technique called the finite-element method (FEM), which is used in engineering design and analysis. There are two versions of FEM. One version is based on the displacement method (stiffness method) and the other is based on the force method (flexibility method). Most commercial FEM computer software is based on the displacement method, which is now described briefly.

A set of linear algebraic equations is created in which the unknowns are the displacements of points called *nodes*. The algebraic equations represent the force

Linear element

Node 1   Node 2

$\longmapsto x$

Quadratic element

Node 1   Node 2   Node 3

$\longmapsto x$

$u(x) = a_0 + a_1 x$

$$u(x) = u_1 \left(\frac{x - x_2}{x_1 - x_2} \right) + u_2 \left(\frac{x - x_1}{x_2 - x_1} \right)$$

$u(x) = u_1 \phi_1(x) + u_2 \phi_2(x)$

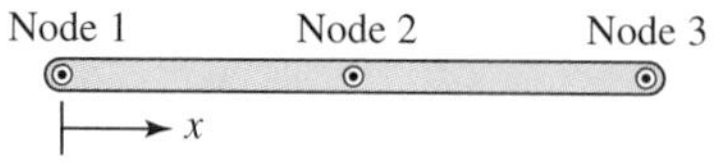

$u(x) = a_0 + a_1 x + a_2 x^2$

$$u(x) = u_1 \left(\frac{x - x_2}{x_1 - x_2} \right)\left(\frac{x - x_3}{x_1 - x_3} \right) + u_2 \left(\frac{x - x_1}{x_2 - x_1} \right)\left(\frac{x - x_3}{x_2 - x_3} \right)$$

$$+ u_3 \left(\frac{x - x_1}{x_3 - x_1} \right)\left(\frac{x - x_2}{x_3 - x_2} \right)$$

$$u(x) = u_1 \phi_1(x) + u_2 \phi_2(x) + u_3 \phi_3(x)$$

Figure 4.71 Linear and quadratic polynomial representations of displacement.

equilibrium equations at the nodes. For example, the displacements of pins (nodes) in a truss would be the unknowns and the equilibrium equations at each joint written in terms of the displacements would be the set of linear algebraic equations. In FEM, however, the equilibrium equations are derived by requiring that the nodal displacement values be such that the potential energy is minimized.[18] Before the equations of the entire body can be created, equations have to be created for small (finite) elements whose assembly represents the body. It is assumed that the displacement in an element can be described by a polynomial. Figure 4.71 shows a linear and a quadratic variation of displacement in a one-dimensional rod.

The constants a_i in the polynomial can be found in terms of the nodal displacement values u_i, as shown in Figure 4.71. The polynomial functions ϕ_i that multiply the nodal displacements are called *interpolation functions* because we can now interpolate the displacement values from the nodal values. Sometimes the same polynomial functions are also used for representing the shapes of the elements. Then the interpolation functions are also referred to as *shape functions*. When the same polynomials represent the displacement and the shape of an element, then the element is called an *isoparametric* element. Figure 4.72 shows some popular elements in two and three

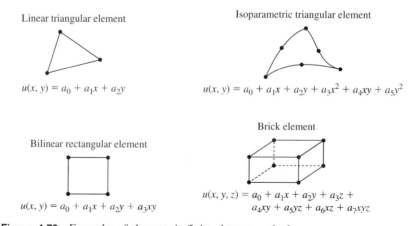

Linear triangular element

$u(x, y) = a_0 + a_1 x + a_2 y$

Isoparametric triangular element

$u(x, y) = a_0 + a_1 x + a_2 y + a_3 x^2 + a_4 xy + a_5 y^2$

Bilinear rectangular element

$u(x, y) = a_0 + a_1 x + a_2 y + a_3 xy$

Brick element

$u(x, y, z) = a_0 + a_1 x + a_2 y + a_3 z +$
$a_4 xy + a_5 yz + a_6 xz + a_7 xyz$

Figure 4.72 Examples of elements in finite-element method.

[18] A body in equilibrium has a minimum potential energy, like a marble in a trough.

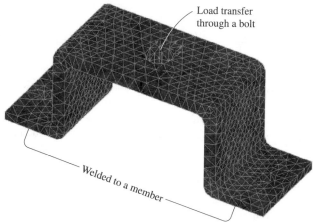

Load transfer
through a bolt

Welded to a member

Figure 4.73 Finite-element mesh of bracket. (Courtesy Professor C. R. Vilmann.)

dimensions. Strains from the displacements can be found by using Equations (2.9*a*) through (2.9*i*). The strains are substituted into potential energy, which is then minimized to generate the algebraic equations.

Figure 4.73 shows a finite-element mesh of a bracket constructed using three-dimensional tetrahedron elements. The bottom of the bracket is welded to another member and is modeled in the finite-element method as points with zero displacements. The load that is transferred through the bolt must be specified before a solution can be found. This load may be measured or estimated by some means.

4.10 CLOSURE

In this chapter we established formulas for deformations and stresses in axial members. We saw that the calculation of stresses and relative deformations requires the calculation of the internal axial force at a section. For statically determinate axial members, the internal axial force can be calculated either by making an imaginary cut and drawing an appropriate free-body diagram, or by drawing an axial force diagram.

In statically indeterminate structures there are more unknowns than there are equilibrium equations. Compatibility equations have to be generated from approximate deformed shapes to solve a statically indeterminate problem. In the displacement method the equilibrium and compatibility equations are written in terms of the deformation of axial members, or in terms of the displacements of points on the structure, and the set of equations is solved. In the force method the equilibrium and compatibility equations are written in terms of internal forces in the axial members, or in terms of the reactions at the support of the structure, and the set of equations is solved.

In Chapter 8 on stress transformation we shall consider problems in which we first find the axial stress using the stress formula in this chapter, and then we find stresses on inclined planes, including planes with maximum shear stress. In Chapter 9 on strain transformation we shall find the axial strain and the strains in the transverse direction due to Poisson's effect. We will then consider strains in different coordinate systems, including coordinate systems in which shear strain is a maximum. In Section 10.1 we shall consider the combined loading problems of axial, torsion, and bending and the design of simple structures that may be determinate or indeterminate.

POINTS AND FORMULAS TO REMEMBER

- Theory is limited to (i) slender members, (ii) regions away from regions of stress concentration, (iii) members in which the variation in cross-sectional areas and external loads is gradual, (iv) members on which axial load is applied such that there is no bending.

$$u = u(x) \quad (4.1) \qquad \text{Small strain } \varepsilon_{xx} = \frac{du(x)}{dx} \quad (4.2) \qquad N = \int_A \sigma_{xx}\, dA \quad (4.4)$$

where u is the axial displacement, which is positive in the positive x direction, ε_{xx} is the axial strain, σ_{xx} is the axial stress, and N is the internal axial force over cross section A.

- Axial strain ε_{xx} is uniform across the cross section.
- Equations (4.2) and (4.4) do not change with material model.
- Formulas below are valid for material that is linear, elastic, isotropic, with no inelastic strains:

  Homogeneous cross-section:

$$\frac{du}{dx} = \frac{N}{EA} \quad (4.8) \qquad \sigma_{xx} = \frac{N}{A} \quad (4.9) \qquad u_2 - u_1 = \frac{N(x_2 - x_1)}{EA} \quad (4.11)$$

  where EA is the axial rigidity of the cross section.

- If N, E, or A change with x, then find deformation by integration of Equation (4.8).
- If N, E, and A do not change between x_1 and x_2, then use Equation (4.11) to find deformation.
- For homogeneous cross sections all external loads must be applied at the centroid of the cross section, and centroids of all cross sections must lie on a straight line.

  Nonhomogeneous cross section:

$$(\sigma_{xx})_i = \frac{NE_i}{\sum_{j=1}^{n} E_j A_j} \quad (4.21) \qquad u_2 - u_1 = \frac{N(x_2 - x_1)}{\sum_{j=1}^{n} E_j A_j} \quad (4.22) \qquad \eta_C = \frac{\sum_{j=1}^{n} \eta_j E_j A_j}{\sum_{j=1}^{n} E_j A_j} \quad (4.23)$$

  where n is the number of materials in a cross section, E_j is the modulus of elasticity of the jth material, A_j is the cross-sectional area of the jth material, $(\sigma_{xx})_i$ is the axial stress in the ith material, η_C is the location of the point through which axial loads must pass for no bending to occur, and η_j is the location of the centroid of area A_j.

$$\text{Structural analysis:} \quad \delta = \frac{NL}{EA} \quad (4.27)$$

  where δ is the deformation in the original direction of the axial bar.

- If N is a tensile force, then δ is elongation. If N is a compressive force, then δ is contraction.
- Degree of static redundancy is the number of unknown reactions minus the number of equilibrium equations.
- If degree of static redundancy is not zero, then we have a statically indeterminate structure.

- Compatibility equations are a geometric relationship between the deformation of bars derived from the deformed shapes of the structure.

- The number of compatibility equations in the analysis of statically indeterminate structures is always equal to the degree of redundancy.

- The direction of forces drawn on the free-body diagram must be consistent with the deformation shown in the deformed shape of the structure.

- The variables necessary to describe the deformed geometry are called degrees of freedom.

- In the displacement method, the displacements of points are treated as unknowns. The number of unknowns is equal to the degrees of freedom.

- In the force method, reaction forces are the unknowns. The number of unknowns is equal to the degrees of redundancy.

CHAPTER FIVE
TORSION OF SHAFTS

5.0 OVERVIEW

Structural members used in transmitting torque from one plane to another are called *shafts*. The transfer of power from the engine to the wheels in an automobile requires many shafts in the power train, some of which are shown in Figure 5.1a. The transfer of torque (power) from the engine to the rotor blades of a helicopter is shown in Figure 5.1b. Lawn mowers, blenders, circular saws, drills, and just about any power equipment in which there is circular motion will have shafts.

The simplest theory for torsion in circular shafts will be developed rigorously following the logic shown in Figure 3.15 but subject to the limitations described in Section 3.13. The formulas from the theory will be used in the design and analysis of statically determinate and indeterminate shafts.

The two major learning objectives are:

1. Understand the theory, its limitations, and its applications for the design and analysis of torsion of circular shafts.

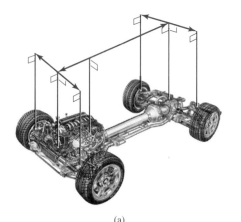

(a)

(b)

Figure 5.1   Transfer of torques between planes.

2. Develop the discipline to visualize the direction of torsional shear stress and the surface on which it acts.

5.1 PRELUDE TO THEORY

Several numerical examples are solved in this section in order to develop the observations that affect the development of the theory of torsion of circular shafts.

- Example 5.1 shows the kinematics of calculating the shear strain in torsion and the application of the logic described in Figure 3.15, using discrete bars attached to a rigid plate.

- Examples 5.2 and 5.3 show how the kinematics of calculating shear strain, which is developed for discrete bars in Example 5.1, is similar to the kinematics of calculating shear strains for continuous circular shafts.

- Example 5.4 shows the impact that the choice of a material model[1] has on the calculation of internal torque. This example also shows that the material model only impacts the stress distribution, leaving all other equations unaffected. Thus the strain distribution, which is a kinematic relationship, is unaffected; the static equivalency equation between shear stress and internal torque is unaffected; and the equilibrium equations relating internal torques to external torques are unaffected. Though we shall develop the simplest theory using Hooke's law, most of the equations that are being developed can still be used after the material model becomes more complex. This is demonstrated for composite shafts in Section 5.4.

EXAMPLE 5.1

Two thin bars of hard rubber (shear modulus $G = 280$ MPa) have cross-sectional areas of 20 mm². The bars are attached to a rigid disc of 20-mm radius (Figure 5.2). Due to the applied torque T, the rigid disc is observed to rotate by an angle of 0.04 rad about the axis of the disc. Determine the applied torque T.

PLAN

We can relate the rotation ($\Delta\phi = 0.04$) of the disc, the radius ($r = 0.02$ m) of the disc, and the length (0.2 m) of the bars to the shear strain in the bars as we did in Example 2.7. Using Hooke's law, we can find the shear stress in each bar. By assuming uniform shear stress in

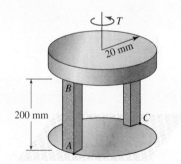

Figure 5.2 Geometry in Example 5.1.

[1]Two material models are considered: (i) a linear elastic homogeneous material cross section, and (ii) a linear elastic laminated material cross section. Example 5.16 on elastic–perfectly plastic material cross sections and Example 5.20 on nonlinear material cross sections are additional examples emphasizing the points made in this paragraph.

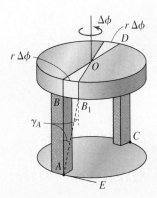

Figure 5.3 Deformed geometry in Example 5.1.

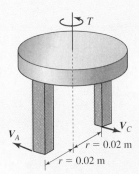

Figure 5.4 Free-body diagram in Example 5.1.

each bar, we can find the shear force. By drawing the free-body diagram of the rigid disc we can find the applied torque T.

Solution

1. *Strain calculations:* We draw the approximated deformed shape of the two bars, as shown in Figure 5.3, and calculate the shear strain as we did in Example 2.7. The two bars have the same length and are located at the same radial distance, hence the shear strain in bar C is the same as that in bar A.

$$BB_1 = 0.02\ \Delta\phi = 0.0008\ \text{m} \tag{E1}$$

$$\tan\gamma_A \approx \gamma_A = \frac{BB_1}{AB} = 0.004 \tag{E2}$$

$$\gamma_C = 0.004 \tag{E3}$$

2. *Stress calculations:* From Hooke's law we can find the shear stresses,

$$\tau_A = G_A\gamma_A = 280\times10^6\times0.004 = 1.12\times10^6\ \text{N/m}^2 \tag{E4}$$

$$\tau_C = G_C\gamma_C = 280\times10^6\times0.004 = 1.12\times10^6\ \text{N/m}^2 \tag{E5}$$

3. *Internal forces:* Assuming uniform shear stresses across the cross section, we obtain the shear forces by multiplying the shear stresses by the cross-sectional area $A = 20\times10^{-6}\ \text{m}^2$,

$$V_A = A_A\tau_A = 1.12\times10^6\times20\times10^{-6} = 22.4\ \text{N} \tag{E6}$$

$$V_C = A_C\tau_C = 1.12\times10^6\times20\times10^{-6} = 22.4\ \text{N} \tag{E7}$$

4. *External torque:* We make imaginary cuts through the bars and draw the free-body diagram of the top part as shown in Figure 5.4. By balancing the moment about the axis of the disc, we obtain the applied torque,

$$T = rV_A + rV_C \tag{E8}$$

ANS. $T = 0.896\ \text{N}\cdot\text{m}$

COMMENTS

1. In Figure 5.3 we approximated the arc BB_1 by a straight line, and we approximated the tangent function by its argument in Equation (E2). These approximations are valid only for small deformations and small strains. The net consequence of these approximations is that the shear strain along length AB_1 is uniform, as can be seen by the angle between any vertical line and line AB_1 at any point along the line.

2. The shear stress is assumed uniform across the cross section because of thin bars, but it is also uniform along the length because of the approximations described in comment 1.

3. The shear stress acts on a surface with outward normal in the direction of the length of the bar, which is also the axis of the disc. The shear force acts in the tangent direction to the circle of radius r. If we label the direction of the axis x, and the tangent direction θ, then as per the notation in Section 1.2, the shear stress is represented by $\tau_{x\theta}$.

4. The summation in Equation (E8) can be rewritten as $\sum_{i=1}^{2} r\tau \, \Delta A_i$, where τ is the shear stress acting at the radius r, and ΔA_i is the cross-sectional area of the ith bar. If we had n bars attached to the disc at the same radius, then the total torque would be given by $\sum_{i=1}^{n} r\tau \, \Delta A_i$. As we increase the number of bars n to infinity, the cross-sectional area ΔA_i will tend to zero (infinitesimal area written as dA) as we try to fit an infinite number of bars on the same circle—resulting in a continuous body with the summation replaced by an integral. In Section 5.1.1 we will formalize the observations in comments 3 and 4.

5. The visualization of a circular shaft as an assembly of bars will be developed further in the next two examples.

EXAMPLE 5.2

A rigid disc of 20-mm diameter is attached to a circular shaft made of hard rubber, as shown in Figure 5.5. The left end of the shaft is fixed into a rigid wall. The rigid disc was rotated counterclockwise by 3.25°. Determine the average shear strain at point A.

PLAN

We can visualize the shaft as made up of infinitesimally thick bars of the type shown in Example 5.1. We relate the shear strain in the bar to the rotation of the disc, as we did in Example 5.1.

Solution　We consider one line on the bar, as shown in Figure 5.6. Point B moves to point B_1. The right angle between AB and AC

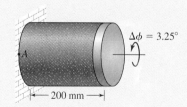

Figure 5.5　Geometry in Example 5.2.

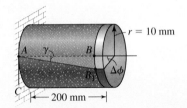

Figure 5.6 Deformed shape in Example 5.2.

changes, and the change represents the shear strain γ. We relate the shear strain and the rotation of the disc as in Example 5.1, as shown,

$$\Delta\phi = \frac{3.25° \pi}{180°} = 0.05672 \text{ rad}$$

$$BB_1 = 10 \, \Delta\phi = 0.5672 \text{ mm}$$

$$\tan \gamma = \gamma = \frac{BB_1}{AB} = \frac{0.5672}{200}$$

ANS. $\gamma = 2836 \, \mu$rad

COMMENTS

1. As in Example 5.1, we assumed that the line AB remains straight. If the assumption were not valid, then the shear strain would vary in the axial direction.

2. The change of right angle that is being measured by the shear strain is the angle between a line in the axial direction and the tangent at any point. If we designate the axial direction x and the tangent direction θ (i.e., use polar coordinates), then the shear strain with subscripts will be $\gamma_{x\theta}$.

3. The value of the shear strain does not depend on the angular position as the problem is axisymmetric.

4. If we start with a rectangular grid overlaid on the shaft, as shown in Figure 5.7a, then each rectangle will deform by the same amount, as shown in Figure 5.7b. Based on the argument of axisymmetry, we will deduce this deformation for any circular shaft under torsion in the next section.

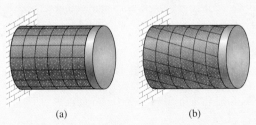

(a)   (b)

Figure 5.7 Deformation of a shaft in torsion. (*a*) Un-deformed shaft. (*b*) Deformed shaft.

EXAMPLE 5.3

Three cylindrical shafts made from hard rubber are securely fastened to rigid discs, as shown in Figure 5.8. The radii of the shaft sections are $r_{AB} = 20$ mm, $r_{CD} = 15$ mm, and $r_{EF} = 10$ mm. If the rigid discs are twisted by the angles shown, determine the average shear strain in each section assuming the lines AB, CD, and EF remain straight.

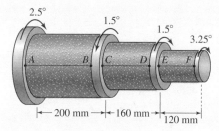

2.5° 1.5° 1.5° 3.25°

A B C D E F

← 200 mm → ← 160 mm → 120 mm

Figure 5.8 Shaft geometry in Example 5.3.

We will solve this problem by two methods. Both methods require the conversion of the rotation of each disc from degrees to radians. Let the leftmost disc be labeled disc 1 and the rightmost disc, disc 4. The rotation of each disc in radians is as follows: $\phi_1 = 0.0436$ rad, $\phi_2 = 0.0262$ rad, $\phi_3 = 0.0262$ rad, and $\phi_4 = 0.0567$ rad.

Method 1

PLAN

Each section of the shaft will undergo the deformation pattern shown in Figure 5.6, but now we need to account for the rotation of the disc at each end. We can analyze each section as we did in Example 5.2. In each section we can calculate the change of angle between the tangent and a line drawn in the axial direction at the point where we want to know the shear strain. We can then determine the sign of the shear strain using the definition of shear strain[2] in Chapter 3.

Solution We draw the approximate deformed shape of section AB in Figure 5.9 and find the shear strain,

$$AA_1 = r_{AB}\phi_1 = 0.872 \text{ mm}$$

$$BB_1 = r_{AB}\phi_2 = 0.524 \text{ mm}$$

$$\tan|\gamma_{AB}| \approx |\gamma_{AB}| = \frac{AA_1 + BB_1}{AB}$$

$$\text{ANS.} \qquad \gamma_{AB} = 6980 \ \mu\text{rad}$$

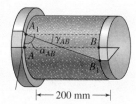

← 200 mm →

Figure 5.9 Calculation of shear strain in section AB by Method 1 in Example 5.3.

---

[2]The shear strain is given by $\gamma = \pi/2 - \alpha$, where α is the final angle. A decrease in angle from a right angle results in positive shear strains.

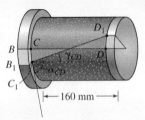

Figure 5.10 Calculation of shear strain in section CD by Method 1 in Example 5.3.

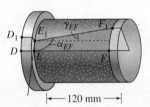

Figure 5.11 Calculation of shear strain in section EF by Method 1 in Example 5.3.

We draw the approximate deformed shape of section CD in Figure 5.10 and find the shear strain,

$$CC_1 = r_{CD}\phi_2 = 0.393 \text{ mm}$$

$$DD_1 = r_{CD}\phi_3 = 0.393 \text{ mm}$$

$$\tan|\gamma_{CD}| \approx |\gamma_{CD}| = \frac{CC_1 + DD_1}{CD}$$

ANS. $\gamma_{CD} = -4913 \ \mu\text{rad}$

We draw the approximate deformed shape of section EF in Figure 5.11 and find the shear strain,

$$EE_1 = r_{EF}\phi_3 = 0.262 \text{ mm}$$

$$FF_1 = r_{EF}\phi_4 = 0.567 \text{ mm}$$

$$\tan|\gamma_{EF}| \approx |\gamma_{EF}| = \frac{FF_1 - EE_1}{EF}$$

ANS. $\gamma_{EF} = -2542 \ \mu\text{rad}$

Method 2

PLAN

We assign a sign to the direction of rotation, calculate the relative deformation of the right disc with respect to the left disc, and analyze the entire shaft.

We draw an approximate deformed shape of the entire shaft, as shown in Figure 5.12. Let the counterclockwise rotation with respect to the x axis be positive and write each angle with the correct sign,

$$\phi_1 = -0.0436 \text{ rad} \qquad \phi_2 = 0.0262 \text{ rad}$$
$$\phi_3 = -0.0262 \text{ rad} \qquad \phi_4 = -0.0567 \text{ rad}$$

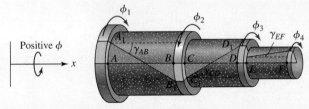

Figure 5.12 Shear strain calculation by Method 2 in Example 5.3.

We compute the relative rotation in each section and multiply the result by the corresponding section radius to obtain the relative movement of two points in a section. We then divide by the length of the section as we did in Example 5.2.

$$\Delta\phi_{AB} = \phi_2 - \phi_1 = 0.0698 \quad \text{ANS. } \gamma_{AB} = \frac{r_{AB}\,\Delta\phi_{AB}}{AB} = 6980 \ \mu\text{rad}$$

$$\Delta\phi_{CD} = \phi_3 - \phi_2 = -0.0524 \quad \text{ANS. } \gamma_{CD} = \frac{r_{CD}\,\Delta\phi_{CD}}{CD} = -4913 \ \mu\text{rad}$$

$$\Delta\phi_{EF} = \phi_4 - \phi_3 = -0.0305 \quad \text{ANS. } \gamma_{EF} = \frac{r_{EF}\,\Delta\phi_{EF}}{EF} = -2542 \ \mu\text{rad}$$

COMMENTS

1. Method 1 is easier to visualize, but the repetitive calculations can be tedious. Method 2 is more mathematical and procedural, but the repetitive calculations are easier. By solving the problems by Method 2 but spending time visualizing the deformation as in Method 1, we can reap the benefits of both methods.

2. We note that the shear strain in each section is directly proportional to the radius and the relative rotation of the shaft and inversely proportional to its length.

5.1.1 Internal Torque

In this section we formalize the observation made in Example 5.1, namely, that the shear stress $\tau_{x\theta}$ can be replaced by an equivalent torque using an integral over the cross-sectional area.

Figure 5.13 shows the shear stress distribution $\tau_{x\theta}$ that is to be replaced by an equivalent internal torque T. Let ρ represent the radial coordinate, that is, the radius of the circle at which the shear stress acts. The moment at the center due to the shear stress on the differential area is $\rho\tau_{x\theta}\,dA$. By integrating over the entire area we obtain the total internal torque at the cross section, as given by the equation

$$T = \int_A \rho \, dV = \int_A \rho\tau_{x\theta} \, dA \tag{5.1}$$

Equation (5.1) is independent of the material model as it represents static equivalency between the shear stress on the entire cross section and the internal torque. If we were to

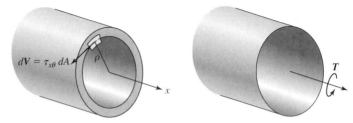

Figure 5.13 Statically equivalent internal torque.

consider a composite shaft cross section or nonlinear material behavior, then it would affect the value and distribution of $\tau_{x\theta}$ across the cross section. But Equation (5.1), relating $\tau_{x\theta}$ and T, would remain unchanged. Examples 5.4 and 5.20 will clarify the discussion in this paragraph.

EXAMPLE 5.4

A homogeneous cross section made of brass and a composite cross section of brass and steel are shown in Figure 5.14. The shear moduli of elasticity for brass and steel are $G_B = 40$ GPa and $G_S = 80$ GPa, respectively. The shear strain in polar coordinates at the cross section was found to be $\gamma_{x\theta} = 0.08\rho$, where ρ is in meters.

(a) Write expressions for $\tau_{x\theta}$ as a function of ρ and plot the shear strain and shear stress distributions across both cross sections.

(b) For each of the cross sections determine the statically equivalent internal torques.

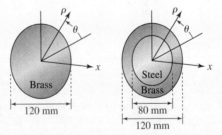

Figure 5.14 Homogeneous and composite cross sections in Example 5.4.

PLAN

(a) Using Hooke's law we can find the shear stress distribution as a function of ρ in each material.

(b) Each of the shear stress distributions can be substituted into Equation (5.1) and the equivalent internal torque obtained by integration.

Solution

(a) From Hooke's law we can write

$$(\tau_{x\theta})_{\text{brass}} = 40 \times 10^9 \times 0.08\rho = 3200\rho \text{ MPa} \qquad \text{(E1)}$$

$$(\tau_{x\theta})_{\text{steel}} = 80 \times 10^9 \times 0.08\rho = 6400\rho \text{ MPa} \qquad \text{(E2)}$$

For the homogeneous cross section the stress distribution is as given in Equation (E1), but for the composite section it switches between Equations (E2) and (E1), depending on the value of ρ. We can write the shear stress distribution for both cross sections as a function of ρ as follows:

Homogeneous cross section:

$$\tau_{x\theta} = 3200\rho \text{ MPa} \qquad 0.00 \leq \rho < 0.06$$

Composite cross section:

$$\tau_{x\theta} = \begin{cases} 6400\rho \text{ MPa} & 0.00 \leq \rho < 0.04 \\ 3200\rho \text{ MPa} & 0.04 < \rho \leq 0.06 \end{cases}$$

The shear strain and the shear stress can now be plotted as a function of ρ, as shown in Figure 5.15.

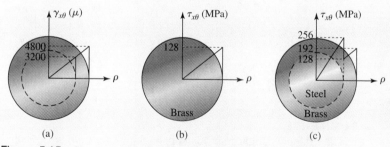

(a)   (b)   (c)

Figure 5.15 Shear strain and shear stress distributions in Example 5.4. (*a*) Shear strain distribution. (*b*) Shear stress distribution in homogeneous cross section. (c) Shear stress distribution in composite cross section.

(b) The differential area dA is the area of a ring of radius ρ and thickness $d\rho$, that is, $dA = 2\pi\rho \, d\rho$. Equation (5.1) can be written as

$$T = \int_0^{0.06} \rho\tau_{x\theta} \times 2\pi\rho \, d\rho \qquad \text{(E3)}$$

Homogeneous cross section: The stress distribution in Example 5.4 for the homogeneous cross section is $\tau_{x\theta} = 3200\rho$ MPa. Substituting this into Equation (E3) and integrating, we obtain the equivalent internal torque,

$$T = \int_0^{0.06} \rho \times 3200\rho \times 10^6 \times 2\pi\rho \ d\rho = 6400\pi \times 10^6 \times \left.\frac{\rho^4}{4}\right|_0^{0.06}$$

ANS. $T = 65.1 \ \text{kN·m}$

Composite cross section: The stress distribution for the composite cross section is $\tau_{x\theta} = 6400\rho$ MPa for $0.00 \le \rho < 0.04$ and $\tau_{x\theta} = 3200\rho$ MPa for $0.04 < \rho \le 0.06$. Writing the integral in Equation (E3) as a sum of two integrals and substituting the stress expressions we can obtain the equivalent internal torque,

$$T = \int_0^{0.06} \rho\tau \times 2\pi\rho \ d\rho = \underbrace{\int_0^{0.04} \rho\tau \times 2\pi\rho \ d\rho}_{T_{\text{steel}}} + \underbrace{\int_{0.04}^{0.06} \rho\tau \times 2\pi\rho \ d\rho}_{T_{\text{brass}}}$$

$$T_{\text{steel}} = \int_0^{0.04} \rho \times 6400\rho \times 10^6 \times 2\pi\rho \ d\rho = 12{,}800\pi \times 10^6 \times \left.\frac{\rho^4}{4}\right|_0^{0.04}$$

$$= 25.7 \ \text{kN·m}$$

$$T_{\text{brass}} = \int_{0.04}^{0.06} \rho \times 3200\rho \times 10^6 \times 2\pi\rho \ d\rho = 6400\pi \times 10^6 \times \left.\frac{\rho^4}{4}\right|_{0.04}^{0.06}$$

$$= 52.3 \ \text{kN·m}$$

$$T = T_{\text{steel}} + T_{\text{brass}} = 25.7 + 52.3$$

ANS. $T = 78 \ \text{kN·m}$

COMMENTS

1. The example demonstrates that although the shear strain varies linearly across the cross section, the shear stress may not. In this example we considered material nonhomogeneity. In a similar manner we can consider other models, such as elastic–perfectly plastic, or material models that have nonlinear stress–strain curves.[3]

2. The material models dictate the shear stress distribution across the cross section, but once the stress distribution is known, Equation (5.1) can be used to find the equivalent internal torque, emphasizing that Equation (5.1) does not depend on the material model.

---

[3]See Section 3.11.

5.1 A pair of 48-in-long bars and a pair of 60-in-long bars are symmetrically attached to a rigid disc at a radius of 2 in at one end and built into the wall at the other end, as shown in Figure P5.1. The shear strain at point A due to a twist of the rigid disc was found to be 3000 μrad. Determine the magnitude of shear strain at point D.

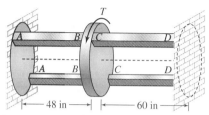

Figure P5.1

5.2 If the four bars in Problem 5.1 are made from a material that has a shear modulus of 12,000 ksi, determine the applied torque T on the rigid disc. The cross sectional areas of all bars are 0.25 in^2.

5.3 If bars AB in Problem 5.1 are made of aluminum with a shear modulus $G_{al} = 4000$ ksi and bars CD are made of bronze with a shear modulus $G_{br} = 6500$ ksi, determine the applied torque T on the rigid disc. The cross-sectional areas of all bars are 0.25 in^2.

5.4 Three pairs of bars are symmetrically attached to rigid discs at the radii shown in Figure P5.4. The discs were observed to rotate by angles $\phi_1 = 1.5°$, $\phi_2 = 3.0°$, and $\phi_3 = 2.5°$ in the direction of the applied torques T_1, T_2, and T_3, respectively. The shear modulus of the bars is 40 ksi and cross-sectional area is 0.04 in^2. Determine the applied torques.

5.5 A circular shaft of radius r and length Δx has two rigid discs attached at each end, as shown in Figure P5.5. If the rigid discs are rotated as shown, determine the shear strain γ at point A in terms of r, Δx, and $\Delta \phi$, assuming that line AB remains straight, where $\Delta \phi = \phi_2 - \phi_1$.

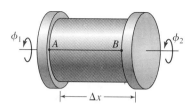

Figure P5.5

5.6 A hollow circular shaft made from hard rubber has an outer diameter of 4 in and an inner diameter of 1.5 in. The shaft is fixed to the wall on the left end and the rigid disc on the right hand is twisted, as shown in Figure P5.6. The shear strain at point A, which is on the outside surface, was found to be 4000 μrad. Determine the shear strain at point C, which is on the inside surface, and the angle of rotation. Assume that lines AB and CD remain straight during deformation.

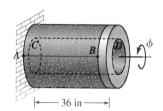

Figure P5.6

5.7 The magnitude of shear strains in the segments of the stepped shaft in Figure P5.7 were found to be $\gamma_{AB} = 3000$ μrad, $\gamma_{CD} = 2500$ μrad, and $\gamma_{EF} = 6000$ μrad. The radius of section AB is 150 mm, of section CD 70 mm, and of section EF 60 mm. Determine the angle by which each of the rigid discs was rotated.

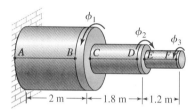

Figure P5.7

A hollow brass shaft ($G_B = 6500$ ksi) and a solid steel shaft ($G_S = 13,000$ ksi) are securely fastened to form a composite shaft, as shown in Figure P5.8. The shear strain $\gamma_{x\theta}$ in polar coordinates at the section and the dimensions of the cross section are as given. In Problems 5.8 through 5.10, determine the equivalent internal torque acting at the cross section.

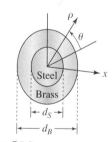

Figure P5.8

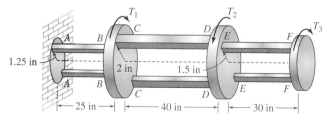

Figure P5.4

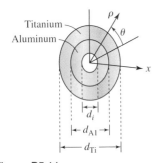

5.8 $\gamma_{x\theta} = 0.001\rho$, where ρ is in inches. Use $d_B = 4$ in and $d_S = 2$ in.

5.9 $\gamma_{x\theta} = -0.0005\rho$, where ρ is in inches. Use $d_B = 6$ in and $d_S = 4$ in.

5.10 $\gamma_{x\theta} = 0.002\rho$, where ρ is in inches. Use $d_B = 3$ in and $d_S = 1$ in.

A hollow titanium shaft ($G_{Ti} = 36$ GPa) and a hollow aluminum shaft ($G_{Al} = 26$ GPa) are securely fastened to form a composite shaft, as shown in Figure P5.11. The shear strain $\gamma_{x\theta}$ in polar coordinates at the section and the dimensions of the cross section are as given.

In Problems 5.11 through 5.13, determine the equivalent internal torque acting at the cross section.

5.11 $\gamma_{x\theta} = 0.05\rho$, where ρ is in meters. Use $d_i = 40$ mm, $d_{Al} = 80$ mm, and $d_{Ti} = 120$ mm.

5.12 $\gamma_{x\theta} = 0.04\rho$, where ρ is in meters. Use $d_i = 50$ mm, $d_{Al} = 90$ mm, and $d_{Ti} = 100$ mm.

5.13 $\gamma_{x\theta} = -0.06\rho$, where ρ is in meters. Use $d_i = 30$ mm, $d_{Al} = 40$ mm, and $d_{Ti} = 50$ mm.

5.2 THEORY

The theory will be developed subject to the limitations described in Section 3.13, namely, (i) the length of the member is significantly greater than the greatest dimension in the cross section; (ii) we are away from the regions of stress concentration; (iii) the variation of external torque or change in cross-sectional areas is gradual except in regions of stress concentration. In addition the following limitation will be imposed; (iv) the cross section is circular. This permits us to use arguments of axisymmetry in deducing deformation.

We assume[4] that external torques are not functions of time, that is, we have a static problem. Figure 5.16 shows a circular shaft that is loaded by external torques T_1 and T_2 at each end and an external distributed torque $t(x)$, which has units of torque per unit length. The radius of the shaft $R(x)$ varies as a function of x. We expect that the internal torque T will be a function of x. ϕ_1 and ϕ_2 are the angles of rotation of the imaginary cross sections at x_1 and x_2, respectively.

The objectives of the theory are:

1. To obtain a formula for the relative rotation $\phi_2 - \phi_1$ in terms of the internal torque T.

2. To obtain a formula for the shear stress $\tau_{x\theta}$ in terms of the internal torque T.

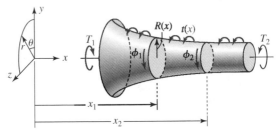

Figure 5.16 Circular shaft.

[4]See Problems 5.45 and 5.46 for dynamic problems.

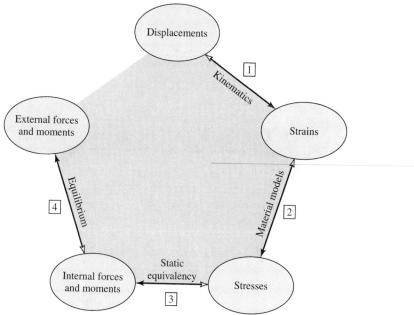

Figure 5.17 Logic in mechanics of materials.

To account for the variations[5] in $t(x)$ and $R(x)$ we will take $\Delta x = x_2 - x_1$ as an infinitesimal distance in which these quantities can be treated as constants. The deformation behavior across the cross section will be approximated. The logic shown in Figure 5.17 and discussed in Section 3.2 will be used to develop the simplest theory for the torsion of circular shafts members. But assumptions will be identified as we move from one step to the next. These assumptions are the points at which complexities can be added to the theory, as described by footnotes associated with the assumptions.

5.2.1 Kinematics

In Example 5.1 the shear strain in a bar was related to the rotation of the disc that was attached to it. In Example 5.2 it was remarked that a shaft could be viewed as an assembly of bars. The assumptions that will let us simulate the behavior of a cross section like that of a rigid plate are:

Assumption 1 Plane sections perpendicular to the axis remain plane during deformation.

Assumption 2 On a cross section, all radial lines rotate by equal angles during deformation.

Assumption 3 Radial lines remain straight during deformation.

Figure 5.18 shows a circular rubber shaft with a grid on the surface that is twisted by hand. The vertical lines, which are the edges of the circle, continue to remain vertical during deformation. This suggests that Assumption 1 would be valid if the interior points deformed in the same manner as the points on the surface.

---

[5]Note that we have imposed the limitation that the variations in $R(x)$ and in $t(x)$ are gradual.

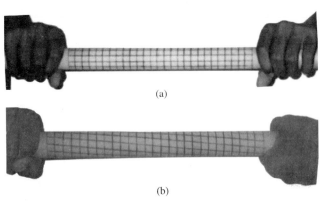

(a)

(b)

Figure 5.18 Torsional deformation. (*a*) Original grid. (*b*) Deformed grid. (Courtesy of Professor J. B. Ligon.)

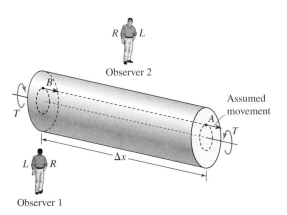

Figure 5.19 Justifying the kinematic assumption, Assumption 1, in circular shafts.

In Figure 5.18, if we consider the various rectangles between two consecutive circles, we notice that the change of shape does not depend on the angular position of the rectangle on the circle. This suggests that the radial lines from the center to the corners of the rectangle on the same circle must be rotating by the same amount, provided the lines are straight in the interior.

The shear strain that is of interest to us is the measure of the angle change between the axial direction and the tangent to the circle. If we use polar coordinates, then the change in angle in which we are interested is between the x and θ directions; in other words, $\gamma_{x\theta}$.

The deformation observed in Figure 5.18 is on the surface of the shaft. Can we extrapolate the surface observations into the interior of the shaft? To answer this question, we consider the analytical justification of Assumptions 1 through 3. The analytical deductions are based on the axisymmetric character of the problem, that is, the applied torque and the geometry are symmetric about the shaft axis, hence deformation must also be symmetric and independent of the angular location of the point.

We will start by assuming that a point moves out of the plane and then show that for circular shafts this deformation is not possible. Consider a small segment of a shaft with torque applied as shown in Figure 5.19. The applied torque is counterclockwise with respect to the outward normal on each end. Thus two observers positioned on opposite sides of the shaft see exactly the same type of shaft and loading, and hence both observers should see the same type of deformation.

In an axisymmetric problem, the deformation of a point on a cross section is independent of its angular location. Thus if a point were to move out of the plane of the cross section, then all points on the circle passing through that point must move out of the plane of the cross section. If all points on the circle passing through A move out of the plane due to the applied torques, as shown in Figure 5.19, then all points on the circle passing through point B must also move in the same direction if no holes are to be created in a small length of the shaft. But now consider the deformation as seen by the two observers:

- Observer 1 sees the points on the right (A) moving *out* of the material while the points on the left (B) are moving *into* the material.

- Observer 2 sees the points on the right (B) moving *into* the material and the points on the left (A) moving *out* of the material.

Both observers see the same shaft and loading but see different deformations, which is clearly not possible. We conclude that no point can move out of the plane of the cross section for a circular shaft. In other words, circular shafts do not warp. Clearly, for noncircular shafts, the arguments of axisymmetry cannot be used. As a consequence noncircular shafts do warp, and it is the accounting of this additional deformation that leads to additional complexities in the theory.[6]

Figure 5.20 shows that all radial lines must rotate by the same angle ϕ because the problem is axisymmetric. But in doing so we have made the implicit assumption that radial lines during deformation remain straight. This assumption is justified as long as we are working with small strains and the radius of the shaft is small compared to its length. We also note that if all lines rotate by equal amounts on the cross section, then ϕ does not change across the cross section and hence can only be a function of x. We record these assumptions and observations.

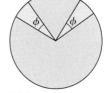

Figure 5.20 Equal rotation of all radial lines.

Assumption 4   Strains are small.

$$\phi = \phi(x) \tag{5.2}$$

Definition 1 ϕ is positive counterclockwise with respect to the x axis.

Assumptions 1 through 3 are analogous to thinking that each cross section in the shaft may be viewed as a rigid disc that rotates about its own axis and we can calculate the shear strain as in Example 5.2. We consider a shaft with radius ρ and length Δx in which the right section with respect to the left section is rotated by an angle $\Delta\phi$, as shown in Figure 5.21,

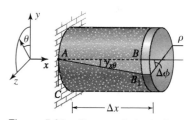

Figure 5.21 Shear strain in torsion.

$$\tan\gamma_{x\theta} \approx \gamma_{x\theta} = \lim_{AB \to 0}\left(\frac{AB}{AC}\right) = \lim_{\Delta x \to 0}\left(\frac{\rho\Delta\phi}{\Delta x}\right)$$

By letting Δx tend to zero we obtain

$$\boxed{\gamma_{x\theta} = \rho\frac{d\phi}{dx}} \tag{5.3}$$

where ρ is the radial coordinate of a point on the cross section. The subscripts x and θ emphasize that the change in angle is between the axial and tangent directions, as

---

[6]See Problem 5.43 for torsion of noncircular shafts.

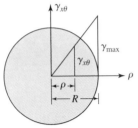

Figure 5.22 Linear variation of shear strain in torsion.

shown in Figure 5.21. The quantity $d\phi/dx$ is called the *rate of twist*. It is a function of x only because ϕ is a function of x only.

Equation (5.3) was derived from purely geometric considerations. If[7] Assumptions 1 through 4 are valid, then Equation (5.3) is independent of the material models. Equation (5.3) shows that the shear strain is a linear function of the radial coordinate ρ and reaches the maximum value γ_{max} at the outer surface ($\rho = R$), as shown in Figure 5.22. Using a similar triangle, Equation (5.4) can be derived as an alternative form of Equation (5.3),

$$\gamma_{x\theta} = \frac{\gamma_{max}\rho}{R} \tag{5.4}$$

5.2.2 Material Model

Our motivation is to develop a simple theory for axial deformation. Thus we make assumptions regarding material behavior that will permit us to use the simplest material model given by Hooke's law.

Assumption 5 Material is linearly elastic.[8]

Assumption 6 Material is isotropic.

Substituting Equation (5.3) into Hooke's law, that is, $\tau = G\gamma$, we obtain

$$\boxed{\tau_{x\theta} = G\rho\frac{d\phi}{dx}} \tag{5.5}$$

Noting that θ is positive in the counterclockwise direction with respect to the x axis, we can represent the shear stress due to torsion on a stress element, as shown in Figure 5.23. Also shown in Figure 5.23 are aluminum and wooden shafts that broke in torsion. The shear stress component that exceeds the shear strength value in aluminum is $\tau_{x\theta}$. The shear strength of wood is weaker along the surface parallel to the grain, which for shafts is in the longitudinal direction. Thus $\tau_{\theta x}$ causes the failure in wooden shafts, as shown. The two failure surfaces highlight the importance of visualizing the torsional shear stress element.

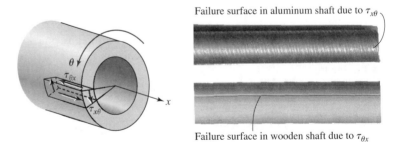

Failure surface in aluminum shaft due to $\tau_{x\theta}$

Failure surface in wooden shaft due to $\tau_{\theta x}$

Figure 5.23 Stress element showing torsional shear stress.

---

[7]If a shaft is made from a very soft rubber-like material that undergoes excessive deformation even for small applied torques, then clearly Assumptions 1 through 4 are inadmissible for these types of soft material, and hence Equations (5.3) and (5.4) are not valid for these materials.
[8]See Problems 5.42 and 5.80 for nonlinear material behavior.

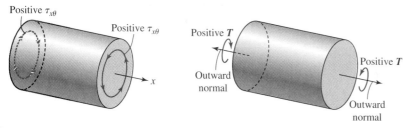

Figure 5.24 Sign convention for positive internal torque.

5.2.3 Internal Torque Sign Convention

The shear stress can be replaced by a statically equivalent internal torque using Equation (5.1). The shear stress $\tau_{x\theta}$ is positive on two surfaces; hence the equivalent internal torque is positive on two surfaces, as shown in Figure 5.24. When we make the imaginary cut to draw the free-body diagram, then the internal torque must be drawn in the positive direction if[9] we want the formulas to give the correct signs.

Definition 2 Internal torque is considered positive if it is counterclockwise with respect to the outward normal to the imaginary cut surface.

5.2.4 Torsion Formulas

Substituting Equation (5.5) into Equation (5.1) and noting that $d\phi/dx$ is a function of x only, we obtain

$$T = \int_A G\rho^2 \frac{d\phi}{dx}\, dA = \frac{d\phi}{dx}\int_A G\rho^2\, dA \qquad (5.6)$$

To simplify, we would like to take G outside the integral, which implies that G cannot change across the cross section. We make the following assumption:

Assumption 7 Material is homogeneous across the cross section.[10]

From Equation (5.6) we obtain

$$T = G\frac{d\phi}{dx}\int_A \rho^2\, dA = GJ\frac{d\phi}{dx}$$

or

$$\boxed{\frac{d\phi}{dx} = \frac{T}{GJ}} \qquad (5.7)$$

---

[9]If the direction of rotation and the direction of shear stress are determined by inspection, then the sign convention can be ignored and formulas used to give the magnitudes of the quantities.
[10]See Section 5.4 on composite shafts, where this assumption is not valid.

where J is the polar moment of inertia for the cross section. For a circular cross section of radius R or diameter D,

$$J = \frac{\pi}{2}R^4 = \frac{\pi}{32}D^4$$

as shown in Example 5.5.

The higher the value of GJ, the smaller will be the deformation ϕ for a given value of the internal torque. Thus the rigidity of the shaft increases with the increase in GJ. A shaft may be made more rigid either by choosing a stiffer material (higher value of G) or by increasing the polar moment of inertia. Example 5.5 shows that for the same amount of material, a hollow shaft will give higher values of J. Example 5.8 brings out the importance of torsional rigidity in design.

Definition 3 The quantity GJ is called *torsional rigidity*.

Substituting Equation (5.7) into Equation (5.5) we obtain

$$\boxed{\tau_{x\theta} = \frac{T\rho}{J}} \qquad (5.8)$$

The quantities T and J do not vary across the cross section. Thus the shear stress varies linearly across the cross section with ρ and reaches a maximum value on the outer surface of the shaft, as shown in Figure 5.25.

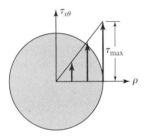

Figure 5.25 Linear variation of torsional shear stress.

To obtain the angle of rotation of the cross section, we can integrate Equation (5.7). Let the angle of rotation of the cross section at x_1 and x_2 be ϕ_1 and ϕ_2, respectively. From Equation (5.7) we obtain

$$\phi_2 - \phi_1 = \int_{\phi_1}^{\phi_2} d\phi = \int_{x_1}^{x_2} \frac{T}{GJ}\, dx \qquad (5.9)$$

where ϕ_1 and ϕ_2 are the rotation of the cross section at x_1 and x_2, respectively. To obtain a simple formula we would like to take the three quantities T, G, and J outside the integral, which means that these quantities should not change with x. To achieve this simplicity we make the following assumptions:

Assumption 8 Material is homogeneous between x_1 and x_2.

Assumption 9 The shaft is not tapered.

Assumption 10 The external (hence internal) torque does not change with x between x_1 and x_2.

In Example 5.9 the change in radius with x makes J a function of x, and Assumption 9 is violated. In Example 5.10 the distributed external torque makes the internal torque T a function of x, and hence Assumption 10 is violated. In both these examples we have to use Equation (5.9) or, equivalently, Equation (5.7) to solve the problems and not the simplified equation given next.

$$\phi_2 - \phi_1 = \frac{T(x_2 - x_1)}{GJ}$$ (5.10)

If two shafts from different materials are joined at some point x_m, or if the cross-sectional area J changes at x_m, or if there is an external torque applied at x_m that would cause the internal torque to change at x_m, then the point x_m should not be between x_1 or x_2. In other words, in Equation (5.10) points x_1 and x_2 must be chosen such that neither T, G, nor J change between these points.

T is an internal torque that has to be determined by making an imaginary cut and drawing a free-body diagram. There are two possible ways in which T may be found as described next and elaborated further in Example 5.6.

1. T is always drawn in the counterclockwise direction with respect to the outward normal of the imaginary cut. The equilibrium equation then is used to get a positive or negative value for T. In other words, we are following the sign convention for internal torque given by Definition 2. Therefore the sign for relative rotation obtained from Equation (5.10) is positive counterclockwise with respect to the x axis, as per Definition 1, and the direction of shear stress can be determined using the subscripts, as was elaborated in Section 1.2.1.

2. T is drawn at the imaginary cut in a direction to equilibrate the external torques. Since inspection is being used in determining the direction of T, the direction of relative rotation in Equation (5.10) and the direction of shear stress $\tau_{x\theta}$ in Equation (5.8) must also be determined by inspection.

5.2.5   Torsional Stresses and Strains

The significant shear stress in the torsion of circular shafts is $\tau_{x\theta}$. All other stress components can be neglected provided the ratio of the length of the shaft to its diameter is on the order of 10 or more. The direction of torsional shear stress in Equation (5.8) can be determined using subscripts provided the sign convention for internal torque is followed when drawing the free-body diagram. Alternatively the direction of shear stress can be determined by inspection, as outlined next.

We consider the problem of showing the torsional shear stress at point A on a shaft. Figure 5.26a shows a segment of a shaft under torsion containing point A. We visualize point A on the left segment and consider the stress element on the left segment. The left segment rotates clockwise in relation to the right segment. This implies that point A, which is part of the left segment, is moving upward on the shaded surface. Hence the shear stress, like friction, on the shaded surface will be downward. Using the symmetry of shear stresses, that is, a pair of symmetric shear stress components points toward or away from the corner, the shear stresses on the rest of the surfaces can be drawn as shown.

Suppose we had considered point A on the right segment of the shaft. In such a case we consider the stress element as part of the right segment, as shown in Figure 5.26b. The

---

[11]An alternative derivation is that if T, G, and J are constant, then the slope $d\phi/dx$ is a constant, and can be represented by the difference formula as $d\phi/dx = (\phi_2 - \phi_1)/(x_2 - x_1) = T/GJ$.

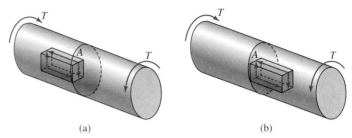

Figure 5.26 Direction of shear stress by inspection.

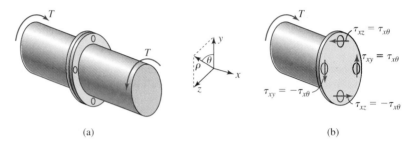

Figure 5.27 Torsional shear stresses.

right segment rotates counterclockwise in relation to the left segment. This implies that point A, which is part of the right segment, is moving down on the shaded surface. Hence the shear stress, like friction, will be upward. Once more using the symmetry of shear stress components, the shear stress on the remaining surfaces can be drawn as shown.

The shaft segments in Figure 5.26 were drawn for the purpose of explanation and are not necessary during visualization. But care must be taken during visualization to clearly identify in one's mind the surface on which the shear stress is being considered, for the direction of the shear stress on the adjoining imaginary surfaces is opposite. Notice, however, that irrespective of the surface on which we visualize the stress element, we obtain the same stress element, as shown in Figure 5.26. This is to be expected because the two stress elements shown represent the same point A.

An alternative view of visualizing torsional shear stress is to think of a coupling at an imaginary section and to visualize the shear stress directions on the bolt surfaces, as shown in Figure 5.27. Once the direction of the shear stress on the bolt surface is visualized, the remaining stress elements can be completed using the symmetry of shear stresses.

After having obtained the torsional shear stress, either by using subscripts or by inspection, we can examine the shear stresses in Cartesian coordinates and obtain the stress components with correct signs, as shown in Figure 5.27b. This process of obtaining stress components in Cartesian coordinates will be important when we consider stress and strain transformation equations in Chapters 8 and 9, where we will relate stresses and strains in different coordinate systems.

The shear strain can be obtained by dividing the shear stress by G, the shear modulus of elasticity.

Consolidate your knowledge

1. Identify five examples of circular shafts from your daily life.

2. With book closed, derive Equations (5.8) and (5.10).

EXAMPLE 5.5

The two shafts shown in Figure 5.28 are of the same material and have the same cross-sectional areas A. Show that the hollow shaft has a higher polar moment of inertia than the solid shaft.

PLAN

We can find the values of R_H and R_S in terms of the cross-sectional area A. We can then substitute these radii in the formulas for polar area moment to obtain the polar area moments in terms of A.

Solution We can calculate the radii R_H and R_S in terms of the cross-sectional area A as

$$A_H = \pi[(2R_H)^2 - R_H^2] = A \qquad R_H^2 = \frac{A}{3\pi} \qquad \text{(E1)}$$

$$A_S = \pi R_S^2 = A \qquad R_S^2 = \frac{A}{\pi} \qquad \text{(E2)}$$

We derive the polar area moment of inertia for a hollow shaft with inside radius R_i and outside radius R_o as

$$J = \int_A \rho^2 \, dA = \int_{R_i}^{R_o} \rho^2 (2\pi\rho) \, d\rho = \left.\frac{\pi}{2}\rho^4\right|_{R_i}^{R_o} = \frac{\pi}{2}(R_o^4 - R_i^4) \qquad \text{(E3)}$$

For the hollow shaft $R_o = 2R_H$ and $R_i = R_H$, whereas for the solid shaft $R_o = R_S$ and $R_i = 0$. Substituting these values into Equation (E3) and using Equations (E1) and (E2), we can obtain the two polar area moments,

$$J_H = \frac{\pi}{2}[(2R_H)^4 - R_H^4] = \frac{15}{2}\pi R_H^4 = \frac{15}{2}\pi\left(\frac{A}{3\pi}\right)^2 = \frac{5A^2}{6\pi}$$

$$J_S = \frac{\pi}{2}R_S^4 = \frac{\pi}{2}\left(\frac{A}{\pi}\right)^2 = \frac{A^2}{2\pi}$$

Thus,

$$\frac{J_H}{J_S} = \frac{5}{3} = 1.67 \qquad \text{(E4)}$$

As $J_H > J_S$ we have proved that the polar moment for the hollow shaft is greater than that of the solid shaft for the same amount of material.

COMMENT

The hollow shaft has a polar moment of inertia of 1.67 times that of the solid shaft for the same amount of material. Equation (5.8) and (5.10) imply that the hollow shaft will have lower stresses and smaller deformation. Alternatively a hollow shaft will require less material (lighter in weight) to obtain the same polar moment of inertia. This reduction in weight is the primary reason why metal shafts are made hollow. Wooden shafts, however, are usually solid as the machining cost does not justify the small saving in weight.

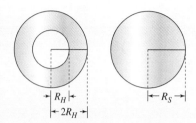

Figure 5.28 Hollow and solid shafts of Example 5.5.

EXAMPLE 5.6

A solid circular steel shaft ($G_s = 12{,}000$ ksi) of variable diameter is acted upon by torques, as shown in Figure 5.29. The diameter of the shaft between wheels A and B and between wheels C and D is 2 in, and the diameter of the shaft between wheels B and C is 4 in. Determine:

(a) The rotation of wheel D with respect to wheel A.

(b) The magnitude of maximum torsional shear stress in the shaft.

(c) The shear stress at point E. Show it on a stress cube.

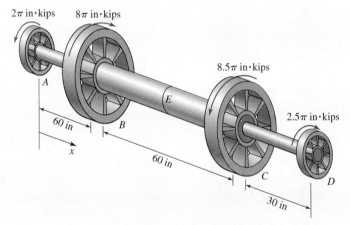

Figure 5.29 Geometry of shaft and loading in Example 5.6.

PLAN

We are required to find $\phi_D - \phi_A$ and τ_{max}. By making imaginary cuts in sections AB, BC, and CD and drawing the free-body diagrams we can find the internal torques in each section.

(a) By using Equation (5.10) we can find the relative rotation in each section. Then by summing the relative rotations we can find $\phi_D - \phi_A$.

(b) By using Equation (5.8) we can find the maximum shear stress in each section, then compare to find the maximum shear stress τ_{max} in the shaft.

(c) In part (b) we found the shear stress in section BC. We can find the direction of the shear stress either using the subscript or intuitively.

Solution   We make an imaginary cut in segment AB in Figure 5.29. We then take the left part after the cut and draw the free-body diagram in Figure 5.30, using the sign convention for internal torque,

$$J_{AB} = \frac{\pi}{32}2^4 = \frac{\pi}{2}$$

$$T_{AB} = -2\pi \text{ in} \cdot \text{kips} \tag{E1}$$

Substituting Equation (E1) into Equation (5.10), we obtain the rotation of the section at B with respect to the section at A. Substituting Equation (E1) into Equation (5.8) and noting that the maximum torsional shear stress in section AB will exist at $\rho = 1$, we obtain

$$\phi_B - \phi_A = \frac{T_{AB}(x_B - x_A)}{G_{AB}J_{AB}} = \frac{-2\pi \times 24}{12,000 \times \pi/2} = -8 \times 10^{-3} \text{ rad} \tag{E2}$$

$$(\tau_{AB})_{max} = \frac{T_{AB}(\rho_{AB})_{max}}{J_{AB}} = \frac{-2\pi \times 1}{\pi/2} = -4 \text{ ksi} \tag{E3}$$

We make an imaginary cut in segment BC in Figure 5.29. We then take the left part after the cut and draw the free-body diagram in Figure 5.31, using the sign convention for internal torque,

$$J_{BC} = \frac{\pi}{32}4^4 = 8\pi$$

$$T_{BC} = 6\pi \text{ in} \cdot \text{kips} \tag{E4}$$

Substituting Equation (E4) into Equation (5.10), we obtain the rotation of the section at C with respect to the section at B. Substituting Equation (E4) into Equation (5.8) and noting that the maximum torsional shear stress in section BC will occur at $\rho = 2$, we obtain

$$\phi_C - \phi_B = \frac{T_{BC}(x_C - x_B)}{G_{BC}J_{BC}} = \frac{6\pi \times 60}{12,000 \times 8\pi} = 3.75 \times 10^{-3} \text{ rad} \tag{E5}$$

$$(\tau_{BC})_{max} = \frac{T_{BC}(\rho_{BC})_{max}}{J_{BC}} = \frac{6\pi \times 2}{8\pi} = 1.5 \text{ ksi} \tag{E6}$$

We make an imaginary cut in segment CD in Figure 5.29. We then take the right part after the cut and draw the free-body diagram in Figure 5.32, using the sign convention for internal torque,

$$J_{CD} = \frac{\pi}{32}2^4 = \frac{\pi}{2}$$

$$T_{CD} = -2.5\pi \text{ in} \cdot \text{kips} \tag{E7}$$

Substituting Equation (E7) into Equation (5.10), we obtain the rotation of the section at D with respect to the section at C. Substituting

Figure 5.30 Free-body diagram of section AB in Example 5.6.

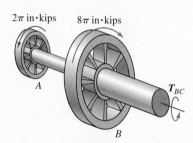

Figure 5.31 Free-body diagram of section BC in Example 5.6.

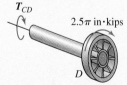

Figure 5.32 Free-body diagram of section CD in Example 5.6.

Equation (E7) into Equation (5.8) and noting that the maximum torsional shear stress in segment CD will occur at $\rho = 1$, we obtain

$$\phi_D - \phi_C = \frac{T_{CD}(x_D - x_C)}{G_{CD}J_{CD}} = \frac{-2.5\pi \times 30}{12{,}000 \times \pi/2} = -12.5 \times 10^{-3} \text{ rad} \quad \text{(E8)}$$

$$(\tau_{CD})_{max} = \frac{T_{CD}(\rho_{CD})_{max}}{J_{CD}} = \frac{-2.5\pi \times 1}{\pi/2} = -5 \text{ ksi} \quad \text{(E9)}$$

(a) Adding Equations (E2), (E5), and (E8), we obtain the relative rotation of the section at D with respect to the section at A,

$$\phi_D - \phi_A = (\phi_B - \phi_A) + (\phi_C - \phi_B) + (\phi_D - \phi_C)$$

$$= (-8 + 3.75 - 12.5)10^{-3} = -16.75 \times 10^{-3}$$

ANS. $\qquad \phi_D - \phi_A = 0.01675 \text{ rad cw}$

(b) By inspection of Equations (E3), (E6), and (E9) we obtain that the magnitude of maximum torsional shear stress is in segment CD,

ANS. $\qquad |\tau_{max}| = 5 \text{ ksi}$

(c) The direction of shear stress can be determined using subscripts or intuitively, as shown in Figure 5.33.

- *Shear stress direction using subscripts:* In Figure 5.33a we note that $\tau_{x\theta}$ in segment BC is +1.5 ksi. The outward normal is in the positive x direction and the force has to be pointed in the positive θ direction (tangent direction), which at point E is downward.

- *Shear stress direction determined intuitively:* Figure 5.33b shows a schematic of segment BC. Consider an imaginary section through E in segment BC. Segment BE tends to rotate clockwise with respect to segment EC. The shear stress will oppose the imaginary clockwise motion of segment BE; hence the direction will be counterclockwise, as shown.

We complete the rest of the stress cube using the fact that a pair of symmetric shear stresses points either toward the corner or away from the corner, as shown in Figure 5.33c.

COMMENTS

1. If we do not follow the sign convention for internal torque and show the internal torque in a direction that counterbalances the external torque, then in the calculation of $\phi_D - \phi_A$ the addition and subtraction must be done manually to account for clockwise and counterclockwise

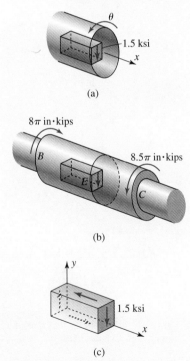

Figure 5.33 Direction of shear stress in Example 5.6.

rotation, as shown in Figure 5.34. Also the shear stress direction must now be determined intuitively.

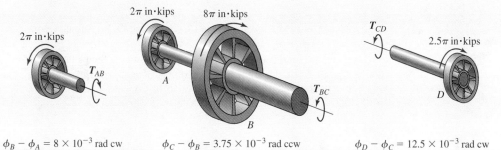

$\phi_B - \phi_A = 8 \times 10^{-3}$ rad cw $\phi_C - \phi_B = 3.75 \times 10^{-3}$ rad ccw $\phi_D - \phi_C = 12.5 \times 10^{-3}$ rad cw

Figure 5.34   Intuitive analysis in Example 5.6.

2. An alternative perspective of the calculation of $\phi_D - \phi_A$ is as follows:

$$\phi_D - \phi_A = \int_{x_A}^{x_D} \frac{T}{GJ}\, dx$$

$$= \underbrace{\int_{x_A}^{x_B} \frac{T_{AB}}{G_{AB}J_{AB}}\, dx}_{\phi_B - \phi_A} + \underbrace{\int_{x_B}^{x_C} \frac{T_{BC}}{G_{BC}J_{BC}}\, dx}_{\phi_C - \phi_B} + \underbrace{\int_{x_C}^{x_D} \frac{T_{CD}}{G_{CD}J_{CD}}\, dx}_{\phi_D - \phi_C}$$

or, written more compactly,

$$\Delta\phi = \sum_i \frac{T_i\, \Delta x_i}{G_i J_i} \tag{5.11}$$

3. Note that $T_{BC} - T_{AB} = 8\pi$ is the magnitude of the applied external torque at the section at B. Similarly $T_{CD} - T_{BC} = -8.5\pi$, which is the magnitude of the applied external torque at the section at C. In other words, the internal torques jump by the value of the external torque as one crosses the external torque from left to right. We will make use of this observation in the next section when plotting the torque diagram.

5.2.6   Torque Diagram

A torque diagram is a plot of the internal torque across the entire shaft. To construct torque diagrams we create a small torsion template to guide us in which direction the internal torque will jump.

Definition 4 A torsion template is an infinitesimal segment of the shaft constructed by making imaginary cuts on either side of a supposed external torque.

Figure 5.35 shows a torsion template. The external torque can be drawn either clockwise or counterclockwise. The ends of the torsion template represent the imaginary cuts just to the left and just to the right of the applied external torque. On these cuts, the internal torques are drawn counterclockwise with respect to the outward normal, according to the sign convention in Definition 2. An equilibrium equation is written, which we will call the *template equation*,

$$T_2 = T_1 - T_{ext}$$

If the external torque on the shaft is in the direction of the assumed torque shown on the template, then value of T_2 is calculated according to the template equation. If the external torque on the shaft is opposite to the direction shown, then T_2 is calculated by changing the sign of T_{ext} in the template equation. Moving across the shaft using the template equation, we can draw the torque diagram, as demonstrated in the next example.

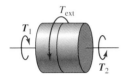

Figure 5.35   Torsion template.

EXAMPLE 5.7

Calculate the rotation of the section at D with respect to the section at A by drawing the torque diagram using the template shown in Figure 5.35.

PLAN

We can start the process of drawing the torque diagram by considering an imaginary extension on the left end. In the imaginary extension the internal torque is zero. Using the template in Figure 5.35 to guide us, we can draw the torque diagram.

Solution   Let *LA* be an imaginary extension on the left side of the shaft, as shown in Figure 5.36. Clearly the internal torque in the imaginary section *LA* is zero, that is, $T_1 = 0$. The torque at A is in the same direction as the torque T_{ext} shown on the template in Figure 5.35. Thus as per the template equation, we subtract the value of the applied torque

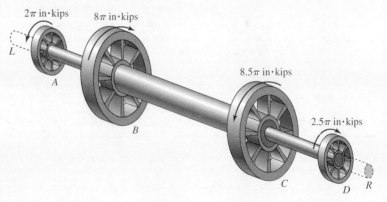

Figure 5.36   Imaginary extensions of the shaft in Example 5.7.

to obtain a value of -2π in·kips for the internal torque T_2 just after wheel A. This is the starting value in the internal torque diagram.

We approach wheel B with an internal torque value of -2π in·kips, that is, $T_1 = -2\pi$ in·kips. The torque at B is in the opposite direction to the torque shown on the template in Figure 5.35. Hence we change the sign of T_{ext} in the template equation, that is, we add 8π in·kips to obtain a value of $+6\pi$ in·kips for the internal torque just after wheel B. We approach wheel C with a value of 6π in·kips and note that the torque at C is in the same direction as that shown on the template. Hence we subtract 8.5π in·kips as per the template equation to obtain -2.5π in·kips for the internal torque just after wheel C. The torque at D is in the same direction as that on the template, and on adding we obtain a zero value in the imaginary extended bar DR as expected, for the shaft is in equilibrium.

From Figure 5.37 the internal torque values in the segments are

$$T_{AB} = -2\pi \qquad T_{BC} = 6\pi \qquad T_{CD} = -2.5\pi$$

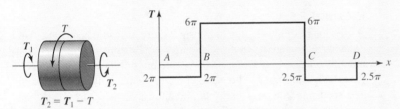

Figure 5.37 Torque diagram in Example 5.7.

Substituting these torque values into Equation (5.11), we obtain the relative rotation of wheel D with respect to wheel A,

$$\phi_D - \phi_A = \frac{T_{AB}(x_B - x_A)}{G_{AB}J_{AB}} + \frac{T_{BC}(x_C - x_B)}{G_{BC}J_{BC}} + \frac{T_{CD}(x_D - x_C)}{G_{CD}J_{CD}}$$

$$= \frac{-2\pi \times 24}{12{,}000 \times \pi/2} + \frac{6\pi \times 60}{12{,}000 \times 8\pi} + \frac{-2.5\pi \times 30}{12{,}000 \times \pi/2}$$

$$= -16.75 \times 10^{-3}$$

> ANS. $\phi_D - \phi_A = 0.01675$ rad cw

COMMENT
We could have created the torque diagram using the template shown in Figure 5.38 and the template equation

$$T_2 = T_1 + T_{ext}$$

We approach wheel A and note that the external torque of 2π is in the opposite direction to that shown on the template in Figure 5.38.

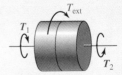

Figure 5.38 Alternative torsion template.

According to the template equation we subtract the value of 2π from the zero value in section LA. Thus our starting value is -2π. As we approach wheel B the internal torque T_1 is -2π, and the applied torque of 8π is in the same direction, as shown on the template. So we add to obtain a value of 6π for the internal torque. We approach section C and note that the applied force is in the opposite direction to the applied torque on the template. Hence we subtract -8.5π to obtain a value of -2.5π for the internal torque. The torque on wheel D is in the same direction as that shown on the template, so we add 2.5π to get a zero value in the extended portion DR. The example shows that the direction of the applied torque T_{ext} on the template is immaterial.

EXAMPLE 5.8

A 1-m-long hollow shaft is to transmit a torque of 400 N·m. The outer diameter of the shaft must be 25 mm to fit existing attachments (Figure 5.39). The relative rotation of the two ends of the shaft is limited to 0.375 rad. The shaft can be made of either titanium alloy or aluminum. The shear modulus of rigidity G, the allowable shear stress τ_{allow}, and the density γ are given in Table 5.1. Determine the maximum inner radius to the nearest millimeter of the lightest shaft that can be used for transmitting the torque.

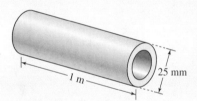

Figure 5.39 Shaft in Example 5.8.

TABLE 5.1   Material properties in Example 5.8

Material	G (GPa)	τ_{allow} (MPa)	γ (Mg/m³)
Titanium alloy	36	450	4.4
Aluminum	28	150	2.8

PLAN

The change in inner radius affects only the polar moment J and no other quantity in Equations (5.8) and (5.10). For each material we can find the minimum polar moment J needed to satisfy the stiffness and strength requirements. Knowing the minimum J for each material we can find the maximum inner radius. We can then find the volume and hence the mass of each material and make our decision on the lighter shaft.

Solution   We note that for both materials $\rho_{max} = 0.0125$ m and $x_2 - x_1 = 1$ m. From Equations (5.8) and (5.10) for titanium alloy we obtain the following limits on J_{Ti}:

$$(\Delta\phi)_{Ti} = \frac{400 \times 1}{36 \times 10^9 \times J_{Ti}} \leq 0.375 \qquad \text{or}$$

$$J_{Ti} \geq 29.63 \times 10^{-9} \text{ m}^4 \tag{E1}$$

$$(\tau_{max})_{Ti} = \frac{400 \times 0.0125}{J_{Ti}} \leq 450 \times 10^6 \qquad \text{or}$$

$$J_{Ti} \geq 11.11 \times 10^{-9} \text{ m}^4 \tag{E2}$$

Using similar calculations for the aluminum shaft we obtain the following limits on J_{Al}:

$$(\Delta\phi)_{Al} = \frac{400 \times 1}{28 \times 10^9 \times J_{Al}} \leq 0.375 \qquad \text{or}$$

$$J_{Al} \geq 38.10 \times 10^{-9} \text{ m}^4 \tag{E3}$$

$$(\tau_{max})_{Al} = \frac{400 \times 0.0125}{J_{Al}} \leq 150 \times 10^6 \qquad \text{or}$$

$$J_{Al} \geq 33.33 \times 10^{-9} \text{ m}^4 \tag{E4}$$

Thus if $J_{Ti} \geq 29.63 \times 10^{-9}$ m^4, it will meet both conditions in Equations (E1) and (E2). Similarly if $J_{Al} \geq 38.10 \times 10^{-9}$ m^4, it will meet both conditions in Equations (E3) and (E4). The internal diameters D_{Ti} and D_{Al} can be found as follows:

$$J_{Ti} = \frac{\pi}{32}(0.025^4 - D_{Ti}^4) \geq 29.63 \times 10^{-9} \quad D_{Ti} \leq 17.3 \times 10^{-3} \text{ m} \quad \text{(E5)}$$

$$J_{Al} = \frac{\pi}{32}(0.025^4 - D_{Al}^4) \geq 38.10 \times 10^{-9} \quad D_{Al} \leq 7.1 \times 10^{-3} \text{ m} \quad \text{(E6)}$$

Rounding downward to the closest millimeter, we obtain

$$D_{Ti} = 17 \times 10^{-3} \text{ m} \tag{E7}$$

$$D_{Al} = 7 \times 10^{-3} \text{ m} \tag{E8}$$

We can find the mass of each material by taking the product of the material density and the volume of a hollow cylinder,

$$M_{Ti} = 4.4 \times 10^6 \times \frac{\pi}{4}(0.025^2 - 0.017^2)1 = 1161 \text{ g} \tag{E9}$$

$$M_{Al} = 2.8 \times 10^6 \times \frac{\pi}{4}(0.025^2 - 0.007^2)1 = 1267 \text{ g} \tag{E10}$$

From Equations (E9) and (E10) we see that the titanium alloy shaft is lighter.

ANS.   A titanium alloy shaft should be used with an inside diameter of 17 mm.

COMMENTS

1. For both materials the stiffness limitation dictated the calculation of the internal diameter, as can be seen from Equations (E1) and (E3).

2. Even though the density of aluminum is lower than that titanium alloy, the mass of titanium is less. Because of the higher modulus of rigidity of titanium alloy we can meet the stiffness requirement using less material than for aluminum.

3. If in Equation (E5) we had 17.95×10^{-3} m on the right side, our answer for D_{Ti} would still be 17 mm because we have to round downward to ensure meeting the less-than sign requirement in Equation (E5).

EXAMPLE 5.9

The radius of a tapered circular shaft varies from $4r$ units to r units over a length of $40r$ units, as shown in Figure 5.40. The radius of the uniform shaft shown is r units. Determine:

(a) The angle of twist of wheel C with respect to the fixed end in terms of T, r, and G.

(b) The maximum shear stress in the shaft.

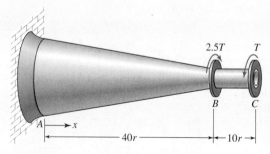

Figure 5.40   Shaft geometry in Example 5.9.

PLAN

(a) Using Equation (5.10), we can find the relative rotation of wheel C with respect to wheel B. But in section AB the radius is changing with x, and hence the polar moment J is changing with x. We can use Equation (5.7) and integrate to obtain the relative rotation of B with respect to A. We can combine the two relative rotations to obtain the relative rotation of C with respect to A.

(b) As per Equation (5.8), the maximum shear stress will exist where radius is minimum (J is minimum) and T is maximum. Thus by inspection, the maximum shear stress will exist on a section just left of B.

Solution We note that R is a linear function of x, which can be found in many ways. Here we find the two constants (intercept and slope) in the equation of a linear function as follows:

$$R(x) = a + bx$$

$$R(x = 0) = 4r = a + b(0) \qquad a = 4r$$

$$R(x = 40r) = r = 4r + b(40r) \qquad b = -0.075$$

$$R(x) = 4r - 0.075x$$

The imaginary cut is made in segment BC. This is the location of the imaginary cut defined by $40r < x < 50r$. To avoid calculating the wall reaction, we take the right part after the cut and draw the free-body diagram in Figure 5.41. From Equation (5.10) we can find the relative rotation of the section at C with respect to the section at B,

Figure 5.41 Free-body diagram of section BC in Example 5.9.

$$J_{BC} = \frac{\pi}{2}r^4$$

$$T_{BC} = T \qquad \text{(E1)}$$

$$\phi_C - \phi_B = \frac{T_{BC}(x_C - x_B)}{G_{BC}J_{BC}} = \frac{T \times 10r}{G(\pi/2)r^4} = \frac{6.366T}{Gr^3} \qquad \text{(E2)}$$

The imaginary cut is made in segment AB. This is the location of the imaginary cut defined by $0 < x < 40r$. To avoid calculating the wall reaction, we take the right part after the cut and draw the free-body diagram in Figure 5.42. We then determine the internal torque in AB,

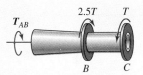

Figure 5.42 Free-body diagram of section AB in Example 5.9.

$$T_{AB} + 2.5T - T = 0 \qquad T_{AB} = -1.5T \qquad \text{(E3)}$$

$$J_{AB} = \frac{\pi}{2}(4r - 0.075x)^4 \qquad \text{(E4)}$$

Substituting Equations (E3) and (E4) into Equation (5.7) and integrating from point A to point B, we can find the relative rotation at the section at B with respect to the section at A,

$$\left(\frac{d\phi}{dx}\right)_{AB} = \frac{T_{AB}}{G_{AB}J_{AB}} = \frac{-1.5T}{G(\pi/2)(4r - 0.075x)^4}$$

or

$$\int_{\phi_A}^{\phi_B} d\phi = \int_{x_A}^{x_B} -\frac{3T}{G\pi(4r - 0.075x)^4} \, dx$$

$$\phi_B - \phi_A = -\frac{3T}{G\pi} \frac{1}{-3} \frac{1}{-0.075} \frac{1}{(4r - 0.075x)^3}\Big|_0^{40r}$$

$$= -\frac{T}{0.075G\pi}\left[\frac{1}{r^3} - \frac{1}{(4r)^3}\right] = -4.178\frac{T}{Gr^3} \qquad (E5)$$

(a) Adding Equations (E2) and (E5), we can find the rotation of the section at C with respect to the section at A as

$$\phi_C - \phi_A = \frac{T}{Gr^3}(6.366 - 4.178)$$

ANS. $\qquad \phi_C - \phi_A = 2.2\frac{T}{Gr^3}$ ccw

(b) Just left of the section at B we have $J_{AB} = \pi r^4/2$ and $\rho_{max} = r$. Substituting these values into Equation (5.8), we obtain the maximum torsional shear stress in the shaft,

$$\tau_{max} = \frac{-1.5Tr}{\pi r^4/2} = -\frac{0.955T}{r^3} \qquad \text{ANS.} \qquad |\tau_{max}| = \frac{0.955T}{r^3}$$

Dimension check: The dimensional consistency[12] of the answer is checked as follows:

$$T \rightarrow O(FL) \qquad r \rightarrow O(L) \qquad G \rightarrow O\left(\frac{F}{L^2}\right)$$

$$\phi \rightarrow O(\ ) \qquad \frac{T}{Gr^3} \rightarrow O\left(\frac{FL}{\frac{F}{L^2}L^3}\right) \rightarrow O(\ ) \rightarrow \text{checks}$$

$$\tau \rightarrow O\left(\frac{F}{L^2}\right) \qquad \frac{T}{r^3} \rightarrow O\left(\frac{FL}{L^3}\right) \rightarrow O\left(\frac{F}{L^2}\right) \rightarrow \text{checks}$$

---

[12]$O(\ )$ represents the dimension of the quantity on the left. F represents dimension for the force. L represents the dimension for length. Thus shear modulus, which has dimension of force (F) per unit area (L^2), is shown as $O(F/L^2)$.

COMMENT

The direction of the maximum shear stress can be determined using subscripts or intuitively, as shown in Figure 5.43.

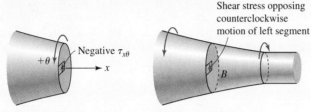

Figure 5.43   Direction of shear stress in Example 5.9.

EXAMPLE 5.10

A uniformly distributed torque of q in·lb/in is applied to an entire shaft, as shown in Figure 5.44. In addition to the distributed torque a concentrated torque of $T = 3qL$ in·lb is applied at section B. Let the shear modulus be G and the radius of the shaft r. In terms of $q, L, G,$ and r, determine:

(a) The rotation of the section at C.

(b) The maximum shear stress in the shaft.

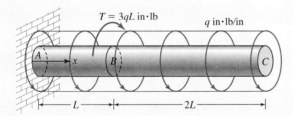

Figure 5.44   Shaft and loading in Example 5.10.

PLAN

(a) The internal torque in segments AB and BC as a function of x must be determined first. Then the relative rotation in each section is found from integrating Equation (5.7).

(b) Since J and ρ_{max} are constant over the entire shaft, the maximum shear stress will exist on a section where the internal torque is maximum. By plotting the internal torque as a function of x we can determine its maximum value.

The imaginary cut is made in segment AB, that is, $0 < x < L$. To avoid calculating the wall reaction, we take the right part after the cut and draw the free-body diagram in Figure 5.45. We replace the distributed torque by an equivalent torque that is equal to the distributed torque intensity multiplied by the length of the cut shaft (the rectangular area). From equilibrium we determine the internal torque in AB,

$$T_{AB} + 3qL - q(3L - x) = 0 \qquad T_{AB} = -qx \qquad \text{(E1)}$$

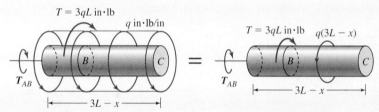

Figure 5.45 Free-body diagram of section AB in Example 5.10.

Substituting Equation (E1) into Equation (5.7) and integrating from point A to point B, we obtain the relative rotation of the section at B with respect to the section at A,

$$\left(\frac{d\phi}{dx}\right)_{AB} = \frac{T_{AB}}{G_{AB}J_{AB}} = \frac{-qx}{G\pi r^4/2} \qquad \int_{\phi_A}^{\phi_B} d\phi = -\int_{x_A=0}^{x_B=L} \frac{2qx}{G\pi r^4}\, dx$$

$$\phi_B - \phi_A = -\frac{qx^2}{G\pi r^4}\bigg|_0^L = -\frac{qL^2}{G\pi r^4} \qquad \text{(E2)}$$

The imaginary cut is now made in segment BC, that is, $L < x < 3L$. To avoid calculating the wall reaction, we take the right part after the cut and draw the free-body diagram in Figure 5.46. We then calculate the internal torque in segment BC,

$$T_{BC} - q(3L - x) = 0 \qquad T_{BC} = q(3L - x) \qquad \text{(E3)}$$

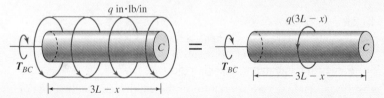

Figure 5.46 Free-body diagram of section BC in Example 5.10.

Substituting Equation (E3) into Equation (5.7) and integrating from point B to point C, we obtain the relative rotation of the section at C with respect to the section at B,

$$\left(\frac{d\phi}{dx}\right)_{BC} = \frac{T_{BC}}{G_{BC}J_{BC}} = \frac{q(3L-x)}{G\pi r^4/2} \qquad \int_{\phi_B}^{\phi_C} d\phi = \int_{x_B=L}^{x_C=3L} \frac{2q(3L-x)}{G\pi r^4}\,dx$$

$$\phi_C - \phi_B = \frac{2q}{G\pi r^4}\left(3Lx - \frac{x^2}{2}\right)\Bigg|_{L}^{3L}$$

$$= \frac{2q}{G\pi r^4}\left[9L^2 - \frac{(3L)^2}{2} - 3L^2 + \frac{L^2}{2}\right] = \frac{4qL^2}{G\pi r^4} \qquad (E4)$$

(a) Adding Equations (E2) and (E4), we obtain the rotation of the section at C with respect to the section at A,

$$\text{ANS.} \qquad \phi_C - \phi_A = \frac{qL^2}{G\pi r^4}(4-1) = \frac{3qL^2}{G\pi r^4} \text{ ccw}$$

(b) We can plot the internal torque as a function of x using Equations (E1) and (E3), as shown in Figure 5.47. The maximum torque will occur on a section just to the right of B. By using Equation (5.8) the maximum torsional shear stress can be calculated,

$$\text{ANS.} \qquad \tau_{max} = \frac{2qLr}{\pi r^4/2} = \frac{3qL}{\pi r^3}$$

Dimension check: The dimensional consistency (see Footnote 12) of the answers is checked as follows:

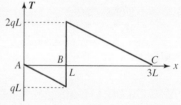

Figure 5.47 Torque diagram in Example 5.10.

$$q \to O\left(\frac{FL}{L}\right) \to O(F) \qquad r \to O(L) \quad L \to O(L) \quad G \to O\left(\frac{F}{L^2}\right)$$

$$\phi \to O(\ ) \qquad\qquad \frac{qL^2}{Gr^4} \to O\left(\frac{FL^2}{(F/L^2)L^4}\right) \to O(\ ) \to \text{checks}$$

$$\tau \to O\left(\frac{F}{L^2}\right) \qquad\qquad \frac{qL}{r^3} \to O\left(\frac{FL}{L^3}\right) \to O\left(\frac{F}{L^2}\right) \to \text{checks}$$

COMMENT

A common mistake is to write the incorrect length of the shaft as a function of x in the free-body diagrams. It should be remembered that the location of the cut is being defined by the variable x, which is measured from the common origin for all segments. Each cut produces two parts, and we are free to choose either part.

*5.2.7 General Approach to Distributed Torque

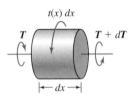

Figure 5.48 Equilibrium of an infinitesimal shaft element.

Distributed torques are usually due to inertial forces or frictional forces acting on the surface of the shaft. The internal torque T becomes a function of x when a shaft is subjected to a distributed external torque, as seen in Example 5.10. If $t(x)$ is a simple function, then we can find T as a function of x by drawing a free-body diagram, as we did in Example 5.10. However, if the distributed torque $t(x)$ is a complex function,[13] it may be easier to use the alternative described in this section.

Consider an infinitesimal shaft element that is created by making two imaginary cuts at a distance dx from each other, as shown in Figure 5.48.

By balancing moments in the axial direction, we obtain

$$(T + dT) + t(x)\ dx - T = 0$$

or

$$\boxed{\frac{dT}{dx} + t(x) = 0} \qquad (5.12)$$

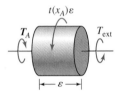

Figure 5.49 Boundary condition on internal torque.

Equation (5.12) represents the equilibrium equation at any section x. It assumes that $t(x)$ is positive counterclockwise with respect to the x axis. The sign of T obtained from Equation (5.12) corresponds to the direction defined by the sign convention in Definition 2. If $t(x)$ is zero in a segment of a shaft, then the internal torque is constant in that segment.

Equation (5.12) can be integrated to obtain the internal torque T. The integration constant can be found by knowing the value of the internal torque T at either end of the shaft. To obtain the value of T at the end of the shaft (say, point A), a free-body diagram is constructed after making an imaginary cut at an infinitesimal distance ε from the end, as shown in Figure 5.49. We then write the equilibrium equation as

$$\lim_{\varepsilon \to 0} [T_{ext} - T_A - t(x_A)\varepsilon] = 0 \qquad \text{or} \qquad T_A = T_{ext}$$

This equation shows that the distributed torque does not affect the boundary condition on the internal torque. The value of the internal torque T at the end of the shaft is equal to the concentrated external torque applied at the end.

EXAMPLE 5.11

The external torque on a drill bit varies linearly to a maximum intensity of q in·lb/in, as shown in Figure 5.50. If the drill bit diameter is d, its length L, and the modulus of rigidity G, determine the relative rotation of the end of the drill bit with respect to the chuck.

PLAN

The relative rotation of section B with respect to section A has to be found. We can substitute the given distributed torque in Equation (5.12)

---

[13]See Problems 5.32, 5.51, and 5.52.

and integrate to find the internal torque as a function of x. We can find the integration constant by using the condition that at section B the internal torque will be zero. We can substitute the internal torque expression into Equation (5.7) and integrate from point A to point B to find the relative rotation of section B with respect to section A.

Solution The distributed torque on the drill bit is counterclockwise with respect to the x axis. Thus we can substitute $t(x) = q(x/L)$ into Equation (5.12) and integrate to obtain

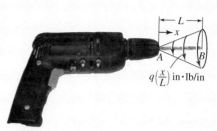

Figure 5.50 Distributed torque on a drill bit in Example 5.11.

$$\frac{dT}{dx} + q\frac{x}{L} = 0 \quad \text{or} \quad T = -q\frac{x^2}{2L} + c \qquad \text{(E1)}$$

where c is an integration constant. At point B, that is, at $x = L$, the internal torque should be zero as there is no concentrated applied torque at B. Substituting this into Equation (E1) we can find the integration constant,

$$T(x = L) = -q\frac{L^2}{2L} + c = 0 \quad \text{or} \quad c = \frac{qL}{2} \qquad \text{(E2)}$$

Substituting Equation (E2) into Equation (E1) and simplifying, we obtain

$$T = \frac{q}{2L}(L^2 - x^2) \qquad \text{(E3)}$$

Substituting Equation (E3) into Equation (5.7), we obtain

$$\frac{d\phi}{dx} = \frac{(q/2L)(L^2 - x^2)}{G\pi d^4/32} \qquad \text{(E4)}$$

Integrating Equation (E4) from point A to point B, we obtain the relative rotation of the section at B with respect to the section at A,

$$\int_{\phi_A}^{\phi_B} d\phi = \frac{16q}{\pi G L d^4} \int_{x_A=0}^{x_B=L} (L^2 - x^2)\, dx$$

or

$$\phi_B - \phi_A = \frac{16q}{\pi G L d^4}\left(L^2 x - \frac{x^3}{3}\right)\Bigg|_0^L$$

$$\text{ANS.} \quad \phi_B - \phi_A = \frac{32qL^2}{3\pi G d^4} \text{ ccw}$$

Dimension check: The dimension of consistency (see Footnote 12) of the answer is checked as follows:

$$q \to O\left(\frac{FL}{L}\right) \to O(F) \qquad d \to O(L) \qquad L \to O(L) \qquad G \to O\left(\frac{F}{L^2}\right)$$

$$\phi \to O(\ ) \qquad\qquad \frac{qL^2}{Gd^4} \to O\left(\frac{FL^2}{(F/L^2)L^4}\right) \to O(\ ) \to \text{checks}$$

COMMENTS

1. There was no free-body diagram drawn to find the internal torque because Equation (5.12), from which we started, is an equilibrium equation that is valid at each and every section of the shaft.

2. We could have obtained the internal torque by integrating Equation (5.12) from L to x as follows:

$$\int_{T_B=0}^{T} dT = -\int_{x_B=L}^{x} t(x)\, dx = -\int_{L}^{x} q\left(\frac{x}{L}\right) dx = \frac{q}{2L}(L^2 - x^2)$$

We could also integrate from x to L, that is,

$$\int_{T}^{T_B=0} dT = -\int_{x}^{x_B=L} t(x)\, dx$$

which will give the same expression for the internal torque. As long as the limits on the integrals on the left and right are consistent, the direction of integration is immaterial.

3. The internal torque can also be found using a free-body diagram. We can make an imaginary cut at some location x and draw the free-body diagram of the right side. The distributed torque represented by $\int_{x}^{L} t(x)\, dx$ is the area of the trapezoid *BCDE*, and this observation can be used for drawing a statically equivalent diagram, as shown in Figure 5.51. Equilibrium then gives us the value of the internal torque as before. We can find the internal torque as shown.

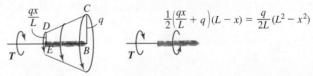

$$\frac{1}{2}\left(\frac{qx}{L} + q\right)(L - x) = \frac{q}{2L}(L^2 - x^2)$$

Figure 5.51 Internal torque by free-body diagram in Example 5.11.

4. The free-body diagram approach in Figure 5.51 is intuitive but more tedious and difficult than the use of Equation (5.12). As the function representing the distributed torque grows in complexity, the attractiveness of the mathematical approach of Equation (5.12) grows correspondingly.

QUICK TEST 5.1

Time: 20 minutes/Total: 20 points

Answer true or false and justify each answer in one sentence. Grade yourself with the answers given in Appendix G.

1. Torsional shear strain and stress vary linearly across a homogeneous cross section.

2. Torsional shear *strain* is maximum at the outermost radius for a homogeneous and a nonhomogeneous cross section.

3. Torsional shear *stress* is maximum at the outermost radius for a homogeneous and a nonhomogeneous cross section.

4. The formula $\tau_{x\theta} = T\rho/J$ can be used for finding the shear stress on a cross section of a tapered shaft.

5. The formula $\phi_2 - \phi_1 = T(x_2 - x_1)/GJ$ can be used for finding the relative rotation of a segment of a tapered shaft.

6. The formula $\tau_{x\theta} = T\rho/J$ can be used for finding the shear stress on a cross section of a shaft subjected to distributed torques.

7. The formula $\phi_2 - \phi_1 = T(x_2 - x_1)/GJ$ can be used for finding the relative rotation of a segment of a shaft subjected to distributed torques.

8. The equation $T = \int_A \rho\tau_{x\theta}\, dA$ cannot be used for nonlinear materials.

9. The equation $T = \int_A \rho\tau_{x\theta}\, dA$ can be used for a nonhomogeneous cross section.

10. Internal torques jump by the value of the concentrated external torque at a section.

PROBLEM SET 5.2

In Problems 5.14 through 5.17 determine the direction of torsional shear stress at points A and B: (a) by inspection; (b) by using the sign convention for internal torque and the subscripts. Report your answer as a positive or negative τ_{xy}.

5.14 (Figure P5.14).

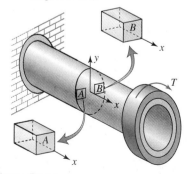

Figure P5.14

5.15 (Figure P5.15).

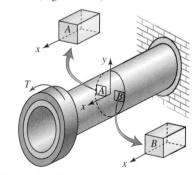

Figure P5.15

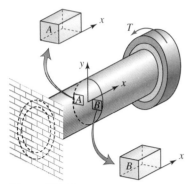

Figure P5.16

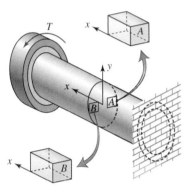

Figure P5.17

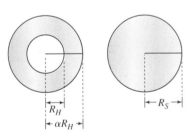

Figure P5.18

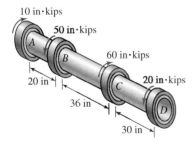

Figure P5.20

5.16 (Figure P5.16).

5.17 (Figure P5.17).

5.18 The two shafts shown in Figure P5.18 have the same cross sectional areas A. Show that the ratio of the polar moment of inertia of the hollow shaft to that of the solid shaft is as follows:

$$\frac{J_{\text{hollow}}}{J_{\text{solid}}} = \frac{\alpha^2 + 1}{\alpha^2 - 1}$$

5.19 Show that for a thin tube of thickness t and centerline radius R the polar moment of inertia can be approximated by $J = 2\pi R^3 t$. By thin tube we imply $t < R/10$.

In Problems 5.20 through 5.23 draw the torque diagrams. Check the values of internal torque by making imaginary cuts and drawing free-body diagrams.

5.20 Draw the torque diagram in Figure P5.20.

5.21 Draw the torque diagram in Figure P5.21.

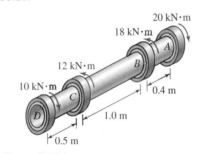

Figure P5.21

5.22 Draw the torque diagram in Figure P5.22.

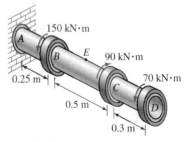

Figure P5.22

5.23 Draw the torque diagram in Figure P5.23.

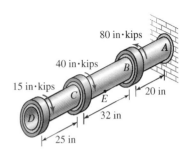

Figure P5.23

5.24 The torsional rigidity of the shaft in Problem 5.20 is 90,000 kips·in². Determine the rotation of the rigid wheel D with respect to the rigid wheel A.

5.25 The torsional rigidity of the shaft in Problem 5.21 is 1270 kN·m². Determine the rotation of the rigid wheel D with respect to the rigid wheel A.

5.26 The shaft in Problem 5.22 is made of steel ($G = 80$ GPa) and has a diameter of 150 mm. Determine: (a) the rotation of the rigid wheel D; (b) the magnitude of the torsional shear stress at point E and show it on a stress cube (Point E is on the top surface of the shaft.); (c) the magnitude of maximum torsional shear *strain* in the shaft.

5.27 The shaft in Problem 5.23 is made of aluminum ($G = 4000$ ksi) and has a diameter of 4 in. Determine: (a) the rotation of the rigid wheel D; (b) the magnitude of the torsional shear stress at point E and show it on a stress cube (Point E is on the bottom surface of the shaft.); (c) the magnitude of maximum torsional shear *strain* in the shaft.

5.28 A solid circular steel shaft BC ($G_s = 12,000$ ksi) is securely attached to two hollow steel shafts AB and CD, as shown in Figure P5.28. Determine: (a) the angle of rotation of the section at D with respect to the section at A; (b) the magnitude of maximum torsional shear stress in the shaft; (c) the

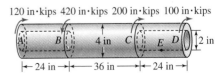

Figure P5.28

torsional shear stress at point E and show it on a stress cube. (Point E is on the inside bottom surface of CD.)

5.29 A steel shaft ($G = 80$ GPa) is subjected to the torques shown in Figure P5.29. Determine: (a) the rotation of section A with respect to the no-load position; (b) the torsional shear stress at point E and show it on a stress cube. (Point E is on the surface of the shaft.)

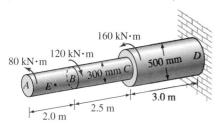

Figure P5.29

Tapered shafts

5.30 The radius of the tapered circular shaft shown in Figure P5.30 varies from 200 mm at A to 50 mm at B. The shear modulus of the material is $G = 40$ GPa. Determine: (a) the angle of rotation of wheel C with respect to the fixed end; (b) the maximum shear strain in the shaft.

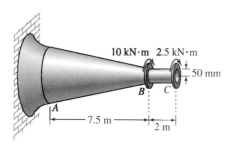

Figure P5.30

5.31 The radius of the tapered shaft shown in Figure P5.31 varies as $R = (r/L)(2L - 0.25x)$.

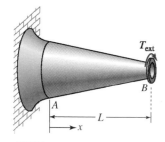

Figure P5.31

In terms of T_{ext}, L, G, and r, determine: (a) the rotation of wheel B; (b) the magnitude of maximum torsional shear stress in the shaft.

Distributed torques

5.32 The external torque on a drill bit varies as a quadratic function to a maximum intensity of q in·lb/in, as shown in Figure P5.32. If the drill bit diameter is d, its length L, and its modulus of rigidity G, determine: (a) the maximum torsional shear stress on the drill bit; (b) the relative rotation of the end of the drill bit with respect to the chuck.

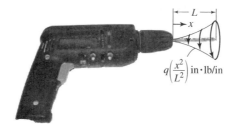

Figure P5.32

5.33 A circular solid shaft is acted upon by torques, as shown in Figure P5.33. Determine the rotation of the rigid wheel A with respect to the fixed end C in terms of q, L, G, and J.

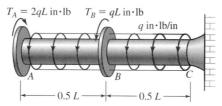

Figure P5.33

Design problems

5.34 A thin steel tube ($G = 12,000$ ksi) of $\frac{1}{8}$-in thickness has a mean diameter of 6 in and a length of 36 in. What is the maximum torque the tube can transmit if the allowable torsional shear stress is 10 ksi and the allowable relative rotation of the two ends is 0.015 rad?

5.35 Determine the maximum torque that can be applied on a 2-in-diameter solid aluminum shaft ($G = 4000$ ksi) if the allowable torsional shear stress is 18 ksi and the relative rotation over 4 ft of the shaft is to be limited to 0.2 rad.

5.36 A hollow steel shaft ($G = 80$ GPa) with an outside radius of 30 mm is to transmit a torque of 2700 N·m. The allowable torsional shear stress is 120 MPa and the allowable relative rotation over 1 m is 0.1 rad. Determine the maximum permissible inner radius to the nearest millimeter.

5.37 A 5-ft-long hollow shaft is to transmit a torque of 200 in·kips. The outer diameter of the shaft must be 6 in to fit existing attachments. The relative rotation of the two ends of the shaft is limited to 0.05 rad. The shaft can be made of steel or aluminum. The shear modulus of elasticity G, the allowable shear stress τ_{allow}, and the specific weight γ are given in Table P5.37. Determine the maximum inner diameter to the nearest $\frac{1}{8}$ in of the lightest shaft that can be used for transmitting the torque and the corresponding weight.

TABLE P5.37

Material	G (ksi)	τ_{allow} (ksi)	γ (lb/in³)
Steel	12,000	18	0.285
Aluminum	4000	10	0.100

Transmission of power[14]

Power transmitted through a shaft is given by the equation

$$P = T\omega = 2\pi f T \qquad (5.13)$$

where T is the torque transmitted, ω is the rotational speed in radians per second, and f is the frequency of rotation in hertz (Hz). Power is reported in units of horsepower in U.S. customary units or in watts. 1 horsepower (hp) is equal to 550 ft·lb/s = 6600 in·lb/s and 1 watt (W) is equal to 1 N·m/s. Using Equation (5.13), solve Problems 5.38 through 5.40.

5.38 A 100-hp motor is driving a pulley and belt system, as shown in Figure P5.38. If

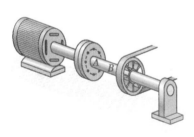

Figure P5.38

[14]Power P is the rate at which work dW/dt is done; and work W done by a constant torque is equal to the product of torque T and angle of rotation ϕ. Noting that $\omega = d\phi/dt$, we obtain Equation (5.13).

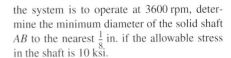

the system is to operate at 3600 rpm, determine the minimum diameter of the solid shaft AB to the nearest $\frac{1}{8}$ in. if the allowable stress in the shaft is 10 ksi.

5.39 The bolts used in the coupling for transferring power in Problem 5.38 have an allowable strength of 12 ksi. Determine the minimum number (> 4) of $\frac{1}{4}$-in-diameter bolts that must be placed at a radius of $\frac{5}{8}$ in.

5.40 A 20-kW motor drives three gears, which are rotating at a frequency of 20 Hz. Gear A next to the motor transfers 8 kW of power. Gear B, which is in the middle, transfers 7 kW of power. Gear C, which is at the far end from the motor, transfers the remaining 5 kW of power. A single solid steel shaft connecting the motors to all three gears is to be used. The steel used has a yield strength in shear of 145 MPa. Assuming a factor of safety of 1.5, what is the minimum diameter of the shaft to the nearest millimeter that can be used if failure due to yielding is to be avoided? What is the magnitude of maximum torsional stress in the segment between gears A and B?

Stretch yourself

5.41 A circular shaft has a constant torsional rigidity GJ and is acted upon by a distributed torque $t(x)$. If at section A the internal torque is zero, show that the relative rotation of the section at B with respect to the rotation of the section at A is given by

$$\phi_B - \phi_A = \frac{1}{GJ}\left[\int_{x_A}^{x_B}(x - x_B)t(x)\,dx\right] \quad (5.14)$$

5.42 A solid circular shaft of radius R and length L is twisted by an applied torque T. The stress–strain relationship for a nonlinear material is given by the power law $\tau = G\gamma^n$. If Assumptions 1 through 4 are applicable, show that the maximum shear stress in the shaft and the relative rotation of the two ends are as follows:

$$\tau_{max} = \frac{T(n+3)}{2\pi R^3}$$

$$\Delta\phi = \left[\frac{(n+3)T}{2\pi GR^{3+n}}\right]^{1/n} L$$

Substitute $n = 1$ in the formulas and show that we obtain the same results as from Equations (5.8) and (5.10).

5.43 The internal torque T and the displacements of a point on a cross section of a noncircular shaft (Figure P5.43) are given by the equations

$$u = \psi(y, z) \frac{d\phi}{dx} \quad (5.15a)$$

$$v = -xz \frac{d\phi}{dx} \quad (5.15b)$$

$$w = xy \frac{d\phi}{dx} \quad (5.15c)$$

$$T = \int_A (y \tau_{xz} - z \tau_{xy}) \, dA \quad (5.16)$$

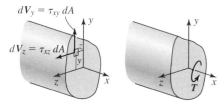

Figure P5.43 Torsion of noncircular shafts.

where u, v, and w are the displacements in the x, y, and z directions, respectively; $d\phi/dx$ is the rate of twist and is considered constant; and $\psi(x, y)$ is called the warping function[15] and describes the movement of points out of the plane of cross section. Using Equations (2.10d) and (2.10f) and Hooke's law, show that the shear stresses for a noncircular shaft are given by

$$\tau_{xy} = G\left(\frac{\partial \psi}{\partial y} - z\right)\frac{d\phi}{dx} \quad (5.17a)$$

$$\tau_{xz} = G\left(\frac{\partial \psi}{\partial z} + y\right)\frac{d\phi}{dx} \quad (5.17b)$$

5.44 Show that for circular shafts, $\psi(x, y) = 0$, the equations in Problem 5.43 reduce to Equation (5.7).

5.45 Consider the dynamic equilibrium of the differential element shown in Figure P5.45, where T is the internal torque, γ is the material density, J is the polar area moment of inertia, and $\partial^2 \phi/\partial t^2$ is the angular acceleration. Show that

$$\frac{\partial^2 \phi}{\partial t^2} = c^2 \frac{\partial^2 \phi}{\partial x^2} \quad \text{where } c^2 = \sqrt{\frac{G}{\gamma}} \quad (5.18)$$

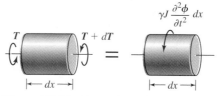

Figure P5.45 Dynamic equilibrium.

5.46 Show by substitution that the following solution satisfies Equation (5.18):

$$\phi = \left(A \cos \frac{\omega x}{c} + B \sin \frac{\omega x}{c}\right)$$
$$\times (C \cos \omega t + D \sin \omega t) \quad (5.19)$$

where A, B, C, and D are constants that are determined from the boundary conditions and the initial conditions and ω is the frequency of vibration.

Computer problems[16]

5.47 A hollow aluminum shaft of 5 ft in length is to carry a torque of 200 in·kips. The inner radius of the shaft is 1 in. If the maximum torsional shear stress in the shaft is to be limited to 10 ksi, determine the minimum outer radius to the nearest $\frac{1}{8}$ in.

5.48 A 4-ft-long hollow shaft is to transmit a torque of 100 in·kips. The relative rotation of the two ends of the shaft is limited to 0.06 rad. The shaft can be made of steel or aluminum. The shear modulus of rigidity G, the allowable shear stress τ_{allow}, and the specific weight γ are given in Table P5.48. The inner radius of the shaft is 1 in. Determine the outer radius of the lightest shaft that can be used for transmitting the torque and the corresponding weight.

TABLE P5.48

Material	G (ksi)	τ_{allow} (ksi)	γ (lb/in³)
Steel	12,000	18	0.285
Aluminum	4000	10	0.100

---

[15]Equations of elasticity show that the warping function satisfies the Laplace equation, $\partial^2 \psi/\partial y^2 + \partial^2 \psi/\partial z^2 = 0$.

[16]See Appendix B for the numerical algorithms you can program either in your calculator, on a spread sheet, or on a computer in any language you are comfortable with.

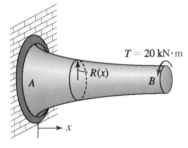

Figure P5.49

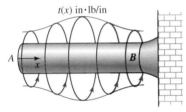

Figure P5.51

5.49 Table P5.49 shows the measured radii of the solid tapered shaft shown in Figure P5.49, at several points along the axis of the shaft. The shaft is made of aluminum ($G = 28$ GPa) and has a length of 1.5 m. Determine: (a) the rotation of the free end with respect to the wall using numerical integration; (b) the maximum shear stress in the shaft.

TABLE P5.49

x (m)	$R(x)$ (mm)
0.0	100.6
0.1	92.7
0.2	82.6
0.3	79.6
0.4	75.9
0.5	68.8
0.6	68.0
0.7	65.9
0.8	60.1
0.9	60.3
1.0	59.1
1.1	54.0
1.2	54.8
1.3	54.1
1.4	49.4
1.5	50.6

5.50 Let the radius of the tapered shaft in Problem 5.49 be represented by the equation $R(x) = a + bx$. Using the data in Table P5.49 determine the constants a and b by the least-squares method and then find

the rotation of the section at B by analytical integration.

5.51 Table P5.51 shows the values of distributed torque at several points along the axis of the solid steel shaft ($G = 12,000$ ksi) shown in Figure P5.51. The shaft has a length of 36 in and a diameter of 1 in. Determine: (a) the rotation of end A with respect to the wall using numerical integration; (b) the maximum shear stress in the shaft.

TABLE P5.51

x (inches)	$t(x)$ (in·lb/in)
0	93.0
3	146.0
6	214.1
9	260.4
12	335.0
15	424.7
18	492.6
21	588.8
24	700.1
27	789.6
30	907.4
33	1040.3
36	1151.4

5.52 Let the distributed torque $t(x)$ in Problem 5.51 be represented by the equation $t(x) = a + bx + cx^2$. Using the data in Table P5.51 determine the constants a, b, and c by the least-squares method and then find the rotation of the section at B by analytical integration.

5.3 STATICALLY INDETERMINATE SHAFTS

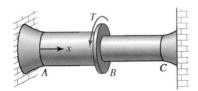

Figure 5.52 Statically indeterminate shaft.

Figure 5.52 shows a statically indeterminate shaft. In statically indeterminate shafts we have two reaction torques, one at the left and the other at the right end of the shaft. But we have only one static equilibrium equation—the sum of all torques in the x direction should be zero. Thus the degree of static redundancy is 1. As in statically indeterminate axial structures, we need to generate a number of compatibility equations equal to the degree of static redundancy.

The continuity of ϕ, and the fact that the sections at the left and right walls have zero rotation, can be used for writing the compatibility equation. The compatibility equation is that the relative rotation of the right wall with respect to the left wall is zero. Once more we can use either the displacement method or the force method.

In the displacement method we can use the rotation of the sections as the unknowns. If torque is applied at several sections along the shaft, then the rotation of each of the sections is treated as an unknown.

In the force method we can use either the reaction torque as the unknown or the internal torques in the sections as the unknowns. Since the degree of static redundancy is 1, the simplest approach is often to work the problem using the left wall (or the right wall) reaction as the unknown variable and solving it using the compatibility equation. The solution procedure is outlined next.

Step 1 Make an imaginary cut in each segment and draw free-body diagrams by taking the left (or right) part if the left (right) wall reaction is carried as the unknown in the problem. Alternatively, draw the torque diagram in terms of the reaction torque.

Step 2 From equilibrium equations for each free-body diagram, obtain the internal torque in terms of the reaction torque.

Step 3 Using Equation (5.10) write the relative rotation of each segment in terms of the reaction torque.

Step 4 Add all the relative rotations. Obtain the rotation of the right wall with respect to the left wall and set it equal to zero to obtain the reaction torque.

Step 5 Once the reaction torque is known, the internal torques can be found from equations obtained in Step 2, and deformation and stresses are calculated.

EXAMPLE 5.12

A solid circular steel shaft ($G_s = 12,000$ ksi, $E_s = 30,000$ ksi) of 4-in diameter is loaded as shown in Figure 5.53. Determine the maximum shear stress in the shaft.

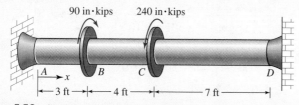

Figure 5.53 Shaft in Example 5.12.

PLAN

We draw a torque diagram in terms of T_A, the reaction torque at A, and obtain the internal torques in terms of T_A. Substituting the internal torques into Equation (5.10), we can find the relative rotation of the right end with respect to the left end for each segment, in terms of T_A. Adding the relative rotations, we obtain the rotation of section D with respect to section A in terms of T_A, equate it to zero, and find T_A. Knowing T_A, the internal torque in each segment can be found. The maximum shear stress will occur in the segment that has the maximum internal torque as the radius of the shaft is the same in all segments.

Solution We can find the polar moment of inertia as

$$J = \frac{\pi 4^2}{32} = 25.13 \text{ in}^4 \tag{E1}$$

and the torsional rigidity as

$$GJ = 301.6 \times 10^3 \text{ kips} \cdot \text{in}^2 \tag{E2}$$

Let T_A, the reaction torque at A, be counterclockwise with respect to the x axis. Using the template shown in Figure 5.54a, we can draw the torque diagram in Figure 5.54b. The internal torques are

$$\boldsymbol{T}_{AB} = -T_A \qquad \boldsymbol{T}_{BC} = -T_A + 90 \qquad \boldsymbol{T}_{CD} = -T_A - 150 \tag{E3}$$

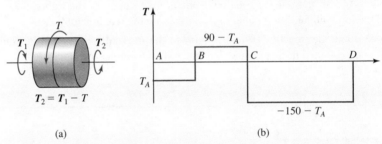

(a) (b)

Figure 5.54 Template and torque diagram in Example 5.12.

Substituting Equations (E2) and (E3) into Equation (5.10), we obtain the relative rotation in each segment,

$$\phi_B - \phi_A = \frac{T_{AB}(x_B - x_A)}{G_{AB}J_{AB}} = \frac{-T_A \times 36}{301.6 \times 10^3} = -0.1194 \times 10^{-3} T_A \tag{E4}$$

$$\phi_C - \phi_B = \frac{T_{BC}(x_C - x_B)}{G_{BC}J_{BC}} = \frac{(-T_A + 90)48}{301.6 \times 10^3}$$
$$= (-0.1592 T_A + 14.32)10^{-3} \tag{E5}$$

$$\phi_D - \phi_C = \frac{T_{CD}(x_D - x_C)}{G_{CD}J_{CD}} = \frac{(-T_A - 150)84}{301.6 \times 10^3}$$
$$= (-0.2785 T_A - 41.78)10^{-3} \tag{E6}$$

Adding Equations (E4), (E5), and (E6), we obtain $\phi_D - \phi_A$. Equating $\phi_D - \phi_A$ to zero, we can find T_A,

$$\phi_D - \phi_A = (-0.1194 T_A - 0.1592 T_A$$
$$+ 14.32 - 0.2785 T_A - 41.78)10^{-3} = 0$$

$$T_A = -49.28 \text{ in} \cdot \text{kips} \tag{E7}$$

Substituting Equation (E7) into Equations (E3), we obtain the internal torques,

$$T_{AB} = 49.28 \text{ in·kips} \qquad T_{BC} = 139.28 \text{ in·kips}$$
$$T_{CD} = -100.72 \text{ in·kips} \tag{E8}$$

$\rho_{max} = 2$ in for all segments and J is the same for all segments. Thus the maximum shear stress will occur in segment BC since the internal torque in BC is maximum. The maximum shear stress can be found using Equation (5.8) as follows:

$$\tau_{max} = \frac{T_{BC}(\rho_{BC})_{max}}{J_{BC}} = \frac{139.3 \times 2}{25.13}$$

ANS. $\qquad \tau_{max} = 11.1 \text{ ksi}$

COMMENTS

1. We can find the internal torques in terms of T_A by making imaginary cuts through AB, BC, and CD and drawing free-body diagrams, as shown in Figure 5.55.

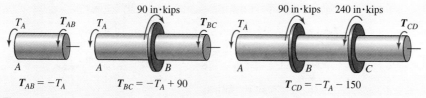

Figure 5.55 Free-body diagrams in Example 5.12.

2. Because the applied torque at C is bigger than that at B, the reaction torques at A and D will be opposite in direction to the torque at C. In other words, the reaction torques at A and D should by clockwise with respect to the x axis. The sign of T_A confirms this intuitive reasoning.

3. Suppose we had started by assuming that the reaction torque at A was counterclockwise in direction with respect to the x axis, and we had drawn the torque diagram. The effect would be to shift the entire torque diagram upward, and we would obtain $T_{AB} = T_A$, $T_{BC} = T_A + 90$, and $T_{CD} = T_A - 150$. Using these internal torques we can calculate $\phi_D - \phi_A$ as before and equate it to zero, to obtain $T_A = 49.28 \text{ in·kips}$ which, on substitution into the internal torque expressions, would yield internal torque values as given by Equations (E8). Thus for the purpose of calculation, the assumed direction of the reaction torques is immaterial, but the answer should be checked intuitively, as described in comment 2.

EXAMPLE 5.13

A solid aluminum shaft ($G_{al} = 27$ GPa) and a solid bronze shaft ($G_{br} = 45$ GPa) are securely connected to a rigid wheel, as shown in Figure 5.56. The shaft has a diameter of 75 mm. The allowable shear stresses in aluminum and bronze are 100 MPa and 120 MPa, respectively. Determine the maximum torque that can act on the wheel.

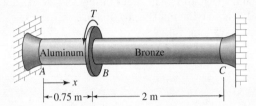

Figure 5.56 Shaft in Example 5.13.

PLAN

We can solve for the maximum shear stress in aluminum and bronze in terms of T and calculate the two limiting values on T to meet the specified conditions. The lower of the two values would meet both conditions and hence is the maximum permissible value of T.

Solution We can find the polar moment of inertia,

$$J = \frac{\pi 0.075^4}{32} = 3.106 \times 10^{-6} \text{ m}^4 \qquad (E1)$$

and the torsional rigidities,

$$G_{AB}J_{AB} = 83.87 \times 10^3 \text{ N·m}^2 \qquad (E2)$$

$$G_{BC}J_{BC} = 139.8 \times 10^3 \text{ N·m}^2 \qquad (E3)$$

Let T_A, the reaction torque at A, be clockwise with respect to the x axis. We can make imaginary cuts in AB and BC and draw the free-body diagrams as shown in Figure 5.57. From equilibrium we can find the internal torques in terms of T_A and T,

$$T_{AB} = T_A \qquad (E4)$$

$$T_{BC} = T_A - T \qquad (E5)$$

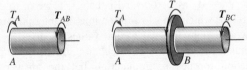

Figure 5.57 Free-body diagrams in Example 5.13.

Substituting Equations (E2) through (E5) into Equation (5.10), we obtain the relative rotation in each segment,

$$\phi_B - \phi_A = \frac{T_{AB}(x_B - x_A)}{G_{AB}J_{AB}} = \frac{T_A \times 0.75}{83.87 \times 10^3} = 8.942 \times 10^{-6}T_A \quad \text{(E6)}$$

$$\phi_C - \phi_B = \frac{T_{BC}(x_C - x_B)}{G_{BC}J_{BC}} = \frac{(T_A - T)2}{139.8 \times 10^3}$$

$$= (14.31T_A - 14.31T)10^{-6} \quad \text{(E7)}$$

Adding Equations (E6) and (E7), we obtain $\phi_C - \phi_A$. Equating $\phi_C - \phi_A$ to zero, we can find T_A in terms of T,

$$\phi_C - \phi_A = (8.942T_A + 14.31T_A - 14.31T)10^{-6} = 0$$

$$T_A = 0.6154T \quad \text{(E8)}$$

Substituting Equation (E8) into Equations (E4) and (E5), we obtain the internal torques,

$$T_{AB} = 0.6154T \qquad T_{BC} = -0.3846T$$

The maximum shear stress in segment AB can be found in terms of T using Equation (5.8) and noting that $\rho_{max} = 0.0375$ mm. Noting that the maximum stress in AB must be less than 100 MPa, we obtain one limit on the external torque T as follows:

$$|(\tau_{AB})_{max}| = \left|\frac{T_{AB}(\rho_{AB})_{max}}{J_{AB}}\right| = \frac{0.6154T \times 0.0375}{3.106 \times 10^{-6}}$$

$$= 7.43T \times 10^3 \le 100 \times 10^6$$

Thus,

$$T \le 13.46 \times 10^3 \text{ N·m} \quad \text{(E9)}$$

By calculating the maximum shear stress in segment BC and noting that it must be less than 120 MPa, we obtain the second limit on T as follows:

$$|(\tau_{BC})_{max}| = \left|\frac{T_{BC}(\rho_{BC})_{max}}{J_{BC}}\right| = \frac{0.3846T \times 0.0375}{3.106 \times 10^{-6}}$$

$$= 4.643T \times 10^3 \le 120 \times 10^6$$

Thus,

$$T \le 25.84 \times 10^3 \text{ N·m} \quad \text{(E10)}$$

The maximum value of T that satisfies Equations (E9) and (E10) is $T_{max} = 13.4$ kN·m.   ANS.

COMMENTS

1. The maximum torque is limited by the maximum shear stress in bronze. If we had a limitation on the rotation of the wheel, then we could easily incorporate it by calculating ϕ_B from Equation (E6) in terms of T.

2. We could have solved the problem by initially assuming that one of the materials has the limiting stress value, say aluminum. We can then do our calculations and find the maximum stress in bronze, which would exceed the limiting value of 120 MPa. We would then resolve the problem. The process, though correct, can become tedious as the number of limitations increases. In the solution process shown, we solve the problem only once, irrespective of the number of limitations, as we defer the decision as to which limitation dictates the maximum value of the torque toward the end.

3. We could have solved this problem by the displacement method. In that case we would carry the rotation of the wheel ϕ_B as the unknown.

*5.4 COMPOSITE SHAFTS

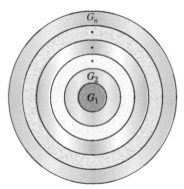

Figure 5.58   Composite shaft cross section.

Shafts with cross sections constructed from more than one material are called *composite shafts*. Clearly, Assumption 7 of material homogeneity across the cross section is no longer valid. Hence the formulas that follow after Assumption 7 are no longer valid. However, we shall see that after accounting for the cross-sectional nonhomogeneity, the derivation process to get the new formulas is the same as before.

We assume that all material cross sections are axisymmetric, as shown in Figure 5.58. We assume that all materials are securely bonded to one another, that is, there is no relative rotation at any interface of the materials. This assumption of being securely bonded is necessary to preserve our kinematic assumptions, Assumptions 1 through 3, that is, there is no warping and all radial lines in a cross section rotate by equal amounts. We further assume that we still have small strains, and all materials are linear, elastic, and isotropic. In other words, Assumptions 1 through 6 are still valid. Thus Equation (5.6), which precedes Assumption 7, is still valid and forms our starting point for deriving the torsional formulas for composite shafts.

Consider the laminated cross section shown in Figure 5.58. Each material has a shear modulus of elasticity G_i that is constant over the material cross-sectional area A_i. Suppose there are n materials in the cross section. The integral in Equation (5.6) can be written as the sum of the integrals over each material,

$$T = \frac{d\phi}{dx}\left(\int_{A_1} G_1\rho^2\,dA + \int_{A_2} G_2\rho^2\,dA + \cdots + \int_{A_n} G_n\rho^2\,dA\right)$$

where G_i is a constant in each integral and can be taken outside the integral. The remaining integral is the polar moment J_i of the ith material. We thus obtain

$$T = \frac{d\phi}{dx}(G_1J_1 + G_2J_2 + \cdots + G_NJ_N)$$

Written more compactly,

$$\boxed{T = \frac{d\phi}{dx}\left(\sum_{j=1}^{n} G_jJ_j\right)} \tag{5.20}$$

Equation (5.20) shows that the torsional rigidity of the composite shaft is the sum of the torsional rigidities of the individual materials. We can write Equation (5.7) for the ith material as $(\tau_{x\theta})_i = G_i\rho\,(d\phi/dx)$, where $(\tau_{x\theta})_i$ is the torsional shear stress in the ith material. Substituting Equation (5.20), we obtain

$$\boxed{(\tau_{x\theta})_i = \frac{G_i\rho T}{\sum_{j=1}^{n} G_jJ_j}} \tag{5.21}$$

We assume that torsional rigidity and the internal torque are not functions of x, between x_1 and x_2. In other words, Assumptions 8 through 10 are applicable to composite shafts. Then $d\phi/dx$ is constant between x_1 and x_2, and we can write $d\phi/dx = (\phi_2 - \phi_1)/(x_2 - x_1)$ in Equation (5.20) to obtain

$$\boxed{\phi_2 - \phi_1 = \frac{T(x_2 - x_1)}{\sum_{j=1}^{n} G_jJ_j}} \tag{5.22}$$

Equations (5.21) and (5.22),[17] which are applicable to composite shafts, now replace Equations (5.8) and (5.10) for the homogeneous cross section. The internal torque T represents the statically equivalent torque over the entire cross section, as it did for the homogeneous cross section. The analysis techniques for finding the internal torque T at a cross section remain the same as before, that is, one may use either free-body diagrams or torque diagrams to find T.

From Equations (5.20) and (5.3) it is clear that the torsional shear strain still varies linearly across the cross section. But the shear stress $(\tau_{x\theta})_i$ in each material, given by Equation (5.21), is a piecewise linear function that changes its slope with each material point at which shear stress is evaluated, as was demonstrated in Example 5.4.

---

[17]If we consider a homogeneous material, then $G_1 = G_2 = \cdots = G_i = \cdots = G_n = G$. Substituting this into Equations (5.21) and (5.22), we obtain Equations (5.8) and (5.10) for homogeneous material, as expected.

EXAMPLE 5.14

A solid steel shaft (G_{st} = 80 GPa), 3 m long, is securely fastened to a hollow bronze shaft (G_{br} = 40 GPa) that is 2 m long, as shown in Figure 5.59. Determine:

(a) The maximum value of torsional shear stress in the shaft.

(b) The rotation of the right end with respect to the wall.

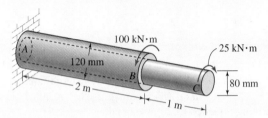

Figure 5.59 Composite shaft in Example 5.14.

PLAN

Segment BC is homogeneous, and Equations (5.8) and (5.10) can be used in this segment to find the relative rotation of section C with respect to B and the maximum shear stress in the segment. Segment AB is made from two materials. We can use Equations (5.21) and (5.22) to find the relative rotation of section B with respect to A and the maximum shear stress in each material in segment AB.

Solution We can find the polar moments as

$$J_{st} = \frac{\pi}{32} 0.08^4 = 4.02 \times 10^{-6} \text{ m}^4$$

$$J_{br} = \frac{\pi}{32}(0.12^4 - 0.08^4) = 16.33 \times 10^{-6} \text{ m}^4$$

and the torsional rigidities as

$$G_{st} J_{st} = 321.6 \times 10^3 \text{ N} \cdot \text{m}^2 \qquad G_{br} J_{br} = 653.2 \times 10^3 \text{ N} \cdot \text{m}^2$$

$$\sum_{j=1}^{2} G_j J_j = G_{st} J_{st} + G_{br} J_{br} = 974.8 \times 10^3 \text{ N} \cdot \text{m}^2$$

We can find the internal torque in BC by drawing the free-body diagram of the right part after making an imaginary cut in BC, as shown in Figure 5.60,

$$T_{BC} = -25 \text{ kN} \cdot \text{m} \qquad (E1)$$

Figure 5.60 Free-body diagram of segment BC in Example 5.14.

Using Equations (5.8) and (5.10), we can find the relative rotation of the section at C with respect to B and the maximum torsional shear stress in BC,

$$\phi_C - \phi_B = \frac{T_{BC}(x_C - x_B)}{G_{st} J_{st}} = \frac{-25 \times 10^3 \times 1}{321.6 \times 10^3} = -0.0781 \text{ rad} \quad (E2)$$

$$(\tau_{BC})_{max} = \frac{T_{BC}(\rho_{BC})_{max}}{J_{st}} = \frac{-25 \times 10^3 \times 0.04}{4.02 \times 10^{-6}} = -250 \times 10^6 \quad (E3)$$

We can find the internal torque in AB by drawing the free-body diagram of the right part after making an imaginary cut in AB, as shown in Figure 5.61,

$$T_{AB} - 100 + 25 = 0$$

$$T_{AB} = 75 \text{ kN·m} \quad (E4)$$

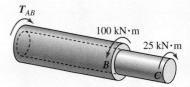

Figure 5.61 Free-body diagram of segment AB in Example 5.14.

Using Equation (5.22), we can find the relative rotation of the section at B with respect to the section at A,

$$\phi_B - \phi_A = \frac{T_{AB}(x_B - x_A)}{\sum_{j=1}^{2} G_j J_j} = \frac{75 \times 10^3 \times 2}{974.8 \times 10^3} = 0.1538 \text{ rad} \quad (E5)$$

Adding Equations (E2) and (E5) we obtain the relative rotation of the section at C with respect to the section at A. Noting that rotation at A is zero, we obtain the rotation of the section at C,

$$\phi_C - \phi_A^{\,0} = 0.1538 - 0.0781 = 0.0757 \text{ rad}$$

$$\text{ANS.} \qquad \phi_C = 0.0757 \text{ rad ccw}$$

Using Equation (5.22), we can find the torsional shear stress at any point on the imaginary cross section in AB,

$$\tau_i = \frac{G_i \rho T_{AB}}{\sum_{j=1}^{2} G_j J_j} = \frac{G_i \rho \times 75 \times 10^3}{974.8 \times 10^3} = 76.9 \times 10^{-3} G_i \rho \quad (E6)$$

In AB the torsional shear stress in steel τ_{st} will be maximum at $\rho = 0.04$ and the torsional shear stress in bronze τ_{br} will be maximum at $\rho = 0.06$. Substituting the values of G and ρ, the maximum torsional shear stress in each material in segment AB can be found,

$$(\tau_{br})_{max} = 76.9 \times 10^{-3} \times 40 \times 10^9 \times 0.06 = 184.6 \times 10^6 \quad (E7)$$

$$(\tau_{st})_{max} = 76.9 \times 10^{-3} \times 80 \times 10^9 \times 0.04 = 246.2 \times 10^6 \quad (E8)$$

Comparing Equations (E3), (E7), and (E8), we see that the maximum torsional shear stress occurs in section *AB*, and its magnitude is $\tau_{max} = 250$ MPa.   ANS.

COMMENTS

1. This example demonstrates that a similar analysis approach can be used for a composite shaft and a homogeneous shaft, provided one takes care to use the appropriate equations for the homogeneous and the composite cross sections.

2. An alternative approach is to view the composite shaft as two homogeneous shafts—each shaft has an internal torque, but the shafts are constraint to rotate by the same amount, as depicted in Figure 5.62. We can write one equation representing the internal torque as the sum of the internal torque in steel and the internal torque in bronze. The second equation we generate by imposing the constraint that an equal length of each shaft rotate by an equal amount, and we write the equation using Equation (5.10). Substituting the torsional rigidities, we obtain a second equation for the unknown internal torques,

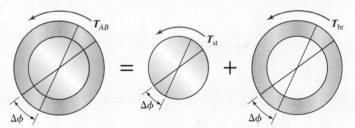

Figure 5.62   Composite shaft as two homogeneous shafts in Example 5.14.

$$T_{AB} = T_{st} + T_{br} = 75 \text{ kN·m} \qquad (E9)$$

$$\Delta\phi = \frac{T_{st}\,\Delta x}{G_{st}J_{st}} = \frac{T_{br}\,\Delta x}{G_{br}J_{br}} \qquad (E10)$$

or

$$T_{st} = 2.03\,T_{br} \qquad (E11)$$

Solving for the internal torques, we obtain $T_{st} = 24.75$ kN·m and $T_{br} = 50.25$ kN·m. Substituting these values into Equation (5.8), we obtain the maximum torsional shear stress in bronze and steel as

$$(\tau_{br})_{max} = \frac{50.25 \times 10^{3} \times 0.06}{16.33 \times 10^{-6}} = 184.6 \times 10^{6}$$

$$(\tau_{st})_{max} = \frac{24.75 \times 10^{3} \times 0.04}{4.02 \times 10^{-6}} = 246.2 \times 10^{6}$$

which are the same values as calculated in Equations (E7) and (E8). Similarly, by substituting $\Delta x = 2$ m and the internal torque values in Equation (E10), we can obtain

$$\phi_B - \phi_A = \frac{24.75 \times 10^3 \times 2}{321.6 \times 10^3} = 0.1538 \text{ rad}$$

which is the same value as given in Equation (E5).

3. The approach outlined in comment 2 is more intuitive than the solution procedure used, but it is also more tedious. As the number of materials in a cross section grows, this approach becomes less attractive. Laminated composite shafts have many layers of different materials, which can be accounted for easily in the use of Equations (5.21) and (5.22).

EXAMPLE 5.15

A composite shaft with a titanium outer layer and an aluminum layer on the inside is to be designed for transmitting a torque of 1000 N·m. The shaft must be 2 m long and must have an outer diameter of 25 mm to fit existing attachments (Figure 5.63). The relative rotation of the two ends of the shaft is limited to 1.5 rad. The shear modulus of rigidity G, the allowable shear stress τ_{allow}, and the density γ are given in Table 5.2. Determine the amount of each material in grams to be used, if the lightest shaft is to be designed.

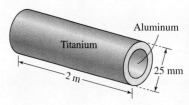

Figure 5.63 Composite shaft in Example 5.15.

TABLE 5.2 **Material properties**

Material	G (GPa)	τ_{allow} (MPa)	γ (Mg/m³)
Titanium alloy	36	450	4.4
Aluminum	28	180	2.8

PLAN

As the specific weight of aluminum is less than that of titanium, the more aluminum we use, the lighter the shaft will be. Thus the problem is to determine the maximum diameter of the aluminum shaft without exceeding the specified limits. We can use Equation (5.21) to determine the maximum torsional shear stress in each shaft and Equation (5.22) to find the relative rotation of the two ends and determine the diameter of the aluminum shaft that satisfies all the limits.

Let the diameter of the aluminum shaft be d mm. We can find the torsional rigidity of the shaft as follows:

$$J_{Ti} = \frac{\pi}{32}(25^4 - d^4)10^{-12} \text{ m}^4 \qquad J_{Al} = \frac{\pi}{32}d^4 \times 10^{-12} \text{ m}^4$$

$$\sum_{j=1}^{2} G_j J_j = G_{Ti}J_{Ti} + G_{Al}J_{Al}$$

$$= 36 \times 10^9 \frac{\pi}{32}(25^4 - d^4)10^{-12} + 28 \times 10^9 \frac{\pi}{32}d^4 \times 10^{-12}$$

$$\sum_{j=1}^{2} G_j J_j = 1380.6 - 0.785d^4 \times 10^{-3} \qquad \text{(E1)}$$

The relative rotation of the two ends should be less than 1.5 rad over the length $x_2 - x_1 = 2$ m. From Equation (5.22) we obtain the following limitation on d:

$$\phi_2 - \phi_1 = \frac{T(x_B - x_A)}{\sum_{j=1}^{2} G_j J_j} = \frac{1000 \times 2}{1380.6 - 0.785d^4 \times 10^{-3}} \leq 1.5$$

or

$$1380.6 - 0.785d^4 \times 10^{-3} \geq \frac{2000}{1.5}$$

or

$$0.785d^4 \times 10^{-3} \leq 1380.6 - \frac{2000}{1.5} \qquad \text{(E2)}$$

$$d \leq 15.7 \text{ mm} \qquad \text{(E3)}$$

The maximum torsional shear stress in titanium will occur at $\rho = 12.5 \times 10^{-3}$ m. From Equation (5.21) we obtain the following limitation on d:

$$\tau_{Ti} = \frac{G_{Ti}\rho T}{\sum_{j=1}^{2} G_j J_j} = \frac{36 \times 10^9 \times 12.5 \times 10^{-3} \times 1000}{1380.6 - 0.785d^4 \times 10^{-3}} \leq 450 \times 10^6$$

or

$$1380.6 - 0.785d^4 \times 10^{-3} \geq \frac{450 \times 10^3}{450}$$

or

$$0.785d^4 \times 10^{-3} \leq 380.6 \qquad \text{(E4)}$$

$$d \leq 26 \text{ mm} \qquad \text{(E5)}$$

The maximum torsional shear stress in aluminum will occur at $\rho = (d/2)10^{-3}$ m. From Equation (5.21) we obtain the following limitation on d:

$$\tau_{Al} = \frac{G_{Al}\rho T}{\sum_{j=1}^{2} G_j J_j} = \frac{28 \times 10^9 (d/2) 10^{-3} \times 1000}{1380.6 - 0.785 d^4 \times 10^{-3}} \leq 180 \times 10^6$$

or

$$1380.6 - 0.785 d^4 \times 10^{-3} \geq \frac{14 d \times 10^3}{180}$$

or

$$0.785 d^4 \times 10^{-3} + 77.8 d \leq 1380.6 \qquad \text{(E6)}$$

To determine the limitation on d from Equation (E6), the roots of the equation would have to be found using a numerical method. Before trying that, we check whether d obtained from Equations (E3) and (E5) would satisfy Equation (E6). As $d = 15.7$ satisfies both Equations (E3) and (E5), we substitute this value into Equation (E6) and obtain $1269.1 \leq 1380.6$, which implies that the limitation on d from Equation (E6) is met by $d = 15.7$. To ensure that we meet the limitation of Equation (E3), we round off the value of d downward to $d = 15$ mm. We multiply the densities of each material by the volume to obtain the amounts of titanium and aluminum to use in the design of the shaft as follows:

$$m_{Ti} = \left[\frac{\pi}{4}(0.025^2 - 0.015^2) \right] 2 \times 4.4 \times 10^6 = 2764.6$$

ANS. $m_{Ti} = 2765$ g

$$m_{Al} = \frac{\pi}{4} 0.015^2 \times 2 \times 2.8 \times 10^6 = 989.6$$

ANS. $m_{Al} = 990$ g

COMMENT

If we constructed a hollow shaft made from titanium only, then the inner diameter to the nearest millimeter would be 8 mm to satisfy the condition on the relative rotation and the maximum shear stress in titanium. The amount of titanium we would then need would be 3877 g. As titanium is significantly more expensive than aluminum, it would result in higher material cost. The total mass of the composite shaft is 3755 g, which is less than the all-titanium shaft mass by 122 g. Thus if the shaft were used in an aircraft, then the all-titanium shaft would result in higher fuel costs.

PROBLEM SET 5.3

Statically indeterminate shafts

A steel shaft (G_{st} = 12,000 ksi) and a bronze shaft (G_{br} = 5600 ksi) are securely connected at B, as shown in Figure P5.53. Using this information solve Problems 5.53 and 5.54.

5.53 Determine the maximum torsional shear stress in the shaft and the rotation of the section at B if the applied torque T = 50 in·kips in Figure P5.53.

5.54 Due to the torque T, the section at B in Figure P5.53 rotates by an amount of 0.02 rad. Determine the maximum torsional shear strain and the applied torque T.

Two hollow aluminum shafts (G = 10,000 ksi) are securely fastened to a solid aluminum shaft and loaded as shown Figure P5.55. Point E is on the inner surface of the shaft. Solve Problems 5.55 and 5.56.

5.55 If T = 300 in·kips in Figure P5.55, determine: (a) the rotation of the section at C with respect to the wall at A; (b) the shear strain at point E.

5.56 If the torsional shear strain at point E in Figure P5.55 is −250 μ, determine the rotation of the section at C and the applied torque T that produced this shear strain.

Two solid circular steel shafts (G_{st} = 80 GPa) and a solid circular bronze shaft (G_{br} = 40 GPa) are securely connected by a coupling at C. A torque of T = 10 kN·m is applied to the rigid wheel B, as shown in Figure P5.57. Solve Problems 5.57 and 5.58.

5.57 If the coupling plates in Figure P5.57 cannot rotate relative to one another, determine the angle of rotation of wheel B due to the applied torque.

5.58 If the coupling plates in Figure P5.57 can rotate relative to one another by 0.5° before engaging, then what will be the angle of rotation of wheel B?

A solid steel shaft (G = 80 GPa) is securely fastened to a solid bronze shaft (G = 40 GPa) that is 2 m long, as shown in Figure P5.59. Solve Problems 5.59 and 5.60.

5.59 If T_{ext} = 10 kN · m in Figure P5.59, determine: (a) the magnitude of maximum torsional shear stress in the shaft; (b) the rotation of the section at 1 m from the left wall.

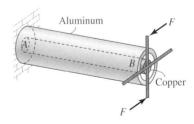

Figure P5.59

5.60 If the section at B in Figure P5.59 rotates by 0.05 rad, determine: (a) the maximum torsional shear strain in the shaft; (b) the applied torque T_{ext}.

An aluminum tube and a copper tube, each having a thickness of 5 mm, are securely fastened to two rigid bars, as shown in Figure P5.61. The bars force the tubes to rotate by equal angles. The two tubes are 1.5 m long and the mean diameters of the aluminum and copper tubes are 125 mm and 50 mm, respectively. The shear moduli for aluminum and copper are G_{al} = 28 GPa and G_{cu} = 40 GPa. Using this information, solve Problems 5.61 and 5.62.

5.61 Under the action of the applied couple section B of the two tubes shown Figure P5.61 rotates by an angle of 0.03 rad. Determine:

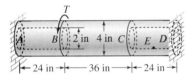

Figure P5.53

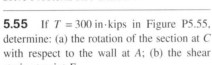

Figure P5.55

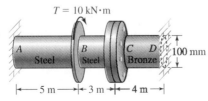

Figure P5.57

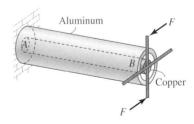

Figure P5.61

(a) the magnitude of maximum torsional shear stress in aluminum and copper; (b) the magnitude of the couple that produced the given rotation.

5.62 The applied couple on the tubes shown in Figure P5.61 is 10 kN·m. Determine: (a) the magnitude of maximum torsional shear stress in aluminum and copper; (b) the rotation of the section at B.

5.63 Two shafts with shear moduli $G_1 = G$ and $G_2 = 2G$ are securely fastened at section B, as shown in Figure P5.63. In terms of T_{ext}, L, G, and d, find the magnitude of maximum torsional shear stress in the shaft and the rotation of the section at B.

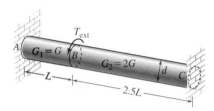

Figure P5.63

5.64 A uniformly distributed torque of q in·lb/in is applied to the entire shaft, as shown in Figure P5.64. In addition to the distributed torque a concentrated torque of $T = 3qL$ in·lb is applied at section B. Let the shear modulus be G and the radius of the shaft r. Determine in terms of q, L, G, and r: (a) the rotation of the section at B; (b) the magnitude of maximum torsional shear stress in the shaft.

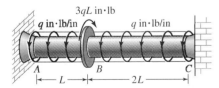

Figure P5.64

Design problems

A steel shaft ($G_{st} = 80$ GPa) and a bronze shaft ($G_{br} = 40$ GPa) are securely connected at B, as shown in Figure P5.65. The magnitude of maximum torsional shear stresses in steel and bronze are to be limited to 160 MPa and 60 MPa, respectively, and the rotation of section B is limited to 0.05 rad. Solve Problems 5.65 and 5.66.

5.65 (a) Determine the maximum allowable torque T to the nearest kN·m that can act on the shaft in Figure P5.65 if the diameter of the shaft is $d = 100$ mm. (b) What are the magnitude of maximum torsional shear stress and the maximum rotation in the shaft corresponding to the answer in part (a)?

5.66 (a) Determine the minimum diameter d of the shaft to the nearest millimeter if the applied torque $T = 20$ kN · m. (b) What are the magnitude of maximum torsional shear stress and the maximum rotation in the shaft corresponding to the answer in part (a)?

The solid steel shaft shown in Figure P5.67 has a shear modulus of elasticity $G = 80$ GPa and an allowable torsional shear stress of 60 MPa. The allowable rotation of any section is 0.03 rad. Solve Problems 5.67 and 5.68.

5.67 The applied torques on the shaft shown in Figure P5.67 are $T_1 = 10$ kN·m and $T_2 = 25$ kN· m. Determine: (a) the minimum diameter d of the shaft to the nearest millimeter; (b) the magnitude of maximum torsional shear stress in the shaft and the maximum rotation of any section.

5.68 The diameter of the shaft shown in Figure P5.67 $d = 80$ mm. Determine the maximum values of the torques T_1 and T_2 to the nearest kN·m that can be applied to the shaft.

Composite shafts

5.69 The composite shaft shown in Figure P5.69 is constructed from aluminum ($G_{al} = 4000$ ksi), bronze ($G_{br} = 6000$ ksi), and steel ($G_{st} = 12,000$ ksi). (a) Determine the rotation of the free end with respect to the wall. (b) Plot the torsional shear strain and the shear stress across the cross section.

Figure P5.65

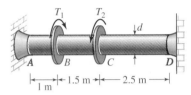

Figure P5.67

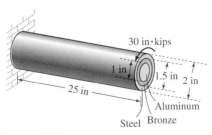

Figure P5.69

5.70 Solve Problem 5.61 using Equations (5.21) and (5.22).

5.71 Solve Problem 5.62 using Equations (5.21) and (5.22).

A cast-iron pipe ($G_{ir} = 70$ GPa) and a copper pipe ($G_{cu} = 40$ GPa) are securely bonded together, as shown in Figure P5.72. The outer diameters of the two pipes are 50 mm and 70 mm and the wall thickness of each pipe is 10 mm. Solve Problems 5.72 and 5.73.

5.72 If $T = 1500$ N·m in Figure P5.72, determine: (a) the magnitude of maximum torsional shear stress in cast iron and copper; (b) the rotation of the section at D with respect to the section at A.

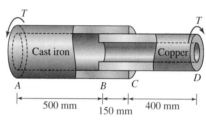

Figure P5.72

5.73 The allowable rotation of section at D with respect to section at A is 0.02 rad. Determine the maximum torque T to the nearest kN · m that can be applied to the shaft in Figure P5.72.

A solid steel shaft ($G = 80$ GPa), 3 m long, is securely fastened to a hollow bronze shaft ($G = 40$ GPa), 2 m long, as shown Figure P5.74. Solve Problems 5.74 and 5.75.

5.74 The applied torque in Figure P5.74 is $T = 10$ kN · m. Determine: (a) the magnitude of maximum torsional shear stress in the shaft; (b) the rotation of the section at 1 m from the left wall.

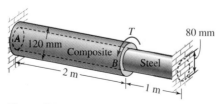

Figure P5.74

5.75 If the section at B in Figure P5.74 rotates by 0.05 rad, determine: (a) the maximum shear strain in the shaft; (b) the applied torque T.

Design problems

5.76 A composite shaft is constructed from aluminum ($G_{al} = 4000$ ksi) and steel ($G_{st} = 12,000$ ksi), as shown in Figure P5.76. The allowable torsional shear stresses are 18 ksi and 24 ksi in aluminum and steel, respectively. Determine: (a) the maximum torque T that can be applied; (b) the rotation of the free end with respect to the wall when the maximum torque is applied.

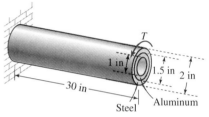

Figure P5.76

5.77 A shaft 1 m long is to be designed to transmit a torque of 3300 N·m. The outside diameter of the shaft must be 40 mm to fit existing attachments. The shaft can be all aluminum, all titanium, or a composite of the two material. The shear modulus of rigidity G, the allowable shear stress τ_{allow}, and the density γ of titanium and aluminum are given in Table P5.77. Determine the diameters to the nearest millimeter of the *lightest* shaft and the corresponding mass.

TABLE P5.77

Material	G (GPa)	τ_{allow} (MPa)	γ (Mg/m³)
Titanium alloy	36	300	4.4
Aluminum	28	180	2.8

Stress concentration

5.78 The allowable shear stress in the stepped shaft shown Figure P5.78 is 17 ksi. Determine the smallest fillet radius that can be used at section B. Use the stress concentration graphs given in Appendix C.

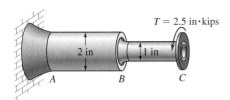

Figure P5.78

5.79 The fillet radius in the stepped shaft shown in Figure P5.79 is 6 mm. Determine the maximum torque that can act on the rigid wheel if the allowable shear stress is 80 MPa and the

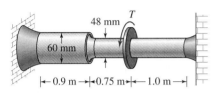

Figure P5.79

modulus of rigidity is 28 GPa. Use the stress concentration graphs given in Appendix C.

Stretch yourself

5.80 A composite circular shaft (Figure P5.80) is made from two nonlinear materials. The stress–strain relationship for material 1 is given by $\tau_1 = G\gamma^{0.5}$, and for material 2 it is given by $\tau_2 = 2G\gamma^{0.5}$. Determine the rotation of the section at B in terms of T, L, G, and R.

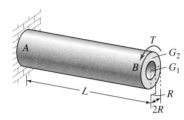

Figure P5.80

---

*5.5 ELASTIC–PERFECTLY PLASTIC CIRCULAR SHAFTS

Figure 5.64 shows the stress–strain curves describing the elastic–perfectly plastic behavior of a material that was discussed in Section 3.11.1. It is assumed that the material behavior is the same for positive and negative stresses and strains.

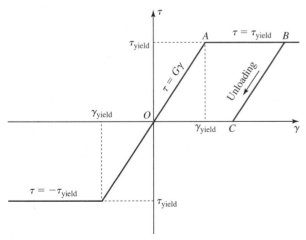

Figure 5.64 Elastic–perfectly plastic material behavior.

Recall the following from Section 3.11.1:

1. Before yield stress the material stress–strain relationship is represented by Hooke's law, and after yield stress the stress is assumed to be constant.

2. To determine the strain (deformation) in the horizontal portion *AB* of the curve we have to use the requirement that deformation be continuous.

3. Unloading (elastic recovery) from a point in the plastic region is along line *BC*, which is parallel to the linear portion of the stress–strain curve *OA*.

We make use of these observations in our analysis that follows.

The torsional shear stress varies linearly across a homogeneous, linear, elastic cross section and is maximum at the outer radius of the shaft. Thus the plastic zone will start from the outside and move inward for a homogeneous material. At the elastic–plastic boundary we know that the shear strain is γ_{yield}. The linear variation of torsional shear strain was derived on the basis of geometry and is unaffected by the fact that we are in the plastic region if the kinematic assumptions[18] are valid. Thus Equation (5.3) can be rewritten as

$$\gamma_{yield} = \rho_y \frac{d\phi}{dx} \tag{5.23}$$

where ρ_y is the radial coordinate of the elastic–plastic boundary. If we assume[19] that ρ_y is not a function of x, then $d\phi/dx$ is a constant and can be replaced by $(\phi_2 - \phi_1)/(x_2 - x_1)$. We obtain the following formula for the relative rotation:

$$\phi_2 - \phi_1 = \frac{\gamma_{yield}}{\rho_y}(x_2 - x_1) \tag{5.24}$$

The shear stress varies linearly from a value of zero at the center to a value of yield stress at the elastic–plastic boundary. After the elastic–plastic boundary the shear stress is equal to the yield stress. The shear stress distribution can be written as

$$\tau = \begin{cases} \dfrac{\tau_{yield}\rho}{\rho_y} & \rho \le \rho_y \\[2mm] \tau_{yield} & \rho \ge \rho_y \end{cases} \tag{5.25}$$

We can find the equivalent internal torque by substituting Equation (5.25) into Equation (5.1) as demonstrated in Example 5.16. We can relate the internal torque to the external torque through equilibrium. There are two kinds of problems we will consider:

1. Problems in which we know ρ_y, such as Examples 5.17 and 5.18.

2. Problems in which we need to find ρ_y, such as Example 5.19.

5.5.1 Residual Shear Stress

By prestressing the shaft into the plastic region we can increase the torque-carrying capacity of the shaft, but now we need to compute the residual stresses in the shaft. During the elastic recovery, that is, unloading along line *BC* in Figure 5.64, the assumption of material linearity is valid and Equation (5.7) can be used for calculating the elastic shear stress. The internal torque for calculating the elastic shear stress corresponds to the torque that was equivalent to the stress distribution for a given location of the elastic–plastic boundary. We subtract the elastic stresses from the elastic–plastic stress distribution and obtain the residual stresses. This is demonstrated in Example 5.17.

---

[18]If strains in the plastic region are very large, then we are violating the assumption of small strains and the analysis is not valid.

[19]This implies that the shaft is not tapered and external torque is not a function of x between x_1 and x_2.

*EXAMPLE 5.16

In the hollow circular shaft shown in Figure 5.65 the shear strain at a section in polar coordinates was found to be $\gamma_{x\theta} = 3\rho \times 10^{-3}$, where ρ is the radial coordinate measured in inches. Assume the shaft is made from elastic–perfectly plastic material, which has a yield stress $\tau_{\text{yield}} = 24$ ksi and a shear modulus $G = 6000$ ksi. Determine the equivalent internal torque acting at the cross section.

PLAN

For the given strain and material model, the shear stress $\tau_{x\theta}$ as a function of ρ was found in Example 3.14. Substituting this stress distribution into Equation (5.1) and integrating across the cross section, we can obtain the equivalent internal torque T.

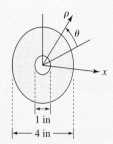

Figure 5.65 Hollow shaft in Example 5.16.

Solution The torsional shear stress $\tau_{x\theta}$ as a function of ρ was found in Example 3.14. It is written here for convenience and plotted in Figure 5.66.

$$\tau_{x\theta} = \begin{cases} 18\rho \ \text{ksi} & 0.5 \le \rho \le 1.333 \\ 24 \ \text{ksi} & 1.333 \le \rho \le 2.0 \end{cases} \qquad \text{(E1)}$$

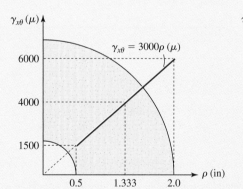 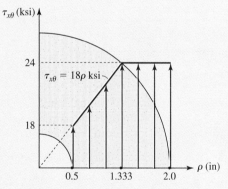

Figure 5.66 Strain and stress distributions in Example 5.16.

The internal torque can be found from Equation (5.1),

$$T = \int_A \rho \tau_{x\theta} \, dA = \underbrace{\int_{A_E} \rho \tau_{x\theta} \, dA}_{T_E} + \underbrace{\int_{A_P} \rho \tau_{x\theta} \, dA}_{T_P} \qquad \text{(E2)}$$

The differential area dA is the area of a ring of radius ρ and thickness $d\rho$, that is, $dA = 2\pi\rho \, d\rho$. Substituting the stress distribution from Equation (E1) into Equation (E2) and performing integration, we obtain the internal torque,

$$T_E = \int_{0.5}^{1.333} \rho \times 18\rho \times 2\pi\rho \; d\rho = 36\pi \left. \frac{\rho^4}{4} \right|_{0.5}^{1.333} = 87.50 \text{ in·kips}$$

$$T_P = \int_{1.333}^{2.0} \rho \times 24 \times 2\pi\rho \; d\rho = 48\pi \left. \frac{\rho^3}{3} \right|_{1.333}^{2.0} = 283.07 \text{ in·kips}$$

$$T = T_E + T_P = 370.6 \text{ in·kips}$$

ANS. $T = 370.6 \text{ in · kips}$

COMMENTS

1. The elastic–plastic material affects the stress distribution, but Equation (5.1), which represents static equivalency, is unaffected.

2. The relationship of the internal torque and the external torque can be established by drawing the appropriate free-body diagram for a particular problem. The relationship of internal and external torques depends on the free-body diagram and on the elasticity or the plasticity of the material.

EXAMPLE 5.17
A 100-mm-diameter solid steel circular shaft ($G = 80$ GPa, $\tau_{\text{yield}} = 160$ MPa) has a plastic zone that is 30 mm deep. Assuming elastic–perfectly plastic material, determine:

(a) The equivalent internal torque at the cross section.

(b) The residual stresses in the cross section on unloading the shaft.

PLAN
The radius of the shaft is 50 mm and the plastic zone starting from outside is 30 mm deep. Thus the elastic–plastic boundary is located at $\rho_y = 20$ mm.

(a) The shear stress varies linearly, starting from zero at $\rho = 0$ and reaching a value of yield stress at $\rho_y = 20$ mm. After 20 mm the shear stress is constant and equal to the yield stress. We can write the equation for the shear stress in terms of ρ and, using Equation (5.1), determine the equivalent internal torque.

(b) Using the internal torque obtained in part (a) and Equation (5.8), we can determine the shear stress distribution during unloading and subtract it from the stress distribution in part (a) to obtain the residual stress distribution.

SOLUTION

(a) The shear stress varies linearly from zero at the center to yield stress at the elastic–plastic boundary. Thus for $0 \leq \rho \leq 0.02$,

$$\tau = \frac{\tau_{\text{yield}}\rho}{\rho_y} = \left(\frac{160 \times 10^6}{20 \times 10^{-3}}\right)\rho = 8000 \times 10^6 \rho$$

Beyond the elastic–plastic boundary the shear stress is equal to the yield stress. We can write the equation for shear stress as

$$\tau = \begin{cases} 8000\rho \text{ MPa} & 0 \leq \rho \leq 0.02 \\ 160 \text{ MPa} & 0.02 \leq \rho \leq 0.05 \end{cases} \tag{E1}$$

Substituting Equation (E1) into Equation (5.1), we can obtain the equivalent internal torque,

$$T = \int_A \rho \tau_{x\theta} \, dA$$

$$= \int_0^{0.02} \rho(8000 \times 10^6 \rho)2\pi\rho \, d\rho + \int_{0.02}^{0.05} \rho(160 \times 10^6)2\pi\rho \, d\rho$$

$$= 16{,}000\pi \times 10^6 \frac{\rho^4}{4}\bigg|_0^{0.02} + 320\pi \times 10^6 \frac{\rho^3}{3}\bigg|_{0.02}^{0.05}$$

$$= (2.01 + 39.2)10^3 \text{ N·m} \tag{E2}$$

ANS. $T = 41.3 \text{ kN·m}$

(b) During unloading we can use Equation (5.8) and obtain the elastic rebound shear stress. The polar moment for the shaft is $J = (\pi/2)0.05^4 = 9.8175 \times 10^{-6}$. Then

$$\tau_{\text{elastic}} = \frac{T\rho}{J} = \frac{41.3 \times 10^3}{9.8175 \times 10^{-6}}\rho = 4207\rho \text{ MPa} \tag{E3}$$

We subtract Equation (E3) from stresses in Equation (E1) to obtain the residual shear stresses, that is, $\tau_{\text{residual}} = \tau - \tau_{\text{elastic}}$,

ANS. $\tau_{\text{residual}} = \begin{cases} 3793\rho \text{ MPa} & 0 \leq \rho \leq 0.02 \\ 160 - 4207\rho \text{ MPa} & 0.02 \leq \rho \leq 0.05 \end{cases} \tag{E4}$

COMMENTS

1. The calculation of residual stress can be shown graphically in Figure 5.67.

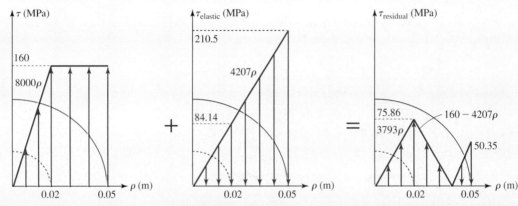

Figure 5.67 Residual stress calculation in Example 5.17.

2. The torque that the shaft could carry without any plastic deformation can be calculated using yield stress as the maximum elastic stress, that is, $160 \times 10^6 = T \times 0.5 / (7.8715 \times 10^{-6})$, or $T = 25.2$ kN·m. With the residual stresses, the maximum torque the shaft can carry without any additional plastic deformation is now 41.3 kN·m, which would be equivalent to a shaft of the same dimensions but made from a material with a yield stress of 210.5 MPa.

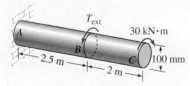

Figure 5.68 Shaft in Example 5.18.

EXAMPLE 5.18

The steel shaft ($G = 80$ GPa, $\tau_{yield} = 160$ MPa) shown in Figure 5.68 has a 30-mm-deep plastic zone in segment AB. Determine:

(a) The magnitude of the applied torque T_{ext}.

(b) The rotation of section C with respect to the wall.

PLAN

The cross section in BC is the same as in Example 5.17, and hence the internal torque in AB is $\boldsymbol{T}_{AB} = 41.3$ kN·m.

(a) We can draw a free-body diagram after making an imaginary cut in AB and determine the external torque T_{ext} by equilibrium.

(b) We can find the rotation of the section at B with respect to the section at A using Equation (5.24). To find the rotation of the section at C with respect to B we first check to see whether segment AB is elastic or parts of it have gone plastic. If it is elastic, then we can use Equation (5.10) to find the relative rotation of C with respect to B and then compute the rotation of the section at C with respect to A.

Solution

(a) We draw the free-body diagram in Figure 5.69 after making an imaginary cut in AB and taking the right part to avoid calculating the wall reaction,

$$T_{ext} = 41.3 + 30 = 71.3 \text{ kN·m}$$

(b) The strain at yield is

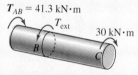

$T_{AB} = 41.3$ kN·m

T_{ext}

30 kN·m

Figure 5.69 Free-body diagram of segment AB in Example 5.18.

$$\gamma_{yield} = \frac{\tau_{yield}}{G} = \frac{160 \times 10^6}{80 \times 10^9} = 2 \times 10^{-3}$$

Using Equation (5.24), we can find the relative rotation of the section at B with respect to the section at A,

$$\gamma_{yield} = \rho_y \frac{d\phi}{dx} = 20 \times 10^{-3} \frac{d\phi}{dx} = 2 \times 10^{-3} \qquad \frac{d\phi}{dx} = 0.1$$

$$\phi_B - \phi_A = 0.1(x_B - x_A) = 0.1 \times 2.5 = 0.25 \qquad \text{(E1)}$$

We draw the free-body diagram in Figure 5.70 after making an imaginary cut in BC and find the internal torque $\boldsymbol{T}_{BC}$ and the maximum torsional shear stress in segment BC from Equation (5.8),

$$J_{BC} = \frac{\pi(0.1)^4}{32} = 9.817(10^{-6}) \text{ m}^4$$

$$\boldsymbol{T}_{BC} = -30 \text{ kN·m} \qquad \text{(E2)}$$

T_{BC} 30 kN·m

C

Figure 5.70 Free-body diagram of segment BC in Example 5.18.

$$\tau_{max} = \frac{T_{BC}\rho_{max}}{J_{BC}} = \frac{-30 \times 10^3 \times 50 \times 10^{-3}}{9.817(10^{-6})} = -153(10^6)$$

The maximum torsional shear stress in BC of 153 MPa does not exceed the yield stress of 160 MPa. Hence we can use Equation (5.10) to find the rotation of the section at C with respect to B,

$$\phi_C - \phi_B = \frac{-30 \times 10^3 \times 2}{80 \times 10^9 \times 9.817 \times 10^{-6}} = -0.0764 \qquad \text{(E3)}$$

Adding Equations (E1) and (E3) we obtain the rotation of the section at C with respect to the section at A. Noting that the rotation at A is zero, we obtain the rotation of the section at C,

$$\phi_C - \overset{0}{\cancel{\phi_A}} = 0.25 - 0.0764 = 0.1736$$

ANS. $\quad \phi_C = 0.1736$ ccw

COMMENTS

1. The example brings out the difference in the calculation of relative rotations of segment ends in the elastic and plastic regions. Segment AB has a plastic zone, and Equation (5.24) is used to find the relative rotation as shown in Equation (E1). Segment BC is in the elastic region, and Equation (5.10) is used to find the relative rotation as shown in Equation (E3).

2. In this example and in Example 5.18, the location of elastic–plastic boundary was given. This simplified the calculations. The next example demonstrates how the elastic–plastic boundary can be found for an applied torque.

EXAMPLE 5.19

A steel shaft ($G = 80$ GPa, $\tau_{\text{yield}} = 160$ MPa) is loaded as shown in Figure 5.71. Determine the location of the elastic–plastic boundary in section AB.

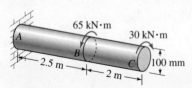

Figure 5.71 Shaft in Example 5.19.

PLAN

We can assume that the elastic–plastic boundary is located at a radial distance ρ_y. We can write the stress distribution in terms of ρ_y and obtain the internal torque in AB in terms of ρ_y, which we can equate to the internal torque obtained from equilibrium.

Solution The shear stress distribution across the cross section can be written as

$$\tau = \begin{cases} \dfrac{\tau_{\text{yield}}\rho}{\rho_y} & 0 \le \rho \le \rho_y \\[2mm] \tau_{\text{yield}} & \rho_y \le \rho \le 0.05 \end{cases} \qquad \text{(E1)}$$

Substituting Equation (E1) into Equation (5.1) to obtain the equivalent internal torque,

$$T_{AB} = \int_A \rho\tau\, dA = \int_0^{\rho_y} \rho\frac{\tau_{\text{yield}}\rho}{\rho_y}2\pi\rho\, d\rho + \int_{\rho_y}^{0.05} \rho\tau_{\text{yield}}2\pi\rho\, d\rho$$

$$= 2\pi\frac{\tau_{\text{yield}}}{\rho_y}\frac{\rho^4}{4}\Big|_0^{\rho_y} + 2\pi\tau_{\text{yield}}\frac{\rho^3}{3}\Big|_{\rho_y}^{0.05}$$

$$= \frac{\pi\tau_{\text{yield}}}{6}[3\rho_y^3 + 4(0.05^3 - \rho_y^3)]$$

or

$$T_{AB} = 83.78(10^6)[0.5(10^{-3}) - \rho_y^3] \qquad \text{(E2)}$$

We make an imaginary cut in AB and draw the free-body diagram in Figure 5.72 to calculate the internal torque,

$$T_{AB} = 65 - 30 = 35 \text{ kN·m} = 35 \times 10^3 \text{ N·m} \qquad \text{(E3)}$$

Figure 5.72 Free-body diagram of segment AB in Example 5.19.

Equating Equations (E2) and (E3) we obtain

$$0.5 \times 10^{-3} - \rho_y^3 = \frac{35 \times 10^3}{83.78 \times 10^6} = 0.4178 \times 10^{-3}$$

$$\rho_y = [(0.5 - 0.4178)10^{-3}]^{1/3} = 0.0435 \text{ m} \qquad \text{(E4)}$$

$$\text{ANS.} \qquad \rho_y = 43.5 \text{ mm}$$

COMMENT
In this example, the value of ρ_y could be found by simply taking a cubic root of a number as shown in Equation (E4). For some problems, it may be necessary to find roots of a cubic equation by a numerical method such as described in Section B.2 of the Appendix.

*EXAMPLE 5.20

Resolve Example 5.16 assuming that the stress–strain relationship is given by the power law (see Section 3.11.3) $\tau = 450\gamma^{0.75}$ ksi.

PLAN

For the given strain and material model, the shear stress $\tau_{x\theta}$ as a function of ρ was found in Example 3.15. Substituting this stress distribution into Equation (5.1) and integrating across the cross section, we can obtain the equivalent internal torque τ.

Solution   The torsional shear stress $\tau_{x\theta}$ as a function of ρ that was found in Example 3.15 is rewritten here for convenience,

$$\tau_{x\theta} = 5.768\rho^{0.75} \text{ ksi} \tag{E1}$$

The internal torque can be found by substituting Equation (E1) and the differential area dA into Equation (5.1). The differential area dA is the area of a ring of radius ρ and thickness $d\rho$, that is, $dA = 2\pi\rho\,d\rho$. Integration yields the internal torque T,

$$T = \int_A \rho\tau_{x\theta}\,dA = \int_{0.5}^{2.0} \rho \times 5.768\rho^{0.75} \times 2\pi\rho\,d\rho = 36.24\int_{0.5}^{2.0} \rho^{2.75}\,d\rho$$

$$T = 36.24\left.\frac{\rho^{3.75}}{3.75}\right|_{0.5}^{2.0} = 9.664(2^{3.75} - 0.5^{3.75})$$

ANS. $T = 129.3$ in·kips

COMMENT

The three Examples 5.4, 5.16, and 5.20 show that the impact of changing the material model affects the stress distribution. But the strain distribution, which is a kinematic relationship, will be unaffected; the static equivalency, Equation (5.1), will be unaffected; and the equilibrium equations relating internal torque to external torque will be unaffected.

*5.6   TORSION OF THIN-WALLED TUBES

The sheet metal skin on a fuselage, the wing of an aircraft, and the shell of a tall building are examples in which a body can be analyzed as a thin-walled tube. By thin wall we imply that the thickness t of the wall is smaller by a factor of at least 10 in comparison to the length b of the biggest line that can be drawn across two points on the cross section, as shown in Figure 5.73. By tube we imply that the length L is at least

$$b > 10t \qquad L > 10b$$

we assume that this thin-walled tube is subjected to torsional moments only.

The walls of the tube are bounded by two free surfaces, and hence by the symmetry of shear stresses, the shear stress in the normal direction τ_{xn} must go to zero on these bounding surfaces, as shown in Figure 5.74. This does not imply that τ_{xn} is zero in the interior. But because the walls are thin, we approximate τ_{xn} to be zero everywhere. The normal stress σ_{xx} would be equivalent to an internal axial force or an internal bending moment. Since there is no external axial force or bending moment, we approximate the value of σ_{xx} as zero.

Figure 5.74 shows that the only nonzero stress component is τ_{xs}, but it can be assumed uniform in the n direction because the tube is thin. We next show that it is also uniform in the s direction by making an imaginary cut along the x direction, as shown in Figure 5.75. The forces in the x direction on the surfaces passing through points A and B balance each other as there is no external axial force,

$$\sum F_x \qquad \tau_A(t_A \, dx) = \tau_B(t_B \, dx)$$

$$\tau_A t_A = \tau_B t_B$$

$$q_A = q_B$$

The quantity $q = \tau_{xs} t$ is called shear flow[20] and has units of force per unit length. Figure 5.75 shows that shear flow is uniform at a given cross section.

We can replace the shear stresses (shear flow) by an equivalent internal torque, as shown in Figure 5.76. The line OC is perpendicular to the line of action of the force dV, which is in the tangent direction to the arc at that point. Noting that the shear flow is a constant, we take it outside the integral sign,

$$T = \oint dT = \oint q(h \, ds) = q \oint 2 \, dA_E = 2q A_E \qquad \text{or} \qquad q = \frac{T}{2A_E}$$

We thus obtain

$$\boxed{\tau_{xs} = \frac{T}{2t A_E}} \qquad (5.26)$$

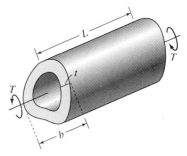

Figure 5.73 Torsion of thin-walled tubes.

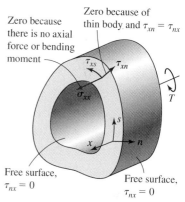

Zero because there is no axial force or bending moment

Zero because of thin body and $\tau_{xn} = \tau_{nx}$

Free surface, $\tau_{nx} = 0$

Free surface, $\tau_{nx} = 0$

Figure 5.74 Deducing stress behavior in thin-walled tubes.

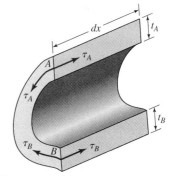

Figure 5.75 Deducing constant shear flow in thin-walled tubes.

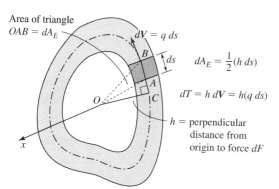

Area of triangle $OAB = dA_E$

$dV = q \, ds$

$dA_E = \frac{1}{2}(h \, ds)$

$dT = h \, dV = h(q \, ds)$

h = perpendicular distance from origin to force dF

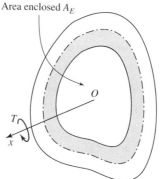

Area enclosed A_E

Figure 5.76 Equivalency of internal torque and shear stress (flow).

---

[20]This terminology is from fluid mechanics, where an incompressible ideal fluid has a constant flow rate in a channel.

where T is the internal torque at the section containing the point at which the shear stress is to be calculated, A_E is the area enclosed by the centerline of the tube, and t is the thickness at the point where the shear stress is to be calculated.

The thickness t can vary with different points on the cross section provided the assumption of thin-walled is not violated. If the thickness varies, then the shear stress will not be constant on the cross section, even though the shear flow is constant.

EXAMPLE 5.21

A semicircular thin tube is subjected to torques as shown in Figure 5.77. Determine:

(a) The maximum torsional shear stress in the tube.

(b) The torsional shear stress at point O. Show the results on a stress cube.

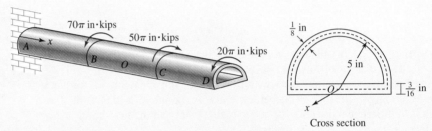

Figure 5.77   Thin-walled tube in Example 5.21.

PLAN

From Equation (5.26) we know that the maximum torsional shear stress will exist in a section where the internal torque is maximum and the thickness minimum. To determine the maximum internal torque, we make cuts in AB, BC, and CD and draw free-body diagrams by taking the right part of each cut to avoid calculating the wall reaction.

Solution   Figure 5.78 shows the free-body diagrams. The right part of each cut is taken to avoid calculating the wall reaction that would appear on the left part of the imaginary cut.

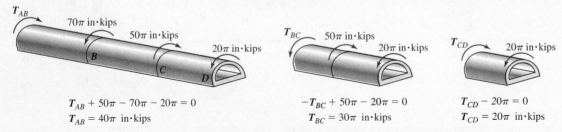

$$T_{AB} + 50\pi - 70\pi - 20\pi = 0$$
$$T_{AB} = 40\pi \text{ in·kips}$$

$$-T_{BC} + 50\pi - 20\pi = 0$$
$$T_{BC} = 30\pi \text{ in·kips}$$

$$T_{CD} - 20\pi = 0$$
$$T_{CD} = 20\pi \text{ in·kips}$$

Figure 5.78   Internal torque calculations in Example 5.21.

(a) The maximum torque is in AB and the minimum thickness is $\frac{1}{8}$ in. The enclosed area is $A_E = \pi 5^2/2 = 12.5\pi$. From Equation (5.26) we obtain

$$\tau_{max} = \frac{40\pi}{12.5\pi \times \frac{1}{8}}$$

ANS. $\tau_{max} = 25.6$ ksi

(b) At point O the internal torque is T_{BC} and $t = \frac{3}{16}$. Thus,

$$\tau_O = \frac{30\pi}{12.5\pi \times \frac{3}{16}}$$

ANS. $\tau_O = 12.8$ ksi

Figure 5.79 shows part of the tube between sections B and C. Segment BO would rotate counterclockwise with respect to segment OC. The shear stress must be opposite to this possible motion and hence is in the clockwise direction, as shown. The direction on the rest of the surface can be drawn using the observation that the symmetric pair of shear stress components either point toward the corner or away from it.

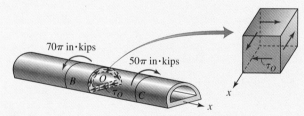

Figure 5.79   Direction of shear stress in Example 5.21.

COMMENT

The shear flow in the cross-section containing point O is a constant over the entire cross-section. The magnitude of torsional shear stress at point O however will be two-thirds that of the value of the shear stress in the circular part of the cross-section because of the variation in thickness.

PROBLEM SET 5.4

Nonlinear response of circular shafts

In Problems 5.81 through 5.84 a solid circular shaft of 3-in diameter has a torsional shear strain at a section in polar coordinates which was found to be $\gamma_{x\theta} = 0.002\rho$, *where* ρ *is the radial coordinate measured in inches. For the given material model in each problem:*

(a) *Write expressions for* $\tau_{x\theta}$ *as a function of* ρ *and plot the shear strain* $\gamma_{x\theta}$ *and shear stress* $\tau_{x\theta}$ *distributions across the cross section.*

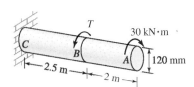

Figure P5.90

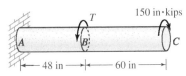

Figure P5.91

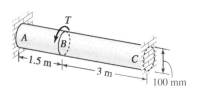

Figure P5.92

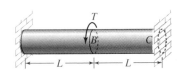

Figure P5.94

(b) Determine the equivalent internal torque acting at the cross section.

5.81 The shaft is made from elastic–perfectly plastic material that has a yield stress $\tau_{yield} = 18$ ksi and a shear modulus $G = 12{,}000$ ksi.

5.82 The shaft is made from a bilinear material that has a yield stress $\tau_{yield} = 18$ ksi and shear moduli $G_1 = 12{,}000$ ksi and $G_2 = 4800$ ksi.

5.83 The shaft material has a stress–strain relationship given by $\tau = 243\,\gamma^{0.4}$ ksi.

5.84 The shaft material has a stress–strain relationship given by $\tau = 12{,}000\gamma - 120{,}000\,\gamma^2$ ksi.

In Problems 5.85 through 5.88 a hollow circular shaft has an inner diameter of 50 mm and an outside diameter of 100 mm. The shear strain at a section in polar coordinates was found to be $\gamma_{x\theta} = 0.2\rho$, where ρ is the radial coordinate measured in meters. For the given material model in the problem:

(a) Write expressions for $\tau_{x\theta}$ as a function of ρ and plot the shear strain $\gamma_{x\theta}$ and shear stress $\tau_{x\theta}$ distributions across the cross section.

(b) Determine the equivalent internal torque acting at the cross section.

5.85 The shaft is made from elastic–perfectly plastic material that has a shear yield stress $\tau_{yield} = 175$ MPa and a shear modulus $G = 26$ GPa.

5.86 The shaft is made from a bilinear material that has a shear yield stress $\tau_{yield} = 175$ MPa and shear moduli $G_1 = 26$ GPa and $G_2 = 14$ GPa.

5.87 The shaft material has a stress–strain relationship given by $\tau = 3435\,\gamma^{0.6}$ MPa.

5.88 The shaft material has a stress–strain relationship given by $\tau = 26{,}000\gamma - 208{,}000\,\gamma^2$ MPa.

5.89 A 5-ft-long hollow shaft with an outside diameter of 4 in and an inside diameter of 2 in is twisted through an angle of 12°. The

shaft material has a yield stress of 24 ksi and a modulus of rigidity of 4000 ksi. Assuming elastic–perfectly plastic material behavior, determine: (a) the magnitude of the applied torque; (b) the residual shear stress when the applied torque is removed; (c) show that the residual stresses are self-equilibrating, that is, produce no internal torque.

5.90 The elastic–perfectly plastic shaft shown in Figure P5.90 has a plastic zone 30 mm deep in segment BC. The yield stress of the shaft is 200 MPa and the shear modulus $G = 80$ GPa. Determine: (a) the magnitude of the applied torque T; (b) the rotation of section A with respect to the wall.

5.91 The 4-in solid elastic–perfectly plastic shaft shown in Figure P5.91 has a yield stress of 30,000 psi and a shear modulus $G = 12{,}000$ ksi. The plastic zone in AB is 0.5 in deep. Determine: (a) the magnitude of the applied torque T; (b) the rotation of section C with respect to the wall.

5.92 The shaft made from elastic–perfectly plastic material shown in Figure P5.92 has a yield stress of 200 MPa and a shear modulus $G = 80$ GPa. The plastic zone in section AB is 25 mm deep. Determine: (a) the torque T; (b) the rotation of the section at B.

5.93 A circular solid shaft of radius R is made from a nonlinear material that has a shear stress–shear strain relationship given by $\tau = K\gamma^{0.4}$. Show that $\tau_{max} = 0.5411(T/R^3)$ and $\phi_2 - \phi_1 = 0.2154(x_2 - x_1)(T/KR^{3.4})^{2.5}$, where τ_{max} is the maximum shear stress at a section, T is the internal torque at the section, and ϕ_1 and ϕ_2 are the rotations of the section at x_1 and x_2, respectively.

5.94 The circular solid shaft of radius R shown in Figure P5.94 is made from a nonlinear material that has a shear stress–shear strain relationship given by $\tau = K\gamma^{0.6}$. Assuming small deformation, determine: (a) the rotation of section B with respect to the left wall; (b) the maximum shear stress in the shaft.

Torsion of thin-walled tubes

5.95 Calculate the magnitude of the maximum torsional shear stress in the cross section shown in Figure P5.95 that is subjected to a torque $T = 100$ in·kips.

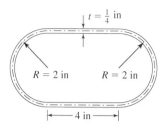

Figure P5.95

5.96 Calculate the magnitude of the maximum torsional shear stress in the cross section shown in Figure P5.96 that is subjected to a torque $T = 900$ N·m.

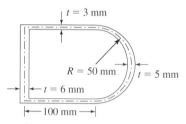

Figure P5.96

5.97 Calculate the magnitude of the maximum torsional shear stress in the cross section shown in Figure P5.97 that is subjected to a torque $T = 15$ kN·m.

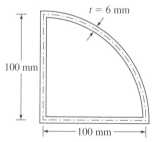

Figure P5.97

In Problems 5.98 through 5.100, a tube of uniform thickness t has a torque T applied to it. Determine the maximum torsional shear stress in terms of t, a, and T.

5.98 In Figure P5.98, determine the maximum torsional shear stress in terms of t, a, and T.

5.99 In Figure P5.99, determine the maximum torsional shear stress in terms of t, a, and T.

5.100 In Figure P5.100, determine the maximum torsional shear stress in terms of t, a, and T.

5.101 The tube of uniform thickness t shown in Figure P5.101 has a torque T applied to it. Determine the maximum torsional shear stress in terms of t, a, b, and T.

5.102 A hexagonal tube of uniform thickness is loaded as shown in Figure P5.102. Determine the magnitude of the maximum torsional shear stress in the tube.

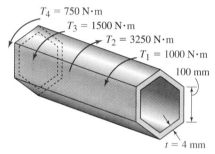

Figure P5.102

5.103 A rectangular tube is loaded as shown in Figure P5.103. Determine the magnitude of the maximum torsional shear stress.

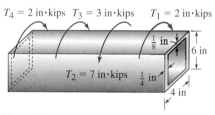

Figure P5.103

5.104 The three tubes shown in Problems 5.98 through 5.100 are to be compared for the maximum torque-carrying capability, assuming that all tubes have the same thickness t, the maximum torsional shear stress in each tube can be τ, and the amount of material used in the cross section of each tube is A. (a) Which shape would you use? (b) What is the percentage torque carried by the remaining two shapes in terms of the most efficient structural shape?

Figure P5.98

Figure P5.99

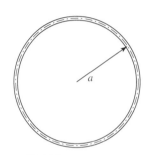

Figure P5.100

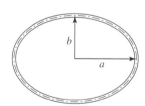

Figure P5.101

*5.7 GENERAL INFORMATION

This section describes the historical development of the torsion of circular and non-circular shafts. It seems very fitting that Coulomb, the man who first differentiated shear stress (see Section 1.3) from normal stress, is the person who initiated the development of the theory of torsion, in which shear stress is the dominant stress component. In 1781 Coulomb started research in electricity and magnetism. To measure the small electrical and magnetic forces, he devised a very sensitive torsion balance in which a weight with a pointer was suspended by a wire. The angular rotation of the wire was measured by the pointer attached to the weight. It is the design of this torsion balance that led Coulomb to investigate the resistance of a wire in torsion. Coulomb assumed that the resistance torque (internal torque T) in a twisted wire is proportional to the angle of twist ϕ and developed a formula for the time period of oscillation of the pendulum when the wire is twisted by a small angle and then set free to oscillate. After validating his formula experimentally, thus confirming his assumption, he proceeded to conduct a parametric study with regard to the length L and the diameter D of the wire and developed the following formula: $T = (\mu D^4 / L) \phi$, where μ is a material constant. If we substitute $d\phi / dx = \phi / L$ and $J = \pi D^4 / 32$ into Equation (5.7) and compare it with Coulomb's formula, we see that Coulomb's material constant is $\mu = \pi G / 32$. Coulomb's formula, though correct, represented an empirical relationship between internal torque and the angle of twist. The analytical development of the theory of torsion of circular shafts is credited to A. Duleau.

A. Duleau, a graduate of the Ecole Polytechnique,[21] was commissioned in 1811 to design a forged iron bridge over the Dordogne river in the French city of Cubzac. He conducted extensive experiments on tension, compression, flexure, torsion, and elastic stability of bars of circular, triangular, elliptical, and rectangular cross sections as there existed few or no data on the behavior of bars of different cross sections under the loading conditions he needed for bridge design. In 1820 he published the results of his study, and in this paper he presented the analytical development of Coulomb's torsion formula by starting with Assumptions 1 and 3, that is, cross sections remain plane and radial lines straight during small twists to circular bars. He also established that these assumptions were not valid for noncircular shafts.

Cauchy, whose contributions to the mechanics of materials we have encountered in several previous chapters, was also interested in the torsion of rectangular bars. Cauchy showed that the cross section of a rectangular bar does not remain plane but warps due to torsional loads.

Saint-Venant, using the observations and conclusions of Coulomb, Duleau, and Cauchy, proposed in 1855 the displacement behavior given in Problem 5.43, and developed torsion formulas for a variety of shapes. This approach of assuming a displacement behavior that incorporates some features based on experience and empirical information but containing sufficient unknowns to satisfy equations of elasticity, is called *Saint-Venant's semi-inverse method*.

Ludwig Prandtl, the originator of boundary-layer theory in fluid mechanics and the inventor of the wind tunnel and its use in airplane design, published a paper in 1903 in

[21]Ecole Polytechnique, one of the early engineering schools, was founded in 1794 and had many pioneers in the field of mechanics of materials among its faculty and student graduates.

which he described the similarity of differential equations that describe the equilibrium of a soap film (membrane) and the equations in torsion derived using Saint-Venant's semi-inverse method. This similarity, called *membrane analogy,* can be used for obtaining torsional rigidities of complex cross sections by conducting experiments on soap films. Torsional rigidities obtained using the membrane analogy are well documented in the literature and in handbooks.

5.8 CLOSURE

In this chapter we established formulas for torsional deformation and stresses in circular shafts. We saw that the calculation of stresses and relative deformation requires the calculation of the internal torque at a section. For statically determinate shafts, the internal torque can be calculated either by making an imaginary cut and drawing an appropriate free-body diagram, or by drawing a torque diagram. In statically indeterminate single-axis shafts, the internal torque expression contains an unknown reaction torque that has to be determined using the compatibility equation. For single-axis shafts the compatibility equation states: the relative rotation of a section at the right wall with respect to the rotation at the left wall is zero.

We also saw that torsional shear stress should be drawn on a stress element for the purpose of understanding the character of torsional shear stress. This drawing of shear stress on a stress element is also important for the purpose of stress and/or strain transformation as will be described in future chapters.

In Chapter 8, on stress transformation, we will consider problems in which we first find torsional shear stress using the stress formula from this chapter and then find stresses on inclined planes, including planes with maximum normal stress. In Chapter 9, on strain transformation, we will find the torsional shear strain and then consider strains in different coordinate systems, including coordinate systems in which the normal strain is maximum. In Section 10.1 we will consider the combined loading problems of axial, torsion, and bending and the design of simple structures that may be determinate or indeterminate.

POINTS AND FORMULAS TO REMEMBER

- Theory is limited to: (1) slender shafts of circular cross sections; and (2) regions away from the neighborhood of stress concentration. The variation in cross sections and external torques is gradual.

$$T = \int_A \rho\tau_{x\theta}\, dA \quad (5.1) \qquad \phi = \phi(x) \quad (5.2) \qquad \text{small strain } \gamma_{x\theta} = \rho\frac{d\phi}{dx} \quad (5.3)$$

where T is the internal torque that is positive counterclockwise with respect to the outward normal to the imaginary cut surface, ϕ is the angle of rotation of the cross section that is positive counterclockwise with respect to the x axis, $\tau_{x\theta}$ and $\gamma_{x\theta}$ are the torsional shear stress and strain in polar coordinates, and ρ is the radial coordinate of the point where shear stress and shear strain are defined.

- Equations (5.1) and (5.4) are independent of material model.
- Torsional shear strain varies linearly across the cross section.
- Torsional shear strain is maximum at the outer surface of the shaft.
- Formulas below are valid for material that is linear, elastic, and isotropic.
- Homogeneous cross section:

$$\frac{d\phi}{dx} = \frac{T}{GJ} \quad (5.7) \qquad \tau_{x\theta} = \frac{T\rho}{J} \quad (5.8) \qquad \phi_2 - \phi_1 = \frac{T(x_2 - x_1)}{GJ} \quad (5.10)$$

where G is the shear modulus of elasticity, and J is the polar moment of the cross section given by $J = (\pi/2)(R_o^4 - R_i^4)$, R_o and R_i being the outer and inner radii of a hollow shaft.

- The quantity GJ is called torsional rigidity.
- If T, G, or J *change* with x, then find the relative rotation of a cross section by integration of Equation (5.7).
- If T, G, and J *do not change* between x_1 and x_2, then use Equation (5.10) to find the relative rotation of a cross section.
- Torsional shear stress varies linearly across the homogeneous cross section, reaching a maximum value on the outer surface of the shaft.
- Nonhomogeneous cross section:

$$\frac{d\phi}{dx} = \frac{T}{\sum_{j=1}^{n} G_j J_j} \qquad (\tau_{x\theta})_i = G_i \frac{\rho T}{\sum_{j=1}^{n} G_j J_j} \qquad \phi_2 - \phi_1 = \frac{T(x_2 - x_1)}{\sum_{j=1}^{n} G_j J_j}$$

where n is the number of materials in a cross section, G_j is the shear modulus of elasticity of the jth material, J_j is the polar moment of the jth material cross section, and $(\tau_{x\theta})_i$ is the torsional shear stress in the ith material.

CHAPTER SIX
SYMMETRIC BENDING OF BEAMS

6.0 OVERVIEW

A beam is any long structural member on which loads act perpendicular to the longitudinal axis. The bookshelf in Figure 6.1a can be modeled as a beam because the length is much greater than its width or thickness, and the weight of the books is perpendicular to the length of the shelf. The horizontal members in the frame of the building in Figure 6.1b can be modeled as beams that transmit the weight of the roof to the vertical columns and are braced for stiffness with axial members. The mast of a ship, the pole of a sign post, the frame of a car, the bulkheads in an aircraft, and the plank of a seesaw are among the countless examples of beams. In most cases the bending loads are coupled to torsional and axial loads. These combined loadings will be considered in Chapter 10. Stresses in symmetric bending are considered in this chapter, whereas the deflection of the beam will be discussed in Chapter 7.

The simplest theory for symmetric bending of beams will be developed rigorously, following the logic described in Figure 3.15, but subject to the limitations described in Section 3.13.

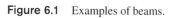

(a)

(b)

Figure 6.1 Examples of beams.

The two major learning objectives are:

1. Understand the theory, its limitations, and its applications for a strength-based design and analysis of symmetric bending of beams.

2. Develop the discipline to visualize the normal and shear stresses in the symmetric bending of beams.

6.1 PRELUDE TO THEORY

Several numerical examples are solved in this section in order to develop the observations that guide the development of the theory of the symmetric bending of beams.

- Example 6.1, using discrete bars welded to a rigid plate, illustrates the kinematics required to calculate the normal strain in bending.

- Example 6.2 shows that the kinematics of calculating normal strain in bending, which are developed for discrete bars in Example 6.1, are similar to the kinematics of calculating normal strains for a continuous beam.

- Example 6.3 shows the application of the logic described in Figure 3.15, which will be used for developing the theory of beam bending.

- Example 6.4 shows the impact that the choice of a material model[1] has on the calculation of the internal bending moment. This example shows that the material model only impacts the stress distribution, leaving all other equations unaffected. Thus the strain distribution, which is a kinematic relationship, is unaffected; the static equivalency equations between stress and internal moment are unaffected; and the equilibrium equations relating internal forces and moments are unaffected. Though we shall develop the simplest theory using Hooke's law, most of the equations that are developed can still be used after the material model becomes more complex. This will be demonstrated for composite beams in Section 6.7.

EXAMPLE 6.1
The left ends of three bars are built into a rigid wall and the right ends are welded to a rigid plate, as shown in Figure 6.2. The undeformed bars are straight and perpendicular to the wall and the rigid plate. Due to the applied moment, the rigid plate is observed to rotate by an angle of 3.5°. If the normal strain in bar 2 is zero, determine the normal strains in bars 1 and 3.

---

[1]Two material models are considered: (i) linear elastic homogeneous material cross section, and (ii) linear elastic laminated material cross section. Example 6.20 on elastic–perfectly plastic material cross sections and Example 6.24* on nonlinear material cross sections are additional examples emphasizing the points in this paragraph.

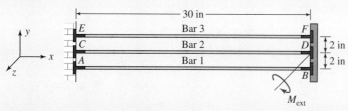

Figure 6.2 Geometry in Example 6.1.

Method 1

PLAN

Draw an approximate deformed shape of the three bars subject to the kinematic restriction: the right angles between the bars and the rigid bodies cannot change during deformation because these bars are welded to rigid bodies. We note that in a circular arc, the tangent to the arc is perpendicular to the radial line at any point. If the bars are approximated as circular arcs and the wall and the rigid plate are in the radial direction, then the kinematic restriction is satisfied by the deformed shape. We can relate the angle subtended by the arc to the length of arc formed by CD, as we did in Example 2.3. Using the deformed geometry, the strains of the remaining bars can be found.

Solution Figure 6.3 shows the deformed bars as circular arcs with the wall and the rigid plate in the radial direction. We know that the length of arc CD_1 is still 30 in since it does not undergo any strain. We can relate the angle subtended by the arc to the length of arc formed by CD, as we did in Example 2.3, and calculate the radius of the arc R as follows:

$$\psi = \left(\frac{3.5°}{180°}\right)\pi = 0.0611 \text{ rad}$$

$$CD_1 = R\psi = 30$$

or

$$R = 491.1 \tag{E1}$$

The distance between points B_1 and D_1 is 2 in, as the distance along the rigid plate does not change during deformation. The radius for arc AB_1 is $R - 2$, as shown in Figure 6.3, and thus the arc length AB_1 can be found and the strain in bar 1 calculated,

$$AB_1 = (R-2)\psi = 29.8778$$

$$\varepsilon_1 = \frac{AB_1 - AB}{AB} = \frac{-0.1222}{30} \tag{E2}$$

$$\text{ANS.} \quad \varepsilon_1 = -4073 \ \mu\text{in/in}$$

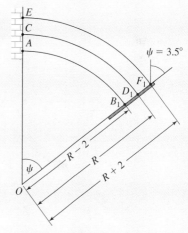

Figure 6.3 Normal strain calculations in Example 6.1.

The strain in bar 3 is similar to that of bar 1,

$$EF_1 = (R+2)\psi = 30.1222$$

$$\varepsilon_3 = \frac{EF_1 - EF}{EF} = \frac{0.1222}{30} \qquad (E3)$$

ANS. $\varepsilon_3 = 4073\ \mu\text{in/in}$

COMMENT

In developing the theory for beam bending, we will assume that the squashing (dimensional change in the y direction) is much smaller than the bending action, hence the normal strain in the y direction is negligible. As a consequence, for kinematic considerations we will view the cross section like a rigid plate that rotates about the z axis. We shall also assume that the cross sections during deformation are nearly perpendicular to the longitudinal direction. This will let us view a line in the longitudinal direction like a bar that is welded to the right plate at a right angle. With these assumptions the kinematic derivation of strains will be as described. In the next example we shall visualize a beam as infinitesimal rods that deform as in this example.

Method 2

PLAN

We can use small-strain approximation and find the deformation component in the horizontal (original) direction for bars 1 and 3, from which the normal strains can be found.

Solution   Figure 6.4 shows the rigid plate in the deformed position. As there is no strain in bar 2, the horizontal displacement of point D is zero, that is, D_1 is at the same location as point D in the undeformed geometry. We can use point D_1 to find the relative displacements of points B and F, knowing that the distance along the rigid plate between these points cannot change.

From the triangle $D_1 F_1 D_2$ the horizontal movement of point F can be related to the distance $D_1 F_1$ on the rigid plate. As these are small strains, we can approximate the sine function by its argument.

$$DF_2 = D_2 F_1 = D_1 F_1 \sin\psi \approx 2\psi = 0.1222 \qquad (E4)$$

The horizontal deformation of bar 3 can next be found and the normal strain in bar 3 calculated,

$$u_3 = DF_2 = 0.1222$$

$$\varepsilon_3 = \frac{u_3}{30} \qquad (E5)$$

ANS. $\varepsilon_3 = 4073\ \mu\text{in/in}$

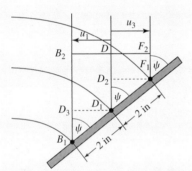

Figure 6.4 Alternate method for normal strain calculations in Example 6.1.

From triangle $D_1D_3B_1$ the horizontal movement of point B can be related to the distance D_1B_1 on the rigid plate,

$$B_2D = D_3D_1 = B_1D_1 \sin \psi \approx 2\psi = 0.1222 \qquad \text{(E6)}$$

Noting that point B moves in the negative x direction, the horizontal deformation of bar 1 can next be found and the normal strain in bar 1 calculated,

$$u_1 = -B_2D$$

$$\varepsilon_1 = \frac{u_1}{30} \qquad \text{(E7)}$$

$$\text{ANS.} \qquad \varepsilon_1 = -4073 \ \mu\text{in/in}$$

COMMENTS

1. Method 1 is intuitive and easier to visualize than Method 2. But Method 2 is computationally simpler. We will use both methods when we develop the kinematics in beam bending in Section 6.2.

2. In Chapter 2 it was observed that small-strain approximation results in a linear relationship between deformation and strain. To demonstrate the simplification of small-strain approximation, suppose that the normal strain of bar 2 was not zero but $\varepsilon_2 = 800 \ \mu\text{in/in}$. What would be the normal strains in bars 1 and 3? One way of solving this new problem would be to compute the radius $R = 491.5$ in. in Equation (E1) and then repeat the steps as before with the new value of R, obtaining $\varepsilon_1 = -3272 \ \mu\text{in/in}$ and $\varepsilon_3 = 4872 \ \mu\text{in/in}$.

   An alternate method would be to use the principle of superposition, which is applicable to linear systems. Imagine that before rotating the rigid plate, we pulled it in the axial direction until we obtained the given normal strain in bar 2. This imaginary axial deformation would change the length by 0.024 in, which can be neglected because it is small compared to 30 in. The total strain is the strain due to axial deformation and the bending strain found before. The net result is $\varepsilon_1 = -4073 + 800 = -3273 \ \mu\text{in/in}$ and $\varepsilon_3 = 4073 + 800 = 4873 \ \mu\text{in/in}$, which is nearly the same as before.

   The foregoing implies that the axial and bending normal strains can be superposed, which suggests that if we use a linear material model, then the axial stresses and the bending stresses can be superposed, as we shall do in Chapter 10 on combined loading.

EXAMPLE 6.2

A beam made from hard rubber is built into a rigid wall at the left end
and attached to a rigid plate at the right end as shown in Figure 6.5. After
rotation of the rigid plate the strain in line AB at $y = 0$ is zero. Determine
the strain in line CD in terms of y and R, where y is the distance of line
CD from line AB, and R is the radius of curvature of line AB.

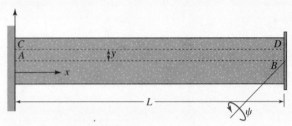

Figure 6.5 Beam geometry in Example 6.2.

PLAN

We visualize the beam as made up of bars, as in Example 6.1, but of
infinitesimal thickness. We consider two such bars, AB and CD, and
analyze the deformations of these two bars as we did in Example 6.1.

Solution   Due to deformation, point B moves to point B_1 and point D
moves to point D_1 as shown in Figure 6.6. We can calculate the strain
in CD as we did in Example 6.1,

$$\varepsilon_{AB} = \frac{AB_1 - AB}{AB} = 0$$

$$AB_1 = AB = R\psi = L \qquad \psi = \frac{L}{R}$$

$$CD_1 = (R - y)\psi = \frac{(R - y)L}{R}$$

$$\varepsilon_{CD} = \frac{CD_1 - CD}{CD} = \frac{(R - y)L/R - L}{L}$$

ANS. $\varepsilon_{CD} = \dfrac{-y}{R}$

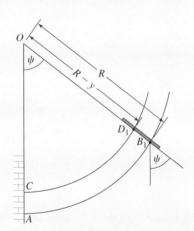

Figure 6.6 Deformed geometry in
Example 6.2.

COMMENTS

1. In Example 6.1, $R = 491.1$ and $y = +2$ for bar 3 and $y = -2$ for bar 1.
On substituting these values into the preceding results, we see that
we obtain the results of Example 6.1.

2. In Section 6.2 we will derive these results based on assumptions that would imply that the cross section of the beam behaves like a rigid plate that can rotate about the z axis.

3. Supposing the strain in AB was ε_{AB}, then the strain in CD can be calculated as we did in comment 2 of Method 2 in Example 6.1, obtaining $\varepsilon_{CD} = \varepsilon_{AB} - y/R$. The strain ε_{AB} is the axial strain, and the remaining component is the normal strain due to bending.

EXAMPLE 6.3

The modulus of elasticity of the bars in Example 6.1 is 30,000 ksi. Each bar has a cross-sectional area $A = \frac{1}{2}$ in^2. Determine the external moment M_{ext} that caused the strains in the bars in Example 6.1.

PLAN

Using Hooke's law, determine the stresses from the strains calculated in Example 6.1. Replace the stresses by equivalent internal axial forces. Draw the free-body diagram of the rigid plate and determine the moment M_{ext}.

Solution

1. *Strain calculations:* The strains in the three bars as calculated in Example 6.1 are

$$\varepsilon_1 = -4073 \ \mu\text{in/in} \qquad \varepsilon_2 = 0 \qquad \varepsilon_3 = 4073 \ \mu\text{in/in} \qquad \text{(E1)}$$

2. *Stress calculations:* From Hooke's law,

$$\sigma_1 = E\varepsilon_1 = 30,000(-4073)10^{-6} = 122.19 \ \text{ksi (C)} \qquad \text{(E2)}$$

$$\sigma_2 = E\varepsilon_2 = 0 \qquad \text{(E3)}$$

$$\sigma_3 = E\varepsilon_3 = 30,000 \times 4073 \times 10^{-6} = 122.19 \ \text{ksi (T)} \qquad \text{(E4)}$$

3. *Internal forces calculations:* The equivalent internal normal forces in each bar can be found as

$$N_1 = \sigma_1 A = 61.095 \ \text{kips (C)} \qquad \text{(E5)}$$

$$N_3 = \sigma_3 A = 61.095 \ \text{kips (T)} \qquad \text{(E6)}$$

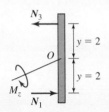

Figure 6.7 Free-body diagram in Example 6.1.

4. *External moment calculations:* From the free-body diagram of the rigid plate (Figure 6.7) we can find the moment M_z,

$$\sum M_O \qquad M_z = N_1(y) + N_3(y) \qquad \text{(E7)}$$

$$\text{ANS.} \qquad M_z = 244.4 \text{ in}\cdot\text{kips}$$

COMMENTS

1. The summation in Equation (E7) can be rewritten as $\sum_{i=1}^{2} y\sigma\Delta A_i$, where σ is the normal stress acting at a distance y from the zero strain bar, and ΔA_i is the cross-sectional area of the ith bar. If we had n bars attached to the rigid plate, then the moment would be given by $\sum_{i=1}^{n} y\sigma\Delta A_i$. As we increase the number of bars n to infinity, the cross-sectional area ΔA_i will tend to zero (infinitesimal area written as dA) as we try to fit an infinite number of bars to the plate—resulting in a continuous body with the summation replaced by an integral.

2. The total axial force in this example is zero because of symmetry. If this were not the case, then the axial force would be given by the summation $\sum_{i=1}^{n} \sigma\Delta A_i$. As in comment 1, this summation would be replaced by an integral when n tends to infinity, as will be shown in Section 6.1.1.

6.1.1 Internal Bending Moment

In this section we formalize the observation made in Example 6.3, namely, the normal stress σ_{xx} can be replaced by an equivalent bending moment using an integral over the cross-sectional area. Figure 6.8 shows the normal stress distribution σ_{xx} that is to be replaced by an equivalent internal bending moment M_z. Let y represent the coordinate

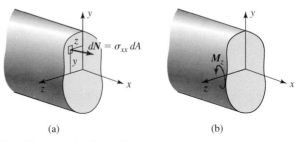

(a)   (b)

Figure 6.8 Statically equivalent internal moment.

at which the normal stress acts. Static equivalency in Figure 6.8 results in the following condition:

$$M_z = -\int_A y\sigma_{xx}\, dA \qquad (6.1)$$

Figure 6.8a suggests that for static equivalency there should be an axial force N and a bending moment about the y axis M_y. However, the requirement of symmetric bending implies that the normal stress σ_{xx} is symmetric about the axis of symmetry, that is, the y axis. Thus M_y is implicitly zero due to the limitation of symmetric bending. The argument of symmetry has no bearing on the axial force N. But our desire to study bending independent of axial loading requires that the stress distribution be such that the internal axial force should be zero. Thus we must explicitly satisfy the following condition:

$$\int_A \sigma_{xx}\, dA = 0 \qquad (6.2)$$

Equation (6.2) implies that the stress distribution across the cross section must be such that there is no net axial force, that is, the compressive force must equal the tensile force on a cross section in bending. If stress is to change from compression to tension, then there must be a location of zero normal stress in bending. This location is called the neutral axis.

Definition 1 The line on the cross section where the bending normal stress is zero is called *neutral axis*.

Equations (6.1) and (6.2) are independent of the material model as these equations represent static equivalency between the normal stress on the entire cross section and the internal moment. If we were to consider a composite beam cross section or a nonlinear material model, then it would affect the value and distribution of σ_{xx} across the cross section, but Equation (6.1) relating σ_{xx} to M_z would remain unchanged. Example 6.4 elaborates further the discussion in this paragraph. The origin of the y coordinate is located at the neutral axis irrespective of the material model. Hence determining the location of the neutral axis is critical in all bending problems. For a homogeneous linearly elastic isotropic material the location of the origin will be discussed in greater detail in Section 6.2.4.

EXAMPLE 6.4

Figure 6.9 shows a homogeneous wooden cross section and a cross section in which the wood is reinforced with steel. The normal strain for both cross sections was found to vary as $\varepsilon_{xx} = -200y\ \mu$. The moduli of elasticity for steel and wood are $E_{\text{steel}} = 30{,}000$ ksi and $E_{\text{wood}} = 8000$ ksi.

(a) Write expressions for normal stress σ_{xx} as a function of y and plot the σ_{xx} distribution for each of the two cross sections shown.

(b) Calculate the equivalent internal moment M_z for each cross section.

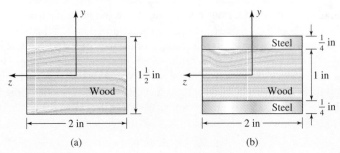

Figure 6.9 Cross sections in Example 6.4. (*a*) Homogeneous. (*b*) Laminated.

PLAN

(a) From the given strain distribution we can find the stress distribution by Hooke's law. We note that the problem is symmetric and stresses in each region will be linear in y.

(b) The integral in Equation (6.1) can be written as twice the integral for the top half since the stress distribution is symmetric about the center. After substituting the stress as a function of y in the integral, the integration can be performed to obtain the equivalent internal moment.

Solution

(a) From Hooke's law we can write

$$(\sigma_{xx})_{\text{wood}} = 8 \times 10^3 (-200y) 10^{-6} = -1.6y \text{ ksi} \qquad \text{(E1)}$$

$$(\sigma_{xx})_{\text{steel}} = 30 \times 10^3 (-200y) 10^{-6} = -6y \text{ ksi} \qquad \text{(E2)}$$

For the homogeneous cross section the stress distribution is given in Equation (E1), but for the laminated case it switches from Equation (E1) to Equation (E2), depending on the value of y. We can write the stress distribution for both cross sections as a function of y as follows:

Homogeneous cross section:

$$\sigma_{xx} = -1.6y \text{ ksi} \qquad -0.75 \le y < 0.75$$

Laminated cross section:

$$\sigma_{xx} = \begin{cases} -6y \text{ ksi} & 0.5 < y \le 0.75 \\ -1.6y \text{ ksi} & -0.5 < y < 0.5 \\ -6y \text{ ksi} & -0.75 \le y < -0.5 \end{cases}$$

The strains and stresses can now be plotted as a function of y, as shown in Figure 6.10.

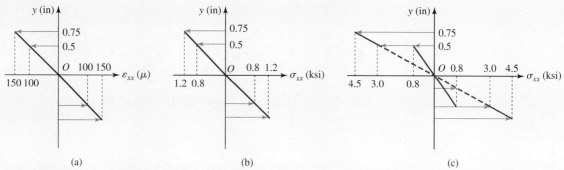

Figure 6.10 Strain and stress distributions in Example 6.4. (*a*) Strain distribution. (*b*) Stress distribution in homogeneous cross section. (*c*) Stress distribution in laminated cross section.

(b) The thickness (dimension in the z direction) is 2. Hence we can write $dA = 2dy$. Noting that the stress distribution is symmetric, we can write the integral in Equation (6.1) as

$$M_z = -\int_{-0.75}^{0.75} y\sigma_{xx}(2dy) = -2\left[\int_{0}^{0.75} y\sigma_{xx}(2dy)\right] \qquad \text{(E3)}$$

Homogeneous cross section: The stress distribution for the homogeneous cross section is $\sigma_{xx} = -1.6y$ ksi. Substituting this into Equation (E3) and integrating, we obtain the equivalent internal moment,

$$M_z = -2\left[\int_{0}^{0.75} y(-1.6y \text{ ksi})(2dy)\right] = 6.4\frac{y^3}{3}\bigg|_{0}^{0.75} = 6.4\frac{0.75^3}{3}$$

ANS. $\qquad M_z = 0.9 \text{ in} \cdot \text{kips}$

Laminated cross section: The stress distribution for the laminated cross section is $\sigma_{xx} = -1.6y$ ksi for $0 \le y < 0.5$ and $\sigma_{xx} = -6y$ ksi for $0.5 < y \le 0.75$. Writing the integral in Equation (E3) as a sum of two integrals and substituting the stress expressions, we can obtain the equivalent internal moment,

$$M_z = -2\left[\int_{0}^{0.5} y(-1.6y)(2dy) + \int_{0.5}^{0.75} y(-6y)(2dy)\right]$$

$$= 4\left(1.6\frac{y^3}{3}\bigg|_{0}^{0.5} + 6\frac{y^3}{3}\bigg|_{0.5}^{0.75}\right)$$

ANS. $\qquad M_z = 2.64 \text{ in} \cdot \text{kips}$

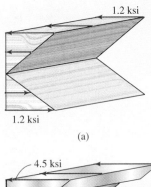

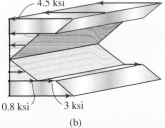

Figure 6.11 Surface stress distributions in Example 6.4. (*a*) Homogeneous cross section. (*b*) Laminated cross section.

COMMENTS

1. The example demonstrates that though the strain varies linearly across the cross section, the stress may not. In this example we considered material nonhomogeneity. In a similar manner we can consider other models, such as elastic–perfectly plastic, or material models that have nonlinear stress–strain curves.

2. Figure 6.11 shows the stress distribution on the surface. The symmetry of stresses about the center results in a zero axial force. The moment can also be found by doubling the moment from one side, as was shown in Equation (E3).

3. We can obtain the equivalent internal moment for a homogeneous cross section by replacing the triangular load by an equivalent load at the centroid of each triangle and then finding the equivalent moment, as shown in Figure 6.12. This approach is very intuitive. However, as the stress distribution becomes more complex, such as in a laminated cross section, or for more complex cross-sectional shapes, this intuitive approach will become very tedious and the generalization represented by Equation (6.1) and the resulting formula will help us simplify the calculations.

4. The relationship of the internal moment and the external loads can be established by drawing the appropriate free-body diagram for a particular problem. The relationship of internal and external moments depends on the free-body diagram and is independent of the material homogeneity.

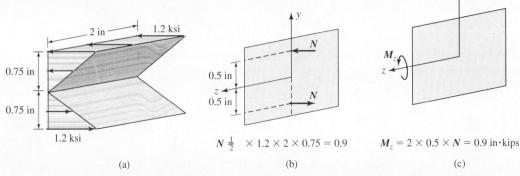

Figure 6.12 Statically equivalent internal moment in Example 6.4.

6.1 The rigid plate that is welded to the two bars in Figure P6.1 is rotated about the z axis, causing the two bars to bend. The normal strains in bars 1 and 2 were found to be $\varepsilon_1 =$ 2000 μin/in and $\varepsilon_2 = -1500$ μin/in. Determine the angle of rotation ψ.

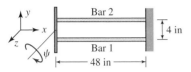

Figure P6.1

6.2 Determine the location h in Figure P6.2 at which a third bar in Problem 6.1 must be placed so that there is no normal strain in the third bar.

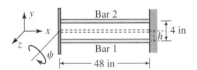

Figure P6.2

6.3 The two rigid plates that are welded to six bars in Figure P6.3 are rotated about the z axis, causing the six bars to bend. The normal strains in bars 2 and 5 were found to be zero. What are the strains in the remaining bars?

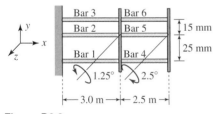

Figure P6.3

6.4 The strains in bars 1 and 3 in Figure P6.4 were found to be $\varepsilon_1 = 800$ μ and $\varepsilon_3 = 500$ μ. Determine the strains in the remaining bars.

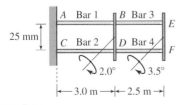

Figure P6.4

6.5 Due to the action of the external moment M_z and force P, the rigid plate shown in Figure P6.5 was observed to rotate by $2°$ and the normal strain in bar 1 was found $\varepsilon_1 =$ 2000 μin/in. Both bars have a cross-sectional area $A = \frac{1}{2}$ in^2 and a modulus of elasticity $E =$ 30,000 ksi. Determine the applied moment M_z and force P.

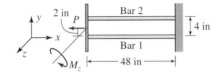

Figure P6.5

6.6 The rigid plate shown in Figure P6.6 was observed to rotate $1.25°$ due to the action of the external moment M_z and the force P. All three bars have a cross-sectional area $A = 100$ mm^2 and a modulus of elasticity $E = 200$ GPa. If the strain in bar 2 was measured as zero, determine the external moment M_z and the force P.

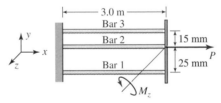

Figure P6.6

6.7 The rigid plates BD and EF in Figure P6.7 were observed to rotate by $2°$ and $3.5°$ in the direction of applied moments. All bars have a cross-sectional area of $A = 125$ mm^2. Bars 1 and 3 are made of steel $E_S = 200$ GPa, and bars 2 and 4 are made of aluminum $E_{al} =$ 70 GPa. If the strains in bars 1 and 3 were found to be $\varepsilon_1 = 800$ μ and $\varepsilon_3 = 500$ μ determine the applied moment M_1 and M_2 and the forces P_1 and P_2 that act at the center of the rigid plates.

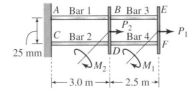

Figure P6.7

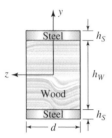

Figure P6.8

Laminated cross sections

In Problems 6.8 through 6.10 steel strips
($E_S = 30,000$ ksi) are securely attached to a
wooden beam ($E_W = 2000$ ksi), as shown in
Figure P6.8. The normal strain ε_{xx} at the cross
section due to bending about the z axis and
the dimensions of the cross section are as
given in each problem. Determine the equiva-
lent internal moment M_z.

6.8 $\varepsilon_{xx} = -100y$ μ, where y is measured in
inches. Use $d = 2$ in, $h_W = 4$ in, and $h_S = \frac{1}{8}$ in.

6.9 $\varepsilon_{xx} = -50y$ μ, where y is measured in
inches. Use $d = 1$ in, $h_W = 6$ in, and $h_S = \frac{1}{4}$ in.

6.10 $\varepsilon_{xx} = 200y$ μ, where y is measured in
inches. Use $d = 1$ in, $h_W = 2$ in, and $h_S = \frac{1}{16}$ in.

In Problems 6.11 through 6.13 the flanges
made from steel ($E_S = 200$ GPa) are securely
attached to the wooden beam ($E_W = 10$ GPa),
as shown in Figure P6.11. The normal strain
ε_{xx} at the cross section due to bending about
the z axis and the dimensions of the cross sec-
tion are as given in each problem. Determine
the equivalent internal moment M_z.

6.11 $\varepsilon_{xx} = -0.012y$, where y is measured in
meters. Use $t_W = 20$ mm, $h_W = 250$ mm, $t_F = 20$ mm, and $d_F = 125$ mm.

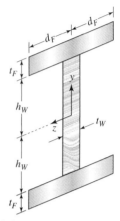

Figure P6.11

6.12 $\varepsilon_{xx} = -0.15y$, where y is measured in
meters. Use $t_W = 10$ mm, $h_W = 50$ mm, $t_F = 10$ mm, and $d_F = 25$ mm.

6.13 $\varepsilon_{xx} = 0.02y$, where y is measured in
meters. Use $t_W = 15$ mm, $h_W = 200$ mm, $t_F = 20$ mm, and $d_F = 150$ mm.

6.2 THEORY

The theory will be developed subject to the limitations described in Section 3.13,
namely: (i) the length of the member is significantly greater than the greatest dimension
in the cross section; (ii) we are away from the regions of stress concentration; (iii) the
variation of external loads or changes in the cross-sectional areas are gradual except in
regions of stress concentration. In addition the following limitations will be imposed:

1. The cross section has a plane of symmetry.[2]
2. The loads are in the plane of symmetry.
3. The load direction does not change with deformation.
4. The external loads are not functions of time,[3] that is, we have a static problem.

Load P_1 in Figure 6.13 would bend the beam as well as twist (rotate) the cross section. Load P_2, which lies in the plane of symmetry, will cause only bending. Thus the limitation of loads applied in the plane of symmetry decouples the bending problem from the

[2]This limitation also separates bending about the z axis from bending about the y axis. See
Problem 6.113.
[3]See Problems 7.33 and 7.34 for dynamic problems.

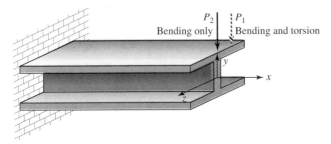

Figure 6.13 Loading in plane of symmetry.

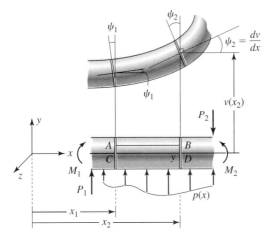

Figure 6.14 Beam segment.

torsion problem.[4] The limitation that the load direction not change with deformation is required to obtain a linear theory and works well as long as the deformations are small.

We assume that Figure 6.14 shows a segment of a beam with $x - y$ as the plane of symmetry. The beam is loaded by transverse forces P_1 and P_2 in the y direction, moments M_1 and M_2 about the z axis, and a transverse distributed force $p(x)$. The distributed force $p(x)$ has units of force per unit length and is considered positive in the positive y direction. Due to external loads, a line on the beam deflects by v in the y direction.

The objectives of the derivation are:

1. To obtain a formula for bending normal stress σ_{xx} and bending shear stress τ_{xy} in terms of the internal moment M_z and the internal shear force V_y
2. To obtain a formula for calculating the beam deflection $v(x)$

To account for the variation[5] of $p(x)$ and the cross-sectional dimensions, we will take $\Delta x = x_2 - x_1$ as infinitesimal distance in which these quantities can be treated as constants. The logic shown in Figure 6.15 and discussed in Section 3.2 will be used to

---

[4]The separation of torsion from bending requires that the load pass through the shear center, which always lies on the axis of symmetry.
[5]Note that we have imposed the limitation that the variation in $p(x)$ and the cross-sectional dimensions be gradual.

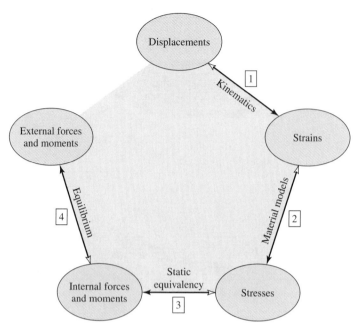

Figure 6.15 Logic in mechanics of materials.

develop the simplest theory for the bending of beams. Assumptions will be identified as we move from one step to the next. These assumptions are the points at which additional complexities can be added to the theory, as described by footnotes associated with the assumptions.

6.2.1 Kinematics

In Example 6.1 we found the normal strains in bars welded to rigid plates rotating about the z axis. The assumptions that will let us simulate the behavior of a cross section like that of the rigid plate are listed, justified with experimental evidence, and the impact of these assumptions on the theory is discussed.

Assumption 1 Squashing, that is, dimensional changes in the y-direction, is significantly smaller than bending.

Assumption 2 Plane sections before deformation remain plane after deformation.

Assumption 3 Plane sections perpendicular to the beam axis remain *nearly* perpendicular after deformation.

Figure 6.16 shows a rubber beam with a grid on its surface. When this beam is bent by hand, notice that lines initially in the y direction continue to remain straight but rotate about the z axis. Also notice that the rotation is such that the original right angle between the x and y directions is nearly preserved during bending. The horizontal distance between the vertical lines does change, showing that there is a normal strain in the x direction. There are some dimensional changes in the y direction, but these appear to be significantly smaller than those in the x direction.

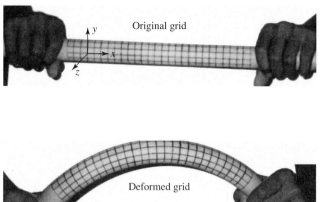

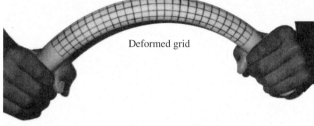

Figure 6.16 Deformation in bending. (Courtesy Professor J. B. Ligon.)

The deformation shown in Figure 6.16 is on the surface of the beam. Can we extrapolate the surface observations into the interior of the beam? To answer this question, and in order to appreciate the impact of these assumptions on the theoretical development that follows, let us reconsider Assumptions 1 through 3.

Assumption 1 implies that the dimensional changes in the cross section are much smaller than the movement of the cross section as a whole. The longer the beam, the better is the validity of Assumption 1. Neglecting dimensional changes in the y direction implies that the normal strain in the y direction is small[6] and can be neglected in the kinematic calculations, that is, $\varepsilon_{yy} = \partial v/\partial y \approx 0$, which implies that v cannot be a function of y, as shown,

$$v = v(x) \tag{6.3}$$

Another way of viewing Assumption 1 is to think that the function v has been approximated as constant in the y direction. This implies is that if we know the curve of one axial line on the beam, then we know how all other lines on the beam bend. The curve described by $v(x)$ is called the elastic curve and will be discussed in detail in the next chapter.

Assumption 2 implies that the axial displacement u varies linearly, as shown in Figure 6.17. In other words, the equation for u is

$$u = u_0 - \psi y$$

where u_0 is the axial displacement at $y = 0$ and ψ is the slope of the plane. We accounted for uniform axial displacement u_0 in Chapter 4. In order to study each problem independently,[7] we will assume $u_0 = 0$.

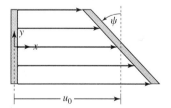

Figure 6.17 Linear variation of axial displacement u.

---

[6]It is accounted for as the Poisson effect, but the normal strain in the y direction is not an independent variable and hence is considered negligible in kinematics.
[7]See Problem 6.111, in which u_0 is carried along with v.

Assumption 3 implies that the shear strain γ_{xy} is nearly zero. γ_{xy} is the measure of change of the angle between the axis of the beam and the perpendicular plane. We cannot use this assumption in building theoretical models of beam bending if shear is important,[8] such as in sandwich beams.[9] But Assumption 3 helps simplify the theory as it eliminates the variable ψ by imposing the constraint that the angle between the axial direction and the cross section be always 90°. This is accomplished by relating ψ to v as described next.

The displacement curve is defined by $v(x)$. As shown in Figure 6.14, the angle of the tangent to the curve $v(x)$ is equal to the rotation of the cross section when Assumption 3 is valid. For small strains, the tangent of an angle can be replaced by the angle itself, that is, $\tan \psi \approx \psi = dv/dx$. Upon substitution into $u = u_0 - \psi y$ and noting that $u_0 = 0$, we obtain

$$u = -y\frac{dv}{dx}(x) \tag{6.4}$$

6.2.2 Strain Distribution

Assumption 4 Strains are small.

Figure 6.18 shows two ways of calculating the normal strain of a line AB at a distance y from a line that is unstrained, line CD, as was discussed in Examples 6.1 and 6.2,

$$AB = CD = CD_1$$

$$\varepsilon_{xx} = \frac{AB_1 - AB}{AB} = \frac{(R-y)\Delta\psi - R\Delta\psi}{R\Delta\psi}$$

or

$$\varepsilon_{xx} = -\frac{y}{R} \tag{6.5a}$$

$$\Delta u = -y \sin \Delta\psi \approx -y\Delta\psi$$

$$\varepsilon_{xx} = \lim_{\Delta x \to 0}\left(\frac{\Delta u}{\Delta x}\right) = -y\frac{d\psi}{dx}$$

or

$$\varepsilon_{xx} = -y\frac{d^2v}{dx^2}(x) \tag{6.5b}$$

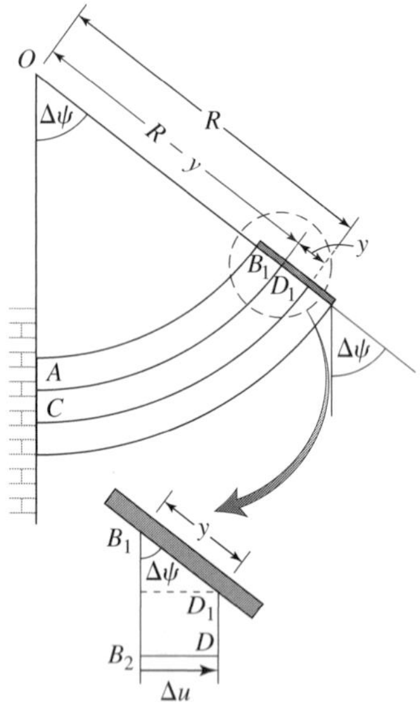

Figure 6.18 Normal strain calculations in symmetric bending.

Equations (6.5a) and (6.5b) show that the bending normal strain ε_{xx} varies linearly with y and has a maximum value at either the top or the bottom of the beam. (d^2v/dx^2) is the curvature of the beam, and its magnitude is equal to $1/R$, where R is the radius of curvature.

---

[8]Such beams are called Timoshenko beams. See Problem 7.32 for accounting for shear.
[9]A sandwich beam cross section consists of two stiff plates with a soft core bonded together. Sandwich beams are common in the design of lightweight structures such as aircraft and boats.

6.2.3 Material Model

Our motivation is to develop a simple theory for bending of symmetric beams. Thus we make assumptions regarding material behavior that will permit us to use the simplest material model given by Hooke's law.

Assumption 5 Material is isotropic.

Assumption 6 Material is linearly elastic.[10]

Assumption 7 There are no inelastic strains.[11]

Substituting Equation (6.5b) into Hooke's law $\sigma_{xx} = E\varepsilon_{xx}$, we obtain

$$\sigma_{xx} = -Ey\frac{d^2v}{dx^2} \tag{6.6}$$

Though the strain is a linear function of y, we cannot say the same for stress. The modulus of elasticity E could change across the cross section, as in laminated structures.

6.2.4 Location of Neutral Axis

Equation (6.6) shows that the stress σ_{xx} across the cross section is a function of y, and that its value must be zero at $y = 0$, that is, the origin of y must be at the neutral axis. But where is the neutral axis on the cross section? In Section 6.1.1 it was noted that the distribution of σ_{xx} is such that the total tensile force cancels the total compressive force on a cross section, as given by Equation (6.2). Substituting Equation (6.6) into Equation (6.2) and noting that $(d^2v/dx^2)(x)$ is a function of x only, whereas the integration is with respect to y and $z(dA = dy\,dz)$, we obtain

$$-\int_A Ey\frac{d^2v}{dx^2}(x)\,dA = -\frac{d^2v}{dx^2}(x)\int_A Ey\,dA = 0$$

We note that (d^2v/dx^2) cannot be zero as that would imply that there is no bending. Therefore the integral must be zero,

$$\int_A yE\,dA = 0 \tag{6.7}$$

Equation (6.7) is used for determining the origin (neutral axis) in composite beams discussed in Section 6.7. But for the simplest theory we would like to take E outside the integral, that is, E cannot change across the cross section. We make the following assumption of homogeneity:

Assumption 8 Material is homogeneous across the cross section[12] of a beam.

---

[10]See Section 6.8 for nonlinear material behavior.

[11]Inelastic strains could be due to temperature, humidity, plasticity, viscoelasticity, etc. See Problem 6.112 for including thermal strains.

[12]See Section 6.7 on composite beams for nonhomogeneous cross sections.

We obtain Equation (6.8) from Equation (6.7),

$$\int_A y \, dA = 0 \qquad (6.8)$$

Equation (6.8) is satisfied if y is measured from the centroid of the cross section, that is, the origin must be chosen to be at the centroid of the cross section constructed from linear elastic, isotropic, homogeneous material. It should be noted that Equation (6.8) is the same as Equation (4.7a) in axial deformation. However, in axial problems it was the requirement that the internal bending moment that generated Equation (4.7a) be zero. Here it is the requirement of zero axial force that generated Equation (6.8). Thus by choosing the origin to be the centroid, we decouple the axial problem from the bending problem.[13]

From Equations (6.6) and (6.8) the following conclusion can be drawn. For cross sections constructed from linear elastic, isotropic, homogeneous material, the bending normal stress σ_{xx} varies linearly, with y reaching a maximum value at the point farthest from the centroid of the cross section. The point farthest from the centroid is the top surface or the bottom surface of the beam.[14] Example 6.5 demonstrates the use of the observations made in this paragraph.

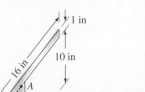

Figure 6.19 T cross section in Example 6.5.

EXAMPLE 6.5

The maximum bending normal strain on a homogeneous steel cross section ($E = 30{,}000$ ksi) was found to be $\varepsilon_{xx} = +1000\ \mu$. Determine the bending normal stress at point A (Figure 6.19).

PLAN

We can find the maximum bending normal stress from Hooke's law and the given strain. We can find the centroid of the cross section where we know that the bending normal stress will be zero. The maximum bending normal stress will be at the point farthest from the centroid, which we expect to be the bottom of the cross section. Knowing the normal stress at two points of linear distribution, we can find the normal stress at point A.

Solution   The centroid η_c of the cross section[15] from the bottom can be found as (Figure 6.20)

$$\eta_c = \frac{\sum_i \eta_i A_i}{\sum_i A_i} = \frac{5 \times 10 \times 1.5 + 10.5 \times 16 \times 1}{10 \times 1.5 + 16 \times 1} = 7.84 \text{ in}$$

---

[13]See Problem 6.111 for the impact when y is not measured from the centroid.

[14]In un-symmetric bending this conclusion is still applicable, but we have to be specific about the location of the point on the top or bottom surface where σ_{xx} reaches its maximum value.

[15]See Appendix A.4 for the calculation of centroids.

Point B is at a distance of 7.84 in from the centroid, whereas point D is at a distance of 3.16 in. Hence the maximum bending normal stress in the given cross section will occur at point B, and its value can be found from Hooke's law as

$$\sigma_B = E\varepsilon_{max} = 30,000 \times 1000 \times 10^{-6} = 30 \text{ ksi}$$

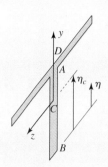

Figure 6.20 Locating the centroid in Example 6.5.

Knowing that the bending normal stress is linear across the cross section, zero at the centroid, and +30 ksi at point B, we can find σ_A at a distance of $y_A = 2.16$ in from the centroid using a triangle, as shown in Figure 6.21,

$$\frac{30}{7.84} = \frac{\sigma_A}{2.16}$$

or

ANS. $\sigma_A = 8.27 \text{ ksi (C)}$

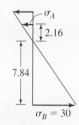

Figure 6.21 Linear stress distribution in Example 6.5.

COMMENT
The stress distribution in Figure 6.21 can be represented as $\sigma_{xx} = -3.82y$ ksi. The equivalent internal moment can be found using Equation (6.1).

6.2.5   Internal Shear Force, Moment, and Sign Conventions

Figure 6.22 shows a cantilever beam loaded with a transverse force P. An imaginary cut is made at section AA, and a free-body diagram is drawn. For equilibrium it is clear that we need an internal shear force V_y, which is possible only if there is a non-zero shear stress τ_{xy}. By Hooke's law this implies that the shear strain γ_{xy} cannot be zero. Assumption 3 implied that shear strain was small but not zero. In beam bending, a check on the validity of the analysis is to compare the maximum shear stress τ_{xy} to the maximum normal stress σ_{xx} for the entire beam. If the two stress components are comparable, then the shear strain cannot be neglected in kinematic considerations

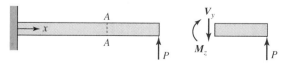

Figure 6.22   Internal forces and moment necessary for equilibrium.

and our theory is not valid. The maximum normal stress σ_{xx} in the beam should be nearly an order of magnitude greater than the maximum shear stress τ_{xy}. The shear force is defined as

$$V_y = \int_A \tau_{xy} \, dA \qquad (6.9)$$

When making an imaginary cut, we create two surfaces. The sign of the stresses at a point does not depend upon which of the two imaginary surfaces we consider. Hence the signs of the internal bending moment M_z and the shear force V_y, which are statically equivalent to the stresses σ_{xx} and τ_{xy}, should not depend on the imaginary surface being considered in the free-body diagram. When we make the imaginary cut to draw the free-body diagram, then M_z and V_y must be drawn in the positive direction if we want the formulas to give the correct signs.[16]

Equation (6.6) shows that σ_{xx} is compressive when the value of y is positive; it is tensile when the value of y is negative. Figure 6.23 shows the positive direction for internal moment M_z.

Definition 2 The direction of positive internal moment M_z on a free-body diagram must be such that it puts a point in the positive y direction into compression.

In Section 1.2.1 we studied the use of subscripts to determine the direction of a stress component, which we can now use to determine the positive direction of τ_{xy}. The equivalent shear force V_y is in the same direction as the shear stress τ_{xy}. Figure 6.24 shows the positive direction[17] for the internal shear force V_y.

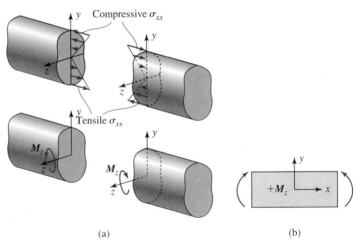

(a)   (b)

Figure 6.23 Sign convention for internal bending moment M_z.

---

[16]If stress directions are determined by inspection, then the sign convention may be ignored and formulas used for determining the magnitude of the stresses. However, to find M_z and V_y as a function of x, as in Example 6.10, and for any theoretical development, sign conventions must be followed.

[17]Some mechanics of materials books use an opposite direction for a positive shear force. This is possible because Equation (6.9) is a definition, and a minus sign can be incorporated into the definition. Unfortunately this shows positive shear force and positive shear stress opposite in direction, causing problems with intuitive understanding.

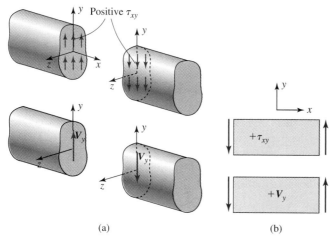

Figure 6.24 Sign convention for internal shear force V_y.

Definition 3 The direction of positive internal shear force V_y on a free-body diagram is in the direction of the positive shear stress on the surface.

The sign conventions for the internal bending moment and the internal shear force are tied to the coordinate system because the sign convention for stresses is tied to the coordinate system. But we are free to choose the directions for our coordinate system. Example 6.6 elaborates this comment further.

The sign convention is used for drawing the internal quantities on the free-body diagram. Equilibrium then gives the proper sign for the internal quantities that can be used in the formulas for deformation and stresses.

EXAMPLE 6.6

Figure 6.25 shows a beam and loading in three different coordinate systems. Determine for the three cases the internal shear force and bending moment at a section 36 in from the free end using the sign conventions described in Figures 6.23 and 6.24.

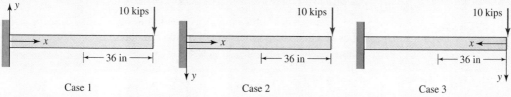

Figure 6.25 Example 6.6 on sign convention.

PLAN

We can first determine the positive direction of the shear force and bending moment for each coordinate system. Then we can make an imaginary cut at 36 in from the free end and take the right-hand part in drawing the free-body diagram. We draw the shear force and bending moment in the positive direction for each of the three cases. By writing equilibrium equations we can then find the values of the shear force and the bending moment.

Solution   We draw three rectangles and the coordinate axes corresponding to each of the three cases, as shown in Figure 6.26. Point A is on the surface that has an outward normal in the positive x direction, and hence the force will be in the positive y direction to produce a positive shear stress. Point B is on the surface that has an outward normal in the negative x direction, and hence the force will be in the negative y direction to produce a positive shear stress. Point C is on the surface where the y coordinate is positive. The moment direction is shown to put this surface into compression.

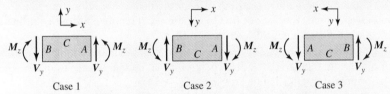

Case 1   Case 2   Case 3

Figure 6.26   Positive shear forces and bending moments in Example 6.6.

We make imaginary cuts 36 in from the right sides and take the right-hand parts for drawing the free-body diagrams as shown in Figure 6.27. On the imaginary cut surfaces we draw the shear forces and bending moments as shown to the left of the rectangles in Figure 6.26.

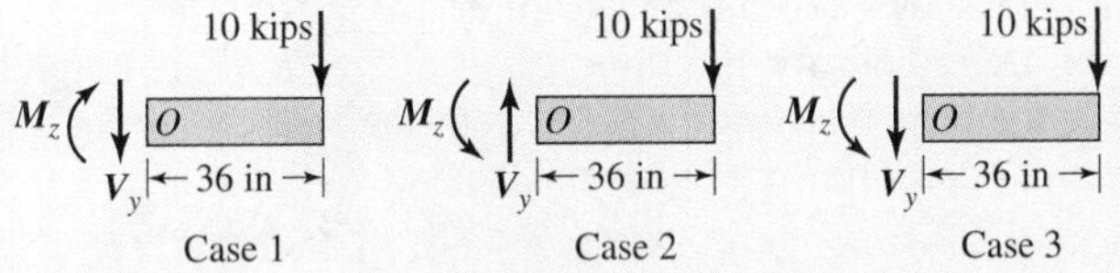

Figure 6.27   Free-body diagrams in Example 6.6.

We balance forces in the y direction to obtain the shear force values and take the moment about point O to obtain the bending moments for each of the three cases as shown,

	Case 1	Case 2	Case 3
$\sum F_y$	$V_y = -10$ kips	$V_y = 10$ kips	$V_y = -10$ kips
$\sum M_O$	$M_z = -360$ in·kips	$M_z = 360$ in·kips	$M_z = 360$ in·kips

COMMENTS

1. In Figure 6.27 we drew the shear force and bending moment directions without consideration of the external force of 10 kips. The equilibrium equations then gave us the correct signs. When we substitute these internal quantities, with the proper signs, into the respective stress formulas, we will obtain the correct signs for the stresses.

2. If we draw the shear force and the bending moment in a direction such that it satisfies equilibrium, then we shall always obtain positive values for the shear force and the bending moment, irrespective of the coordinate system. In such cases the sign for the stresses will have to be determined intuitively, and the stress formulas should be used only for determining the magnitude. To reap the benefit of both approaches, it is recommended that the internal quantities be drawn using the sign convention, and the answers be checked intuitively.

3. All three cases show that the shear force is acting upward and the bending moment is counterclockwise, which are the directions for equilibrium.

6.2.6 Flexure Formulas

The formulas for curvature d^2v/dx^2 and bending normal stress are derived in this section. Substituting σ_{xx} from Equation (6.7) into Equation (6.1) and noting that d^2v/dx^2 is a function of x only, while integration is with respect to y and z ($dA = dy\,dz$), we obtain

$$M_z = \int_A Ey^2 \frac{d^2v}{dx^2}\,dA = \frac{d^2v}{dx^2}\int_A Ey^2\,dA \qquad (6.10)$$

If material is homogeneous, as per Assumption 8, then we can take E outside the integral and note that the remaining integral $I_{zz} = \int_A y^2\,dA$ is the second area moment of inertia about the z axis. We obtain the following moment curvature formula from Equation (6.10):

$$\boxed{M_z = EI_{zz} \frac{d^2v}{dx^2}} \qquad (6.11)$$

The higher the value of EI_{zz}, the smaller will be the deformation (curvature) of the beam. Thus the rigidity of the beam increases with the increase in EI_{zz}. A beam can be made more rigid either by choosing a stiffer material (a higher value of E) or by choosing a cross sectional shape that has a large area moment of inertia (see Example 6.7).

Definition 4 The quantity EI_{zz} is called the *bending rigidity* of a beam cross section.

Solving for d^2v/dx^2 in Equation (6.11) and substituting into Equation (6.6), we obtain the equation

$$\sigma_{xx} = -\frac{M_z y}{I_{zz}} \qquad (6.12)$$

which is called *bending* stress formula or *flexure* stress formula.

In Equation (6.12) the subscript z emphasizes that the bending occurs about the z axis. If bending occurs about the y axis, then y and z in Equation (6.12) are interchanged, as elaborated in Section 10.1 on combined loading.

In Equation (6.12) I_{zz} is always positive. Thus for positive M_z, the stress σ_{xx} will be compressive at points where y is positive, which is consistent with our sign convention in Definition 2. M_z is an internal bending moment that has to be determined by making an imaginary cut, drawing a free-body diagram, and writing an equilibrium equation. There are two possible ways in which M_z may be found as described next and elaborated further in Example 6.8.

1. On a free-body diagram M_z is always drawn according to the sign convention in Definition 2. The equilibrium equation then is used to get a positive or negative value for M_z. Positive values of stress σ_{xx} from Equation (6.12) are tensile and negative values of σ_{xx} are compressive.

2. M_z is drawn at the imaginary cut in a direction to equilibrate the external loads. Since inspection is being used in determining the direction of M_z, Equation (6.12) should only be used to determine the magnitude. The tensile and compressive nature of σ_{xx} must be determined by inspection.

Consolidate your knowledge

1. Identity five examples of beams from your daily life.
2. With the book closed, derive Equations (6.11) and (6.12).

EXAMPLE 6.7
The two square beam cross sections shown in Figure 6.28 have the same material cross-sectional area A. Show that the hollow cross section has a higher area moment of inertia about the z axis than the solid cross section.

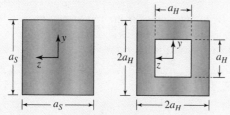

Figure 6.28   Cross sections in Example 6.7.

PLAN

We can find the lengths a_S and a_H in terms of the cross-sectional area A. Then we can find the area moments of inertia in terms of A and compare.

Solution   The lengths a_S and a_H in terms of area can be found as

$$A_S = a_S^2 = A \qquad\qquad a_S = \sqrt{A}$$

$$A_H = (2a_H)^2 - a_H^2 = 3a_H^2 = A \qquad a_H = \sqrt{\frac{A}{3}}$$

Let I_S and I_H represent the area moments of inertia about the z axis for the solid cross section and the hollow cross section, respectively. We can find I_S and I_H in terms of area A,

$$I_S = \frac{1}{12}a_S a_S^3 = \frac{1}{12}A^2$$

$$I_H = \frac{1}{12}(2a_H)(2a_H)^3 - \frac{1}{12}a_H a_H^3 = \frac{15}{12}a_H^4 = \frac{15}{12}\left(\frac{A}{3}\right)^2 = \frac{5}{36}A^2$$

The ratio of the two area moments of inertia is $I_H/I_S = 5/3 = 1.67$, which proves that the hollow cross section has a higher area moment of inertia.

COMMENTS

1. The hollow cross section has a higher area moment of inertia for the same cross-sectional area. From Equations (6.11) and (6.12) this implies that the hollow cross section will have lower stresses and deformation. Alternatively a hollow cross section will require less material (lighter in weight) to obtain the same area moment of inertia. This observation plays a major role in the design of beam shapes. Figure 6.29 shows some typical steel beam cross sections used in structures. Notice that in each case material from the region near the centroid is removed. Cross sections so created are thin near the centroid. This thin region near the centroid is called the *web,* whereas the wide material near the top or bottom is referred to as the *flange.* Appendix E has tables showing the geometric properties of some structural steel members.

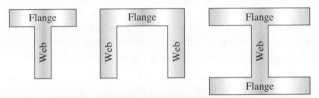

Figure 6.29 Metal beam cross sections.

2. We know that the bending normal stress is zero at the centroid and maximum at the top and/or bottom surfaces. By taking material near the centroid, where it is not severely stressed, to the top or bottom surface, where stress is maximum, we are trying to use material where it does the most good in terms of carrying load. This phenomenological explanation is an alternative explanation to the design of the cross sections shown in Figure 6.29. It is also the motivation in sandwich beam design, where two stiff panels are separated by softer and lighter core material.

3. Wooden beams are usually rectangular as machining costs do not offset any saving in weight.

EXAMPLE 6.8

An S180 × 30 steel beam is loaded and supported as shown in Figure 6.30. Determine:

(a) The bending normal stress at a point A that is 20 mm above the bottom of the beam.

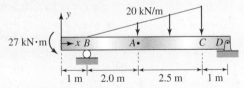

Figure 6.30 Beam in Example 6.8.

(b) The maximum compressive bending normal stress in a section 0.5 m from the left end.

PLAN

From Appendix E we can find the cross section, the centroid, and the moment of inertia. Knowing the centroid, we can determine the y coordinate of point A and the y coordinate of the point farthest from the centroid. Thus in Equation (6.12) the only undetermined quantity is the internal moment. We first find the reaction at the left support and then make an imaginary cut through A. Using the sign convention, we draw the internal moment and the shear force. We next write the equilibrium equation and determine the internal moment. Using Equation (6.12) we determine the bending normal stress at point A and the maximum bending normal stress in the section.

Solution From Appendix E we obtain the centroid and the area moment of inertia of the cross section of $S180 \times 30$ following Figure 6.31,

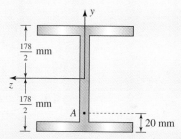

Figure 6.31 $S178 \times 30$ cross section in Example 6.8.

$$I_{zz} = 17.65 \times 10^6 \text{ mm}^4 \tag{E1}$$

$$y_A = -\left(\frac{178}{2} - 20\right) = -69 \text{ mm} \tag{E2}$$

$$y_{max} = \pm\frac{178}{2} = \pm 89 \text{ mm} \tag{E3}$$

The coordinates of point A can be found as shown Figure 6.31. The maximum flexure normal stress will occur at the top or at the bottom of the cross section. We draw the free-body diagram of the entire beam as shown in Figure 6.32 and replace the distributed load by a statically equivalent load at the centroid of the triangle. We then calculate R_B, by equilibrium of moment about point D.

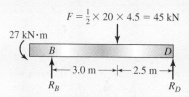

Figure 6.32 Free-body diagram of entire beam in Example 6.8.

$$R_B \times 5.5 - 27 - 45 \times 2.5 = 0$$

$$R_B = 25.36 \text{ kN} \tag{E4}$$

The intensity of the distributed load acting on the beam at point A can be found from similar triangles, as shown in Figure 6.33,

$$\frac{p_A}{2} = \frac{20}{4.5} \quad \text{or} \quad p_A = 8.89 \text{ kN/m} \tag{E5}$$

We make an imaginary cut through point A in Figure 6.30 and draw the internal bending moment and the shear force using our sign convention. We also replace that portion of the distributed load acting at left of A by an equivalent force to obtain the free-body diagram shown

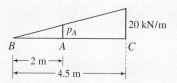

Figure 6.33 Intensity of distributed force at point A in Example 6.8.

in Figure 6.34. By equilibrium of moment at point A we obtain the internal moment,

$$M_A + 27 - 25.36 \times 2 + 8.89 \times \frac{2}{3} = 0$$

$$M_A = 17.8 \text{ kN·m} \qquad \text{(E6)}$$

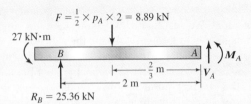

Figure 6.34 Internal moment calculations in part (a) of Example 6.8.

(a) From Equations (6.12), (E1), (E2), and (E6) we obtain the bending normal stress at point A,

$$\sigma_A = -\frac{M_A y_A}{I_{zz}} = -\frac{17.8 \times 10^3 (-69) 10^{-3}}{17.65 \times 10^{-6}}$$

ANS. $\sigma_A = 69.6 \text{ MPa (T)}$

(b) We make an imaginary cut at 0.5 m from the left and draw the internal bending moment and the shear force using the sign convention. Balancing the moment at the right end, we obtain the internal moment as shown in Figure 6.35,

$$M_{0.5} = -27 \text{ kN·m} \qquad \text{(E7)}$$

With the moment being negative in Equation (E7), the maximum value of y must be in the negative y direction for Equation (6.12) to yield a negative normal stress value, that is, a compressive value. Substituting $y = -88.9 \times 10^{-3}$ m, corresponding to the bottom of the beam, the moment from Equation (E7), and I_{zz} into Equation (6.12), we obtain

$$\sigma_{0.5} = -\frac{(-27) \times 10^3 (-89) 10^{-3}}{17.65 \times 10^{-6}}$$

ANS. $\sigma_{0.5} = 136.1 \text{ MPa (C)}$

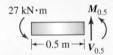

Figure 6.35 Internal moment calculations in part (a) of Example 6.8.

COMMENTS

1. In Figure 6.35 the internal moment was drawn according to the sign convention. But suppose we drew it in the opposite direction so that it equilibrates the applied moment. Now we are *not* following the

sign convention. Thus by inspection we must determine that the bottom surface of the beam would be in compression. But how would inspection work for stress at point A? The next comment answers this question.

2. An intuitive check on the answer can be conducted by drawing an approximate deformed shape of the beam, as shown in Figure 6.36. We start by first drawing the approximate shape of the bottom surface (or the top surface). At the left end the beam deflects downward due to the applied moment. At the support point B the deflection must be zero. Noting that the slope of the beam must be continuous (otherwise a corner will be formed), we note that the beam has to deflect upward as one crosses B. Now the externally distributed load pushes the beam downward. Eventually the beam will deflect downward, and finally it must have zero deflection at the support point D. After drawing the bottom surface, the top surface is drawn parallel to it.

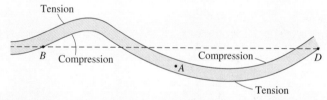

Figure 6.36   Approximate deformed shape of beam in Example 6.8.

By inspection of Figure 6.36 we see that point A is in the region where the bottom surface is in tension and the top surface in compression. If point A were closer to the inflection point, then we would have greater difficulty in assessing the situation, which once more emphasizes the point that intuitive checks are valuable but the conclusions must be viewed with caution.

PROBLEM SET 6.2

Second area moments of inertia

6.14   A solid and a hollow square beam have the same cross-sectional area A, as shown in Figure P6.14. Show that the ratio of the second area moment of inertia for the hollow beam I_H to that of the solid beam I_S is

$$\frac{I_H}{I_S} = \frac{\alpha^2 + 1}{\alpha^2 - 1}$$

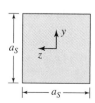
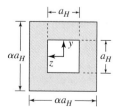

Figure P6.14

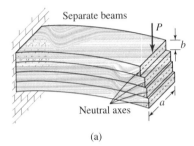

Separate beams

P

b

Neutral axes

a

(a)

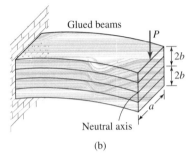

Glued beams

P

$2b$

$2b$

a

Neutral axis

(b)

Figure P6.15

60° 60°

a

Figure P6.16

a

a

Figure P6.17

a

Figure P6.18

What is the ratio of the section moduli for the two cross sections?

6.15 Figure P6.15a shows four separate wooden strips that bend independently about the neutral axis passing through the centroid of each strip. Figure 6.15b shows the four strips glued together and bending as a unit about the centroid of the glued cross section. (a) Show that $I_G = 16I_S$, where I_G is the area moment of inertia for the glued cross section and I_S is the total area moment of inertia of the four separate beams. (b) Also show that $\sigma_G = \sigma_S/4$, where σ_G and σ_S are the maximum bending normal stresses at any cross section for the glued and separate beams, respectively.

In Problems 6.16 through 6.18, the cross sections of the beams shown are constructed from thin sheet metal of thickness t. Assume that the thickness t ≪ a and obtain the second area moments of inertia about an axis passing through the centroid in terms of a and t.

6.16 Using Figure P6.16 solve the problem as described in the text above.

6.17 Using Figure P6.17 solve the problem as described in the text above.

6.18 Using Figure P6.18 solve the problem as described in the text above.

6.19 The same amount of material is used for constructing the cross sections shown in Figures P6.16, P6.17, and P6.18. Let the maximum bending normal stresses be σ_T, σ_S, and σ_C for the triangular, square, and circular cross sections, respectively. For the same moment-carrying capability determine the proportional ratio of the maximum bending normal stresses, that is, $\sigma_T: \sigma_S: \sigma_C$. What is the proportional ratio of the section moduli?

Normal stress and strain variations across a cross section

In Problems 6.20 through 6.22, the cross section of a beam with a coordinate system that has its origin at the centroid C is shown in each figure. The normal strain at point A due to bending about the z axis as well as the modulus of elasticity are as given in each problem.

(a) Plot the stress distribution across the cross section.

(b) Determine the maximum bending normal stress in the cross section.

(c) Determine the equivalent internal bending moment M_z by using Equation (6.1).

(d) Determine the equivalent internal bending moment M_z by using Equation (6.12).

6.20 $\varepsilon_{xx} = 200\ \mu$; $E = 8000$ ksi.

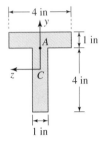

4 in

y

A

1 in

z

C

4 in

1 in

Figure P6.20

6.21 $\varepsilon_{xx} = -200\ \mu$; $E = 200$ GPa.

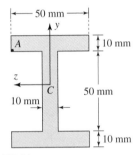

50 mm

y

A

10 mm

z

C

50 mm

10 mm

10 mm

Figure P6.21

6.22 $\varepsilon_{xx} = 300\ \mu$; $E = 30{,}000$ ksi.

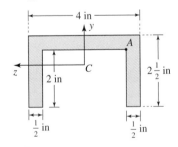

4 in

y

A

z

C

2 in

$2\frac{1}{2}$ in

$\frac{1}{2}$ in

$\frac{1}{2}$ in

Figure P6.22

In Problems 6.23 through 6.25, the cross section of a beam with a coordinate system that has its origin at the centroid C is shown for each figure. The internal moment at a beam cross section is given in each problem. Determine the bending normal stresses at points A, B, and D.

6.23 $M_z = 20$ in·kips (Figure P6.23).

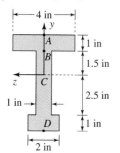

Figure P6.23

6.24 $M_z = 10$ kN·m (Figure P6.24).

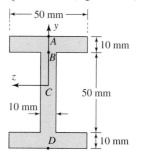

Figure P6.24

6.25 $M_z = -15$ kN·m (Figure P6.25).

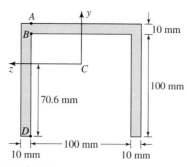

Figure P6.25

Sign convention

In Problems 6.26 through 6.28, a beam and loading in three different coordinate systems is shown. Determine the internal shear force and bending moment at the section containing point A for the three cases shown using the sign convention described in Section 6.2.5.

6.26 Using Figure P6.26 solve the problem as described in text above.

6.27 Using Figure P6.27 solve the problem as described in text above.

6.28 Using Figure P6.28 solve the problem as described in text above.

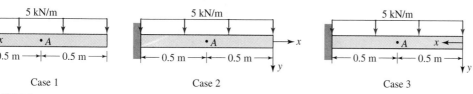

Figure P6.26

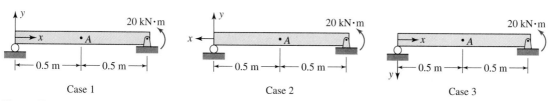

Figure P6.27

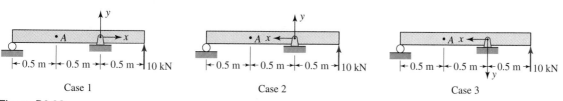

Figure P6.28

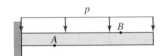

Figure P6.29

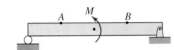

Figure P6.30

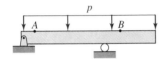

Figure P6.31

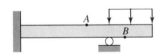

Figure P6.32

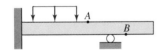

Figure P6.33

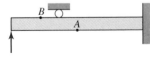

Figure P6.34

Sign of stress by inspection

In Problems 6.29 through 6.34, for the beam and loading shown, draw an approximate deformed shape of the beam. By inspection determine whether the bending normal stress is tensile or compressive at points A and B.

6.29 Using Figure P6.29 solve the problem as described in text above.

6.30 Using Figure P6.30 solve the problem as described in text above.

6.31 Using Figure P6.31 solve the problem as described in text above.

6.32 Using Figure P6.32 solve the problem as described in text above.

6.33 Using Figure P6.33 solve the problem as described in text above.

6.34 Using Figure P6.34 solve the problem as described in text above.

Bending normal stress and strain calculations

6.35 A W150 × 24 steel beam is simply supported over a length of 4 m and supports a distributed load of 2 kN/m. At the midsection of the beam, determine: (a) the bending normal stress at a point 40 mm above the bottom surface; (b) the maximum bending normal stress.

6.36 A W10 × 30 steel beam is simply supported over a length of 10 ft and supports a distributed load of 1.5 kips/ft. At the midsection of the beam, determine: (a) the bending normal stress at a point 3 in below the top surface; (b) the maximum bending normal stress.

6.37 An S12 × 35 steel cantilever beam has a length of 20 ft. At the free end a force of 3 kips acts downward. At the section near the built-in end, determine: (a) the bending normal stress at a point 2 in above the bottom surface; (b) the maximum bending normal stress.

6.38 An S250 × 52 steel cantilever beam has a length of 5 m. At the free end a force of 15 kN acts downward. At the section near the built-in end, determine: (a) the bending normal stress at a point 30 mm below the top surface; (b) the maximum bending normal stress.

In Problems 6.39 through 6.41, the beam, the loading, and the cross section of the beam are as shown. Determine the bending normal stress at point A and the maximum bending normal stress in the section containing point A.

6.39 (Figure P6.39).

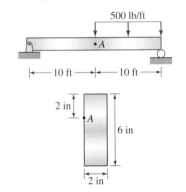

Figure P6.39

6.40 (Figure P6.40).

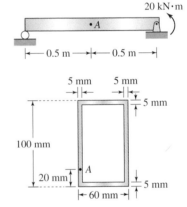

Figure P6.40

6.41 (Figure P6.41).

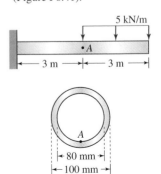

Figure P6.41

In Problems 6.42 through 6.44, the beam, the loading, and the cross section of the beam are as shown. Determine the bending normal stress at point A and the maximum bending normal stress in the section containing point A.

6.42 Using Figure P6.42 solve the problem as described in the text above.

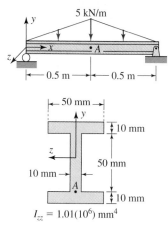

$I_{zz} = 1.01(10^6)$ mm^4

Figure P6.42

6.43 Using Figure P6.43 solve the problem as described in the text above.

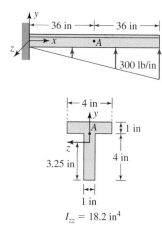

$I_{zz} = 18.2$ in^4

Figure P6.43

6.44 Using Figure P6.44 solve the problem as described in the text above.

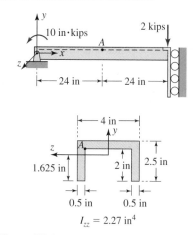

$I_{zz} = 2.27$ in^4

Figure P6.44

In Problems 6.45 and 6.46, a wooden rectangular beam (E = 10 GPa), its loading, and its cross section are as shown in Figure P6.45.

6.45 If the distributed force $w = 5$ kN/m in Figure P6.45, determine the normal strain ε_{xx} at point A.

6.46 The normal strain at point A in Figure P6.45 was measured as $\varepsilon_{xx} = -600$ μ. Determine the distributed force w that is acting on the beam.

In Problems 6.47 and 6.48, a wooden beam (E = 8000 ksi), its loading, and its cross section are as shown in Figure P6.47.

6.47 If the applied load $P = 6$ kips in Figure P6.47, determine the normal stress ε_{xx} at point A.

6.48 The normal strain at point A in Figure P6.47 was measured as $\varepsilon_{xx} = -250$ μ. Determine the load P.

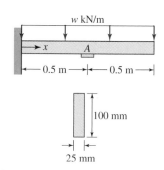

Figure P6.45

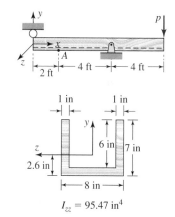

$I_{zz} = 95.47$ in^4

Figure P6.47

6.3 SHEAR AND MOMENT BY EQUILIBRIUM

Equilibrium equations at a point on the beam are differential equations relating the distributed force p, the shear force V_y, and the bending moment M_z. These differential equations can be integrated analytically or graphically to obtain V_y and M_z as a function of x. This will permit us to determine the maximum values of V_y and M_z, and

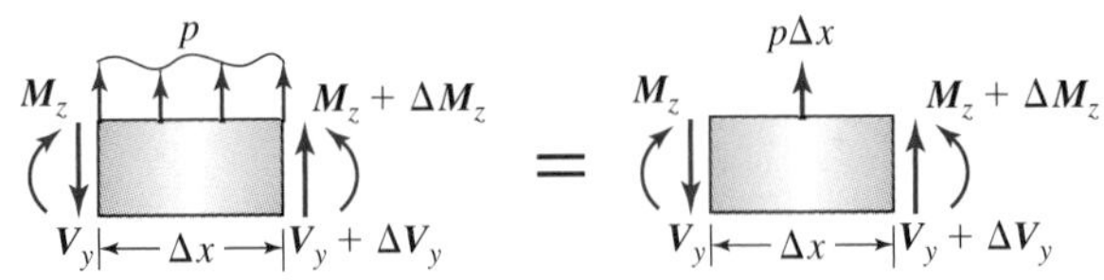

Figure 6.37 Differential beam element.

hence the maximum values of the bending normal stress from Equation (6.12) and the maximum bending shear stress as discussed in Section 6.6. Also, M_z written as a function of x will be needed when integrating Equation (6.11) to find the deflection of the beam. This will be discussed in Chapter 7.

Consider a small element Δx of the beam, as shown at left in Figure 6.37. Recall that a positive distributed force p acts in the positive y direction, as shown in Figure 6.14. Internal shear forces and the internal moment change as one moves across the element, as shown. By replacing the distributed force by an equivalent force, we obtain the diagram on the right of Figure 6.37.

By equilibrium of forces in the y direction, we obtain

$$-V_y + (V_y + \Delta V_y) + p\Delta x = 0 \qquad \text{or} \qquad \frac{\Delta V_y}{\Delta x} = -p$$

As $\Delta x \to 0$, we obtain

$$\boxed{\frac{dV_y}{dx} = -p} \qquad (6.13)$$

By equilibrium of moment in the z direction about an axis passing through the right side, we obtain

$$-M_z + (M_z + \Delta M_z) + V_y\Delta x + (p\Delta x)\frac{\Delta x}{2} = 0 \qquad \text{or} \qquad \frac{\Delta M_z}{\Delta x} + \frac{p\Delta x}{2} = -V_y$$

As $\Delta x \to 0$, we obtain

$$\boxed{\frac{dM_z}{dx} = -V_y} \qquad (6.14)$$

Equations (6.13) and (6.14) are differential equilibrium equations that are applicable at every point on the beam, except where V_y and M_z are discontinuous. In Example 6.10 we shall see that V_y and M_z are discontinuous at the points where concentrated (point) external forces or moments are applied. We shall consider two methods for finding V_y and M_z as a function of x:

1. We can integrate Equation (6.13) to obtain V_y and then integrate Equation (6.14) to obtain M_z. The integration constants can be found from the values of V_y and M_z at the end of the beam, as illustrated in Example 6.9.

2. We can make an imaginary cut at some location defined by the variable x, draw the free-body diagram, and determine V_y and M_z in terms of x by writing equilibrium equations. We can check our results by substituting the expressions of V_y and M_z in Equations (6.13) and (6.14), respectively.

The approach to finding V_y and M_z by integration is a general approach, which is particularly useful if p is represented by a complicated function. But for uniform and linear variations of p the free-body diagram method is simpler. Example 6.9 compares the two methods, and Example 6.10 elaborates the use of the free-body diagram approach further.

EXAMPLE 6.9

Two models of wind pressure on a light pole are shown in Figure 6.38.
Find V_y and M_z as a function of x for the two distributions shown.
Neglect the weights of the light and the pole.

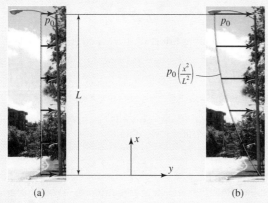

(a)   (b)

Figure 6.38   Light pole in Example 6.9. (*a*) Uniform distribution. (*b*) Quadratic distribution.

PLAN

For uniform distribution we can find V_y and M_z as a function of x by
making an imaginary cut at a distance x from the bottom and drawing
the free-body diagram of the top part. For the quadratic distribution we
can first integrate Equation (6.13) to find V_y and then integrate Equation (6.14) to find M_z. To find the integration constants, we can construct a free-body diagram of infinitesimal length at the top ($x = L$) and
obtain the boundary conditions on V_y and M_z. Using boundary conditions and integrated expressions, we can obtain V_y and M_z as a function
of x for the quadratic distribution.

Solution

1. *Uniform distribution:* We can make an imaginary cut at location x and
   draw the free-body diagram of the top part, as shown in Figure 6.39.

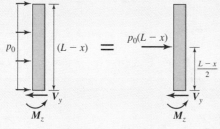

Figure 6.39   Shear force and bending moment by free-body diagram.

We then replace the distributed load by an equivalent force and write the equilibrium equations,

$$V_y = p_0(L - x) \tag{E1}$$

$$M_z = p_0(L - x)\left(\frac{L-x}{2}\right) = \frac{p_0}{2}(x^2 - 2xL + L^2) \tag{E2}$$

Check: Differentiating Equations (E1) and (E2) we obtain

$$\frac{dV_y}{dx} = -p_0 = -p, \qquad \frac{dM_z}{dx} = -p_0(L - x) = -V_y$$

which is consistent with Equations (6.13) and (6.14).

2. *Quadratic distribution:* Substituting $p = p_0(x^2/L^2)$ into Equation (6.13) and integrating, we obtain

$$V_y = -\left(\frac{p_0}{3L^2}\right)x^3 + C_1 \tag{E3}$$

Substituting Equation (E3) into Equation (6.14) and integrating, we obtain

$$M_z = \frac{p_0}{3L^2}\left(\frac{x^4}{4}\right) - C_1 x + C_2 \tag{E4}$$

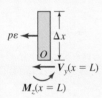

Figure 6.40 Boundary conditions on shear force and bending moments.

We make an imaginary cut at a distance Δx from the top and draw the free-body diagram shown in Figure 6.40. By equilibrium of forces in the y direction and letting Δx tend to zero we obtain a condition on the shear force. By equilibrium of moment at point O and letting Δx tend to zero we obtain a condition on the bending moment,

$$\lim_{\Delta x \to 0} [V_y(x = L) + p\Delta x] = 0 \qquad \text{or} \qquad V_y(x = L) = 0 \tag{E5}$$

$$\lim_{\Delta x \to 0}\left[M_z(x = L) + \frac{p\Delta x^2}{2}\right] = 0 \qquad \text{or} \qquad M_z(x = L) = 0 \tag{E6}$$

Substituting $x = L$ into Equation (E3) and using the condition (E5) we obtain $C_1 = p_0 L/3$. Thus from Equation (E3) we obtain

$$V_y = -\frac{p_0}{3L^2}x^3 + \frac{p_0 L}{3} \tag{E7}$$

$$\text{ANS.} \qquad V_y = \frac{p_0}{3L^2}(L^3 - x^3)$$

Substituting $x = L$ into Equation (E4), and $C_1 = p_0 L/3$, we obtain from Equation (E6)

$$\frac{p_0}{3L^2}\left(\frac{L^4}{4}\right) - \frac{p_0 L}{3}(L) + C_2 = 0 \qquad \text{or} \qquad C_2 = \frac{p_0 L^2}{4}$$

Thus from Equation (E4) we obtain

$$M_z = \frac{p_0 x^4}{12L^2} - \frac{p_0}{3}xL + \frac{p_0 L^2}{4} \tag{E8}$$

ANS. $\quad M_z = \dfrac{p_0}{12L^2}(x^4 - 4xL^3 + 3L^4)$

COMMENTS

1. Suppose that for the uniform distribution we integrate Equation (6.13) after substituting $p = p_0$. We would obtain $V_y = -p_0 x + C_3$. On substituting this into Equation (6.14) and integrating, we would obtain $M_z = p_0(x^2/2) - C_3 x + C_4$. Note that the boundary conditions on V_y and M_z given by Equations (E5) and (E6) are independent of the distributed load as Δx tends to zero. Substituting $x = L$ in the expressions of V_y and M_z and equating the results to zero, we obtain $C_3 = p_0 L$ and $C_4 = p_0 L^2/2$. Substituting these in the expressions of V_y and M_z we obtain Equations (E1) and (E2).

2. The free-body diagram approach is simpler than the integration approach for uniform distribution for two reasons. First, we did not have to perform any integration to obtain the equivalent load $p_0 L$ or to determine its location when we constructed the free-body diagram in Figure 6.39. Second, we do not have to impose zero boundary conditions on the shear force and bending moments at $x = L$ because these conditions are implicitly included in the free-body diagram in Figure 6.39.

3. The free-body diagram approach would present difficulties for the quadratic distribution, as we would need to find the equivalent load and its location—both would involve the same integrals as those obtained from Equations (6.13) and (6.14). Thus for simple distributions the free-body diagram approach is preferred, whereas the integration approach is better for more complex loading.

EXAMPLE 6.10

(a) Write the equations for the internal shear force V_y and the internal bending moments M_z as a function of x for the entire beam shown in Figure 6.41.

(b) Determine the values of V_y and M_z just before and after point B.

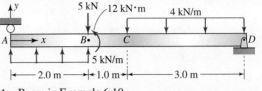

Figure 6.41 Beam in Example 6.10.

PLAN

By considering the free-body diagram of the entire beam we can determine the reactions at supports A and D.

(a) The loading changes at points B and C. Thus shear force and bending moment will be represented by different functions in AB, BC, and CD. In segments AB and BC we will take the left part after making the imaginary cut for the free-body diagram. In segment CD we can take the right part after making the imaginary cut as it has fewer loads than the left part. We can use Equations (6.13) and (6.14) to check our answers.

(b) By substituting $x = 2$ in the expressions for V_y and M_z in segment AB we can find the values just before B, and by substituting $x = 2$ in segment BC we find the values just after B.

Solution

(a) We replace the distributed loads by equivalent forces and draw the free-body diagram of the entire beam in Figure 6.42. We then have

$$\sum M_D \qquad R_A(6) - 10(5) + 12 + 5(4) + 12(1.5) = 0$$

or

$$R_A = 0 \qquad\qquad\qquad (E1)$$

$$\sum F_y \qquad -R_A + 10 - 5 - 12 + R_D = 0$$

or

$$R_D = 7 \text{ kN} \qquad\qquad\qquad (E2)$$

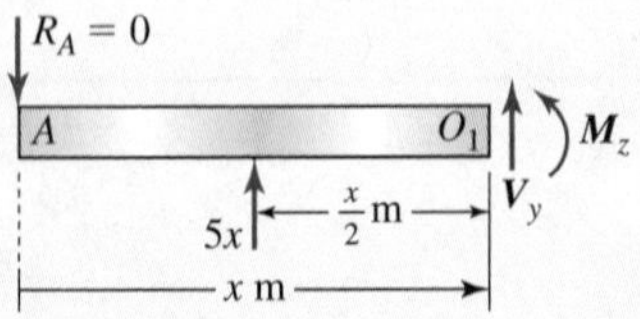

1. *Segment AB*, $0 \le x < 2$: We make an imaginary cut at some location x in segment AB. We take the left part of the cut and draw the free-body diagram after replacing the distributed force over the distance x by a statically equivalent force, as shown in Figure 6.43. We write the equilibrium equations to obtain V_y and M_z as a function of x,

$$\sum F_y \qquad V_y + 5x = 0 \qquad \text{or} \qquad \text{ANS.} \quad V_y = -5x \text{ kN} \quad (E3)$$

$$\sum M_{O_1} \qquad M_z - 5x\left(\frac{x}{2}\right) = 0 \qquad \text{or} \qquad \text{ANS.} \quad M_z = \frac{5}{2}x^2 \text{ kN·m} \quad (E4)$$

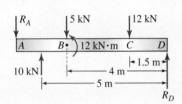

Figure 6.42 Free-body diagram of entire beam in Example 6.10.

Figure 6.43 Imaginary cut in AB in Example 6.10.

Check: Differentiating Equations (E3) and (E4) we obtain

$$\frac{dV_y}{dx} = -5 = -p, \qquad \frac{dM_z}{dx} = 5x = -V_y$$

which are consistent with Equations (6.13) and (6.14), respectively.

2. *Segment BC, 2 < x < 3:* We make an imaginary cut at some location *x* in segment *BC*. We take the left part of the cut and draw the free-body diagram after replacing the distributed force by a statically equivalent force, as shown in Figure 6.44. We write the equilibrium equations to obtain V_y and M_z as a function of *x*,

$$\sum F_y \qquad V_y + 10 - 5 = 0 \quad \text{or} \quad \text{ANS.} \quad V_y = -5 \text{ kN} \quad (E5)$$

$$\sum M_{O_2} \qquad M_z - 10(x - 1) + 12 + 5(x - 2) = 0 \qquad \text{or}$$

$$\text{ANS.} \qquad M_z = 5x - 12 \text{ kN·m} \qquad (E6)$$

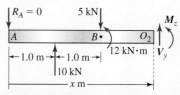

Figure 6.44 Imaginary cut in *BC* in Example 6.10.

Check: Differentiating Equations (E5) and (E6) we obtain

$$\frac{dV_y}{dx} = 0 = -p, \qquad \frac{dM_z}{dx} = 5 = -V_y$$

which are consistent with Equations (6.13) and (6.14), respectively.

3. *Segment CD, 3 < x < 6:* We make an imaginary cut at some location *x* in segment *CD*. We take the right part of the cut and note that left part is *x* m long and the right part hence is 6 − *x* m long. We draw the free-body diagram after replacing the distributed force by a statically equivalent force, as shown in Figure 6.45. We write the equilibrium equations to obtain V_y and M_z as a function of *x*,

$$\sum F_y \qquad V_y + 4(6 - x) - 7 = 0 \qquad \text{or}$$

$$\text{ANS.} \qquad V_y = 4x - 17 \text{ kN} \qquad (E7)$$

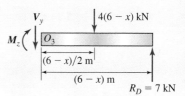

Figure 6.45 Imaginary cut in *CD* in Example 6.10.

$$\sum M_{O_3} \qquad M_z + 4(6 - x)\left(\frac{6 - x}{2}\right) - 7(6 - x) = 0 \qquad \text{or}$$

$$\text{ANS.} \qquad M_z = -2x^2 + 17x - 30 \text{ kN·m} \qquad (E8)$$

Check: Differentiating Equations (E7) and (E8) we obtain

$$\frac{dV_y}{dx} = 4 = -p, \qquad \frac{dM_z}{dx} = -4x + 17 = -V_y$$

which are consistent with Equations (6.13) and (6.14), respectively.

(b) Substituting $x = 2$ into Equations (E3) and (E4) we obtain the values of V_y and M_z just before point B,

ANS. $V_y(2^-) = -10$ kN, $M_z(2^-) = +10$ kN·m

where the superscripts − refer to just before $x = 2$. Substituting $x = 2$ into Equations (E5) and (E6) we obtain the values of V_y and M_z just after point B,

ANS. $V_y(2^+) = -5$ kN, $M_z(2^+) = -2$ kN·m

where the superscripts + refer to just after $x = 2$.

COMMENTS

1. In Figures 6.43 and 6.44 the left part after the imaginary cut was taken and the distance from A was labeled x. In Figure 6.45 the right part of the imaginary cut was taken and the distance from the right end was labeled $(6 - x)$. These free-body diagrams emphasize that x defines the location of the imaginary cut, irrespective of the part used in drawing the free-body diagram.

   Furthermore, the distance (coordinate) x is always measured from the same point in all free-body diagrams, which in this problem is point A.

2. We note that $V_y(2^+) - V_y(2^-) = 5$ kN, which is the magnitude of the applied external force at point B. Similarly, $M_z(2^+) - M_z(2^-) = -12$ kN·m, which is the magnitude of the applied external moment at point B. This emphasizes that the external point force causes a jump in internal shear force, and the external point moment causes a jump in the internal bending moment.

   We will make use of these observations in the next section in plotting the shear force—bending moment diagrams.

3. We can obtain V_y and M_z in each segment by integrating Equations (6.13) and (6.14). But to determine the integration constants in each segment, we will have to derive the conditions on V_y and M_z by making use of the observation that shear force and bending moment values jump by the value of applied force and moment, respectively. Thus in the current form the method of integration would not be simpler than the free-body approach.

4. As part of studying the calculation of beam deflection, Section 7.4 introduces a method based on the integration approach, which eliminates drawing free-body diagrams for each segment to account for jumps in the loading. But to understand the method we need the concept of "discontinuity functions," also called "singularity functions."

6.4 SHEAR AND MOMENT DIAGRAMS

Shear and moment diagrams are plots of internal shear force and internal bending moment as a function of x. By looking at these plots we can immediately see the maximum values of the shear force and the bending moment as well as the location of these maximum values. One way of making these plots is to determine the shear force and bending moment as a function of x, as was done in Section 6.3, and plot the results. However, for simple loadings there exists an easier alternative, which is presented in this section. We first discuss how the distributed forces are accounted, then how to account for the point forces and moments.

6.4.1 Distributed Force

The graphical technique described in this section is based on the interpretation that an integral represents an area under the curve described by the integrand. The minus signs[18] in Equations (6.13) and (6.14) lead to positive areas being subtracted and negative areas being added. To overcome this problem of flip-flop of sign in the graphical procedure, we introduce $V = -V_y$. Let V_1 and V_2 be the values of V at x_1 and x_2, respectively. Let M_1 and M_2 be the of values of M_z at x_1 and x_2, respectively. Equations (6.13) and (6.14) can be written in terms of V as $dV/dx = p$ and $dM_z/dx = V$, which upon integration, yields

$$V_2 = V_1 + \int_{x_1}^{x_2} p \; dx \tag{6.15}$$

$$M_2 = M_1 + \int_{x_1}^{x_2} V \; dx \tag{6.16}$$

The key idea is to recognize that the values of the integrals in Equations (6.15) and (6.16) are the areas under the load curve p and the curve defining V, respectively. If we know V_1 and M_1, then by adding or subtracting the areas under the respective curves, we can find V_2 and M_2. We then move to point 2, where we now know the shear force and bending moment, and consider it as point 1 for the next segment of the beam. Moving in this bootstrap manner we go across the beam accounting for the distributed forces.

Shear force curve

Recollect that p is positive in the positive y direction. Thus in Figure 6.46a and b, $p = +w$, and from Equation (6.15) we obtain $V_2 = V_1 + w(x_2 - x_1)$. Similarly, in Figure 6.46c and d, $p = -w$, and from Equation (6.15) we obtain $V_2 = V_1 - w(x_2 - x_1)$. The term $w(x_2 - x_1)$ is the area of the rectangle and represents the magnitude of the integral in Equation (6.15). The line joining the values of V_1 and V_2 is a straight line

---

[18]This is a consequence of trying to stay mathematically consistent but not violating the intuitive understanding that the directions of shear force and shear stress are the same. See footnote 17.

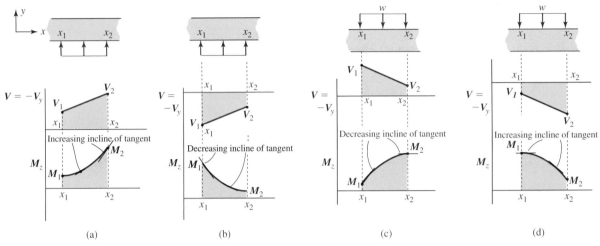

Figure 6.46 Shear and moment diagrams for uniformly distributed load.

because the integral of a constant function will result in a linear function. Another way to remember this is the following observation. The value of V_2 relative to V_1 will move in the direction of the distributed force.

Bending moment curve

The integral in Equation (6.16) represents the area under the curve defining V, that is, the areas of the trapezoids shown by the shaded regions in Figure 6.46. In Figure 6.46a and c, V is positive and we add the area to M_1 to get M_2. In Figure 6.46b and d, V is negative and we subtract the area from M_1 to get M_2. As V is linear between x_1 and x_2, the integral in Equation (6.16) will generate a quadratic function. But what would be the curvature of the moment curve, concave or convex? To answer this question, we note that the derivative of the moment curve, that is, the slope of the tangent, is equal to the value on shear force diagram. To avoid some ambiguities associated with the sign[19] of a slope, we consider the inclination of the tangent to the moment curve, that is, $|dM_z/dx| = |V|$. If the magnitude of V is increasing, the inclination of the tangent to the moment curve must increase, as shown in Figure 6.46a and d. If the magnitude of V is decreasing, the inclination of the tangent to the moment curve must decrease, as shown in Figure 6.46b and c. An alternative approach to getting the curvature of the moment curve is to note that if we substitute Equation (6.14) into Equation (6.13), we obtain $d^2M_z/dx^2 = p$. If p is positive, then the curvature of the moment curve is positive, hence the curve is concave, as shown in Figure 6.46a and b. If p is negative, then the curvature of the moment curve is negative, and the curve is convex, as shown in Figure 6.46c and d. We record the discussion of this paragraph as the following *curvature rule* for

---

[19]We avoid statements such as "increasing negative slope," which has the ambiguity whether the slope became more negative or less negative. Similarly there is ambiguity in "decreasing negative slope."

quadratic M_z curves:

> The curvature of the M_z curve must be such that the incline of the tangent to the M_z curve must increase (or decrease) as the magnitude of V increases (or decreases).
>
> <div align="center">or</div> (6.17)
>
> The curvature of the moment curve is concave if p is positive, and convex if p is negative.

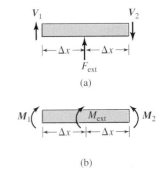

Figure 6.47 Beam templates. (*a*) Shear force template. (*b*) Moment template.

6.4.2 Point Force and Moments

In comment 2 of Example 6.10 it was observed that the values of the internal shear force and the bending moment jump as one crosses an applied point force and moment, respectively. In Section 4.1.8 on axial force diagrams and in Section 5.2.6 on torque diagrams we used a template to give us the correct direction of the jump. We use the same idea here.

A template is a small segment (Δx tends to zero in Figure 6.47) of a beam on which the external moment M_{ext} and an external force F_{ext} are a drawn. The directions of F_{ext} and M_{ext} are arbitrary. The ends at $+\Delta x$ and $-\Delta x$ from the applied external force and moment represent the imaginary cut just to the left and just to the right of the applied external forces and moments. On these cuts the internal shear force and the internal bending moment are drawn. Equilibrium equations are written for this $2\Delta x$ segment of the beam to obtain the template equations.

Shear force template:

$$V_2 = V_1 + F_{ext}$$

Moment template:

$$M_2 = M_1 + M_{ext}$$

The moment equation does not contain the moment due to the forces because these moments will go to zero as Δx goes to zero.

Shear force template

Notice that the internal forces V_1 and V_2 are drawn opposite to the direction of positive internal shear forces, as per the definition $V = -V_y$, which is an additional artifact of the procedure to remember. To avoid this, we note that the sign of F_{ext} is the same as the direction in which V_2 will move relative to V_1. In the future we will not draw the shear force template but use the following observation: V will jump in the direction of the external point force. This observation is similar to the one we made regarding distributed forces and can be confirmed to be independent of the orientation of the coordinate system.

Moment template

On the moment template, the internal moments are drawn according to our sign convention, discussed in Section 6.2.5. Unlike the observation about the jump in V, there is no single observation that is valid for all coordinate systems. Thus the moment template must be drawn and the corresponding template equation used as follows.

If the external moment on the beam is in the direction of the assumed moment M_{ext} on the template, then the value of M_2 is calculated according to the template equation. If the external moment on the beam is opposite to the direction of M_{ext} on the template, then M_2 is calculated by changing the sign of M_{ext} in the template equation.

6.4.3 Construction of Shear and Moment Diagrams

The procedure for constructing shear and moment diagrams is outlined in this section and explained using Figure 6.48.

Step 1 Determine the reaction forces and moments.

The free-body diagram for the entire beam is drawn, and the reaction forces and moments are calculated at the supports at A and B, as shown in Figure 6.48.

Step 2 Draw and label the vertical axes for V and M_z along with the units to be used.

Labeling the axes along with units can prevent the common mistake of unit mixup when using formulas. It is also recommended that $V = -V_y$ be shown on the axis to remind ourselves that the positive and negative values read from the plots are for V, whereas the formula that will be developed in Section 6.6 for the bending shear stress will be in terms of V_y.

Step 3 Draw the beam with all forces and moments. At each change of loading draw a vertical line.

The vertical lines define the segments of the beam between two points x_1 and x_2 where the values of shear force and moment will be calculated. The vertical lines also

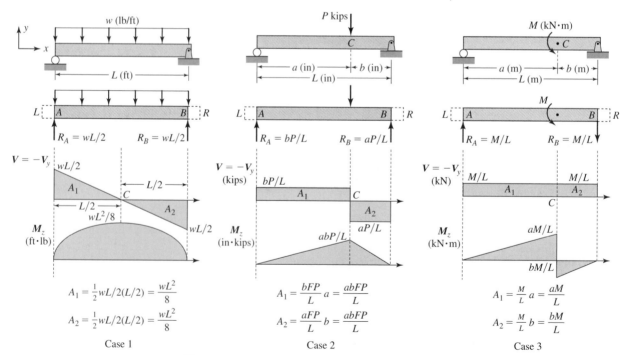

Figure 6.48 Construction of shear and moment diagrams.

represent points where V and M_z values may jump, such as at point C in cases 2 and 3 in Figure 6.48.

Step 4 Consider imaginary extensions on the left and right ends of the beam. V and M_z are zero in these imaginary extensions.

In the imaginary left extension, LA at the beams shown in Figure 6.48, V_1 and M_1 are zero, and we can start our process at this segment. Which way to jump as we cross point A (the start of the beam) can now be treated like any other point on the beam at which there is a point force and/or point moment. At the right imaginary extension BR, the values of the shear force and bending moment must return to zero, providing a check on our solution procedure.

Shear force diagram

Step 5 If there is a point force, then jump the value of V in the direction of the point force.

Just before point A in Figure 6.48, $V_1 = 0$ as we are in the imaginary extension. As we cross point A, the value of V jumps upward (positive) by the value of the reaction force R_A, which is in the upward direction.

At point C in case 2 we jump in the direction of P, which is pointed downward, that is, we subtract P from the value of V_1. In other words, $V_2 = bP/L - P = (b-L)P/L = -aP/L$ just after point C, as shown.

The reaction force R_B is upward in cases 1 and 2, so we add the value of R_B to V_1. In case 3 R_B is downward, so we subtract the value of R_B from V_1. As expected in all cases, we return to a zero value for force V in the imaginary extension BR.

Step 6 Compute the area under the curve of the distributed load. Add the area to the value of V_1 if p is positive, and subtract it if p is negative, to obtain the value of V_2.

In case 1 the area under the distributed force is wL and p is negative. Therefore we subtract wL from the value of V just after A $(+wL/2)$ to get the value of V just before B $(-wL/2)$.

Step 7 Repeat Steps 5 and 6 until you reach the imaginary extension at the right of the beam. If the value of V is not zero in the imaginary extension, then check Steps 5 and 6 for each segment of the beam.

For the three simple cases considered in Figure 6.48, this step is not required.

Step 8 Draw additional vertical lines at any point where the value V is zero. Determine the location of these points by using geometry.

The points where V is zero represent the location of the maximum or minimum values of the bending moment because $dM_z/dx = 0$ at these points. In case 1 $V = 0$ at point C. The location can be found by using similar triangles.

Step 9 Calculate the areas under the V curve and between two adjacent vertical lines.

Areas A_1 and A_2 can be found and recorded as shown in Figure 6.48.

Moment diagram

Step 10 If there is a point moment, then use the moment template and the template equation to determine the direction of the jump.

In case 3 there is a point moment at point C. Comparing the direction of the moment at C to that in the template in Figure 6.47, we conclude that $M_{ext} = -M$. Just before C, $M_1 = aM/L$. As per the template equation, $M_2 = aM/L - M = (a - L)M/L = -bM/L$, which is the value just after C.

In all three cases there is no point moment at A, hence our starting value is zero. If there were a point moment at A, we would use the moment template and the template equation to determine the starting value as we move from the imaginary segment to just right of A.

Step 11 To move from the right of one vertical line to the left of the next vertical line, add the areas under the V curve if V is positive, and subtract the areas if V is negative. Draw the curve according to the curvature rule in Equation (6.17).

In all three cases the area A_1 is positive and we add the value of the area to the value of the moment at point A to obtain the moment just before C. In cases 1 and 2 the area A_2 is negative, hence we subtract the value of A_2 from the moment value just after C to get a zero value just before B. In case 3 A_2 is positive and we add the value of A_2 to the moment value just after C.

In cases 2 and 3 the V curve is constant in each segment, hence the M_z curve is linear in each segment. In case 1 the V curve is linear, hence the M_z curve is quadratic and we need to determine the curvature of the curve. The inclination of the tangent to the M_z curve at A is nonzero, and it decreases to zero at C, that is, the inclination of the tangent decreases as the magnitude of V decreases. Similarly as we move from C to B, we note that the inclination of the tangent increases from zero to a nonzero value, which is consistent with the increasing magnitude of V.

Alternatively in case 1 $p = -w$, hence the curvature of the moment curve is convex.

Step 12 Repeat Steps 10 and 11 until you reach the imaginary extension on the right of the beam. If the value of M_z is not zero in the imaginary extension, then check Steps 10 and 11 for each segment of the beam.

This procedure is applied and elaborated in Examples 6.11 and 6.12.

6.5 BEAM DESIGN ISSUES

We address two issues in this section. The first issue relates to choosing a standard (commonly manufactured) beam cross section that will be cheapest to use. The second issue relates to determining the maximum tensile compressive bending normal stress.

6.5.1 Section Modulus

In design of steel beams the tensile strength and the compressive strength are usually assumed equal. The magnitude of the maximum bending normal stress is calculated using Equation (6.12), which can be written as $\sigma_{max} = M_{max} y_{max}/I_{zz}$, where M_{max} is the magnitude of the maximum internal bending moment, and y_{max} the distance of the point farthest from the neutral axis. The moment of inertia I_{zz} and y_{max} depend on the geometry of the cross section. Rather than using two variables in trying to determine the best geometric shape for use in a particular design, a variable called section modulus S, is defined and the equation for the maximum bending normal stress is written as

$$S = \frac{I_{zz}}{y_{max}} \qquad \sigma_{max} = \frac{M_{max}}{S} \qquad (6.18)$$

For steel beams of standard shapes the section modulus S is tabulated in Appendix E and used in design as elaborated in the Example 6.12.

6.5 BEAM DESIGN ISSUES | **379**

6.5.2 Maximum Tensile and Compressive Bending Normal Stresses

In Section 3.1 it was observed that a brittle material usually ruptures when the maximum tensile normal stress exceeds the ultimate tensile stress of the material. Cracks in material also propagate due to tensile stress.[20] Adhesively bonded material debonds due to the tensile normal stress called *peel stress*. Thus it is possible that a structure designed for maximum normal stress will fail when the maximum tensile stress that triggers the failure is less in magnitude than the maximum compressive stress. This must be properly accounted for in design. Similarly, failure may occur due to a maximum compressive normal stress that is less than the maximum tensile normal stress. This may happen because of a phenomenon called buckling, which is discussed in Chapter 11.

Thus in beam design it may be necessary to determine the maximum tensile bending normal stress and the maximum compressive bending normal stress. These two stress values may be different when the top and the bottom of the beam are at different distances from the neutral axis of the cross section. Since both the M_z value and the y value affect the sign of the bending normal stress in Equation (6.12), stresses must be checked at four points:

- On the top and bottom surfaces on the cross-section location where M_z is a maximum positive value.

- On the top and bottom surfaces on the cross-section location where M_z is a maximum negative value.

Example 6.11 elaborates this issue.

EXAMPLE 6.11
A loaded beam and cross section are shown in Figure 6.49.

(a) Draw the shear force and bending moment diagrams for the beam shown and determine the maximum shear force and bending moment.

(b) Determine the maximum tensile and compressive bending normal stress in the beam.

PLAN
(a) We can determine the reaction force and moment at wall C and follow the procedure for drawing shear and moment diagrams described in Section 6.4.3.

---

[20]See Section 10.3 for additional details.

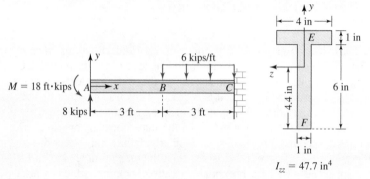

Figure 6.49 Beam and loading in Example 6.11.

(b) We can find σ_{xx} from Equation (6.12) at points E and F at those
cross sections where the M_z value is maximum positive and maximum negative. From these four values we can find the maximum
tensile and compressive bending normal stresses.

Solution

(a) We draw the shear force and bending moment diagram as per the
procedure outlined in Section 6.4.3.

Step 1: From the free-body diagram shown in Figure 6.50 we can
determine the value of the reaction force R_w by equilibrium of
forces in the y direction. By balancing the moment at point C we
can determine the reaction moment M_w. The values of these reactions are as follows:

$$R_w = 10 \text{ kips} \tag{E1}$$
$$M_w = 3 \text{ ft·kips} \tag{E2}$$

Step 2: We draw and label the axes for V and M_z and record the
units.

Step 3: The beam is shown in Figure 6.50 with all forces and
moments acting on it. Vertical lines at points A, B, and C are
shown as drawn.

Step 4: We draw imaginary extensions LA and CR to the beam.

1. *Shear force diagram in Figure 6.50*

Steps 5, 6, 7: In segment LA the shear force is zero, hence $V_1 = 0$.
The 8-kips force at A is upward, so we jump to a value of
$V_2 = +8$ kips just to the right of point A. $p = 0$ in segment AB,
hence the value of V remains at 8 kips just before B. Since there is
no point force at B, there is no jump in V at B.

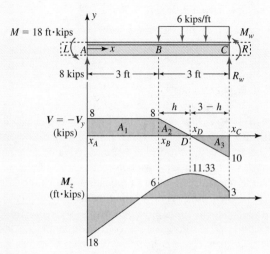

Figure 6.50   Shear and moment diagrams in Example 6.11.

In segment BC the area under the distributed load is 18 kips. As p is negative, we subtract the area from 8 kips to get a value of -10 kips just before C. We join it by a straight line as p is uniform in BC.

The reaction force R_w is upward, so we add the value to -10 kips to get a zero value just after C, confirming the correctness of our solution.

Step 8: At point D, where $V = 0$, we draw another vertical line. To find the location of point D, we use the two similar triangles on either side of point D to get

$$\frac{8}{h} = \frac{10}{3-h} \qquad (E3)$$

Solving Equation (E3) we obtain the value of h,

$$h = 1.333 \qquad (E4)$$

Step 9: We calculate areas A_1, A_2, and A_3,

$$A_1 = 8 \times 3 = 24 \qquad (E5)$$

$$A_2 = \frac{1}{2}8 \times h = 5.33 \qquad (E6)$$

$$A_3 = \frac{1}{2}10(3-h) = 8.33 \qquad (E7)$$

2. *Moment diagram in Figure 6.50*

Steps 10, 11, 12: In segment LA the bending moment is zero, hence $M_1 = 0$. Comparing the 18-ft·kips couple at point A with M_{ext} in the moment template in Figure 6.47, we obtain $M_{ext} = -18$ ft·kips. Hence from the template equation $M_2 = -18$ ft·kips just to the right of point A.

The area A_1 is positive, so we add its value to -18 ft·kips to obtain $M_2 = +6$ ft·kips just before B. As V was constant in AB, we join the moments at points A and B by a straight line, as shown.

The area A_2 is positive, so we add its value to $+6$ ft·kips to obtain $M_2 = +11.33$ ft·kips just before D. As V is linear between B and D, the integral will result in a quadratic function. The magnitude of the shear force is decreasing, hence the incline of the tangent to the moment curve must decrease as we move from point B toward point D, resulting in the convex curve shown between B and D. Alternatively p is negative between B and D, hence the curve is convex.

The area A_3 is negative, so we subtract its value from 11.33 ft·kips to obtain $+3$ ft·kips just before C. As V is linear between D and C, the integral will result in a quadratic function. Since the magnitude of the shear force is increasing, hence the incline of the tangent to the moment curve must increase as we move from point D toward C, resulting in the convex curve shown between D and C.

Comparing the moment M_w at C with M_{ext} in the template in Figure 6.47, we obtain $M_{ext} = -M_w = -3$ ft·kips. Hence from the template equation $M_2 = 0$ just to the right of point C, that is, in the imaginary segment CR the moment is zero as expected, confirming the correctness of our construction.

From Figure 6.50 we see that the maximum values of V and M_z are -10 kips and -18 ft·kips, respectively. Recollect that, $V = -V_y$. Thus the maximum values of the shear force and the bending moment are

$$\text{ANS.} \qquad (V_y)_{max} = 10 \text{ kips} \qquad \text{(E8)}$$

$$\text{ANS.} \qquad (M_z)_{max} = -18 \text{ ft·kips} \qquad \text{(E9)}$$

(b) The maximum positive moment occurs at D ($M_D = +11.33$ ft·kips $= 136$ in·kips) and the maximum negative moment occurs at A ($M_A = -18$ ft·kips $= -216$ in·kips). We can evaluate the bending normal stress at points E ($y_E = +2.6$ in) and F ($y_F = -4.4$ in) on the cross sections at A and D using Equation (6.12) as follows:

- On cross section A, at point E, the bending normal stress is

$$\sigma_{AE} = -\frac{-216 \times 2.6}{47.7} = +11.8 \text{ ksi}$$

- On cross section A, at point F, the bending normal stress is

$$\sigma_{AF} = -\frac{-216(-4.4)}{47.7} = -19.9 \text{ ksi}$$

- On cross section D, at point E, the bending normal stress is

$$\sigma_{DE} = -\frac{136 \times 2.6}{47.7} = -7.4 \text{ ksi}$$

- On cross section D, at point F, the bending normal stress is

$$\sigma_{DF} = -\frac{136(-4.4)}{47.7} = +12.6 \text{ ksi}$$

From these results it is clear that the maximum tensile bending normal stress occurs at point F on the cross section at D, whereas the maximum compressive bending normal stress occurs at point F on the cross section at A, and the values are as follows:

ANS. $\quad \sigma_{DF} = 12.6 \text{ ksi (T)} \qquad \sigma_{AF} = 19.9 \text{ ksi (C)}$

COMMENTS

1. It should be emphasized that writing the steps is for the purpose of explanation. All the calculations needed to draw the shear and moment diagrams are shown by Equations (E3) through (E7).

2. From Equation (6.14) we know that at the points at which the shear force is zero, the slope of the moment curve is zero, that is, the bending moment will have a local maximum or a local minimum at such points. In Figure 6.50 we see that at point D there is a local maximum in the bending moment. But the maximum moment in the beam is at point A, where a point moment is applied.

3. In calculating the maximum tensile and compressive bending normal stresses we evaluated four possible points. But if we were determining the magnitude of the maximum bending normal stress only, then we would need to evaluate one point only, namely, where the moment is maximum and where y is maximum. In other words, we would have found the stress at point F on the cross section at A.

4. Suppose we use the template shown in Figure 6.51 to determine the direction of the jump in the moment. In LA, $M_1 = 0$ just before A. Comparing the 18-ft·kips couple at point A with M_{ext} in the

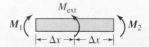

Figure 6.51 Alternative template.

template in Figure 6.51, we obtain $M_{ext} = 18\text{-ft·kips}$. Hence from the template equation

$$M_2 = M_1 - M_{ext}$$

and Figure 6.51, $M_2 = -18\ \text{ft·kips}$ just to the right of point A the same as before. It can be verified that using the template in Figure 6.51 for jumps at point C will result in a zero moment value in the imaginary segment CR, as before.

This shows that the direction of M_{ext} on the template is immaterial. Thus there is no need to memorize the template, which can be drawn before starting on the shear and moment diagrams.

EXAMPLE 6.12

Consider the beam shown in Figure 6.41. Select the lightest W- or S-shaped beams from these given in Appendix E if the allowable bending normal stress is 53 MPa in tension or compression.

PLAN

We can draw the shear and moment diagrams using the procedure described in Section 6.4.3. From the moment diagram we can find the maximum moment. Using the allowable bending normal stress of 53 MPa and Equation (6.18), we can find the minimum sectional modulus. Using Appendix E, we can make a list of the beams for which the sectional modulus is just above the one we determined and choose the lightest beam we can use.

Solution

Step 1: The reaction forces at points A and D were determined in Example 6.10.

Step 2: The beam with all forces and moments acting on it is shown in Figure 6.52. Vertical lines at points A, B, C, and D are shown as drawn.

Step 3: We draw imaginary extensions LA and DR to the beam.

Step 4: We label the axes for V and M_z and record the units.

1. *Shear force diagram in Figure 6.52*

Steps 5, 6, 7: In segment LA, $V_1 = 0$. Since R_A is zero, there is no jump at A and we start our diagram at zero.

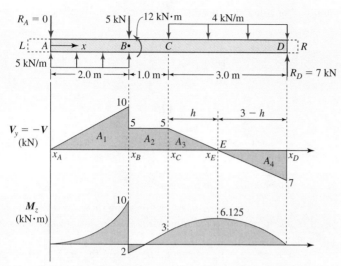

Figure 6.52 Shear and moment diagrams in Example 6.12.

In segment AB, $p = +5$ kN/m. Hence we add the area of $5 \times 2 = 10$ kN to obtain $V_2 = 10$ kN just before point B and draw a straight line between the values of V at A and B, as shown in Figure 6.52. At B the point force of 5 kN is downward, thus we jump downward by 5 kN to obtain $V_2 = 5$ kN just after point B.

In segment BC, $p = 0$; hence the value of V does not change until point C.

In segment CD, $p = -4$ kN/m. Hence we subtract the area of $4 \times 3 = 12$ kN to obtain $V_2 = -7$ kN just before point D and draw a straight line between the values of V at C and D, as shown in Figure 6.52. The reaction force at D is upward, so we jump upward by 7 kN to obtain $V_2 = 0$ kN just after point D, that is, in the imaginary segment DR the shear force is zero as expected, confirming the correctness of our construction.

Step 8: At point E, where $V_y = 0$, we draw another vertical line. To find the location of point E, we use the two similar triangles on either side of point E to get

$$\frac{5}{h} = \frac{7}{3-h} \tag{E1}$$

Solving Equation (E1) we obtain the value of h,

$$h = 1.25 \tag{E2}$$

Step 9: We calculate the areas A_1 through A_4,

$$A_1 = \frac{1}{2}10 \times 2 = 10 \tag{E3}$$

$$A_2 = 5 \times 1 = 5 \tag{E4}$$

$$A_3 = \frac{1}{2}5h = 3.125 \tag{E5}$$

$$A_4 = \frac{1}{2}7(3-h) = 6.125 \tag{E6}$$

2. *Bending moment diagram in Figure 6.52*

Steps 10, 11, 12: In segment *LA* the bending moment is zero, hence $M_1 = 0$. As there is no point moment at *A*, we start our moment diagram at zero.

As *V* is positive in segment *AB*, we add the area A_1 to obtain $M_2 = +10$ kN·m just before *B*. As *V* is linear in *AB*, the integral will result in a quadratic function between *A* and *B*. As the magnitude of the shear force is increasing, the incline of the tangent to the moment curve must increase as we move from point *A* toward point *B*, resulting in the concave curve shown between *A* and *B*. Alternatively, in *AB* *p* is positive, hence the moment curve is concave.

Comparing the moment 12 kN·m at *B* with M_{ext} in the template in Figure 6.47, we obtain $M_{\text{ext}} = -12$ kN·m. Hence from the template equation $M_2 = 10 - 12 = -2$ kN·m just to the right of point *B*.

As *V* is positive in segment *BC*, we add the area A_2 to obtain the value of $M_2 = +3$ kN·m just before *C*. As *V* is constant between *B* and *C*, the integral will result in a linear function, so we draw a straight line between *B* and *C*.

As *V* is positive in segment *CE*, we add the area A_3 to obtain the value of $M_2 = +6.125$ kN·m just before *E*. As *V* is linear between *C* and *E*, the integral will result in a quadratic function. As the magnitude of the shear force is decreasing, the incline of the tangent to the moment curve must also decrease as we move from point *C* toward *E*, resulting in the convex curve between *C* and *E*, as shown. Alternatively, in *CE* *p* is negative, hence the moment curve is convex.

As *V* is positive in *ED*, we add the area A_4 to obtain the value of $M_2 = 0$ just before *D*. As *V* is linear between *E* and *D*, the integral will result in a quadratic function. As the magnitude of the shear force is increasing, the incline of the tangent to the moment curve must also increase as we move from point E toward D, resulting in the convex curve between E and D, as shown. Alternatively, in ED p is negative, hence the moment curve is convex.

As there is no point moment at D, there will be no jump in the moment at D. Hence we obtain a zero value for the moment in the imaginary segment DR as expected, confirming the correctness of our construction.

From the moment diagram in Figure 6.52 the maximum moment is $M_{max} = 10$ kN·m. Noting that the allowable bending normal stress is 53 MPa, Equation (6.18) yields

$$\sigma_{max} = \frac{10 \times 10^3}{S} \leq 53 \times 10^6 \qquad \text{or} \qquad S \geq 188.7 \times 10^3 \text{ mm}^3$$

$$\text{or} \qquad S \geq 188.7 \times 10^3 \text{ mm}^3 \qquad (E7)$$

From Appendix D we obtain the following list of W- and S-shaped beams that have a section modulus close to that given in Equation (E7):

$$W150 \times 29.8 \qquad S = 219 \times 10^3 \text{ mm}^3$$

$$W200 \times 22.5 \qquad S = 194.2 \times 10^3 \text{ mm}^3$$

$$S200 \times 27.4 \qquad S = 236 \times 10^3 \text{ mm}^3$$

$$S180 \times 30 \qquad S = 198.3 \times 10^3 \text{ mm}^3$$

ANS.   We select W200 × 22.5 as it is the lightest beam with a mass of only 22.5 kg/m.

COMMENTS

1. The use of the section modulus in selecting the beam cross section from a set of standard shapes is simplified, as demonstrated by this example. But the concept of section modulus can also be used in nonstandard shapes.

2. Suppose we used the template shown in Figure 6.51 for determining the direction of jump in internal moments. Just before B, $M_1 = +10$ kN·m. Comparing the moment 12 kN·m at B with M_{ext} in the template in Figure 6.47, we obtain $M_{ext} = 12$ kN·m. Hence from the template equation $M_2 = -2$ kN·m just to the right of point B—the same as before—demonstrating once more that the directions of the external forces and the moment on the template are immaterial.

3. It should again be emphasized that writing the steps is for the purpose of explanation. All the calculations needed for drawing the shear and moment diagrams are given by Equations (E1) through (E6). From here on the shear and moment diagrams will be drawn without additional explanations.

QUICK TEST 6.1   Time: 20 minutes/Total: 20 points

Answer true or false and justify each answer in one sentence. Grade yourself with the answers given in Appendix G. Assume linear elastic, homogeneous material unless stated otherwise.

1. If you know the geometry of the cross section and the bending normal strain at one point on a cross section, then the bending normal strain can be found at any point on the cross section.

2. If you know the geometry of the cross section and the maximum bending normal stress on a cross section, then the bending normal stress at any point on the cross section can be found.

3. A rectangular beam with a 2-in × 4-in cross section should be used with the 2-in side parallel to the bending (transverse) forces.

4. The best place to drill a hole in a beam is through the centroid.

5. In the formula $\sigma_{xx} = -M_z y / I_{zz}$, y is measured from the bottom of the beam.

6. The formula $\sigma_{xx} = -M_z y / I_{zz}$ can be used for finding the normal stress on a cross section of a tapered beam.

7. The equations $\int_A \sigma_{xx}\, dA = 0$ and $M_z = -\int_A y\sigma_{xx}\, dA$ cannot be used for nonlinear materials.

8. The equation $M_z = -\int_A y\sigma_{xx}\, dA$ can be used for nonhomogeneous cross sections.

9. The internal shear force jumps by the value of the applied transverse force as one crosses it from left to right.

10. The internal bending moment jumps by the value of the applied concentrated moment as one crosses it from left to right.

Equilibrium of shear force and bending moment

In Problems 6.49 through 6.56, (a) write the equations for shear force and bending moments as a function of x for the entire beam; (b) show that your results satisfy Equations (6.13) and (6.14).

6.49 Using Figure P6.49 solve the problem as described in text above.

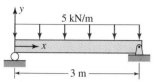

Figure P6.49

6.50 Using Figure P6.50 solve the problem as described in text above.

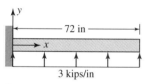

Figure P6.50

6.51 Using Figure P6.51 solve the problem as described in text above.

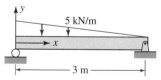

Figure P6.51

6.52 Using Figure P6.52 solve the problem as described in text above.

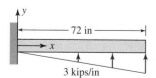

Figure P6.52

6.53 Using Figure P6.53 solve the problem as described in text above.

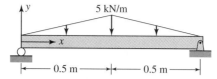

Figure P6.53

6.54 Using Figure P6.54 solve the problem as described in text above.

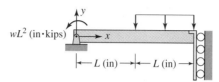

Figure P6.54

6.55 Using Figure P6.55 solve the problem as described in text above.

6.56 Using Figure P6.56 solve the problem as described in text above.

6.57 Consider the beam shown in Figure P6.57. (a) Write the shear force and moment equations as a function of x in segments AB and BC. (b) Show that your results satisfy Equations (6.13) and (6.14). (c) What are the shear force and bending moment values just before and just after point B?

6.58 Consider the beam shown in Figure P6.57. (a) Write the shear force and moment equations as a function of x in segments CD and DE. (b) Show that your results satisfy Equations (6.13) and (6.14). (c) What are the shear force and bending moment values just before and just after point D?

6.59 Consider the beam shown in Figure P6.59. (a) Write the shear force and moment equations as a function of x in segments AB and BC. (b) Show that your results satisfy Equations (6.13) and (6.14). (c) What are the shear force and bending moment values just before and just after point B?

6.60 Consider the beam shown in Figure P6.59. (a) Write the shear force and

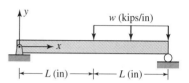

Figure P6.55

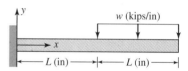

Figure P6.56

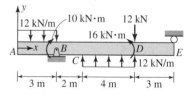

Figure P6.57

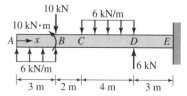

Figure P6.59

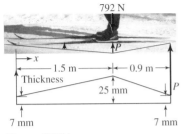

Figure P6.61

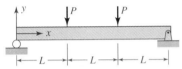

Figure P6.62

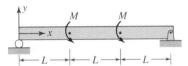

Figure P6.63

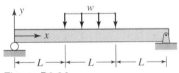

Figure P6.66

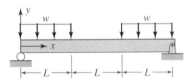

Figure P6.67

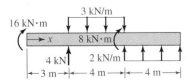

Figure P6.70

moment equations as a function of x in segments CD and DE. (b) Show that your results satisfy Equations (6.13) and (6.14). (c) What are the shear force and bending moment values just before and just after point D?

6.61 During skiing, the weight of a person is often all on one ski. The ground reaction $p(x)$, when the weight is on one ski, is modeled as shown in Figure P6.61. (a) Find shear force and bending moment as a function of x across the ski. (b) The ski is 50 mm wide and the thickness of the ski varies as shown. Determine the maximum bending normal stress. Use of spread sheet recommended.

Shear and moment diagrams

6.62 Draw the shear and moment diagrams for the beam and loading shown in Figure P6.62.

6.63 Draw the shear and moment diagrams for the beam and loading shown in Figure P6.63.

6.64 For the beam shown in Figure P6.49, draw the shear force bending moment diagram. Determine the values of maximum shear force and bending moment.

6.65 For the beam shown in Figure P6.50, draw the shear force bending moment diagram. Determine the values of maximum shear force and bending moment.

In Problems 6.66 and 6.67, draw the shear and moment diagrams for the beam and loading shown in each problem. Determine the values of maximum shear force and bending moment.

6.66 Using Figure P6.66 solve the problem as described in text above.

6.67 Using Figure P6.67 solve the problem as described in text above.

6.68 Draw the shear and moment diagrams for the beam shown in Figure P6.57 and determine the values of maximum shear force and bending moment.

6.69 Draw the shear and moment diagrams for the beam shown in Figure P6.59 and determine the values of maximum shear force and bending moment.

In Problems 6.70 through 6.75, draw the shear and moment diagrams and determine

the values of maximum shear force and bending moment in each problem.

6.70 Using Figure P6.70 solve the problem as described in text above.

6.71 Using Figure P6.71 solve the problem as described in text above.

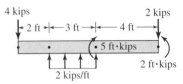

Figure P6.71

6.72 Using Figure P6.72 solve the problem as described in text above.

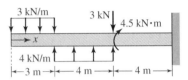

Figure P6.72

6.73 Using Figure P6.73 solve the problem as described in text above.

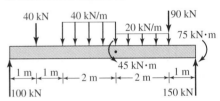

Figure P6.73

6.74 Using Figure P6.74 solve the problem as described in text above.

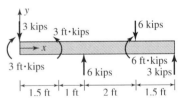

Figure P6.74

6.75 Using Figure P6.75 solve the problem as described in text above.

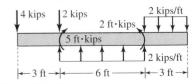

Figure P6.75

Design problems

6.76 A beam and its loading, and cross section are as shown Figure P6.76. Determine the intensity w of the distributed load if the maximum bending normal stress is limited to 10 ksi (C) and 6 ksi (T). The second area moment of inertia is $I_{zz} = 47.73$ in^4.

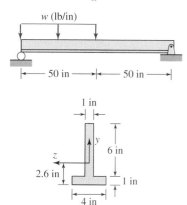

Figure P6.76

6.77 Two pieces of lumber are glued together to form the beam shown in Figure P6.77. Determine the intensity w of the distributed load if the maximum tensile bending normal stress in the glue is limited to 800 psi (T) and the maximum bending normal stress in wood is limited to 1200 psi.

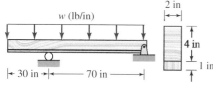

Figure P6.77

6.78 The beam shown in Figure P6.54 has a load $w = 25$ lb/in and $L = 72$ in. Select the lightest W- or S-shaped beam from Appendix E if the allowable bending normal stress is 21 ksi in tension and compression.

6.79 The beam shown in Figure P6.55 has a load $w = 0.4$ kips/in and $L = 48$ in. Select the lightest W- or S-shaped beam from Appendix E if the allowable bending normal stress is 16 ksi in tension and compression.

6.80 The beam shown in Figure P6.56 has a load $w = 0.15$ kips/in and $L = 48$ in. Select

the lightest W- or S-shaped beams from Appendix E if the allowable bending normal stress is 21 ksi in tension and compression.

6.81 Consider the beam shown in Figure P6.57. Select the lightest W- or S-shaped beam from Appendix E if the allowable bending normal stress is 180 MPa in tension and compression.

6.82 Consider the beam shown in Figure P6.59. Select the lightest W- or S-shaped beam from Appendix E if the allowable bending normal stress is 225 MPa in tension and compression.

6.83 The wind pressure on a signpost is approximated as a uniform pressure, as shown Figure P6.83. A similar signpost is to be designed using a hollow square steel beam for the post. The outer dimension of the square is to be 12 in. If the allowable bending normal stress is 24 ksi and the pressure $p = 33$ lb/ft^2, determine the inner dimension of the lightest hollow beam to the nearest $\frac{1}{8}$ in.

Stress concentration

6.84 The allowable bending normal stress in the stepped circular beam shown in Figure P6.84 is 200 MPa and $P = 200$ N. Determine the smallest fillet radius that can be used at section B.

6.85 The allowable bending normal stress in the stepped circular beam shown in Figure P6.85 is 48 ksi. Determine the maximum intensity of the distributed load w assuming the fillet radius is: (a) 0.3 in; (b) 0.5 in. Use the stress concentration graphs given in Appendix C.

Fatigue

6.86 The fillet radius is 5 mm in the stepped aluminum circular beam shown in Figure P6.84. What should be the peak value of the cyclic load P to ensure a service life of one-half million cycles? Use the S–N curve shown in Figure 3.34.

6.87 The beam in Figure P6.85 is made from a steel alloy that has the S–N curve shown in Figure 3.34. The peak intensity of the cyclic distributed load is $w = 80$ lbs/in and the fillet radius is 0.36 in. What is the predicted service life of the beam?

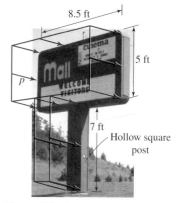

Figure P6.83

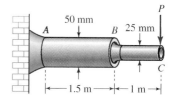

Figure P6.84

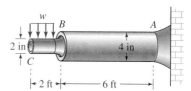

Figure P6.85

6.6 SHEAR STRESS IN THIN SYMMETRIC BEAMS

In Section 6.2.5 it was observed that the maximum bending shear stress has to be nearly an order of magnitude less than the maximum bending normal stress for our theory to be valid. But shear stress plays an important role in bending, particularly when beams are constructed by joining a number of beams together to increase stiffness. In this section we develop a theory that can be used for calculating the bending shear stress.

Figure 6.53*a* shows the bending of four separate wooden strips and Figure 6.53*b* shows the bending of the four wooden strips after the strips have been glued together. In Figure 6.53*a* each wooden strip slides relative to the other in the axial direction. But in Figure 6.53*b* the relative sliding is prevented by the shear resistance of the glue, that is, the shear stress in the glue. One may thus hypothesize that in any beam there will be shear stresses on imaginary surfaces parallel to the axis of the beam.

Notice that the beam in Figure 6.53*a* has significantly more curvature (it bends more) than that in Figure 6.53*b*, even though the forces exerted in both cases are approximately the same. This phenomenon of increasing stiffness[21] at the expense of introducing shear stress is exploited in the design of lightweight structures. The flanges are designed for carrying most of the normal stress in bending and the webs are designed for carrying most of the shear stress (see also Figure 6.29). In sandwich beams two stiff panels are separated by a soft core material (see also footnote 9). The stiff panels are designed to carry the normal stress and the soft core is designed to carry the shear stress.

6.6.1 Shear Stress Direction

Before developing formulas for the shear stress in bending, it is worthwhile to understand the character of these shear stresses and to be able to determine the direction of the shear stress on a cross section by inspection.

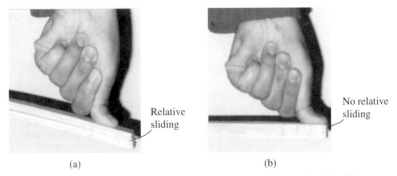

Relative
sliding

No relative
sliding

(a)   (b)

Figure 6.53 Effect of shear stress in bending. (*a*) Separate beams. (*b*) Glued beams.

---

[21]The separate beams bend about the centroid of each strip whereas the glued cross section bends about the centroid of the glued unit. Thus the area moment of inertia for the glued unit is significantly greater than the total of the separate beams. See Problem 6.15.

Consider the beam in Figure 6.54a. The beam is constructed by gluing five pieces of wood together. Due to the bending load P, a normal stress distribution across the cross section will develop as shown in Figure 6.54b. The moment M_z varies along the length of the beam. Thus from Equation (6.12) we know that the magnitude of the normal stress σ_{xx} will change along the length of the beam. From the evidence of the photographs in Figure 6.53 we know that shear stress will exist at each glued surface to resist the relative sliding of the wood strips. If we take a small element Δx of strips 3 and 5, we obtain Figure 6.54c and d. On the glued surface between wooden strips 2 and 3 there will be a shear stress τ_{zx} as the outward normal of the surface is in the z direction and the internal shear force is in the x direction. On the glued surface between wooden strips 4 and 5 there will be a shear stress τ_{yx} as the outward normal of the surface is in the y direction and the internal shear force is in the x direction. On the small element Δx the equivalent shear force from these shear stresses must balance the change in the equivalent normal axial force, as shown Figure 6.54e and f. Thus the shear stress in bending must balance the variations in the normal stress σ_{xx} along the length of the beam.[22]

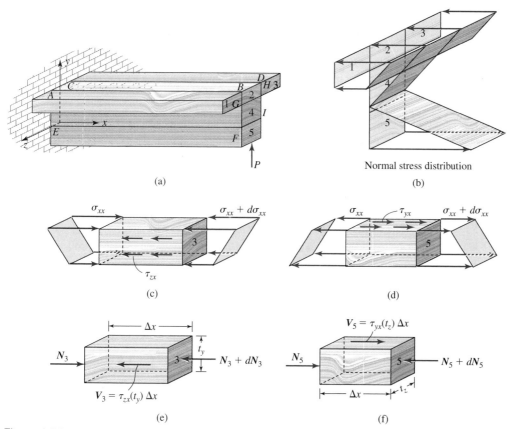

Figure 6.54 Shear stress on different surfaces in bending.

[22]From the field of elasticity it is known that in the absence of body forces, the equilibrium at a point requires: $\partial\sigma_{xx}/\partial x + \partial\tau_{yx}/\partial y + \partial\tau_{zx}/\partial z = 0$ (see Problem 1.98). Thus if σ_{xx} varies with x, then τ_{yx} (or τ_{xy}) must vary with y, and τ_{zx} (or τ_{xz}) must vary with z. See Problem 6.114 for additional details.

The preceding shows that shear stress develops on surfaces cut parallel to the axis of the beam. But from the symmetry of shear stresses $\tau_{xy} = \tau_{yx}$ and $\tau_{xz} = \tau_{zx}$. These stresses, τ_{xy} and τ_{xz}, are on the cross sections perpendicular to the axis of the beam. We know from Equation (6.10) that on the cross section of the beam the resultant of the shear stress τ_{xy} distribution is the shear force V_y. Thus the direction (sign) of τ_{xy} should be the same as that of V_y. But the shear force V_z that would be statically equivalent to τ_{xz} must be zero, as there is no external force in the z direction. This means that τ_{xz} must reverse sign (direction) on the cross section if the net force from it is zero. We also know that the y axis is the axis of symmetry, and the loading is in the plane of symmetry. Therefore all stresses including τ_{xz} must be symmetric about the y axis. In other words, the shear stress τ_{xz} will reverse its direction as one crosses the y axis on the cross section. This sometimes implies that the shear stress τ_{xz} will be zero at the y axis.

Consider now a circular cross section that is glued together from nine wooden strips, as shown in Figure 6.55a. Once more shear stresses will develop along each glued surface to resist relative sliding between two adjoining wooden strips, and the shear stress value must balance the change in axial force due to the variation in σ_{xx}. The outward normal of the surface will be in a different direction for each glued surface on which we consider the shear stress. If we define a tangential coordinate s that is in the direction of the tangent to the centerline of the cross section, then the outward normal to the glued surface will be in the s direction and the shear stress will be τ_{sx}. Once more by the symmetry of shear stresses, $\tau_{xs} = \tau_{sx}$. At a point if the s direction and the y direction are the same, then τ_{xs} will equal $\pm\tau_{xy}$. If the s direction and the z direction are the same at a point, then τ_{xs} will equal $\pm\tau_{xz}$.

It should be noted that in Figure 6.54e and f and in Figure 6.55b the shear force that balances the change in the axial force N is shown on only one surface. The surface on the other end of the free-body diagram is always assumed to be a free surface, that is, the shear stress is zero on these other surfaces. The origin of the s coordinate is chosen to be one of the free surfaces and will be used in the next section in developing shear stress formulas. In a beam cross section the top and bottom and the side surfaces are always assumed to be surfaces on which shear stress is zero.

In Figure 6.54e and f and in Figure 6.55b we notice that the shear force expression contains the product of the shear stress and the thickness t of the cross section at that point. This product is called *shear flow* and is denoted by q as follows:

$$q = \tau_{xs}t \tag{6.19}$$

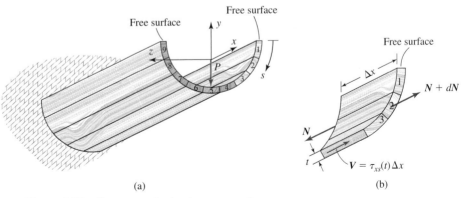

(a)   (b)

Figure 6.55   Shear stress in circular cross section.

The units of the shear flow q are force per unit length. The terminology is from fluid flow in channels, but it is used extensively to discuss shear stresses in thin cross sections, probably because of the visual image of a flow it conveys in discussing shear stress directions, as elaborated further in Example 6.13.

The shear flow along the centerline of the cross section is drawn in such a direction as to satisfy the following rules:

1. The resultant force in the y direction is in the same direction as V_y.

2. The resultant force in the z direction is zero.

3. It is symmetric about the y axis. This requires that shear flow change direction as one crosses the y axis on the centerline. Sometimes this will imply that shear stress is zero at the point(s) where the centerline intersects the y axis.

EXAMPLE 6.13

Assuming a positive shear force V_y, sketch the direction of the shear flow along the centerline on the thin cross sections shown in Figure 6.56.

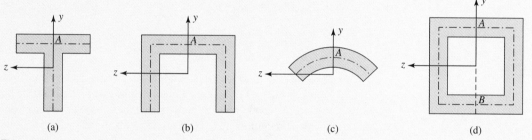

| (a) | (b) | (c) | (d) |

Figure 6.56   Cross sections in Example 6.13.

PLAN

With the outward normal of the cross section in the positive x direction, the positive shear force V_y will be in the positive y direction according to the sign convention in Section 6.2.5. At point A the centerline intersects the y axis. Thus on either side of A the shear flow will be in the opposite direction. In the fourth cross section the shear flow will also change direction at point B. We can determine the direction of the flow in each cross section to satisfy the rules described at the end of Section 6.6.1.

Solution

(a) On the cross section shown in Figure 6.57a the shear flow (shear stress) from C to A will be in the positive y direction as V_y on the cross section is in the positive y direction. At point A in the flange the flow will break in two and go in opposite directions, as shown in Figure 6.57a. The resultant force due to shear flow from A to D will cancel the force due to shear flow from A to E, satisfying the condition of zero resultant force in the z direction and the condition of symmetric flow about the y axis.

(b) On the cross section shown in Figure 6.57b the shear flow from C to E and from D to F will be in the positive y direction to satisfy the condition of symmetry about the y axis and to have the same direction as V_y. In the flange the two flows will approach point A from opposite directions. The resultant force due to shear flow from A to E will cancel the force due to shear flow from A to F, satisfying the condition of zero resultant force in the z direction and the condition of symmetric flow about the y axis.

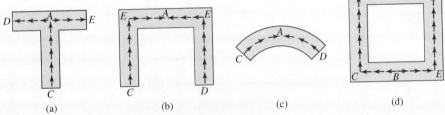

Figure 6.57 Shear flow in Example 6.13.

(c) On the cross section shown in Figure 6.57c the shear flows from points C and D will approach point A in opposite directions. This ensures the condition of symmetry, and the condition of zero force in the z direction is met. The y components of the two shear flows are in the positive y direction, satisfying the condition that shear flow result in positive V_y.

(d) The shear flow from C to D and the shear flow from E to F have to be in the positive y direction to satisfy the condition of symmetry about the y axis and to have the same direction as V_y. At points A and B the shear flows must change direction to ensure symmetric shear flows about the y axis. The force from the shear flows in BC and DA will cancel the force from the shear flows in BE and FA, ensuring the condition of a zero force in the z direction.

COMMENTS

1. It should be noted that the shear flow (shear stress) is zero at the following points because these points are on the free surface: points C, D, and E in Figure 6.57a; points C and D in Figure 6.57b and c.

2. At point A in Figure 6.57b, c, and d the shear flow will be zero, but it will not be zero in Figure 6.57a, which is a conclusion that requires use of the formulas that will be developed in the next section (see Example 6.15). But using the analogy of fluid flow we can appreciate the conclusion in the following manner. In Figure 6.57a

the shear flows at point A in branches AD and AE add up to the value of shear flow at point A in branch CA. With no other branch at point A in Figure 6.57*b*, *c*, and *d* the values of the shear flow are equal and opposite, which is only possible if the value of shear flow is zero.

3. The term *flow* invokes an image that helps in visualizing the direction of shear stress.

4. By examining the direction of the stress components in the Cartesian system, we can determine whether a stress component is positive or negative τ_{xy} or τ_{xz}, as shown in Figure 6.58. Note that τ_{xy} is positive in all cases, which is a consequence of having to be in the direction of the shear force V_y. But τ_{xz} can be positive or negative, depending on the location of the point.

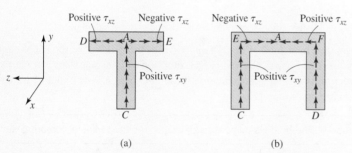
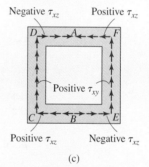

(a) (b) (c)

Figure 6.58 Directions and signs of stress components in Example 6.13.

6.6.2 Bending Shear Stress Formula

The discussion in the previous section and in Example 6.13 highlights that the bending shear stress is τ_{xy} in the web, τ_{xz} in the flange, and for symmetric curvilinear cross sections it depends on the location of the point. To develop a single formula that is applicable to all situations requires the definition of a tangential coordinate s that is in the direction of the tangent to the centerline of the cross section and starts from a free surface. In this section we derive the formula for bending shear stress τ_{xs}.

Consider a differential element of a wooden beam with circular cross section as shown in Figure 6.59*a*. Consider the shear stress acting on the surface between wooden pieces 3 and 4. There are two possible free-body diagrams we can consider, which are shown in Figure 6.59*b* and *c*. The axial force N_s (or N_{s*}) acting on the part of cross section A_s (or A_{s*}) varies because of the variation of the bending stress σ_{xx} along the length of the beam. Assuming the shear stress does not change across the

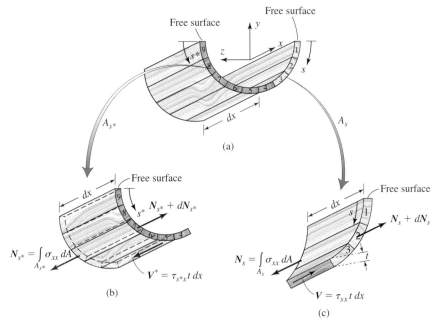

Figure 6.59 Differential element of beam for shear stress calculations.

thickness, the shear force V (or V^*) is equal to the product of the shear stress multiplied by the area $t\,dx$, as shown. The assumption of constant shear stress in the thickness direction is a good approximation if the thickness is small, that is, we are making the following assumption.

Assumption 9 The beam is thin perpendicular to the centerline of the cross section.

Balancing forces in Figure 6.59c, we obtain

$$N_s + dN_s - N_s + \tau_{sx} t\, dx = 0$$

or

$$\tau_{sx} t = -\frac{dN_s}{dx} = -\frac{d}{dx}\int_{A_s}\sigma_{xx}\, dA \qquad (6.20)$$

In Equation (6.20) identification of the area A_s is critical. So we formally define it next.

Definition 5 Area A_s is the area between the free surface and the point where the shear stress is being evaluated.

Substituting Equation (6.12) into Equation (6.20) and noting that the moment M_z and the area moment of inertia I_{zz} do not vary over the cross section, we obtain

$$\tau_{sx} t = \frac{d}{dx}\left(\frac{M_z}{I_{zz}}\int_{A_s} y\, dA\right) = \frac{d}{dx}\left(\frac{M_z Q_z}{I_{zz}}\right) \qquad (6.21)$$

where Q_z is referred to as the first moment of the area A_s and is defined as

$$\boxed{Q_z = \int_{A_s} y\, dA} \qquad (6.22)$$

We make the following assumption.

Assumption 10 The beam is not tapered.

Assumption 10 implies that I_{zz} and Q_z are not a function of x, and these quantities can be taken outside the derivative sign. We obtain $\tau_{sx}t = (Q_z/I_{zz})dM_z/dx$. Substituting Equation (6.14), the relation between shear force V_y and moment M_z, we obtain the formula $\tau_{sx}t = -Q_zV_y/I_{zz}$ or

$$\boxed{\tau_{sx} = \tau_{xs} = -\frac{V_yQ_z}{I_{zz}t}} \tag{6.23}$$

In Equation (6.23) the shear force V_y can be found either by equilibrium or by drawing the shear force diagram. Also t is the thickness at the point where the shear stress is being found, and I_{zz} is known from the geometry of the cross section. Thus it is the calculation of Q_z that is the new learning objective.

6.6.3 Calculation of Q_z

From Equation (6.22) we note that Q_z is the first moment of the area A_s about the z axis. Figure 6.60 shows the area A_s between the top free surface and the point at which the shear stress is being found (line s–s). The integral in Equation (6.22) is the numerator in the definition of the centroid of the area A_s. Analogous to the moment due to a force, the first moment of an area can be found by placing the area A_s at its centroid and finding the moment about the neutral axis, that is, Q_z is the product of area A_s and the distance of the centroid of the area A_s from the neutral axis, as shown in Figure 6.60. Alternatively, Q_z can be found by using the bottom surface as the free surface, shown as Q_{z*} in Figure 6.60.

At the top surface, which is a free surface, the value of Q_z is zero as the area A_s is zero. When we reach the bottom surface after starting from the top, the value of Q_z is once more zero because $A_s = A$, and from Equation (6.8), $\int_A y\, dA = 0$. If Q_z starts with a zero value at the top and ends with a zero value at the bottom, then it must reach a maximum value somewhere on the cross section. To answer the question where Q_z reaches a maximum value, consider the change in Q_z as the line s–s moves downward in Figure 6.60. As the line moves downward toward the neutral axis, the moment of the area Q_z increases as we add the moments from the additional areas. When the line s–s crosses the neutral axis, then the new additional area after the neutral axis produces a negative moment because the centroid of this area is in the negative y direction. In other words, Q_z increases up to the neutral axis, then it starts decreasing. Thus Q_z is maximum at the neutral axis. From Equation (6.23) the implication of the last statement is: bending shear stress is maximum at the neutral axis of a cross section. Note that to find the maximum shear stress in the beam, we first need to find the maximum shear force V_y, and at that location we find the shear stress at the neutral axis.

We can write $A = A_s + A_{s*}$ in Equation (6.8) and write the integral as $\int_{A_s} y\, dA + \int_{A_{s*}} y\, dA = 0$ to obtain the following condition: $Q_z + Q_{z*} = 0$. The equation implies that Q_z and Q_{z*} will have the same magnitude but opposite signs. Thus if we used Q_z or Q_{z*} in Equation (6.23), we would get the same magnitude of the shear stress, but

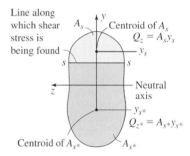

Figure 6.60 Calculation of Q_z.

which would give the correct sign (direction)?[23] The answer is that both will give the correct sign provided the following point is remembered: the s direction in Equation (6.23) is measured from the free surface used in the calculation of Q_z.

We can find the magnitude and the direction of the bending shear stress in two ways:

1. Use Equation (6.23) to find the magnitude of the shear stress. Use the rules described in Section 6.6.1 to determine the direction of the shear stress.

2. Follow the sign convention described in Section 6.2.5 for determining the shear force V_y. The shear stress is found from Equation (6.23) and the direction of the shear stress is determined using the subscripts, as was elaborated in Section 1.2.1.

6.6.4 Shear Flow Formula

The formula for shear flow can be obtained by substituting Equation (6.23) into Equation (6.19) to get

$$q = -\frac{V_y Q_z}{I_{zz}} \qquad (6.24)$$

Equation (6.24) can be used in two ways. It can be used for finding the magnitude of the shear flow at a point, and the direction of shear flow can be found by inspection following the rules described in Section 6.6.1. Alternatively, the sign convention for the shear force V_y is followed and the shear flow is determined from Equation (6.24). A positive value of shear flow implies that the flow is in the positive s direction, where s is measured from the free surface used in the calculation of Q_z.

One of the applications of Equation (6.24) is the determination of the spacing between mechanical fasteners used in holding strips of two or more beams together. Nails or screws are examples of mechanical fasteners that are used in wooden beams. Nuts and bolts or rivets are examples of mechanical fasteners used in metal beams. Figure 6.61 shows two strips of beams held together by a row of mechanical fasteners. Suppose the fasteners are spaced at intervals Δs, and each fastener can support a shear force V_F. Then the row of fasteners can support an average shear force per unit length of $V_F / \Delta s$, which can be approximated as the shear flow in the beam, that is, $q \approx V_F / \Delta s$.

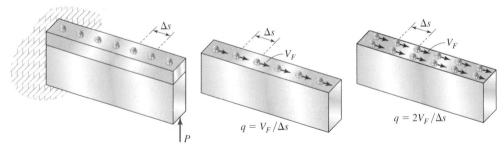

Figure 6.61 Spacing in mechanically fastened beams.

---

[23]In many mechanics of materials textbooks the shear stress formula gives only the magnitude of the shear stress correctly. The correct sign (direction) has to be found by inspection. In this book inspection as well as subscripts in the formulas will be used for determining the correct direction of shear stress.

Thus once we know the shear flow from Equation (6.24), we can find the spacing in a row of fasteners as $\Delta s \approx V_F/q$. If there is more than one row of fasteners holding two pieces of wood together, then each row of fasteners can carry an average shear flow of $V_F/\Delta s$, and thus the total shear flow carried by two rows is $2V_F/\Delta s$, which is then approximated by the shear flow in the beam, that is, $q = 2V_F/\Delta s$. Thus once we know the shear flow from Equation (6.24), we can use it to determine either the spacing between the fasteners if the shear force that the fasteners can support is known, or if the spacing is known, then we can find the shear force carried by each fastener. Example 6.17 further elaborates on this discussion.

6.6.5   Bending Stresses and Strains

In symmetric bending about the z axis, the significant stress components in Cartesian coordinates are σ_{xx} and τ_{xy} in the web and σ_{xx} and τ_{xz} in the flange. We can find σ_{xx} from Equation (6.12), but from Equation (6.23) we get τ_{xs}. How do we get τ_{xy} or τ_{xz} from τ_{xs}? There are two alternatives.

1. Follow the sign convention for the shear force, given by Definition 3, in determining V_y. Using Equation (6.23), get τ_{xs}. Note that the positive s direction is from the free surface to the point where the shear stress is found. Draw the stress cube using the argument of subscripts as described in Section 1.2.1. Now look at the shear stress in the Cartesian coordinates and determine the direction and sign of the stress component (τ_{xy} or τ_{xz}).

2. Use Equation (6.23) for finding the magnitude of τ_{xs}, and determine the direction of the shear stress by inspection, as described in Section 6.6.1. Draw the stress cube. Now look at the shear stress in the Cartesian coordinates and determine the direction and sign of the stress component (τ_{xy} or τ_{xz}).

In beam bending problems there are four possible stress elements, as shown in Figure 6.62. At the top and bottom surfaces of the beam the bending shear stress τ_{xy} is zero and the bending normal stress σ_{xx} is maximum at a cross section. The state of stress at the top and bottom is shown on in Figure 6.62a. No arrows are shown in the figures as the normal stress could be tensile or compressive. At the neutral axis σ_{xx} is zero and τ_{xy} is maximum in a cross section, as shown by the stress element in Figure 6.62b. At any point on the web σ_{xx} and τ_{xy} are nonzero, whereas at any point in the flange σ_{xx} and τ_{xz} are nonzero, as shown in Figure 6.62c and d.

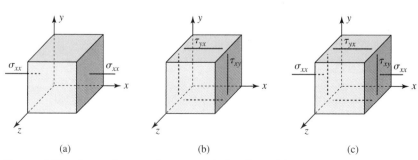
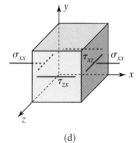

(a)   (b)   (c)   (d)

Figure 6.62   Stress elements in symmetric bending of beams. (a) Top or bottom. (b) Neutral axis. (c) Any point on web. (d) Any point on flange.

From the generalized Hooke's law given by Equations (3.12a) through (3.12f) we obtain the strains as follows:

$$\varepsilon_{xx} = \frac{\sigma_{xx}}{E} \qquad \varepsilon_{yy} = -\frac{\nu\sigma_{xx}}{E} = -\nu\varepsilon_{xx} \qquad \varepsilon_{zz} = -\frac{\nu\sigma_{xx}}{E} = -\nu\varepsilon_{xx}$$

$$\gamma_{xy} = \frac{\tau_{xy}}{G} \qquad \gamma_{xz} = \frac{\tau_{xz}}{G}$$

(6.25)

The normal strains in the y and z directions are due to the Poisson effect.

Consolidate your knowledge   With book closed, derive Equation (6.23).

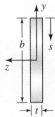

Figure 6.63   Cross section in Example 6.14.

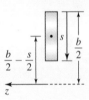

Figure 6.64   Calculation of Q_z in Example 6.14.

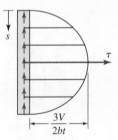

Figure 6.65   Shear stress distribution in Example 6.14.

EXAMPLE 6.14

A positive shear force V acts on the thin rectangular cross section shown in Figure 6.63. Determine the shear stress τ_{xs} due to bending about the z axis as a function of s and sketch it.

PLAN

We can find Q_z by taking the first moment of the area between the top surface and the surface located at an arbitrary point s. By substituting Q_z as a function of s in Equation (6.23), we can obtain τ_{xs} as a function of s.

Solution   We can draw the area A_s between the top surface and some arbitrary location s in Figure 6.64 and determine the first moment about the z axis to find Q_z,

$$Q_z = st\left(\frac{b}{2} - \frac{s}{2}\right) \tag{E1}$$

Substituting Equation (E1) and the area moment of inertia $I_{zz} = tb^3/12$ into Equation (6.23), we obtain

ANS. $$\tau_{xs} = -\frac{Vst\left(\dfrac{b}{2} - \dfrac{s}{2}\right)}{\left(\dfrac{tb^3}{12}\right)t} = \frac{-6Vs(b-s)}{b^3t} \tag{E2}$$

Noting that we obtain a negative sign for the shear stress, the direction of the shear stress has to be in the negative s direction on a surface with a normal in the positive x direction, as shown in Figure 6.65.

COMMENTS

1. Figure 6.65 shows that the shear stress is zero at the top ($s = 0$) and the bottom ($s = b$) and is maximum at the neutral axis, as expected. The maximum bending stress at a cross section can be written as $\tau_{max} = 1.5V/A$, where A is the cross-sectional area.

2. The shear force is in the positive y direction. Hence the shear stress on the cross section should be in the positive direction, as shown in Figure 6.65.

3. We note that the s direction is in the negative y direction. Hence $\tau_{xy} = -\tau_{xs}$, which is confirmed by the direction of shear stress in Figure 6.65.

4. Substituting $\tau_{xy} = -\tau_{xs} = 6Vs(b-s)/b^3t$ into Equation (6.9) and noting that $dA = t \, ds$, we obtain by integration

$$V_y = \int_0^b \frac{6Vs(b-s)}{b^3t} t \, ds = \frac{6V}{b^3}\left(\frac{bs^2}{2} - \frac{s^3}{3}\right)\Bigg|_0^b = V$$

which once more confirms our results.

EXAMPLE 6.15

A positive shear force $V_y = 30$ N acts on the thin cross sections shown in Figure 6.66 (not drawn to scale). Determine the shear flow along the centerlines and sketch it.

PLAN

V_y and I_{zz} are known in Equation (6.24). Hence the shear flow along the centerline will be determined if Q_z is determined along the centerline.

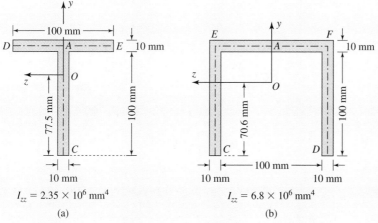

Figure 6.66 Cross sections in Example 6.15.

Noting that the cross section is symmetric about the y axis, the shear
flow needs to be found only on one side of the y axis.

Solution

(a) Figure 6.67 shows the areas A_s that can be used for finding the shear
flows in DA and CA of the cross section in Figure 6.66a. The parameters s_1 and s_2 are defined from the free surface to the point where
the shear flow is to be found. The distance from the centroid of the
areas A_s to the z axis can be found and Q_z calculated for each case:

$$Q_1 = s_1 \times 0.01(0.105 - 0.0775) = 0.275s_1 \times 10^{-3} \text{ m}^3 \qquad \text{(E1)}$$

$$Q_2 = s_2 \times 0.01[-(0.0775 - s_2/2)] = -(0.775s_2 - 5s_2^2)10^{-3} \text{ m}^3$$
$$\text{(E2)}$$

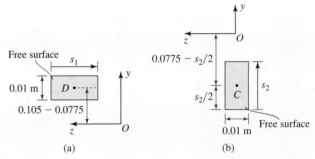

Figure 6.67 Calculation of Q_z in part (a) of Example 6.15.

Substituting V_y, I_{zz}, and Equations (E1) and into Equation (6.24),
the shear flow in DA and CA of the cross section in Figure 6.66a
can be found:

$$q_1 = -\frac{30 \times 0.275s_1 \times 10^{-3}}{2.35 \times 10^{-6}} \text{ N/m} = -3.51s_1 \text{ kN/m} \qquad \text{(E3)}$$

$$q_2 = -\frac{30[-(0.775s_2 - 5s_2^2)]10^{-3}}{2.35 \times 10^{-6}} \text{ N/m}$$

$$= 9.89s_2 - 63.83s_2^2 \text{ kN/m} \qquad \text{(E4)}$$

The shear flow q_1 is negative, implying that the direction of the
flow is opposite to the direction of s_1. The values of q_1 can be calculated from Equation (E3) and plotted as shown in Figure 6.68a.
By symmetry the flow in AE can also be plotted. The values of q_2

are positive between C and A, implying the flow is in the direction of s_2. The values of q_2 can be calculated from Equation (E4) and plotted as shown in Figure 6.68a.

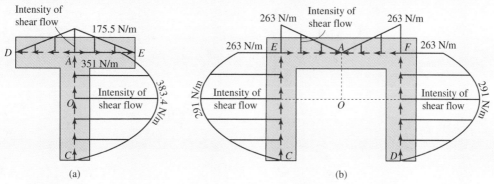

Figure 6.68 Shear flows on cross sections in Example 6.15.

(b) Figure 6.69 shows the areas A_s that can be used for finding the shear flows in CE and EA of the cross section in Figure 6.66b. In CE the parameter s can vary between points C and E. The same parameter s can be used for EA, but now its value is restricted between points E and A. The distance from the centroid of the areas A_s to the z axis can be found and Q_z calculated for each case.

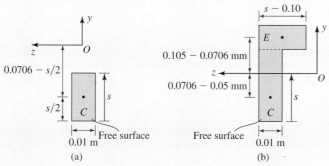

Figure 6.69 Calculation of Q_z in part (b) of Example 6.15.

For $0 \le s \le 0.10$, that is, between C and E,

$$Q_1 = s \times 0.01[-(0.0706 - s/2)] = -(0.706s - 5s^2)10^{-3} \text{ m}^3 \quad (E5)$$

For $0.10 \leq s \leq 0.16$, that is, between E and F,

$$Q_2 = 0.01 \times 0.1[-(0.0706 - 0.05)]$$
$$+ (s - 0.10)0.01(0.105 - 0.0706)$$
$$= (0.344s - 0.055)10^{-3} \text{ m}^3 \tag{E6}$$

Substituting V_y, I_{zz}, and Equations (E5) and (E6) into Equation (6.24), the shear flows in CE and EA of the cross section shown in Figure 6.66b can be found:

$$q_1 = -\frac{30[-(0.706s - 5s^2)]10^{-3}}{2.35 \times 10^{-6}} \text{ N/m}$$
$$= 9.016s - 63.83s^2 \text{ kN/m} \tag{E7}$$

$$q_2 = -\frac{30(0.344s - 0.055)10^{-3}}{2.35 \times 10^{-6}} \text{ N/m}$$
$$= 0.702 - 4.388s \text{ kN/m} \tag{E8}$$

In Equations (E7) and (E8) the shear flows q_1 and q_2 are positive, implying that the direction of the flows is in the direction of positive s. The values of q_1 and q_2 can be calculated from Equation (E7) and plotted as shown in Figure 6.68a. Using symmetry, the flows in DF and FA can also be plotted as shown in Figure 6.68a.

COMMENTS

1. In Example 6.13 the direction of flow was determined by inspection, whereas in this example it was determined using the formulas. A comparison of Figures 6.68 and 6.57a and b shows the same results. Thus inspection could be used as a check on the result of the formulas. Alternatively, the formulas could be used for determining the magnitude of the shear flow (stress) and inspection to determine the direction of the shear flow.

2. In Figure 6.68a the flow value at point A in CA is 351 N/m, which is the sum of the flows in AD and AE. In Figure 6.68b the flows in EA and FA approach A from opposite directions, and hence the flow at A is zero. Also notice that the flow at E in CE is the same as in EA in Figure 6.68b. Thus the behavior of shear flow is similar to that of fluid flow in a channel.

3. Figure 6.68 shows that the shear flow in the flanges varies linearly, whereas the shear flow in the web varies quadratically and its maximum value is at the neutral axis.

EXAMPLE 6.16

A beam is loaded as shown in Figure 6.70. The cross section of the beam is shown on the right and has an area moment of inertia $I_{zz} = 40.83$ in⁴.

(a) Determine the maximum bending normal and shear stresses.

(b) Determine the bending normal and shear stresses at point D on a section just to the right of support A. Point D is just below the flange.

(c) Show the results of parts (a) and (b) on stress cubes.

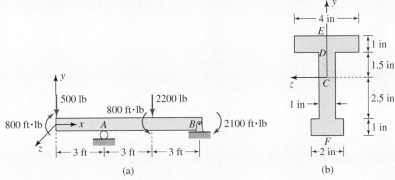

Figure 6.70 Beam and loading in Example 6.16.

PLAN

We can draw the shear force and bending moment diagrams and determine the maximum bending moment M_{max}, the maximum shear force $(V_y)_{max}$, and the value of the bending moment M_A and the shear force $(V_y)_A$ just to the right of support A. Using Equations (6.12) and (6.23) we can determine the required stresses and show the results on a stress cube.

Solution By considering the free-body diagram of the entire beam we can find the reaction forces at A and B and draw the shear force and bending moment diagrams in Figure 6.71. The areas under the shear force curve are

$$A_1 = 500 \times 3 = 1500$$

$$A_2 = 1000 \times 3 = 3000$$

$$A_3 = 1200 \times 3 = 3600$$

From the diagrams we can find the maximum shear force and moment, as well as the values of shear force and moment just to the right of support A,

$$(V_y)_{max} = 1200 \text{ lb} \tag{E1}$$

$$M_{max} = 2300 \text{ ft} \cdot \text{lb} \tag{E2}$$

$$(V_y)_A = -1000 \text{ lb} \tag{E3}$$

$$M_A = -700 \text{ ft} \cdot \text{lb} \tag{E4}$$

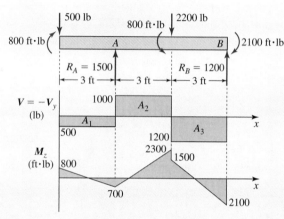

Figure 6.71 Shear force and bending moment diagrams in Example 6.16.

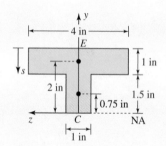

Figure 6.72 Calculation of Q_z at neutral axis in Example 6.16.

(a) Point F is the point farthest away from the neutral axis. Hence the maximum bending normal stress will occur at point F. Substituting $y_F = -3.5$ and Equation (E2) into Equation (6.12), we obtain

$$\sigma_{max} = -\frac{2300 \times 12(-3.5)}{40.83} = 2365.7 \tag{E5}$$

ANS. $\sigma_{max} = 2366 \text{ psi (T)}$

The maximum bending shear stress will occur at the neutral axis in the section where V_y is maximum. We can draw the area A_s between the top surface and the neutral axis (NA) (Figure 6.72) and determine the first moment about the z axis to find Q_z,

$$Q_{NA} = 4 \times 1 \times 2 + 1.5 \times 1 \times 0.75 = 9.125 \text{ in}^3 \tag{E6}$$

Substituting Equations (E1) and (E6) into Equation (6.23), we obtain

$$(\tau_{xs})_{max} = -\frac{1200 \times 9.125}{40.83 \times 1} = -268.2 \tag{E7}$$

ANS. $(\tau_{xs})_{max} = -268 \text{ psi}$

(b) Substituting $y_D = 1.5$ in and Equation (E4) into Equation (6.12), we obtain the value of the normal stress at point D on a section just to the right of A,

$$(\sigma_{xx})_D = -\frac{-700 \times 12 \times 1.5}{40.83} = 308.6 \qquad \text{(E8)}$$

ANS. $(\sigma_{xx})_D = 309$ psi (T)

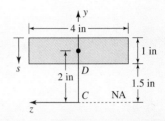

Figure 6.73 Calculation of Q_z at D in Example 6.16.

We can draw the area A_s between the free surface at the top and point D (Figure 6.73) and find Q_z at D,

$$Q_D = 4 \times 1 \times 2 = 8 \text{ in}^3 \qquad \text{(E9)}$$

Substituting Equations (E3) and (E9) into Equation (6.23), we obtain the value of the shear stress at point D on a section just to the right of A,

$$(\tau_{xs})_D = -\frac{-1000 \times 8}{40.83 \times 1} = 196 \qquad \text{(E10)}$$

ANS. $(\tau_{xs})_D = 196$ psi

(c) In Figures 6.72 and 6.73 the coordinate s is in the opposite direction to y at points D and the neutral axis. Hence at both these points $\tau_{xy} = -\tau_{xs}$ and from Equations (E7) and (E10), we obtain

$$(\tau_{xy})_{max} = 268 \text{ psi} \qquad (\tau_{xy})_D = -196 \text{ psi}$$

We can show these results along with the normal stress values in Equations (E1) and (E5) on the stress elements in Figure 6.74.

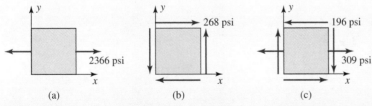

(a)   (b)   (c)

Figure 6.74 Stress elements in Example 6.16. (*a*) Maximum bending normal stress. (*b*) Maximum bending shear stress. (*c*) Bending and normal shear stresses at point D just to the right of A.

COMMENTS

1. The maximum value of V is -1200 kN, but $V = -V_y$. Hence the maximum value of V_y is a positive value, as given in Equation (E1). Similarly the sign is accounted for in Equation (E3).

2. V_y is positive in Equation (E1), thus we expect $(\tau_{xy})_{max}$ to be positive. Just after support A the shear force V_y is negative, thus we expect that $(\tau_{xy})_D$ will be negative, as shown in Figure 6.74.

3. Note that the maximum bending shear stress in the beam given by Equation (E7) is nearly an order of magnitude smaller than the maximum bending normal stress given by Equation (E5), which is consistent with the requirement for validity of our beam theory, as was remarked in Section 6.2.5. If in some problem the maximum bending shear stress were nearly the same as the maximum bending normal stress, then that would indicate that the assumptions of beam theory are not valid and the theory needs to be modified to account for shear stress.

EXAMPLE 6.17

A wooden cantilever box beam is to be constructed by nailing four pieces of lumber in one of the two ways shown in Figure 6.75. The allowable bending normal and shear stresses in the wood are 750 psi and 150 psi, respectively. The maximum force that the nails can support is 100 lb. Determine the maximum value of load P, the spacing of the nails to the nearest half inch, and the preferred nailing method.

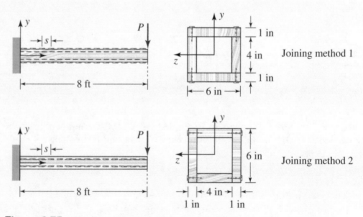

Figure 6.75 Wooden beams in Example 6.17.

PLAN

The maximum bending normal and shear stresses for both beams are the same and can be found in terms of P. These maximum values can be compared to the allowable stress values, and the limiting value on force P can be found. The shear flow at the junction of the wood pieces can be found using Equation (6.24). The spacing of the nails for each joining method can be found by dividing the allowable force in the nail by the shear flow. The method that gives the greater spacing between the nails is better as fewer nails will be needed.

Solution We can draw the shear force and bending moment diagrams for the beams as shown in Figure 6.76 and calculate the maximum shear force and movement,

$$(V_y)_{max} = -P \text{ lb} \tag{E1}$$

$$M_{max} = -8\,P \text{ ft·lb} = -96\,P \text{ in·lb} \tag{E2}$$

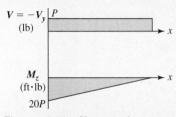

Figure 6.76 Shear and moment diagrams in Example 6.17.

The area moment of inertia about the z axis is

$$I_{zz} = \frac{1}{12}6 \times 6^3 - \frac{1}{12}4 \times 4^3 = 86.67 \text{ in}^4 \tag{E3}$$

Substituting Equations (E2) and (E3) and $y_{max} = \pm 3$ in, we can find the magnitude of the maximum bending normal stress from Equation (6.12) in terms of P. Using the allowable normal stress as 750 psi, we can obtain one limiting value on P,

$$\sigma_{max} = \left| \frac{M_{max} y_{max}}{I_{zz}} \right| = \frac{96P \times 3}{86.67} = 3.323P \le 750 \quad \text{or} \quad P \le 225.7 \text{ lb} \tag{E4}$$

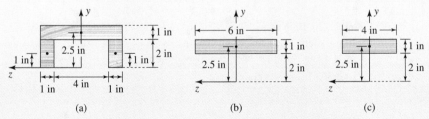

(a)   (b)   (c)

Figure 6.77 Calculation of Q_z in Example 6.17. (a) Q_z at neutral axis. (b) Q_z at nails in joining method 1. (c) Q_z at nails in joining method 2.

Figure 6.77a shows the calculation of Q_z at the neutral axis,

$$Q_{NA} = 6 \times 1 \times 2.5 + 2(2 \times 1 \times 1) = 19 \text{ in}^3 \tag{E5}$$

The area A_s is between the top surface and the neutral axis. We also note that at the neutral axis, the thickness perpendicular to the centerline is $t = 1 + 1 = 2$ in. Substituting Equations (E1), (E3), and (E5) and $t = 2$ into Equation (6.23), we can obtain the magnitude of the bending shear stress in terms of P. Using the allowable shear stress as 150 psi, we can obtain another limiting value on P,

$$\tau_{max} = \left| \frac{P \times 19}{86.67 \times 2} \right| = 0.1096 P \le 150 \qquad \text{or} \qquad P \le 1368 \text{ lb} \quad \text{(E6)}$$

Comparing Equations (E4) and (E6), we determine the maximum value of force,

$$P_{max} = 225.7 \text{ lb}$$

To find the shear flow on the surface joined by the nails, we make imaginary cuts through the nails and draw the area A_s, as shown in Figure 6.77b and c, and find Q_z for each joining method,

$$Q_1 = 6 \times 1 \times 2.5 = 15 \text{ in}^3 \quad \text{(E7)}$$

$$Q_2 = 4 \times 1 \times 2.5 = 10 \text{ in}^3 \quad \text{(E8)}$$

From Equations (E1) and (E4) we obtain the shear force as $V_y = -225.7$ lb. Substituting this value along with Equations (E3), (E7), and (E8) into Equation (6.24), we obtain the magnitude of the shear flow for each joining method,

$$q_1 = \left| \frac{V_y Q_1}{I_{zz}} \right| = \left| \frac{225.7 \times 15}{86.67} \right| = 39.06 \text{ lb/in} \quad \text{(E9)}$$

$$q_2 = \left| \frac{V_y Q_2}{I_{zz}} \right| = \left| \frac{225.7 \times 10}{86.67} \right| = 26.04 \text{ lb/in} \quad \text{(E10)}$$

This shear flow is to be carried by two rows of nails for each of the joining methods. Thus each row resists half of the flow. Using this fact, we can find the spacing between the nails,

$$\frac{100}{\Delta s_1} = \frac{q_1}{2} = \frac{39.06}{2} \qquad \text{or} \qquad \Delta s_1 = \frac{2 \times 100}{39.06} = 5.1 \text{ in} \quad \text{(E11)}$$

$$\frac{100}{\Delta s_2} = \frac{q_2}{2} = \frac{26.04}{2} \qquad \text{or} \qquad \Delta s_2 = \frac{2 \times 100}{26.04} = 7.7 \text{ in} \quad \text{(E12)}$$

As $\Delta s_2 > \Delta s_1$, fewer nails will be used in joining method 2. Rounding downward to the nearest half inch, the answer is:

ANS. Use joining method 2 with a nail spacing of 7.5 in.

COMMENTS

1. In this particular example only the magnitudes of the stresses were important, the sign did not play any role. This shall not always be the case, particularly in later chapters when we consider combined loading and stresses on different planes.

2. From visualizing the imaginary cut surface of the nails we observe that the shear stress component in the nails is τ_{yx} in joining method 1 and τ_{zx} in joining method 2.

3. In Section 6.6.1 it was observed that the shear stresses in bending develop to balance the changes in axial force due to σ_{xx}. The shear stresses in the nails balance σ_{xx}, which acts on a greater area in joining method 1 (6 in wide) than in joining method 2 (4 in wide). This is reflected in the higher value of Q_z, which led to a higher value of shear flow for joining method 1 than for joining method 2, as shown by Equations (E9) and (E10).

4. The observations in comment 3 are valid as long as σ_{xx} is the same for both joining methods at any location. If I_{zz} and y_{max} were different, then it is possible to arrive at a different answer. See Problem 6.104.

PROBLEM SET 6.4

Bending normal and shear stresses

In Problems 6.88 through 6.94, assume a positive shear force V_y. (a) Sketch the direction of the shear flow along the centerline on the thin cross sections shown. (b) At points A, B, C, and D, determine whether the stress component is τ_{xy} or τ_{xz} and whether it is positive or negative.

6.88 Using Figure P6.88 solve the problem as described in text above.

6.89 Using Figure P6.89 solve the problem as described in text above.

6.90 Using Figure P6.90 solve the problem as described in text above.

6.91 Using Figure P6.91 solve the problem as described in text above.

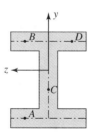

Figure P6.88

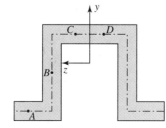

Figure P6.89

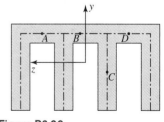

Figure P6.90

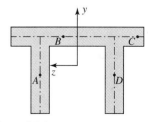

Figure P6.91

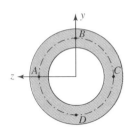

Figure P6.92

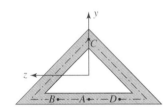

Figure P6.93

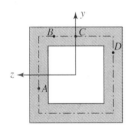

Figure P6.94

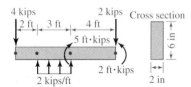

Figure P6.95

6.92 Using Figure P6.92 solve the problem as described in text above.

6.93 Using Figure P6.93 solve the problem as described in text above.

6.94 Using Figure P6.94 solve the problem as described in text above.

6.95 Determine the magnitude of the maximum bending normal stress and bending shear stress in the beam shown in Figure P6.95.

6.96 For the beam, loading, and cross section shown in Figure P6.96, determine: (a) the magnitude of the maximum bending normal stress, and shear stress; (b) the bending normal stress and the bending shear stress at point A. Point A is just below the flange on the cross section just right of the 4 kN force. Show your result on a stress cube. The area moment of inertia for the beam was calculated to be $I_{zz} = 3.6 \times 10^6$ mm^4.

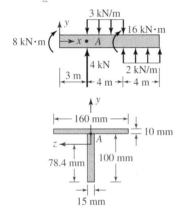

Figure P6.96

6.97 For the beam, loading, and cross section shown in Figure P6.97, determine: (a) the magnitude of the maximum bending normal stress and shear stress; (b) the bending normal stress and the bending shear stress at point A. Point A is on the cross section 2 m from the right end. Show your result on a stress cube. The area moment of inertia for the beam was calculated to be $I_{zz} = 453 \times 10^6$ mm^4.

6.98 Two pieces of lumber are nailed together as shown in Figure P6.98. The nails are uniformly spaced 10 in apart along the length. Determine the average shear force in each nail in segments AB and BC.

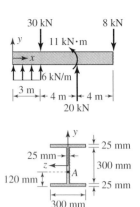

Figure P6.97

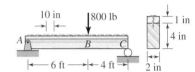

Figure P6.98

6.99 A cantilever beam is constructed by nailing three pieces of lumber, as shown in Figure P6.99. The nails are uniformly spaced at intervals of 75 mm. Determine the average shear force in each nail.

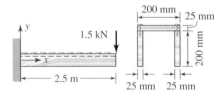

Figure P6.99

6.100 A cantilever beam is constructed by nailing three pieces of lumber, as shown in Figure P6.100. The nails are uniformly spaced at intervals of 75 mm. (a) Determine the shear force in each nail. (b) Which is the better nailing method, the one shown in Problem 6.99 or the one in this problem?

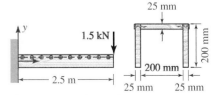

Figure P6.100

Design problems

6.101 The planks in a park bench are made from recycled plastic and are bolted to concrete supports, as shown in Figure P6.101. For the purpose of design the front plank is modeled as a simply supported beam that carries all the weight of two individuals. Assume that each person has a mass 100 kg and the weight acts at one-third the length of the plank, as shown. The allowable bending normal stress for the recycled plastic is 10 MPa and allowable bending shear stress is 2 MPa. The width d of the planks that can be manufactured is in increments of 2 cm, from 12 to 20 cm. To design the lightest bench, determine the corresponding values of the thickness t to the closest centimeter for the various values of d.

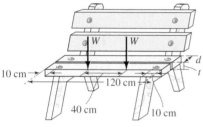

Figure P6.101

6.102 Two pieces of wood are glued together to form a beam, as shown in Figure P6.102. The allowable bending normal and shear stresses in wood are 3 ksi and 1 ksi, respectively. The allowable bending normal and shear stresses in the glue are 600 psi (T) and 250 psi, respectively. Determine the maximum moment M_{ext} that can be applied to the beam.

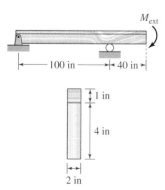

Figure P6.102

6.103 A wooden cantilever beam is to be constructed by nailing two pieces of lumber together, as shown in Figure P6.103. The allowable bending normal and shear stresses in the wood are 7 MPa and 1.5 MPa, respectively. The maximum force that the nail can support is 300 N. Determine the maximum value of load P to the nearest Newton and the spacing of the nails to the nearest centimeter.

6.104 A wooden cantilever box beam is to be constructed by nailing four 1-in × 6-in pieces of lumber in one of the two ways shown in Figure P6.104. The allowable bending normal and shear stresses in the wood are 750 psi and 150 psi, respectively. The maximum force that a nail can support is 100 lb. Determine the maximum value of load P to the nearest pound, the spacing of the nails to the nearest half inch, and the preferred nailing method.

Joining method 1

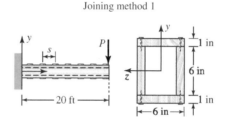

Joining method 2

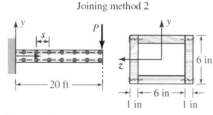

Figure P6.104

Historical problems[24]

6.105 Leonardo da Vinci conducted experiments on simply supported beams and drew the following conclusion: "If a beam 2 braccia long (L) supports 100 libbre (W), a beam 1 braccia long ($L/2$) will support 200 libbre ($2W$). As many times as the shorter length is contained in the longer (L/α), so many times more weight (αW) will it support than the longer one."

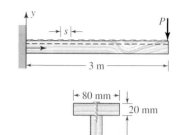

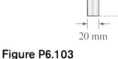

Figure P6.103

[24]See Section 6.9 for additional details.

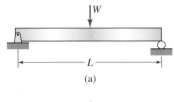

(a)

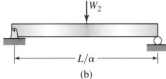

(b)

Figure P6.105

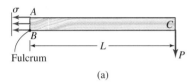

(a)

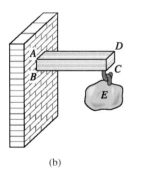

(b)

Figure P6.106 Galileo's beam experiment.

Prove this statement to be true by considering the two simply supported beams in Figure P6.105 and showing that $W_2 = \alpha W$ for the same allowable bending normal stress.

6.106 Galileo believed that the cantilever beam shown in Figure P6.106 would break at point B, which he considered to be a fulcrum point of a lever, with AB and BC as the two arms. He believed that the material resistance (stress) was uniform across the cross section. Show that the stress value σ that Galileo obtained from Figure P6.106 is three times larger than the bending normal stress predicted by Equation (6.12).

6.107 Galileo concluded that the bending moment due to the beam's weight increases as the square of the length at the built-in end of a cantilever beam. Show that Galileo's statement is correct by deriving the bending moment at the built-in end in the cantilever beam in terms of specific weight γ, cross-sectional area A, and beam length L.

6.108 In the simply supported beam shown in Figure P6.108. Galileo determined that the bending moment is maximum at the applied load and its value is proportional to the product ab. He then concluded that to break the beam with the smallest load P, the load should be placed in the middle. Prove Galileo's conclusions by drawing the shear force and bending moment diagrams and finding the value of the maximum bending moment in terms of P, a, and b. Then show that this value is largest when $a = b$.

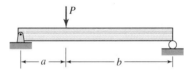

Figure P6.108

6.109 Mariotte, in an attempt to correct Galileo's strength prediction, hypothesized that the stress varied in proportion to the distance from the fulcrum, point B in Figure P6.106, that is, it varied linearly from point B. Show that the maximum bending stress value obtained by Mariotte is twice that predicted by Equation (6.12).

Stretch yourself

6.110 A beam is acted upon by a distributed load $p(x)$. Let M_A and V_A represent the internal bending moment and the shear force at A. Show that the internal moment at B is given by

$$M_B = M_A - V_A(x_B - x_A)$$

$$+ \int_{x_A}^{x_B} (x_B - x)p(x)\, dx \quad (6.26)$$

6.111 The displacement in the x direction in a beam cross section is given by $u = u_0(x) - y(dv/dx)(x)$. Assuming small strains and linear, elastic, isotropic, homogeneous material with no inelastic strains, show that

$$N = EA\frac{du_0}{dx} - EAy_c\frac{d^2v}{dx^2}$$

$$M_z = -EAy_c\frac{du_0}{dx} + EI_{zz}\frac{d^2v}{dx^2}$$

where y_c is the y coordinate of the centroid of the cross section measured from some arbitrary origin, A is the cross-sectional area, I_{zz} is the area moment of inertia about the z axis, and N and M_z are the internal axial force and the internal bending moment. Note that if y is measured from the centroid of the cross section, that is, $y_c = 0$, then the axial and bending problems decouple. In such a case show that $\sigma_{xx} = N/A - M_zy/I_{zz}$.

6.112 Show that the bending normal stresses in a homogeneous, linearly elastic, isotropic symmetric beam subject to a temperature change $\Delta T(x, y)$ is given by the equation

$$\sigma_{xx} = -\frac{M_zy}{I_{zz}} + \frac{M_Ty}{I_{zz}} - E\alpha\Delta T(x, y) \quad (6.27)$$

where $M_T = E\alpha\int_A y\Delta T(x, y)\, dA$, α is the coefficient of thermal expansion, and E is the modulus of elasticity.

6.113 In unsymmetrical bending of beams, under the assumption of plane sections remaining plane and perpendicular to the beam axis, the displacement u in the x direction can be shown to be given by $u = -y\, dv/dx - z\, dw/dx$, where y and z are measured from the centroid of the cross section, and v and w are the deflections of the beam in the y and z directions, respectively. Assume small strain, a linear, elastic, isotropic, homogeneous material, and no inelastic strain. Using Equations (1.5b) and (1.5c), show that

$$M_z = EI_{zz}\frac{d^2v}{dx^2} + EI_{yz}\frac{d^2w}{dx^2}$$

$$M_y = EI_{yz}\frac{d^2v}{dx^2} + EI_{yy}\frac{d^2w}{dx^2}$$ (6.28)

and

$$\sigma_{xx} = -\left(\frac{I_{yy}M_z - I_{yz}M_y}{I_{yy}I_{zz} - I_{yz}^2}\right)y - \left(\frac{I_{zz}M_y - I_{yz}M_z}{I_{yy}I_{zz} - I_{yz}^2}\right)z$$ (6.29)

Note that if either y or z is a plane of symmetry, then $I_{yz} = 0$. From Equation (6.28) this implies that the moment about the z axis causes deformation in the y direction only and the moment about the y axis causes deformation in the z direction only. In other words, the bending problems about the y and z axes are decoupled.

6.114 The equation $\partial\sigma_{xx}/\partial x + \partial\tau_{yx}/\partial y = 0$ was derived in Problem 1.98. Into this equation, substitute Equations (6.12) and (6.14) and integrate with y to show that

$$\tau_{yx} = \frac{6V_y(b^2/4 - y^2)}{b^3 t}$$

See Figure P6.114.

Computer problems

6.115 A cantilever, hollow circular aluminum beam of 5-ft length is to support a load of 1200 lb. The inner radius of the beam is 1 in. If the maximum bending normal stress is to be limited to 10 ksi, determine the minimum outer radius of the beam to nearest $\frac{1}{16}$ in.

6.116 Table P6.116 shows the values of the distributed loads at several points along the axis of the rectangular beam shown in Figure P6.116. Determine the maximum bending normal and shear stresses in the beam.

TABLE P6.116 Data for Problem 6.116

x (ft)	$p(x)$ (lb/ft)
0	275
1	348
2	398
3	426
4	432
5	416
6	377
7	316
8	233
9	128
10	0

6.117 Let the distributed load $p(x)$ in Problem 6.116 be represented by the equation $p(x) = a + bx + cx^2$. Using the data in Table P6.116, determine the constants a, b, and c by the least-squares method. Then find the maximum bending moment and the maximum shear force by analytical integration and determine the maximum bending normal and shear stresses.

Figure P6.114

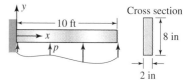

Figure P6.116

Beam cross sections constructed from more than one material are referred to as composite beams. A wooden beam stiffened with steel plates, iron bars inserted in concrete, or a laminated structure of composite material[25] are examples of composite beams. Clearly, Assumption 8 of material homogeneity across the cross section is no longer valid. Hence the formulas that follow Assumption 8 are no longer valid, but as

---

[25]See Section 3.12.3.

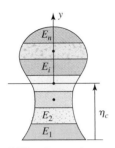

Figure 6.78 Composite cross section.

we shall see, after accounting for the cross-sectional nonhomogeneity, the derivation process to get the new formulas is the same as before.

Figure 6.78 shows a laminated composite cross section. We assume that all material cross sections are symmetric about the y axis and the loading is still in the plane of symmetry. We assume that the materials are securely bonded to each other, that is, there is no relative movement at any interface. This assumption of secure bonding is necessary to preserve our kinematic assumptions, Assumptions 1 through 3. We further assume that we still have small strains, that all materials are linear, elastic, isotropic, and that there are no inelastic strains. In other words, Assumptions 1 through 7 are still valid. Thus Equations (6.7) and (6.10), which were developed before the use of the assumption of material homogeneity, are still valid and form our starting point for deriving the formulas for composite beams. Equation (6.7) will yield the location of the origin (neutral axis), whereas Equation (6.10) can be developed to give the flexure formulas.

6.7.1 Normal Bending Stress in Composite Beams

Consider the laminated cross section shown in Figure 6.78. Each material has a modulus of elasticity E_i, which is constant over the material cross-sectional area A_i. Suppose there are n materials in the cross section. The integral in Equation (6.10) can be written as the sum of the integrals over all materials,

$$M_z = \frac{d^2v}{dx^2}\left(\int_{A_1} E_1 y^2 \, dA + \int_{A_2} E_2 y^2 \, dA + \cdots + \int_{A_n} E_n y^2 \, dA\right)$$

E_i is a constant in each integral and can be taken outside the integral. The remaining integral is the second area moment of inertia $(I_{zz})_i$ of the ith material. We thus obtain

$$M_z = \frac{d^2v}{dx^2}[E_1(I_{zz})_1 + E_2(I_{zz})_2 + \cdots + E_n(I_{zz})_n]$$

Written more compactly,

$$M_z = \frac{d^2v}{dx^2}\left[\sum_{j=1}^{n} E_j(I_{zz})_j\right] \tag{6.30}$$

Equation (6.30) shows that the bending rigidity of the composite beam is the sum of the bending rigidities of all materials. Substituting Equation (6.30) into Equation (6.6) we obtain

$$\sigma_{xx} = -EyM_z \Big/ \left[\sum_{j=1}^{n} E_j(I_{zz})_j\right] \tag{6.31}$$

where E in the numerator is the modulus of elasticity at the point at which the stress σ_{xx} is being evaluated. We can rewrite Equation (6.31) for the ith material as

$$(\sigma_{xx})_i = -E_i y M_z \Big/ \left[\sum_{j=1}^{n} E_j(I_{zz})_j\right] \tag{6.32}$$

where $(\sigma_{xx})_i$ is the bending normal stress in the ith material. Equations (6.30) and (6.28),[26] which are applicable to composite beams, now replace Equations (6.11) and (6.12) for the homogeneous cross section. The internal bending moment M_z represents the statically equivalent moment over the entire cross section, as it did for the homogeneous cross section. The analysis techniques for finding the internal bending moment M_z at a cross section remain the same as before, that is, one may use either the equilibrium method or the shear and moment diagrams to find M_z.

From Equations (6.30) and (6.5b) it is clear that the bending normal *strain* still varies linearly across the composite cross section. But the normal stress in each material $(\sigma_{xx})_i$, given by Equation (6.32), is a piecewise linear function, which changes it slope with each material point at which the normal stress is evaluated, as was shown in Example 6.4.

6.7.2 Location of Neutral Axis in Composite Beams

In the discussion of the location of the origin for homogeneous beams it was observed that we choose the origin such that the total axial force on a cross section is zero, which resulted in decoupling the axial problem from the bending problem. The location of the origin for a nonhomogeneous section is determined from Equation (6.7), which is identical to Equation (4.24). Hence the result in Example 4.11, which gives the location of the origin for multiple materials, is valid here also and is rewritten for convenience,

$$\eta_c = \left(\sum_{j=1}^{n} \eta_j E_j A_j \right) \bigg/ \left(\sum_{j=1}^{n} E_j A_j \right) \qquad (6.33)$$

where η_j is the location of the centroid of the jth material as measured from a common datum line.

6.7.3 Bending Shear Stress in Composite Beams

Equation (6.20), which relates shear and normal stresses in a beam, was derived from the equilibrium of a beam element and is independent of the material model. It is assumed that the kinematic assumptions remain valid, as is the case for our composite beams. Substituting Equation (6.31) into Equation (6.20) and noting that the moment and the bending rigidity do not change across the cross section, we obtain

$$\tau_{sx} t = -\frac{d}{dx} \int_{A_s} \frac{-E y M_z}{\sum_{j=1}^{n} E_j (I_{zz})_j} \, dA = \frac{d}{dx} \left[\frac{M_z}{\sum_{j=1}^{n} E_j (I_{zz})_j} \int_{A_s} E y \, dA \right] = \frac{d}{dx} \left[\frac{M_z Q_{comp}}{\sum_{j=1}^{n} E_j (I_{zz})_j} \right]$$

$$(6.34)$$

where

$$Q_{comp} = \int_{A_s} E y \, dA \qquad (6.35)$$

---

[26]If we consider a homogeneous material, then $E_1 = E_2 = \cdots = E_i = \cdots = E_n = E$. Substituting this into Equations (6.30) and (6.32), we obtain Equations (6.11) and (6.12) for homogeneous material, as expected.

We assume here that the beam is not tapered and the modulus of elasticity does not change with x, which implies that Q_{comp} and the bending rigidity in Equation (6.34) do not change with x and can be taken outside the derivative sign. Then, using Equation (6.14), we obtain

$$\tau_{sx}t = \left[\frac{Q_{comp}}{\sum_{j=1}^{n}E_j(I_{zz})_j}\right]\frac{dM_z}{dx} = -\left[\frac{Q_{comp}V_y}{\sum_{j=1}^{n}E_j(I_{zz})_j}\right]$$

or

$$\tau_{sx} = \tau_{xs} = -\frac{Q_{comp}V_y}{\left[\sum_{j=1}^{n}E_j(I_{zz})_j\right]t} \tag{6.36}$$

In Equation (6.36) Q_{comp} is the new quantity that we need to evaluate. Let n_s represent the number of materials in A_s, that is, n_s denotes the number of materials between the free surface and where we are evaluating the shear stress. The integral in Equation (6.35) can be written as the sum of integrals over all materials,

$$Q_{comp} = \int_{A_1} E_1 y\, dA + \int_{A_2} E_2 y\, dA + \cdots + \int_{A_{n_s}} E_{n_s} y\, dA$$

where E_j is a constant in each integral and can be taken outside the integral. The remaining integral, designated as $(Q_z)_j$, is the first area moment of the jth material about the neutral axis. Thus we obtain

$$Q_{comp} = E_1(Q_z)_1 + E_2(Q_z)_2 + \cdots + E_{n_s}(Q_z)_{n_s}$$

or, written more compactly,

$$Q_{comp} = \sum_{j=1}^{n_s} E_j(Q_z)_j \tag{6.37}$$

Here $(Q_z)_j$ is a geometric property independent of the material and can be found as was discussed in Section 6.6.3. Then Equation (6.37) can be used to find Q_{comp}. If $n_s = n$, that is, if we consider the entire cross section, then from Equation (6.7), $Q_{comp} = 0$. Thus, once more, like Q_z for homogeneous materials, Q_{comp} starts and ends with a zero value as we move from one free surface and reach the other free surface and is maximum at the neutral axis. Thus the bending shear stress for composite beams is maximum on a cross section at the neutral axis.

6.7.4 Cross-Section Transformation Method

For composite beam cross sections made from materials with rectangular cross sections an alternative approach is possible, which often looks simpler, particularly if the number of materials involved is small. The process of solving beam problems by the cross section transformation method is described first and justified immediately afterward.

The dimensions of the cross section in the z direction (bending axis) are transformed using the equation $\tilde{z} = z(E_j/E_{ref})$, where E_{ref} is the modulus of elasticity of

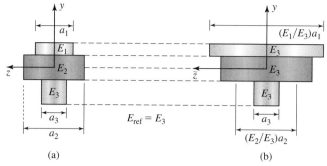

Figure 6.79   Transformation of cross section. (*a*) Original composite cross section. (*b*) Transformed homogeneous cross section.

any material on the cross section that is used as the reference material, and $\tilde{z}$ is the transformed dimension. Figure 6.79 shows a cross section made from three materials. Using material 3 as the reference material, the dimensions in the z direction are transformed using the ratio of the modulus of elasticity of the material to the reference material. The transformed cross section is considered homogeneous with the modulus of elasticity of the reference material. The centroid, the second area moment of inertia $\tilde{I}_{zz}$, and the first area moment $\tilde{Q}_z$ are found for the transformed cross section like for any homogeneous material. The following formulas are then used for finding the bending normal and shear stresses:

$$(\sigma_{xx})_i = -\left(\frac{E_i}{E_{ref}}\right)\frac{M_z y}{\tilde{I}_{zz}} \tag{6.38}$$

$$\tau_{sx} = \tau_{xs} = -\left(\frac{V_y \tilde{Q}_z}{\tilde{I}_{zz} t}\right) \tag{6.39}$$

where $\tilde{I}_{zz}$ and $\tilde{Q}_z$ are the second area moment of inertia and the first area moment about the axis through the centroid of the transformed homogeneous cross section, and t is the thickness perpendicular to the centerline of the original composite cross section. Note that the location of the centroid for the original composite cross section and the transformed homogeneous cross section will be the same, as the dimensions in the y direction are not being changed.

To justify this approach, we note that the differential area element can be written as $dA = dy\,dz$, which will now transform to $dA = (E_{ref}/E_j)dy\,d\tilde{z} = (E_{ref}/E_j)\,d\tilde{A}$, where $d\tilde{A}$ is the differential area element in the transformed cross section. Thus the second moment of inertia transforms as

$$(I_{zz})_j = \int_{A_j} y^2\,dA = \frac{E_{ref}}{E_j}\int_{\tilde{A}_j} y^2\,d\tilde{A} \quad \text{or} \quad (I_{zz})_j = \frac{E_{ref}}{E_j}(\tilde{I}_{zz})_j$$

The total bending rigidity thus can be written as

$$\sum_{j=1}^{n} E_j(I_{zz})_j = \sum_{j=1}^{n} E_j\left(\frac{E_{ref}}{E_j}\right)(\tilde{I}_{zz})_j = E_{ref}\sum_{j=1}^{n}(\tilde{I}_{zz})_j = E_{ref}\tilde{I}_{zz} \tag{6.40}$$

Substituting into Equation (6.32), we obtain Equation (6.38). Similarly the first moment of the area will transform as shown,

$$(Q_z)_j = \int_{A_j} y \, dA = \frac{E_{ref}}{E_j} \int_{\tilde{A}_j} y \, d\tilde{A} = \frac{E_{ref}}{E_j} (\tilde{Q}_z)_j$$

Substituting into Equation (6.37), we obtain

$$Q_{comp} = \sum_{j=1}^{n_s} E_j \left(\frac{E_{ref}}{E_j} \right) (\tilde{Q}_z)_j = E_{ref} \sum_{j=1}^{n_s} (\tilde{Q}_z)_j = E_{ref} \tilde{Q}_z \qquad (6.41)$$

Substituting Equations (6.40) and (6.41) into Equation (6.36), we obtain Equation (6.39). The reason the dimension t in Equation (6.39) is not transformed is because if the point is in the flange, then the perpendicular direction to the centerline is the y direction, which is not being transformed. Had we restricted ourselves to points in the web, where the perpendicular direction to the centerline is the z direction and transformed the t dimension, then the shear stress formula in the ith material would be

$$(\tau_{xs})_i = -\frac{E_i}{E_{ref}} \frac{V_y \tilde{Q}_z}{\tilde{I}_{zz} \tilde{t}}$$

This formula reflects a similar modification as the normal stress formula and the dimension $\tilde{t}$ is the dimension perpendicular to the centerline in the transformed homogeneous cross section provided the point is in the web. We will use Equation (6.39), which is valid at all points.

The advantage of the method of this section is that the analysis process is the same as that for homogeneous cross sections. The disadvantages are that it will become tedious if either the number of materials in the cross section is large or the material cross sections are curvilinear instead of rectangles. Example 6.18 demonstrates this method along with the direct use of Equations (6.32) and (6.36).

EXAMPLE 6.18

An aluminum strip (E_{al} = 10,000 ksi) is securely fastened to a wooden beam (E_w = 1600 ksi) and loaded as shown in Figure 6.80.

(a) On a section just to the right of support A, determine and plot the bending normal and shear stresses across the cross section.

(b) Determine the maximum bending normal and shear stresses in aluminum and wood.

Method 1

PLAN

The beam and loading are the same as in Example 6.16, hence we know the values of shear force and moment just to the right of support A, as well as the maximum shear force and bending moment. Using Equation (6.33), the location of the neutral axis (origin) can be determined, I_{zz} for

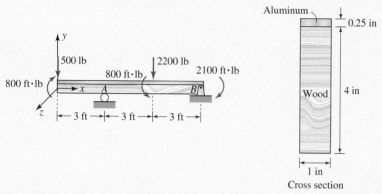

Figure 6.80 Beam and loading in Example 6.18.

each material found, the bending rigidity for the cross section calculated, and Q_{comp} as a function of y determined using Equation (6.37).

Solution

(a) Using Equations (6.32) and (6.36), the bending normal stress and the shear stress as a function of y can be found and plotted.

(b) The maximum bending normal stress in a material will exist at points that are farthest from the neutral axis in the material, whereas the maximum shear stress will exist on points in the material that are closest to the neutral axis.

From Equations (E1) through (E4) in Example 6.16 the following values of shear force and bending moment are known:

$$(V_y)_{\text{max}} = 1200 \text{ lb} \qquad (\text{E1})$$

$$M_{\text{max}} = 2300 \text{ ft·lb} \qquad (\text{E2})$$

$$(V_y)_A = -1000 \text{ lb} \qquad (\text{E3})$$

$$M_A = -700 \text{ ft·lb} \qquad (\text{E4})$$

Using Equation (6.33) and Figure 6.81, we can determine the location of the neutral axis (origin),

$$\eta_c = \frac{E_w A_w \eta_w + E_{\text{al}} A_{\text{al}} \eta_{\text{al}}}{E_w A_w + E_{\text{al}} A_{\text{al}}}$$

$$= \frac{1600 \times 4 \times 1 \times 2 + 10{,}000 \times 1 \times 0.25 \times 4.125}{1600 \times 4 \times 1 + 10{,}000 \times 1 \times 0.25}$$

$$= 2.597 \text{ in} \qquad (\text{E5})$$

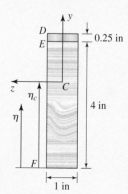

Figure 6.81 Location of origin in Example 6.18.

Using the parallel axis theorem we can find the area moment of inertia about the z axis,

$$(I_{zz})_{al} = \frac{1}{12}1 \times 0.25^3 + 0.25(2.597 - 4.125)^2 = 0.585 \text{ in}^4 \quad \text{(E6)}$$

$$(I_{zz})_w = \frac{1}{12}1 \times 4^3 + 4(2.597 - 2)^2 = 6.759 \text{ in}^4 \quad \text{(E7)}$$

We can find the total bending rigidity,

$$\sum E_j(I_{zz})_j = E_w(I_{zz})_w + E_{al}(I_{zz})_{al}$$

$$= 1600 \times 6.759 + 10,000 \times 0.585$$

$$= 16.664 \times 10^3 \text{ kips} \cdot \text{in}^2 \quad \text{(E8)}$$

Figure 6.82 shows the calculation of $(Q_{comp})_{al}$ using Equation (6.37). We consider the material between the top surface (free surface) and any point y in aluminum,

$$(Q_{comp})_{al} = E_{al}(Q_z)_{al}$$

$$= 10,000(1.653 - y)1\left(\frac{1.653 + y}{2}\right)$$

$$= 5000(1.653^2 - y^2) \quad \text{(E9)}$$

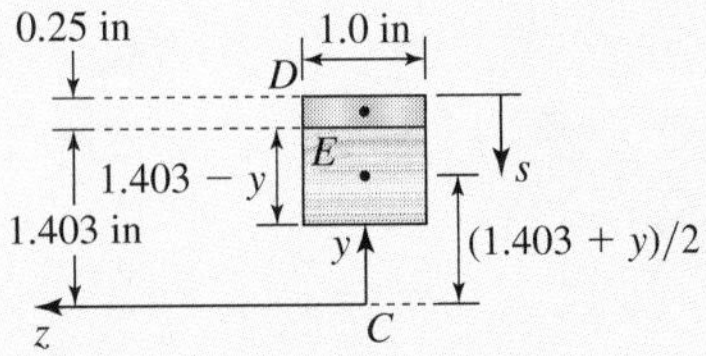

Figure 6.83 shows the calculation of $(Q_{comp})_w$. Once more we consider the material between the top surface and any point y in wood. From Equation (6.37) there are two terms, one over aluminum and the other over wood. The contribution from the aluminum can be found by substituting $y = 1.403$ in Equation (E9),

$$(Q_{comp})_w = [E_{al}(Q_z)_{al}]\big|_{y=1.403} + E_w(Q_z)_w$$

$$= 5000(1.653^2 - 1.403^2) + 1600(1.403 - y)1\left(\frac{1.403 + y}{2}\right)$$

$$= 3820 + 800(1.403^2 - y^2) \quad \text{(E10)}$$

(a) The bending normal stress as a function of y on a section just to the right of support A can be found by substituting Equations (E4) and (E8) into Equation (6.32) to obtain

$$(\sigma_{xx})_{al} = -E_{al}y\frac{M_A}{\sum E_j(I_{zz})_j} = -10,000y\frac{(-700)12}{16.664 \times 10^3}$$

$$= 5041y \text{ psi} \quad \text{(E11)}$$

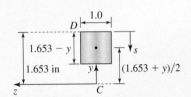

Figure 6.82 Calculation of Q_{comp} in aluminum.

Figure 6.83 Calculation of Q_{comp} in wood.

$$(\sigma_{xx})_w = -E_w y \frac{M_A}{\sum E_j(I_{zz})_j} = -1600y \frac{(-700)12}{16.664 \times 10^3}$$

$$= 806.5y \text{ psi} \tag{E12}$$

We can write Equations (E11) and (E12) as

$$\sigma_{xx} = \begin{cases} 5041y \text{ psi}, & -2.597 \le y < 1.403 \\ 806.5y \text{ psi}, & 1.403 < y \le 1.65 \end{cases} \tag{E13}$$

The bending shear stress τ_{xs} in each material, as a function of y on a section just to the right of support A, can be found by substituting Equations (E3), (E9), and (E10) into Equation (6.36). The directions y and s are opposite in Figures 6.82 and 6.83. Hence $\tau_{xs} = -\tau_{xy}$ in each material and can be found as

$$(\tau_{xs})_{al} = -(\tau_{xy})_{al} = -\frac{(Q_{comp})_{al}(V_y)_A}{\left[\sum E_j(I_{zz})_j\right]t}$$

$$= -\frac{5000(1.653^2 - y^2)(-1000)}{16.664 \times 10^3 \times 1}$$

$$= -300(1.653^2 - y^2) \tag{E14}$$

$$(\tau_{xy})_w = -(\tau_{xy})_w = -\frac{(Q_{comp})_w(V_y)_A}{\left[\sum E_j(I_{zz})_j\right]t}$$

$$= -\frac{[3820 + 800(1.403^2 - y^2)](-1000)}{16.664 \times 10^3 \times 1}$$

$$= -[229.2 + 48.01(1.403^2 - y^2)] \tag{E15}$$

We can write Equations (E14) and (E15) as

$$\tau_{xy} = \begin{cases} -300(1.653^2 - y^2) \text{ psi}, & -2.597 \le y < 1.403 \\ -[229.2 + 48.01(1.403^2 - y^2)] \text{ psi}, & 1.403 < y \le 1.653 \end{cases}$$

$$\tag{E16}$$

Equations (E13) and (E16) can be plotted as a function of y across the cross section, as shown in Figure 6.84.

(b) The maximum bending normal stress in aluminum will be at point D, and in wood it will be at point F as in each material these points are farthest from the neutral axis. Substituting Equations (E2) and (E8)

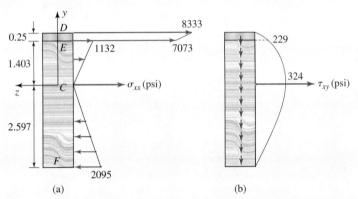

Figure 6.84 Stress distribution across cross section just to the right of support at A in Example 6.18.

and $y_D = 1.653$ and $y_F = -2.597$ into Equation (6.32), we obtain

$$(\sigma_{max})_{al} = -E_{al} y_D \frac{M_{max}}{\sum E_j (I_{zz})_j}$$

$$= -10,000 \times 1.653 \frac{2300 \times 12}{16.664 \times 10^3} = -27,378 \text{ psi}$$

ANS. $(\sigma_{max})_{al} = 27.4 \text{ ksi (C)}$

$$(\sigma_{max})_w = -E_w y_F \frac{M_{max}}{\sum E_j (I_{zz})_j}$$

$$= -1600(-2.597) \frac{2300 \times 12}{16.664 \times 10^3} = 6882 \text{ psi}$$

ANS. $(\sigma_{max})_w = 6.9 \text{ ksi (T)}$

The maximum bending shear stress in aluminum will be at point E, and in wood it will be at point C as in each material these points are closest to the neutral axis. Substituting Equations (E1), (E9), and (E10) and $y_E = 1.403$ and $y_C = 0$ into Equation (6.36), we obtain

$$(\tau_{max})_{al} = \left| \frac{(Q_{comp})_{al}|_{y=1.403} (V_y)_{max}}{\left[\sum E_j (I_{zz})_j \right] t} \right|$$

$$= \frac{5000(1.653^2 - 1.403^2)1200}{16.664 \times 10^3 \times 1}$$

ANS. $(\tau_{max})_{al} = 275 \text{ psi}$

$$(\tau_{max})_w = \left| \frac{(Q_{comp})_w\big|_{y=0}(V_y)_{max}}{\left[\sum E_j(I_{zz})_j\right]t} \right|$$

$$= -\frac{[3820 + 800(1.403^2 - 0^2)]1200}{16.664 \times 10^3 \times 1}$$

ANS. $(\tau_{max})_w = 388$ psi

COMMENTS

1. The internal forces and moments in Example 6.16 and in this example are the same, even though the cross section geometry and materials are very different. This once more emphasizes that the equilibrium equations that relate external forces to internal forces are independent of the material model.

2. The maximum bending shear stress in each material is an order of magnitude less than the maximum bending normal stress, which is consistent with the requirement for validity of our beam theory, as was remarked in Section 6.2.5. But as the difference between the moduli of elasticity E reaches orders of magnitude (as in sandwich beams[27]), the normal and shear stresses for the material with the smaller E become nearly of the same order and the theory is no longer valid.

3. Figure 6.84 shows that the bending normal stress σ_{xx} is discontinuous but the bending shear stress τ_{xy} is continuous at the aluminum–wood interface, even though τ_{xy} is represented by two different functions in Equation (E16). If we examine Figure 6.83 we note that the calculation of Q_{comp} in wood is equal to $(Q_{comp})_{al}$ at the interface plus the additional value from wood. The additional value of Q_{comp} from the wood will be zero at the interface, implying that Q_{comp} is a continuous function at the interface, even though it is represented by different functions in each material. As the rest of the terms in Equation (6.36) are the same for both materials, the continuity of shear stress at the interface reflects the continuity of Q_{comp}.

Method 2

PLAN

Using aluminum as the reference material, we note that the ratio $E_{wood}/E_{ref} = 0.16$. Thus the dimension of the wood parallel to the $\tilde{z}$ axis becomes $1 \times 0.16 = 0.16$ in. We can now find the centroid and $\tilde{I}_{zz}$ for the transformed cross section. We can also find $\tilde{Q}_z$ as a function of y.

---

[27]See Footnote 9 in Section 6.2.1.

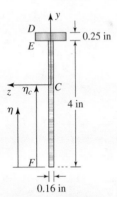

Figure 6.85 Transformed cross
section in Example 6.18.

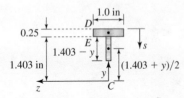

Figure 6.86 Calculation of $\tilde{Q}_z$ in
wood.

(a) Using Equations (6.34) and (6.39), the bending normal stress and
the shear stress as a function of y can be found and plotted.

(b) The maximum bending normal stress in a material will exist in a
material at points that are farthest from the neutral axis, whereas
the maximum shear stress will exist on points in the material that
are closest to the neutral axis.

Solution

(a) We can draw the transformed cross section using aluminum as
the reference material, as shown in Figure 6.85, and calculate the
centroid,

$$\eta_c = \frac{4 \times 0.16 \times 2 + 1 \times 0.25 \times 4.125}{4 \times 0.16 + 1 \times 0.25} = 2.597 \text{ in} \quad (E17)$$

The area moment of inertia $\tilde{I}_{zz}$ for the transformed cross section
can be found as

$$\tilde{I}_{zz} = \frac{1}{12}1 \times 0.25^3 + 0.25(2.597 - 4.125)^2 + \frac{1}{12}0.16 \times 4^3$$

$$+ 0.16 \times 4(2.597 - 2)^2$$

$$= 1.6664 \text{ in}^4 \quad (E18)$$

Substituting Equations (E4) and (E18) into Equation (6.38), we
obtain the normal stress in aluminum and wood as function of y,

$$(\sigma_{xx})_{al} = -\left(\frac{E_{al}}{E_{ref}}\right)\frac{M_A y}{\tilde{I}_{zz}} = -1y\frac{(-700)12}{1.6664} = 5041y \text{ psi} \quad (E19)$$

$$(\sigma_{xx})_w = -\left(\frac{E_w}{E_{ref}}\right)\frac{M_A y}{\tilde{I}_{zz}} = -0.16y\frac{(-700)12}{1.6664} = 806.5y \text{ psi} \quad (E20)$$

Equations (E19) and (E20) are the same as Equations (E11) and
(E12). Hence the variation of σ_{xx} across the cross section is given
by Equation (E13) and the plot is as given in Figure 6.84.
We can use Figure 6.82 to calculate $\tilde{Q}_z$ in aluminum as the
dimensions are unchanged in the transformed cross section,

$$\tilde{Q}_{al} = (1.653 - y)1\frac{1.653 + y}{2} = \frac{1.653^2 - y^2}{2} \quad (E21)$$

Figure 6.86 shows the calculation of $\tilde{Q}_z$ in wood. We consider the
material between the top surface and any point y in the wood. The
contribution from the aluminum can be found by substituting

$y = 1.403$ into Equation (E21),

$$\tilde{Q}_w = \frac{1.653^2 - 1.403^2}{2} + (1.403 - y)0.16\frac{1.403 + y}{2}$$

$$= 0.3820 + 0.08(1.403^2 - y^2) \qquad (E22)$$

The bending shear stress τ_{xs} in each material, as a function of y, on a section just to the right of support A can be found by substituting $t = 1$ and Equations (E3), (E21), and (E22) into Equation (6.39). The directions y and s are opposite in Figures 6.82 and 6.86. Hence $\tau_{xs} = -\tau_{xy}$ in each material and can be found as

$$(\tau_{xs})_{al} = -(\tau_{xy})_{al} = -\left(\frac{(V_y)_A \tilde{Q}_{al}}{\tilde{I}_{zz}t}\right)$$

$$= -\frac{-1000(1.653^2 - y^2)/2}{1.6664 \times 1} = -300(1.653^2 - y^2) \quad (E23)$$

$$(\tau_{xy})_w = -(\tau_{xy})_w = -\left(\frac{(V_y)_A \tilde{Q}_w}{\tilde{I}_{zz}t}\right)$$

$$= -\frac{-1000[0.3820 + 0.08(1.403^2 - y^2)]}{1.6664 \times 1}$$

$$= -[229.2 + 48.01(1.403^2 - y^2)] \qquad (E24)$$

Equations (E23) and (E24) are the same as Equations (E14) and (E15). Hence the plot of the shear stress is as given in Figure 6.84.

(b) The maximum bending normal stress in aluminum will be at point D and in wood it will be at point F as in each material these points are farthest from the neutral axis. Substituting Equations (E2) and (E18) and $y_D = 1.653$ and $y_F = -2.597$ into Equation (6.38), we obtain

$$(\sigma_{max})_{al} = -\left(\frac{E_{al}}{E_{ref}}\right)\frac{M_{max}y_D}{\tilde{I}_{zz}}$$

$$= -1\frac{2300 \times 12 \times 1.653}{1.6664} = -27{,}378 \text{ psi}$$

$$\text{ANS.} \qquad (\sigma_{max})_{al} = 27.4 \text{ ksi (C)}$$

$$(\sigma_{max})_w = -\left(\frac{E_w}{E_{ref}}\right)\frac{M_{max}y_F}{\tilde{I}_{zz}}$$

$$= -1.6\frac{2300 \times 12(-2.597)}{1.6664} = 6882 \text{ psi}$$

$$\text{ANS.} \qquad (\sigma_{max})_w = 6.9 \text{ ksi (T)}$$

The maximum bending shear stress in aluminum will be at point E and in wood it will be at point C as in each material these points are closest to the neutral axis. Substituting Equations (E1), (E21), and (E22) and $y_E = 1.403$ and $y_C = 0$ into Equation (6.39), we obtain

$$(\tau_{max})_{al} = \left| \frac{(V_y)_{max} \tilde{Q}_{al}}{\tilde{I}_{zz} t} \right| = \frac{1200(1.653^2 - 1.403^2)/2}{1.6664 \times 1}$$

ANS. $(\tau_{max})_{al} = 275$ psi

$$(\tau_{max})_w = \left| \frac{(V_y)_{max} \tilde{Q}_w}{\tilde{I}_{zz} t} \right| = \frac{1200[0.3820 + 0.08(1.403^2 - 0^2)]}{1.6664 \times 1}$$

ANS. $(\tau_{max})_w = 388$ psi

COMMENTS

1. As in Method 1, the calculation of internal forces and moments is independent of the material nonhomogeneity across the cross section.

2. The chief advantage of Method 2 are the simplified equations. The calculation of $\tilde{I}_{zz}$ in Equation (E18) is simpler than computing the total bending rigidity $\sum E_j (\tilde{I}_{zz})_j$ in Equation (E8) of Method 1. Similarly finding $\tilde{Q}_{al}$ and $\tilde{Q}_w$ in Equations (E21) and (E22) is simpler than the calculation of $(Q_{comp})_{al}$ and $(Q_{comp})_w$ in Equations (E9) and (E10) in Method 1. But as mentioned earlier, Method 2 cannot be used for nonrectangular cross sections.

EXAMPLE 6.19

In reinforced concrete beams, the tension-carrying capacity of the concrete is ignored because of low tensile strength and unpredictability to cracking on the tension side. We incorporate this information to solve the following problem.

A reinforced concrete beam is constructed by embedding circular steel rods of 20-mm diameter, as shown in Figure 6.87. The maximum resisting (internal) moment in the beam was determined as 40 kN·m. Determine the maximum bending normal stress in the concrete and steel. Use the moduli of elasticity for steel and concrete as $E_s = 200$ GPa and $E_c = 28$ GPa, respectively.

PLAN

We can use a modulus of elasticity of zero to simulate a zero tensile capacity of concrete. In other words, we view the cross section as if it were made from three materials: material 1 would be concrete from the top to the neutral axis with a modulus of elasticity of 28 GPa, material 2 would be concrete below the neutral axis with a zero modulus of elasticity, and material 3 would be steel with a modulus of elasticity of 200 GPa. Assuming that the neutral axis is η_c from the top, an equation in η_c can be written using Equation (6.33) and solved. Once the location of the neutral axis is determined, the maximum bending normal stress in each material can be determined using Equation (6.32).

Solution   We assume that the neutral axis is η_c from the top, as shown in Figure 6.88. The area of concrete in compression is $A_1 = 0.15\eta_c$. The total area of steel is $A_3 = 3\pi \times 0.01^2 = 0.942 \times 10^{-3}$ m². Substituting these values and $E_1 = E_c = 28$ GPa, $E_2 = 0$, and $E_3 = E_s = 200$ GPa into Equation (6.33), we obtain a quadratic equation in η_c,

$$
\begin{aligned}
\eta_c &= \frac{E_1 A_1 \eta_1 + E_2 A_2 \eta_2 + E_3 A_3 \eta_3}{E_1 A_1 + E_2 A_2 + E_3 A_3} \\[2mm]
&= \frac{28 \times 0.15\,\eta_c(\eta_c/2) + 0 + 200 \times 0.942 \times 10^{-3} \times 0.35}{28 \times 0.15\,\eta_c + 0 + 200 \times 0.942 \times 10^{-3}}
\end{aligned}
$$

or

$$
\eta_c^2 + 0.0897\,\eta_c - 0.0314 = 0 \tag{E1}
$$

Solving for the positive root in the quadratic Equation (E1), we obtain

$$
\eta_c = 0.1379 \text{ m} \tag{E2}
$$

Using the parallel-axis theorem, we can find the area moments of inertia about the z axis,

$$
\begin{aligned}
(I_{zz})_1 &= \frac{1}{12}0.15 \times 0.1379^3 + 0.15 \times 0.1379\left(\frac{0.1379}{2}\right)^2 \\[2mm]
&= 131.1 \times 10^{-6} \text{ m}^4 \tag{E3}
\end{aligned}
$$

$$
\begin{aligned}
(I_{zz})_3 &= 3\left[\frac{\pi}{4}0.01^4 + \pi \times 0.01^2(0.35 - 0.1379)^2\right] \\[2mm]
&= 42.42 \times 10^{-6} \text{ m}^4 \tag{E4}
\end{aligned}
$$

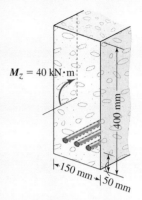

Figure 6.87 Reinforced concrete beam cross section.

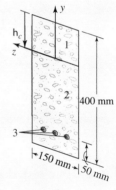

Figure 6.88 Neutral axis calculation in Example 6.19.

We can find the total bending rigidity as

$$\sum E_j(I_{zz})_j = E_1(I_{zz})_1 + E_2(I_{zz})_2 + E_3(I_{zz})_3$$

$$= 28 \times 10^9 \times 131.1 \times 10^{-6} + 0 + 200 \times 10^9 \times 42.42 \times 10^{-6}$$

$$= 12.15 \times 10^6 \ \text{N·m}^2 \tag{E5}$$

The maximum compressive bending stress in concrete will occur at the top surface. Substituting $y = \eta_c = 0.1379$ m, $M_z = 40$ kN·m, $E_1 = 28$ GPa, and Equation (E5) into Equation (6.32), we obtain

$$(\sigma_{max})_c = -E_1 y \frac{M_z}{\sum E_j(I_{zz})_j}$$

$$= -28 \times 10^9 \times 0.1379 \frac{40 \times 10^3}{12.15 \times 10^6} = -12.7 \times 10^6 \ \text{N/m}^2 \tag{E6}$$

ANS. $(\sigma_{max})_c = 12.7$ MPa (C)

The maximum tensile bending stress in steel will occur at the lowest point on the circular rod. Substituting $y = -(0.35 + 0.01 - \eta_c) = -0.2221$ m, $M_z = 40$ kN·m, $E_3 = 200$ GPa, and Equation (E5) into Equation (6.32), we obtain

$$(\sigma_{max})_s = -E_3 y \frac{M_z}{\sum E_j(I_{zz})_j} = -200 \times 10^9 (-0.2221) \frac{40 \times 10^3}{12.15 \times 10^6}$$

$$= 146.2 \times 10^6 \ \text{N/m}^2 \tag{E7}$$

ANS. $(\sigma_{max})_s = 146.2$ MPa (T)

COMMENT

If we analyze this problem assuming the cross section is made from two materials, then the location of the neutral axis from Equation (6.33) can be found as $\eta_c = 0.2153$ m. The corresponding bending rigidity will be $\sum E_j(I_{zz})_j = 26.2 \times 10^6$ N·m^2. The maximum normal stress in concrete in compression will be $(\sigma_{max})_c = 9.42$ MPa (C) and the maximum normal stress in steel will be $(\sigma_{max})_s = 44.2$ MPa (T). Both these stress values are smaller than those obtained in Equations (E6) and (E7). Thus by neglecting the concrete capacity to support stresses in tension we make a conservative approximation in our design.

*6.8 ELASTIC–PERFECTLY PLASTIC BEAMS

Figure 6.89 shows the elastic–perfectly plastic material stress–strain curve. It is assumed that the material has the same behavior in tension and in compression.

Recall the following from Section 3.11.1:

1. Before yield stress, the material stress–strain relationship is represented by Hooke's law, and after yield stress, the stress is assumed to be constant.

2. To determine the strain (deformation) in the horizontal portion AB of the curve, we have to use the requirement that deformation be continuous.

3. Unloading (elastic recovery) from a point in the plastic region is along line BC, which is parallel to the linear portion of the stress–strain curve OA shown in Figure 6.89.

We make use of these observations in our analysis that follows. Consider a symmetric cross section of a beam that is not only symmetric about the y axis but also about the z axis, as shown in Figure 6.90a. Points A and B will reach yield stress simultaneously at the same load. As the load increases, the plastic zone will move inward in a symmetric manner, and the neutral axis will continue to be at the centroid of the cross section. Let the location of the elastic–plastic boundary be given by the distance a from the neutral axis. Assuming we know a, we can write the stress distribution as a function of y and use Equation (6.1) to find the equivalent internal moment

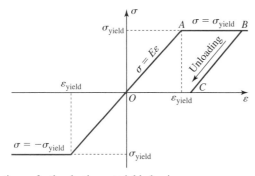

Figure 6.89 Elastic–perfectly plastic material behavior.

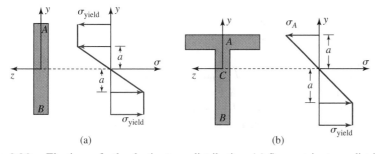

Figure 6.90 Elastic–perfectly plastic stress distribution. (a) Symmetric stress distribution. (b) Asymmetric stress distribution.

at the cross section. Alternatively, if the internal moment is known and we need to find the location of the elastic–plastic boundary, then we can write the stress distribution in terms of a, calculate the internal moment in terms of a, and equate it to the given moment value and find a.

Now consider a cross section that is not symmetric about the z axis, such as shown in Figure 6.90b. Point B is farther away than point A is from the elastic neutral axis at C. Thus point B will reach yield stress before point A as the loads are increased. The plastic zone will start from point B and move inward. In order to satisfy Equation (6.2), that is, the total axial force on the cross section is zero, the neutral axis will shift from the centroid. So with cross sections that are unsymmetrical with respect to the z axis we will need to find the location of the neutral axis also. Once more we will start by writing the stress distribution in terms of a across the cross section. We will first determine the location of the neutral axis in terms of a by using Equation (6.2). Once we know the location of the neutral axis, the equivalent internal moment can be determined using Equation (6.1).

Definition 6 The moment at which the maximum bending normal stress just reaches the yield stress is called *elastic moment* and will be designated by M_e.

Definition 7 The internal moment for which the entire cross section becomes fully plastic is called *plastic moment* and will be designated by M_p.

Definition 8 The ratio of the plastic moment to the elastic moment is called the *shape factor* for the cross section and will be designated by f,

$$f = \frac{M_p}{M_e} \tag{6.42}$$

As the name suggests, the shape factor depends only on the shape of the cross section and is independent of the material from which the beam is made. When the cross section becomes fully plastic, the beam can no longer support any load at the cross section and the corresponding moment represents the collapse moment. For an elastic design the shape factor thus reflects the margin of safety.

EXAMPLE 6.20
The normal strain at a cross section in a beam due to bending about the z axis was found to vary as $\varepsilon_{xx} = -0.0125y$, with y measured in meters (Figure 6.91). Assume the beam is made from elastic–perfectly plastic material that has a yield stress $\sigma_{yield} = 250$ MPa in tension and compression, and a modulus of elasticity $E = 200$ GPa. Determine the equivalent internal moment.

PLAN
For the given strain and material model, the normal stress σ_{xx} as a function of y was found in Example 3.16. Substituting this stress distribution into Equation (6.1) and integrating across the cross section, we can obtain the equivalent internal moment M_z.

Solution The normal stress σ_{xx} as a function of y that was found in Example 3.16 is rewritten here for convenience and plotted in Figure 6.92,

$$\sigma_{xx} = \begin{cases} -250 \text{ MPa}, & 0.1 \leq y \leq 0.3 \\ -2500y \text{ MPa}, & -0.1 \leq y \leq 0.1 \\ 250 \text{ MPa}, & -0.3 \leq y \leq -0.1 \end{cases} \qquad \text{(E1)}$$

The internal moment M_z can be calculated by substituting Equation (E1) into Equation (6.1). The area dA changes as we move across the cross section, as shown in Figure 6.92. The integral in Equation (6.1) must account for changes in stress distribution as well as changes in the differential area. We also note that the stress distribution is symmetric. Thus we could calculate the moment from the top half and double it to obtain the total equivalent moment as shown,

$$M_z = -\int_A y\sigma_{xx} \, dA = -2\left(\int_0^{0.1} y\sigma_{xx} \, dA + \int_{0.1}^{0.25} y\sigma_{xx} \, dA + \int_{0.25}^{0.3} y\sigma_{xx} \, dA \right)$$

$$\text{(E2)}$$

Substituting for σ_{xx} and dA in Equation (E2), we obtain

$$M_z = -2 \times 10^6 \left[\int_0^{0.1} y(-125y) \, dy + \int_0^{0.25} y(-12.5) \, dy \right.$$

$$\left. + \int_{0.25}^{0.3} y(-62.5) \, dy \right]$$

$$= -2 \times 10^6 \left[(-125)\frac{y^3}{3} \Big|_0^{0.1} + (-12.5)\frac{y^2}{2} \Big|_{0.1}^{0.25} + (-62.5)\frac{y^2}{2} \Big|_{0.25}^{0.3} \right]$$

$$= 83.33 \times 10^3 + 656.25 \times 10^3 + 1718.75 \times 10^3$$

$$\text{ANS.} \qquad M_z = 2458 \text{ kN·m}$$

COMMENTS

1. The total internal axial force is zero because the tensile stress behavior above the centroid is symmetrically duplicated by the compressive stress behavior below the centroid.

2. The elastic–plastic material affects the stress distribution but Equation (6.1), which represents static equivalency, is unaffected.

3. The relationship of the internal moment and the external loads can be established by drawing the appropriate free-body diagram for a particular problem. The relationship of internal and external moments depends on the free-body diagram and is independent of the material model.

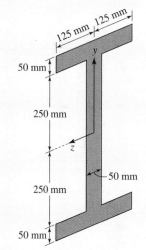

Figure 6.91 Beam cross section in Example 6.20.

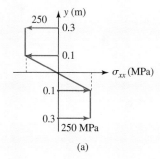

(a)

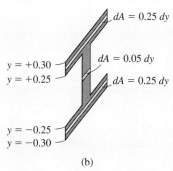

(b)

Figure 6.92 Variation of stress and differential area across cross section in Example 6.20.

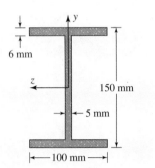

Figure 6.93 Cross section in Example 6.21.

EXAMPLE 6.21

The internal moment at a section of the beam shown in Figure 6.93 was found to be 21 kN·m. The beam material has a yield stress of 200 MPa. Determine the location of the elastic–plastic boundary.

PLAN

The neutral axis will pass through the centroid even as the plastic zone grows. Let the distance of the elastic–plastic boundary from the centroid be a. As a starting point we can assume the elastic–plastic boundary a to be located in the web, that is, we assume $a < 69$ mm. We can write the stress expression across the cross section in terms of a and the coordinate y. Substituting the stress expression into Equation (6.2) we find the internal moment M_z in terms of a and equate it to 21 kN to determine a. For integration we note that the area dA changes as we move from the flange to the web.

We draw the stress distribution in Figure 6.94, assuming the elastic–plastic boundary to be located in the web, and record the value of y at each point where either stress or area changes.

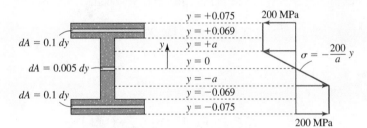

Figure 6.94 Stress distribution in Example 6.21.

The stress σ and dA in each interval can be written as follows:

$$\sigma = -200 \text{ MPa} \qquad dA = 0.100 \ dy \qquad 0.069 \leq y \leq 0.075$$

$$\sigma = -200 \text{ MPa} \qquad dA = 0.005 \ dy \qquad a \leq y \leq 0.069$$

$$\sigma = -\frac{200}{a}y \text{ MPa} \qquad dA = 0.005 \ dy \qquad -a \leq y \leq a \qquad \text{(E1)}$$

$$\sigma = 200 \text{ MPa} \qquad dA = 0.005 \ dy \qquad -0.069 \leq y \leq -a$$

$$\sigma = 200 \text{ MPa} \qquad dA = 0.100 \ dy \qquad -0.075 \leq y \leq -0.069$$

We note that the moment from the bottom half ($y < 0$) will be the same as that from the top half ($y > 0$). Thus we can find the moment from only the top half and double it. From Equation (6.1) we find the moment in terms of a,

$$M_z = -2 \times 10^6 \left[\int_0^a y \left(-\frac{200}{a} y \right) 0.005 \, dy + \int_a^{0.069} y(-200)0.005 \, dy \right.$$

$$\left. + \int_{0.069}^{0.075} y(-200)0.100 \, dy \right]$$

or

$$M_z = 400 \times 10^6 \left[\frac{0.005}{a} \left(\frac{y^3}{3} \right) \Big|_0^a + 0.005 \left(\frac{y^2}{2} \right) \Big|_a^{0.069} + 0.1 \left(\frac{y^2}{2} \right) \Big|_{0.069}^{0.075} \right]$$

$$= 22.04 \times 10^3 - \frac{a^2}{3} \times 10^6 \qquad \text{(E2)}$$

Equating the moment in Equation (E2) to 21 kN, we can find a,

$$22.04 \times 10^3 - \frac{a^2}{3} \times 10^6 = 21 \times 10^3 \qquad \text{or} \qquad a = 0.0558 \text{ m}$$

$$\text{ANS.} \qquad a = 55.8 \text{ mm}$$

COMMENT
Since $a < 69$ mm, our assumption about the location of the elastic–plastic boundary is correct.

EXAMPLE 6.22
Determine the shape factor for the beam cross section shown in Figure 6.95. The centroid of the cross section and the moment of inertia about the axis passing through the centroid are as given.

PLAN
Point B is farther from the neutral axis than point A, and hence the elastic stress will be maximum at point B. We can find the elastic moment by using Equation (6.12) and equating the normal stress to yield stress. To find the plastic moment we can assume that the neutral axis is at a distance a from the bottom. We can write the stress distribution in terms

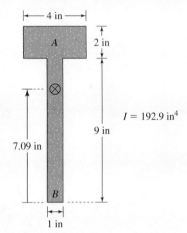

Figure 6.95 Cross section in Example 6.22.

of a and the coordinate y. Using Equation (6.2) we can find the value of a. Using Equation (6.1) we can find the plastic moment.

From Equation (6.12) we find the bending normal stress at point B and equate it to the yield stress to obtain the elastic moment,

$$\sigma_{yield} = -\frac{M_e(-7.09)}{192.9} \qquad \text{or} \qquad M_e = 27.21\,\sigma_{yield} \qquad (E1)$$

We draw the stress distribution in Figure 6.96, assuming the location of the neutral axis at a distance a from the bottom, and record the value of y at each point where either stress or area changes.

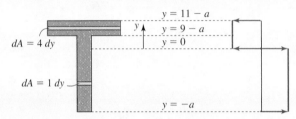

Figure 6.96 Stress distribution in Example 6.22.

The stress σ and dA in each interval can be written as

$$\sigma = -\sigma_{yield} \qquad dA = 4\,dy \qquad 9 - a \le y \le 11 - a$$
$$\sigma = -\sigma_{yield} \qquad dA = dy \qquad 0 \le y \le 9 - a \qquad (E2)$$
$$\sigma = \sigma_{yield} \qquad dA = dy \qquad -a \le y \le 0$$

Substituting Equation (E2) into Equation (6.2), we can find the value of a,

$$\int_{-a}^{0} \sigma_{yield}\,dy + \int_{0}^{9-a} -\sigma_{yield}\,dy + \int_{9-a}^{11-a} -\sigma_{yield}\,4\,dy$$
$$= \sigma_{yield}[a - (9 - a) - 4 \times 2] = 0$$

or

$$a = 8.5 \text{ in} \qquad (E3)$$

Substituting Equations (E2) and (E3) into Equation (6.1) we obtain

$$M_p = -\int_A y\sigma_{xx}\,dA$$
$$= -\left[\int_{-a}^{0} y(\sigma_{yield})\,dy + \int_{0}^{9-a} y(-\sigma_{yield})\,dy + \int_{9-a}^{11-a} y(-\sigma_{yield})4\,dy\right]$$

or

$$M_p = -\sigma_{\text{yield}}\left[\frac{y^2}{2}\Big|_{-8.5}^{0} - \left(\frac{y^2}{2}\right)\Big|_{0}^{0.5} - 4\left(\frac{y^2}{2}\right)\Big|_{0.5}^{2.5}\right] = 48.25\,\sigma_{\text{yield}} \quad (E4)$$

Substituting Equations (E1) and (E4) into Equation (6.42), the shape factor can be found,

$$f = \frac{M_p}{M_e} = \frac{48.25\,\sigma_{\text{yield}}}{27.21\,\sigma_{\text{yield}}}$$

ANS. $f = 1.77$

COMMENT
σ_{yield} in M_p and M_e cancel and the shape factor depends only upon the geometry of the cross section.

EXAMPLE 6.23
A beam of elastic–perfectly plastic material has a yield stress of 48 ksi. If point A on the beam cross section shown in Example 6.22 just reaches the yield stress, determine: (a) the location of the neutral axis; (b) the moment required to produce the given state of stress.

PLAN
We draw the stress distribution across the cross section, as shown in Figure 6.97, assuming that the neutral axis is in the web at a distance a from the top, and record the value of y at each point where either stress or area changes.

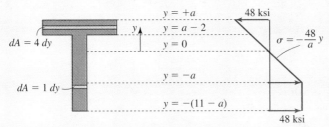

Figure 6.97 Stress distribution in Example 6.23.

We can write the stress expression and the value of dA for each interval,

$$\sigma = -\frac{48}{a}y \qquad dA = 4\,dy \qquad a - 2 \le y \le a$$

$$\sigma = -\frac{48}{a}y \qquad dA = dy \qquad -a \le y \le a - 2 \qquad \text{(E1)}$$

$$\sigma = 48 \text{ ksi} \qquad dA = dy \qquad -(11 - a) \le y \le -a$$

We can substitute Equation (E1) into Equation (6.2) and obtain the equation for determining the value of a,

$$\int_A \sigma_{xx}\,dA = \int_{-(11-a)}^{-a} 48\,dy + \int_{-a}^{a-2} -\frac{48}{a}y\,dy + \int_{a-2}^{a} -\frac{48}{a}y4\,dy = 0$$

or

$$48\left[y\Big|_{-(11-a)}^{-a} - \frac{y^2}{2a}\Big|_{-a}^{a-2} - 4\frac{y^2}{2a}\Big|_{a-2}^{a} \right] = 0 \quad \text{or} \quad 10a - 4a^2 + 12 = 0$$

$$\text{(E2)}$$

The two roots of the quadratic equation, Equation (E2) are $a = 3.386$ and $a = -0.886$. Only the positive root is admissible. Thus the location of neutral axis is

$$\text{ANS.} \qquad a = 3.386 \text{ in} \qquad \text{(E3)}$$

We can substitute Equations (E1) and (E3) into Equation (6.1) and obtain the equivalent internal moment,

$$M_z = -\left[\int_{-(11-a)}^{-a} y48\,dy + \int_{-a}^{a-2} y\left(-\frac{48}{a}y\right) dy + \int_{a-2}^{a} y\left(-\frac{48}{a}y\right)4\,dy \right]$$

or

$$M_z = -48\left[\frac{y^2}{2}\Big|_{-7.614}^{-3.386} - \frac{y^3}{3 \times 3.386}\Big|_{-3.386}^{1.386} - 4\frac{y^3}{3 \times 3.386}\Big|_{1.386}^{3.386} \right] = 1995.6$$

$$\text{ANS.} \qquad M_z = 1996 \text{ in}\cdot\text{kips}$$

*EXAMPLE 6.24

Resolve Example 6.20 assuming that the stress–strain relationship is given by the power law (see Section 3.11.3) $\sigma = 9000\varepsilon^{0.6}$ MPa in tension and compression.

PLAN

For the given strain and material model, the normal stress σ_{xx} as a function of y was found in Example 3.17. Substituting this stress distribution into Equation (6.1) and integrating across the cross section we can obtain the equivalent internal moment M_z.

Solution The normal stress σ_{xx} as a function of y that was found in Example 3.17 is rewritten here for convenience and plotted in Figure 6.98,

$$
\sigma_{xx} = \begin{cases} E\varepsilon_{xx}^{0.6}, & \varepsilon_{xx} \ge 0 \\ -E(-\varepsilon_{xx})^{0.6}, & \varepsilon_{xx} \le 0 \end{cases} \qquad \sigma_{xx} = \begin{cases} 649.2(-y)^{0.6}, & y \le 0 \\ -649.2y^{0.6}, & y \ge 0 \end{cases} \quad \text{(E1)}
$$

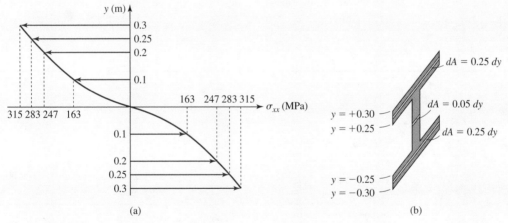

Figure 6.98 Variation of stress and differential area across the cross section in Example 6.24.

The internal moment M_z can be found by substituting the stress expressions of Equation (E1) into Equation (6.1). We note that the stresses are symmetric in tension and compression, and hence we can compute the moment from the top half and double it,

$$
M_z = -\int_A y\sigma_{xx}\, dA = -2\left(\int_0^{0.25} y\sigma_{xx}0.05\, dy + \int_{0.25}^{0.3} y\sigma_{xx}0.25\, dy \right)
$$

Substituting for σ_{xx},

$$M_z = -2\left[\int_0^{0.25} y(-649.2y^{0.6})0.05 \; dy + \int_{0.25}^{0.3} y(-649.2y^{0.6})0.25 \; dy\right]$$

or

$$M_z = -2\times 10^6\left[-32.46\frac{y^{2.6}}{2.6}\Big|_0^{0.25} + (-162.3)\frac{y^{2.6}}{2.6}\Big|_{0.25}^{0.3}\right]$$

$$= 679.3 \times 10^3 + 2059.8 \times 10^3$$

ANS. $M_z = 2739 \text{ kN·m}$

COMMENT
Examples 6.4, 6.20, and 6.24 show that changing the material model affects the stress distribution. But the strain distribution, which is a kinematic relationship, will be unaffected, the static equivalency equations, Equations (6.1) and (6.2), will be unaffected, and the equilibrium equations relating internal forces and moments are unaffected.

PROBLEM SET 6.5

Composite beams

In Problems 6.118 through 6.120, the cross section of a composite beam with a coordinate system that has an origin at C is shown. The normal strain at point A due to bending about the z axis and the modulus of elasticity of each material are as given in each problem. In solving the problems use the fact that ε_{xx} varies linearly with y.

(a) *Plot the stress distribution across the cross section.*

(b) *Determine the maximum bending normal stress in each material.*

(c) *Determine the equivalent internal bending moment M_z by using Equation (6.1).*

(d) *Determine the equivalent internal bending moment M_z by using Equation (6.31).*

6.118 $\varepsilon_{xx} = -200 \; \mu$, $E_1 = 8000$ ksi, $E_2 = 200$ ksi.

6.119 $\varepsilon_{xx} = -200 \; \mu$, $E_1 = 200$ GPa, $E_2 = 70$ GPa.

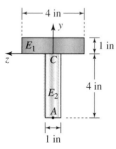

Figure P6.118

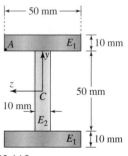

Figure P6.119

6.120 $\varepsilon_{xx} = 300\ \mu$, $E_1 = 30,000$ ksi, $E_2 = 20,000$ ksi.

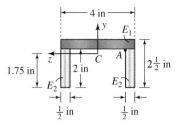

Figure P6.120

6.121 A simply supported 3-m-long beam has a uniformly distributed load of 10 kN/m over the entire length of the beam. If the beam has the composite cross section shown in Figure P6.121, determine the maximum bending normal stress in each of the three materials. Use $E_{al} = 70$ GPa, $E_w = 10$ GPa, and $E_s = 200$ GPa.

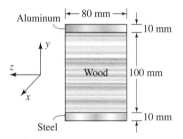

Figure P6.121

6.122 A wooden rod ($E_w = 2000$ ksi) and a steel strip ($E_s = 30,000$ ksi) are fastened securely to rigid plates, as shown in Figure P6.122. Determine (a) maximum

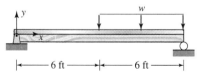

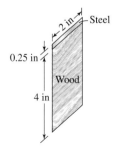

Figure P6.122

---

intensity of the load w if the allowable bending normal stresses in steel and wood are 20 ksi and 4 ksi, respectively; (b) the magnitude of maximum shear stress in the beam.

6.123 A steel strip ($E_s = 200$ GPa) is attached to an aluminum beam ($E_{al} = 70$ GPa) to form a composite cantilever beam, as shown in Figure P6.123. Determine (a) the maximum bending normal and shear stresses in steel and aluminum; (b) the magnitude of maximum shear stress in the beam.

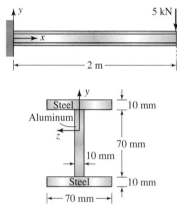

Figure P6.123

6.124 A steel ($E_{steel} = 200$ GPa) tube of outside diameter of 240 mm is attached to a brass ($E_{brass} = 100$ GPa) tube to form the cross section in Figure P6.124. Determine (a) the maximum bending normal stress in steel and brass; (b) the magnitude of the maximum shear stress in the beam.

Reinforced concrete beams

In Problems 6.125 through 6.127, neglect the capacity of concrete to support stresses in tension.

6.125 A reinforced concrete beam is constructed by embedding circular steel rods of 20-mm diameter, as shown in Figure P6.125. The maximum resisting (internal) moment in the beam was determined as 30 kN · m. Determine the maximum bending normal stress in the concrete and steel. Use the moduli of elasticity for steel and concrete as $E_s = 200$ GPa and $E_c = 28$ GPa, respectively.

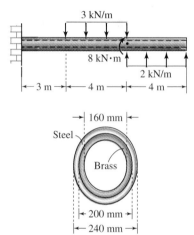

Figure P6.124

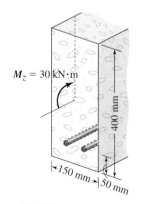

Figure P6.125

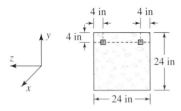

Figure P6.126

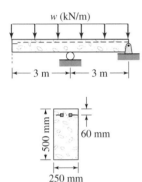

Figure P6.127

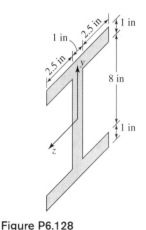

Figure P6.128

6.126 A reinforced concrete beam is constructed by embedding 2-in × 2-in square steel rods. The reinforced concrete beam is used as a 10-ft cantilever beam, with a force of 36 kips applied at the free end in the negative y direction (Figure P6.126). Determine the maximum compressive bending normal stress in the concrete. Use the moduli of elasticity for steel and concrete as $E_s = 25000$ ksi and $E_c = 4500$ ksi, respectively.

6.127 A reinforced concrete beam is constructed by embedding 20-mm × 20-mm square steel rods and loaded and supported as shown in Figure P6.127. The allowable tensile bending normal stress is 160 MPa and the allowable compressive bending normal stress for concrete is 20 MPa. Determine the maximum intensity of the distributed load w. Use the moduli of elasticity for steel and concrete as $E_s = 200$ GPa and $E_c = 28$ GPa, respectively.

Nonlinear beams

6.128 The bending normal stress on the cross-section shown in Figure P6.128 is given by

$$\sigma_{xx} = \begin{cases} 24 \text{ ksi}, & -5 \le y < -4 \\ -6y \text{ ksi}, & -4 < y < 4 \\ -24 \text{ ksi}, & 4 < y \le 5 \end{cases}$$

Determine the equivalent internal bending moment M_z.

In Problems 6.129 through 6.132, a hollow rectangular beam has the dimensions shown in Figure P6.129. The normal strain due to bending about the z axis was found to vary as

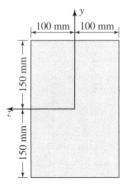

Figure P6.129

$\varepsilon_{xx} = -0.01y$, *with y measured in meters. For the given material model in the problem: determine the internal moment that would produce the given strain.*

6.129 The beam is made from elastic–perfectly plastic material that has a yield stress $\sigma_{yield} = 250$ MPa and a modulus of elasticity $E = 200$ GPa.

6.130 The beam is made from a bilinear material that has a yield stress $\sigma_{yield} = 250$ MPa and moduli of elasticity $E_1 = 200$ GPa and $E_2 = 80$ GPa.

6.131 The beam material has a stress–strain relationship given by $\sigma = 952\varepsilon^{0.2}$ MPa.

6.132 The beam material has a stress–strain relationship given by $\sigma = 200\varepsilon - 2000\varepsilon^2$ MPa.

In Problems 6.133 through 6.136, a wide-flange beam has the dimensions shown in Figure P6.133. The normal strain due to bending about the z axis was found to vary as $\varepsilon_{xx} = -0.4y \times 10^{-3}$, with y measured in inches. For the given material model in the problem: determine the internal moment that would produce the given strain.

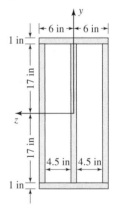

Figure P6.133

6.133 The beam is made from elastic–perfectly plastic material that has a yield stress $\sigma_{yield} = 40$ ksi and a modulus of elasticity $E = 10,000$ ksi.

6.134 The beam is made from a bilinear material that has a yield stress $\sigma_{yield} = 40$ ksi and moduli of elasticity $E_1 = 10,000$ ksi and $E_2 = 2000$ ksi.

6.135 The beam material has a stress–strain relationship given by $\sigma = 70\varepsilon^{0.1}$ ksi.

6.136 The beam material has a stress–strain relationship given by $\sigma = 10,000\varepsilon - 90,000\varepsilon^2$ ksi.

6.137 A beam made from elastic–perfectly plastic material has a yield stress of 30 ksi, as shown in Figure P6.137. Determine the internal bending moment if the elastic–plastic boundary is 0.5 in from the top and bottom.

Figure P6.137

6.138 Determine the shape factor for the rectangular cross section shown in Figure P6.137.

6.139 An elastic–perfectly plastic material has a yield stress $\sigma_{yield} = 40$ ksi. Point A in Figure P6.139 is at yield stress due to bending of the beam. Determine: (a) the location of the neutral axis assuming it is in the web; (b) the applied moment that produced the state of stress.

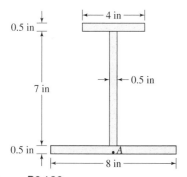

Figure P6.139

6.140 Determine the shape factor for the cross section shown in Figure P6.139.

6.141 A uniformly loaded simply supported beam is made of elastic–perfectly plastic material that has a yield stress of 30 ksi. The beam has the hollow square cross section shown in Figure P6.141. If point A is at yield stress, determine the intensity w of the uniform load.

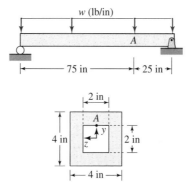

Figure P6.141

6.142 A beam of elastic–perfectly plastic material has a yield stress of 50 ksi and a cross section as shown in Figure P6.142. If point A just reaches yield stress, determine: (a) the location of the neutral axis; (b) the applied moment that produced the state of stress.

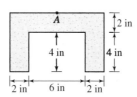

Figure P6.142

6.143 The hollow square beam shown in Figure P6.143 is made from a material that has a stress–strain relation given by $\sigma = K\varepsilon^{0.4}$. Assume the same behavior in tension and in compression. In terms of K, L, a, and M_{ext} determine the bending normal strain and stress at point A.

6.144 The stress–strain curve in tension for a material is given by $\sigma = K\varepsilon^{0.5}$. For the rectangular cross section shown in Figure P6.144

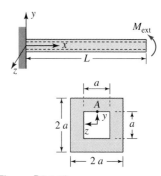

Figure P6.143

Figure P6.144

show that the bending normal stress is given by

$$
\sigma_{xx} = \begin{cases} \dfrac{-5\sqrt{2}}{bh^2}\left(\dfrac{y}{h}\right)^{0.5} M_z, & y > 0 \\[3mm] \dfrac{5\sqrt{2}}{bh^2}\left(-\dfrac{y}{h}\right)^{0.5} M_z, & y < 0 \end{cases}
$$

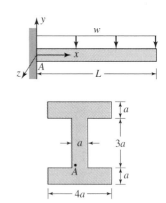

6.145 The cantilever beam shown in Figure P6.145 is made from a material that has a stress–strain relation in tension given by $\sigma = K\varepsilon^{0.7}$. In terms of K, w, a, and L, determine the fiber stress at point A. Point A is just above the bottom flange.

Figure P6.145

*6.9 GENERAL INFORMATION

The initial development in strength of materials and related fields came from the study of the strength of beams. Mistakes regarding the location of the neutral axis and the stress distribution across the cross section influenced significant amounts of earlier work. The driving force was that the fracture loads on a beam as predicted by the theory did not correlate well with the experimental values. But as you read the history, keep in mind that the pioneers' struggle in the dark was made difficult by the fact that near fracture, the stress–strain relationship is nonlinear, which influences the stress distribution and the location of the neutral axis.

The earliest known work on beams is by Leonardo da Vinci (1452–1519). In addition to his statements on simply supported beams, which are described in Problem 6.105, he correctly concluded that in a cantilever untapered beam the cross section farthest from the built-in end would deflect the most. But it was Galileo's work that had the greatest early influence on the study of beams.

Galileo (1564–1642) was born in Pisa, Italy. In 1581 he enrolled at Pisa University to study medicine. But it was the work of Euclid, Archimedes, and Leonardo da Vinci that attracted him to mathematics and mechanics. In 1589 he was given a professorship of mathematics at Pisa University, where he conducted his famous experiments on falling bodies, and the field of dynamics was born. His conclusions about falling bodies were in complete disagreement with the popular Aristotelian mechanics. He paid the price for his views, for the proponents of Aristotelian mechanics made his stay at Pisa University untenable, and he left the university in early 1592. But by the end of 1592 he was appointed a professor of mathematics at the University of Padua. During this period he discovered his interest in astronomy. Based on sketchy reports he built himself a telescope. On seeing Jupiter moons in 1610 he found evidence for his belief in the Copernican theory that the earth was not the center of the universe. In 1616 Copernicus was condemned by the Church, and the Inquisition warned Galileo to limit himself to the physical world and leave theology to the Church. In 1632 he published his views on the Copernican theory under the mistaken belief that the new pope, Maffeo Barberini, who was Galileo's admirer, would be more tolerant in his views. Galileo was condemned by the Inquisition and was put under house arrest for the last eight years of his life. During this period he wrote the book called *Two New Sciences,* in which he describes his various works in mechanics, including work on mechanics of materials. We have seen his contribution toward the early concept of stress in Section 1.3, here we discuss briefly his contributions on the bending of beams.

Figure P6.106 shows Galileo's illustration of the bending test from which he drew a number of correct conclusions, two of which are described in Problems 6.107 and 6.108. His other correct conclusions included the following. (i) A beam whose width is greater than its thickness would offer greater resistance standing on its edge than lying flat—the area moment of inertia will be greater. (ii) The resisting moment (strength of the beam) increases as the cube of the radius for circular beams—the section modulus increases as the cube of the radius. (iii) The cross-sectional dimensions must increase at a greater rate than the length for constant strength cantilever beam bending due to its own weight. These conclusions have been used in the design of beams since Galileo's time. But as we saw in Problem 6.106, Galileo's prediction on the load-carrying capacity of beams was incorrect because the stress distribution and the location of the neutral axis were incorrect.

An important correction to the stress distribution is credited to Mariotte (1620–1684). Mariotte became interested in the strength of beams while trying to design pipes for supplying water to the palace of Versailles. His experiments with wooden and glass beams convinced him that Galileo's load predictions were greatly exaggerated. Thus he developed his own theory, which incorporated linear elasticity, and concluded that the stress distribution was linear, with a zero stress value at the bottom of the beam. Mariotte's predicted values did not correlate well with the experimental values either. He argued that beams loaded over a long time would have failure loads closer to his predicted values which, while true, is not the correct explanation for the discrepancy. As we saw in Problem 6.109, the cause of the discrepancy lay in an incorrect assumption about the location of the neutral axis. This incorrect location of the neutral axis influenced the study of beams by many pioneers, including Navier and Jacob Bernoulli. We saw some of the contributions of Navier in Section 1.3 and will discuss Jacob Bernoulli's contributions in Chapter 7 on beam deflection. But a curious design philosophy developed during this time. Engineers would use Galileo's theory in designing beams of brittle material such as stone, but would use Mariotte's theory for wooden beams.

Parent (1666–1716) was born in Paris. He obtained a degree in law on the insistence of his parents, but he never practiced it because he wanted to do mathematics. Starting with Mariotte's stress formula he first found that it was not applicable to circular cross sections of beams. Later he proved that the linear stress distribution across a rectangular cross section had to be such that the zero stress point must be at the center, provided the material behavior was elastic. Unfortunately Parent's work was published in a journal that he edited and published, and not in the more widely read journal published by the French Academy. More than half a century later Coulomb, whose contribution we saw in Section 5.7, unaware of Parent's work, independently deduced the more general statement for deriving the location of the neutral axis. Coulomb showed that the stress distribution had to be such that the net axial force was zero, and this statement was independent of material behavior. In other words, Coulomb correctly deduced Equation (6.2) and noted that it was independent of the material model. Saint-Venant rigorously examined the implication of the kinematic assumptions, Assumptions 1 through 3, and demonstrated that these are met exactly only if the beam is subjected to couples only, with no transverse force, that is, the shear force must be zero.

The shear stresses in beams did not receive much attention in the earlier development of beam theory. In Section 1.3 it was mentioned that the concept of shear stress was only developed in 1781 by Coulomb, but Coulomb believed that shear was only important in short beams. Vicat's experiment in 1833 with short beams gave ample evidence of the importance of shear. Jourawaski, a Russian railroad engineer, was working in 1844 on building a railroad between St. Petersburg and Moscow. A 180-ft-long bridge had to be built over the river Werebia, and Jourawaski had to use thick wooden beams. These thick beams were failing along the length of the fibers, which were in the longitudinal direction. Jourawaski realized the importance of shear in long beams and developed the theory we studied in Section 6.6.

Thus starting with Galileo, it took nearly 250 years of struggle to understand the correct nature of stresses in beam bending. Other historical developments related to beam theory will be discussed in Section 7.6.

6.10 CLOSURE

In this chapter we established formulas for calculating normal and shear stresses in beams under symmetric bending. We saw that the calculation of bending stresses requires the calculation of the internal bending moment and the shear force at a section. We considered only statically determinate beams, for which the internal shear force and bending moment diagrams can be found either by making an imaginary cut and drawing an appropriate free-body diagram or by drawing a shear force–bending moment diagram. The free-body diagram is preferred if stresses are to be found at a specified cross section, but shear force–moment diagrams are the better choice if maximum bending normal or shear stress is to be found in the beam.

The shear force–bending moment diagrams can be drawn by using the graphical interpretation of integrals as the area under the curve defined by the integrand. Alternatively internal shear force and bending moments could be found as a function of the x coordinate along the beam and plotted. Finding the bending moment as a function of x is important in the next chapter, where the moment–curvature relationship established in this chapter is integrated. Once we know how to find the deflection in a beam, problems of statically indeterminate beams can be solved.

We also saw that the bending normal and shear stresses should be drawn on a stress element for the purpose of understanding the character of bending stresses. The correct direction of the stresses can be obtained by inspection in many cases and is the quicker method. Alternatively, by following the sign convention for drawing the internal shear force and bending moment on free-body diagrams and using the subscripts, the correct direction of the bending stresses can be obtained. It should be emphasized that shear–moment diagrams yield the values and the sign of the shear force and the bending moment according to our sign convention. Hence both methods of determining the direction of stresses should be understood. The drawing of bending stresses on a stress element is also important for the purpose of stress and/or strain transformation, as will be described later.

In Chapter 8 on stress transformation we will consider problems in which we first find bending stresses using the stress formulas in this chapter and then find stresses on inclined planes, including planes with maximum normal and shear stress. In Chapter 9 on strain transformation we will find the bending strains and then consider strains in different coordinate systems, including coordinate systems in which the normal and shear strains are maximum. In Section 10.1 we will consider the combined loading problems of axial, torsion, and bending and the design of simple structures that may be determinate or indeterminate.

POINTS AND FORMULAS TO REMEMBER

- Theory is limited to: (1) slender beams; (2) regions away from the neighborhood of stress concentration; (3) gradual variation in cross section and external loads; (4) loads acting in the plane of symmetry in the cross section; (5) no change in direction of loading during bending.

$$M_z = -\int_A y\sigma_{xx}\, dA \quad (6.1) \qquad u = -y\frac{dv}{dx}, \quad v = v(x) \quad (6.4)$$

$$\text{small strain, } \varepsilon_{xx} = -\frac{y}{R} = -y\frac{d^2v}{dx^2} \quad (6.5a, b)$$

where M_z is the internal bending moment that is drawn on the free-body diagram to put a point with positive y coordinate in compression, u and v are the displacements in the x and y directions, respectively, σ_{xx} and ε_{xx} are the bending (flexure) normal stress and strain, y is the coordinate measured from the neutral axis to the point where normal stress and normal strain are defined, and d^2v/dx^2 is the curvature of the beam.

- The normal bending strain ε_{xx} is a linear function of y.
- The normal bending strain ε_{xx} will be maximum at either the top or the bottom of the beam.
- Equations (6.1), (6.5a), and (6.5b) are independent of the material model.
- Formulas below are valid for material that is linear, elastic, isotropic, with no inelastic strains.
- Homogeneous cross section:

$$M_z = EI_{zz}\frac{d^2v}{dx^2} \quad (6.11) \qquad \sigma_{xx} = -\frac{M_z y}{I_{zz}} \quad (6.12)$$

where y is measured from the centroid of the cross section, and I_{zz} is the second area moment about the z axis passing through the centroid.

- EI_{zz} is the bending rigidity of a beam cross section.
- Normal stress σ_{xx} in bending varies linearly with y on a homogeneous cross section.
- Normal stress σ_{xx} is zero at the centroid ($y = 0$) and maximum at the point farthest from the centroid for a homogeneous cross section.
- The shear force V_y will jump by the value of the applied external force as one crosses it from left to right.
- M_z will jump by the value of the applied external moment as one crosses it from left to right.

$$V_y = \int_A \tau_{xy}\, dA \quad (6.9) \qquad \tau_{xs} = -\frac{V_y Q_z}{I_{zz}t} \quad (6.23)$$

where Q_z is the first moment of the area A_s about the z axis passing through the centroid, t is the thickness perpendicular to the centerline, A_s is the area between the free surface and the line at which the shear stress is being found, and the coordinate s is measured from the free surface used in computing Q_z.

- Direction of shear flow on a cross section must be such that: (1) the resultant force in the y direction is in the same direction as V_y; (2) the resultant force in the z direction is zero; (3) it is symmetric about the y axis.
- Q_z is zero at the top and bottom surfaces and is maximum at the neutral axis.
- Shear stress is maximum at the neutral axis of a cross section in symmetric bending of beams.
- Bending strains:

$$\varepsilon_{xx} = \frac{\sigma_{xx}}{E} \qquad \varepsilon_{yy} = -\frac{\nu\sigma_{xx}}{E} = -\nu\varepsilon_{xx} \qquad \varepsilon_{zz} = -\frac{\nu\sigma_{xx}}{E} = -\nu\varepsilon_{xx} \qquad \gamma_{xy} = \frac{\tau_{xy}}{G} \qquad \gamma_{xz} = \frac{\tau_{xz}}{G} \quad (6.25)$$

CHAPTER SEVEN
DEFLECTION OF SYMMETRIC BEAMS

7.0 OVERVIEW

Figure 7.1a shows a diving board bending under the weight of the diver at the end of the board. The diving board must have adequate flexibility to provide the spring force that the divers can use to launch themselves into a dive. The bridge in Figure 7.1b shows beams that must provide enough stiffness to resist large deflections due to the weight of the traffic. Adequate flexibility or stiffness can be incorporated into beam design if we can find the beam deflection, which is the topic of this chapter.

As shall be seen, the deflection of a beam can be obtained by integrating either a second-order or a fourth-order differential equation. A differential equation, together

(a)

(b)

Figure 7.1 Examples of beam deflection.

with all the conditions necessary to solve for the integration constants is called a boundary-value problem. The solution of the boundary-value problem gives the deflection of the beam at any location x along the length of the beam.

The learning objective in this chapter is how to formulate and solve the boundary-value problem for the deflection of a beam at any point.

7.1 SECOND-ORDER BOUNDARY-VALUE PROBLEM

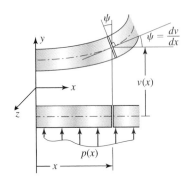

Figure 7.2 Beam deflection.

In Chapter 6 on the symmetric bending of beams, we established that if we can find the deflection in the y direction of one point on the cross section, then we know the deflection of all points on the cross section. In other words, the deflection at a cross section is independent of the y and z coordinates. However, the deflection can be a function of x, as shown in Figure 7.2. The deflected curve represented by $v(x)$ is called *elastic curve*. The deflection function $v(x)$ can be found by integrating Equation (6.11) twice, provided we can find the internal moment as a function of x, as we did in Section 6.3. Equation (6.11) is rewritten here for convenience:

$$M_z = EI_{zz}\frac{d^2v}{dx^2} \tag{7.1}$$

The deflection $v(x)$ can be obtained by integrating Equation (7.1), which is a second-order differential equation that will generate two integration constants. These integration constants are determined from boundary conditions, as discussed in Section 7.1.1.

As one moves across the beam it is possible that the applied load changes, resulting in different functions of x that represent the internal moment M_z. In such cases there are as many differential equations as there are functions representing the moment M_z. Each additional differential equation generates additional integration constants. These additional integration constants are determined from continuity (compatibility) equations obtained by considering the point where the functional representation of the moment changes character. The continuity conditions are discussed in Section 7.1.2.

Definition 1 The mathematical statement listing all the differential equations and all the conditions necessary for solving for $v(x)$ is called the *boundary-value problem* for the beam deflection.

7.1.1 Boundary Conditions

The integration of Equation (7.1) will result in v and dv/dx. Thus the conditions that we are seeking are on v and/or dv/dx. Figure 7.3 shows three types of support and the associated boundary conditions.

Note that for a second-order differential equation we need two boundary conditions. If on one end there is only one boundary condition, such as in Figure 7.3b or c, then the remaining boundary condition must come from the other end. If we are in

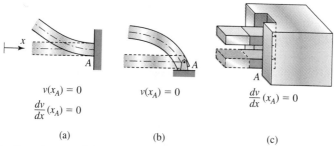

$$v(x_A) = 0$$
$$\frac{dv}{dx}(x_A) = 0$$

$$v(x_A) = 0$$

$$\frac{dv}{dx}(x_A) = 0$$

(a)   (b)   (c)

Figure 7.3   Boundary conditions for second-order differential equations. (*a*) Built-in end. (*b*) Simple support. (*c*) Smooth slot.

doubt about a boundary condition at a support then, often, the doubt can be resolved by drawing an approximate deformed shape of the beam.

7.1.2   Continuity Conditions

Suppose that because of change in the applied loading, the internal moment M_z in a beam is represented by one function to the left of x_j and another function to the right of x_j. Then there are two second-order differential equations, which on integration will produce two different displacement functions, one for each side of x_j, which will contain a total of four integration constants. Two of these four integration constants can be determined from the boundary conditions, as discussed in Section 7.1.1. The remaining two constants will have to be determined from conditions at x_j. Figure 7.4 shows that a discontinuous displacement at x_j implies a broken beam and a discontinuous slope at x_j implies that a beam is kinked at x_j.

Assuming that the beam neither breaks nor kinks, then the displacement functions must satisfy the following conditions:

$$v_1(x_j) = v_2(x_j) \tag{7.2a}$$

$$\frac{dv_1}{dx}(x_j) = \frac{dv_2}{dx}(x_j) \tag{7.2b}$$

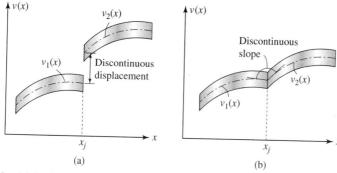

Figure 7.4   (*a*) Broken beam. (*b*) Kinked beam.

where v_1 and v_2 are the displacement functions to the left and right of x_j. The conditions[1] given by Equations (7.2) are the *continuity conditions*, also known as *compatibility conditions* or *matching conditions*.

- Example 7.1 demonstrates the formulation and solution of a boundary-value problem with one second-order differential equation and the associated boundary conditions.

- Example 7.2 demonstrates the formulation and solution of a boundary-value problem with two second-order differential equations, the associated boundary conditions, and the continuity conditions.

- Example 7.3 demonstrates the formulation only of a boundary-value problem with multiple second-order differential equations, the associated boundary conditions, and the continuity conditions.

- Example 7.4 demonstrates the formulation and solution of a boundary-value problem with variable area moment of inertia, that is, I_{zz} is a function of x.

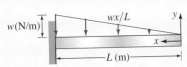

Figure 7.5 Beam and loading in Example 7.1.

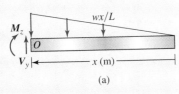

(a)

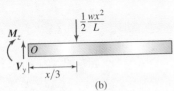

(b)

Figure 7.6 Free-body diagram in Example 7.1.

EXAMPLE 7.1

A beam has a linearly varying distributed load, as shown in Figure 7.5. Determine:

(a) The equation of the elastic curve in terms of E, I, w, L, and x.

(b) The maximum intensity of the distributed load if the maximum deflection is to be limited to 20 mm.

Use $E = 200$ GPa, $I = 600 \times 10^6$ mm^4, and $L = 8$ m.

PLAN

(a) We can make an imaginary cut at an arbitrary location x and take the right part for drawing the free-body diagram. Using equilibrium equations, the moment M_z can be written as a function of x. By integrating Equation (7.1) and using the boundary conditions that deflection and slope at $x = L$ are zero, we can find $v(x)$.

(b) The maximum deflection for this problem will occur at the free end and can be found by substituting $x = 0$ in the $v(x)$ expression. By requiring that $|v_{max}| \leq 0.02$ m, we can find w_{max}.

Solution

(a) Figure 7.6 shows the free-body diagram of the right part after making an imaginary cut at some location x. Internal moment and shear forces are drawn according to the sign convention discussed in Section 6.2.5. The distributed force is replaced by an equivalent

---
[1]An alternate form for writing Equations (7.2a) and (7.2b) is $v(x_j^-) = v(x_j^+)$ and $v'(x_j^-) = v'(x_j^+)$, where the minus and plus refer to just left and just right of x_j, and the prime refers to the first derivative.

force, and the internal moment is found by balancing the moment about point O,

$$\sum M_O \qquad M_z = -\frac{1}{2}\frac{wx^2}{L}\left(\frac{x}{3}\right) = -\frac{1}{6}\frac{wx^3}{L} \qquad (E1)$$

Substituting Equation (E1) into Equation (7.1) and noting the zero slope and deflection at the built-in end, the boundary-value problem can be stated as follows:

- Differential equation:

$$EI_{zz}\frac{d^2v}{dx^2} = -\frac{1}{6}\frac{wx^3}{L} \qquad (E2)$$

- Boundary conditions:

$$v(L) = 0 \qquad (E3)$$

$$\frac{dv}{dx}(L) = 0 \qquad (E4)$$

Equation (E2) can be integrated to obtain

$$EI_{zz}\frac{dv}{dx} = -\frac{1}{24}\frac{wx^4}{L} + c_1 \qquad (E5)$$

Substituting $x = L$ in Equation (E5) and using Equation (E4), the constant c_1 can be found and substituted into Equation (E5) to obtain a new expression for the slope,

$$-\frac{1}{24}\frac{wL^4}{L} + c_1 = 0 \qquad \text{or} \qquad c_1 = \frac{wL^3}{24}$$

$$EI_{zz}\frac{dv}{dx} = -\frac{1}{24}\frac{wx^4}{L} + \frac{wL^3}{24} \qquad (E6)$$

Equation (E6) can be integrated to obtain

$$EI_{zz}v = -\frac{1}{120}\frac{wx^5}{L} + \frac{wL^3}{24}x + c_2 \qquad (E7)$$

Substituting $x = L$ in Equation (E7) and using Equation (E3), the constant c_2 can be found,

$$-\frac{1}{120}\frac{wL^5}{L} + \frac{wL^3}{24}L + c_2 = 0 \qquad \text{or} \qquad c_2 = -\frac{wL^4}{30} \qquad (E8)$$

The deflection expression can be obtained by substituting Equation (E8) into Equation (E7) and simplifying,

$$v(x) = -\frac{w}{120EI_{zz}L}(x^5 - 5L^4x + 4L^5) \qquad \text{(E9)}$$

Dimension check: We note that all terms in the parentheses of Equation (E9) have the dimension of length to the power of five, that is, $O(L^5)$. Thus Equation (E9) is dimensionally homogeneous. But we can also check whether the left-hand side and any one term of the right-hand side has the same dimension,

$$w \rightarrow O\left(\frac{F}{L}\right) \qquad x \rightarrow O(L) \qquad E \rightarrow O\left(\frac{F}{L^2}\right) \qquad I_{zz} \rightarrow O(L^4)$$

$$v \rightarrow O(L) \qquad \frac{wx^5}{EI_{zz}L} \rightarrow O\left(\frac{(F/L)L^5}{(F/L^2)O(L^4)L}\right) \rightarrow O(L) \rightarrow \text{checks}$$

(b) By inspection it can be seen that the maximum deflection for this problem will occur at the free end. Substituting $x = 0$ in Equation (E9), we obtain $v_{max} = -wL^4/30EI_{zz}$. The minus sign indicates that the deflection is in the negative y direction, as expected. Substituting the given values of the variables and requiring that the magnitude of the deflection $|v_{max}| \leq 0.02$ m, the value of w_{max} can be found,

$$|v_{max}| = \frac{w_{max}L^4}{30EI_{zz}} = \frac{w_{max}8^4}{30(200 \times 10^9)(600 \times 10^{-6})} \leq 0.02$$

or

$$w_{max} \leq 17.58 \times 10^3 \text{ N/m}$$

$$\text{ANS.} \qquad w_{max} = 17.5 \text{ kN/m}$$

COMMENTS

1. From calculus we know that the maximum of a function occurs at the point where the slope of the function is zero. But the slope at $x = L$, where the deflection is maximum, is not zero. This is because the function $v(x)$ is monotonic, that is, a continuously increasing (decreasing) function. For monotonic functions the maximum (or minimum) always occurs at the end of the interval. It was the monotonic character of the function that we intuitively recognized in our statement that the maximum deflection occurs at the free end.

2. If the dimension check showed that some term did not have the proper dimension, then we would backtrack, check each equation for dimensional homogeneity, and identify the error.

EXAMPLE 7.2

For the beam and loading shown in Figure 7.7, determine: (a) the equation of the elastic curve in terms of E, I, L, P, and x; (b) the maximum deflection in the beam.

PLAN

(a) Due to the load P at B the internal moment will be represented by different functions in AB and BC, which can be found by making imaginary cuts and drawing a free-body diagram. We can write the two differential equations using Equation (7.1), the two boundary conditions of zero deflection at A and C, and the two continuity conditions at B. The boundary-value problem can be solved to obtain the elastic curve.

(b) In each section we can set the slope to zero and find the roots of the equation that will give the location of zero slope. We can substitute the location values in the elastic curve equation derived in part (a) to determine the maximum deflection in the beam.

Solution

(a) The free-body diagram of the entire beam can be drawn, and the reaction at A found as $R_A = 3P/2$ upward, and the reaction at C as $R_C = P/2$ downward. Make an imaginary cut in AB, that is, the location of the imaginary cut is defined by $0 \le x < L$. The free-body diagram can be drawn using the left part, as shown in Figure 7.8. The internal moment can be found by balancing the moment balanced at point O_1,

$$\sum M_{O_1} \qquad M_1 + 2PL - R_A x = 0$$

or

$$M_1 = \frac{3}{2}Px - 2PL \qquad \text{(E1)}$$

Make an imaginary cut in BC, that is, the location of the imaginary cut is defined by $L \le x < 2L$. The free-body diagram can be drawn by taking the left part (see comment at end of example), as shown in Figure 7.9. The internal moment can be found by balancing the moment balanced at point O_2,

$$\sum M_{O_2} \qquad M_2 + 2PL - R_A x + P(x - L) = 0$$

or

$$M_2 = \frac{3}{2}Px - 2PL - P(x - L) \qquad \text{(E2)}$$

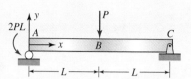

Figure 7.7 Beam and loading in Example 7.2.

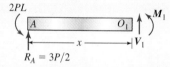

Figure 7.8 Imaginary cut in AB in Example 7.2.

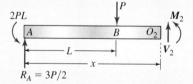

Figure 7.9 Imaginary cut in BC in Example 7.2.

Check: The internal moment must be continuous at *B* as there is
no external point moment at *B*. Substituting $x = L$ in Equations (E1)
and (E2), we find $M_1 = M_2$ at $x = L$.

The boundary-value problem can be stated using Equation (7.1),
(E1), and (E2), the zero deflection at points *A* and *C*, and the continuity conditions at *B* as follows:

- Differential equations:

$$EI_{zz}\frac{d^2v_1}{dx^2} = \frac{3}{2}Px - 2PL, \qquad\qquad 0 \le x < L \qquad \text{(E3)}$$

$$EI_{zz}\frac{d^2v_2}{dx^2} = \frac{3}{2}Px - 2PL - P(x - L), \qquad L \le x < 2L \qquad \text{(E4)}$$

- Boundary conditions:

$$v_1(0) = 0 \qquad\qquad \text{(E5)}$$

$$v_2(2L) = 0 \qquad\qquad \text{(E6)}$$

- Continuity conditions:

$$v_1(L) = v_2(L) \qquad\qquad \text{(E7)}$$

$$\frac{dv_1}{dx}(L) = \frac{dv_2}{dx}(L) \qquad\qquad \text{(E8)}$$

Integrating Equations (E3) and (E4) we obtain

$$EI_{zz}\frac{dv_1}{dx} = \frac{3}{4}Px^2 - 2PLx + c_1 \qquad\qquad \text{(E9)}$$

$$EI_{zz}\frac{dv_2}{dx} = \frac{3}{4}Px^2 - 2PLx - \frac{P}{2}(x - L)^2 + c_2 \qquad \text{(E10)}$$

Substituting $x = L$ in Equations (E9) and (E10) and using Equation (E8), we obtain

$$\frac{3}{4}PL^2 - 2PL^2 + c_1 = \frac{3}{4}PL^2 - 2PL^2 - 0 + c_2 \qquad \text{or} \qquad c_1 = c_2$$
$$\text{(E11)}$$

Substituting Equation (E11) into Equation (E10) and integrating
Equations (E9) and (E10), we obtain

$$EI_{zz}v_1 = \frac{1}{4}Px^3 - PLx^2 + c_1x + c_3 \qquad\qquad \text{(E12)}$$

$$EI_{zz}v_2 = \frac{1}{4}Px^3 - PLx^2 - \frac{P}{6}(x - L)^3 + c_1x + c_4 \qquad \text{(E13)}$$

Substituting $x = L$ in Equations (E12) and (E13) and using Equation (E7), we obtain

$$\frac{1}{4}PL^3 - PL^3 + c_1L + c_3 = \frac{1}{4}PL^3 - PL^3 - 0 + c_1L + c_4$$

or

$$c_3 = c_4 \tag{E14}$$

Substituting $x = 0$ in Equation (E12) and using Equation (E5), we obtain

$$c_3 = 0 \tag{E15}$$

and from Equation (E14) we obtain

$$c_4 = 0 \tag{E16}$$

Substituting $x = 2L$ and Equation (E16) into Equation (E13) and using Equation (E6), we obtain

$$\frac{1}{4}P(2L)^3 - PL(2L)^2 - \frac{P}{6}(L)^3 + c_1(2L) = 0$$

or

$$c_1 = \frac{13}{12}PL^2 \tag{E17}$$

Substituting Equations (E15), (E16), and (E17) into Equations (E12) and (E13) and simplifying, we obtain

$$v_1(x) = \frac{P}{12EI_{zz}}(3x^3 - 12Lx^2 + 13L^2x), \qquad 0 \le x < L \tag{E18}$$

$$v_2(x) = \frac{P}{12EI_{zz}}[3x^3 - 12Lx^2 + 13L^2x - 2(x-L)^3], \quad L \le x < 2L \tag{E19}$$

Dimension check: All terms in parentheses are dimensionally homogeneous as all have the dimensions of length cubed. But we can also check whether the left-hand side and any one term of the right-hand side have the same dimension as follows:

$$P \to O(F) \qquad x \to O(L) \qquad E \to O\!\left(\frac{F}{L^2}\right) \qquad I_{zz} \to O(L^4)$$

$$v \to O(L) \qquad \frac{Px^3}{EI_{zz}} \to O\!\left(\frac{FL^3}{(F/L^2)L^4}\right) \to O(L) \to \text{checks}$$

(b) Let dv_1/dx be zero at $x = x_1$. Differentiating Equation (E18) [or substituting Equation (E17) into Equation (E9) and simplifying], we obtain

$$\frac{P}{12EI_{zz}}(9x_1^2 - 24Lx_1 + 13L^2) = 0$$

or

$$9x_1^2 - 24Lx_1 + 13L^2 = 0 \tag{E20}$$

The roots of the quadratic equation in Equation (E20) are $x_1 = 1.91L$ and $x_1 = 0.756L$. Equation (E18), hence Equation (E20), is valid only in the range from 0 to L. Thus the admissible root is $x_1 = 0.756L$. Substituting this root into Equation (E18), the following is obtained:

$$v_1(0.756L) = \frac{P}{12EI_{zz}}(3 \times 0.756L^3 - 12L \times 0.756L^2$$

$$+ 13L^2 \times 0.756L) = \frac{0.355PL^3}{EI_{zz}} \tag{E21}$$

To find the maximum deflection in BC, assume dv_2/dx to be zero at $x = x_2$. Differentiating Equation (E19) [or substituting Equation (E17) into Equation (E10) and simplifying], we obtain

$$\frac{P}{12EI_{zz}}[9x_2^2 - 24Lx_2 + 13L^2 - 6(x_2 - L)^2] = 0$$

or

$$3x_2^2 - 12Lx_2 + 7L^2 = 0 \tag{E22}$$

The roots of the quadratic equations in Equation (E22) are $x_2 = 0.709L$ and $x_2 = 3.29L$. Both roots are outside the range of L to $2L$ and hence are inadmissible. Thus in this problem the slope is zero only at $0.756L$, and the maximum deflection is given by Equation (E21),

$$\text{ANS.} \quad v_{max} = \frac{0.355PL^3}{EI_{zz}}$$

COMMENT
When we made the imaginary cut in BC, we took the left part for drawing the free-body diagram. Had we taken the right part, we would have

obtained the moment expression $M_2 = \frac{1}{2}Px - PL$, which is the simplified form of Equation (E2). We can start with this moment expression
and obtain our results by integration and use of conditions as shown.
The values of the integration constants will be different, there will be
slightly more algebra, but the final result will be the same. The form of
the moment expression used in the example made use of the observation that the continuity conditions are at $x = L$ and the terms in power of
$(x - L)$ will be zero, resulting in less algebra and simplified relations of
the constants given by Equations (E11) and (E14).

EXAMPLE 7.3

Write the boundary-value problem for solving the deflection at any
point x of the beam shown in Figure 7.10. Do not integrate or solve.

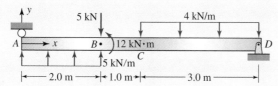

Figure 7.10 Beam and loading in Example 7.3.

PLAN

The moment expressions in each interval were found in Example 6.10.
Substituting these moment expressions into Equation (7.1), the differential equations can be written. We can also write the zero-deflection
conditions at points A and D and the continuity conditions at points B
and C to complete the boundary-value problem statement.

Solution   From Equations (E4), (E6), and (E8) in Example 6.10, we
have $M_1 = 5x^2/2$ in AB, $M_2 = 5x - 12$ in BC, and $M_3 = -2x^2 + 17x - 30$ in
CD. Substituting these expressions into Equation (7.1) and writing the
zero-displacement boundary conditions at A and D and the continuity

conditions at B and C, we obtain the boundary-value problem statement as follows:

- Differential equations:

$$EI_{zz}\frac{d^2 v_1}{dx^2} = \frac{5}{2}x^2, \qquad\qquad 0 < x < 2 \qquad \text{(E1)}$$

$$EI_{zz}\frac{d^2 v_2}{dx^2} = 5x - 12, \qquad\qquad 2 < x < 3 \qquad \text{(E2)}$$

$$EI_{zz}\frac{d^2 v_3}{dx^2} = -2x^2 + 17x - 30, \qquad 3 < x < 6 \qquad \text{(E3)}$$

- Boundary conditions:

$$v_1(0) = 0 \qquad\qquad \text{(E4)}$$

$$v_3(6) = 0 \qquad\qquad \text{(E5)}$$

- Continuity conditions:

$$v_1(2) = v_2(2) \qquad\qquad \text{(E6)}$$

$$\frac{dv_1}{dx}(2) = \frac{dv_2}{dx}(2) \qquad\qquad \text{(E7)}$$

$$v_2(3) = v_3(3) \qquad\qquad \text{(E8)}$$

$$\frac{dv_2}{dx}(3) = \frac{dv_3}{dx}(3) \qquad\qquad \text{(E9)}$$

COMMENTS

1. Equations (E1), (E2), and (E3) are three differential equations of order 2. Integrating these three differential equations would result in six integration constants. We have two boundary conditions and four continuity conditions. A properly formulated boundary-value problems will always have *exactly* the right number of conditions needed to solve a problem.

2. In Example 7.3 there were two differential equations and the resulting algebra was tedious. This example has three differential equations, which will make the algebra even more tedious. Fortunately there is a method, called discontinuity method, discussed in Section 7.4, which reduces the algebra by introducing functions that will let us write all three differential equations as a single equation and implicitly satisfy the continuity conditions during integration.

EXAMPLE 7.4

A cantilever beam with variable width $b(x)$ is shown in Figure 7.11. Determine the maximum deflection in terms of P, b_L, t, L, and E.

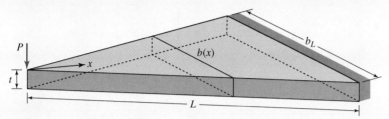

Figure 7.11   Variable-width beam in Example 7.4.

PLAN

$b(x)$ as a linear function of x can be found, and the area moment of inertia for a rectangular cross section can be calculated as a function of x. The bending moment as a function of x can also be found and substituted into Equation (7.1) to obtain the differential equation. The zero deflection and slope at $x = L$ are the boundary conditions necessary to solve the boundary-value problem for the elastic curve. The maximum deflection will be at $x = 0$ and can be found from the equation of the elastic curve.

Solution   Noting that $b(x)$ is a linear function of x that passes through the origin and has a value of b_L at $x = L$, we obtain $b(x) = b_L x / L$ and the area moment of inertia as

$$I_{zz} = \frac{b(x)t^3}{12} = \left(\frac{b_L t^3}{12L}\right)x \qquad \text{(E1)}$$

An imaginary cut can be made at some location x, the left part can be taken, and the free-body diagram drawn as shown in Figure 7.12. Balancing the moment at point O, the internal bending moment can be found,

$$M_z = -Px \qquad \text{(E2)}$$

Substituting Equations (E1) and (E2) into Equation (7.1), the differential equation can be obtained. Writing the deflection and slope as zero at $x = L$, the formulation of the boundary-value problem is completed as follows:

- Differential equation:

$$\frac{d^2v}{dx^2} = \frac{M_z}{EI_{zz}} = \frac{-Px}{E(b_L t^3 / 12L)x} = -\frac{12PL}{Eb_L t^3} \qquad \text{(E3)}$$

Figure 7.12   Free-body diagram in Example 7.4.

- Boundary conditions:

$$v(L) = 0 \tag{E4}$$

$$\frac{dv}{dx}(L) = 0 \tag{E5}$$

Integrating Equation (E3),

$$\frac{dv}{dx} = -\frac{12PL}{Eb_Lt^3}x + c_1 \tag{E6}$$

Substituting Equation (E6) into Equation (E5), we obtain

$$0 = -\frac{12PL}{Eb_Lt^3}L + c_1 \quad \text{or} \quad c_1 = \frac{12PL^2}{Eb_Lt^3} \tag{E7}$$

Substituting Equation (E7) into Equation (E6) and integrating, we obtain

$$v = -\frac{12PL}{Eb_Lt^3}\left(\frac{x^2}{2}\right) + \frac{12PL^2}{Eb_Lt^3}x + c_2 \tag{E8}$$

Substituting Equation (E8) into Equation (E4), we obtain

$$0 = -\frac{12PL}{Eb_Lt^3}\left(\frac{L^2}{2}\right) + \frac{12PL^2}{Eb_Lt^3}L + c_2 \quad \text{or} \quad c_2 = -\frac{6PL^3}{Eb_Lt^3} \tag{E9}$$

The maximum deflection will occur at the free end. Substituting $x = 0$ and Equation (E9) into Equation (E8), we obtain

$$\text{ANS.} \qquad v_{max} = -\frac{6PL^3}{Eb_Lt^3} \tag{E10}$$

COMMENTS

1. The beam taper must be gradual as per the limitation on the theory described in Section 6.2.
2. We can calculate the maximum bending normal stress in any section by substituting $y = t/2$ and Equations (E1) and (E2) into Equation (6.12), to obtain

$$\sigma_{max} = \left| -Px\frac{t/2}{(b_Lt^3/12L)x} \right| = \frac{6PL}{b_Lt^2} \tag{E11}$$

Equation (E11) shows that the maximum bending normal stress is a constant throughout the beam. Such beams are called *constant-strength beams* and are used in many designs were reduction in weight is a serious consideration. One such design is elaborated in comment 3.

3. In a leaf spring each leaf is considered an independent beam that bends about its own neutral axis because there is no restriction to sliding (see Problem 6.15). The variable-width beam is designed for constant strength, and b_L is found using Equation (E11). The width b_L is then divided into n parts, as shown in Figure 7.13a. Except for the main leaf A, all other leaf dimensions are found by taking the one-half leaf width on either side of the main leaf. In the assembled spring, the distance in each leaf from the applied load P is the same as in the original variable-width beam shown in Figure 7.11. Hence each leaf has the same allowable strength at all points. Defining $\bar{b}$ as the width of each leaf and $\bar{L}$ as the total length of the spring, $L = \bar{L}/2$, Equations (E10) and (E11) can be rewritten as

$$\delta = \frac{3P\bar{L}^3}{4nE\bar{b}\,t^3}$$

$$\sigma_{\max} = \frac{3P\bar{L}}{n\bar{b}\,t^2}$$

(7.3)

and used in design as was done in Example 3.8.

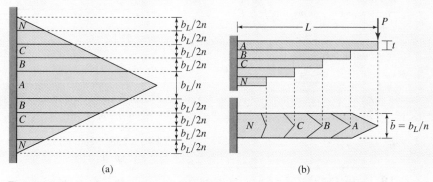

(a)   (b)

Figure 7.13 Explanation of leaf spring design.

PROBLEM SET 7.1

Figure P7.1

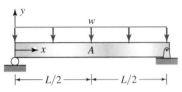

Figure P7.2

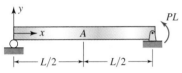

Figure P7.3

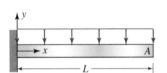

Figure P7.4

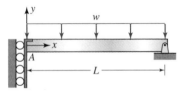

Figure P7.5

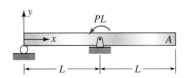

Figure P7.8

Second-order boundary-value problems

In Problems 7.1 through 7.6, in terms of w, P, L, E, and I, determine: (a) the equation of the elastic curve; (b) the deflection of the beam at point A.

7.1 Using Figure P7.1 solve the problem as described in text above.

7.2 Using Figure P7.2 solve the problem as described in text above.

7.3 Using Figure P7.3 solve the problem as described in text above.

7.4 Using Figure P7.4 solve the problem as described in text above.

7.5 Using Figure P7.5 solve the problem as described in text above.

7.6 Using Figure P7.6 solve the problem as described in text above.

Figure P7.6

In Problems 7.7 and 7.8 determine the deflection at point A in terms of w, P, L, E, and I.

7.7 Using Figure P7.7 solve the problem as described in text above.

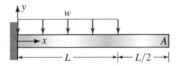

Figure P7.7

7.8 Using Figure P7.8 solve the problem as described in text above.

In Problems 7.9 through 7.12, in terms of w, L, E, and I, determine: (a) the equation of the elastic curve; (b) the deflection at x = L

7.9 Using Figure P7.9 solve the problem as described in text above.

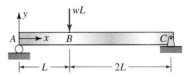

Figure P7.9

7.10 Using Figure P7.10 solve the problem as described in text above.

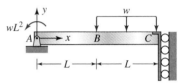

Figure P7.10

7.11 Using Figure P7.11 solve the problem as described in text above.

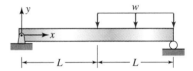

Figure P7.11

7.12 Using Figure P7.12 solve the problem as described in text above.

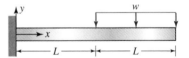

Figure P7.12

In Problems 7.13 and 7.14, write the boundary-value problem for determining the deflection of the beam at any point x. Assume EI is constant. Do not integrate or solve.

7.13 Using Figure P7.13 solve the problem as described in text above.

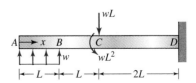

Figure P7.13

7.14 Using Figure P7.14 solve the problem as described in text above.

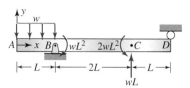

Figure P7.14

Variable area moment of inertia

7.15 A cantilever beam with variable depth $h(x)$ and constant width b is shown in Figure P7.15. The beam is to have a constant strength σ. In terms of b, L, E, x, and σ, determine: (a) the variation of $h(x)$; (b) the maximum deflection.

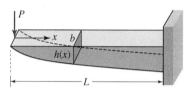

Figure P7.15

7.16 A cantilever tapered circular beam with variable radius $R(x)$ is shown in Figure P7.16. The beam is to have a constant strength σ. In terms of L, E, x, and σ, deter-

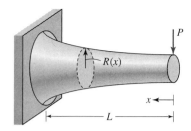

Figure P7.16

mine: (a) the variation of $R(x)$; (b) the maximum deflection.

7.17 For the tapered beam shown in Figure P7.17, determine the maximum bending normal stress and the maximum deflection in terms of E, w, b, h_0, and L.

7.18 For the tapered circular beam shown in Figure P7.18, determine the maximum bending normal stress and the maximum deflection in terms of E, P, d_0, and L.

7.19 The 2-in × 8-in wooden beam of rectangular cross section shown in Figure P7.19 is braced at the support using 2-in × 1-in wooden pieces. The modulus of elasticity of wood is 2000 ksi. Determine the maximum bending normal stress and the maximum deflection.

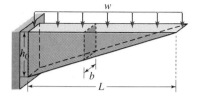

Figure P7.17

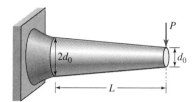

Figure P7.18

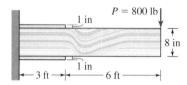

Figure P7.19

7.2 FOURTH-ORDER BOUNDARY-VALUE PROBLEM

We were able to solve for the deflection of a beam in Section 7.1 using second-order differential equations because we could find M_z as a function of x. In statically indeterminate beams, the internal moment determined from static equilibrium will contain some unknown reaction(s) in the moment expression. Also, if the distributed load p is not uniform or linear but a more complicated function, then finding the internal moment M_z as a function of x may be difficult. In either case it may be preferable to start from an alternate equation, which is obtained by substituting Equation (7.1) into Equation (6.13), that is, into $dM_z/dx = -V_y$, and substituting the result into Equation (6.14), that is, $dV_y/dx = -p$, to obtain

$$V_y = -\frac{d}{dx}\left(EI_{zz}\frac{d^2v}{dx^2}\right) \tag{7.4}$$

$$\frac{d^2}{dx^2}\left(EI_{zz}\frac{d^2v}{dx^2}\right) = p \tag{7.5}$$

If the bending rigidity EI_{zz} is constant, then it can be taken outside the differentiation. However, if the beam is tapered, then I_{zz} is a function of x and the form given here has to be used.

7.2.1 Boundary Conditions

The deflection $v(x)$ can be obtained by integrating Equation (7.5), which is a fourth-order differential equation that will generate four integration constants. To determine these four integration constants, four boundary conditions are needed. The integration of Equation (7.5) will yield V_y of Equation (7.4), which on integration would yield M_z of Equation (7.1), which on integration would yield v and dv/dx. Thus boundary conditions could be imposed on any of the four quantities v, dv/dx, M_z, and V_y. In order to understand how these conditions are determined, we generalize a principle discussed in statics for determining the reaction force and/or moments. Recall that in drawing free-body diagrams the following principles are used for determining reaction forces and moments at the supports:[2]

- If a point cannot move in a given direction, then a reaction force opposite to the direction acts at that support point.
- If a line cannot rotate about an axis in a given direction, then a reaction moment opposite to the direction acts at that support.

Consider the cantilever beam with an arbitrarily varying distributed load shown in Figure 7.14a. We make an imaginary cut very close to the support at A (infinitesimal distance Δx) and draw the free-body diagram as shown in Figure 7.14b. The internal shear force and the internal moment are drawn according to our sign convention. Notice that the distributed force is not shown because as Δx goes to zero, the contribution of the distributed force will drop out from the equilibrium equations. By equilibrium we obtain $V_y(0) = -R_A$ and $M_z(0) = -M_A$. Thus if a point cannot move, that is, the deflection v is zero at a point, then the shear force is not known because the reaction force is not known. Similarly if a line cannot rotate around an axis passing through a point, that is, dv/dx is zero, then the internal moment is not known because the reaction moment is not known. The reverse is equally true. Consider the free-body diagram constructed after making an imaginary cut at an infinitesimal distance from end B, as shown Figure 7.14c. By equilibrium we obtain $V_y(L) = 0$ and $M_z(L) = 0$, but the free end can deflect and rotate by any amount that is dictated by the loading. Thus when we specify a value of shear force, then we cannot specify displacement, and when we specify a value of internal moment at a point, then we cannot

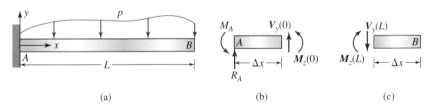

(a) (b) (c)

Figure 7.14 Example demonstrating grouping of boundary conditions.

---

[2]See Appendix A.1.2.

specify rotation. Thus there are two sets into which the four quantities v, dv/dx, M_z, and V_y are grouped for the purpose of determining the boundary conditions.

- *Group 1:* At a boundary point either the deflection v can be specified or the internal shear force V_y can be specified, but not both.

- *Group 2:* At a boundary point either the slope dv/dx can be specified or the internal bending moment M_z can be specified, but not both.

In order to generate four boundary conditions, two conditions are specified at each end of the beam. One condition is chosen from each group. Stated succinctly, the boundary conditions at each end of the beam are

- *Group 1:* $\qquad\qquad\qquad v \qquad$ or $\qquad V_y$

and $\qquad\qquad\qquad\qquad\qquad\qquad\qquad\qquad\qquad\qquad\qquad$ (7.6)

- *Group 2:* $\qquad\qquad\qquad \dfrac{dv}{dx} \qquad$ or $\qquad M_z$

From Figure 7.14c we concluded that the shear force and the bending moment at the free end were zero. This conclusion can be reached by inspection without drawing a free-body diagram. If at the end there were a point force or a point moment, then clearly the magnitude of the shear force would equal the point force and the magnitude of the internal moment would equal the point moment. Again, we can reach this conclusion without drawing a free-body diagram. But to get the correct sign of V_y and M_z we need a free-body diagram with the internal quantities drawn according to our sign convention. We address the issue in Section 7.2.3.

7.2.2 Continuity and Jump Conditions

If there is a point force or a point moment at x_j, or if the distributed force is given by different functions on the left and right of x_j, then, again, the displacement will be represented by different functions on the left and right of x_j. Thus we have two fourth-order differential equations which, upon integration, will generate eight integration constants requiring eight conditions:

- Four conditions are the boundary conditions discussed in Section 7.2.1.

- Two additional conditions are the continuity conditions at x_j discussed in Section 7.1.2.

- The remaining two conditions are the equilibrium equations on V_y and M_z at x_j.

The equilibrium conditions on V_y and M_z at x_j are jump conditions due to a point force or a point moment to be discussed in the next section.

7.2.3 Use of Template in Boundary Conditions or Jump Conditions

We discussed the concept of templates in drawing shear–moment diagrams in Section 6.4.2. Here we discuss it in determining the boundary conditions on V_y and M_z and jumps in these internal quantities due to a point force or a point moment.

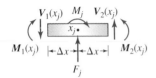

Figure 7.15 Template at x_j.

Recollect that a template is a small segment of a beam on which a point moment M_j and a point force F_j are a drawn (Δx tends to zero in Figure 7.15). The directions of F_j and M_j are arbitrary. F_j and M_j could be applied or reactive forces and moments. The ends at $+\Delta x$ and $-\Delta x$ represent the imaginary cut just to the left and just to the right of the point forces and point moments. On these imaginary cuts the internal shear force and the internal bending moment are drawn according to our sign convention, as discussed in Section 6.2.5. Equilibrium equations are written for this $2\Delta x$ segment of the beam to obtain the template equations,

$$V_2(x_j) - V_1(x_j) = -F_j$$

$$M_2(x_j) - M_1(x_j) = M_j$$

The moment equation does not contain the moment due to the forces because these moments will go to zero as Δx goes to zero.

If the point force on the beam is in the direction of F_j shown on the template, the template equation for force is used as given. If the point force on the beam is opposite to the direction of F_j shown on the template, then the template equation is used by changing the sign of F_j. The template equation for the moment is used in a similar fashion.

If x_j is a *left* boundary point, then there is no beam left of x_j; hence V_1 and M_1 are zero and we obtain the boundary conditions on V_y and M_z from V_2 and M_2.

If x_j is a *right* boundary point, then there is no beam right of x_j; hence V_2 and M_2 are zero and we obtain the boundary conditions on V_y and M_z from V_1 and M_1.

If x_j is in between the ends of the beam, then the jump in shear force and internal moment is calculated using the template equations.

An alternative to the templates is to draw free-body diagrams after making imaginary cuts at an infinitesimal distance from the point force and writing equilibrium equations. Example 7.5 demonstrates the use of the template. Examples 7.6 and 7.7 demonstrate the use of free-body diagrams to determine the boundary conditions or the jump in internal quantities.

EXAMPLE 7.5

The bending rigidity of the beam shown in Figure 7.16 is 135×10^6 lb·in^2 and the displacements of the beam in segments AB and BC are as given:

$$v_1 = 5(x^3 - 20x^2)10^{-6} \text{ in}, \qquad 0 \le x \le 20$$

$$v_2 = 10(x^3 - 30x^2 + 200x)10^{-6} \text{ in}, \qquad 20 \le x \le 40$$

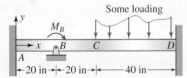

Figure 7.16 Beam in Example 7.5.

Determine: (a) the reactions at the left wall at A; (b) the reaction force at B and the applied moment M_B.

PLAN

By differentiating the given displacement functions and using Equations (7.1) and (7.4), the bending moment M_z and the shear force V_y can be found in segments AB and BC.

(a) Using the template in Figure 7.15, the reactions at A can be found from the values of V_y and M_z at $x = 0$.

(b) Using the template in Figure 7.15, the reaction force and the applied moment at B can be determined from the values of V_y and M_z before and after $x = 20$.

Solution   The shear force calculation requires the third derivative of the displacement functions. The functions v_1 and v_2 can be differentiated three times, as shown:

$$\frac{dv_1}{dx} = 5(3x^2 - 40x)10^{-6} \tag{E1}$$

$$\frac{d^2v_1}{dx^2} = 5(6x - 40)10^{-6} \tag{E2}$$

$$\frac{d^3v_1}{dx^3} = 5 \times 6 \times 10^{-6} = 30 \times 10^{-6} \tag{E3}$$

$$\frac{dv_2}{dx} = 10(3x^2 - 60x + 200)10^{-6} \tag{E4}$$

$$\frac{d^2v_2}{dx^2} = 10(6x - 60)10^{-6} \tag{E5}$$

$$\frac{d^3v_2}{dx^3} = 10 \times 6 \times 10^{-6} = 60 \times 10^{-6} \tag{E6}$$

Using Equations (7.1), (E2), and (E5), the internal moment can be found:

$$M_{z_1} = 135 \times 10^6 \times 5(6x - 40)10^{-6} = 675(6x - 40) \text{ in·lb} \tag{E7}$$

$$M_{z_2} = 135 \times 10^6 \times 10(6x - 60)10^{-6} = 1350(6x - 60) \text{ in·lb} \tag{E8}$$

Using Equations (7.4), (E3), and (E6), the shear force can be found:

$$V_{y_1} = 135 \times 10^6 \times 30 \times 10^{-6} = 4050 \text{ lb} \tag{E9}$$

$$V_{y_2} = 135 \times 10^6 \times 60 \times 10^{-6} = 8100 \text{ lb} \tag{E10}$$

The internal moment and shear force at A can be found by substituting $x = 0$ into Equations (E7) and (E9),

$$M_{z_1}(0) = 675(-40) = -27{,}000 \text{ in·lb} \qquad \text{(E11)}$$

$$V_{y_1}(0) = 4050 \text{ lb} \qquad \text{(E12)}$$

The internal moment and shear force just before and after B can be found by substituting $x = 20$ into Equations (E7) through (E10)

$$M_{z_1}(20) = 675(6 \times 20 - 40) = 54{,}000 \text{ in·lb} \qquad \text{(E13)}$$

$$M_{z_2}(20) = 1350(6 \times 20 - 60) = 81{,}000 \text{ in·lb} \qquad \text{(E14)}$$

$$V_{y_1}(20) = 4050 \text{ lb} \qquad \text{(E15)}$$

$$V_{y_2}(20) = 8100 \text{ lb} \qquad \text{(E16)}$$

Figure 7.17 shows the free-body diagram of the entire beam. It also shows the template of Figure 7.15 for convenience.

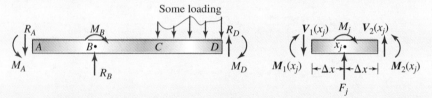

Figure 7.17 Free-body diagram of entire beam in Example 7.5.

If we compare the reaction force at A to F_j and the reaction moment to M_j in Figure 7.15, we obtain $F_j = -R_A$ and $M_j = -M_A$. As point A is the left end of the beam, $V_1(x_j)$ and $M_1(x_j)$ are zero on the template and $V_2(x_j) = V_{y_1}(0)$ and $M_2(x_j) = M_{z_1}(0)$. From the template equation we obtain $R_A = V_{y_1}(0)$ and $M_A = -M_{z_1}(0)$. Substituting Equations (E11) and (E12), we obtain

ANS. $\qquad R_A = 4050 \text{ lb} \qquad M_A = 27{,}000 \text{ in·lb}$

If we compare the reaction force at B to F_j and the applied moment to M_j in Figure 7.15, we obtain $F_j = R_B$ and $M_j = M_B$. Substituting for $x_j = 20$ and using Equations (E13) through (E16), we obtain R_B and M_B,

$$V_{y_2}(20) - V_{y_1}(20) = -R_B \qquad \text{or} \qquad \text{ANS.} \quad R_B = -3600 \text{ lb}$$

$$M_{z_2}(20) - M_{z_1}(20) = M_B \qquad \text{or} \qquad \text{ANS.} \quad M_B = 27{,}000 \text{ in·lb}$$

COMMENTS

1. An alternative to the use of the template is to draw a free-body diagram after making imaginary cuts at an infinitesimal distance from the point forces, as shown in Figure 7.18. The internal forces and moments must be drawn according to our sign convention. By writing equilibrium equations the required quantities can be found.

2. The free-body diagram of the entire beam in Figure 7.17 is not necessary. From the use of the template equations the force F_j and the moment M_j with the correct signs can be found. If F_j and M_j are positive, then R_B and M_B will be in the direction shown on the template. If these quantities are negative, then the direction is opposite. The figures were drawn for the purpose of explanation.

3. This problem demonstrates the basic principles used in determining the conditions on shear force and bending moment and relating these internal quantities to the reaction forces and moments. These basic principles will be used in subsequent examples in which the displacement functions will have to be determined first.

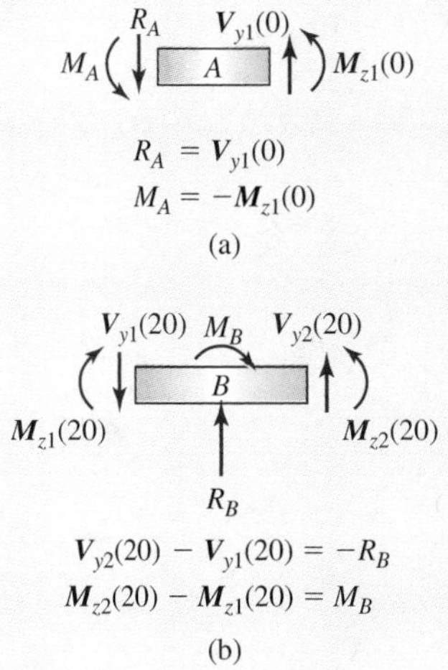

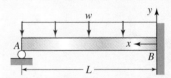

$$R_A = V_{y1}(0)$$
$$M_A = -M_{z1}(0)$$
(a)

$$V_{y2}(20) - V_{y1}(20) = -R_B$$
$$M_{z2}(20) - M_{z1}(20) = M_B$$
(b)

Figure 7.18 Alternative to template.

EXAMPLE 7.6

In terms of E, I, w, L, and x, determine: (a) the elastic curve; (b) the reaction force at A in Figure 7.19.

Method 1: Fourth-order differential equation approach

PLAN

(a) Substituting $p = -w$ in Equation (7.5), the fourth-order differential equation can be written. The two boundary conditions at A are zero deflection and zero moment and the two boundary conditions at B are zero deflection and zero slope. We can solve the boundary-value problem and obtain the elastic curve.

(b) We can draw a free-body diagram after making an imaginary cut just to the right of A and relate the reaction force to the shear force. We can find the shear force at point A by substituting $x = L$ in the solution obtained in part (a).

Figure 7.19 Beam and loading in Example 7.6.

Solution

(a) Noting that the distributed force is in the negative y direction, we can substitute $p = -w$ into Equation (7.5) and write the differential equation. The deflection at A is zero but the slope depends on the loading. Hence from Equation (7.6) the condition on the moment has to be specified. Since there is no applied moment, we conclude that the internal moment at A is zero. The deflection and slope at point B are zero. The boundary-value problem statement can be written as follows:

- Differential equation:

$$\frac{d^2}{dx^2}\left(EI_{zz}\frac{d^2v}{dx^2}\right) = -w \qquad \text{(E1)}$$

- Boundary conditions:

$$v(0) = 0 \qquad \text{(E2)}$$

$$\frac{dv}{dx}(0) = 0 \qquad \text{(E3)}$$

$$v(L) = 0 \qquad \text{(E4)}$$

$$EI_{zz}\frac{d^2v}{dx^2}(L) = 0 \qquad \text{(E5)}$$

Integrating Equation (E1) twice,

$$\frac{d}{dx}\left(EI_{zz}\frac{d^2v}{dx^2}\right) = -wx + c_1 \qquad \text{(E6)}$$

$$EI_{zz}\frac{d^2v}{dx^2} = -\frac{wx^2}{2} + c_1x + c_2 \qquad \text{(E7)}$$

Substituting Equation (E7) into Equation (E5), we obtain

$$c_1L + c_2 = \frac{wL^2}{2} \qquad \text{(E8)}$$

Integrating Equation (E7), we obtain

$$EI_{zz}\frac{dv}{dx} = -\frac{wx^3}{6} + c_1\frac{x^2}{2} + c_2x + c_3 \qquad \text{(E9)}$$

Substituting Equation (E9) into Equation (E3), we obtain

$$c_3 = 0 \qquad \text{(E10)}$$

Integrating Equation (E9) after substituting Equation (E10), we obtain

$$EI_{zz}v = -\frac{wx^4}{24} + c_1\frac{x^3}{6} + c_2\frac{x^2}{2} + c_4 \qquad \text{(E11)}$$

Substituting Equation (E11) into Equation (E2), we obtain

$$c_4 = 0 \tag{E12}$$

Substituting Equations (E12) and (E11) into Equation (E4), we obtain

$$\frac{c_1 L^3}{6} + \frac{c_2 L^2}{2} = \frac{wL^4}{24} \tag{E13}$$

Solving Equations (E8) and (E13) simultaneously, we obtain

$$c_1 = \frac{5wL}{8} \tag{E14}$$

$$c_2 = -\frac{wL^2}{8} \tag{E15}$$

Substituting Equations (E12), (E14), and (E15) into Equation (E11) and simplifying, we obtain the elastic curve,

ANS. $$v(x) = -\frac{w}{48EI_{zz}}(2x^4 - 5Lx^3 + 3L^2x^2) \tag{E16}$$

Dimension check: We note that all terms in parentheses on the right-hand side of Equation (E16) have the dimension of length to the power of 4, that is, $O(L^4)$. Thus Equation (E16) is dimensionally homogeneous. But we can also check whether the left-hand side and any one term of the right-hand side have the same dimension, as follows:

$$w \to O\left(\frac{F}{L}\right) \qquad x \to O(L) \qquad E \to O\left(\frac{F}{L^2}\right) \qquad I_{zz} \to O(L^4)$$

$$v \to O(L) \qquad \frac{wx^4}{EI_{zz}} \to O\left(\frac{(F/L)L^4}{(F/L^2)O(L^4)}\right) \to O(L) \to \text{checks}$$

(b) We make an imaginary cut just to the right of point A (infinitesimal distance) and draw the free-body diagram of the left part using the sign convention in Section 6.2.5 (Figure 7.20). By force equilibrium in the y direction we can relate the shear force at A to the reaction force at A,

$$R_A = V_A = V_y(L) \tag{E17}$$

From Equations (7.4), (E6), and (E14) the shear force is given as

$$V_y(x) = -\frac{d}{dx}\left(EI_{zz}\frac{d^2v}{dx^2}\right) = wx - \frac{5wL}{8} \tag{E18}$$

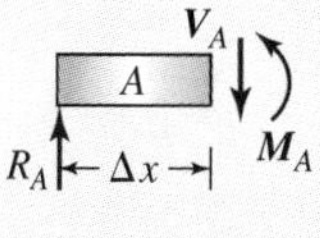

Figure 7.20 Infinitesimal equilibrium element at A in Example 7.6.

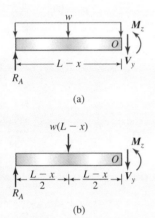

Figure 7.21 Free-body diagram in Example 7.6.

Substituting Equation (E18) into Equation (E17), we obtain the reaction at A as

ANS. $\quad R_A = \dfrac{3wL}{8}$

Method 2: Second-order differential equation approach

PLAN
We can make an imaginary cut at some arbitrary location x and use the left part to draw the free-body diagram. The moment expression will contain the reaction force at A as an unknown. The second-order differential equation, Equation (7.1), would generate two integration constants, leading to a total of three unknowns. The three conditions we need are that the displacement at A is zero and displacement and slope at B are zero. Solving the boundary-value problem, we can obtain the elastic curve and the unknown reaction force at A.

Solution We make an imaginary cut at a distance x from the right wall and take the left part of length $L - x$ to draw the free-body diagram shown in Figure 7.21 using the sign convention for internal quantities discussed in Section 6.2.5.

Balancing the moment at point O, we obtain the moment expression,

$$M_z - R_A(L - x) + w\frac{(L - x)^2}{2} = 0$$

or

$$M_z = R_A(L - x) - \frac{w}{2}(L^2 + x^2 - 2Lx) \tag{E19}$$

Substituting Equation (E1) into Equation (7.1) and writing the boundary conditions, we obtain the following boundary-value problem:

• Differential equation:

$$EI_{zz}\frac{d^2v}{dx^2} = R_A(L - x) - \frac{w}{2}(L^2 + x^2 - 2Lx) \tag{E20}$$

• Boundary conditions:

$$v(0) = 0 \tag{E21}$$

$$\frac{dv}{dx}(0) = 0 \tag{E22}$$

$$v(L) = 0 \tag{E23}$$

Integrating Equation (E20), we obtain

$$EI_{zz}\frac{dv}{dx} = R_A\left(Lx - \frac{x^2}{2}\right) - \frac{w}{2}\left(L^2x + \frac{x^3}{3} - Lx^2\right) + c_1 \quad \text{(E24)}$$

Substituting Equation (E24) into Equation (E22), we obtain

$$c_1 = 0 \quad \text{(E25)}$$

Integrating Equation (E24) after substituting Equation (E25), we obtain

$$EI_{zz}v = R_A\left(\frac{Lx^2}{2} - \frac{x^3}{6}\right) - \frac{w}{2}\left(\frac{L^2x^2}{2} + \frac{x^4}{12} - \frac{Lx^3}{3}\right) + c_2 \quad \text{(E26)}$$

Substituting Equation (E26) into Equation (E21), we obtain

$$c_2 = 0 \quad \text{(E27)}$$

Substituting Equations (E26) and (E27) into Equation (E23), we obtain

$$R_A\left(\frac{L^3}{2} - \frac{L^3}{6}\right) - \frac{w}{2}\left(\frac{L^4}{2} + \frac{L^4}{12} - \frac{L^4}{3}\right) = 0 \quad \text{or} \quad \text{ANS.} \quad R_A = \frac{3wL}{8}$$

$$\text{(E28)}$$

Substituting Equations (E27) and into Equation (E26) and simplifying, we obtain

$$\text{ANS.} \qquad v(x) = -\frac{w}{48EI_{zz}}(2x^4 - 5Lx^3 + 3L^2x^2) \quad \text{(E29)}$$

COMMENTS

1. Method 2 has less algebra than Method 1 and should be used whenever possible.

2. Suppose that in drawing the free-body diagram for calculating the internal moment, we had taken the right-hand part. Then we would have the wall reaction force and the wall reaction moment in the moment expression, that is, two unknowns rather than one. In such a case we would have to eliminate one of the unknowns using the static equilibrium equation for the entire beam. In other words, in statically indeterminate problems, the internal moment should contain a number of unknown reactions equal to the degree of static redundancy.

3. The moment boundary condition given by Equation (E5) in Method 1 is implicitly satisfied in Equation (E19). We can confirm this by substituting $x = L$ in Equation (E19).

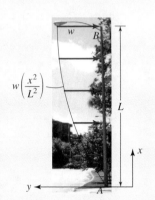

Figure 7.22 Beam and loading in Example 7.7.

EXAMPLE 7.7

A light pole is subjected to a wind pressure that varies as a quadratic function, as shown in Figure 7.22. In terms of E, I, w, L, and x, determine: (a) the deflection at the free end; (b) the ground reactions.

PLAN

(a) Though the problem is statically determinate, finding the moment as a function of x by static equilibrium is difficult. We can use the fourth-order differential equation, Equation (7.5). The four boundary conditions are: the deflection and slope at A are zero, and the moment and shear force at B are zero. We can solve the boundary-value problem and determine the elastic curve. By substituting $x = L$ in the elastic curve equation, we can obtain the deflection at the free end.

(b) By making an imaginary cut just above point A, we can relate the internal shear force and the internal moment at point A to the reactions at A. By sub stituting $x = 0$ in the moment and shear force expressions, we can obtain the shear force and moment values at point A.

Solution Noting that the distributed force is in the negative y direction, the differential equation can be written using Equation (7.5). The four boundary conditions discussed in the plan can also be written to complete the boundary-value problem statement as follows.

- Differential equation:

$$\frac{d^2}{dx^2}\left(EI_{zz}\frac{d^2 v}{dx^2}\right) = -w\left(\frac{x^2}{L^2}\right) \tag{E1}$$

- Boundary conditions:

$$v(0) = 0 \tag{E2}$$

$$\frac{dv}{dx}(0) = 0 \tag{E3}$$

$$EI_{zz}\frac{d^2 v}{dx^2}\bigg|_{x=L} = 0 \tag{E4}$$

$$\frac{d}{dx}\left(EI_{zz}\frac{d^2 v}{dx^2}\right)\bigg|_{x=L} = 0 \tag{E5}$$

Integrating Equation (E1), we obtain

$$\frac{d}{dx}\left(EI_{zz}\frac{d^2v}{dx^2}\right) = -\frac{wx^3}{3L^2} + c_1 \tag{E6}$$

Substituting Equation (E6) into Equation (E5), we obtain

$$c_1 = \frac{wL}{3} \tag{E7}$$

Substituting Equation (E7) into Equation (E6) and integrating, we obtain

$$EI_{zz}\frac{d^2v}{dx^2} = -\frac{wx^4}{12L^2} + \frac{wL}{3}x + c_2 \tag{E8}$$

Substituting Equation (E8) into Equation (E4), we obtain

$$c_2 = -\frac{wL^2}{4} \tag{E9}$$

Substituting Equation (E9) into Equation (E8) and integrating, we obtain

$$EI_{zz}\frac{dv}{dx} = -\frac{wx^5}{60L^2} + \frac{wLx^2}{6} - \frac{wL^2x}{4} + c_3 \tag{E10}$$

Substituting Equation (E10) into Equation (E3), we obtain

$$c_3 = 0 \tag{E11}$$

Substituting Equation (E11) into Equation (E10) and integrating, we obtain

$$EI_{zz}\frac{dv}{dx} = -\frac{wx^6}{360L^2} + \frac{wLx^3}{18} - \frac{wL^2x^2}{8} + c_4 \tag{E12}$$

Substituting Equation (E12) into Equation (E2), we obtain

$$c_4 = 0 \tag{E13}$$

Substituting Equation (E13) into Equation (E12) and simplifying, we obtain

$$v(x) = -\frac{w}{360EI_{zz}L^2}(x^6 - 20L^3x^3 + 45L^4x^2) \tag{E14}$$

Dimension check: We note that all terms in parentheses on the right-hand side of Equation (E14) have the dimension of length to the power of 6, that is, $O(L^6)$. Thus Equation (E14) is dimensionally homogeneous. But we can also check whether the left-hand side and any one term of the right-hand side have the same dimension as follows:

$$w \rightarrow O\left(\frac{F}{L}\right) \qquad x \rightarrow O(L) \qquad E \rightarrow O\left(\frac{F}{L^2}\right) \qquad I_{zz} \rightarrow O(L^4)$$

$$v \rightarrow O(L) \qquad \frac{wx^6}{EI_{zz}L^2} \rightarrow O\left(\frac{(F/L)L^6}{(F/L^2)L^4L^2}\right) \rightarrow O(L) \rightarrow \text{checks}$$

(a) Substituting $x = L$ into Equation (E14), we obtain the deflection at the free end as

$$\text{ANS.} \qquad v(L) = -\frac{13wL^4}{180EI_{zz}}$$

(b) We make an imaginary cut just to the right of point A $(\varepsilon \rightarrow 0)$ and take the left part to draw the free-body diagram shown in Figure 7.23. Balancing forces and moments we can relate the reaction force R_A and the reaction moment M_A to the internal shear force and the internal bending moment at point A,

$$M_z(0) = -M_A \tag{E15}$$

$$V_y(0) = -R_A \tag{E16}$$

Substituting Equations (E7) and (E6) into Equation (7.4) and Equations (E9) and (E8) into Equation (7.1), we can obtain the shear force and bending moment expressions,

$$M_z(x) = -\frac{wx^4}{12L^2} + \frac{wL}{3}x - \frac{wL^2}{4} \tag{E17}$$

$$V_y(x) = \frac{wx^3}{3L^2} - \frac{wL}{3} \tag{E18}$$

Substituting Equations (E17) and (E18) into Equations (E15) and (E16), we obtain the reaction force and the reaction moment,

$$\text{ANS.} \qquad R_A = \frac{wL}{3} \text{ upward} \qquad M_A = \frac{wL^2}{4} \text{ ccw}$$

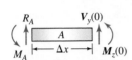

Figure 7.23 Infinitesimal equilibrium element at A in Example 7.7.

COMMENTS

1. The directions of R_A and M_A can be checked by inspection to be correct, as these are the directions necessary for equilibrium of the externally distributed force.

2. In drawing the free-body diagram in Figure 7.23, the reaction force R_A and the reaction moment M_A can be drawn in any direction, but the internal quantities V_y and M_z must be drawn according to the sign convention in Section 6.2.5. Irrespective of the direction in which R_A and M_A are drawn, the final answer will be as given. The sign in the equilibrium equations, Equations (E15) and (E16), will account for the assumed directions of the reactions.

PROBLEM SET 7.2

Fourth-order boundary-value problems

7.20 The displacement in the y direction in segment AB, shown in Figure P7.20, was found to be $v(x) = (20x^3 - 40x^2)10^{-6}$ in. If the bending rigidity is 135×10^6 lb·in², determine the reaction force and the reaction moment at the wall at A.

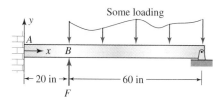

Figure P7.20

7.21 The displacement in the y direction in section AB, shown in Figure P7.21, is given by $v_1(x) = -3(x^4 - 20x^3)10^{-6}$ in and in BC by $v_2(x) = -8(x^2 - 100x + 1600)10^{-3}$ in. If the bending rigidity is 135×10^6 lb·in², determine: (a) the reaction force at B and the applied moment M_B; (b) the reactions at the wall at A.

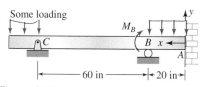

Figure P7.21

In Problems 7.22 and 7.23, determine the elastic curve and the reaction(s) at A in terms of E, I, P, w, and x.

7.22 Using Figure P7.22 solve the problem as described in text above.

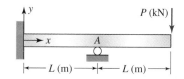

Figure P7.22

7.23 Using Figure P7.23 solve the problem as described in text above.

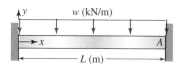

Figure P7.23

7.24 Determine the slope at $x = L$ and the reaction moment at the left wall in terms of E, I, w, and L (Figure P7.24).

7.25 Determine the deflection and the moment reaction at $x = L$ in terms of E, I, w, and L (Figure P7.25).

7.26 Determine the deflection and the slope at $x = L$ (Figure P7.26) in terms of E, I, w, and L.

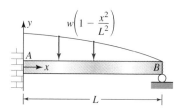

Figure P7.24

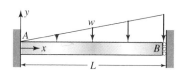

Figure P7.25

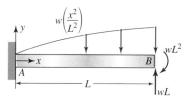

Figure P7.26

7.27 A linear spring that has a spring constant K is attached at the end of a beam, as shown in Figure P7.27. In terms of w, E, I, L, and K, write the boundary-value problem but do not integrate or solve.

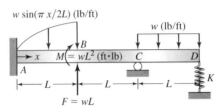

Figure P7.27

Historical problems[3]

7.28 The beam and loading shown in Figure P7.28 was the first statically indeterminate beam for which a solution was obtained by Navier. Show that Navier's solution for the reaction at A given here is correct,

$$R_A = \frac{Pa^2(3L-a)}{2L^3}$$

where $L = a + b$.

7.29 Jacob Bernoulli incorrectly assumed that the neutral axis was tangent to the concave side of the curve in Figure P7.29 and obtained the following relationship:

$$\frac{Ebh^3}{3}\left(\frac{1}{R}\right) = Px$$

where R is the radius of curvature of the beam at any location x. Derive this equation based on Bernoulli's assumption and show that it is incorrect by a factor of 4. (*Hint:* Follow the process in Section 6.1 and take the moment about point B.)

7.30 Clebsch considered a beam loaded by several concentrated forces P_i placed at a location x_i, as shown in Figure P7.30. He obtained the second-order differential equation between the concentrated forces. By integration he obtained the slope and deflection as given and concluded that all C_i's are equal and all D_i's are equal. Show that the conclusion of Clebsch

---

[3]See Section 7.6 for additional details.

is correct. For $x_i \le x \le x_{i+1}$

$$EI\frac{d^2v}{dx^2} = Rx - \sum_{j=1}^{i} P_j(x - x_j)$$

$$EI\frac{dv}{dx} = R\frac{x^2}{2} - \sum_{j=1}^{i} P_i\frac{(x-x_i)^2}{2} + C_i$$

$$EIv = R\frac{x^3}{6} - \sum_{j=1}^{i} P_i\frac{(x-x_i)^3}{6} + C_i x + D_i$$

Stretch yourself

7.31 A beam resting on an elastic foundation has a distributed spring force that depends on the deflections at a point acting as shown in Figure P7.31. Show that the differential equation governing the deflection of the beam is

$$\frac{d^2}{dx^2}\left(EI_{zz}\frac{d^2v}{dx^2}\right) + kv = p \qquad (7.7)$$

where k is the foundation modulus, that is, spring constant per unit length.

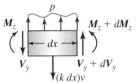

Figure P7.31 Elastic foundation effect.

7.32 To account for shear, the assumption of planes remaining perpendicular to the axis of the beam (Assumption 3 in Section 6.2) is dropped, and it is assumed that the plane rotates by the angle ψ from the vertical. This yields the following displacement equations:

$$u = -y\,\psi(x) \qquad v = v(x)$$

The rest of the derivation[4] is as before. Show that the following equations apply:

$$\frac{d}{dx}\left[GA\left(\frac{dv}{dx} - \psi\right)\right] = -p$$

$$\frac{d}{dx}\left(EI_{zz}\frac{d\psi}{dx}\right) = -GA\left(\frac{dv}{dx} - \psi\right) \qquad (7.8)$$

---

[4]Use Equations (2.10a) and (2.10d) to get ε_{xx} and γ_{xy}. Use Hooke's law, the static equivalency equations [Equations (6.1) and (6.9)], and the equilibrium equations [Equations (6.13) and (6.14)].

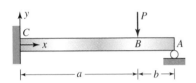

Figure P7.28

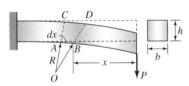

Figure P7.29

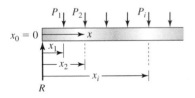

Figure P7.30

where A is the cross-sectional area and G is the shear modulus of elasticity. Beams governed by these equations are called *Timoshenko beams*.

7.33 Figure P7.33 shows a differential element of a beam that is free to vibrate, where ρ is the material density, A is the cross-sectional area, and $\partial^2 v/\partial t^2$ is the linear acceleration. Show that the equation for dynamic equilibrium is given by

$$\frac{\partial^2 v}{\partial t^2} + c^2 \frac{\partial^4 v}{\partial x^4} = 0 \qquad (7.9)$$

where $c = \sqrt{EI_{zz}/\rho A}$.

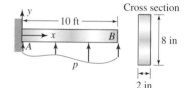

Figure P7.33   Dynamic equilibrium.

7.34 Show by substitution that the following solution satisfies Equation (7.9):

$$v(x, t) = G(x)H(t)$$

$$G(x) = A\cos\omega x + B\sin\omega x + C\cosh\omega x + D\sinh\omega x$$

$$H(t) = E\cos(c\omega^2)t + D\sin(c\omega^2)t$$

7.35 By substitution show that the following deflection solution satisfies the fourth order boundary value problem of the cantilever beam shown in Figure P7.35.

$$v(x) = \frac{1}{6EI}\Big[R_A x^3 + 3M_A x^2$$

$$+ \int_0^x (x - x_1)^3 p(x_1)\, dx_1 \Big] \qquad (7.10)$$

where $R_A = -\int_0^L p(x_1)\, dx_1$ and $M_A = \int_0^L x_1 p(x_1)\, dx_1$.

Computer problems

7.36 Table P7.36 shows the value of distributed load at several point along the axis of a rectangular beam. Determine the slope and deflection at the free end using. Use modulus of elasticity as 2000 ksi.

TABLE P7.36   Data in Problem 7.36

x (ft)	$p(x)$ (lb/ft)	x (ft)	$p(x)$ (lb/ft)
0	275	6	377
1	348	7	316
2	398	8	233
3	426	9	128
4	432	10	0
5	416		

7.37 For the beam and loading given in Problem 7.35, determine the slope and deflection at the free end in the following manner. First represent the distributed load by $p(x) = a + bx + cx^2$ and, using the data in Table P7.36, determine constants a, b, and c by the least-squares method. Then using fourth-order differential equations solve the boundary-value problem. Use the modulus of elasticity as 2000 ksi.

7.38 Table P7.38 shows the measured radii of a solid tapered circular beam at several points along the axis, as shown in Figure P7.38. The beam is made of aluminum ($E = 28$ GPa) and has a length of 1.5 m. Determine the slope and deflection at point B.

TABLE P7.38   Data for Problem 7.38

x (m)	$R(x)$ (mm)	x (m)	$R(x)$ (mm)
0.0	100.6	0.8	60.1
0.1	92.7	0.9	60.3
0.2	82.6	1.0	59.1
0.3	79.6	1.1	54.0
0.4	75.9	1.2	54.8
0.5	68.8	1.3	54.1
0.6	68.0	1.4	49.4
0.7	65.9	1.5	50.6

7.39 Let the radius of the tapered beam in Problem 7.38 be represented by the equation $R(x) = a + bx$. Using the data in Table P7.38, determine constants a and b by the least-squares method and then find slope and deflection at point B by analytical integration.

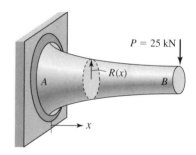

Figure P7.35

Figure P7.38

*7.3 SUPERPOSITION

The assumptions and limitations that were imposed in deriving the simplest theory for beam bending ensured that we have a linear theory. As a consequence, the differential equations governing beam deflection, that is, Equation (7.1) or Equation (7.5), are linear differential equations, and hence the principle of superposition can be applied to beam deflection.

The leftmost beam in Figure 7.24 is loaded with a uniformly distributed load w and a concentrated load P_1. The superposition principle says that the deflection of a beam with uniform load w and point force P_1 is equal to the sum of the deflections calculated by considering each load separately, as shown on the right two beams in Figure 7.24. Though the example in Figure 7.24 demonstrates the principle of superposition, there is no intrinsic gain in calculating the deflection of each load separately and adding to find the final answer. But if the solutions to basic cases are tabulated, as in Table 7.1, then the principle of superposition becomes a very useful tool to obtain results quickly. Thus the maximum deflection of the beam on the left can be found using the results of cases 1 and 3. Comparing the loading of the two beams in Figure 7.24 to those shown for cases 1 and 3, we note that $P = -P_1 = -wL$ and $p_0 = -w$, $a = L$, and $b = 0$. Substituting these values into v_{max} given in Table 7.1 and adding, we obtain

$$v_{max} = -\frac{(wL)L^3}{3EI} - \frac{wL^4}{8EI} = -\frac{11wL^4}{24EI}$$

Another very useful application of superposition is when solving for the deflection of statically indeterminate beams. Consider a beam built in at one end and simply supported at the other end with a uniformly distributed load, as shown in Figure 7.25. The support at A can be replaced by a reaction force, and once more the total loading can be shown as the sum of two individual loads, as shown at right in Figure 7.25. Comparing the loading of the two beams in Figure 7.24 to those shown for cases 1 and 3 in Table 7.1, we note that $P = R_A$ and $p_0 = -w$, $a = L$, and $b = 0$ and that v_{max} is at point A in both cases. Substituting these values into v_{max} given in Table 7.1 and adding, we obtain

$$v_A = \frac{R_A L^3}{3EI} - \frac{wL^4}{8EI}$$

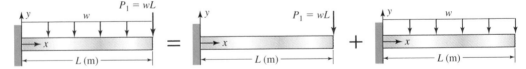

Figure 7.24 Example of superposition principle.

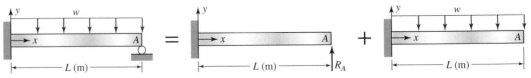

Figure 7.25 Example of use of superposition principle in solving statically indeterminate beam deflection.

But the deflection at A must be zero in the original beam. Thus we can solve for the reaction force as $R_A = 3wL/8EI$. Now the solution of $v(x)$ given in Table 7.1 can be superposed to obtain

$$v(x) = \frac{R_A x^2}{6EI}(3L - x) + \frac{(-w)x^2}{24EI}(x^2 - 4Lx + 6L^2)$$

Substituting for R_A and simplifying, the solution for the elastic curve is

$$v(x) = \frac{wx^2(-2x^2 + 5Lx - 3L^2)}{48EI}$$

TABLE 7.1 Deflections and slopes of beams[a]

Case	Beam and Loading	Maximum Deflection and Slope	Elastic Curve
1		$v_{max} = \dfrac{Pa^2}{6EI}(2a + 3b)$ $\theta_{max} = \dfrac{Pa^2}{2EI}$	$v = \dfrac{Px^2}{6EI}(3a - x)$ for $0 \le x \le a$ $v = \dfrac{Pa^2}{6EI}(3x - a)$ for $x \ge a$
2		$v_{max} = \dfrac{Ma(a + 2b)}{2EI}$ $\theta_{max} = \dfrac{Ma}{EI}$	$v = \dfrac{Mx^2}{2EI}$ for $0 \le x \le a$ $v = \dfrac{Ma}{2EI}(2x - a)$ for $x \ge a$
3		$v_{max} = \dfrac{p_0 a^3(3a + 4b)}{24EI}$ $\theta_{max} = \dfrac{p_0 a^3}{6EI}$	$v = \dfrac{p_0 x^2}{24EI}(x^2 - 4ax + 6a^2)$ for $0 \le x \le a$ $v = \dfrac{p_0 a^3}{24EI}(4x - a)$ for $x \ge a$
4		$v_{max} = \dfrac{PL^3}{48EI}$ $\theta_{max} = \dfrac{PL^2}{16EI}$	$v = \dfrac{Px}{48EI}(3L^2 - 4x^2)$ for $0 \le x \le \dfrac{L}{2}$
5		$v_{max} = \dfrac{5p_0 L^4}{384EI}$ $\theta_{max} = \dfrac{p_0 L^3}{24EI}$	$v = \dfrac{p_0 x}{24EI}(x^3 - 2Lx^2 + L^3)$
6		$v_{max} = \dfrac{ML^2}{9\sqrt{3}EI}$ @ $x = 0.4226L$ $\theta_1 = \dfrac{ML}{3EI}$ $\theta_2 = \dfrac{ML}{6EI}$	$v = \dfrac{Mx}{6EIL}(x^2 - 3Lx + 2L^2)$

[a]These equations can be used for composite beams by replacing the bending rigidity EI by the sum of bending rigidities $\sum E_i I_i$.

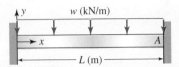

Figure 7.26 Beam in Example 7.8.

EXAMPLE 7.8

Using the principle of superposition and Table 7.1, determine for the beam shown in Figure 7.26: (a) the reactions at A; (b) the maximum deflection.

PLAN

(a) The wall at A can be replaced by a force reaction R_A and a moment reaction M_A. Thus the beam would be a cantilever beam with a uniformly distributed load, a point force at the end, and a point moment at the end, corresponding to the first three cases in Table 7.1. Superposing the slope and deflection values from Table 7.1 and equating the result to zero would generate two equations in the two unknowns R_A and M_A that can be solved to obtain the reactions at A.

(b) From the symmetry of the problem, we can conclude that the maximum deflection will occur at the center. Substituting $x = L/2$ in the elastic curve equation of Table 7.1 and adding the results, we can find the maximum deflection of the beam.

Solution

(a) The right wall at A can be replaced by a reaction force and a reaction moment, as shown at left in Figure 7.27. The total loading on the beam can be considered as the sum of the three loadings shown at right in Figure 7.27.

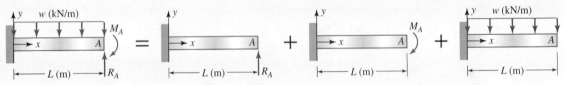

Figure 7.27 Superposition of three loadings in Example 7.8.

Comparing the three beam loadings in Figure 7.27 to that shown for cases 1 through 3 in Table 7.1, we obtain $P = R_A$, $M = -M_A$, and $p = -w$, $a = L$, and $b = 0$. Noting that v_{max} and θ_{max} shown in Table 7.1 for the cantilever beam occur at point A, we can substitute the load values and superpose to obtain the deflection v_A and the slope at θ_A. Noting that at the wall at A the deflection v_A and the slope at θ_A must be zero, we obtain two simultaneous equations in R_A and M_A,

$$v_A = \frac{R_A L^3}{3EI} + \frac{(-M_A)L^2}{2EI} + \frac{(-w)L^4}{8EI} = 0$$

or

$$8R_A L - 12M_A = 3wL^2 \qquad \text{(E1)}$$

$$\theta_A = \frac{R_A L^2}{2EI} + \frac{(-M_A)L}{EI} + \frac{(-w)L^3}{6EI} = 0$$

or

$$3R_A L - 6M_A = wL^2 \qquad \text{(E2)}$$

Equations (E1) and (E2) can be solved to obtain

$$\text{ANS.} \qquad R_A = \frac{wL}{2} \qquad M_A = \frac{wL^2}{12} \qquad \text{(E3)}$$

(b) The maximum deflection would occur at the center of the beam. Substituting $x = L/2$, $P = R_A = wL/2$, $M = -M_A = -wL/12$, and $p = -w$ in the equation of the elastic curve for cases 1 through 3 in Table 7.1 and superposing the solution, we obtain

$$v_{max} = v\left(\frac{L}{2}\right) = \frac{(wL/2)(L/2)^2}{6EI}\left(3L - \frac{L}{2}\right) + \frac{-(wL^2/12)(L/2)^2}{2EI}$$
$$+ \frac{(-w)(L/2)^2}{24EI}\left[\left(\frac{L}{2}\right)^2 - 4L\left(\frac{L}{2}\right) + 6L^2\right]$$

or

$$v_{max} = \frac{5wL^4}{96EI} - \frac{wL^4}{96EI} - \frac{17wL^4}{384EI}$$

$$\text{ANS.} \qquad v_{max} = -\frac{wL^4}{384EI}$$

COMMENTS

1. All terms in Equations (E1) and (E2) have the same dimension, as they should. If this were not the case, then the equations arrived at using superposition and the subsequent simplifications must be examined carefully to ensure dimensional homogeneity.

2. By symmetry we know that the reaction forces at each wall must be equal; hence the value of the reaction forces should be $wL/2$, as calculated in Equation (E3).

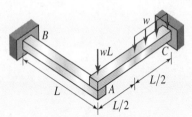

Figure 7.28 Two cantilever beams
in Example 7.9.

EXAMPLE 7.9

The end of one cantilever beam rests on the end of another cantilever
beam, as shown in Figure 7.28. Both beams have length L and bending
rigidity EI. Determine the deflection at A and the wall reactions at B
and C in terms of w, L, E, and I.

PLAN

The two beams can be separated by putting an unknown force R_A that
is equal but opposite in direction on each beam at point A. Using case 1
in Table 7.1 for beam AB, the deflection at A can be found in terms of
R_A. Using cases 1 and 3 for beam AC, the deflection at A can be found
by superposition. By equating the deflection at A for the two beams,
the force R_A can be found.

(a) Once R_A is known, the deflection at A is found from the equation
written for beam AB.

(b) The reactions at B and C can be found using equilibrium equations
on each beam's free-body diagram.

Solution

(a) The assembly of the beams shown in Figure 7.28 can be represented by two beams with a force R_A that acts in equal but opposite
directions, as shown in Figure Figure 7.29a and b. The loading on
the beam in Figure 7.29b can be represented as the sum of the two
loadings shown in Figure 7.29c and d.

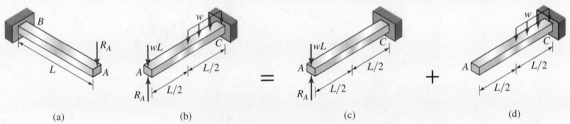

| (a) | (b) | | (c) | | (d) |

Figure 7.29 Analysis of beam assembly by superposition in Example 7.9.

Comparing the beam of Figure 7.29a to that shown in case 1
in Table 7.1, we obtain $P = -R_A$, $a = L$, and $b = 0$. Noting that v_{max}
in case 1 occurs at A, the deflection at A can be written as

$$v_A = \frac{(-R_A)L^2}{6EI}2L = -\frac{R_A L^3}{3EI} \qquad \text{(E1)}$$

Comparing the beam in Figure 7.29c to that of case 1 in Table 7.1,
we obtain $P = R_A - wL$, $a = L$, and $b = 0$. Comparing the beam in

Figure 7.29*d* to that of case 3 in Table 7.1, we obtain $p_0 = -w$, $a = L/2$, and $b = L/2$. Noting that v_{max} for both cases occurs at A, by superposition the deflection at A can be written as

$$v_A = \frac{(R_A - wL)L^2}{6EI}2L + \frac{(-w)(L/2)^3(3L/2 + 4L/2)}{24EI}$$

$$= \frac{(R_A - wL)L^3}{3EI} - \frac{7wL^4}{384EI} \qquad (E2)$$

Equating Equations (E1) and (E2), the reaction R_A can be found,

$$-\frac{R_A L^3}{3EI} = \frac{(R_A - wL)L^3}{3EI} - \frac{7wL^4}{384EI} \qquad \text{or} \qquad R_A = \frac{135wL}{256} \quad (E3)$$

Substituting Equation (E3) into Equation (E1), we obtain the deflection at A as

$$\text{ANS.} \qquad v_A = -\frac{45wL^4}{256EI}$$

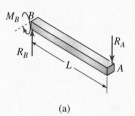

(a)

(b) The reactions at the wall can be found from the free-body diagrams of each beam, as shown in Figure 7.30. By balancing the forces in the y direction and the moments about B in Figure 7.30*a*, the reactions at B can be found,

$$\text{ANS.} \qquad R_B = R_A = \frac{135wL}{256}$$

$$M_B - R_A L = 0 \qquad \text{or} \qquad \text{ANS.} \qquad M_B = \frac{135wL^2}{256}$$

By balancing the forces in the y direction and the moments about C in Figure 7.30*b*, the reactions at C can be found,

$$R_C + R_A - wL - \frac{wL}{2} = 0 \qquad \text{or} \qquad \text{ANS.} \qquad R_C = \frac{249wL}{256}$$

$$M_C + R_A L - wL(L) - \frac{wL}{2}\left(\frac{L}{4}\right) = 0 \qquad \text{or} \qquad \text{ANS.} \qquad M_C = \frac{153wL^2}{256}$$

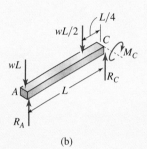

(b)

Figure 7.30 Free-body diagrams in Example 7.9.

COMMENT

This example demonstrates how the analysis and design of structures can be significantly simplified by using the principle of superposition. Handbooks now document an extensive number of cases for which beam deflections are known. These cases can be used in a wide variety of beam assemblies. But to develop a list of formulas (as in Table 7.1) requires a knowledge of methods such as those described in Sections 7.1 and 7.2.

*7.4 DEFLECTION BY DISCONTINUITY FUNCTIONS

The representation of the distributed load p, or moment $\boldsymbol{M}_z$, by different functions for different parts of the beam, and the determination of the integration constants to satisfy the continuity conditions and equilibrium conditions at the junctions x_j are tedious and algebraically intensive tasks, which may be unavoidable for a complicated distributed loading function. But for a large class of engineering problems, where the distributed loads are either constant or varying linearly, there is an alternative that avoids the algebraic tedium. The method is based on the concept of discontinuity functions, which is elaborated in this section.

7.4.1 Discontinuity Functions

Consider a distributed load p and an equivalent load $P = p\varepsilon$, as shown in Figure 7.31. Suppose we now let the intensity of the distributed load increase continuously to infinity, while we decrease the length over which the distributed force is applied to zero in such a manner that the area $p\varepsilon$ remains a finite quantity. Then we obtain a concentrated force P applied at $x = a$. Mathematically this is stated as

$$P = \lim_{p \to \infty} \lim_{\varepsilon \to 0} (p\varepsilon)$$

Rather than write the limit operations to represent a concentrated force, the following notation is used for representing the concentrated force: $P\langle x - a\rangle^{-1}$.

The function $\langle x - a\rangle^{-1}$ is called the *Dirac delta function*, or delta function. The delta function is zero except in an infinitesimal region near a. As x tends towards a, the delta function tends to infinity, but the area under the function is equal to 1. Mathematically the delta function is defined as

$$\langle x - a\rangle^{-1} = \begin{cases} 0, & x \neq a \\ \infty, & x \to a \end{cases} \qquad \int_{a-\varepsilon}^{a+\varepsilon} \langle x - a\rangle^{-1} \, dx = 1 \qquad (7.11)$$

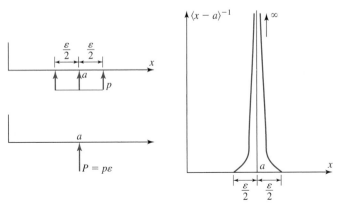

Figure 7.31 Delta function.

Step function   Ramp function

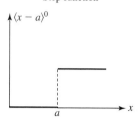
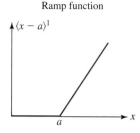
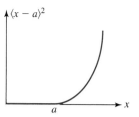

Figure 7.32 Discontinuity functions.

Now consider the following integral of the delta function:

$$\int_{-\infty}^{x} \langle x - a \rangle^{-1} \, dx$$

The lower limit of minus infinity emphasizes that the point is before a. If $x < a$, then in the interval of integration, the delta function is zero at all points; hence the integral value is zero. If $x > a$, then the integral can be written as the sum of three integrals,

$$\int_{-\infty}^{a-\varepsilon} \langle x - a \rangle^{-1} \, dx + \int_{a-\varepsilon}^{a+\varepsilon} \langle x - a \rangle^{-1} \, dx + \int_{a+\varepsilon}^{x} \langle x - a \rangle^{-1} \, dx$$

The first and third integrals are zero because the delta function is zero at all points in the interval of integration, whereas the second integral is equal to 1 as per Equation (7.11). Thus the integral $\int_{-\infty}^{x} \langle x - a \rangle^{-1} \, dx$ is zero before a and one after a. It is called the *step function* and is represented by the notation $\langle x - a \rangle^{0}$ (see Figure 7.32):

$$\langle x - a \rangle^{0} = \int_{-\infty}^{x} \langle x - a \rangle^{-1} \, dx = \begin{cases} 0, & x < a \\ 1, & x > a \end{cases} \tag{7.12}$$

Now consider the integral of the step function,

$$\int_{-\infty}^{x} \langle x - a \rangle^{0} \, dx$$

If $x < a$, then in the interval of integration the step function is zero at all points. Hence the integral value is zero. If $x > a$, then we can write the integral as the sum of two integrals,

$$\int_{-\infty}^{a} \langle x - a \rangle^{0} \, dx + \int_{a}^{x} \langle x - a \rangle^{0} \, dx$$

The first integral is zero because the step function is zero at all points in the interval of integration, whereas the second integral value is $x - a$. The integral $\int_{-\infty}^{x} \langle x - a \rangle^{0} \, dx$ is called the *ramp function*. It is represented by the notation $\langle x - a \rangle^{1}$ and is shown in Figure 7.32. Proceeding in this manner we can define an entire class of functions, which are represented mathematically as follows:

$$\langle x - a \rangle^{n} = \begin{cases} 0, & x \leq a \\ (x - a)^{n}, & x > a \end{cases} \tag{7.13}$$

We can also generate the following integral formula from Equation (7.13):

$$\int_{-\infty}^{x} \langle x - a \rangle^{n} \, dx = \frac{\langle x - a \rangle^{n+1}}{n + 1}, \qquad n \geq 0 \tag{7.14}$$

We define one more function, called the *doublet function*. It is represented by the notation $\langle x - a \rangle^{-2}$ and is defined mathematically as

$$\langle x - a \rangle^{-2} = \begin{cases} 0, & x \neq a \\ \infty, & x \to a \end{cases} \qquad \int_{-\infty}^{x} \langle x - a \rangle^{-2} \, dx = \langle x - a \rangle^{-1} \qquad (7.15)$$

The delta function $\langle x - a \rangle^{-1}$ and the doublet function $\langle x - a \rangle^{-2}$ become infinite at $x = a$, that is, they are singular at $x = a$ and are referred to as *singularity functions*.

Definition 2 The entire class of functions $\langle x - a \rangle^{n}$ for positive and negative n are called *discontinuity functions*.

The discontinuity functions are zero if the argument is negative. By differentiating Equations (7.12), (7.14), and (7.15) we can obtain the following formulas:

$$\frac{d\langle x - a \rangle^{-1}}{dx} = \langle x - a \rangle^{-2} \qquad \frac{d\langle x - a \rangle^{0}}{dx} = \langle x - a \rangle^{-1}$$

$$\frac{d\langle x - a \rangle^{n}}{dx} = n\langle x - a \rangle^{n-1}, \qquad n \geq 1$$

$$(7.16)$$

7.4.2 Use of Discontinuity Functions

Before proceeding to develop a method for solving for the elastic curve using discontinuity functions, we discuss the process by which the internal moment M_z and the distributed force p can be written using the discontinuity functions. We will develop the procedure using a simple example of a cantilever beam subject to different types of loading, as shown in Figure 7.33. Then we will generalize the procedure to more general loading and types of support.

When we make an imaginary cut before $x = a$ in the cantilever beams shown in Figure 7.33, the internal moment M_z will be zero. If the imaginary cut is made after

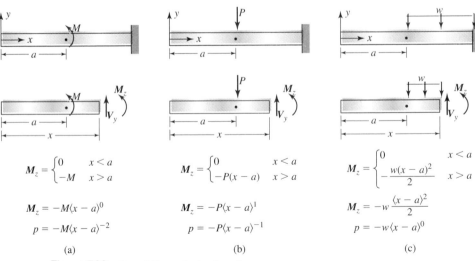

$$M_z = \begin{cases} 0 & x < a \\ -M & x > a \end{cases} \qquad M_z = \begin{cases} 0 & x < a \\ -P(x-a) & x > a \end{cases} \qquad M_z = \begin{cases} 0 & x < a \\ -\dfrac{w(x-a)^2}{2} & x > a \end{cases}$$

$$M_z = -M\langle x-a \rangle^0 \qquad\qquad M_z = -P\langle x-a \rangle^1 \qquad\qquad M_z = -w\frac{\langle x-a \rangle^2}{2}$$

$$p = -M\langle x-a \rangle^{-2} \qquad\qquad p = -P\langle x-a \rangle^{-1} \qquad\qquad p = -w\langle x-a \rangle^0$$

(a)   (b)   (c)

Figure 7.33 Use of discontinuity functions.

$x = a$, then the internal moment M_z will not be zero and can be determined using a free-body diagram. Once the moment expression is known, then it can be rewritten using the discontinuity functions. This moment expression can be used to find displacement using the second-order differential equation, Equation (7.1). However, if the fourth-order differential equation, Equation (7.5), has to be solved, then the expression of the distributed force p is needed. Now the distributed force can be obtained from the moment expression using the identity that is obtained by substituting Equation (6.14) into Equation (6.13), namely, $d^2M_z/dx^2 = p$. By using Equation (7.16) we can obtain the distributed force expression from the moment expression, as shown in Figure 7.33.

If the loading is the distributed load, as in the beam in Figure 7.33c, then the expression for the distributed load can be obtained directly, without the free-body diagram, and the moment expression obtained by integrating twice. For the concentrated force and moment also, it is not difficult to recognize the type of discontinuity function that will be used in the representation. The difficulty lies in obtaining the correct sign in the expression for the internal moment M_z and the distributed force p. We shall overcome this problem by using a template to guide us. A template is created by making an imaginary cut beyond the applied load. On the imaginary cut the internal moment is drawn according to the sign convention discussed in Section 6.2.5. A moment equilibrium equation is written. If the applied load is in the assumed direction on the template, then the sign used is the sign in the moment equilibrium equation. If the direction of the applied load is opposite to that on the template, then the sign in the equilibrium equation is changed. The beams shown in Figure 7.33 are like templates for the given coordinate systems.

EXAMPLE 7.10

Write the moment and distributed force expressions using discontinuity functions for the three templates shown in Figure 7.34.

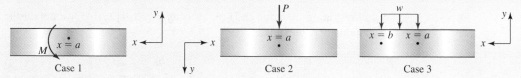

Figure 7.34   Three cases of Example 7.10.

PLAN

For cases 1 and 2 we can make an imaginary cut after $x = a$ and draw the shear force and bending moment according to the sign convention in Section 6.2.5. By equilibrium we can obtain the moment expression and rewrite it using discontinuity functions. By differentiating twice, we can obtain the distributed force expression. For case 3 we can write the expression for the distributed force using discontinuity functions and integrate twice to obtain the moment expression.

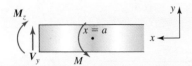

Figure 7.35 Case 1 in Example 7.10.

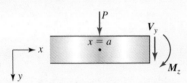

Figure 7.36 Case 2 in Example 7.10.

Solution

Case 1 (Figure 7.35): We make an imaginary cut at $x > a$ and draw the free-body diagram using the sign convention in Section 6.2.5. We then obtain the moment expression,

$$M_z = M \qquad \text{(E1)}$$

Equation (E1) is valid only after $x > a$. Using the step function we can write the moment expression, and by differentiating twice as per Equation (7.16) we obtain

ANS. $\quad M_z = M\langle x - a\rangle^0 \qquad p = M\langle x - a\rangle^{-2}$

Case 2 (Figure 7.36): We make an imaginary cut at $x > a$ and draw the free-body diagram using the sign convention in Section 6.2.5. We then obtain the moment expression,

$$M_z = P(x - a) \qquad \text{(E2)}$$

Equation (E2) is valid only after $x > a$. Using the ramp function we can write the moment expression, and by differentiating twice as per Equation (7.16) we can obtain the distributed force,

ANS. $\quad M_z = P\langle x - a\rangle^1 \qquad p = P\langle x - a\rangle^{-1}$

Case 3: The distributed force is in the negative y direction. Its start can be represented by the step function at $x = a$. The end of the distributed force can also be represented by a step function using a sign opposite to that used at the start,

$$p = -w\langle x - a\rangle^0 + w\langle x - b\rangle^0 \qquad \text{(E3)}$$

Integrating Equation (E3) twice using Equation (7.14), we obtain the moment expression,

ANS. $\quad M_z = -\frac{w}{2}\langle x - a\rangle^2 + \frac{w}{2}\langle x - b\rangle^2 \qquad \text{(E4)}$

COMMENTS

1. The three cases shown could be part of a beam with more complex loading. But the contribution for each of the loads would be calculated as shown in the example.

2. In obtaining Equation (E4) we did not write integration constants, as all we are seeking at this stage are expressions for use in the calculation of displacements. However, when we shall be integrating for displacements, we will write integration constants that will be determined from boundary conditions.

3. In case 3 we did not have to draw the free-body diagram. This is an advantage in problems where the distributed load changes character over the length of the beam. Thus even for statically determinate beams it may be advantageous to start with the fourth-order differential equation, rather than the second-order differential equation.

EXAMPLE 7.11

Using discontinuity functions, determine the equation of the elastic curve in terms of E, I, L, P, and x for the beam shown in Figure 7.7.

PLAN

Two templates can be created, one for an applied moment and one for the applied force. Using the templates as a guide, the moment expression in terms of discontinuity functions can be written. The second-order differential equation, Equation (7.1), can be written and solved using the zero deflection boundary conditions at A and C to obtain the elastic curve.

Solution   Figure 7.37 shows two templates. By balancing the moment about point O, the moment expressions for the two templates can be written as

$$M_z = M\langle x - a\rangle^0 \qquad M_z = F\langle x - a\rangle^1$$

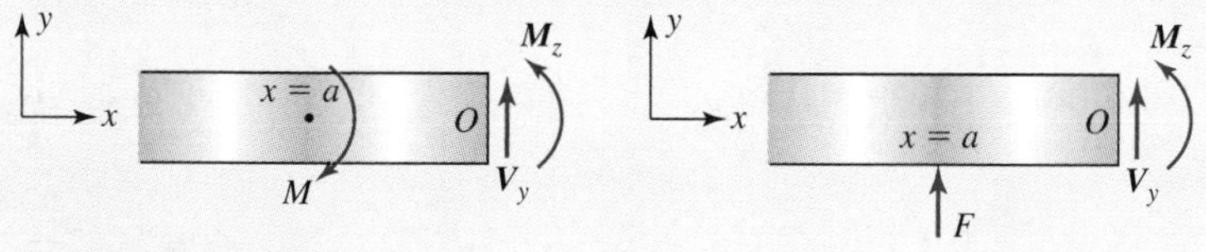

Figure 7.37   Templates for Example 7.11.

Figure 7.38 shows the free-body diagram of the beam. By balancing the moment at C the reaction at A can be found as $R_A = 3P/2$. We can write the moment expressions using the templates in Figure 7.37 to guide us. The reaction force is in the same

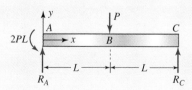

Figure 7.38   Free-body diagram in Example 7.11.

direction as the force in the template. Hence the term in the moment expression will have the same sign as shown in the template equation,

$$M_z = \frac{3P}{2}\langle x \rangle^1 - 2PL\langle x \rangle^0 - P\langle x - L \rangle^1 \qquad \text{(E1)}$$

The applied moment at point A has an opposite direction to that shown in the template in Figure 7.37, hence the term in the moment expression in Equation (E1) will have a negative sign to that shown in the template equation. The force P at B has an opposite sign to that shown on the template, and hence the term in the moment expression will have a negative sign, as shown in Equation (E1).

Substituting Equation (E1) into Equation (7.1) and writing the zero deflection conditions at A and C, we obtain the following boundary-value problem:

- Differential equation:

$$EI_{zz}\frac{d^2v}{dx^2} = \frac{3P}{2}\langle x \rangle^1 - 2PL\langle x \rangle^0 - P\langle x - L \rangle^1, \qquad 0 \le x < 2L \quad \text{(E2)}$$

- Boundary conditions:

$$v(0) = 0 \qquad \text{(E3)}$$
$$v(2L) = 0 \qquad \text{(E4)}$$

Integrating Equation (E2) twice using Equation (7.14), we obtain

$$EI_{zz}\frac{dv}{dx} = \frac{3P}{4}\langle x \rangle^2 - 2PL\langle x \rangle^1 - \frac{P}{2}\langle x - L \rangle^2 + c_1 \qquad \text{(E5)}$$

$$EI_{zz}v = \frac{P}{4}\langle x \rangle^3 - PL\langle x \rangle^2 - \frac{P}{6}\langle x - L \rangle^3 + c_1 x + c_2 \qquad \text{(E6)}$$

Substituting Equation (E6) into Equation (E3), the constant c_2 can be found,

$$\frac{P}{4}\langle 0 \rangle^3 - PL\langle 0 \rangle^2 - \frac{P}{6}\langle -L \rangle^3 + c_2 = 0 \qquad \text{or} \qquad c_2 = 0 \quad \text{(E7)}$$

Substituting Equation (E6) into Equation (E4), the constant c_1 can be found,

$$\frac{P}{4}\langle 2L \rangle^3 - PL\langle 2L \rangle^2 - \frac{P}{6}\langle L \rangle^3 + c_1(2L) = 0 \qquad \text{or} \qquad c_1 = \frac{13}{12}PL^2 \quad \text{(E8)}$$

Substituting Equations (E7) and (E8) into Equation (E6), we obtain the elastic curve,

ANS. $\quad v = \dfrac{P}{12EI_{zz}}[3\langle x \rangle^3 - 12L\langle x \rangle^2 - 2\langle x - L \rangle^3 + 13L^2 x] \qquad \text{(E9)}$

Dimension check: All terms in brackets are dimensionally homogeneous as all have the dimensions of length cubed. But we can also check whether the left-hand side and any one term of the right-hand side have the same dimension,

$$P \rightarrow O(F) \qquad x \rightarrow O(L) \qquad E \rightarrow O\left(\frac{F}{L^2}\right) \qquad I_{zz} \rightarrow O(L^4)$$

$$v \rightarrow O(L) \qquad \frac{Px^3}{EI_{zz}} \rightarrow O\left(\frac{FL^3}{(F/L^2)L^4}\right) \rightarrow O(L) \rightarrow \text{checks}$$

COMMENTS

1. Comparing the boundary-value problem in this example with that of Example 7.2, we note the following: (i) There is only one differential equation here representing the two differential equations of Example 7.2. (ii) There are no continuity equations at $x = L$ as there were in Example 7.2. The net impact of these two features is a significant reduction in the algebra in this example compared to the algebra in Example 7.2.

2. Equation (E9) represents the two equations of the elastic curve in Example 7.2. We note that $\langle x - L \rangle^3 = 0$ for $0 \leq x < L$. Hence Equation (E9) can be written as $v(x) = P(3x^3 - 12Lx^2 + 13L^2x)/12EI_{zz}$, which is same as Equation (E18) in Example 7.2. For $L \leq x < 2L$, the term $\langle x - L \rangle^3 = (x - L)^3$. Hence Equation (E9) can be written as $v(x) = P[3x^3 - 12Lx^2 + 13L^2x - 2(x - L)^3]/12EI_{zz}$, which is same as Equation (E19) in Example 7.2.

EXAMPLE 7.12

A beam with a bending rigidity $EI = 42{,}000$ N·m² is shown in Figure 7.39. Determine: (a) the deflection at point B; (b) the moment and shear force just before and after B.

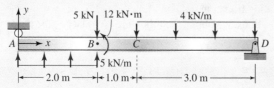

Figure 7.39 Beam and loading in Example 7.12.

PLAN

The coordinate system in this example is the same as in Example 7.11, and hence we can use the templates in Figure 7.37. Differentiating the template equations twice, we obtain the template equation for the distributed forces, and we write the distributed force expression in terms of discontinuity functions. Using Equation (7.5) and the boundary conditions at A and D, we can write the boundary-value problem and solve it to obtain the elastic curve.

(a) Substituting $x = 2$ m in the elastic curve, we can obtain the deflection at B.

(b) Substituting $x = 2.5$ in the shear force expression, we can obtain the shear force value.

Solution

(a) The templates of Example 7.11 are repeated in Figure 7.40. The moment expression is differentiated twice to obtain the template equations for the distributed force as shown,

$$M_z = M\langle x - a\rangle^0 \qquad M_z = F\langle x - a\rangle^1$$

$$p = M\langle x - a\rangle^{-2} \qquad p = F\langle x - a\rangle^{-1}$$

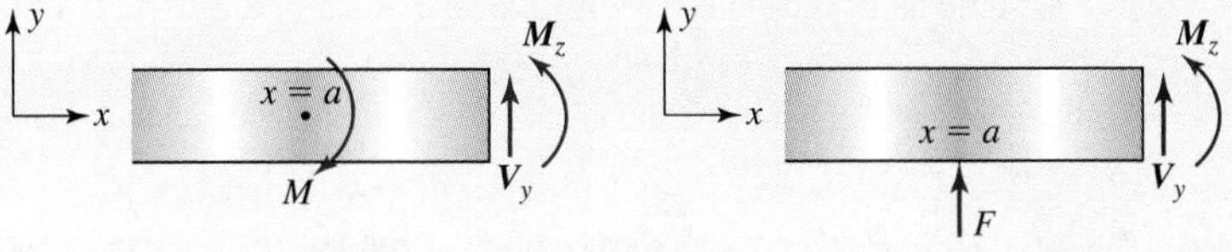

Figure 7.40 Templates for Example 7.12.

We note that the distributed force in segment AB is positive, starts at zero, and ends at $x = 2$. The distributed force in segment CD is negative, starts at $x = 3$, and is over the rest of the beam. Using the template equations and Figure 7.40, we can write the distributed force expression,

$$p = 5\langle x\rangle^0 - 5\langle x - 2\rangle^0 - 4\langle x - 3\rangle^0 - 5\langle x - 2\rangle^{-1} - 12\langle x - 2\rangle^{-2} \tag{E1}$$

Substituting Equation (E1) into Equation (7.5) and writing the boundary conditions, we obtain the following boundary-value problem:

- Differential equation:

$$\frac{d^2}{dx^2}\left(EI_{zz}\frac{d^2v}{dx^2}\right) = 5\langle x\rangle^0 - 5\langle x - 2\rangle^0 - 4\langle x - 3\rangle^0 - 5\langle x - 2\rangle^{-1}$$

$$- 12\langle x - 2\rangle^{-2}, \qquad 0 \le x < 6 \tag{E2}$$

- Boundary conditions:

$$v(0) = 0 \tag{E3}$$

$$EI_{zz}\frac{d^2v}{dx^2}(0) = 0 \tag{E4}$$

$$v(6) = 0 \tag{E5}$$

$$EI_{zz}\frac{d^2v}{dx^2}(6) = 0 \tag{E6}$$

Integrating Equation (E2) twice, we obtain

$$\frac{d}{dx}\left(EI_{zz}\frac{d^2v}{dx^2}\right) = 5\langle x\rangle^1 - 5\langle x-2\rangle^1 - 4\langle x-3\rangle^1$$
$$- 5\langle x-2\rangle^0 - 12\langle x-2\rangle^{-1} + c_1 \tag{E7}$$

$$EI_{zz}\frac{d^2v}{dx^2} = \frac{5}{2}\langle x\rangle^2 - \frac{5}{2}\langle x-2\rangle^2 - 2\langle x-3\rangle^2 - 5\langle x-2\rangle^1$$
$$- 12\langle x-2\rangle^0 + c_1x + c_2 \tag{E8}$$

Substituting Equation (E8) into Equation (E3), we obtain

$$c_2 = 0 \tag{E9}$$

Substituting Equation (E8) into Equation (E6), we obtain

$$\frac{5}{2}\langle 6\rangle^2 - \frac{5}{2}\langle 4\rangle^2 - 2\langle 3\rangle^2 - 5\langle 4\rangle^1 - 12\langle 4\rangle^0 + c_1(6) = 0 \quad c_1 = 0 \tag{E10}$$

Substituting Equations (E9) and (E10) into Equation (E8) and integrating twice, we obtain

$$EI_{zz}\frac{dv}{dx} = \frac{5}{6}\langle x\rangle^3 - \frac{5}{6}\langle x-2\rangle^3 - \frac{2}{3}\langle x-3\rangle^3$$
$$- \frac{5}{2}\langle x-2\rangle^2 - 12\langle x-2\rangle^1 + c_3 \tag{E11}$$

$$EI_{zz}v = \frac{5}{24}\langle x\rangle^4 - \frac{5}{24}\langle x-2\rangle^4 - \frac{2}{12}\langle x-3\rangle^4$$
$$- \frac{5}{6}\langle x-2\rangle^3 - 6\langle x-2\rangle^2 + c_3x + c_4 \tag{E12}$$

Substituting Equation (E12) into Equation (E3), we obtain

$$c_4 = 0 \qquad \text{(E13)}$$

Substituting Equation (E12) into Equation (E4), we obtain

$$\frac{5}{24}\langle 6 \rangle^4 - \frac{5}{24}\langle 4 \rangle^4 - \frac{2}{12}\langle 3 \rangle^4 - \frac{5}{6}\langle 4 \rangle^3 - 6\langle 4 \rangle^2 + c_3(6) = 0$$

$$c_3 = -\frac{323}{36} = -8.97 \qquad \text{(E14)}$$

Substituting Equations (E13) and (E14) into Equation (E12) and simplifying, we obtain the elastic curve,

$$v = \frac{1}{72EI_{zz}}[15\langle x \rangle^4 - 15\langle x - 2 \rangle^4 - 12\langle x - 3 \rangle^4 - 60\langle x - 2 \rangle^3$$

$$- 432\langle x - 2 \rangle^2 - 646x] \qquad \text{(E15)}$$

Substituting $x = 2$ into Equation (E15), we obtain the deflection at point B,

$$v(2) = \frac{1}{72 \times 42 \times 10^3}[15\langle 2 \rangle^4 - 15\langle 0 \rangle^4 - 12\langle -1 \rangle^4 - 60\langle 0 \rangle^3$$

$$- 432\langle 0 \rangle^2 - 646(6)]$$

ANS. $v(2) = -1.2$ mm

(b) As per Equation (7.1), the moment M_z can be found from Equation (E8). As per Equation (7.4), the shear force V_y is the negative of the expression given in Equation (E9). Noting that the constants c_1 and c_2 are zero, we obtain the following expressions for M_z and V_y:

$$M_z(x) = \left[\frac{5}{2}\langle x \rangle^2 - \frac{5}{2}\langle x - 2 \rangle^2 - 2\langle x - 3 \rangle^2 - 5\langle x - 2 \rangle^1 \right.$$

$$\left. - 12\langle x - 2 \rangle^0 \right] \text{ kN} \cdot \text{m} \qquad \text{(E16)}$$

$$V_y(x) = [-5\langle x \rangle^1 + 5\langle x - 2 \rangle^1 + 4\langle x - 3 \rangle^1 + 5\langle x - 2 \rangle^0$$

$$+ 12\langle x - 2 \rangle^{-1}] \text{ kN} \qquad \text{(E17)}$$

Point B is at $x = 2$. Just after point B, that is, at $x = 2-$, all terms except the first term in Equations (E16) and (E17) are zero. Hence we obtain

$$M_z(2-) = 10 \text{ kN} \cdot \text{m} \qquad V_y(2-) = -10 \text{ kN}$$

Just after point B, that is, at $x = 2+$, the step function $\langle x - 2 \rangle^0$ is equal to 1. Hence this term along with the first term are the non-zero terms in Equations (E16) and (E17). Hence we obtain

$$M_z(2+) = 10 - 12 \text{ kN} \cdot \text{m} \qquad \text{(E18)}$$

$$\text{ANS.} \qquad M_z(2+) = -2 \text{ kN} \cdot \text{m}$$

$$V_y(2+) = -10 + 5 \text{ kN} \qquad \text{(E19)}$$

$$\text{ANS.} \qquad V_y(2+) = -5 \text{ kN}$$

COMMENT

We note that $M_z(2+) - M_z(2-) = -12 \text{ kN} \cdot \text{m}$ and $V_y(2+) - V_y(2-) = 5 \text{ kN} \cdot \text{m}$, which are the values of the applied moment and applied shear force. Thus the jump in the internal shear force and internal moment difference is captured by the step function.

*7.5 AREA-MOMENT METHOD

The utility of the area-moment method is fully realized if the deflection or the slope of the beam is to be found at one specific point only. The method is based on graphical interpretation of the integrals that are generated when the differential equation, Equation (7.1), is integrated.

Equation (7.1) can be written as

$$\frac{d}{dx} v'(x) = \frac{M_z}{EI_{zz}}$$

where $v'(x) = dv(x)/dx$ represents the slope of the elastic curve. Integrating the equation from any point A to any other point x, the following can be obtained:

$$\int_{v'(x_A)}^{v'(x)} dv'(x) = \int_{x_A}^{x} \frac{M_z}{EI_{zz}} dx_1$$

or, written alternatively,

$$v'(x) = v'(x_A) + \int_{x_A}^{x} \frac{M_z}{EI_{zz}} dx_1 \qquad (7.17)$$

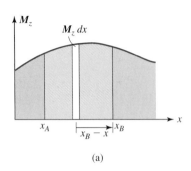

(a)

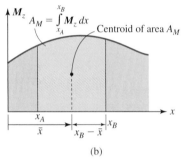

(b)

Figure 7.41 Graphical interpretation of integrals in area-moment method.

Integrating Equation (7.17) between point A and any point x, the following is obtained:

$$v(x) = v(x_A) + v'(x_A)(x - x_A) + \int_{x_A}^{x}\left(\int_{x_A}^{x}\frac{M_z}{EI_{zz}}\,dx_1\right)dx \qquad (7.18)$$

The last integral[5] can be written as

$$v(x) = v(x_A) + v'(x_A)(x - x_A) + \int_{x_A}^{x}(x - x_1)\frac{M_z}{EI_{zz}}\,dx_1 \qquad (7.19)$$

Assume EI_{zz} is a constant for the beam. Using Equations (7.17) and (7.19), the slope and the deflection at point B can be written,

$$v'(x_B) = v'(x_A) + \frac{1}{EI_{zz}}\int_{x_A}^{x_B}M_z\,dx \qquad (7.20)$$

$$v(x_B) = v(x_A) + v'(x_A)(x_B - x_A) + \frac{1}{EI_{zz}}\int_{x_A}^{x_B}(x_B - x)\,M_z\,dx \qquad (7.21)$$

The integral in Equation (7.20) can be interpreted as the area under the bending moment curve, as shown in Figure 7.41. The moment diagram can be constructed as discussed in Section 6.4. Thus if the slope $v'(x_A)$ at point A is known, then by adding the area under the moment curve, the slope at point B can be found. The area A_M will be considered positive if the moment curve is in the upper plane and negative if it is in the lower plane.

From Figure 7.41a, it is seen that the integral in Equation (7.21) is the first moment of the area under the moment curve about point B. This first moment of the area can be found by taking the distance of the centroid from B and multiplying by the area,

$$\int_{x_A}^{x_B}(x_B - x)M_z\,dx = (x_B - \bar{x})A_M$$

With this interpretation the deflection of B can be found from Equation (7.21). Table A.2 in the Appendix lists the areas and the centroids of the areas under various curves. These values can be used in calculating the integrals in Equations (7.20) and (7.21).

Consider the cantilever beam in Figure 7.42a and the associated bending moment diagram. At point A the slope and the deflection at A are zero. Hence $v'(x_A) = 0$ and $v(x_A) = 0$ in Equations (7.20) and (7.21). The area A_M, representing the integral in Equation (7.20), is $-PL(L)/2$. Thus the slope at B is $v'(x_B) = -PL^2/2EI$. The distance of the centroid from B, that is, $x_B - \bar{x} = 2L/3$, hence $(x_B - \bar{x})A_M = (2L/3)(-PL^2/2)$, is the value of the integral equation, Equation (7.21). Thus the deflection at B is $v(x_B) = -PL^3/3EI$.

Now consider the simply supported beam and the associated bending moment in Figure 7.42b. The value of the slope is not known at any point on the beam. Thus before the deflection and slope at B can be determined, the slope at A must be found. The deflection at A is zero. Treating the slope at A as an unknown constant, the

---

[5]By integrating by parts it can be shown that

$$\int_{x_A}^{x}\left[\int_{x_A}^{x}f(x_1)dx_1\right]dx = \int_{x_A}^{x}(x - x_1)f(x_1)\,dx_1$$

By letting $f(x_1) = M_z/EI_{zz}$, Equation (7.19) can be obtained from Equation (7.18).

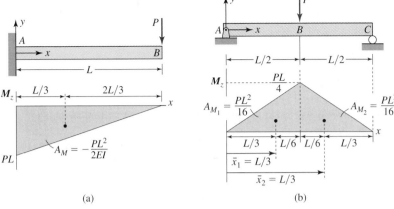

Figure 7.42   Application of area-moment method.

deflection at C can be written using Equation (7.21) and equated to zero to obtain the slope at A,

$$v(x_C) = v(x_A) + v'(x_A)(x_C - x_A) + \frac{1}{EI}[A_{M_1}(x_C - \bar{x}_1) + A_{M_2}(x_C - \bar{x}_2)] = 0$$

or

$$v'(x_A)(L) + \frac{1}{EI}\left[\frac{PL^2}{16}\left(\frac{2L}{3}\right) + \frac{PL^2}{16}\left(\frac{L}{3}\right)\right] = 0 \qquad \text{or} \qquad v'(x_A) = -\frac{PL^2}{16EI}$$

Using Equation (7.21) once more, the deflection at point B can be found,

$$v(x_B) = v(x_A) + v'(x_A)(x_B - x_A) + \frac{1}{EI}[A_{M_1}(x_B - \bar{x}_1)]$$

$$= -\frac{PL^2}{16EI}\left(\frac{L}{2}\right) + \frac{1}{EI}\left[\frac{PL^2}{16}\left(\frac{L}{6}\right)\right] = -\frac{PL^3}{48}$$

EXAMPLE 7.13

In terms of E, I, w, and L, determine the deflection and slope at point B for the beam and loading shown in Figure 7.43.

PLAN

The reaction forces at A and C can be found and then the shear–moment diagram can be drawn as discussed in Section 6.4. The area under the moment curve and the location of the centroids can then be determined. Noting that the deflection at A is zero, the deflection at C can be written in terms of the unknown slope at A using Equation (7.21). Equating the deflection at C to zero, the slope at A can be found. Slope and deflection at B can now be found using Equations (7.20) and (7.21), respectively.

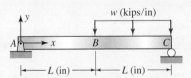

Figure 7.43   Beam and loading in Example 7.13.

Solution From the free-body diagram of the entire beam, the reaction forces at the supports can be found and the shear–moment diagram drawn, as shown in Figure 7.44. The moment curve in region BC is a quadratic and the areas under the curves can be written as the sum of the three areas in order to use the formulas for the areas and centroids given in Table 1.2 in the Appendix. Using Table 1.2 the areas and the centroid locations can be found,

$$A_1 = \frac{L}{2}\left(\frac{wL^2}{4}\right) = \frac{wL^3}{8}$$

$$A_2 = \frac{L}{4}\left(\frac{wL^2}{4}\right) = \frac{wL^3}{16}$$

$$A_3 = \frac{2}{3}\left(\frac{L}{4}\right)\left(\frac{wL^2}{32}\right) = \frac{wL^3}{192}$$

$$A_4 = \frac{2}{3}\left(\frac{3L}{4}\right)\left(\frac{9wL^2}{32}\right) = \frac{9wL^3}{64}$$

$$\bar{x}_1 = \frac{2}{3}(L)$$

$$\bar{x}_2 = L + \frac{L}{8} = \frac{9L}{8}$$

$$\bar{x}_3 = L + \frac{5}{8}\left(\frac{L}{4}\right) = \frac{37L}{32}$$

$$\bar{x}_4 = 2L - \frac{5}{8}\left(\frac{3L}{4}\right) = \frac{49L}{32}$$

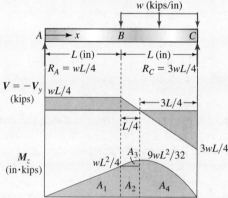

Figure 7.44 Shear–moment diagram in Example 7.13.

The deflection at C can be written as

$$v(x_C) = v(x_A) + v'(x_A)(x_C - x_A) + \frac{1}{EI}[A_1(x_C - \bar{x}_1)$$
$$+ A_2(x_C - \bar{x}_2) + A_3(x_C - \bar{x}_3) + A_4(x_C - \bar{x}_4)] \quad \text{(E1)}$$

The deflections at the support are zero. Substituting $v(x_C) = 0$ and $v(x_A) = 0$ and the values of the areas and centroids in Equation (E1), the slope at A can be found,

$$v'(x_A)(2L) + \frac{1}{EI}\left[\frac{wL^3}{8}\left(2L - \frac{2L}{3}\right) + \frac{wL^3}{16}\left(2L - \frac{9L}{8}\right) + \frac{wL^3}{192}\left(2L - \frac{37L}{32}\right)\right.$$
$$\left. + \frac{9wL^3}{64}\left(2L - \frac{49L}{32}\right)\right] = 0$$

or

$$v'(x_A) = -\frac{7wL^3}{48EI} \quad \text{(E2)}$$

The deflection at B can be written as

$$v(x_B) = v(x_A) + v'(x_A)(x_B - x_A) + \frac{1}{EI}[A_1(x_B - \bar{x}_1)] \quad \text{(E3)}$$

Substituting the calculated values into Equation (E3), the deflection at B can be found,

$$v(x_B) = -\frac{7wL^3}{48EI}(L) + \frac{1}{EI}\left(\frac{wL^3}{8}\right)\left(L - \frac{2L}{3}\right)$$

$$\text{ANS.} \quad v(x_B) = -\frac{5wL^4}{48EI}$$

COMMENTS

1. The example demonstrates the two important uses of the area moment method, namely, for finding reactions in indeterminate beams and for finding slopes and deflection at a point in the beam.

2. If the elastic curve needs to be determined for an indeterminate beam, we can use the area moment method to determine the reactions and then use the second-order differential equation to solve the problem. But if this approach is to have any computational advantage over using the fourth-order differential equations, then it will if the moment diagram can be drawn quickly by inspection.

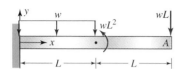

Figure P7.42

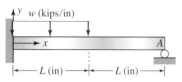

Figure P7.43

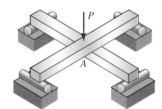

Figure P7.44

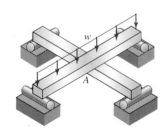

Figure P7.45

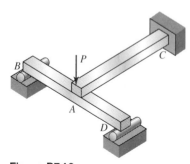

Figure P7.46

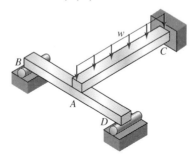

Figure P7.47

PROBLEM SET 7.3

Superposition

7.40 Determine the deflection at the free end of the beam shown in Figure P7.12.

7.41 Determine the reaction force at support A in Figure P7.22.

7.42 Determine the deflection at point A on the beam shown in Figure P7.42 in terms of w, L, E, and I.

7.43 Determine the reaction force and the slope at A for the beam shown in Figure P7.43, using superposition.

7.44 Two beams of length L and bending rigidity EI, shown in Figure P7.44, are simply supported at the ends and are in contact at the center. Determine the deflection at the center in terms of P, L, E, and I.

7.45 Two beams of length L and bending rigidity EI, shown in Figure P7.45, are simply supported at the ends and are in contact at the center. Determine the deflection at the center in terms of w, L, E, and I.

7.46 A cantilever beam's end rests on the middle of a simply supported beam, as shown in Figure P7.46. Both beams have length L and bending rigidity EI. Determine the deflection at A and the reactions at the wall at C in terms of P, L, E, and I.

7.47 A cantilever beam's end rests on the middle of a simply supported beam, as shown in Figure P7.47. Both beams have length L and bending rigidity EI. Determine the deflection at A and the reactions at the wall at C in terms of w, L, E, and I.

7.48 The end of one cantilever beam rests on the end of another cantilever beam, as shown in Figure P7.48. Both beams have length L and bending rigidity EI. Determine the deflection at A and the reactions at the wall at C in terms of w, L, E, and I.

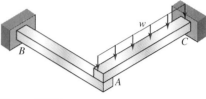

Figure P7.48

Discontinuity functions

7.49 Solve Problem 7.9 using discontinuity functions.

7.50 Solve Problem 7.10 using discontinuity functions.

7.51 Solve Problem 7.11 using discontinuity functions.

7.52 Solve Problem 7.12 using discontinuity functions.

7.53 (a) Solve for the elastic curve for the beam and loading shown in Figure P7.13. (b) Determine the slope and deflection at point C.

7.54 Solve Problem 7.22 using discontinuity functions.

Area-moment method

7.55 Using the area-moment method, determine the deflection in the middle for the beam shown in Figure P7.2.

7.56 Using the area-moment method, determine the deflection in the middle of the beam shown in Figure P7.3.

7.57 Using the area-moment method, determine the deflection and slope at the free end of the beam shown in Figure P7.4.

7.58 Using the area-moment method, determine the slope at $x = 0$ and deflection at $x = L$ of the beam shown in Figure P7.6.

7.59 Using the area-moment method, determine the slope at $x = 0$ and deflection at $x = L$ of the beam shown in Figure P7.9.

7.60 Using the area-moment method, determine slope at $x = 0$ and deflection at $x = L$ of the beam shown in Figure P7.10.

7.61 Using the area-moment method, determine the slope at the free end of the beam shown in Figure P7.12.

*7.6 GENERAL INFORMATION

The development of the theory regarding the strength of beams described in Section 6.9 was relatively intuitive compared to the theory regarding the deflection of beams described here. The very term "elastic curve" for describing the deflection of a beam reflects the early impact of mathematicians.

In 1666 Isaac Newton of England developed a mathematical tool called the fluxional method, but it did not receive much attention. Gottfried Wilhelm Leibniz of Germany, developed in 1675 nearly the same method independent of Newton, which he called differential calculus. In continental Europe the notation and method of Leibniz became very popular. Members of the Bernoulli family were in the forefront of finding applications for this new mathematical tool. One of the applications they considered was the determination of the elastic curve.

Jacob Bernoulli (1646–1716) and John Bernoulli (1667–1748) were two brothers who won acclaim for their mathematical work, which the French Academy of Science recognized by making them members in 1699. Daniel Bernoulli (1700–1782), whose contributions in hydrodynamics are renowned, and Leonard Euler (1707–1782), whose name is associated with the buckling theory, to be introduced in Chapter 11, were John Bernoulli's son and pupil. Daniel Bernoulli and Leonard Euler were also among the early pioneers in the development of the theory of the elastic curve. Jacob Bernoulli started with Mariotte's incorrect assumption that the neutral axis was tangent to the bottom (concave side) of the curve in a cantilever beam and obtained a relationship between the curvature of the beam at any point and the applied load, as described in Problem 7.29, which was incorrect only in the value of the constant for the bending rigidity. Euler, on the suggestion of Daniel Bernoulli, approached the same problem from the view of minimizing the strain energy in a beam, and he obtained the correct relationship between moment and curvature. Euler called the constant relating moment and curvature "moment of stiffness" rather than bending rigidity, and he was the first to recognize that this constant had to be determined experimentally. In Section 3.12.1 we saw that Thomas Young made a similar comment concerning axial rigidity much later than Euler, and was recognized by having the modulus of elasticity named after him. Such are the quirks of history.

Navier (1785–1836), whose work on the concept of stress we saw in Section 1.3, was the first to solve for the deflection of statically indeterminate beams by carrying the extra unknown reactions in the second-order differential equation and determining these reactions using conditions on deflection and slopes at the support (see Problem 7.28). Phillips (1821–1889), during his work on railroads, developed the theory of leaf springs, some of which is described in Example 7.4. It is one of the very early demonstrations of the use of mechanics of materials to produce answers to engineering design problems. Saint-Venant, whose work we have seen in several chapters, was first to realize that the deflection at the free end of a cantilever beam due to a force at the free end can be found without formally integrating the differential equations. This was the beginning of the area-moment method we studied in Section 7.5. A. Clebsch (1833–1872), in his book on elasticity, which was published in 1862, considered the issue of the deflection of a beam under concentrated forces, described in Problem 7.30, which eventually evolved into the discontinuity method discussed in Section 7.4. The discontinuity functions were formally introduced by the English mathematician W. H. Macaulay in 1919.

The development of an understanding of the normal stress in bending was very intuitive; shear stress in bending was guided by experimental evidence; and beam deflection by mathematics. Thus the development of the beam theory highlights the need and importance of intuition, experimental evidence, and mathematical formalization in the understanding of nature.

7.7 CLOSURE

In this chapter we saw several methods for determining the deflection of beams. The second-order boundary-value problem is the preferred approach, but may not be usable if the distributed loads on the beam are complicated functions or if the available information on the distributed load is a set of numerical values measured experimentally. The discontinuity function method should be used if the beam loading changes in a discrete manner across the beam because a single differential equation can represent the loading on the entire beam. The area-moment method is a graphical technique that can yield quick solutions of beam deflection and slope at a point, *if* the moment diagram can be constructed easily. The superposition method is a versatile design tool that can be used for solving problems of determinate and indeterminate beams provided the beam deflection and slope values are available for many basic cases, such as in a handbook.

This chapter also concludes the second major part of the book. A synopsis of the theories described in Chapters 4 through 7 for one-dimensional structural elements is given in Table 7.2. The table highlights the essential elements common to all theories. Using these theories we can now obtain deformation, strains, and stresses at a point in one-dimensional structural elements. In the next three chapters we will use this information in many ways. The paragraph that follows gives a brief overview of the uses of the information in the next three chapters.

In order to determine whether a structure would break under a given load, we need failure theories, which we will study in Chapter 10. But to use failure theories we will need to know how to determine maximum normal stress and maximum shear stress at a point from the stresses obtained from our one-dimensional theories, which is the topic of discussion in Chapter 8 on stress transformation. Only experiments can render the final verdict on our design based on the one-dimensional theory. One of the most popular experimental techniques involves the measurement of strains using strain gages. But this experimental technique of strain gages requires an understanding of the relationship between the strains obtained from one-dimensional theory and the strains in any given direction, which is the topic of discussion in Chapter 9 on strain transformation. Chapter 10 on design and failure represents the culmination of the first nine chapters. In Chapter 10 we study stresses and strains in structural elements subject to combined axial, torsional, and bending loads and address issues related to the design and failure of structures and machine elements.

TABLE 7.2 Synopsis of one-dimensional structural theories

	Axial (Rods)	**Torsion (Shafts)**	**Symmetric Bending (Beams)**
Displacements/ deformation	$u(x, y, z) = u(x)$ $v = 0 \quad w = 0$	$u = 0 \quad v = 0 \quad w = 0$ $\phi(x, y, z) = \phi(x)$	$u(x, y, z) = -y\dfrac{dv}{dx}$ $v = v(x) \quad w = 0$
Strains	$\varepsilon_{xx} = \dfrac{du}{dx}$	$\gamma_{x\theta} = \rho\dfrac{d\phi}{dx}$	$\varepsilon_{xx} = -y\dfrac{d^2v}{dx^2}$
Stresses	$\sigma_{xx} = E\varepsilon_{xx} = E\dfrac{du}{dx}$	$\tau_{x\theta} = G\gamma_{x\theta} = \rho\dfrac{d\phi}{dx}$	$\sigma_{xx} = E\varepsilon_{xx} = -Ey\dfrac{d^2v}{dx^2} \quad \tau_{xy} \neq 0$
Internal forces and moments	$N = \displaystyle\int_A \sigma_{xx}\, dA$	$T = \displaystyle\int_A \rho\tau_{x\theta}\, dA$	$\displaystyle\int_A \sigma_{xx}\, dA = 0$ ...Locates neutral axis $M_z = -\displaystyle\int_A y\sigma_{xx}\, dA \quad V_y = \displaystyle\int_A \tau_{xy}\, dA$
Sign convention	$+N$	$+T$	$+V_y \qquad +M_z$
Homogeneous cross section Stress formulas	$\sigma_{xx} = \dfrac{N}{A}$	$\tau_{x\theta} = \dfrac{T\rho}{J}$	$\sigma_{xx} = -\dfrac{M_z y}{I_{zz}}$ $\tau_{sx} = \tau_{xs} = -\dfrac{V_y Q_z}{I_{zz}t}$
Deformation formulas	$\dfrac{du}{dx} = \dfrac{N}{EA}$ $u_2 - u_1 = \dfrac{N(x_2 - x_1)}{EA}$ EA = axial rigidity	$\dfrac{d\phi}{dx} = \dfrac{T}{GJ}$ $\phi_2 - \phi_1 = \dfrac{T(x_2 - x_1)}{GJ}$ GJ = torsional rigidity	$\dfrac{d^2v}{dx^2} = \dfrac{M_z}{EI_{zz}}$ $v = \int\left(\int\dfrac{M_z}{EI_{zz}}\,dx\right)dx + C_1 x + C_2$ EI_{zz} = bending rigidity
Composite cross section Stress formulas	$(\sigma_{xx})_i = \dfrac{NE_i}{\displaystyle\sum_{j=1}^{n} E_j A_j}$	$(\tau_{x\theta})_i = \dfrac{G_i \rho T}{\displaystyle\sum_{j=1}^{n} G_j J_j}$	$(\sigma_{xx})_i = \dfrac{-E_i y M_z}{\displaystyle\sum_{j=1}^{n} E_j (I_{zz})_j}$ $\tau_{sx} = \tau_{xs} = \dfrac{-Q_{comp} V_y}{\left[\displaystyle\sum_{j=1}^{n} E_j (I_{zz})_j\right]t}$
Deformation formulas	$\dfrac{du}{dx} = \dfrac{N}{\displaystyle\sum_{j=1}^{n} E_j A_j}$ $u_2 - u_1 = \dfrac{N(x_2 - x_1)}{\displaystyle\sum_{j=1}^{n} E_j A_j}$	$\dfrac{d\phi}{dx} = \dfrac{T}{\displaystyle\sum_{j=1}^{n} G_j J_j}$ $\phi_2 - \phi_1 = \dfrac{T(x_2 - x_1)}{\displaystyle\sum_{j=1}^{n} G_j J_j}$	$\dfrac{d^2v}{dx^2} = \dfrac{M_z}{\displaystyle\sum_{j=1}^{n} E_j (I_{zz})_j}$ $v = \int\left[\int\dfrac{M_z}{\displaystyle\sum_{j=1}^{n} E_j (I_{zz})_j}\,dx\right]dx + C_1 x + C_2$

POINTS AND FORMULAS TO REMEMBER

- The deflected curve of a beam represented by $v(x)$ is called *elastic curve*.

$$M_z = EI_{zz}\frac{d^2v}{dx^2} \quad (7.1) \qquad V_y = -\frac{d}{dx}\left(EI_{zz}\frac{d^2v}{dx^2}\right) \quad (7.4) \qquad \frac{d^2}{dx^2}\left(EI_{zz}\frac{d^2v}{dx^2}\right) = p \quad (7.5)$$

  where v is the deflection of the beam at any x and is positive in the positive y direction; M_z is the internal bend-ing moment; V_y is the internal shear force; p is the distributed force on the beam and is positive in the positive y direction; EI_{zz} is the bending rigidity of the beam; and d^2v/dx^2 is the curvature of the beam.

- The mathematical statement listing all the differential equations and all the conditions necessary for solving for $v(x)$ is called *boundary-value problem* for beam deflection.

- Boundary conditions for second-order differential equations:

Built-in end at x_A	$v(x_A) = 0$	$\dfrac{dv}{dx}(x_A) = 0$
Simple support at x_A	$v(x_A) = 0$	
Smooth slot at x_A	$\dfrac{dv}{dx}(x_A) = 0$	

- Continuity conditions at x_j:

$$v_1(x_j) = v_2(x_j) \qquad \frac{dv_1}{dx}(x_j) = \frac{dv_2}{dx}(x_j)$$

- Boundary conditions for fourth-order differential equations are determined at each boundary point by specify-ing: (v or V_y) and (dv/dx or M_z)   (7.6)

- In fourth-order boundary-value problems, at each point x_j where the differential equation changes, the conti-nuity conditions and equilibrium conditions must be specified.

- In the discontinuity function method a single differential equation and conditions on deflection and slopes at support describe the complete boundary-value problem.

- Discontinuity functions:

$$\langle x - a \rangle^{-n} = \begin{cases} 0 & x \neq a \\ \infty & x \to a \end{cases} \qquad \langle x - a \rangle^n = \begin{cases} 0 & x \leq a \\ (x-a)^n & x > a \end{cases}$$

- Differentiation formulas:

$$\frac{d\langle x - a \rangle^{-1}}{dx} = \langle x - a \rangle^{-2} \qquad \frac{d\langle x - a \rangle^0}{dx} = \langle x - a \rangle^{-1} \qquad \frac{d\langle x - a \rangle^n}{dx} = n\langle x - a \rangle^{n-1} \quad n \geq 1$$

- Integration formulas:

$$\int_{-\infty}^{x} \langle x - a \rangle^{-2}\, dx = \langle x - a \rangle^{-1} \qquad \int_{-\infty}^{x} \langle x - a \rangle^{-1}\, dx = \langle x - a \rangle^0 \qquad \int_{-\infty}^{x} \langle x - a \rangle^n dx = \frac{\langle x - a \rangle^{n+1}}{n + 1} \quad n \geq 0$$

- Area-moment method is a graphical technique that can yield quick solutions of beam deflection and slope at a point, *if* the moment diagram can be constructed easily.

$$v'(x_B) = v'(x_A) + \frac{1}{EI_{zz}} \underbrace{\int_{x_A}^{x_B} M_z \, dx}_{A_M} \qquad (7.20)$$

$$v(x_B) = v(x_A) + v'(x_A)(x_B - x_A) + \frac{1}{EI_{zz}} \underbrace{\int_{x_A}^{x_B} (x_B - x) M_z \, dx}_{(x_B - \bar{x}) A_M} \qquad (7.21)$$

- The superposition method is a versatile design tool that can be used for solving problems of determinate and indeterminate beams provided the beam deflection and slope values are available for many basic cases, such as in a handbook.

CHAPTER EIGHT
STRESS TRANSFORMATION

8.0 OVERVIEW

Figure 8.1 shows failure surfaces of aluminum and cast iron members under axial and torsional loads. Why do different materials under similar loading produce different failure surfaces? If we had a combined loading of axial and torsion, then what would be the failure surface, and which stress component would cause the failure? The answer to this question is critical for the successful design of solid members

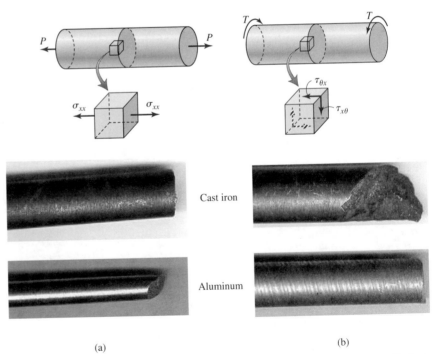

Cast iron

Aluminum

(a)   (b)

Figure 8.1 Failure surfaces. (*a*) Axial load. (*b*) Torsional load. (Specimens courtesy Professor J. B. Ligon.)

that are subjected to combined axial, torsional, and bending loads. In Chapter 10 we will study combined loading and failure theories that relate maximum normal and shear stresses to material strength. In this chapter we develop procedures and equations that transform stress components from one coordinate system to another at a *given point*.

Stress transformation can also be viewed as relating stresses on different planes that pass through a point. The outward normals of the planes form the axes of a coordinate system. Thus relating stresses on different planes is equivalent to relating stresses in different coordinate systems. We will use both viewpoints in this chapter of stress transformation.

The two major learning objectives of this chapter are:

1. Learn the equations and procedures of relating stresses (on different planes) in different coordinate systems at a point.

2. Develop the ability to visualize planes passing through a point on which stresses are given or are being found, in particular the planes of maximum normal stress and maximum shear stress.

8.1 PRELUDE TO THEORY: THE WEDGE METHOD

In this chapter we will study three methods of stress transformation. The wedge method, described in this section, is used to derive stress transformation equations that are used in the next section. The stress transformation equations then are manipulated to generate a graphical procedure called Mohr's circle, which is described in Section 8.3.

We define two coordinate systems that will be used in this chapter.

Definition 1 The fixed reference coordinate system in which the entire problem is described is called *global coordinate system*.

Definition 2 A coordinate system that can be fixed at any point on the body and has an orientation that is defined with respect to the global coordinate system is called *local coordinate system*.

Relating internal forces and moments to external forces and moments is usually done in the global coordinate system. These internal quantities are used to obtain stresses in the global coordinate system, such as axial normal stress, torsional shear stress, and bending normal and shear stresses. In all two-dimensional problems in this book, the local coordinate system will be the n, t, z coordinate system.

- The n direction will be the direction of the *outward normal* to the plane on which we are finding the stresses.

- The z direction is identical to the z direction of the global coordinate system.

- The tangent t direction can be found from the right-hand rule, as shown in Figure 8.2, that is, just as x cross y yields z, in a similar manner n cross t yields z. With the thumb of the right hand pointed in the known z direction, the curl of the fingers is from the known n direction toward the t direction.

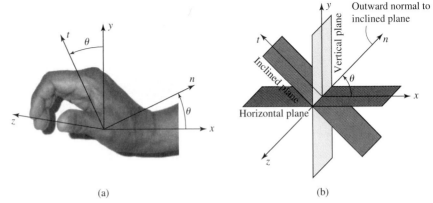

Figure 8.2 Local and global coordinate systems.

Alternatively, the positive t direction can be found as follows. Curl the fingers of the right hand from the z direction toward the n direction, then the positive t direction is given by the direction of the thumb. With the n direction as positive in the outward normal direction, positive shear stress τ_{nt} is in the positive tangent direction and negative τ_{nt} will be in the negative tangent direction.

In this section we restrict ourselves to plane stress problems. We will consider only those inclined planes that can be obtained by rotation about the z axis, as shown in Figure 8.2b.

The steps in the wedge method are described next and elaborated by application in Examples 8.1 and 8.2.

Step 1: A stress cube with the plane on which stresses are to be found, or are given, is constructed.

Step 2: A wedge is constructed from the following three planes:

   1. A vertical plane that has an outward normal in the x direction.
   2. A horizontal plane that has an outward normal in the y direction.
   3. The specified inclined plane on which we either seek stresses or the stresses are given.

Establish a local n, t, z coordinate system using the outward normal of the inclined plane as the n direction. All the known and unknown stresses are shown on the wedge. The diagram so constructed will be called a *stress wedge*.

Step 3: Multiply the stress components by the area of the planes on which the stress components are acting, to obtain forces acting on that plane. The wedge with the forces drawn will be referred to as the *force wedge*.

Step 4: Balance forces in *any* two directions to determine the unknown stresses. We can write equilibrium equations on the force wedge because the wedge represents a point on a body that is in equilibrium.

Step 5: Check the answer intuitively. This is accomplished by considering each stress component individually. By inspection, we decide whether the stress component will produce tensile or compressive normal stress on the incline and whether it will produce positive or negative shear stress on the incline.

EXAMPLE 8.1

A steel beam on a bridge was repaired by welding along a line that is 35° to the axis of the beam. The normal stress near the bottom of the beam is estimated using beam theory and is shown on the stress cube. Determine the normal and shear stress on the plane containing the weld line.

PLAN

Step 1 of the procedure outlined in this section is complete, as shown in Figure 8.3. We follow the remaining steps to solve the problem.

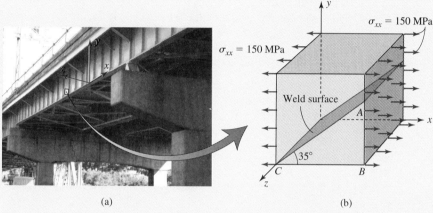

(a)   (b)

Figure 8.3   Stress cube at a point on a bridge.

Solution

Step 2: We construct a wedge from a horizontal plane, a vertical plane, and an inclined plane, as shown in Figure 8.4a. The outward normal to the inclined plane is drawn, and knowing the positive z direction, we establish the positive t direction using the right-hand rule

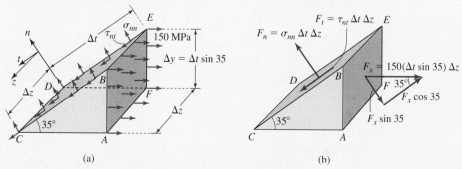

(a)   (b)

Figure 8.4   Example 8.1. (*a*) Stress wedge. (*b*) Force wedge.

for the n, t, z coordinates. On the inclined plane we can show the normal stress σ_{nn} and the shear stress τ_{nt}. From triangle ABC we note that $\Delta y = \Delta t \sin 35$.

Step 3: We multiply the stresses σ_{nn} and τ_{nt} by the area of the incline $BCDE$ to obtain the forces in the n and t directions, respectively. Similarly, we multiply the stress of 150 MPa by the area of the plane $ABEF$ to obtain the force in the x direction. These forces are shown on the force wedge in Figure 8.4b.

Step 4: As the unknowns are in the n and t directions, we balance the forces in the n and t directions. The components of force F_x in the n and t directions are shown on the force wedge in Figure 8.4b. Balancing forces in the n direction, we obtain

$$\sigma_{nn} \Delta t \, \Delta z - \underbrace{(150 \, \Delta t \sin 35 \, \Delta z)}_{F_x} \sin 35 = 0 \qquad \text{(E1)}$$

or

$$(\sigma_{nn} - 150 \sin 35 \sin 35) \Delta t \, \Delta z = (\sigma_{nn} - 49.35) \Delta t \, \Delta z = 0 \qquad \text{(E2)}$$

$$\text{ANS.} \qquad \sigma_{nn} = 49.3 \text{ MPa (T)}$$

In a similar manner, balancing the forces in the t direction, we obtain

$$\tau_{nt} \Delta t \, \Delta z - (150 \, \Delta t \sin 35 \, \Delta z) \cos 35 = 0 \qquad \text{(E3)}$$

or

$$(\tau_{nt} - 150 \sin 35 \cos 35) \Delta t \, \Delta z = (\tau_{nt} - 70.48) \Delta t \, \Delta z = 0 \quad \text{(E4)}$$

$$\text{ANS.} \qquad \tau_{nt} = 70.5 \text{ MPa}$$

Step 5: We check the answer using intuitive arguments. The surface ABC in Figure 8.3 tends to move away from the rest of the cube. Hence the material resistance opposing it results in a tensile stress, as seen. A more visual way is to imagine the inclined plane in Figure 8.3 as a glued surface. Due to the action of σ_{xx}, the two surfaces on either side of the glue are being pulled apart; hence the glue will be put into tension. In a similar manner the wedge ABC due to σ_{xx} will tend to slide upward relative to the rest of the cube; hence the material resistance (like friction) will be downward, resulting in a positive shear stress as seen.

COMMENTS

1. In both Equation (E2) and Equation (E4) the dimensions Δt and Δz were common factors and did not affect the final answer. In other words, the dimensions of the stress cube are immaterial. This is not surprising as the stress cube is a visualization aid that is symbolically representing a point. Only the relative orientation of the plane is important.

2. The stress cube in Figure 8.2 and the stress and force wedges in Figure 8.3 can be represented in two dimensions, as shown in Figure 8.5. These are easier to draw and work with. But once more it must be emphasized that stress is a distributed force and not a vector as depicted in Figure 8.5. Force balance can only be achieved on the force wedge.

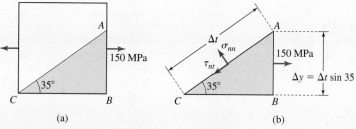

Figure 8.5 Wedge method in Example 8.1. (*a*) Stress cube. (*b*) Stress wedge. (*c*) Force wedge.

3. In constructing the stress wedge we took the lower wedge. The dimensions of the stress cube are immaterial as it represents a point. But the orientation of the planes is important. An alternative approach is to take the upper wedge, as shown in Figure 8.6.

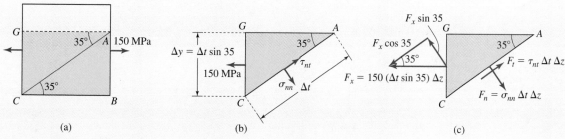

Figure 8.6 Alternative approach in Example 8.1. (*a*) Stress cube. (*b*) Stress wedge. (*c*) Force wedge.

4. Some writing could be saved by taking $\Delta t = 1$ and $\Delta z = 1$, as these terms always drop out. But the geometric visualization may become more difficult in the process.

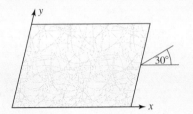

Figure 8.7 Stresses in lamina in Example 8.2.

EXAMPLE 8.2

Fibers are oriented at 30° to the x axis in a lamina of a composite[1] plate, as shown Figure 8.7. Stresses at a point in the lamina were found by the finite-element method[2] as

$$\sigma_{xx} = 30 \text{ MPa (T)}$$

$$\sigma_{yy} = 60 \text{ MPa (C)}$$

$$\tau_{xy} = 50 \text{ MPa}$$

In order to assess the strength of the interface between the fiber and the resin, determine the normal and shear stresses on the plane containing the fiber.

PLAN

As per Step 1 of the procedure outlined in this section, we can draw the stress cube with an plane inclined at 30° and then follow the remaining steps of the procedure.

Solution

Step 1: We draw a stress cube in two dimensions for the given state of stress, and the plane inclined at 30° counterclockwise to the x axis, as shown in Figure 8.8a.

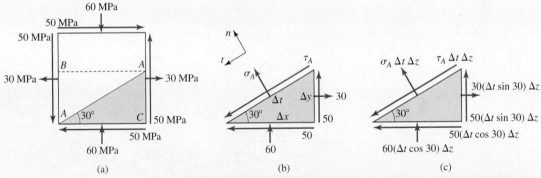

Figure 8.8 (*a*) Stress cube. (*b*) Stress wedge. (*c*) Force wedge in Example 8.2.

Step 2: We can choose wedge *ACA* or wedge *ABA* as a stress wedge. Figure 8.8b shows the stress wedge *ACA* with a local *n, t, z* coordinate system.

Step 3: We assume the length of the inclined plane to be Δt. From geometry we see that $\Delta x = \Delta t \cos 30$ and $\Delta y = \Delta t \sin 30$. If we assume that the

---

[1]See Section 3.12.3 for a brief description of composite materials.
[2]See Section 4.9 for a brief description of the finite-element method.

dimension of the cube out of the paper is Δz, we get the following areas: inclined plane $\Delta t\, \Delta z$, vertical plane $\Delta y\, \Delta z$, and horizontal plane $\Delta x\, \Delta z$. The stresses are converted into forces by multiplying by the area of the plane, and a force wedge is drawn as shown in Figure 8.8c.

Step 4: We can balance forces in any two directions. We choose to balance forces in the n and t directions as the unknowns are in the n and t directions. Figure 8.9 shows the resolution of the forces in the x, y coordinates to n, t coordinates.

By balancing forces in the n direction we obtain

$$\sigma_A(\Delta t\, \Delta z) - \underbrace{[30(\Delta t \sin 30)\Delta z - 50(\Delta t \cos 30)\Delta z]}_{F_x} \sin 30$$

$$+ \underbrace{[60(\Delta t \cos 30)\Delta z + 50(\Delta t \sin 30)\,\Delta z]}_{F_y} \cos 30 = 0$$

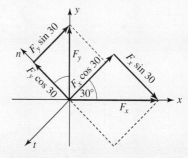

Figure 8.9 Resolution of force components.

Noting that the inclined area $\Delta t\, \Delta z$ is common to all terms and can be removed from the equation, we can find the normal stress,

$$\sigma_A - (30 \sin 30 - 50 \cos 30)\sin 30$$
$$+ (60 \cos 30 + 50 \sin 30)\cos 30 = 0$$

or

ANS. $\sigma_A = 80.8$ MPa (C)

By balancing forces in the t direction, the shear stress can be calculated in a similar manner,

$$\tau_A(\Delta t\, \Delta z) - \underbrace{[30(\Delta t \sin 30)\Delta z - 50(\Delta t \cos 30)\Delta z]}_{F_x} \cos 30$$

$$- \underbrace{[60(\Delta t \cos 30)\Delta z + 50(\Delta t \sin 30)\Delta z]}_{F_y} \sin 30 = 0$$

$$\tau_A - (30 \sin 30 - 50 \cos 30)\cos 30$$
$$- (60 \cos 30 + 50 \sin 30)\sin 30 = 0$$

or

ANS. $\tau_A = 14.0$ MPa

Step 5: We can check the answers intuitively. Consider each stress component individually and visualize the inclined plane as a glue line. The

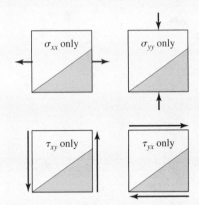

Figure 8.10 Intuitive check.

rectangles shown in Figure 8.10 are for purposes of explanation. One can go through the arguments mentally without drawing these rectangles.

Figure 8.10 shows that the right surface (wedge) and the left surface will move:

- Apart due to σ_{xx}—therefore put the glue in *tension*
- Into each other due to σ_{yy}—therefore put the glue in *compression*
- Into each other due to τ_{xy}—therefore put the glue in *compression*
- Into each other due to τ_{yx}—therefore put the glue in *compression*.

Thus the normal stress in the glue (on the inclined plane) is expected to be in compression, which is consistent with our answer.

Figure 8.10 shows that the right surface (shaded wedge), with respect to the left surface, will slide:

- Upward due to σ_{xx}—therefore the shaded wedge will have a *positive* (downward) shear stress
- Upward due to σ_{yy}—therefore the right wedge will have a *positive* (downward) shear stress
- Upward due to τ_{xy}—therefore the right wedge will have a *positive* (downward) shear stress
- Downward due to τ_{yx}—therefore the right wedge will have a *negative* (upward) shear stress.

Thus the shear stress on the incline is expected to be positive, which is consistent with our answer.

COMMENTS

1. In the intuitive check, three of the components gave one answer, whereas the fourth gave an opposite answer. What happens if two intuitive deductions are positive and two intuitive deductions are negative and the stress components are nearly equal in magnitude? The question emphasizes that intuitive reasoning is a quick and important check on results, but one must be cautious with the conclusions.

2. We could have balanced forces in the x and y directions, in which case we have to find the x and y components of the normal and tangential forces on the force wedge. After removing the common factors $\Delta t\,\Delta z$ we would obtain

$$\sigma_A \sin 30 + \tau_A \cos 30 = -50\cos 30 + 30\sin 30 = -28.30$$

$$\sigma_A \cos 30 - \tau_A \sin 30 = -50\sin 30 - 60\cos 30 = -76.96$$

Solving these two equations we will obtain the values of σ_A and τ_A as before. By balancing forces in the n, t directions we generated one equation per unknown but did extra computation in finding components of forces in the n and t directions. By balancing forces in the x and y directions we did less work finding the components of forces, but we did extra work in solving simultaneous equations. This shows that the important point is to balance forces in any two directions, and the direction chosen for balancing the forces is a matter of preference.

3. Figure 8.9 is useful in reducing the algebra when forces are balanced in the n and t directions. But you may prefer to resolve components of individual forces, as shown in Figure 8.11, and then write the equilibrium equations. The method is a little more tedious, but has the advantage that the intuitive check can be conducted as one writes the equilibrium equations as follows.

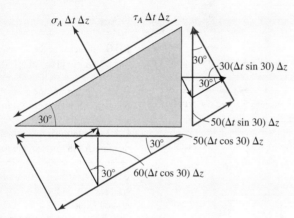

Figure 8.11   Alternative force resolution.

The normal stress σ_A on the incline will be:

- Tensile due to σ_{xx}
- Compressive due to σ_{yy}
- Compressive due to τ_{xy}
- Compressive due to τ_{yx}.

As σ_{xx} is the smallest stress component, it is not surprising that the total result is a compressive normal stress on the inclined plane.

The shear stress on the incline will be:

- Positive due to σ_{xx}
- Positive due to σ_{yy}
- Positive due to τ_{xy}
- Negative due to τ_{yx}.

We expect the net result to be positive shear stress on the incline.

PROBLEM SET 8.1

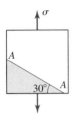

Figure P8.1

Figure P8.2

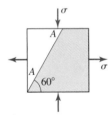

Figure P8.3

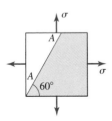

Figure P8.4

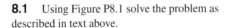

Figure P8.5

In Problems 8.1 through 8.9, one could say that the normal stress on the incline AA is in tension, in compression, or cannot be determined by inspection. Similarly we could say that the shear stress on the incline AA is positive, negative, or cannot be determined by inspection. Choose the correct answers for normal and shear stresses on the incline AA by inspection. Assume the coordinate z is perpendicular to this page and toward you.

8.1 Using Figure P8.1 solve the problem as described in text above.

8.2 Using Figure P8.2 solve the problem as described in text above.

8.3 Using Figure P8.3 solve the problem as described in text above.

8.4 Using Figure P8.4 solve the problem as described in text above.

8.5 Using Figure P8.5 solve the problem as described in text above.

8.6 Using Figure P8.6 solve the problem as described in text above.

Figure P8.6

8.7 Using Figure P8.7 solve the problem as described in text above.

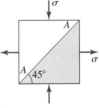

Figure P8.7

8.8 Using Figure P8.8 solve the problem as described in text above.

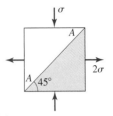

Figure P8.8

8.9 Using Figure P8.9 solve the problem as described in text above.

Figure P8.9

8.10 Determine the normal and shear stresses on plane *AA* in Problem 8.1 for $\sigma = 10$ ksi.

8.11 Determine the normal and shear stresses on plane *AA* in Problem 8.4 for $\sigma = 10$ ksi.

8.12 Determine the normal and shear stresses on plane *AA* in Problem 8.6 for $\tau = 10$ ksi.

8.13 Determine the normal and shear stresses on plane *AA* in Problem 8.7 for $\sigma = 60$ MPa.

8.14 Determine the normal and shear stresses on plane *AA* in Problem 8.9 for $\tau = 60$ MPa.

In Problems 8.15 through 8.18, a shaft is adhesively bonded along the seam as shown. By inspection determine whether the adhesive will be in tension or in compression.

8.15 Using Figure P8.15 solve the problem as described in text above.

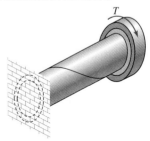

Figure P8.15

8.16 Using Figure P8.16 solve the problem as described in text above.

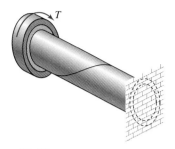

Figure P8.16

8.17 Using Figure P8.17 solve the problem as described in text above.

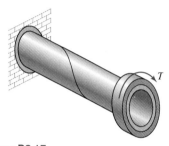

Figure P8.17

8.18 Using Figure P8.18 solve the problem as described in text above.

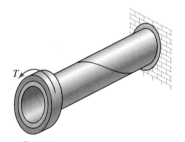

Figure P8.18

In Problems 8.19 through 8.21, determine the normal and shear stresses on plane AA.

8.19 Using Figure P8.19 solve the problem as described in text above.

8.20 Using Figure P8.20 solve the problem as described in text above.

8.21 Using Figure P8.21 solve the problem as described in text above.

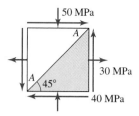

Figure P8.19

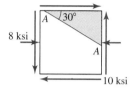

Figure P8.20

8.22 The stresses at a point in plane stress are $\sigma_{xx} = 45$ MPa (T), $\sigma_{yy} = 15$ MPa (T), and $\tau_{xy} = -20$ MPa. Determine the normal and shear stresses on a plane passing through the point at 28° counterclockwise to the x axis.

8.23 The stresses at a point in plane stress are $\sigma_{xx} = 45$ MPa (T), $\sigma_{yy} = 15$ MPa (C), and $\tau_{xy} = -20$ MPa. Determine the normal and shear stresses on a plane passing through the point at 38° clockwise to the x axis.

8.24 The stresses at a point in plane stress are $\sigma_{xx} = 10$ ksi (C), $\sigma_{yy} = 20$ ksi (C), and $\tau_{xy} = 30$ ksi. Determine the normal and shear stresses on a plane passing through the point that is 42° counterclockwise to the x axis.

8.25 A cast-iron shaft of 25-mm diameter fractured along a surface that is 45° to the axis of the shaft. The shear stress τ due to torsion is as shown in Figure P8.25. If the ultimate normal stress for the brittle cast-iron material is 330 MPa (T), determine the torque that caused the fracture.

Design problems

8.26 In a wooden structure a member was adhesively bonded along a plane 40° to the horizontal plane, as shown in Figure P8.26. Due to a load P on the structure, the stresses at a point on the bonded plane were estimated as shown, where P is in lb. If the adhesive strength in tension is 500 psi and its strength

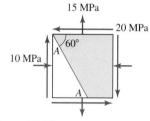

Figure P8.21

Figure P8.25

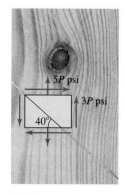

Figure P8.26

in shear is 200 psi, determine the maximum permissible load the structure can support without breaking the adhesive joint.

Stretch yourself

In three dimensions, the area of the inclined plane A can be related to the areas of the surfaces of the stress cube using the direction cosines of the outward normals,

$$n_x = \cos \theta_x \qquad A_x = n_x A$$
$$n_y = \cos \theta_y \qquad A_y = n_y A$$
$$n_z = \cos \theta_z \qquad A_z = n_z A$$

as shown in Figure P8.27. These relationships can be used to convert the stress wedge into a force wedge. Using this information, solve Problems 8.27 and 8.28. (Hint: A component of a vector in a given direction can be found by taking the dot (scalar) product of the vector with a unit vector in the given direction.)

8.27 The stresses at a point are $\sigma_{xx} = 8$ ksi (T), $\sigma_{yy} = 12$ ksi (T), and $\sigma_{zz} = 8$ ksi (C).

Determine the normal stress on a plane that has outward normals at 60°, −60°, and 45° to the x, y, and z directions, respectively.

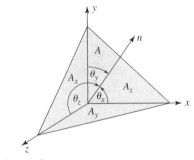

Figure P8.27

8.28 The stresses at a point are $\tau_{xy} = 125$ MPa and $\tau_{xz} = -150$ MPa. Determine the normal stress on a plane that has outward normals at 72.54°, 120°, and 35.67° to the x, y, and z directions, respectively.

8.2 STRESS TRANSFORMATION BY METHOD OF EQUATIONS

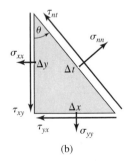

Equations that relate the stresses in the Cartesian coordinate system to the stresses on an arbitrary inclined plane are developed in this section by using the wedge method of Section 8.1. We once more consider only those planes that can be obtained by rotating about the z axis, as shown in Figure 8.2. The outward normal to the inclined plane makes an angle θ with the x axis. The angle θ is considered positive counterclockwise from the x axis, as shown in Figure 8.12.

From triangle *OAC* in Figure 8.12*a* we deduce that the angle *OAC* is 90° − θ. From triangle *OAB* we conclude that the angle *OBA* is θ. The stress wedge *OAB* is drawn as shown in Figure 8.12*b*. Multiplying the stresses by the areas of the plane, the force wedge is constructed, as shown in Figure 8.13.

Figure 8.12 (*a*) Stress cube. (*b*) Stress wedge.

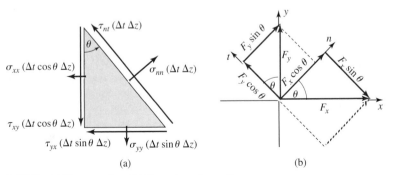

Figure 8.13 (*a*) Force wedge. (*b*) Force transformation.

By equilibrium of forces on the force wedge, we obtain

$$\sum F_n \quad \sigma_{nn} \Delta t\,\Delta z + \underbrace{(-\sigma_{xx} \cos\theta\,\Delta t\,\Delta z - \tau_{yx}\sin\theta\,\Delta t\,\Delta z)}_{F_x}\cos\theta$$

$$+ \underbrace{(-\sigma_{yy}\sin\theta\,\Delta t\,\Delta z - \tau_{xy}\cos\theta\,\Delta t\,\Delta z)}_{F_y}\sin\theta = 0$$

$$\sum F_t \quad \tau_{nt}\Delta t\,\Delta z - \underbrace{(-\sigma_{xx}\cos\theta\,\Delta t\,\Delta z - \tau_{yx}\sin\theta\,\Delta t\,\Delta z)}_{F_x}\sin\theta$$

$$+ \underbrace{(-\sigma_{yy}\sin\theta\,\Delta t\,\Delta z - \tau_{xy}\cos\theta\,\Delta t\,\Delta z)}_{F_y}\cos\theta = 0$$

After noting that $\Delta t\,\Delta z$ is a common factor, these equations simplify to

$$\boxed{\sigma_{nn} = \sigma_{xx}\cos^2\theta + \sigma_{yy}\sin^2\theta + 2\tau_{xy}\sin\theta\cos\theta} \qquad (8.1)$$

$$\boxed{\tau_{nt} = -\sigma_{xx}\cos\theta\sin\theta + \sigma_{yy}\sin\theta\cos\theta + \tau_{xy}(\cos^2\theta - \sin^2\theta)} \qquad (8.2)$$

In Equations (8.1) and (8.2) a positive angle θ is measured from the x axis in the counterclockwise direction to the n direction. Equations (8.1) and (8.2) can be written in terms of double angles of 2θ,[3]

$$\sigma_{nn} = \frac{\sigma_{xx} + \sigma_{yy}}{2} + \frac{\sigma_{xx} - \sigma_{yy}}{2}\cos 2\theta + \tau_{xy}\sin 2\theta \qquad (8.3)$$

$$\tau_{nt} = -\frac{\sigma_{xx} - \sigma_{yy}}{2}\sin 2\theta + \tau_{xy}\cos 2\theta \qquad (8.4)$$

8.2.1 Maximum Normal Stress

In Section 3.1 it was observed that a brittle material usually ruptures when the maximum tensile normal stress exceeds the ultimate tensile stress of the material. Cracks in the material also propagate due to tensile stress.[4] Adhesively bonded material debonds due to tensile normal stress, which is called *peel stress.* Similarly, failure may occur due to a maximum compressive normal stress because of the phenomenon called *buckling,* which is discussed in Chapter 11. In this section we develop equations for maximum tensile and compressive normal stresses.

In Equation (8.3) the stresses σ_{xx}, σ_{yy}, and τ_{xy} are assumed known. Thus Equation (8.3) expresses σ_{nn} as a function of θ. From calculus we know that the maximum or minimum of a function exists where the first derivative is zero. Let $\theta = \theta_p$ be the angle of the outward normal of the plane on which the maximum or minimum normal stress exists. By

---

[3]$\cos^2\theta = (1 + \cos 2\theta)/2$, $\sin^2\theta = (1 - \cos 2\theta)/2$, $\cos^2\theta - \sin^2\theta = \cos 2\theta$, and $\cos\theta\sin\theta = (\sin 2\theta)/2$.

[4]See Section 10.3 for additional details.

differentiating Equation (8.3) we obtain

$$\frac{d\sigma_{nn}}{d\theta}\bigg|_{\theta=\theta_p} = -2\frac{(\sigma_{xx}-\sigma_{yy})\sin 2\theta}{2} + 2\tau_{xy}\cos 2\theta\bigg|_{\theta=\theta_p} = 0$$

or

$$\tan 2\theta_p = \frac{2\tau_{xy}}{\sigma_{xx}-\sigma_{yy}} \tag{8.5}$$

We note that $\tan(180 + 2\theta_p) = \tan 2\theta_p$. Thus there are two angles—180° apart—which satisfy Equation (8.5), as shown in Figure 8.14. The plus sign is to be taken with subscript 1 and the minus sign with subscript 2. Then

$$R = \sqrt{\left(\frac{\sigma_{xx}-\sigma_{yy}}{2}\right)^2 + \tau_{xy}^2}$$

$$\sin 2\theta_{1,2} = \pm\tau_{xy}/R$$

$$\cos 2\theta_{1,2} = \pm\frac{\sigma_{xx}-\sigma_{yy}}{2}/R$$

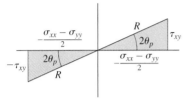

Figure 8.14 Two angles of principal planes.

Defining $\theta_1 = \theta_p$ and $\theta_2 = 90 + \theta_p$ as the two angles that satisfy Equation (8.5), let the normal and shear stresses on planes with outward normals in the θ_1 and θ_2 directions be given by σ_1, τ_1, and σ_2, τ_2, respectively. Using sines and cosines of $2\theta_1$ and $2\theta_2$, as shown in Figure 8.14, and substituting these quantities into Equations (8.3) and (8.4), we obtain

$$\sigma_{1,2} = \frac{\sigma_{xx}+\sigma_{yy}}{2} \pm \sqrt{\left(\frac{\sigma_{xx}-\sigma_{yy}}{2}\right)^2 + \tau_{xy}^2} \tag{8.6}$$

$$\tau_{1,2} = 0 \tag{8.7}$$

where $\sigma_{1,2}$ represents the two stresses σ_1 and σ_2 with the plus sign to be taken with σ_1 and the minus sign with σ_2. Equation (8.7) shows that the planes on which σ_1 and σ_2 act are planes with zero shear stress.

Definition 3 Planes on which the shear stresses are zero are called *principal planes.*

Definition 4 The normal direction to the principal planes is referred to as the principal direction or *principal axis.*

Definition 5 The angles the principal axis makes with the global coordinate system are called *principal angles.*

Definition 6 Normal stress on a principal plane is called *principal stress.*

Definition 7 The greatest principal stress is called *principal stress 1.*

In Definition 7 both the magnitude and the sign are considered in determining the greatest principal stress. A stress of −2 MPa is greater than −10 MPa. Alternatively, if normal stresses are shown on an axis with negative values to the left of the origin and positive values to the right, then the rightmost normal stress is principal stress 1 denoted by σ_1.

The stresses in Equation (8.6) represent the maximum or minimum normal stress at a point. This implies that principal stresses are the maximum and minimum normal stresses at a point. Furthermore, the plane of principal stress 1 (θ_1) is 90° away from the plane of principal stress 2 (θ_2). In other words, principal planes are orthogonal.

Definition 8 Only θ defining principal axis 1 is reported in describing the principal coordinate system in two-dimensional problems. Counterclockwise rotation from the x axis is defined as positive.

We can find σ_{tt} by substituting $90 + \theta$ in place of θ into Equation (8.1),

$$\sigma_{tt} = \sigma_{xx} \sin^2 \theta + \sigma_{yy} \cos^2 \theta - 2\tau_{xy} \cos \theta \sin \theta \qquad (8.8)$$

Adding Equation (8.1), Equation (8.8), and the principal stresses in Equation (8.6), we obtain

$$\boxed{\sigma_{nn} + \sigma_{tt} = \sigma_{xx} + \sigma_{yy} = \sigma_1 + \sigma_2} \qquad (8.9)$$

Equation (8.9) shows that the sum of the normal stresses in an orthogonal coordinate system at a point does not depend on the orientation of the coordinate system. In other words, the sum of the normal stresses is invariant with the coordinate transformation.

Equation (8.5) will give us either θ_1 or θ_2. Thus it is not clear whether the principal angle found from Equation (8.5) is associated with σ_1 or σ_2. The problem can be resolved by the following procedure.

Procedure for Determining Principal Angle 1 and Principal Stresses

Step 1: Find θ_p from Equation (8.5).

Step 2: Substitute θ_p in Equation (8.1) to find a principal stress.

Step 3: Find the other principal stress from Equation (8.9).

Step 4: Using Definition 7, decide which of the two principal stresses is principal stress 1.

Step 5: If the stress obtained from substituting θ_p into Equation (8.1) yields principal stress 1, then we report θ_p as principal angle 1 θ_1, otherwise we subtract (or add) 90° from θ_p and report the result as principal angle 1.

Step 6: Use Equation (8.6) as a check on the results.

From the definition of plane stress,[5] the plane with the outward normal in the z direction has zero shear stress. Therefore this plane is a principal plane and the normal stress σ_{zz} is a principal stress of zero value. In plane strain, the shear stresses with subscript z are also zero, as shown in Figure 3.24. Hence in plane strain also σ_{zz} is the third principal stress, but it is not zero. Using Figure 3.24, we can summarize the discussion in this paragraph as follows:

$$\sigma_3 = \sigma_{zz} = \begin{cases} 0 & \text{plane stress} \\ v(\sigma_{xx} + \sigma_{yy}) = v(\sigma_1 + \sigma_2) & \text{plane strain} \end{cases} \qquad (8.10)$$

---

[5]See Section 1.2.2 for a definition of plane stress, and Section 2.5.1 for a definition of plane strain. See Section 3.5 for the difference between plane stress and plane strain.

The value of the third principal stress affects the maximum shear stress at a point, as shall be seen in the next two sections.

8.2.2 In-Plane Maximum Shear Stress

Ductile materials usually yield when the maximum shear stress exceeds the yield stress. Lap joints in which loads are transferred from one member to another through shear are designed on the basis of the shear strength of the adhesive used in bonding the members. In this section we develop equations for maximum shear stress.

In determining the maximum shear stress from Equation (8.2) we are considering only planes that can be obtained from rotation about the z axis, as shown in Figure 8.2. Thus we are not considering all possible planes that may pass through the point. We use the following definition to differentiate the maximum shear stress obtained from Equation (8.2) from the absolute maximum shear stress at a point.

Definition 9 The maximum shear stress on a plane that can be obtained by rotating about the z axis is called *in-plane maximum shear stress.*

Let $\theta = \theta_s$ be the plane at which the in-plane maximum shear stress exists. By differentiating Equation (8.2) we get

$$\left.\frac{d\tau_{nt}}{d\theta}\right|_{\theta=\theta_s} = -2\frac{(\sigma_{xx} - \sigma_{yy})\cos 2\theta}{2} - 2\tau_{xy}\sin 2\theta\bigg|_{\theta=\theta_s} = 0$$

or

$$\tan 2\theta_s = -\frac{\sigma_{xx} - \sigma_{yy}}{2\tau_{xy}} \tag{8.11}$$

Once more, two angles can satisfy Equation (8.11). Letting $\bar{\theta}_1 = \theta_s$ and $\bar{\theta}_2 = 90 + \theta_s$, then

$$R = \sqrt{\left(\frac{\sigma_{xx} - \sigma_{yy}}{2}\right)^2 + \tau_{xy}^2}$$

$$\sin 2\bar{\theta}_{1,2} = \mp\frac{\sigma_{xx} - \sigma_{yy}}{2}/R$$

$$\cos 2\bar{\theta}_{1,2} = \pm\tau_{xy}/R$$

Let τ_{12} and τ_{21} be the shear stresses on the two planes defined by the angles $\bar{\theta}_1$ and $\bar{\theta}_2$. We can find the sines and cosines of $2\bar{\theta}_1$ and $2\bar{\theta}_2$, as shown in Figure 8.15, and substitute these quantities into Equations (8.1) and (8.2) to obtain

$$\sigma_{av} = \frac{\sigma_{xx} + \sigma_{yy}}{2} \tag{8.12}$$

$$|\tau_p| = \sqrt{\left(\frac{\sigma_{xx} - \sigma_{yy}}{2}\right)^2 + \tau_{xy}^2} = \left|\frac{\sigma_1 - \sigma_2}{2}\right| \tag{8.13}$$

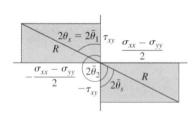

Figure 8.15 Two angles of maximum shear stress planes.

where τ_p is the in-plane maximum shear stress obtained from the magnitude of the equation $\tau_{12} = -\tau_{21} = R$.

From Equations (8.5) and (8.11) we can obtain

$$\tan 2\theta_s = \frac{-1}{\tan 2\theta_p} = \tan(90 + 2\theta_p)$$

Therefore $\theta_s = 45 + \theta_p$. In other words, maximum in-plane shear stress exists on two planes, each of which is 45° away from the principal planes.

8.2.3 Maximum Shear Stress

By default, maximum shear stress implies absolute maximum shear stress at a point, as defined:

Definition 10 The maximum shear stress at a point is the *absolute maximum shear stress* that acts on any plane passing through the point.

In the previous section we saw that as we rotate the coordinate system about the z axis (the third principal axis), the shear stress varies from a zero value, at a principal plane, to a maximum value given by Equation (8.10) on a plane that is 45° to a principal plane. Will this observation also be true if we rotate about principal axis 1 or 2? The answer is yes, because there is no distinction between the three principal planes passing through a point. We consider each of the three rotations and show all the possibilities on the stress cube in Figures 8.16 through 8.18.

Rotation about principal axis 1 (Figure 8.16) will produce the maximum shear stress on a plane that is 45° to principal planes 2 and 3, given by the equation

$$\tau_{23} = -\tau_{32} = \frac{\sigma_2 - \sigma_3}{2}$$

Rotation about principal axis 2 (Figure 8.17) will produce maximum shear stress on a plane that is 45° to principal planes 1 and 3, given by the equation

$$\tau_{31} = -\tau_{13} = \frac{\sigma_3 - \sigma_1}{2}$$

Rotation about principal axis 3 (Figure 8.18) will produce maximum shear stress on a plane that is 45° to principal planes 1 and 2, given by the equation

$$\tau_{21} = -\tau_{12} = \frac{\sigma_1 - \sigma_2}{2}$$

The maximum shear stress at a point is the largest in magnitude of the three values obtained from Figures 8.16 through 8.18. It is written in a convenient form as

$$\tau_{\max} = \left| \max\left(\frac{\sigma_1 - \sigma_2}{2}, \frac{\sigma_2 - \sigma_3}{2}, \frac{\sigma_3 - \sigma_1}{2} \right) \right| \qquad (8.14)$$

Equation (8.14) shows that the maximum shear stress value depends on principal stress 3. Equation (8.10) shows that the value of principal stress 3 depends on whether a plane stress or plane strain exists. In other words, the maximum shear stress value may be different in plane stress and in plane strain.

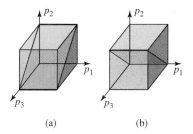

Figure 8.16 Planes of maximum shear obtained by rotating about principal axis 1.

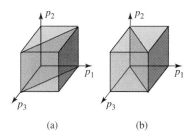

Figure 8.17 Planes of maximum shear obtained by rotating about principal axis 2.

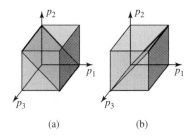

Figure 8.18 Planes of maximum shear obtained by rotating about principal axis 3.

EXAMPLE 8.3

Solve Example 8.2 using Equations (8.1) and (8.2). Also determine the principal stresses, principal angle 1, and the maximum shear stress at the point.

PLAN

(a) We can determine the angle of the outward normal to the inclined plane containing the fiber. We can then use Equations (8.1) and (8.2) to find the normal and shear stresses on the plane.

(b) We can find the principal angle from Equation (8.5) and substitute into Equation (8.1) to find one of the principal stresses. We can then use Equation (8.9) to find the other principal stresses. Using Definition 7, we determine principal stress 1. Principal stress 3 is zero because the point is in plane stress.

(c) We can find the maximum shear stress from Equation (8.14).

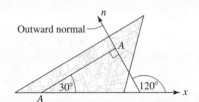

Figure 8.19 Outward normal to plane in Example 8.3.

Solution

(a) The plane AA containing the fiber is at an angle of 30° from the x axis. Hence the direction of the outward normal is $\theta = 120°$, as shown in Figure 8.19. Substituting in Equations (8.1) and (8.2), we obtain

$$\sigma_A = 30 \cos^2 120 + (-60) \sin^2 120 + 2 \times 50 \sin 120 \cos 120$$

$$= 7.5 - 45 - 43.301 = -80.801$$

$$\tau_A = -30 \sin 120 \cos 120 + (-60) \sin 120 \cos 120$$

$$+ 50(\cos^2 120 - \sin^2 120) = 13.97$$

or

ANS. $\sigma_A = 80.8$ MPa (C) $\tau_A = 14.0$ MPa

(b) We can find the principal angle from Equation (8.5),

$$\theta_p = \frac{1}{2} \arctan\left(\frac{50}{[30 - (-60)]/2}\right) = \frac{1}{2} \arctan\left(\frac{50}{45}\right) = 24.01° \tag{E1}$$

We can substitute the principal angle into Equation (8.1) to obtain one of the principal stresses,

$$\sigma_p = 30 \cos^2 24.01 + (-60) \sin^2 24.01$$

$$+ 2 \times 50 \sin 24.01 \cos 24.01$$

$$= 25.03 - 9.93 + 37.16 = 52.26 \tag{E2}$$

If we note that $\sigma_{xx} + \sigma_{yy} = 30 - 60 = -30$, we obtain from Equations (8.9) and (E2) the other principal stress as $-30 - 52.26 = -82.26$.

Thus the principal stress in Equation (E2) is principal stress 1, and the angle in Equation (E1) is principal angle 1. We thus report our answers as

ANS. $\sigma_1 = 52.3$ MPa (T) $\sigma_2 = 82.3$ MPa (C)

ANS. $\sigma_3 = 0$ $\theta_1 = 24.0°$ ccw

(c) The maximum shear stress at the point is half the maximum difference between the principal stresses, as per Equation (8.14), which in this problem is between σ_1 and σ_2,

$$\tau_{max} = \frac{52.26 - (-82.26)}{2} = 67.26$$

ANS. $\tau_{max} = 67.3$ MPa

COMMENTS

1. We can check our calculations of σ_1 and σ_2 using Equation (8.6) as shown,

$$\sigma_1 = \frac{30 + (-60)}{2} + \sqrt{\left[\frac{30 - (-60)}{2}\right]^2 + 50^2}$$

$$= -15 + 67.26 = 52.26 \quad Checks$$

$$\sigma_2 = -15 - 67.26 = -82.26 \quad Checks$$

2. In finding normal stress σ_A and shear stress τ_A on the inclined plane, we substituted $\theta = 120°$ as the angle of the outward normal. It can be checked that if we substituted $\theta = 300°$, $\theta = -60°$, or $\theta = -240°$ into Equations (8.1) and (8.2), we would obtain the same values of σ_A and τ_A. This is illustrated in Figure 8.20. A plane passing through a point has two sides. The stresses on either side are the same, and hence the outward normal direction to either side can be used for computing normal and shear stresses on the plane. The direction of the outward normal can be measured by going counterclockwise (positive direction) or by going clockwise (negative direction) from the x axis. Equations (8.1) and (8.2) reflect this observation, as substitution of

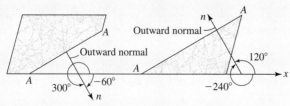

Figure 8.20   Different values of θ in Example 8.3.

$(\theta + 180°)$, $(\theta - 180°)$, $(\theta + 360°)$, or $(\theta - 360°)$ in place of θ in Equations (8.1) and (8.2) results in the same expressions for the two equations. In other words, the values of the stresses on a plane through a point are unique and depend on the orientation of the plane only and not on how its orientation is described or measured.

3. If the point were in plane strain on a material with a Poisson ratio of $\frac{1}{3}$, then the third principal stress would be $\sigma_3 = v(\sigma_1 + \sigma_2) = -10$ MPa. Thus in this problem $\sigma_1 > \sigma_3 > \sigma_2$, and hence for this problem the maximum shear stress would be unaffected. But if the third principal stress value were not in between principal stresses 1 and 2, then by Equation (8.13) the maximum stress value would be affected.

QUICK TEST 8.1   Time: 15 minutes/Total: 20 points

Each question is worth 2 points. Use the solutions given in Appendix G to grade yourself.

In Questions 1 through 3, what is the value of θ you would substitute in the stress transformation equations to find the normal and shear stresses on plane AA?

1.   **2.**   **3.**

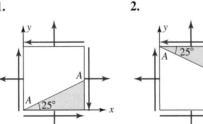 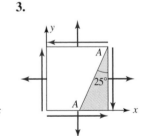

4.   At a point in plane stress, the principal stresses from the equations were found to be 5 ksi (T) and 20 ksi (C). What is the value of principal stress 1?

5.   At a point in plane stress, the principal stresses from the equations were found to be 5 ksi (C) and 20 ksi (C). What is the value of principal stress 1?

6.   At a point in plane stress, the principal stresses from the equations were found to be 5 ksi (T) and 20 ksi (T). What is the value of the maximum shear stress at that point?

7. At a point in plane stress, the principal stresses from the equations were found to be 5 ksi (T) and 20 ksi (C). What is the value of the maximum shear stress at that point?

8. At a point in plane stress, the principal stresses from the equations were found to be 5 ksi (C) and 20 ksi (C). What is the value of the maximum shear stress at that point?

In Questions 9 and 10, the angle θ_p from $\tan 2\theta_p = 2\tau_{xy}/(\sigma_{xx} - \sigma_{yy})$ was found to be $-35°$. On substituting this value into Equation (8.1), the normal stress was found to be 100 MPa (T). If the other principal stress is as given, then what is the value of principal angle 1?

9. 125 MPa (T).

10. 125 MPa (C).

8.3 STRESS TRANSFORMATION BY MOHR'S CIRCLE

To develop this graphical technique, we square Equations (8.3) and (8.4) and add the result to eliminate θ and obtain the following equation:

$$\left(\sigma_{nn} - \frac{\sigma_{xx} + \sigma_{yy}}{2}\right)^2 + \tau_{nt}^2 = \left(\frac{\sigma_{xx} - \sigma_{yy}}{2}\right)^2 + \tau_{xy}^2 \tag{8.15}$$

Consider the following equation of a circle: $(x - a)^2 + y^2 = R^2$. Comparing the equation of the circle with Equation (8.15), we see that Equation (8.15) represents a circle with a center that has coordinates $(a, 0)$ and radius R, where $a = (\sigma_{xx} + \sigma_{yy})/2$ and $R = \sqrt{[(\sigma_{xx} - \sigma_{yy})/2]^2 + \tau_{xy}^2}$. The circle is called *Mohr's circle* for stress. The coordinates of each point on the circle are the stresses (σ_{nn}, τ_{nt}). These are the normal and shear stresses on an arbitrarily oriented plane that is passing through the point at which the stresses σ_{xx}, σ_{yy}, and τ_{xy} are specified. Thus:

- Each point on Mohr's circle represents a unique plane that passes through the point at which the stresses are specified.
- The coordinates of the point on Mohr's circle are the normal and shear stresses on the plane represented by the point.

8.3.1 Construction of Mohr's Circle

The steps in the construction of Mohr's circle are described first and justified later.

Step 1: Show the stresses σ_{xx}, σ_{yy}, and τ_{xy} on a stress cube and label the vertical plane V and the horizontal plane H, as shown in Figure 8.21.

Step 2: Write the coordinates of points V and H as $V(\sigma_{xx}, \tau_{xy}\curvearrowright)$ and $H(\sigma_{yy}, \tau_{yx}\curvearrowright)$. The rotation arrow next to the shear stresses corresponds to the rotation of the cube caused by the set of shear stresses on planes V and H.

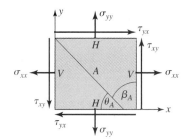

Figure 8.21 Stress cube for construction of Mohr's circle.

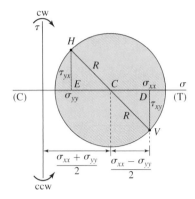

Figure 8.22 Construction of Mohr's circle.

Step 3: Draw the horizontal axis with the tensile normal stress to the right and the compressive normal stress to the left, as shown in Figure 8.22. Draw the vertical axis with the clockwise direction of shear stress up and the counterclockwise direction of rotation down.

Step 4: Locate points V and H and join the points by drawing a line. Label the point at which line VH intersects the horizontal axis as C.

Step 5: With C as the center and CV or CH as the radius, draw Mohr's circle.

The steps described in the construction of Mohr's circle can be justified as follows. The two triangles VCD and HCE are identical because

- angle VCD = angle HCE
- right angle CDV = right angle CEH
- side HE = side DV from the symmetry of shear stresses.

Thus side CE = side CD. In other words, C is the midpoint of DE, and the coordinates of the centerpoint C are the mean values of the coordinates of points D and E, that is, $((\sigma_{xx} + \sigma_{yy})/2, 0)$, which represents the center of the circle, as in Equation (8.15). By noting that the length of side CD is the difference between the coordinates of D and C, we obtain the radius of the circle from the Pythagorean theorem as $\sqrt{[(\sigma_{xx} - \sigma_{yy})/2]^2 + \tau_{xy}^2}$, which is consistent with Equation (8.15).

An important point to remember is the differentiation made in Step 2 between τ_{xy} and τ_{yx}. Equations (8.3) and (8.4) tell us that the stresses on different planes are related by twice the angle between the planes. The vertical plane V and the horizontal plane H are 90° apart on the stress cube. Thus these planes must be 180° apart on Mohr's circle, as each point on Mohr's circle represents a unique plane. This implies that if the vertical plane V is in the upper half of our coordinate system, then the horizontal plane H should be in the lower half to maintain the 180° difference on Mohr's circle. If we use the conventional method of using the upper half-plane for positive values of shear stress and the lower half-plane for negative values of shear stress, then V and H will both be either in the upper half or in the lower half because the shear stresses $\tau_{xy} = \tau_{yx}$. By associating the clockwise and counterclockwise rotation, we can satisfy the requirement that the horizontal plane and the vertical plane on Mohr's circle be 180° apart. The arguments in this paragraph emphasize the following points:

1. Angles between planes on a stress cube are doubled when plotted on Mohr's circle.

2. The sign of shear stress cannot be determined directly from Mohr's circle, as the only information from Mohr's circle is that the shear stress causes the plane to rotate clockwise or counterclockwise.

8.3.2 Principal Stresses from Mohr's Circle

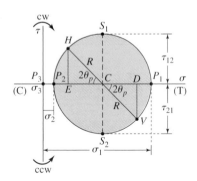

Figure 8.23 Principal stresses and in-plane maximum shear.

Figure 8.23 shows that the shear stresses are zero at points P_1 and P_2. By Definition 3, these points represent planes that are the principal planes. The normal stresses at these points are principal stresses, and these principal stresses can be found by inspection as the average normal stress plus or minus the radius. By Definition 7, the stress at point P_1 is principal stress 1. In other words, principal stress 1 is the rightmost normal stress on Mohr's circle.

The angle between lines CV and CP_1 is labeled $2\theta_p$ because all angles on Mohr's circle are double the actual angle between planes. The value of angle $2\theta_p$ can be found from the known dimensions of triangle *VCD* or triangle *HCE*, and we confirm the relationship given in Equation (8.5). The angle θ_1 is the angle between line *CV* and CP_1 in Figure 8.23 $\theta_p = \theta_1$. To determine the sign of θ_1 we first record the direction in which we move from point *V* to point P_1 on Mohr's circle. If the direction of rotation is counterclockwise, then the sign of θ_1 is positive. If the direction of rotation is clockwise, then the sign of θ_1 is negative.

Inspection of Figure 8.23 confirms the observation that the maximum and the minimum normal stresses will be principal stresses. The observation that the principal planes are orthogonal is also obvious from Figure 8.23, as points P_1 and P_2, which represent principal planes, are at 180° on Mohr's circle. That the coordinate of the center of the circle is the mean value of the normal stresses of any two points that are on a diameter of the circle confirms the observation that the sum of normal stresses is invariant with the coordinate transformation.

8.3.3 Maximum In-Plane Shear Stress

The maximum in-plane shear stress will exist on the plane represented by points S_1 and S_2, and its value is the radius of Mohr's circle, which is consistent with Equation (8.13). Points S_1 and S_2 are at 90° from points P_1 and P_2 on Mohr's circle, which is consistent with the earlier observation that the maximum in-plane shear stress exists on two planes which are at 45° to the principal planes.

8.3.4 Maximum Shear Stress

The circle in Figure 8.23 is the in-plane Mohr's circle, as the coordinate axes *n* and *t* are always in the *xy* plane. The in-plane circle represent all the planes that are obtained by rotating about principal axis 3. Let point P_3 represent principal plane 3. For plane stress problems, point P_3 coincides with the origin, as shown in Figure 8.24.

We can draw two more circles, one between P_1 and P_3 and the second between P_2 and P_3, as shown in Figure 8.24. These two circles represent rotation about principal axis 2 and principal axis 1, respectively, and are termed out-of-plane circles. The three circles together represent the complete state of stress at a point. The maximum shear stress at a point is the radius of the biggest circle. This observation is also valid for plane strain. The difference is that the value of σ_3 will have to be found by using Equation (8.10), plotted on the horizontal axis, and labeled P_3.

8.3.5 Stresses on an Inclined Plane

The stresses on an inclined plane are found by first locating the plane on Mohr's circle and then determining the coordinates of the point representing the plane. This is achieved as follows.

Step 1: Draw the inclined plane on the stress cube and label it *A*, as shown in Figure 8.21.

Step 2: Locate the inclined plane on Mohr's circle as will be described later and label it *A*, as shown in Figure 8.25.

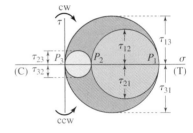

Figure 8.24 Maximum shear stress in plane stress.

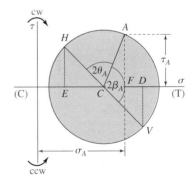

Figure 8.25 Stresses on inclined plane.

Step 3: Calculate the coordinates of point *A*.

Step 4: Determine the sign of shear stress.

There are two alternatives in Step 2.

1. On the stress cube in Figure 8.21, the inclined plane *A* is at an angle θ_A from the horizontal plane in the clockwise direction. Start from line *CH* on Mohr's circle in Figure 8.25, then rotate by an angle $2\theta_A$ in the clockwise direction and draw the line *CA*. Line *CA* represents plane *A*.

2. On the stress cube, the inclined plane *A* is at an angle β_A from the vertical plane in the counterclockwise direction. Starting from plane *CV* we rotate by an angle $2\beta_A$ in the counterclockwise direction and draw the line *CA*, which represents plane *A*.

We note from the stress cube that $\theta_A + \beta_A = 90°$ and from Mohr's circle we see that $2(\theta_A + \beta_A) = 180°$. This once more confirms that each point on Mohr's circle represents a unique plane, and it is immaterial how we locate that point on the circle.

Step 3 is the reverse of Step 2 in the construction of Mohr's circle and is a simple problem in geometry. Angle *FCA* can be found from the known angles. Radius *CA* of the circle is known, and lengths *FA* and *CF* can be found from triangle *FCA*. The coordinates of point *A* are (σ_A, τ_A). The direction of rotation is recorded as clockwise because point *A* is in the upper plane in Figure 8.25. If point *A* had been in the lower plane, we would have recorded a counterclockwise rotation with the shear stress.

To determine the sign of shear stress, we start by drawing the shear stress such that the inclined plane *A* rotates in the same direction as was recorded with the coordinates in Step 3. A local coordinate system is established, and if the shear stress is in the positive tangent direction, then it is positive. The two possibilities are shown in Figure 8.26. In both cases the shear stress is positive.

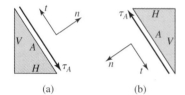

(a)   (b)

Figure 8.26 Sign of shear stress on an incline.

8.3.6 Principal Stress Element

The principal stress element is a visualization aid used in the prediction of failure surfaces. Potential failure surfaces are the planes on which maximum normal or maximum shear stress acts. In other words, principal planes and the plane of maximum shear are the potential failure surfaces.

Definition 11 A *principal stress element* shows stresses on a wedge constructed from the principal planes and the plane of maximum shear stress.

In the steps to be described the stress cubes are drawn for purposes of explanation and are not required to be drawn once the method of construction is understood.

Step 1: Rotate the coordinate axis by an angle θ_1 and the stress cube along with it. The vertical plane rotates to principal plane 1 and the point *V* rotates to point P_1 on Mohr's circle. The horizontal plane rotates to principal plane 2 and the point *H* rotates to point P_2 on Mohr's circle.

Step 2: Draw a line at 45° to the principal planes P_1 and P_2, representing the plane of maximum shear stress. Label the plane S_1 or S_2, whichever is consistent with the label shown on Mohr's circle. This is the wedge of the principal elements.

Step 3: Show principal stress 1 on plane P_1 and principal stress 2 on plane P_2. On the inclined plane show the maximum in-plane shear stress in the clockwise (or counterclockwise) direction if the inclined plane corresponds to the point in the upper (lower) half of Mohr's circle. Also show the center (average) normal stress value on the inclined plane.

Point V rotates to point P_1 and point H rotates to point P_2 on Mohr's circle, as shown in Figure 8.23. The corresponding rotation of the stress cube is shown in Figure 8.27.

We can check the consistency of the labeling on the stress cube and on Mohr's circle in the following manner. Plane S_1 is at $45°$ counterclockwise to plane P_1 in Figure 8.27. By rotating in the counterclockwise direction by $90°$ on Mohr's circle in Figure 8.23, we rotate from point P_1 to point S_1. The stress wedge obtained is shown in Figure 8.28a. Plane S_2 is at $45°$ clockwise to plane P_1 in Figure 8.27. By rotating in the clockwise direction by $90°$ on Mohr's circle in Figure 8.23, we rotate from point P_1 to point S_2. The stress wedge obtained is shown in Figure 8.28b. The stress wedges shown in Figure 8.28 are called principal elements.

Figure 8.29 shows Mohr's circle and the principal element associated with the axial loading of a circular bar. Cast iron, being a brittle material, fails from maximum tensile stress, that is, due to principal stress 1, and the failure surface is the principal plane 1. Aluminum, being a ductile material, fails from maximum shear stress, and the failure surface is the plane of maximum shear S_2. Local imperfections dictated that the failure surface was S_2 rather than S_1, which from our theory is equally likely. An explanation of the failure surfaces due to torsion shown in Figure 8.1 is left to the student in Problem 8.35.

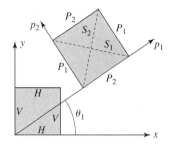

Figure 8.27 Principal planes and planes of maximum in-plane shear stress.

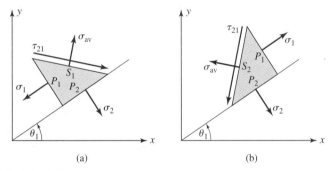

(a)   (b)

Figure 8.28 Principal elements.

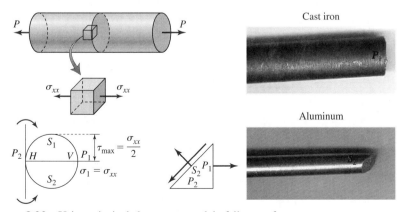

Figure 8.29 Using principal elements to explain failure surface.

EXAMPLE 8.4

For each of the given states of stress plot the normal stress and the shear stress on a plane versus θ—the angle of the outward normal of the plane—draw Mohr's circle for each state of stress, on each diagram identify the planes at $\theta_A = 30°$, $\theta_B = 75°$, $\theta_D = 105°$, and $\theta_E = 150°$.

(a) The uniaxial stress state is $\sigma_{xx} = \sigma_0$, and all other stress components are zero.

(b) The state of pure shear is $\tau_{xy} = \tau_0$, and all other stress components are zero.

PLAN

We can substitute the given states of stress into Equations (8.1) and (8.2) to obtain σ_{nn} and τ_{nt} as a function of θ and plot them. For each state of stress we can draw the stress cube, write the coordinates of planes V and H, and draw Mohr's circle. Starting from point V on Mohr's circle we can rotate by twice the angle in the counterclockwise direction to get the various points on the circle.

Solution

(a) Substituting the stress components for the uniaxial stress states into Equations (8.1) and (8.2), we obtain

$$\sigma_{nn} = \sigma_0 \cos^2 \theta \qquad \text{or} \qquad \frac{\sigma_{nn}}{\sigma_0} = \cos^2 \theta \qquad \text{(E1)}$$

$$\tau_{nt} = -\sigma_0 \sin \theta \cos \theta \qquad \text{or} \qquad \frac{\tau_{nt}}{\sigma_0} = -\sin \theta \cos \theta \qquad \text{(E2)}$$

Equations (E1) and (E2) can be plotted as shown in Figure 8.30a. We can also draw the stress cube showing uniaxial tension and record the coordinates of points V and H (Figure 8.30b). With no shear stress, the two points V and H are on the horizontal axis forming the diameter of Mohr's circle with the center at C and radius $R = \sigma_0/2$. Starting from point V on Mohr's circle we rotate counterclockwise by twice the angle θ to get the inclined planes, as shown in Figure 8.30c.

(b) Substituting the stress components for state of pure shear in Equations (8.1) and (8.2), we obtain

$$\sigma_{nn} = 2\tau_0 \sin \theta \cos \theta \qquad \text{or} \qquad \frac{\sigma_{nn}}{\tau_0} = 2 \sin \theta \cos \theta \qquad \text{(E3)}$$

$$\tau_{nt} = \tau_0(\cos^2 \theta - \sin^2 \theta) \qquad \text{or} \qquad \frac{\tau_{nt}}{\tau_0} = \cos^2 \theta - \sin^2 \theta \qquad \text{(E4)}$$

Equations (E3) and (E4) can be plotted as shown in Figure 8.31. We can also draw the stress cube showing pure shear and record the coordinates of points V and H. With normal stress, the two

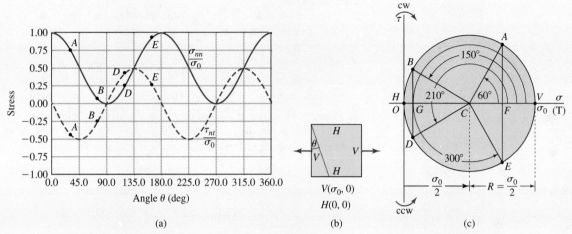

Figure 8.30   Stresses on inclined plane in uniaxial state of stress.

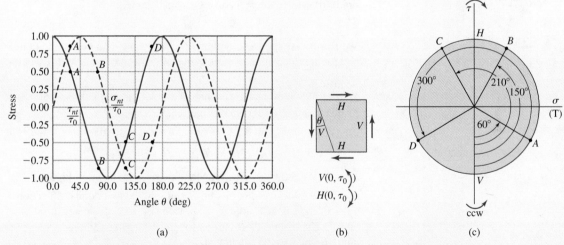

Figure 8.31   Stresses on inclined plane in state of pure shear stress.

points V and H are on the vertical axis forming the diameter of
Mohr's circle with center at C and radius $R = \tau_0$. Starting from
point V on Mohr's circle we rotate counterclockwise by twice the
angle θ to get the inclined planes, as shown in Figure 8.31.

COMMENTS

1. The example shows the relationship of planes on a graph and on a Mohr's circle, emphasizing that angles double when plotted on a Mohr's circle.

2. Figure 8.30 shows that in axial problems, the plane of maximum shear is 45° to the axis of the axial member and the value of maximum shear stress is half the value the axial normal stress at that point.

3. Figure 8.31 shows in torsion problems the principal planes are 45° to the axis of the shaft. The magnitude of principal stresses is same as the magnitude of torsional shear stress at that point.

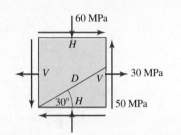

Figure 8.32 Stress cube in Example 8.5.

EXAMPLE 8.5
(a) Solve Example 8.2 using Mohr's circle. Determine (b) the principal stresses, and principal angle 1, (c) the maximum shear stress at the point (d) show the results on a principal element.

PLAN

We can follow the steps outlined for the construction of Mohr's circle and for the calculation of various quantities, as presented in this section.

Solution

Step 1: We draw the stress cube and label the vertical and horizontal planes V and H, as shown in Figure 8.32.

Step 2: From Figure 8.32 we note the coordinates of points V and H as

$$V(30, 50 \circlearrowright) \qquad H(-60, 50 \circlearrowleft)$$

Step 3: We draw the axes for Mohr's circle, as shown in Figure 8.33.

Step 4: We locate points V and H and join the two points. The coordinates of the center C are the mean value of the coordinates of points A and B. The distance AC can be found, from which the radius and the angle θ_p can then be calculated,

$$a = \frac{30 + (-60)}{2} = -15$$

$$R = \sqrt{AC^2 + AH^2} = 67.27$$

$$\tan 2\theta_p = \frac{AH}{AC} = 1.1111$$

$$2\theta_p = 48.01°$$

(a) The plane D is 30° counterclockwise from the horizontal plane in the stress cube in Figure 8.32. We rotate by twice the angle (i.e., by 60°) from line CH on Mohr's circle in Figure 8.33 and draw the line CD. The coordinates (σ_D, τ_D) of point D are calculated from geometry,

$$\beta = 60 - 2\theta_p = 11.99 \qquad \sigma_D = -(15 + R\cos\beta) = -80.8$$

$$\tau_D = R\sin\beta = 13.97$$

To determine the sign of τ_D, we draw the plane and show the direction of shear stress such that it causes the plane to rotate counterclockwise to be consistent with the fact that point D on Mohr's circle was in the lower half of the plane. We establish a local coordinate system and determine that the shear stress has a positive sign (Figure 8.34). The results are

ANS. $\qquad \sigma_D = 80.8$ MPa (C) $\qquad \tau_D = 14.0$ MPa

(b) The principal stresses are the coordinates of points P_1 and P_2,

$$\sigma_1 = -15 + R = 52.27 \qquad \sigma_2 = -15 - R = -82.27$$

$$\theta_p = \frac{48.02}{2} = 24.005°$$

The principal stresses and principal angle 1 for the problem are

ANS. $\qquad \sigma_1 = 52.3$ MPa (T) $\qquad \sigma_2 = 82.3$ MPa (C)

ANS. $\qquad \sigma_3 = 0 \qquad \theta_1 = 24.0°$ ccw

(c) The circle between P_1 and P_3 and the one between P_2 and P_3 are both inscribed within the in-plane circle shown in Figure 8.33. Thus in this problem the in-plane maximum shear stress and the absolute maximum shear stress at the given point are the same,

ANS. $\qquad \tau_{max} = 67.3$ MPa

(d) Figure 8.35 shows the principal element.

COMMENTS

1. The Mohr's circle method looks longer than the method of equations because of the explanation needed for the geometric constructions. However, computationally the difference between the two methods is small. The advantage of the method of equations is that the equations can be programmed and solved by computer. The advantage of using Mohr's circle is that it helps in the intuitive understanding of stress transformation.

2. The maximum shear stress shown in Figure 8.35 is positive, as can be deduced by establishing a local n, t coordinate system.

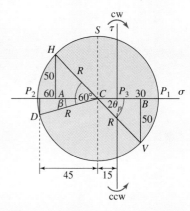

Figure 8.33 Mohr's circle in Example 8.5.

Figure 8.34 Sign of shear stress in Example 8.5.

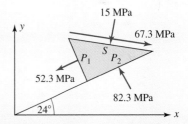

Figure 8.35 Principal element in Example 8.5.

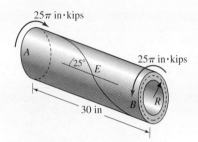

25π in·kips

A

25°

E

25π in·kips

30 in

B

R

Figure 8.36 Geometry of shaft and loading in Example 8.6.

EXAMPLE 8.6

A 30-in-long thin cylindrical tube is to transmit a torque of 25π in·kips. The tube is to be fabricated by butt welding a $\frac{1}{16}$-in-thick steel plate ($G = 12,000$ ksi) along a spiral seam, as shown in Figure 8.36. Buckling considerations limit the allowable stress in steel to 10 ksi in compression. The allowable shear stress in the weld is 12 ksi, and the allowable tensile stress in the weld is 20 ksi. Stiffness considerations limit the relative rotation of the two ends to 3°. Determine the minimum outer radius of the tube to the nearest $\frac{1}{16}$ in.

PLAN

We are required to find R to satisfy four limitations. By inspection we see that the weld material would be put into compression, and hence we can ignore the constraint on the maximum tensile stress in the weld. We can use the thin-tube approximation for computing J in terms of R, as given in Problem 5.19. For this simple loading we can determine the internal torque by inspection as $T = +25\pi$ in·kips. We can find $\phi_B - \phi_A$ in terms of R using Equation (5.10) and find one limit on R. Since the tube is thin, we can further assume that the torsional shear stress will not vary significantly from the inside to the outside, and hence we can evaluate it at the centerline radius. We can find the torsional shear stress in terms of R using Equation (5.8). We can then find the maximum compressive stress in steel and the shear stress in the seam using either Mohr's circle or the method of equations and find two other limits on R. We choose R that satisfies all limits and round it upward to the nearest $\frac{1}{16}$ value.

Solution   For thin tubes, from Problem 5.19, we have $J = 2\pi R^3 t$. Substituting $t = \frac{1}{16}$ we obtain

$$J = \frac{\pi R^3}{8} \tag{E1}$$

Substituting Equation (E1), $T = 25\pi$ in·kips, $G = 12,000$ ksi, and $L = 30$ in into Equation (5.10), we obtain the rotation of the section at B with respect to the section at A, which should be less than $3° = 0.0524$ rad. We thus find one limit on R,

$$\phi_B - \phi_A = \frac{25\pi \times 30}{12,000\,\pi R^3/8} = \frac{0.5}{R^3} \leq 0.0524 \qquad \text{or} \qquad R \geq 2.12 \tag{E2}$$

Substituting Equation (E1), $T = 25\pi$ in·kips, and $\rho = R$ into Equation (5.8),

$$\tau_{x\theta} = \frac{25\pi R}{\pi R^3/8} = \frac{200}{R^2} \text{ ksi} \tag{E3}$$

The direction of shear stress at point E can be determined using subscripts or intuitively, as shown in Figure 8.37. A two-dimensional representation of the stress cube is shown in Figure 8.37c. The directions of shear stress on the other surfaces are determined using the fact that pairs shear stresses either point toward the corner or away from the corner, as shown in Figure 8.37c.

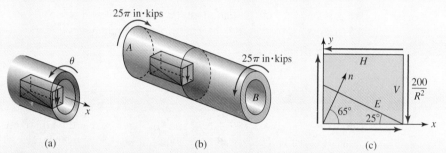

Figure 8.37   Direction of shear stress in Example 8.6.

We can determine the maximum compressive stress and the shear stress on the inclined plane using either Mohr's circle for stress or the method of equations.

Mohr's circle method: We record the coordinates of point V as $V(0, 200/R^2 \searrow)$ and the coordinates of point H as $H(0, 200/R^2 \nwarrow)$ and draw Mohr's circle as shown in Figure 8.38. We determine principal stress 2,

$$\sigma_2 = \frac{200}{R^2} \text{ ksi (C)} \qquad \text{(E4)}$$

We then locate point E on Mohr's circle and determine the shear stress,

$$|\tau_E| = \frac{200}{R^2} \sin 40 = \frac{128.56}{R^2} \text{ ksi} \qquad \text{(E5)}$$

Method of equations: We note from Figure 8.37c that $\tau_{xy} = -200/R^2$ ksi, the normal stresses are zero, and the angle of the normal to the inclined plane is 65°. Substituting this information into Equations (8.6) and (8.2), we obtain principal stress 2,

$$\sigma_2 = 0 - \sqrt{0 + \left(\frac{100}{R^2}\right)^2} = \frac{200}{R^2} \text{ ksi (C)} \qquad \text{(E4)}$$

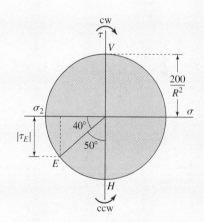

Figure 8.38 Stresses on the inclined plane using Mohr's circle.

and the shear stress on the inclined plane,

$$\tau_E = (0)\cos 65 \sin 65 + (0)\sin 65 \cos 65$$

$$+ \left(-\frac{200}{R^2}\right)(\cos^2 65 - \sin^2 65) \qquad \text{or}$$

$$|\tau_E| = \frac{128.56}{R^2} \text{ ksi} \qquad \text{(E5)}$$

We now consider the limitation on the compressive stress in steel and the shear stress in the weld and find two other limitations on R,

$$\sigma_2 = \frac{200}{R^2} \le 10 \qquad \text{or} \qquad R \ge 4.472 \qquad \text{(E6)}$$

$$|\tau_E| = \frac{128.56}{R^2} \le 12 \qquad \text{or} \qquad R \ge 3.273 \qquad \text{(E7)}$$

Comparing Equations (E2), (E6), and (E7) we see that the minimum value of R that will satisfy all three conditions is 4.472, given by Equation (E6). Rounding upward to the closest $\frac{1}{16}$ value we obtain the value of the centerline circle radius as

$$\text{ANS.} \qquad R = 4\frac{1}{2} \text{ in}$$

COMMENTS

1. Consider the error due to the approximation of J using the thin-tube formula. The outer radius for the tube is $R_o = R + t/2 = 4\frac{17}{32}$, and the inner radius of the tube is $R_i = R - t/2 = 4\frac{15}{32}$. Thus the value of exact $J = \pi(R_o^4 - R_i^4)/2$ would be 35.786 in^4. The value of approximate $J = 2\pi R^3 t$ for thin tubes is 35.785, a percentage difference from exact J of 0.003%, which is negligible for any engineering calculation.

2. Consider the approximation of uniform shear stress in the tube. If we substitute $\rho = 4\frac{17}{32}$ into Equation (5.8) we will obtain a value of 9.945 ksi at the outer surface. If we substitute $\rho = 4\frac{1}{2}$ into Equation (5.8), we will obtain a value of 9.876 ksi at the centerline, for a difference of 0.69%, which is also negligible.

3. If we do not use thin-tube approximation, then we will have to find roots of nonlinear equations requiring numerical methods to solve (see Problem 8.59). The thin-tube approximation can be used if $t < R/10$.

4. In this problem the direction (sign of τ_{xy}) of shear stress is important only to the extent of recognizing that the weld is subjected to compressive stress and not tensile stress. The magnitude of the shear stress in the weld is unaffected by the direction (sign of τ_{xy}) of shear stress—this will not be true in combined loading problems.

EXAMPLE 8.7

A T-section beam is constructed by gluing two pieces of wood together, as shown in Figure 8.39. The maximum stress in the glue joint is to be limited to 2 MPa in tension and the maximum shear stress is to be limited to 1.7 MPa. Determine the maximum value for load w.

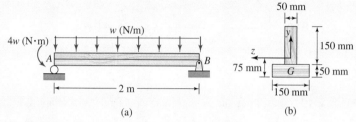

Figure 8.39 Beam and loading in Example 8.7.

PLAN

We are given that principal stress 1 in glue cannot exceed 2 MPa and the maximum shear stress in glue cannot exceed 1.7 MPa. We can draw the shear force and bending moment diagrams in terms of w and find the bending normal stress in glue and the bending shear stress in glue in the sections where the moment M_z is maximum and the shear force V_y is maximum. We can draw stress cubes at the various sections and find principal stress 1 and the maximum shear stress in terms of w. Using the limiting values we can find the value of w.

Solution We can find the reaction forces at A and B by considering the free-body diagram of the entire beam and draw the shear force and bending moment diagrams as shown in Figure 8.40. The maximum shear force and bending moment are given by

$$(V_y)_{max} = -3w \qquad (E1)$$

$$M_{max} = -4w \qquad (E2)$$

As the maximum bending moment and the maximum shear force exist in the section at A, the maximum principal flexural normal stress in glue and the maximum shear stress in glue will also exist in the section at A. The area moment of inertia of the cross section can be calculated as $I_{zz} = 53.125 \times 10^6$ mm^4. Substituting $y_G = -25$ mm and Equation (E2) into Equation (6.12), we obtain the flexural normal stress at G as

$$\sigma_G = -\frac{-4w(-25)10^{-3}}{53.125 \times 10^{-6}} = -1882w \qquad (E3)$$

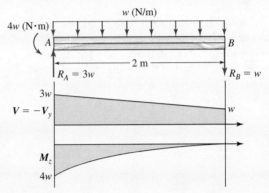

Figure 8.40 Shear force and bending moment diagrams in Example 8.7.

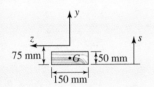

Figure 8.41 Calculation of Q_z for a point in Example 8.7.

We can draw the area A as the area between the bottom surface and point G, as shown in Figure 8.41, and find Q_z,

$$Q_G = 150 \times 50(-50) = -375 \times 10^3 \text{ mm}^3 \quad \text{(E4)}$$

Substituting Equations (E1) and (E4) into Equation (6.23), we obtain the shear stress,

$$\tau_{xs} = -\frac{-3w(-375)10^{-6}}{53.125 \times 10^{-6} \times 50 \times 10^{-3}} = -423.5w \quad \text{(E5)}$$

We can draw the stress cube and show on it the stresses in Equations (E3) and (E5). Note that with s positive upward, the shear stress τ_{xs} on the surface with the outward normal in the positive x direction will be downward to reflect the negative sign in Equation (E5). Alternatively, the sign of the shear stress τ_{xy} is negative as the shear force V_y is negative. We can draw Mohr's circle as shown in Figure 8.42. The radius R is given by

$$R = \sqrt{(941w)^2 + (423.5w)^2} = 1032w \quad \text{(E6)}$$

Principal stress 1 σ_1 can be found and the limiting value on σ_1 yields one limit on w,

$$\sigma_1 = -941w + 1032w = 91w \leq 2 \times 10^6 \quad \text{or} \quad w \leq 22{,}000 \quad \text{(E7)}$$

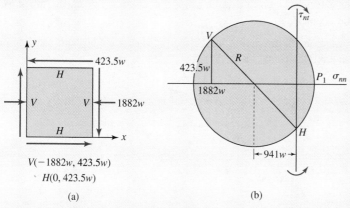

$V(-1882w, 423.5w)$

$H(0, 423.5w)$

(a)

(b)

Figure 8.42 Mohr's circle in Example 8.7.

The maximum shear stress τ_{max} is the radius of the circle, and the limiting value on τ_{max} yields the other limit on w,

$$\tau_{max} = 1032w = 1.7 \times 10^6 \qquad \text{or} \qquad w \leq 1647 \qquad \text{(E8)}$$

Comparing Equations (E7) and (E8) we conclude that the maximum permissible value of w is

$$\text{ANS.} \qquad w_{max} = 1647 \text{ N/m}$$

COMMENTS

1. The maximum normal stress in glue is $\sigma_2 = -941w - 1032w = -1973w$, which in magnitude is nearly 20 times greater than σ_1 but it is not even considered because it is compressive and does not affect the failure of glue in this problem.

2. The maximum bending normal stress in wood is at the top of the beam at section A and its value is $\sigma_{xx} = -9411.8w$. The maximum bending shear stress is at the neutral axis and its value is $\tau_{xy} = -441.2w$. At the top of the beam the only nonzero stress is σ_{xx}, thus from Figure 8.30, the maximum shear stress is $\tau_{max} = \sigma_{xx}/2 = -4705.9w$ which is an order of magnitude greater than the of maximum bending shear stress. If we had to consider shear strength failure of wood then we would use the maximum value of $4705.9w$ in our calculation.

3. Comment 2 emphasizes the difference between maximum stresses in a material and maximum bending stresses. Comment 1 emphasizes that it is not the magnitude of the maximum stress but the type of stress that causes failure in a material that is important in design.

QUICK TEST 8.2

Time: 20 minutes/Total: 20 points

Each question is worth 2 points. Use the solutions given in Appendix G to grade yourself.

In Questions 1 through 3, associate the stress cubes with the appropriate Mohr's circle given:

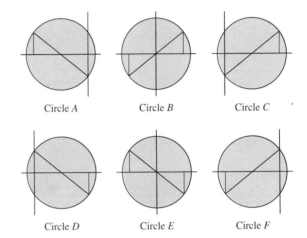

Circle *A*   Circle *B*   Circle *C*

Circle *D*   Circle *E*   Circle *F*

1.   **2.**   **3.**

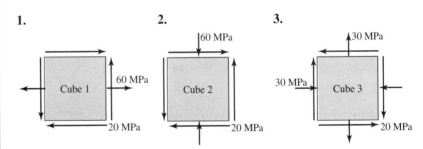

In Questions 4 through 6, Mohr's circles correspond to a plane state of stress. Determine the two possible values of principal angle 1 θ_1 in each question.

4.   **5.**   **6.**

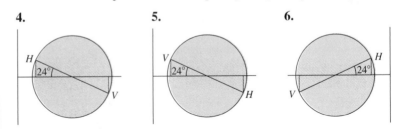

In Questions 7 through 9, Mohr's circle corresponds to the state of stress shown. Associate plane *A* on the stress cube with the corresponding Mohr's circles showing plane *A*, which are given:

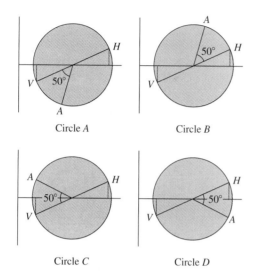

Circle *A*   Circle *B*

Circle *C*   Circle *D*

7.

8.

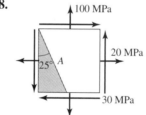

9.

10. Plane *E* passes through a point that has the state of stress given in Question 7. The normal and shear stresses on plane *E* were found to be $\sigma_E = 90$ MPa (T) and $\tau_E = -40$ MPa. What are the normal stress and the shear stress on the plane that is 90° counterclockwise from plane *E*?

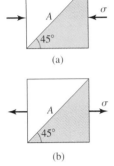

(a)

(b)

Figure P8.32

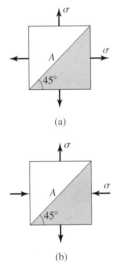

(a)

(b)

Figure P8.33

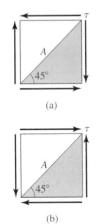

(a)

(b)

Figure P8.34

PROBLEM SET 8.2

8.29 Show that Equations (8.6) and (8.7) are correct by substituting the values of sines and cosines following Figure 8.14 into Equations (8.3) and (8.4).

8.30 Show that Equations (8.12) and (8.13) are correct by substituting the appropriate sines and cosines following Figure 8.15 into Equations (8.3) and (8.4).

8.31 Derive Equation (8.6) by starting from Equation (8.8).

In Problems 8.32 through 8.34, draw Mohr's circle and determine the normal and shear stresses on plane A.

8.32 Using Figure P8.32 solve the problem as described in text above.

8.33 Using Figure P8.33 solve the problem as described in text above.

8.34 Using Figure P8.34 solve the problem as described in text above.

8.35 Explain the failure surfaces due to torsion that are shown in Figure 8.1.

8.36 Solve Problem 8.19 by the method of equations.

8.37 Solve Problem 8.19 by Mohr's circle.

8.38 Solve Problem 8.20 by the method of equations.

8.39 Solve Problem 8.20 by Mohr's circle.

8.40 Solve Problem 8.21 by the method of equations.

8.41 Solve Problem 8.21 by Mohr's circle.

In Problems 8.42 through 8.44, in a thin body (plane stress) the stresses in the xy plane are as shown on each stress element. Determine: (a) the normal and shear stresses on plane A; (b) the principal stresses at the point; (c) the maximum shear stress at the point. (d) Draw the principal element.

8.42 Using Figure P8.42 solve the problem as described in text above.

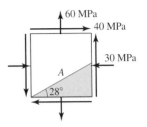

Figure P8.42

8.43 Using Figure P8.43 solve the problem as described in text above.

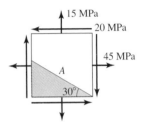

Figure P8.43

8.44 Using Figure P8.44 solve the problem as described in text above.

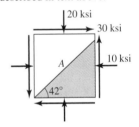

Figure P8.44

In Problems 8.45 through 8.47, in a thick body (plane strain) the stresses in the xy plane are as shown on each stress element. The Poisson's ratio of the material is $v = 0.3$. Determine: (a) the normal and shear stresses on plane A; (b) the principal stresses at the point; (c) the maximum shear stress at the point. (d) Draw the principal element.

8.45 Using Figure P8.45 solve the problem as described in text above.

8.46 Using Figure P8.46 solve the problem as described in text above.

8.47 Using Figure P8.47 solve the problem as described in text above.

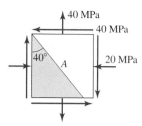

Figure P8.45

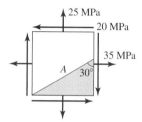

Figure P8.46

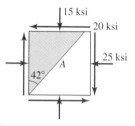

Figure P8.47

In Problems 8.48 through 8.50, the strains at a point and the material properties are as given in each problem. Assuming plane stress, determine the principal stresses, principal angle 1, and the maximum shear stress at the point. (See Problems 3.70–3.72.)

8.48 $\varepsilon_{xx} = 500 \ \mu$ $\varepsilon_{yy} = 400 \ \mu$
$\gamma_{xy} = -300 \ \mu$ $E = 200$ GPa
$\nu = 0.32$

8.49 $\varepsilon_{xx} = -3000 \ \mu$ $\varepsilon_{yy} = 1500 \ \mu$
$\gamma_{xy} = 2000 \ \mu$ $E = 70$ GPa
$G = 28$ GPa

8.50 $\varepsilon_{xx} = -800 \ \mu$ $\varepsilon_{yy} = -1000 \ \mu$
$\gamma_{xy} = -500 \ \mu$ $E = 30,000$ ksi
$\nu = 0.3$

In Problems 8.51 through 8.53, the difference in the principal stresses $\sigma_1 - \sigma_2$ and the principal direction 1 θ_1 from the x axis were measured by photoelasticity[6] at several points and are as given in each problem. The sum of the principal stresses $\sigma_1 + \sigma_2$ was found from elasticity[7] and is also given. Assuming plane stress, determine the stresses σ_{xx}, σ_{yy}, and τ_{xy} at the point.

8.51 $\sigma_1 - \sigma_2 = 10$ ksi, $\theta_1 = -15°$, and $\sigma_1 + \sigma_2 = 6$ ksi.

8.52 $\sigma_1 - \sigma_2 = 3$ ksi, $\theta_1 = +25°$, and $\sigma_1 + \sigma_2 = -17$ ksi.

8.53 $\sigma_1 - \sigma_2 = 5$ ksi, $\theta_1 = +35°$, and $\sigma_1 + \sigma_2 = 5$ ksi.

8.54 A broken 2-in × 6-in wooden bar was glued together as shown in Figure P8.54. Determine the normal and shear stresses in the glue.

8.55 Determine the normal and shear stresses in the seam of the shaft passing through point A at an angle $\theta = 60°$ to the axis of a solid shaft of 2-in diameter, as shown in Figure P8.55.

In Problems 8.56 and 8.57, if the applied force $P = 1.8$ kN, determine the principal stresses and the maximum shear stress at points A, B, and C, which are on the surface of the beam.

8.56 Using Figure P8.56 solve the problem as described in text above.

8.57 Using Figure P8.57 solve the problem as described in text above.

Design problems

8.58 A thin tube of $\frac{1}{8}$-in thickness has a mean diameter of 6 in. What is the maximum torque the tube can transmit if the allowable normal stress in compression is 10 ksi?

---

[6]See Section 8.4.1 for additional information on photoelasticity.
[7]Equations of elasticity show that $(\partial^2/\partial x^2)(\sigma_1 + \sigma_2) + (\partial^2/\partial y^2)(\sigma_1 + \sigma_2) = 0$. This differential equation can be solved numerically or analytically with the appropriate boundary conditions.

Figure P8.54

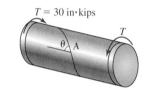

Figure P8.55

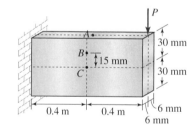

Figure P8.56

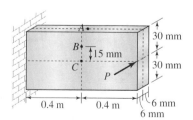

Figure P8.57

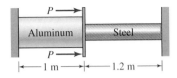

Figure P8.60

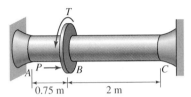

Figure P8.61

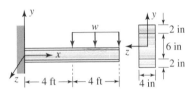

Figure P8.62

Figure P8.64

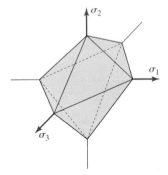

Figure P8.68 Eight octahedral planes.

8.59 Solve Example 8.6 again, now without the thin-tube approximation.

8.60 An aluminum rod (E_{al} = 70 GPa) and a steel rod (E_s = 210 GPa) are securely fastened to a rigid plate that does not rotate during the application of load P, as shown in Figure P8.60. The diameters of the aluminum and steel rods are 20 mm and 10 mm, respectively. The allowable shear stresses in aluminum and steel are 120 MPa and 150 MPa. Determine the maximum force P that can be applied to the rigid plate.

8.61 Two pieces of solid shaft of 75-mm diameter are securely connected to a rigid wheel, as shown in Figure P8.61. The shaft material has a modulus of shear rigidity G = 80 GPa and an allowable normal stress in tension or compression of 90 MPa. Determine the maximum torque T that can act on the wheel.

8.62 A cantilever beam is constructed by gluing three pieces of timber, as shown in Figure P8.62. The allowable shear stress in the adhesive is 300 psi and the allowable tensile stress is 200 psi. The allowable tensile or compressive stress in wood is 2000 psi. Determine the maximum intensity of the distributed load w.

8.63 Determine the thickness of a steel plate required for a thin cylindrical boiler with a centerline diameter of 2.5 m, if the maximum tensile stress is not to exceed 100 MPa and the maximum shear stress is not to exceed 60 MPa, when the pressure in the boiler is 1800 kPa.

8.64 A thin cylindrical tank is fabricated by butt welding a $\frac{1}{2}$-in-thick plate, as shown in Figure P8.64. The centerline diameter of the tank is 4 ft. The maximum tensile stress of the plate cannot exceed 30 ksi. The normal and shear stresses in the weld are limited to 25 ksi and 18 ksi, respectively. What is the maximum pressure the tank can hold?

Stretch yourself

8.65 By multiplying the matrices show that the following matrix equations are the same as Equations (8.1), (8.2), and (8.8):

$$[\sigma]_{nt} = [T]^T[\sigma][T] \qquad (8.16)$$

where

$$[\sigma]_{nt} = \begin{bmatrix} \sigma_{nn} & \tau_{nt} \\ \tau_{tn} & \sigma_{tt} \end{bmatrix}$$

$$[T] = \begin{bmatrix} \cos\theta & -\sin\theta \\ \sin\theta & \cos\theta \end{bmatrix}$$

$$[\sigma] = \begin{bmatrix} \sigma_{xx} & \tau_{xy} \\ \tau_{yx} & \sigma_{yy} \end{bmatrix}$$

and $[T]^T$ represents the transpose of the matrix $[T]$. The matrix $[T]$ is the transformation matrix that relates the x and y coordinates to the n and t coordinates.

8.66 Show that the eigenvalues of the matrix $[\sigma]$ are the principal stresses given by Equation (8.8).

8.67 Using the wedge shown in Figure P8.27, show that the normal stress on an inclined plane is related to the stresses in Cartesian coordinates by the equation

$$\sigma_{nn} = \sigma_{xx}n_x^2 + \sigma_{yy}n_y^2 + \sigma_{zz}n_z^2 + 2\tau_{xy}n_xn_y$$
$$+ 2\tau_{yz}n_yn_z + 2\tau_{zx}n_zn_x \qquad (8.17)$$

8.68 Figure P8.68 show eight (octal) planes that make equal angles with the principal planes. These planes are called *octahedral planes*. Though the signs of the direction cosines change with each plane, the magnitude of the direction cosines is the same for all eight planes, that is, $|n_x| = 1/\sqrt{3}$, $|n_y| = 1/\sqrt{3}$, and $|n_z| = 1/\sqrt{3}$. The normal stress and the shear stress on the octahedral planes σ_{oct} and τ_{oct} are given by

$$\sigma_{oct} = \frac{\sigma_1 + \sigma_2 + \sigma_3}{3} \qquad (8.18)$$

$$\tau_{oct} = \frac{1}{3}\sqrt{(\sigma_1 - \sigma_2)^2 + (\sigma_2 - \sigma_3)^2 + (\sigma_3 - \sigma_1)^2} \qquad (8.19)$$

Using Equation (8.17) obtain Equation (8.18).

Computer problems

8.69   On a machine component made of steel ($E = 30{,}000$ ksi, $G = 11{,}600$ ksi) the following strains were found: $\varepsilon_{xx} = [100(2x + y) + 50]\,\mu$, $\varepsilon_{yy} = -100(2x + y)\,\mu$, and $\gamma_{xy} = 200(x - 2y)\,\mu$. Assuming plane stress, determine the principal stresses, principal angle 1, and the maximum shear stress every $30°$ on a circle of radius 1 around the origin. Use a spreadsheet or write a computer program for the calculations.

8.70   The stresses around a hole in a very large plate subject to a uniform stress σ are given in polar coordinates:

$$\sigma_{rr} = \frac{\sigma}{2}\left(1 - \frac{a^2}{r^2}\right) - \frac{\sigma}{2}\left(1 - \frac{4a^2}{r^2} + \frac{3a^4}{r^4}\right)\cos 2\theta$$

$$(8.20a)$$

$$\sigma_{\theta\theta} = \frac{\sigma}{2}\left(1 + \frac{a^2}{r^2}\right) + \frac{\sigma}{2}\left(1 + \frac{3a^4}{r^4}\right)\cos 2\theta$$

$$(8.20b)$$

$$\tau_{r\theta} = \frac{\sigma}{2}\left(1 + \frac{2a^2}{r^2} - \frac{3a^4}{r^4}\right)\sin 2\theta \qquad (8.20c)$$

On a ship deck with a manhole having a diameter of 2 ft, it was estimated that $\sigma = 10$ ksi (Figure P8.70). Calculate the principal stresses every $15°$ at a radius of 18 in. Use a spreadsheet or write a computer program for the calculations.

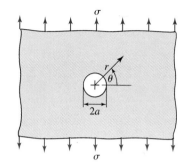

Figure P8.70

QUICK TEST 8.3

Time: 15 minutes/Total: 20 points

Answer true or false. If false, give the correct explanation. Each question is worth 2 points. Use the solution given in Appendix G to grade yourself.

1.   In plane stress there are two principal stresses and in plane strain there are three principal stresses.

2.   Principal planes are always orthogonal.

3.   For a given state of stress at a point, the principal stresses depend on the material.

4.   Depending on the coordinate system used for finding stresses at a point, the values of the stress components differ. Hence the principal stress at that point will depend on the coordinate system in which the stresses were found.

5.   Planes of maximum shear stress are always at $90°$ to principal planes.

6.   The sum of the normal stresses in an orthogonal coordinate system is independent of the orientation of the coordinate system.

7.   If principal stress 1 is tensile and principal stress 2 is compressive, then the in-plane maximum shear stress and the maximum shear stress are the same for plane *stress* problems.

8.   If principal stress 1 is tensile and principal stress 2 is compressive, then the in-plane maximum shear stress and the maximum shear stress are the same for plane *strain* problems.

9.   Two planes passing through a point can be represented by the same point on Mohr's circle.

10.   Two points on Mohr's circle can represent the same plane.

*8.4 GENERAL INFORMATION

Photoelasticity is an experimental method for deducing stress information from observing the effects on light as it passes through a transparent material that is stressed. Several definitions and ideas related to the transmission of light are needed to explain photoelasticity, as described in the next section.

8.4.1 Photoelasticity

The color of a light depends on the frequency of the light wave. White light is a mixture of lights at different frequencies. *Monochromatic* light is a light of a single frequency. Light waves can lie in different planes at a point (that is, the vibrations of the particles are in different planes). *Polarized* light is light with one plane of vibrations. A *polarizer* is a material that will permit only those light rays to pass through that have a plane of vibration parallel to the polarizing axis. If two polarizers are used, then the second polarizer is called an *analyzer*. If the polarizer and the analyzer have the polarizing axis arranged perpendicular to each other, then no light will pass through and a *dark field* will be produced. If the polarizer and the analyzer have the polarizing axis parallel to each other, then all the light that passes through the first polarizer will pass through the analyzer and a *light field* will be produced.

The velocity of light depends on the material through which it is passing. Usually the velocity of light in a material is the same in all directions (that is, the behavior is isotropic). However, there are transparent materials which when stressed have polarizing axes that are at right angles to each other. The velocities of light along these two polarizing axes are different. Materials in which light velocities are different in the two polarizing axes are called *birefringe* materials. When light passes through a birefringe material, then the ray along one axis takes a longer time to pass through it than the ray through the other axis. In other words, the ray along the slower axis reaches the same magnitude as the faster ray after a time Δt. The time Δt is called *retardation* time.

Figure 8.43 shows light originating from a monochromatic source. Light in which the plane of vibration is parallel to the vertical axis passes through the first polarizer. The components of the light passing through the polarizer are found in the direction of the two axes of the birefringe material. The component of light along the slower axis is retarded by time Δt. The second polarizer will only pass that light that is parallel to the horizontal axis. Thus the observer sees the resultant of the horizontal components of the two rays emerge from the birefringe material.

The light the observer sees depends on two variables—the angle θ of the birefringe fast axis with respect to the analyzer axis, and the retardation time Δt. What makes photoelasticity possible was first observed by Maxwell in 1857:

1. The principal axes of stresses in a birefringe material are the fast and slow axes, with the fast axis corresponding to principal stress 1.

2. The retardation time is proportional to the difference of principal stresses.

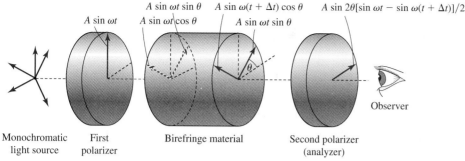

$A \sin \omega t$

$A \sin \omega t \sin \theta$ $A \sin \omega (t + \Delta t) \cos \theta$ $A \sin 2\theta [\sin \omega t - \sin \omega (t + \Delta t)]/2$

$A \sin \omega t \cos \theta$ $A \sin \omega t \sin \theta$

Observer

Monochromatic   First   Birefringe material   Second polarizer
light source   polarizer   (analyzer)

Figure 8.43 Transmission of light in photoelasticity.

Suppose we start with a light field (polarizer and analyzer axes are parallel) and note that $\sin(n\pi)$ is zero. We conclude that the observer will not see light (dark spots) where the angle θ is equal to $n\pi/2$. Lines that connect these dark spots are called *isoclinic lines*. The isoclinic lines thus give us the direction of principal stress 1 at different points. If we start with a dark field, then isoclinics correspond to lines joining points of maximum transmission of light (that is, where $\sin 2\theta = 1$).

If we start with a light field, the observer will also not see any light if the term in brackets in Figure 8.43 equals zero. This will happen at those points at which $\Delta t = 2n\pi/\omega$. Lines connecting these points are called *fringes*. As mentioned earlier, Δt is related to the difference in principal stresses $\sigma_1 - \sigma_2$. Thus the fringes yield the values of $\sigma_1 - \sigma_2$. Figure 8.44 shows these fringes in a disc subjected to diametrically opposite compressive forces.

By using different combinations of the orientation of the polarizer axis and the analyzer axis, and the relative orientations of the axis of birefringe material it is possible to obtain different isoclinic lines and different fringes. Photographs are taken for each isoclinic line and fringe, and a composite photograph that shows all isoclinic lines and fringes is made. To describe the state of plane stress completely, three pieces of information are needed. Photoelasticity yields only two—the orientation of principal axis 1 and the difference of principal stresses. On a free surface we know that one of the principal stresses is zero. Thus photoelasticity will give us a complete state of stress for a point on a free surface. In the interior if we know the sum of the principal stresses, then we can obtain the complete state of stress at that point. To obtain the sum of the principal stresses, analytical, numerical, other experimental techniques, as well as hybrids of these methods are used.

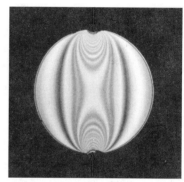

Figure 8.44 Photoelastic fringes showing principal stress difference. (Courtesy Professor I. Miskioglu.)

8.5 CLOSURE

In this chapter we studied the relationship of stresses in different coordinate systems and methods to determine the maximum tensile normal stress, maximum compressive normal stress, and maximum shear stress. In Chapter 10 we shall study various failure theories, including maximum-normal-stress and maximum-shear-stress theories, and we will use these theories for the design and failure analysis of simple structures and machines.

POINTS AND FORMULAS TO REMEMBER

- Stress transformation equations relate stresses *at a point* in different coordinate systems.

$$\sigma_{nn} = \sigma_{xx} \cos^2 \theta + \sigma_{yy} \sin^2 \theta + 2\tau_{xy} \sin \theta \cos \theta \quad (8.1)$$

$$\tau_{nt} = -\sigma_{xx} \cos \theta \sin \theta + \sigma_{yy} \sin \theta \cos \theta + \tau_{xy}(\cos^2 \theta - \sin^2 \theta) \quad (8.2)$$

where σ_{xx}, σ_{yy}, and τ_{xy} are the stresses in x, y, z coordinate system, σ_{nn}, σ_{tt}, and τ_{nt} are the stresses in n, t, z coordinate system, θ is measured from the x axis in the counterclockwise direction to the n direction.

- The values of stresses on a plane through a point are unique and depend on the orientation of the plane only and not on how its orientation is described or measured.
- Planes on which the shear stresses are zero are called *principal planes*.
- Principal planes are orthogonal.
- The normal direction to the principal planes is referred to as the principal direction or *principal axis*.
- The angles the principal axis makes with the global coordinate system are called *principal angles*.
- Normal stress on a principal plane is called *principal stress*.
- The greatest principal stress is called *principal stress 1*.
- Principal stresses are the maximum and minimum normal stresses at a point.
- The maximum shear stress on a plane that can be obtained by rotating about the z axis is called *in-plane maximum shear stress*.
- The maximum shear stress at a point is the *absolute* maximum shear stress that is on any plane passing through the point.
- Maximum in-plane shear stress exists on two planes which are at $45°$ to the principal planes.

$$\tan 2\theta_p = \frac{2\tau_{xy}}{\sigma_{xx} - \sigma_{yy}} \quad (8.5) \qquad \sigma_{1,2} = \frac{\sigma_{xx} + \sigma_{yy}}{2} \pm \sqrt{\left(\frac{\sigma_{xx} - \sigma_{yy}}{2}\right)^2 + \tau_{xy}^2} \quad (8.6)$$

$$|\tau_p| = \left|\frac{\sigma_1 - \sigma_2}{2}\right| \quad (8.13)$$

where θ_p is the angle to either principal plane 1 or 2, σ_1 and σ_2 are the principal stresses, τ_p is the in-plane maximum shear stress.

$$\sigma_{nn} + \sigma_{tt} = \sigma_{xx} + \sigma_{yy} = \sigma_1 + \sigma_2 \quad (8.9)$$

-

$$\sigma_3 = \sigma_{zz} = \begin{cases} 0, & \text{plane stress} \\ v(\sigma_{xx} + \sigma_{yy}), & \text{plane strain} \end{cases} \quad (8.10)$$

$$\tau_{max} = \left|\max\left(\frac{\sigma_1 - \sigma_2}{2}, \frac{\sigma_2 - \sigma_3}{2}, \frac{\sigma_3 - \sigma_1}{2}\right)\right| \quad (8.14)$$

- Each point on Mohr's circle represents a unique plane that passes through the point at which the stresses are specified.

- The coordinates of the point on Mohr's circle are the normal and shear stresses on the plane represented by the point.

- Angles between planes on a stress cube are doubled when plotted on Mohr's circle.

- The sign of shear stress cannot be determined directly from Mohr's circle, as the only information from Mohr's circle is that the shear stress causes the plane to rotate clockwise or counterclockwise.

- The maximum shear stress at a point is the radius of the biggest circle.

- A principal stress element shows stresses on a wedge constructed from principal planes and the plane of maximum shear stress.

CHAPTER NINE
STRAIN TRANSFORMATION

9.0 OVERVIEW

Figure 9.1 shows strain gages attached to the surface of a composite plate on which a bending moment is being applied. Strain gages are the most popular strain-measuring devices. All our stress and strain formulas have been developed in Cartesian coordinates. Thus for the purpose of analysis or experimental verification of our design, we need to transform the strains measured with the strain gages to strains in Cartesian coordinates. In this chapter we study the equations and methods used for strain transformation.

Ideas, definitions, and equations in strain transformation are very similar to those in stress transformation. This similarity will be used in the development of the theory of strain transformation. But there are also several differences. Care must be taken to account for these differences in developing a successful understanding of the strain transformation equations and methods. The major learning objective of this chapter is:

- Learn the equations and procedures of relating strains at a point in different coordinate systems.

(a)   (b)

Figure 9.1 Measurement of strains using strain gages. (Courtesy Professor I. Miskioglu.)

9.1 PRELUDE TO THEORY

In the wedge method of stress transformation, the key idea was to convert stresses into forces, that is, to convert a second-order tensor into a vector. The motivation for the conversion was that we know vector arithmetic but are not familiar with tensor arithmetic. We adopt the same strategy here for strain transformation. By multiplying a strain component by the length of a line we obtain deformation, which is a vector quantity. Using small-strain approximation, we can find the component of deformation in a given direction. Section 9.1.1 elaborates this process of strain transformation.

Principal strains, like principal stresses, are maximum and minimum normal strains. Section 9.1.2 describes a process of determining *approximate* principal strain directions by inspection.

9.1.1 Line Method

In this section we restrict ourselves to plane strain problems, which were described in Section 2.5.1. In plane strain all strains with subscript z are assumed to be zero. We further assume that the strains in the global Cartesian coordinate system, that is, ε_{xx}, ε_{yy}, and γ_{xy}, are known at a point. Define a right-handed local coordinate system n, t, z, as shown in Figure 9.2. As in stress transformation, only those coordinate systems that can be obtained by rotation about the z axis are considered. Our objective is to find ε_{nn}, ε_{tt}, and γ_{nt}.

Recollect that normal strains are a measure of change in the length of a line, whereas shear strains are a measure of change in the angle between two lines. The change in angle between two lines can be found if the angle of rotation of each line can be found. Thus by finding the change in length and rotation of two lines representing the axis of a coordinate system, we can find the strains in that coordinate system. This process is now formally described.

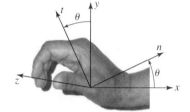

Figure 9.2 Global and local coordinate system.

Given a strain component in a global coordinate system, the normal and shear strains in the local coordinate system can be found by following these steps.

Step 1: View the n and t directions as two separate lines and determine the deformation and rotation of each line as described in the steps that follow.

Step 2: Construct a rectangle with a diagonal in the direction of the line.

Step 3: Relate the length of the diagonal to the lengths of the rectangle's sides.

Step 4: Calculate the deformation due to the given strain component and draw the deformed shape.

In drawing the deformed shapes, assume one side of the rectangle as fixed and apply the entire deformation on the opposite side. This is permissible as the rectangle represents a point and the deformation is a relative movement between two points and we take the two points on the two opposite sides of the rectangle.

Step 5: Find the deformation and rotation of the diagonal using small-strain approximations.

Recollect from Section 2.4 that small-strain approximation can be used for strains less than 0.01 (1%). Small normal strains are calculated by using the deformation component in the original direction of the line element regardless of the orientation of the deformed line element. In small shear strain γ calculations the following approximation may be used for the trigonometric functions: $\tan \gamma \approx \gamma$, $\sin \gamma \approx \gamma$, and $\cos \gamma \approx 1$.

Step 6: Calculate the normal strains by dividing the deformation by the length of the diagonal.

Step 7: Calculate the change of angle from the rotation of the lines in the n and t directions. As per Defintion 4 in Chapter 2, if the angle between the n and t directions decreases, then the shear strain is positive, if the angle increases, then the shear strain is negative.

This procedure must be repeated for each strain component in the x, y, z coordinate system. This makes the line method repetitive and tedious. Therefore we will consider problems with only one nonzero strain component to learn the principle used in developing the equations of strain transformation.

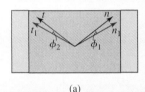

(a)

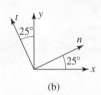

(b)

Figure 9.3 Movement of n and t lines in Example 9.1.

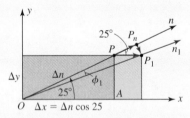

Figure 9.4 Deformation calculations in the n direction rectangles in Example 9.1.

EXAMPLE 9.1

At a point, the only nonzero strain component is $\varepsilon_{xx} = 200\ \mu$. Determine the strain components in the n, t coordinate system that is rotated $25°$ counterclockwise to the x axis.

PLAN

We follow the procedure described in this section.

Solution

Step 1: View the axes of the n, t coordinate system as two lines, as shown in Figure 9.3a. Due to the normal strain in the x direction, the lines in the n and t directions deform to n_1 and t_1, as shown in Figure 9.3b. The deformation and the rotation of each line are calculated next.

Calculations in the n direction

Step 2: We can draw a rectangle with a diagonal in the n direction, as shown in Figure 9.4.

Step 3: Let the length of the diagonal be Δn and the size Δx by Δy. We can relate Δx to Δn using triangle OAP, as shown in Figure 9.4.

Step 4: Let point P move to point P_1 due to strain ε_{xx}. The deformed shape can be drawn as shown in Figure 9.4. Using small-strain approximation, the component PP_1 in the n direction can be found. A perpendicular line from P_1 to the n direction is drawn. We are trying to find PP_n, the deformation in the n direction, and ϕ_1, the rotation of the line OP.

Step 5: The distance PP_1 can be found as the strain times the length of the rectangle. Substituting for Δx, we obtain

$$PP_1 = \varepsilon_{xx}\Delta x = 200 \times 10^{-6} \times \Delta n \cos 25 = 181.3\Delta n \times 10^{-6} \quad \text{(E1)}$$

As line PP_1 is parallel to line OA, angle P_nPP_1 is $25°$. The sides PP_n and P_nP_1 can be calculated from triangle P_nPP_1,

$$PP_n = PP_1 \cos 25 = 164.3\Delta n \times 10^{-6} \quad \text{(E2)}$$

$$P_nP_1 = PP_1 \sin 25 = 76.6\Delta n \times 10^{-6} \quad \text{(E3)}$$

The angle ϕ_1 can be found from triangle P_nOP_1 as

$$OP_n = OP + PP_n = \Delta n(1 + 164.3 \times 10^{-6}) \approx \Delta n$$

$$\tan\phi_1 \approx \phi_1 = \frac{P_nP_1}{OP_n} = 76.6 \times 10^{-6} \text{ rad} \tag{E4}$$

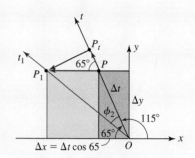

Figure 9.5 Deformation calculation in the *t* direction in Example 9.1.

Calculations in the t direction

Step 2: We can draw a rectangle with a diagonal representing the *t* direction, as shown in Figure 9.5.

Step 3: Let the length of the diagonal be Δt and the size Δx by Δy. We can relate Δx and Δt using triangle *OAP*, as shown in Figure 9.5.

Step 4: Let point *P* move to point P_1 due to strain ε_{xx}. Using small-strain approximation, the component PP_1 in the *t* direction has to be found. A perpendicular line from P_1 to the *t* direction is drawn. We are trying to find PP_t, the deformation in the *t* direction, and ϕ_2, the rotation of the line *OP*.

Step 5: The distance PP_1 can be found as the strain times the length of the rectangle. Substituting for Δx, we obtain

$$PP_1 = \varepsilon_{xx}\Delta x = 200 \times 10^{-6} \times \Delta t \cos 65 = 84.5\Delta t \times 10^{-6} \tag{E5}$$

As line PP_1 is parallel to line *OA*, angle P_nPP_1 is 65°. We can calculate sides PP_t and P_tP_1 from triangle P_tPP_1,

$$P_tP_1 = PP_1 \sin 65 = 76.58\Delta t \times 10^{-6} \tag{E6}$$

$$PP_t = PP_1 \cos 65 = 35.7\Delta t \times 10^{-6} \tag{E7}$$

We can calculate the angle ϕ_2 from triangle P_tOP_1 as

$$OP_t = OP + PP_t = \Delta t(1 + 35.7 \times 10^{-6}) \approx \Delta t$$

$$\tan\phi_2 \approx \phi_2 = \frac{P_tP_1}{OP_t} = 76.6 \times 10^{-6} \tag{E8}$$

Step 6: The normal strain in the *n* and *t* directions can be found by dividing the deformation by the length of the diagonal, that is, $\varepsilon_{nn} = PP_n/\Delta n$ and $\varepsilon_{tt} = PP_t/\Delta t$. Substituting Equations (E3) and (E7) we obtain

$$\text{ANS.} \quad \varepsilon_{nn} = 164.3 \ \mu \quad \varepsilon_{tt} = 35.7 \ \mu$$

Step 7: From Figure 9.3 the shear strain is $\gamma_{nt} = -(\phi_1 + \phi_2)$, the minus sign reflecting the fact that the angle between the *n* and *t* directions increased. Substituting Equations (E4) and (E8), we obtain the shear strain,

$$\text{ANS.} \qquad \gamma_{nt} = -153.2 \ \mu\text{rad}$$

COMMENT

In Figure 9.4 the displacement of point P to P_1 is shown in the positive x direction, whereas in Figure 9.5 the displacement is shown in the negative x direction. But notice that both rectangles show elongation to reflect positive ε_{xx}. However, in showing deformation (relative movement) we held the left side fixed in Figure 9.4, whereas in Figure 9.5 we held the right side fixed to show the deformation. Both rectangles represent the same point. Once more this emphasizes the difference between displacements and deformations.

9.1.2 Visualizing Principal Strain Directions

The visualization process described here will be used for conducting intuitive checks on the results of principal directions obtained by the method of equation and by Mohr's circle, described in Sections 9.2 and 9.3, respectively. In strain transformation, analogous to the terms used in the description of stress transformation, the following terms can be defined.

Definition 1 The *principal directions* are the coordinate axes in which the shear strain is zero.

Definition 2 The angles the principal axes make with the global coordinate system are called *principal angles*.

Definition 3 Normal strains in the principal directions are called *principal strains*.

Definition 4 The greatest principal strain is called *principal strain 1* (ε_1).

By greatest principal strains we refer to the magnitude and the sign of the principal strain. Thus a strain of $-600 \ \mu$ is greater than one of $-1000 \ \mu$.

As in stress transformation, the principal strain directions are the directions of maximum or minimum normal strain. If we start with a circle at a given point in plane strain, then the circle will deform into an ellipse with the major axis in the direction of maximum normal strain (principal strain 1) and the minor axis in the direction of minimum normal strain (principal strain 2). We make use of this observation to estimate the direction of the principal strains within a 45° quadrant. The following are the steps in the visualization process:

Step 1: Visualize or draw a square with a circle drawn inside.

Step 2: Visualize or draw the deformed shape of the square due to *only* normal strains.

The deformed shape will be a rectangle with longer and shorter sides, depending on the relative values of the normal strains ε_{xx} and ε_{yy}. The circle within the square has now become an ellipse with the major axis either along the x direction or along the y direction.

Step 3: Visualize or draw the deformed shape of the rectangle due to the shear strain.

The rectangle will deform into a rhombus and the ellipse inside would have rotated such that the major axis is in the direction of the longer diagonal of the rhombus. The major axis can rotate at most 45° from its orientation in Step 2. The major axis represents principal direction 1 and the minor axis represents principal direction 2.

Step 4: Using the eight 45° sectors shown in Figure 9.6, report the orientation of principal direction 1. Also report principal direction 2 as two sectors counterclockwise from the sector reported for principal direction 1.

As in stress transformation, principal directions 1, 2, and 3 form a right-handed coordinate system. The z direction is the third principal direction. Once principal direction 1 is determined, the right-hand rule places principal direction 2 at two sectors (90°) counterclockwise from it.

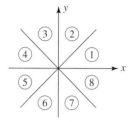

Figure 9.6 Eight sectors in which the principal axis will lie.

EXAMPLE 9.2

At a point in plane strain, the strain components are $\varepsilon_{xx} = 200\ \mu$, $\varepsilon_{yy} = 500\ \mu$, and $\gamma_{xy} = 600\ \mu$. Estimate the orientation of the principal directions and report your results using the sectors shown in Figure 9.6.

PLAN

We will follow the steps outlined in this section.

Solution   The drawings in Figure 9.7 are shown for the purpose of explanation. It is not necessary to draw these figures as these can be visualized.

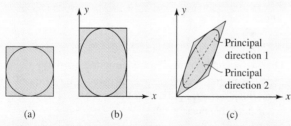

Figure 9.7 Deformation in Example 9.2. (*a*) Undeformed shape. (*b*) Deformation due to normal strains. (*c*) Additional deformation due to shear strain.

Step 1: We draw a circle inside a square, as shown in Figure 9.7*a*.

Step 2: As $\varepsilon_{yy} > \varepsilon_{xx}$, the extension in the y direction is greater than that in the x direction. The square becomes a rectangle and the circle becomes an ellipse, as shown Figure 9.7*b*.

Step 3: As γ_{xy} is positive, the angle between the x and y axes must decrease and we obtain the rhombus shown in Figure 9.7*c*.

Step 4: By inspection (see comment 1) the two solutions to this problem are:

- Principal axis 1 is in sector 2 and principal axis 2 is in sector 4.

or

- Principal axis 1 is in sector 6 and principal axis 2 is in sector 8.

COMMENTS

1. In Figure 9.7*b* the major axis is along the *y* axis. This major axis can rotate at most 45° clockwise or counterclockwise, as dictated by the shear strain. Thus principal axis 1 will be either in sector 2 or in sector 6, according to Figure 9.7*c*.

2. We will always obtain two answers for principal angle 1 as we did in stress transformation. Both answers are correct and either can be reported.

EXAMPLE 9.3

At a point in plane strain, the strain components are $\varepsilon_{xx} = -200 \ \mu$, $\varepsilon_{yy} = -400 \ \mu$, and $\gamma_{xy} = -300 \ \mu$. Estimate the orientation of the principal directions and report your results using the sectors shown in Figure 9.6.

PLAN

This time we will visualize but not draw any deformed shapes.

Solution

Step 1: We visualize a square with a circle.

Step 2: Due to normal strains, the contraction in the *y* direction is greater than that in the *x* direction. Hence the rectangle will have a longer side in the *x* direction, that is, the major axis is along the *x* axis, before we consider the shear strain.

Step 3: As the shear strain is negative, the angle will increase. The major axis will rotate clockwise, and it will lie either in sector 8 or in sector 4.

Step 4: Principal axis 1 is either in sector 8 or in sector 4. The solution is:

- Principal axis 1 is in sector 8 and principal axis 2 is in sector 2.

or

- Principal axis 1 is in sector 4 and principal axis 2 is in sector 6.

Line method

In Problems 9.1 through 9.4, determine the rotation of line OP and the normal strain in the direction OP due to the given strain component in each problem.

9.1 $\varepsilon_{xx} = 500\ \mu$ in Figure P9.1.

Figure P9.1

9.2 $\varepsilon_{yy} = -400\ \mu$ in Figure P9.2.

Figure P9.2

9.3 $\gamma_{xy} = 300\ \mu$ in Figure P9.3.

Figure P9.3

9.4 $\gamma_{yx} = 300\ \mu$ in Figure P9.4.

Figure P9.4

In Problems 9.5 through 9.7, at a point, the only nonzero strain component is as given in each problem. Determine the strain components in the n, t coordinate system shown in Figure P9.5.

9.5 $\varepsilon_{xx} = -400\ \mu$.

9.6 $\varepsilon_{yy} = 600\ \mu$.

9.7 $\gamma_{xy} = -500\ \mu$.

9.8 At a point in plane strain the strain components are $\varepsilon_{xx} = -400\ \mu$, $\varepsilon_{yy} = 600\ \mu$, and $\gamma_{xy} = -500\ \mu$. Using the results obtained in Problems 9.5, 9.6, and 9.7, determine the strain components in the n, t coordinate system shown in Figure P9.5.

In Problems 9.9 through 9.11, at a point, the only nonzero strain component is as given in each problem. Determine the strain components in the n, t coordinate system shown in Figure P9.9.

9.9 $\varepsilon_{xx} = -600\ \mu$.

9.10 $\varepsilon_{yy} = -800\ \mu$.

9.11 $\gamma_{xy} = \gamma_{xy} = 500\ \mu$.

9.12 At a point in plane strain the strain components are $\varepsilon_{xx} = -600\ \mu$, $\varepsilon_{yy} = -800\ \mu$, and $\gamma_{xy} = 500\ \mu$. Using the results obtained in Problems 9.9, 9.10, and 9.11, determine the strain components in the n, t coordinate system shown in Figure P9.9.

In Problems 9.13 through 9.15, at a point, the only nonzero strain component is as given in each problem. Determine the strain components in the n, t coordinate system shown in Figure P9.13.

9.13 $\varepsilon_{xx} = 600\ \mu$.

9.14 $\varepsilon_{yy} = \varepsilon_{yy} = 600\ \mu$.

9.15 $\gamma_{xy} = \gamma_{xy} = 600\ \mu$.

9.16 At a point in plane strain the strain components are $\varepsilon_{xx} = 600\ \mu$, $\varepsilon_{yy} = 600\ \mu$, and $\gamma_{xy} = 600\ \mu$. Using the results obtained in Problems 9.13, 9.14, and 9.15, determine the strain

Figure P9.5

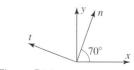

Figure P9.9

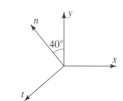

Figure P9.13

components in the n, t coordinate system shown in Figure P9.13.

Visualization of principal axis

In Problems 9.17 through 9.21, the state of strain at a point in plane strain is as given in each problem. Estimate the orientation of the principal directions and report your results using the sectors shown in Figure 9.6.

9.17 $\varepsilon_{xx} = -400 \ \mu$, $\varepsilon_{yy} = 600 \ \mu$, $\gamma_{xy} = -500 \ \mu$.

9.18 $\varepsilon_{xx} = -600 \ \mu$, $\varepsilon_{yy} = -800 \ \mu$, $\gamma_{xy} = 500 \ \mu$.

9.19 $\varepsilon_{xx} = 800 \ \mu$, $\varepsilon_{yy} = 600 \ \mu$, $\gamma_{xy} = -1000 \mu$.

9.20 $\varepsilon_{xx} = 0$, $\varepsilon_{yy} = 600 \ \mu$, $\gamma_{xy} = -500 \ \mu$.

9.21 $\varepsilon_{xx} = -1000 \ \mu$, $\varepsilon_{yy} = -500\mu$, $\gamma_{xy} = 700 \ \mu$.

9.2 METHOD OF EQUATIONS

In this section we develop strain transformation equations using the line method.[1] Assume the strains in the global Cartesian coordinate system, that is, ε_{xx}, ε_{yy}, and γ_{xy}, are known at a point in plane strain. Our objective is to find the strains in the local coordinate system shown in Figure 9.2, that is, ε_{nn}, ε_{tt}, and γ_{nt}. As in stress transformation, only coordinate systems that can be obtained by rotation about the z axis are considered here.

Definition 5 The angle θ describing the orientation of the local coordinate system to the global coordinate system is defined as positive measured from the x axis in the counterclockwise direction.

We will consider one strain component at a time and follow the steps outlined in Section 9.1.1 to determine the deformation and rotation of a line in the n direction. By substituting $90 + \theta$ in place of θ in the expressions obtained for the deformation and rotation in the n direction, the strain and rotation of the line in the t direction can be obtained.

Calculations for ε_{xx} acting alone

Step 1: View the axes of the n, t coordinate system as two lines, as shown in Figure 9.8a. Due to the normal strain in the x direction, the lines in the n and t directions deform to n_1

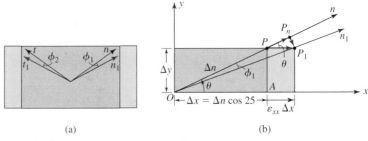

    (a)   (b)

Figure 9.8 Strain transformation with ε_{xx} only.

---

[1]See Problems 9.78 through 9.80 for an alternative derivation of the strain transformation equations.

and t_1, as shown. The deformation and the rotation of each line are calculated as shown next.

9.2 METHOD OF 567
EQUATIONS

Step 2: Draw a rectangle with a diagonal at an angle θ, as shown in Figure 9.8*b*. The diagonal represents the *n* direction.

Step 3: Let the length of the diagonal be Δn and let the length of the rectangle be Δx. We can relate Δx and Δn using triangle *OAP*, as shown in Figure 9.8*b*.

Step 4: Let point *P* move to point P_1 due to strain ε_{xx}. Draw an exaggerated deformed shape. Using small-strain approximation, we need to find the component PP_n in the *n* direction, which we can do by dropping a line from P_1 perpendicular to the *n* direction. Using geometry, we can find PP_n, the deformation in the *n* direction, and ϕ_1, the rotation of the line *OP*.

Step 5: The deformation PP_1 is the strain times the length of the rectangle. Substituting Δx, we obtain $PP_1 = \varepsilon_{xx} \Delta x = \varepsilon_{xx} \Delta n \cos \theta$. As line PP_1 is parallel to line *OA*, the angle $P_n PP_1$ is θ, and we can calculate sides PP_n from triangle $P_n PP_1$ as $PP_n = PP_1 \cos \theta$. Substituting for PP_1 we obtain

$$PP_n = (\varepsilon_{xx} \cos^2 \theta) \Delta n$$

From triangle $P_n OP_1$,

$$\tan \phi_1 = \frac{P_n P_1}{OP_n}$$

The side $P_n P_1$ can be found from triangle $P_n PP_1$ as $P_n P_1 = PP_1 \sin \theta$. Substituting for PP_1, we obtain

$$P_n P_1 = (\varepsilon_{xx} \sin \theta \cos \theta) \Delta n$$

Now $OP_n = OP + PP_n = OP(1 + PP_n/OP) = OP(1 + \varepsilon_{nn})$. For small strain, $\varepsilon_{nn} \ll 1$ and hence can be neglected, obtaining $OP_n = OP = \Delta_n$. For small strain the tangent function can be approximated by its argument. With these two approximations, we obtain the rotation ϕ_1,

$$\tan \phi_1 \approx \phi_1 = \frac{P_n P_1}{OP_n} \approx \frac{P_n P_1}{OP} = \frac{(\varepsilon_{xx} \sin \theta \cos \theta) \Delta n}{\Delta n}$$

or

$$\phi_1 = \varepsilon_{xx} \sin \theta \cos \theta$$

Step 6: The normal strain in the *n* direction can be found as $\varepsilon_{nn}^{(1)} = PP_n/\Delta n$. The superscript 1 is to differentiate the strain calculated from ε_{xx} only from that calculated from ε_{yy} and γ_{xy}. Substituting for PP_n, we obtain

$$\varepsilon_{nn}^{(1)} = \varepsilon_{xx} \cos^2 \theta \qquad (9.1)$$

We note that the *t* axis is a line like the *n* axis, but at an angle of $90 + \theta$ instead of θ. We can obtain the normal strain in the *t* direction and the rotation of the *t* axis by substituting $90 + \theta$ in place of θ in Equation (9.1),

$$\varepsilon_{tt}^{(1)} = \varepsilon_{xx} \cos^2 (\theta + 90) = \varepsilon_{xx} \sin^2 \theta \qquad (9.2)$$

Step 7: We can obtain the rotation of the *t* axis by substituting $90 + \theta$ in place of θ in the expression for ϕ_1 to obtain ϕ_2,

$$\phi_2 = \left| \varepsilon_{xx} \sin(\theta + 90) \cos(\theta + 90) \right| = \varepsilon_{xx} \sin\theta \, \cos\theta$$

The angle between the *n* and *t* directions increases, as seen from the rectangle in Figure 9.8a. This implies that the shear strain will be negative and is given as

$$\gamma_{nt}^{(1)} = -(\phi_1 + \phi_2) = -2\varepsilon_{xx} \sin\theta \, \cos\theta \tag{9.3}$$

Calculations for ε_{yy} acting alone

The preceding calculations can be repeated for ε_{yy}. The calculations for Steps 1–5 are shown in Figure 9.9. Based on small strain, we once more approximate $OP_n \cong OP$ and $\tan\phi_1 \approx \phi_1$ to obtain

$$\tan\phi_1 \approx \phi_1 = \frac{P_nP_1}{OP_n} \approx \frac{P_nP_1}{OP} = \varepsilon_{yy} \sin\theta \, \cos\theta$$

Step 6: The normal strain in the *n* direction can be found as $\varepsilon_{nn}^{(2)} = PP_n/\Delta n$. Substituting for PP_n, we obtain

$$\varepsilon_{nn}^{(2)} = \varepsilon_{yy} \sin^2\theta \tag{9.4}$$

The normal strain in the *t* direction is obtained by substituting $90 + \theta$ in place of θ into Equation (9.4),

$$\varepsilon_{tt}^{(2)} = \varepsilon_{yy} \sin^2(\theta + 90) = \varepsilon_{yy} \cos^2\theta \tag{9.5}$$

Step 7: The rotation of the *t* axis can be obtained by substituting $90 + \theta$ in place of θ in the expression for ϕ_1,

$$\phi_2 = \left| \varepsilon_{yy} \sin(\theta + 90) \cos(\theta + 90) \right| = \varepsilon_{yy} \sin\theta \, \cos\theta$$

The angle between the *n* and *t* directions decreases, as seen from the rectangle in Figure 9.9a. This implies that the shear strain will be positive and is given as

$$\gamma_{nt}^{(2)} = \phi_1 + \phi_2 = 2\varepsilon_{yy} \sin\theta \, \cos\theta \tag{9.6}$$

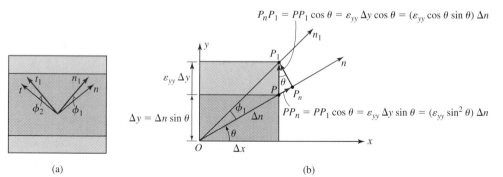

(a)   (b)

Figure 9.9 Strain transformation with ε_{yy} only.

Calculations for γ_{xy} acting alone

The preceding calculations can be repeated for γ_{xy}. The calculations for Steps 1–5 are shown in Figure 9.10. Based on small strain, we once more approximate $OP_n \cong OP$ and $\tan \phi_1 \approx \phi_1$ to obtain

$$\tan \phi_1 \approx \phi_1 = \frac{P_n P_1}{OP_n} \approx \frac{P_n P_1}{OP} = \gamma_{xy} \sin^2 \theta$$

Step 6: The normal strain in the n direction can be found as $\varepsilon_{nn}^{(3)} = PP_n / \Delta n$. Substituting for PP_n, we obtain

$$\varepsilon_{nn}^{(3)} = \frac{PP_n}{\Delta n} = \gamma_{xy} \sin \theta \cos \theta \qquad (9.7)$$

The normal strain in the t direction is obtained by substituting $90 + \theta$ in place of θ into Equation (9.7),

$$\varepsilon_{tt}^{(3)} = \gamma_{xy} \sin(\theta + 90) \cos(\theta + 90) = -\gamma_{xy} \sin \theta \cos \theta \qquad (9.8)$$

Step 7: The rotation of the t axis can be obtained by substituting $90 + \theta$ in place of θ in the expression for ϕ_1,

$$\phi_2 = \left| \gamma_{xy} \cos^2(\theta + 90) \right| = \gamma_{xy} \cos^2 \theta$$

From Figure 9.10a it is seen that the movement of the line in the n direction to n_1 increases the initial angle and the movement of the line in the t direction to t_1 decreases the initial angle. The final angle between the n_1 and t_1 directions is $\pi/2 + \phi_1 - \phi_2$. Thus from the definition of shear strain in Chapter 2, we obtain

$$\gamma_{nt}^{(3)} = \phi_1 - \phi_2 = \gamma_{xy}(\cos^2 \theta - \sin^2 \theta) \qquad (9.9)$$

Total strains

As we are working with small strains, we have a linear system, and the total strain in the n and t directions is the superposition of the strains due to the individual components. That is,

$$\varepsilon_{nn} = \varepsilon_{nn}^{(1)} + \varepsilon_{nn}^{(2)} + \varepsilon_{nn}^{(3)} \qquad \varepsilon_{tt} = \varepsilon_{tt}^{(1)} + \varepsilon_{tt}^{(2)} + \varepsilon_{tt}^{(3)} \qquad \gamma_{nt} = \gamma_{nt}^{(1)} + \gamma_{nt}^{(2)} + \gamma_{nt}^{(3)}$$

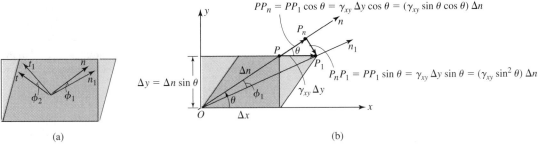

Figure 9.10 Strain transformation with γ_{xy} only.

We obtain the following equations:

$$\varepsilon_{nn} = \varepsilon_{xx} \cos^2 \theta + \varepsilon_{yy} \sin^2 \theta + \gamma_{xy} \sin \theta \cos \theta \tag{9.10}$$

$$\varepsilon_{tt} = \varepsilon_{xx} \sin^2 \theta + \varepsilon_{yy} \cos^2 \theta - \gamma_{xy} \sin \theta \cos \theta \tag{9.11}$$

$$\gamma_{nt} = -2\varepsilon_{xx} \sin \theta \cos \theta + 2\varepsilon_{yy} \sin \theta \cos \theta + \gamma_{xy}(\cos^2 \theta - \sin^2 \theta) \tag{9.12}$$

Equations (9.10), (9.11), and (9.12) are similar to the stress transformation equations, Equations (8.1), (8.2), and (8.8), with the difference that the coefficient of the shear strain term is half the coefficient of shear stress term. This difference is due to the fact that we are using engineering strain instead of tensor strain.[2] With this difference accounted for, we can rewrite the results from stress transformation for strain transformation as described next.

9.2.1 Principal Strains

Noting that the coefficient of shear strain in strain transformation equations is half the coefficient of shear stress in stress transformation equations, we obtain [see Equation (8.5)] the principal angle θ_p,

$$\tan 2\theta_p = \frac{\gamma_{xy}}{\varepsilon_{xx} - \varepsilon_{yy}} \tag{9.13}$$

The principal strains [analogous to Equation (8.6)] can be obtained from the following equation:

$$\varepsilon_{1,2} = \frac{\varepsilon_{xx} + \varepsilon_{yy}}{2} \pm \sqrt{\left(\frac{\varepsilon_{xx} - \varepsilon_{yy}}{2}\right)^2 + \left(\frac{\gamma_{xy}}{2}\right)^2} \tag{9.14}$$

In Equation (9.14) it is to be understood that $\varepsilon_{1,2}$ represents the two strains ε_1 and ε_2. The plus sign is to be taken with ε_1 and the minus sign with ε_2. Like principal stresses, the principal strains correspond to the maximum and minimum normal strains at a point.

Adding Equations (9.10) and (9.11) and the principal strains in Equation (9.14), we obtain

$$\varepsilon_{nn} + \varepsilon_{tt} = \varepsilon_{xx} + \varepsilon_{yy} = \varepsilon_1 + \varepsilon_2 \tag{9.15}$$

Equation (9.15) shows that the sum of the normal strains at any point in an orthogonal coordinate system does not depend on the orientation of the coordinate system.

Definition 6 The angle of principal axis 1 from the x axis is only reported in describing the principal coordinate system in two-dimensional problems. Counterclockwise rotation from the x axis is defined as positive.

---

[2]An alternative is to let $\gamma_{xy} = 2\varepsilon_{xy}$ and $\gamma_{nt} = 2\varepsilon_{nt}$ in Equations (9.10) through (9.12), where it is understood that ε_{xy} is the tensor shear strain and γ_{xy} is the engineering shear strain. In such a case the equations of stress and strain transformation have identical forms.

There are two values of θ_p that satisfy Equation (9.13), separated by 90°. The direction θ_1 corresponding to ε_1 is 90° from the direction θ_2 corresponding to ε_2. In other words, principal directions are orthogonal. It is not clear whether the principal angle found from Equation (9.13) is associated with ε_1 or ε_2. A simple way to resolve this problem is to substitute θ_p into Equation (9.10) and obtain one principal strain, then use Equation (9.15) to find the other principal strain. Using Definition 4, we can decide which of the two principal strains is principal strain 1. If the strain obtained from substituting θ_p into Equation (9.10) yields principal strain 1, then we report θ_p as principal angle 1, otherwise we subtract (or add) 90° from θ_p and report the result as principal angle 1.

We will use Equation (9.14) as a check on our results rather than using it for the calculation of principal strain. In plane strain, the shear strains with subscript z are zero. Therefore, by Definitions 1 and 3, the z direction is a principal direction and the normal strain ε_{zz} is a principal strain of zero value.

In plain stress the shear strains with subscript z are zero, making ε_{zz} the third principal strain, but ε_{zz} is not zero, as shown in Figure 3.24, but equal to $v(\sigma_{xx} + \sigma_{yy})$. If we add Equations (3.12a) and (3.12b) for plane stress problems we obtain $\sigma_{xx} + \sigma_{yy} = E[(\varepsilon_{xx} + \varepsilon_{yy})/(1 - v)]$. Thus[3]

$$\varepsilon_{zz} = -\frac{v}{1 - v}(\varepsilon_{xx} + \varepsilon_{yy})$$

We can write the third principal strain, using Equation (9.15), as

$$\varepsilon_3 = \begin{cases} 0, & \text{plane strain} \\ -\dfrac{v}{1 - v}(\varepsilon_{xx} + \varepsilon_{yy}) = -\dfrac{v}{1 - v}(\varepsilon_1 + \varepsilon_2), & \text{plane stress} \end{cases} \qquad (9.16)$$

The value of the third principal strain affects the maximum shear strain at a point, as described next.

9.2.2 Maximum Shear Strain

As in stress transformation, we differentiate between in-plane maximum shear strain and absolute maximum shear strain using the following definitions.

Definition 7 The maximum shear strain in coordinate systems that can be obtained by rotating about the z axis is called *in-plane maximum shear strain.*

Definition 8 The maximum shear strain at a point is the absolute maximum shear strain that can be obtained in a coordinate system by considering rotation about all three axes.

After accounting for the fact that the coefficient of shear strain in strain transformation equations is half the coefficient of shear stress in stress transformation equations, we obtain the maximum shear strain as

$$\frac{\gamma_{max}}{2} = \left| \max\left(\frac{\varepsilon_1 - \varepsilon_2}{2}, \frac{\varepsilon_2 - \varepsilon_3}{2}, \frac{\varepsilon_3 - \varepsilon_1}{2} \right) \right| \qquad (9.17)$$

---

[3]See Equation (3.16).

The in-plane maximum shear strain γ_p is related to the difference between principal strains 1 and 2 and is given by

$$\left|\frac{\gamma_p}{2}\right| = \left|\frac{\varepsilon_1 - \varepsilon_2}{2}\right| \qquad (9.18)$$

Equation (9.17) shows that the value of the maximum shear strain depends on the value of principal strain 3. Equation (9.16) shows that the value of principal strain 3 depends on the plane stress or plane strain problem. As in stress transformation, the maximum shear strain exists in two coordinate systems that are at 45° to the principal coordinate system.

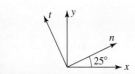

Figure 9.11

EXAMPLE 9.4

At a point in plane strain, the strain components are $\varepsilon_{xx} = 200\ \mu$, $\varepsilon_{yy} = 1000\ \mu$, and $\gamma_{xy} = -600\ \mu$. Determine:

(a) The principal strains and principal angle 1.

(b) The maximum shear strain.

(c) The strain components in a coordinate system that is rotated 25° counterclockwise, as shown in Figure 9.11.

PLAN

(a) Using Equation (9.13), we can find θ_p. We can substitute θ_p into Equation (9.10) and find one of the principal strains. Using Equation (9.15) we find the other principal strain and, using Definition 4, decide which is principal strain 1.

(b) We can find the maximum shear strain using Equation (9.17).

(c) We can find the strains in the n and t coordinates by substituting $\theta = 25°$ in Equations (9.10), (9.11), and (9.12).

Solution

(a) From Equation (9.13) we obtain the principal angle,

$$\tan 2\theta_p = \frac{-600}{200 - 1000} = 0.75 = \tan 36.87$$

$$\theta_p = 18.43° \qquad \text{or} \qquad 18.43° \text{ ccw} \qquad (E1)$$

Substituting θ_p into Equation (9.10), we obtain one of the principal strains,

$$\varepsilon_p = 200 \cos^2 18.43 + 1000 \sin^2 18.43$$

$$+ (-600) \sin 18.43 \cos 18.43 = 100\ \mu \qquad (E2)$$

Now $\varepsilon_{xx} + \varepsilon_{yy} = 1200\ \mu$. From Equation (9.15) we obtain the other principal strain as $1200 - 100 = 1100\ \mu$, which is greater than the principal strain in Equation (E2). Thus $1100\,\mu$ is principal strain 1, and principal angle 1 is obtained by adding (or subtracting) $90°$ from Equation (E1). As the point is in plane strain, the third principal strain is zero. We report our results as

ANS. $\quad \varepsilon_1 = 1100\,\mu \qquad \varepsilon_2 = 100\,\mu \qquad \varepsilon_3 = 0$

ANS. $\quad \theta_1 = 108.4°\ \text{ccw} \qquad \text{or} \qquad 71.6°\ \text{cw} \qquad\qquad$ (E3)

We can check the principal strain values using Equation (9.14),

$$\varepsilon_{1,2} = \frac{200 + 1000}{2} \pm \sqrt{\left(\frac{200 - 1000}{2}\right)^2 + \left(\frac{-600}{2}\right)^2}$$

$$= 600 \pm 500 \qquad \text{Checks}$$

Intuitive check: We can check the orientation of principal axis 1, using the intuitive approach described in Section 9.1.2. We visualize a circle in a square, as shown in Figure 9.12a. As $\varepsilon_{yy} > \varepsilon_{xx}$, the rectangle will become longer in the y direction than in the x direction and the circle will become an ellipse with major axis along the y direction, as shown in Figure 9.12b. As $\gamma_{xy} < 0$, the angle between the x and y directions will decrease. The rectangle will become a rhombus and the major axis of the ellipse will rotate counterclockwise from the y axis. Hence we expect principal axis 1 to be in either the third sector or the seventh sector, confirming the result given in Equation (E3).

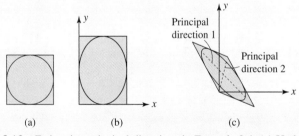

(a)   (b)   (c)

Figure 9.12 Estimating principal directions in Example 9.4. (*a*) Undeformed shape. (*b*) Deformation due to normal strains. (*c*) Additional deformation due to shear strain.

(b) We can find the maximum shear strain from Equation (9.17), that is, the maximum difference is between ε_1 and ε_3, thus the maximum shear strain is

$$\text{ANS.} \qquad \gamma_{max} = 1100 \; \mu$$

(c) Substituting $\theta = 25°$ in Equations (9.10), (9.11), and (9.12), we obtain

$$\varepsilon_{nn} = 200\cos^2 25 + 1000 \sin^2 25 + (-600) \sin 25 \cos 25$$

or

$$\text{ANS.} \qquad \varepsilon_{nn} = 113.1 \; \mu$$

$$\varepsilon_{tt} = 200 \sin^2 25 + 1000 \cos^2 25 - (-600) \sin 25 \cos 25$$

or

$$\text{ANS.} \qquad \varepsilon_{tt} = 1086.9 \; \mu$$

$$\gamma_{nt} = -2(200) \sin 25 \cos 25 + 2(1000) \sin 25 \cos 25$$
$$+ (-600)(\cos^2 25 - \sin^2 25)$$

or

$$\text{ANS.} \qquad \gamma_{nt} = 227.2 \; \mu$$

We can use Equation (9.15) to check our results. We note that $\varepsilon_{nn} + \varepsilon_{tt} = 1200$, which is the same value as for $\varepsilon_{xx} + \varepsilon_{yy}$, confirming the accuracy of our results.

COMMENTS

1. It can be checked that if we substitute $\theta = 25° + 180° = 205°$ or $\theta = 25° - 180° = -155°$ in Equations (9.10), (9.11), and (9.12), we will obtain the same values for ε_{nn}, ε_{tt}, and γ_{nt} as in part (c). In other words, adding or subtracting 180° from the angle θ in Equations (9.10), (9.11), and (9.12) does not affect the results, which emphasizes that the strain at a point in a given direction (coordinate system) is unique and does not dependent upon how the orientation of the line is described or measured.

2. If the point were in plane stress on a material with a Poisson's ratio of $\frac{1}{3}$, then the third principle strain would be $\varepsilon_3 = -(v/1 - v)(\varepsilon_{xx} + \varepsilon_{yy}) = -600 \; \mu$ and the maximum shear strain would be $\gamma_{max} = 1700 \; \mu$ which is different than the value we obtained in part (b) for plane strain.

EXAMPLE 9.5

For the wooden cantilever beam shown in Figure 9.13 determine at point A: (a) the principal strains and the angle of first principal direction θ_1; (b) the maximum shear strain. Use the modulus of elasticity $E = 12.6$ GPa and Poisson's ratio $\nu = 0.3$.

PLAN

The bending stresses σ_{xx} and τ_{xy} at point A can be found using Equations (6.12) and (6.23), respectively. Using Hooke's law, the strains ε_{xx}, ε_{yy}, and γ_{xy} can be found. Using Equation (9.13), θ_p can be found and substituted into Equation (9.10) to obtain one of the principal strains. Using Equation (9.15), we find the other principal strain and, using Definition 4, decide which is principal strain 1. The maximum shear strain can be found using Equation (9.17).

Solution

Bending stress calculations: The area moment of inertia I_{zz} and the first moment Q_z of the area A_s shown in Figure 9.14a can be calculated,

$$I_{zz} = \frac{12 \times 60^3}{12} = 0.216 \times 10^6 \text{ mm}^4 = 0.216 \times 10^{-6} \text{ m}^4 \quad \text{(E1)}$$

$$Q_z = 12 \times 15(15 + 7.5) = 4.050 \times 10^3 \text{ mm}^3$$
$$= 4.050 \times 10^{-6} \text{ m}^3 \quad \text{(E2)}$$

Recollect that A_s is the area between the free surface and the parallel line passing through point A, where shear stress is to be found.

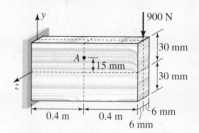

Figure 9.13 Beam and loading in Example 9.5.

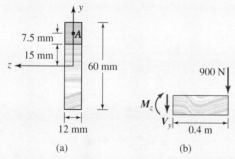

(a)   (b)

Figure 9.14 Calculation of geometric and internal quantities.

Figure 9.14b shows the free-body diagram of the right part of the beam after making the imaginary cut through point A in Figure 9.13.

The shear force V_y and the bending moment M_z are drawn according to our sign convention. By balancing forces and moment we obtain

$$V_y = -900 \text{ N} \tag{E3}$$

$$M_z = -0.4 \times 900 = -360 \text{ N·m} \tag{E4}$$

Substituting Equations (E1), (E4), and $y_A = 0.015$ m into Equation (6.12), we obtain the bending normal stress,

$$\sigma_{xx} = -\frac{M_z y}{I_{zz}} = -\frac{-360 \times 0.015}{0.216 \times 10^{-6}} = 25 \times 10^6 \text{ N/m}^2 \tag{E5}$$

By visualizing the beam deformation, we expect σ_{xx} to be tensile.

Substituting Equations (E1), (E2), (E3), and $t = 0.012$ m into Equation (6.23), we obtain the magnitude of τ_{xy}. Noting that τ_{xy} must have the same sign as V_y, we obtain the sign of τ_{xy} (see Section 6.6.5) as given by

$$|\tau_{xy}| = \left|\frac{V_y Q_z}{I_{zz} t}\right| = \left|\frac{-900 \times 4.050 \times 10^{-6}}{0.216 \times 10^{-6} \times 0.012}\right| = 1.41 \times 10^6 \text{ N}$$

or

$$\tau_{xy} = -1.41 \times 10^6 \text{ N} \tag{E6}$$

Bending strain calculations: The shear modulus of elasticity can be found from $G = E/2(1 + v)$. Substituting $E = 12.6$ GPa and $v = 0.3$, we obtain $G = 4.85$ GPa. The only two nonzero stress components are given by Equations (E5) and (E6). Using the generalized Hooke's law [or Equation (6.25)], we obtain the bending strains,

$$\varepsilon_{xx} = \frac{\sigma_{xx}}{E} = \frac{25 \times 10^6}{12.6 \times 10^9} = 1.984 \times 10^{-3} = 1984 \ \mu \tag{E7}$$

$$\varepsilon_{yy} = -\frac{v \sigma_{xx}}{E} = -v\varepsilon_{xx} = -0.3 \times 1984 = -595.2 \ \mu \tag{E8}$$

$$\gamma_{xy} = \frac{\tau_{xy}}{G} = \frac{-1.41 \times 10^6}{4.85 \times 10^9} = -0.2907 \times 10^{-3} = -290.7 \ \mu \tag{E9}$$

(a) *Stress transformation calculations:* From Equation (9.13) we obtain the principal angle,

$$\tan 2\theta_p = \frac{-290.7}{1984 - (-595.2)} = -0.1124 = -\tan 6.41$$

or

$$\theta_p = -3.21° \tag{E10}$$

Substituting θ_p into Equation (9.10) we obtain one of the principal strains,

$$\varepsilon_p = 1984 \cos^2(-3.21) + (-595.2)\sin^2(-3.21)$$

$$+ (-290.7)\sin(-3.21)\cos(-3.21) = 1992 \; \mu \quad \text{(E11)}$$

Now $\varepsilon_{xx} + \varepsilon_{yy} = 1389 \; \mu$. From Equation (9.15) we obtain the other principal strain as $1389 - 1992 = -603 \; \mu$, which is less than the principal strain in Equation (E11). Thus $1992 \; \mu$ is principal strain 1, and principal angle 1 is obtained from Equation (E10). The third principal strain will be the same as the second principal strain. We report our results as

ANS. $\varepsilon_1 = 1992 \; \mu \qquad \varepsilon_2 = -603 \; \mu \qquad \varepsilon_3 = -603 \; \mu$

$\theta_1 = 3.21° \; \text{cw}$ $\qquad\qquad\qquad$ (E12)

Check: We can check the principal strain values using Equation (9.14),

$$\varepsilon_{1,2} = \frac{1984 + (-595.2)}{2} \pm \sqrt{\left(\frac{1984 - (-595.2)}{2}\right)^2 + \left(\frac{-290.7}{2}\right)^2}$$

$$= 694.4 \pm 1297.7$$

or

$$\varepsilon_1 = 1992.1 \qquad \varepsilon_2 = -603.3 \qquad \textit{Checks}$$

Intuitive check: We can check the orientation of principal axis 1 using the intuitive approach described in Section 9.1.2. We visualize a circle in a square, as shown in Figure 9.15a. As $\varepsilon_{xx} > \varepsilon_{yy}$ the rectangle will become longer in the x direction than in the y direction and the circle will become an ellipse with its major axis along the x direction, as shown in Figure 9.15b. As $\gamma_{xy} < 0$, the angle

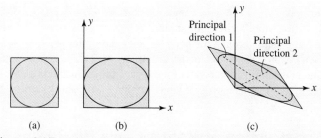

(a) $\qquad\qquad$ (b) $\qquad\qquad\qquad$ (c)

Figure 9.15 Estimating principal directions in Example 9.5. (*a*) Undeformed shape. (*b*) Deformation due to normal strains. (*c*) Additional deformation due to shear strain.

between the x and y directions will increase. The rectangle will become a rhombus and the major axis of the ellipse will rotate clockwise from the x axis, as shown in Figure 9.15c. Hence we expect principal axis 1 to be in the eighth sector (or the fourth sector), confirming the result given in Equations (E12).

(b) We can find the maximum shear strain from Equation (9.17), that is, the difference between ε_1 and ε_2 (or ε_3). Thus the maximum shear strain is

$$\text{ANS.} \qquad \gamma_{max} = 2595 \, \mu$$

COMMENT

The example demonstrates the synthesis of the theory of symmetric bending of beams and the theory of strain transformation. A similar synthesis can be elaborated for axial and torsion members.

9.3 MOHR'S CIRCLE

To develop this graphical technique we eliminate θ from Equations (9.10) and (9.12) after writing in terms of double angles, as we did in stress transformation, to obtain

$$\left(\varepsilon_{nn} - \frac{\varepsilon_{xx} + \varepsilon_{yy}}{2}\right)^2 + \left(\frac{\gamma_{nt}}{2}\right)^2 = \left(\frac{\varepsilon_{xx} - \varepsilon_{yy}}{2}\right)^2 + \left(\frac{\gamma_{xy}}{2}\right)^2 \qquad (9.19)$$

Comparing Equation (9.19) with that for a circle, $(x - a)^2 + y^2 = R^2$, we see that Equation (9.19) represents a circle with a center that has coordinates $(a, 0)$ and radius R, where

$$a = \frac{\varepsilon_{xx} + \varepsilon_{yy}}{2} \qquad R = \sqrt{\left(\frac{\varepsilon_{xx} - \varepsilon_{yy}}{2}\right)^2 + \left(\frac{\gamma_{xy}}{2}\right)^2} \qquad (9.20)$$

The circle is called Mohr's circle for strain. Each point on Mohr's circle represents a unique direction passing through the point at which the strains are specified. The coordinates of each point on the circle are the strains $(\varepsilon_{nn}, \gamma_{nt}/2)$. These represent the normal strain of a line in the n direction and half the shear strain, which represents the rotation of the line passing through the point at which strains ε_{xx}, ε_{yy}, and γ_{xy} are specified.

9.3.1 Construction of Mohr's Circle for Strains

The construction of Mohr's circle for strain is very similar to Mohr's circle for stress. There are two important differences. (i) In stress transformation we talked about planes, here we talk about directions. The directions are the outward normals of the planes. This difference is elaborated in Step 1 of the procedure to construct Mohr's circle that follows.

(ii) The vertical axis is shear strain divided by 2. All values of shear strain that are plotted on Mohr's circle or calculated from Mohr's circle must account for the fact that the vertical coordinate is shear strain divided by 2.

The steps in the construction of Mohr's circle for strain are as follows.

Step 1: Draw a square with a shape deformed due to shear strain γ_{xy}. Label the intersection of the vertical plane and the x axis as V and the intersection of the horizontal plane and the y axis as H, as shown in Figure 9.16.

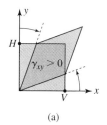

Unlike stress transformation, where V and H represented planes, here V and H refer to directions. The outward normal to the vertical plane is the x axis, and V is the label associated with it. Similarly the outward normal to the horizontal plane is the y axis, which is represented by point H.

Step 2: Write the coordinates of points V and H,

$$V(\varepsilon_{xx}, \gamma_{xy}/2\!\downarrow) \qquad H(\varepsilon_{yy}, \gamma_{xy}/2\!\uparrow), \qquad \text{for } \gamma_{xy} > 0$$

The arrow of rotation along side the shear strains corresponds to the rotation of the line on which the point lies, as shown in Figure 9.16. The construction of the deformed shape in Step 2 for the purpose of determining the direction of rotation of the x and y axes is a critical step for a successful construction of Mohr's circle. The rotation of the lines differentiate γ_{xy} from γ_{yx} for the purpose of plotting.

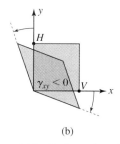

Step 3: Draw the horizontal axis to represent the normal strain, with extensions (E) to the right and contractions (C) to the left, as shown in Figure 9.17a. Draw the vertical axis to represent half the shear strain, with clockwise rotation of a line in the upper plane and counterclockwise rotation of a line in the lower plane.

This step emphasizes the following point. The value of shear strain read from Mohr's circle does not tell us whether shear strain is positive or negative, but it shows that the shear strain will cause a line in a given direction to rotate clockwise or counterclockwise. This point is further elaborated in Section 9.3.2.

Step 4: Locate points V and H and join the points by drawing a line. Label the point at which line VH intersects the horizontal axis as C.

Step 5: The horizontal coordinate of point C is the average normal strain. Distance CE can be found from the coordinates of points E and C and the radius R calculated

Figure 9.16 Deformed cube for construction of Mohr's circle.

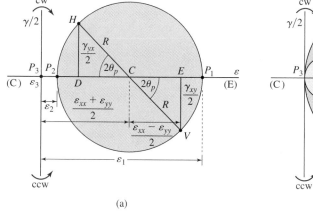

(a)

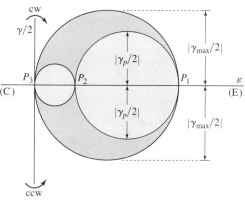

(b)

Figure 9.17 Mohr's circle for strains.

using the Pythagorean theorem. With C as the center and CV or CH as the radius, draw Mohr's circle.

Step 6: Calculate the principal strains by finding the coordinates of points P_1 and P_2 in Figure 9.17a.

Step 7: Calculate principal angle θ_p from either triangle VCE or triangle DCH. Find the angle between lines CV and CP_1 if θ_1 is different from θ_p.

In Figure 9.17a, θ_p and θ_1 have the same value, but this may not always be the case, as elaborated in Example 9.6. The angle θ_1 is the angle measured from the x axis, which is represented by point V on Mohr's circle, and principal direction 1 represented by point P_1.

Step 8: Check your answer for θ_1 intuitively using the visualization technique of Section 9.1.2.

Step 9: The in-plane maximum shear strain $\gamma_p/2$ equals R, the radius of the in-plane circle shown in Figure 9.17a. To find the absolute maximum shear strain, locate point P_3 at the value of the third principal strain and draw two more circles between P_1 and P_3 and between P_2 and P_3, as shown in Figure 9.17b. The maximum shear strain at a point is found from the radius of the largest circle.

For plane strain P_3 is at the origin, as shown in Figure 9.17b. But for plane stress, the third principal strain must be found from Equation (9.16) and located before drawing the remaining two circles. Notice that the radii of the circles yield half the value of the maximum shear strain.

9.3.2 Strains in a Specified Coordinate System

The strains in a specified coordinate system are found by first locating the coordinate directions on Mohr's circle and then determining the coordinates of the point representing the directions. This is achieved as follows.

Step 10: Draw the Cartesian coordinate system and the specified coordinate system along with a square in each coordinate system, representing the undeformed state. Label points V, H, N, and T to represent the four directions, as shown in Figure 9.18a.

Step 11: Points V and H on Mohr's circle are known. Point N on Mohr's circle is located by starting from point V and rotating by $2\theta_V$ in the same direction, as shown in

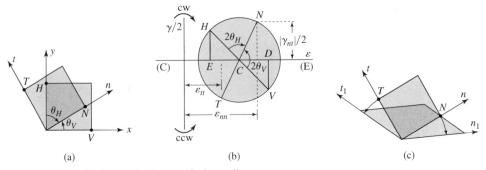

(a)   (b)   (c)

Figure 9.18 Strains in specified coordinate system.

It should be emphasized that we could start from point H on Mohr's circle and reach point N by rotating $2\theta_H$, as shown in Figure 9.18b. In Figure 9.18a, $\theta_H + \theta_V = 90°$, and in Figure 9.18b we see that $2\theta_H + 2\theta_V$ is 180°, which once more emphasizes that each point on Mohr's circle represents a unique direction, and it is immaterial how one reaches it.

Step 12: Calculate the coordinates of points N and T.

This is the reverse of Step 2 in the construction of Mohr's circle and is a problem in geometry. As per Figure 9.18b, the coordinates of points N and T are

$$N(\varepsilon_{nn}, \gamma_{nt}/2 \downarrow) \qquad T(\varepsilon_{tt}, \gamma_{nt}/2 \uparrow)$$

The rotation of the line at point N is clockwise, as it is in the upper plane, whereas the rotation of the line at point T is counterclockwise, as it is in the lower plane in Figure 9.18b.

Step 13: Determine the sign of the shear strain.

To draw the deformed shape we rotate the n coordinate in the direction shown for point N in Step 3. Similarly, we rotate the t coordinate in the direction shown for point T in Step 3, as illustrated in Figure 9.18c. The angle between the n and t directions increases, hence the shear strain γ_{nt} is negative according to Definition 4 in Chapter 2.

EXAMPLE 9.6

At a point in plane strain, the strain components are $\varepsilon_{xx} = 200\ \mu$, $\varepsilon_{yy} = 1000\ \mu$, and $\gamma_{xy} = -600\ \mu$. Using Mohr's circle, determine:

(a) The principal strains and principal angle 1.

(b) The maximum shear strain.

(c) The strain components in a coordinate system that is rotated 25° counterclockwise, as shown in Figure 9.19.

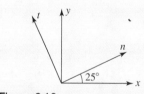

Figure 9.19

PLAN

We can follow the steps outlined for the construction of Mohr's circle and for the calculation of the various quantities as outlined in this section.

Solution

Step 1: The shear strain is negative, hence the angle between the x and y axes should increase. We draw the deformed shape of a square due to shear strain γ_{xy}. We label the intersection of the vertical plane and the x axis as V and the intersection of the horizontal plane and the y axis as H, as shown in Figure 9.20.

Step 2: Using Figure 9.20, we can write the coordinates of points V and H,

$$V(200, 300 \downarrow) \qquad H(1000, 300 \uparrow)$$

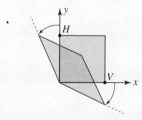

Figure 9.20 Deformed cube for construction of Mohrs's circle in Example 9.6.

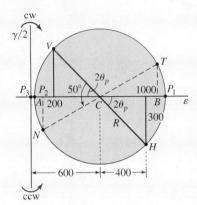

Figure 9.21 Mohr's circle in Example 9.6.

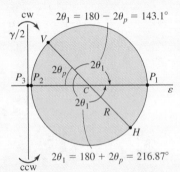

Figure 9.22 Two values of principal angle 1.

Step 3: We draw the axes for Mohr's circle as shown in Figure 9.21.

Step 4: Locate points V and H and join the points by drawing a line.

Step 5: Point C, the center of Mohr's circle, is midway between points A and B, that is, at 600 μ. The distance BC can thus be found as 400 μ, as shown in Figure 9.21. From the Pythagorean theorem we can find the radius R,

$$R = \sqrt{CB^2 + BH^2} = \sqrt{400^2 + 300^2} = 500 \qquad \text{(E1)}$$

Step 6: The principal strains are the coordinates of points P_1 and P_2 in Figure 9.21. By adding CP_1, that is, the radius, to the coordinate of point C, we can obtain the principal strains as $\varepsilon_1 = 600 + 500 = 1100$ and $\varepsilon_2 = 600 - 500 = 100$. Noting that for plane strain the third principal strain is zero,

ANS. $\qquad \varepsilon_1 = 1100 \ \mu \qquad \varepsilon_2 = 100 \ \mu \qquad \varepsilon_3 = 0 \qquad \text{(E2)}$

Step 7: Using triangle BCH we can find the principal angle θ_p,

$$\tan 2\theta_p = \frac{BH}{BC} = \frac{300}{400} \qquad \text{or} \qquad 2\theta_p = 36.87° \qquad \text{(E3)}$$

Principal angle 1 can be found from θ_p as shown in Figure 9.22,

ANS. $\qquad \theta_1 = 71.6° \text{ cw} \qquad \text{or} \qquad \theta_1 = 108.4° \text{ ccw} \qquad \text{(E4)}$

Step 8: Intuitive check: We visualize a circle in a square, as shown in Figure 9.23a. As $\varepsilon_{yy} > \varepsilon_{xx}$ the rectangle will become longer in the y direction than in the x direction, and the circle will become an ellipse with the major axis along the y direction, as shown in Figure 9.23b. As $\gamma_{xy} < 0$, the angle between the x and y directions will increase. The rectangle will become a rhombus, and the major axis of the ellipse will rotate counterclockwise from the y axis, as shown in Figure 9.23c.

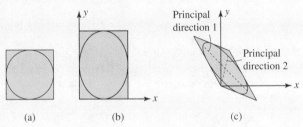

Figure 9.23 Estimating principal directions in Example 9.6. (*a*) Undeformed shape. (*b*) Deformation due to normal strains. (*c*) Additional deformation due to shear strain.

Hence we expect principal axis 1 to be either in the third sector or in the seventh sector, confirming the result given in Equation (E4).

Step 9: The circles between P_1 and P_2 and between P_2 and P_3 will be inscribed in the circle between P_1 and P_3. Thus the maximum shear strain at the point can be determined from the circle between P_1 and P_3,

$$\frac{\gamma_{max}}{2} = \frac{\varepsilon_1 - \varepsilon_3}{2} = \frac{1100}{2} \qquad \text{or} \qquad \text{ANS.} \qquad \gamma_{max} = 1100 \ \mu$$

Step 10: We can draw the Cartesian coordinate system and the specified coordinate system with a square representing the undeformed state. Label points *V, H, N,* and *T* to represent the four directions, as shown in Figure 9.24.

Step 11: Starting from point *V* on Mohr's circle, we rotate by 50° counterclockwise and obtain point *N* on Mohr's circle in Figure 9.21. Similarly by starting from point *H* and rotating by 50° counterclockwise, we obtain point *T* on Mohr's circle in Figure 9.21.

Step 12: Angle *ACN* and angle *BCT* can be found as $50 - 2\theta_p = 13.13°$. From triangle *ACN* in Figure 9.21, the coordinates of point *N* can be found,

$$\varepsilon_{nn} = 600 - 500 \cos 13.13 = 113.1$$

$$\gamma_{nt}/2 = 500 \sin 13.13 = 113.58 \ \text{↘} \qquad (E2)$$

From triangle *BCT*, the coordinates of point *T* can be found,

$$\varepsilon_{tt} = 600 + 500 \cos 13.13 = 1086.9$$

$$\gamma_{nt}/2 = 500 \sin 13.13 = 113.58 \ \text{↗} \qquad (E3)$$

Step 13: In Figure 9.24 line *ON* rotates in the counterclockwise direction to ON_1, as per Equation (E5), and line *OT* rotates in the clockwise direction to OT_1, as per Equation (E6). Angle N_1OT_1 is less than angle *NOT,* hence the shear strain in the *n, t* coordinate system is positive. Thus the strain results are

ANS. $\qquad \varepsilon_{nn} = 113.1 \ \mu \qquad \varepsilon_{tt} = 1086.9 \ \mu \qquad \gamma_{nt} = 227.2 \ \mu$

COMMENT

Example 9.4 and this example solve the same problem. But unlike the method of equations used in Example 9.4, this example shows that we do not need any equation in solving the problem by Mohr's circle. Once Mohr's circle is constructed, the problem of strain transformation becomes a problem of geometry.

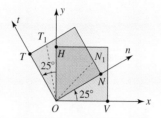

Figure 9.24 *n, t* coordinate system in Example 9.6.

QUICK TEST 9.1

Time: 15 minutes/Total: 20 points

Grade yourself with the answers given in Appendix G. Each question is worth two points.

In Questions 1 through 3, associate the strain states with the appropriate Mohr's circle given.

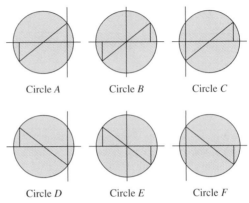

Circle A Circle B Circle C

Circle D Circle E Circle F

1. $\varepsilon_{xx} = -600\ \mu$, $\varepsilon_{yy} = 0$, and $\gamma_{xy} = -600\ \mu$.
2. $\varepsilon_{xx} = 0$, $\varepsilon_{yy} = 600\ \mu$, and $\gamma_{xy} = 600\ \mu$.
3. $\varepsilon_{xx} = 300\ \mu$, $\varepsilon_{yy} = -300\ \mu$, and $\gamma_{xy} = -600\ \mu$.

In Questions 4 and 5, the Mohr's circles corresponding to the states of strain $\varepsilon_{xx} = -500\ \mu$, $\varepsilon_{yy} = 1100\ \mu$, and $\gamma_{xy} = -1200\ \mu$ are shown. Identify the circle you would use to find the strains in the n, t coordinate system in each question.

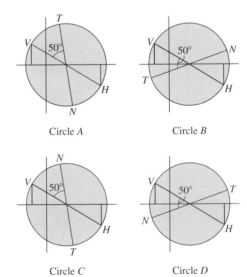

Circle A Circle B

Circle C Circle D

4.

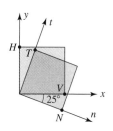

5.

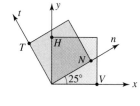

In Questions 6 and 7, the Mohr's circles for a state of strain are given. Determine the two possible values of principal angle 1 (θ_1) in each question.

6.

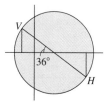

7.

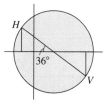

In Questions 8 through 10, the Mohr's circles for points in plane strain are given. Report principal strain 1 and maximum shear strain in each question.

8.

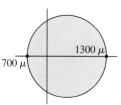

9.

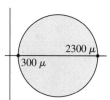

10.

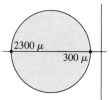

9.4 GENERALIZED HOOKE'S LAW IN PRINCIPAL COORDINATES

In Section 3.4 it was observed that the generalized Hooke's law is valid for any orthogonal coordinate system. We have seen that the principal coordinates for stresses and strains are orthogonal.

It has been shown mathematically and confirmed experimentally that for isotropic materials[4] the principal directions for strains are the same as the principal

---

[4]See Example 9.9, where the principal directions for stresses and strains are different because the material is orthotropic.

directions for stresses. We can write the generalized Hooke's law relating principal stresses to principal strains as follows:

$$\varepsilon_1 = \frac{\sigma_1 - \nu(\sigma_2 + \sigma_3)}{E} \tag{9.21a}$$

$$\varepsilon_2 = \frac{\sigma_2 - \nu(\sigma_3 + \sigma_1)}{E} \tag{9.21b}$$

$$\varepsilon_3 = \frac{\sigma_3 - \nu(\sigma_1 + \sigma_2)}{E} \tag{9.21c}$$

Note that there are no equations for shear stresses and shear strains, as both these quantities are zero in the principal coordinate system.

Now that we know that, at a point, principal axis 1 for stresses and strains is the same for isotropic material, we can extend our intuitive check to stress transformation. This can be done by viewing σ_{xx}, σ_{yy}, and τ_{xy} as analogous to ε_{xx}, ε_{yy}, and γ_{xy} for the purpose of the visualization procedure outlined in Section 9.1.2.

EXAMPLE 9.7

The stresses $\sigma_{xx} = 4$ ksi (T), $\sigma_{yy} = 10$ ksi (C), and $\tau_{xy} = 4$ ksi were calculated at a point on a free surface of an isotropic material. Determine: (a) the orientation of principal axis 1 for stresses, using Mohr's circle for stress; (b) the orientation of principal axis 1 for strains, using Mohr's circle for strain. Use the following material constants: $E = 7500$ ksi, $G = 3000$ ksi, and $\nu = 0.25$.

PLAN

By substituting the stresses and material constants into the generalized Hooke's law in Cartesian coordinates, the strains ε_{xx}, ε_{yy}, and γ_{xy} can be found. We can draw Mohr's circle for stress to find principal direction 1 for stress, and we can draw Mohr's circle for strain to find principal direction 1 for strain.

Solution As the point is on a free surface, the state of stress is plane stress; hence $\sigma_{zz} = 0$. Substituting the stresses and the material constants into Equations (3.12a), (3.12b), and (3.12d), we obtain

$$\varepsilon_{xx} = \frac{\sigma_{xx}}{E} - \frac{\nu}{E}\sigma_{yy} = \frac{4}{7500} - \frac{0.25}{7500}(-10) = 0.867 \times 10^{-3} = 867\ \mu \tag{E1}$$

$$\varepsilon_{yy} = \frac{\sigma_{yy}}{E} - \frac{\nu}{E}\sigma_{xx} = \frac{-10}{7500} - \frac{0.25}{7500}4 = -1.467 \times 10^{-3} = -1467\ \mu \tag{E2}$$

$$\gamma_{xy} = \frac{\tau_{xy}}{G} = \frac{4}{3000} = 1.333 \times 10^{-3} = 1333\ \mu \tag{E3}$$

(a) We draw the stress cube and record the coordinates of planes V and H,

$$V(4, 4\blacktriangle) \qquad H(-10, 4\blacktriangleright) \tag{E4}$$

We then draw Mohr's circle for stress, as shown in Figure 9.25a. The angle θ_p can be found from triangle BCH (or ACV) and is given by

$$\tan 2\theta_p = \frac{4}{7} \quad \text{or} \quad \theta_p = 14.87° \quad \text{(E5)}$$

$$\text{ANS.} \quad \theta_1 = 14.87° \text{ ccw}$$

For this example $\theta_1 = \theta_p$ and we obtain the result for the orientation of principal axis 1.

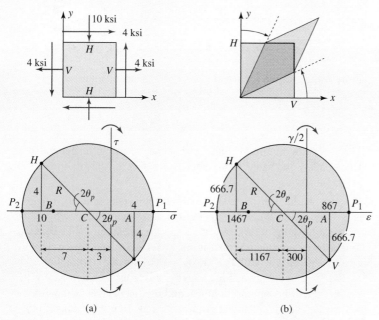

Figure 9.25 Mohr's circles in Example 9.7. (*a*) Stress. (*b*) Strain.

(b) As γ_{xy} is positive, the angle between the x and y coordinates decreases, as shown by the deformed shape in Figure 9.25b. Noting that the vertical coordinate is $\gamma/2$, we record the coordinates of points V and H,

$$V(867, 666.7\downarrow) \quad H(-1467, 666.7\uparrow) \quad \text{(E6)}$$

We then draw Mohr's circle for strain, as shown in Figure 9.25b. The angle θ_p can be found from triangle BCH (or ACV) and is given by

$$\tan 2\theta_p = \frac{666.7}{1167} \quad \text{or} \quad \theta_p = 14.87° \quad \text{(E7)}$$

$$\text{ANS.} \quad \theta_1 = 14.87° \text{ ccw}$$

For this example $\theta_1 = \theta_p$ and we obtain the result for the orientation of principal axis 1.

COMMENTS

1. The example highlights that for isotropic materials the principal axes for stresses and strains are the same.

2. The principal stresses can be found from Mohr's circle for stress as

$$\sigma_1 = -3 + 8.06 = 5.06 \text{ ksi} \qquad \sigma_2 = -3 - 8.06 = -11.06 \text{ ksi}$$

Noting that $\sigma_3 = 0$ because of the plane stress state, we obtain the principal strains from Equations (9.21a) and (9.21b),

$$\varepsilon_1 = \frac{5.06 - 0.25(-11.06)}{7500} = 1044 \ \mu$$

$$\varepsilon_2 = \frac{-11.06 - 0.25 \times 5.06}{7500} = -1644 \ \mu$$

From Mohr's circle for strain we obtain the same values,

$$\varepsilon_1 = -300 + 1344 = 1044 \ \mu$$

$$\varepsilon_2 = -300 - 1344 = -1644 \ \mu$$

The preceding highlights that the sequence of using the generalized Hooke's law and Mohr's circle does not affect the calculation of the principal strains.

3. We can conduct an intuitive check on the orientation of principal axis 1 for strain. We visualize a circle in a square, as shown in this example (Figure 9.26). As $\varepsilon_{xx} > \varepsilon_{yy}$, the rectangle will become longer in the x direction than in the y direction, and the circle will become an ellipse with its major axis along the x direction. As $\gamma_{xy} > 0$, the angle between the x and y directions will decrease. The rectangle will become a rhombus, and the major axis of the ellipse will rotate counterclockwise from the x axis. Hence we expect principal axis 1 to be either in the first sector or in the fifth sector of Figure 9.6. The result given in Equation (E7) puts principal axis 1 in sector 1, which is one of our intuitive answers.

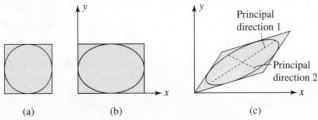

Figure 9.26 Estimating principal directions in Example 9.7. (*a*) Undeformed shape. (*b*) Deformation due to normal strains. (*c*) Additional deformation due to shear strain.

EXAMPLE 9.8

For an isotropic materials show that $G = E/2(1 + v)$.

PLAN

We can start with a state of pure shear and find principal stresses in terms of shear stress τ and principal strains in terms of shear strain γ. Using Equation (9.21a) we can relate principal strain 1 to principal stresses and obtain a relationship between shear stress τ and shear strain γ. This relationship will have only E and v in it, and comparing this to the relationship $\tau = G\gamma$, we can obtain the relationship between E, v, and G.

Solution We start assuming that all stress components except $\tau_{xy} = \tau$ are zero in the Cartesian coordinate system. We draw the stress cube and Mohr's circle in Figure 9.27a and find the principal stresses in terms of τ,

$$\sigma_1 = +\tau \tag{E1}$$

$$\sigma_2 = -\tau \tag{E2}$$

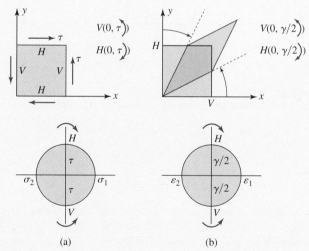

Figure 9.27 Mohr's circles for pure shear in Example 9.8. (a) Stress. (b) Strain.

We then start with all strains except $\gamma_{xy} = \gamma$ as zero and using Mohr's circle in Figure 9.27b, we find the principal strains,

$$\varepsilon_1 = +\gamma/2 \tag{E3}$$

$$\varepsilon_2 = -\gamma/2 \tag{E4}$$

Noting that $\sigma_3 = 0$, we substitute Equations (E1), (E2), and (E3) into Equation (9.21a) to obtain

$$\frac{\gamma}{2} = \frac{\tau - v(-\tau + 0)}{E} = \frac{1 + v}{E}\tau \quad \text{or} \quad \tau = \frac{E}{2(1 + v)}\gamma \quad (E5)$$

Comparing Equation (E5) to $\tau = G\gamma$ we obtain $G = E/2(1 + v)$.

COMMENTS

1. Principal axes 1 in Mohr's circles for stress and for strain are seen to be at 90° counterclockwise from plane V. This implies that for isotropic materials the principal direction for stresses is the same as the principal direction for strains.

2. The state of pure shear can be produced by applying tensile stress in one direction (σ_1) and a compressive stress of equal magnitude in a perpendicular direction (σ_2). Then on a 45° plane a state of pure shear will be seen.

EXAMPLE 9.9

The stresses $\sigma_{xx} = 4$ ksi (T), $\sigma_{yy} = 10$ ksi (C), and $\tau_{xy} = 4$ ksi were calculated at a point on a free surface of an orthotropic composite material. An orthotropic material has the following stress–strain relationship at a point in plane stress:

$$\varepsilon_{xx} = \frac{\sigma_{xx}}{E_x} - \frac{v_{yx}}{E_y}\sigma_{yy} \qquad \varepsilon_{yy} = \frac{\sigma_{yy}}{E_y} - \frac{v_{xy}}{E_x}\sigma_{xx}$$

$$\gamma_{xy} = \frac{\tau_{xy}}{G_{xy}} \qquad \frac{v_{yx}}{E_y} = \frac{v_{xy}}{E_x} \tag{9.22}$$

Determine: (a) the orientation of principal axis 1 for stresses using Mohr's circle for stress; (b) the orientation of principal axis 1 for strains using Mohr's circle for strain. Use the following values for the material constants: $E_x = 7500$ ksi, $E_y = 2500$ ksi, $G_{xy} = 1250$ ksi, and $v_{xy} = 0.3$.

PLAN

By substituting the stresses and material constants into Equation (9.22) strains ε_{xx}, ε_{yy}, and γ_{xy} can be found. We can draw Mohr's circle for stress to find principal direction 1 for stress and we can draw Mohr's circle for strain to find principal direction 1 for strain.

Solution From $v_{yx}/E_y = v_{xy}/E_x$, we obtain

$$v_{yx} = \frac{E_y v_{xy}}{E_x} = \frac{2500}{7500}0.3 = 0.1 \qquad \text{(E1)}$$

Substituting the stresses and the material constants into Equation (9.22), we obtain

$$\varepsilon_{xx} = \frac{\sigma_{xx}}{E_x} - \frac{v_{yx}}{E_y}\sigma_{yy} = \frac{4}{7500} - \frac{0.1}{2500}(-10) = 0.933 \times 10^{-3} = 933\,\mu \qquad \text{(E2)}$$

$$\varepsilon_{yy} = \frac{\sigma_{yy}}{E_y} - \frac{v_{xy}}{E_x}\sigma_{xx} = \frac{-10}{2500} - \frac{0.3}{7500}4 = -4.160 \times 10^{-3} = -4160\,\mu \qquad \text{(E3)}$$

$$\gamma_{xy} = \frac{\tau_{xy}}{G_{xy}} = \frac{4}{1250} = 3.200 \times 10^{-3} = 3200\,\mu \qquad \text{(E4)}$$

(a) We draw the stress cube and record the coordinates of points V and H. We then draw Mohr's circle for stress, as shown in Figure 9.28a, and

$$\tan 2\theta_p = \frac{4}{7} \qquad \text{or} \qquad \text{ANS.} \qquad \theta_p = 14.87° \text{ ccw} \qquad \text{(E5)}$$

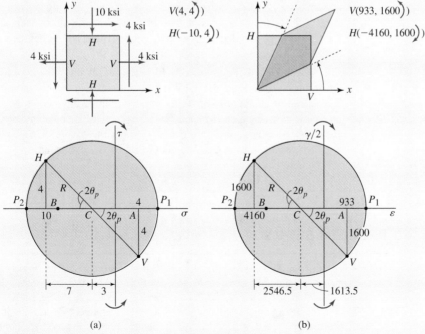

Figure 9.28 Mohr's circles in Example 9.9. (*a*) Stress. (*b*) Strain.

(b) As γ_{xy} is positive, the angle between the x and y coordinates decreases, as shown by the deformed shape in Figure 9.28*b*. Noting that the vertical coordinate is $\gamma/2$, we record the coordinates of points V and H. We then draw Mohr's circle for strain, as shown in Figure 9.28*b*. The angle θ_p can be found from triangle *BCH* (or *ACV*) and is given by

$$\tan 2\theta_p = \frac{1600}{2546.5} \qquad \text{or} \qquad \text{ANS.} \qquad \theta_p = 16.1° \text{ ccw} \quad (E6)$$

For this example $\theta_1 = \theta_p$ and we obtain the result for the orientation of principal axis 1.

COMMENTS

1. The stress state in this example is the same as in Example 9.7. In Example 9.7 we concluded that for isotropic materials the principal directions for stresses and strains are the same. Equations (E5) and (E6) show that for orthotropic materials the principal directions for stresses and strains are different.

2. In Example 9.7, if we change the material constants for the isotropic material, then the stress values will be different, but the result for the principal angle for stress will not change. If we change the material constants for orthotropic materials, then we change not only the stress values but we may also change the principal angle for stress because we may change the degree of orthotropicness, that is, the degree of difference in the material constants in the x and y directions.

3. The preceding two comments highlight some of the reasons why the intuitive experience, usually based on isotropic materials, can be misleading when working with composite materials, which generally are not isotropic. In such cases mathematical rigor can provide answers which, once confirmed by experiment, can form a new knowledge base for the development of intuitive understanding.

PROBLEM SET 9.2

9.22 Starting from Equation (9.10), show that maximum or minimum normal strain will exist in the direction of θ_p, as given by Equation (9.13). (*Hint:* See similar derivation in stress transformation.)

9.23 Show that the values of the maximum and minimum normal strains are given by Equation (9.14). (*Hint:* See similar derivation in stress transformation.)

9.24 Show that angle θ_p as given by Equation (9.13) is the principal angle, that is, shear strain is zero in a coordinate system that is at an angle θ_p to the Cartesian coordinate system. (*Hint:* See similar derivation in stress transformation.)

9.25 Show that the coordinate system of maximum in-plane shear strain is 45° to the principal coordinate system. (*Hint:* See similar derivation in stress transformation.)

9.26 Show that the maximum in-plane shear strain is given by Equation (9.18). (*Hint:* See similar derivation in stress transformation.)

9.27 Starting from Equations (9.10) and (9.12), obtain the expression of Mohr's circle given by Equation (9.19). (*Hint:* See similar derivation in stress transformation.)

9.28 Solve Problem 9.5 by the method of equations.

9.29 Solve Problem 9.5 by Mohr's circle.

9.30 Solve Problem 9.6 by the method of equations.

9.31 Solve Problem 9.6 by Mohr's circle.

9.32 Solve Problem 9.7 by the method of equations.

9.33 Solve Problem 9.7 by Mohr's circle.

In Problems 9.34 through 9.37, at a point in plane strain, the strain components in the x, y coordinate system are as given. Determine:

(a) The principal strains and principal angle 1.

(b) The maximum shear strain.

(c) The strain components in the n, t coordinate system.

9.34 $\varepsilon_{xx} = -400\ \mu$, $\varepsilon_{yy} = 600\ \mu$, and $\gamma_{xy} = -500\ \mu$ (Figure P9.34).

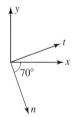

Figure P9.34

9.35 $\varepsilon_{xx} = -600\ \mu$, $\varepsilon_{yy} = -800\ \mu$, and $\gamma_{xy} = 500\ \mu$ (Figure P9.35).

Figure P9.35

9.36 $\varepsilon_{xx} = 250\ \mu$, $\varepsilon_{yy} = 850\ \mu$, and $\gamma_{xy} = 1600\ \mu$ (Figure P9.36).

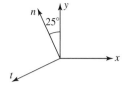

Figure P9.36

9.37 $\varepsilon_{xx} = -1200\ \mu$, $\varepsilon_{yy} = -2400\ \mu$, and $\gamma_{xy} = -1000\ \mu$ (Figure P9.37).

In Problems 9.38 through 9.41, at a point in plane strain, the strain components in the n, t coordinate system are as given. Determine:

(a) The principal strains.

(b) The maximum shear strain.

(c) The strain components in the x, y coordinate system.

9.38 $\varepsilon_{nn} = 2000\ \mu$, $\varepsilon_{tt} = -800\ \mu$, and $\gamma_{nt} = 750\ \mu$ (Figure P9.38).

9.39 $\varepsilon_{nn} = -2000\ \mu$, $\varepsilon_{tt} = -800\ \mu$, and $\gamma_{nt} = -600\ \mu$ (Figure P9.39).

9.40 $\varepsilon_{nn} = 350\ \mu$, $\varepsilon_{tt} = 700\ \mu$, and $\gamma_{nt} = 1400\ \mu$ (Figure P9.40).

9.41 $\varepsilon_{nn} = -3600\ \mu$, $\varepsilon_{tt} = 2500\ \mu$, and $\gamma_{nt} = -1000\ \mu$ (Figure P9.41).

In Problems 9.42 through 9.45, the principal strains ε_1 and ε_2 and the direction of principal direction 1 θ_1 from the x axis are given. Determine strains ε_{xx}, ε_{yy}, and γ_{xy} at the point.

9.42 $\varepsilon_1 = 1200\ \mu$, $\varepsilon_2 = 300\ \mu$, and $\theta_1 = 27.5°$.

9.43 $\varepsilon_1 = 900\ \mu$, $\varepsilon_2 = -600\ \mu$, and $\theta_1 = -20°$.

9.44 $\varepsilon_1 = -200\ \mu$, $\varepsilon_2 = -2000\ \mu$, and $\theta_1 = 105°$.

9.45 $\varepsilon_1 = 1400\ \mu$, $\varepsilon_2 = -600\ \mu$, and $\theta_1 = -75°$.

Generalized Hooke's law in principal coordinates

In Problems 9.46 through 9.48, in a thin body (plane stress) the stresses in the xy plane are

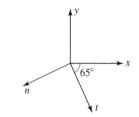

Figure P9.37

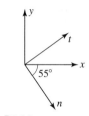

Figure P9.38

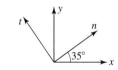

Figure P9.39

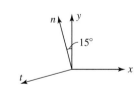

Figure P9.40

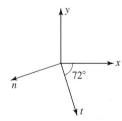

Figure P9.41

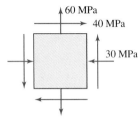

Figure P9.46

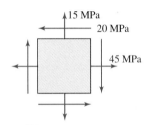

Figure P9.47

as shown on each stress element. The modulus of elasticity E and Poisson's ratio v are given in each problem. Determine:

(a) The principal strains and principal angle 1 at the point.

(b) The maximum shear strain at the point.

9.46 $E = 70$ GPa and $v = 0.25$ (Figure P9.46).

9.47 $E = 70$ GPa and $v = 0.25$ (Figure P9.47).

9.48 $E = 30,000$ ksi and $v = 0.28$ (Figure P9.48).

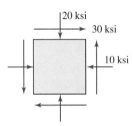

Figure P9.48

In Problems 9.49 through 9.51, in a thick body (plane strain) the stresses in the xy plane are as shown on each stress element. The modulus of elasticity E and Poisson's ratio v are given in each problem. Determine:

(a) The principal strains and principal angle 1 at the point.

(b) The maximum shear strain at the point.

9.49 $E = 70$ GPa and $v = 0.25$ (Figure P9.49).

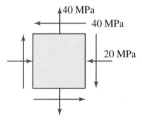

Figure P9.49

9.50 $E = 70$ GPa and $v = 0.25$ (Figure P9.50).

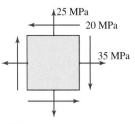

Figure P9.50

9.51 $E = 30,000$ ksi and $v = 0.28$ (Figure P9.51).

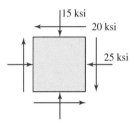

Figure P9.51

Orthotropic materials

In Problems 9.52 and 9.53, an orthotropic material has the following properties: $E_x = 7500$ ksi, $E_y = 2500$ ksi, $G_{xy} = 1250$ ksi, and $v_{xy} = 0.25$. Use Equations (9.22) to solve each problem.

9.52 Determine the principal directions for stresses and strains at a point on a free surface where the following strains were measured: $\varepsilon_{xx} = -400$ μ, $\varepsilon_{yy} = 600$ μ, and $\gamma_{xy} = -500$ μ.

9.53 Determine the principal directions for stresses and strains at a point on a free surface where the following stresses were computed: $\sigma_{xx} = 10$ ksi (T), $\sigma_{yy} = 7$ ksi (C), and $\tau_{xy} = 5$ ksi.

In Problems 9.54 and 9.55, an orthotropic material has the following properties: $E_x = 50$ GPa, $E_y = 18$ GPa, $G_{xy} = 9$ GPa, and $v_{xy} = 0.25$. Use Equations (9.22) to solve each problem.

9.54 Determine the principal directions for stresses and strains at a point on a free surface where the following strains were measured: $\varepsilon_{xx} = 800$ μ, $\varepsilon_{yy} = 200$ μ, and $\gamma_{xy} = 300$ μ.

9.55 Determine the principal directions for stresses and strains at a point on a free surface where the following stresses were computed: $\sigma_{xx} = 70$ MPa (C), $\sigma_{yy} = 49$ MPa (C), and $\tau_{xy} = -30$ MPa.

9.5 STRAIN GAGES

Strain gages are strain-measuring devices based on the fact that the wire resistance changes with changes in the length of the wire due to deformation. Since strain causes a length change, the change in resistance can be correlated to the strain in the wire by conducting an experiment. By bonding a wire to a stressed part, we can *assume* that the deformation of the wire is the same as that of the material. Hence by measuring changes in the resistance of a wire we can get the strains in the material. Strain gages are a sophisticated implementation of this technique.

Strain gages are usually manufactured by etching a thin foil of material, as shown in Figure 9.29. The back and forth pattern is to increase the sensitivity of the gage by providing a long length of wire in a very small area. Strain gages can be as small in length as $\frac{8}{1000}$ in, which for many engineering calculations is equivalent to measuring strain at a point.

Since we are measuring changes in the length of a wire, a strain gage measures only normal strains directly and not shear strains. In this section it will be shown how shear strains are calculated from the measured normal strains. Because of the finite sizes of strain gages, strain gages give an average value of strain at a point. To protect the strain gage from damage, no force is applied on its top. Hence strain gages are bonded to a free surface, that is, measurements take place in plane stress. We record the following observations:

1. Strain gages measure only normal strains directly.

2. Strain gages are bonded to a free surface, that is, the strains are in a state of plane stress and not plane strain.

3. Strain gages measure average strain at a point.

In plane stress there are three *independent*[5] strain components ε_{xx}, ε_{yy}, and γ_{xy}. To determine these, we need three observations at a point. In other words, we need to find normal strains in three directions. The assembly of three strain gages used for finding three normal strains, shown in Figure 9.30, is called *strain rosette*. To determine strains ε_{xx}, ε_{yy}, and γ_{xy} from strain gage readings, Equation (9.10) is used,

$$\varepsilon_a = \varepsilon_{xx} \cos^2 \theta_a + \varepsilon_{yy} \sin^2 \theta_a + \gamma_{xy} \sin \theta_a \cos \theta_a \tag{9.23a}$$

$$\varepsilon_b = \varepsilon_{xx} \cos^2 \theta_b + \varepsilon_{yy} \sin^2 \theta_b + \gamma_{xy} \sin \theta_b \cos \theta_b \tag{9.23b}$$

$$\varepsilon_c = \varepsilon_{xx} \cos^2 \theta_c + \varepsilon_{yy} \sin^2 \theta_c + \gamma_{xy} \sin \theta_c \cos \theta_c \tag{9.23c}$$

Equations (9.23) can be solved for the three unknowns ε_{xx}, ε_{yy}, and γ_{xy} as θ_a, θ_b, and θ_c are known. It should be remembered that angles are positive in the counterclockwise sense from the x axis in Equations (9.23).

The angles at which strain gages are attached are chosen to reduce the algebra in the calculation of ε_{xx}, ε_{yy}, and γ_{xy}. Figure 9.31 shows two popular choices of angles in a strain rosette. Notice in Figure 9.31*b* that angle θ_c can be 120° or −60° (or 300° or −240°). This emphasizes the fact that Equation (9.10) does not change if 180° is added to or subtracted from angle θ. (See Problem 9.56.) An alternative explanation is that normal strain is a measure of the deformation of a line—deformation is the

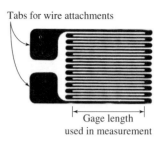

Tabs for wire attachments

Gage length
used in measurement

Figure 9.29 Typical strain gage.

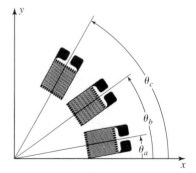

Figure 9.30 Strain rosette.

---

[5]But there are four nonzero strain components.

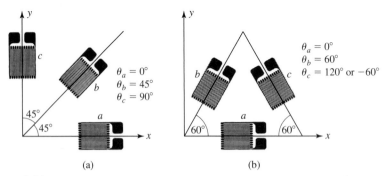

$\theta_a = 0°$
$\theta_b = 45°$
$\theta_c = 90°$

$\theta_a = 0°$
$\theta_b = 60°$
$\theta_c = 120°$ or $-60°$

(a)   (b)

Figure 9.31 Strain rosettes. (*a*) 45°. (*b*) 60°.

relative movement of two points on a line; hence the value does not depend on whether the two points on the line have positive or negative coordinates. We record the observation made in this paragraph as follows.

A change in strain gage orientation by ±180° makes no difference in the strain values.

Once strains ε_{xx}, ε_{yy}, and γ_{xy} are found, then the principal strains can be found. Principal stresses can be found next, if needed, by using the generalized Hooke's law in principal coordinates. Alternatively stresses σ_{xx}, σ_{yy}, and τ_{xy} may be found first from the generalized Hooke's law, and then the principal stresses can be found. But it is important to remember that the point where strains are being measured is in plane stress. Hence $\sigma_{zz} = 0$. The strain in the z direction is the third principal strain and can be found from Equation (9.16).

EXAMPLE 9.10

Strains $\varepsilon_a = 900$ μin/in, $\varepsilon_b = 200$ μin/in, and $\varepsilon_c = 700$ μin/in were recorded by the three strain gages shown in Figure 9.32 at a point on the free surface of a material that has a modulus of elasticity $E = 30,000$ ksi and a Poisson ratio $\nu = 0.3$. Determine the principal stresses, principal angle 1, and the maximum shear stress at the point.

Method 1

PLAN

We note that $\varepsilon_a = \varepsilon_{xx}$. We can find strains ε_{yy} and γ_{xy} from the two equations obtained by substituting $\theta_b = +60°$ and $\theta_c = -60°$ into Equation (9.10). We can then find principal strains 1 and 2 and principal angle 1 by using either Mohr's circle or the method of equations. Principal strain 3 can be found from Equations (9.16), and the maximum shear strain from the radius of the biggest circle. Using the generalized Hooke's law in principal coordinates, the principal stresses can be found.

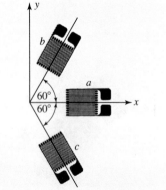

Figure 9.32 Strain rosette in Example 9.10.

Solution

Strain gages: The strain in the x direction is given by the strain gage a reading. Thus

$$\varepsilon_{xx} = 900 \ \mu \qquad \text{(E1)}$$

Substituting $\theta_b = +60°$ and $\theta_c = -60°$ into Equation (9.10), we obtain the following two equations:

$$\varepsilon_b = 900 \cos^2 60 + \varepsilon_{yy} \sin^2 60 + \gamma_{xy} \sin 60 \cos 60 = 200$$

or

$$0.75\varepsilon_{yy} + 0.433\gamma_{xy} = -25 \qquad \text{(E2)}$$

or

$$\varepsilon_c = 900 \cos^2 (-60) + \varepsilon_{yy} \sin^2 (-60) + \gamma_{xy} \sin (-60) \cos (-60) = 700$$

or

$$0.75\varepsilon_{yy} - 0.433\gamma_{xy} = 475 \qquad \text{(E3)}$$

Solving Equations (E2) and (E3), we obtain

$$\varepsilon_{yy} = 300 \ \mu \qquad \gamma_{xy} = -577.4 \ \mu \qquad \text{(E4)}$$

Mohr's circle for strain: We find the principal strains using Mohr's circle for strain shown in Figure 9.33.

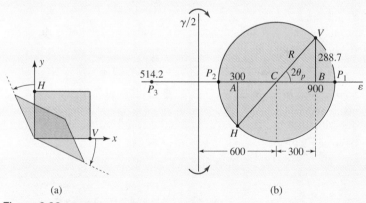

Figure 9.33 Mohr's circle in Example 9.10.

Step 1: The shear strain is negative, hence the angle between the x and y axes should increase. We draw the deformed shape of a square due to shear strain γ_{xy}. We label the intersection of the vertical plane and the x axis as V and the intersection of the horizontal plane and the y axis as H, as shown in Figure 9.33a.

Step 2: Using Figure 9.33a, we write the coordinates of points V and H as

$$V(900, 288.7 \curvearrowright) \qquad H(300, 288.7 \curvearrowleft) \qquad \text{(E5)}$$

Step 3: We draw the axes for Mohr's circle, as shown in Figure 9.33*b*.

Step 4: Locate points *V* and *H* and join the points by drawing a line.

Step 5: Point *C*, the center of Mohr's circle, is midway between points *A* and *B*, that is, at 600 μ. Distance *BC* can thus be found as 300 μ, as shown in Figure 9.33*b*. From the Pythagorean theorem we can find the radius *R*,

$$R = \sqrt{CB^2 + BV^2} = \sqrt{300^2 + 288.7^2} = 416.4 \qquad \text{(E6)}$$

Step 6: The principal strains are the coordinates of points P_1 and P_2 in Figure 9.33*b*,

$$\varepsilon_1 = 600 + 416.4 = 1016.4 \qquad \varepsilon_2 = 600 - 416.4 = 183.6 \quad \text{(E7)}$$

As the point is on a free surface, the state is in plane stress. Hence the third principal strain from Equation (9.16) is

$$\varepsilon_3 = \varepsilon_{zz} = -\frac{0.3}{1-0.3}(900 + 300) = -514.2\,\mu \qquad \text{(E8)}$$

Step 7: Using triangle *BCH* we can find the principal angle θ_p,

$$\cos 2\theta_p = \frac{CB}{CV} = \frac{300}{416.4} \qquad \text{(E9)}$$

From Figure 9.33*b* we see that $\theta_1 = \theta_p$, and the direction is clockwise, as reported in the equations

$$2\theta_p = 43.9° \qquad \theta_1 = \theta_p = 21.9° \text{ cw} \qquad \text{(E10)}$$

Step 8: Intuitive check: We visualize a circle in a square, as in Figure 9.34*a*. As $\varepsilon_{xx} > \varepsilon_{yy}$, the rectangle will become longer in the *x* direction than in the *y* direction, and the circle will become an ellipse with its major axis along the *x* direction, as shown in Figure 9.34*b*.

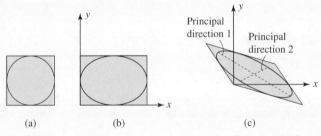

(a) (b) (c)

Figure 9.34 Estimating principal directions in Example 9.10. (*a*) Undeformed shape. (*b*) Deformation due to normal strains. (*c*) Additional deformation due to shear strain.

As $\gamma_{xy} < 0$, the angle between the x and y directions will increase. The rectangle will become a rhombus and the major axis of the ellipse will rotate clockwise from the x axis, as shown in Figure 9.34c. Hence we expect principal axis 1 to be either in the eighth sector or in the fifth sector, confirming the result given in Equation (E10).

Step 9: Locating point P_3, which corresponds to the third principal strain in Figure 9.33b, we note that the circle between P_1 and P_3 will be a bigger circle than between P_2 and P_3, and between P_1 and P_2. Thus the maximum shear strain at the point can be determined from the circle between P_1 and P_3,

$$\frac{\gamma_{max}}{2} = \frac{\varepsilon_1 - \varepsilon_3}{2} = 765.3 \qquad \gamma_{max} = 1531\,\mu$$

Hooke's law: For plane stress $\sigma_3 = 0$. From Equations (9.21a) and (9.21b) we obtain

$$\varepsilon_1 = \frac{\sigma_1 - \nu\sigma_2}{30,000} = 1016 \times 10^{-6} \qquad \sigma_1 - 0.3\sigma_2 = 30.48 \text{ ksi} \quad \text{(E11)}$$

$$\varepsilon_2 = \frac{\sigma_2 - \nu\sigma_1}{30,000} = 184 \times 10^{-6} \qquad \sigma_2 - 0.3\sigma_1 = 5.52 \text{ ksi} \quad \text{(E12)}$$

Solving Equations (E11) and (E12), we obtain $\sigma_1 = 35.31$ ksi and $\sigma_2 = 16.11$ ksi. For isotropic materials the principal direction for stresses and strains is the same. We report the principal stresses and principal angle 1 as

ANS. $\quad \sigma_1 = 35.3 \text{ ksi (T)} \qquad \sigma_2 = 16.1 \text{ ksi (T)} \qquad \sigma_3 = 0 \quad \theta_1 = 21.9° \text{ cw}$

The shear modulus of elasticity is

$$G = \frac{E}{2(1 + \nu)} = 11{,}538 \text{ ksi} \qquad \text{(E13)}$$

The maximum shear stress can be found from Hooke's law as

$$\tau_{max} = G\gamma_{max} = 11{,}538 \times 1531 \times 10^{-6} = 17.66 = 17.7 \text{ ksi} \qquad \text{(E14)}$$

Check: We can also find the maximum shear stress as half the maximum difference between principal stresses, that is, from Equation (8.14),

ANS. $$\tau_{max} = \frac{35.3 - 0}{2} = 17.65 \text{ ksi}$$

confirming our previous result.

COMMENT

This example combines three concepts: use of strain gages to find strain components in Cartesian coordinates, use of Mohr's circle for finding principal strains, and use of Hooke's law in principal coordinates for finding principal stresses.

Method 2

PLAN

We can find ε_{xx}, ε_{yy}, and γ_{xy} from the values of the strains recorded by the strain gages, as we did in Method 1. We can use Hooke's law in Cartesian coordinates to find σ_{xx}, σ_{yy}, and τ_{xy}. Using Mohr's circle for stress (or the method of equations), we can then find the principal stresses, principal angle 1, and the maximum shear stress.

Solution

Strain gages: From Equations (E1) and (E4),

$$\varepsilon_{xx} = 900\,\mu \qquad \varepsilon_{yy} = 300\,\mu \qquad \gamma_{xy} = -577.4\,\mu \qquad \text{(E15)}$$

Hooke's law: We note that for plane stress $\sigma_{zz} = 0$. Using Equations (3.12a) and (3.12b), we can write

$$\varepsilon_{xx} = \frac{\sigma_{xx} - v\sigma_{yy}}{30,000} = 900 \times 10^{-6} \quad \text{or} \quad \sigma_{xx} - 0.3\sigma_{yy} = 27\,\text{ksi} \quad \text{(E16)}$$

$$\varepsilon_{yy} = \frac{\sigma_{yy} - v\sigma_{xx}}{30,000} = 300 \times 10^{-6} \quad \text{or} \quad \sigma_{yy} - 0.3\sigma_{xx} = 9\,\text{ksi} \quad \text{(E17)}$$

Solving Equations (E16) and (E17), we obtain $\sigma_{xx} = 32.63$ ksi and $\sigma_{yy} = 18.79$ ksi. From Equations (3.12d) and (E13) we obtain the shear stress as

$$\tau_{xy} = G\gamma_{xy} = 11,538(-577.4)10^{-6} = -6.66\,\text{ksi} \qquad \text{(E18)}$$

Mohr's circle for stress

Step 1: We draw the stress cube and label the vertical plane V and the horizontal plane H, as shown in Figure 9.35a.

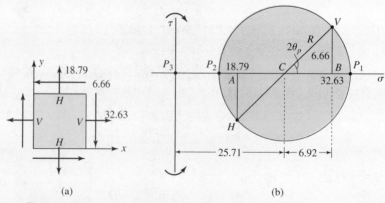

(a)　　　　　　　　　(b)

Figure 9.35　Mohr's circle in Example 9.10.

Step 2: Using Figure 9.35*a*, we write the coordinates of points V and H as

$$V(32.63, 6.66 \downarrow) \qquad H(18.79, 6.66 \uparrow) \qquad \text{(E19)}$$

Step 3: We draw the axes for Mohr's circle, as shown in Figure 9.35*b*.

Step 4: Locate points V and H and join the points by drawing a line.

Step 5: Point C, the center of Mohr's circle, is midway between points A and B, that is, at 25.71 ksi. Distance BC can thus be found as 6.92 ksi, as shown in Figure 9.35*b*. From the Pythagorean theorem we can find the radius R,

$$R = \sqrt{CB^2 + BV^2} = \sqrt{6.92^2 + 6.66^2} = 9.60 \qquad \text{(E20)}$$

Step 6: The principal stresses are the coordinates of points P_1 and P_2 in Figure 9.35*b*. As the point is on free surface, the state is in plane stress. Hence the third principal stress is zero,

$$\sigma_1 = 25.71 + 9.6 = 35.31 \qquad \sigma_2 = 25.71 - 9.6 = 16.11 \qquad \sigma_3 = 0$$
$$\text{(E21)}$$

Step 7: Using triangle BCH we can find the principal angle θ_p

$$\cos 2\theta_p = \frac{CB}{CV} = \frac{6.92}{9.6} \qquad \text{(E22)}$$

From Figure 9.35*b* we see that $\theta_1 = \theta_p$ and the direction is clockwise, as reported in the equation

$$2\theta_p = 43.9° \qquad \theta_1 = \theta_p = 21.9° \text{ cw} \qquad \text{(E23)}$$

We report the principal stresses and principal angle 1 as

ANS. $\qquad \sigma_1 = 35.3 \text{ ksi (T)} \qquad \sigma_2 = 16.1 \text{ ksi (T)} \qquad \sigma_3 = 0 \qquad \theta_1 = 21.9° \text{ cw}$

Step 8: The biggest circle will be between P_1 and P_3. The maximum shear stress is the radius of this circle and can be calculated as

ANS. $\qquad \tau_{max} = \dfrac{35.3 - 0}{2} = 17.65 = 17.7 \text{ ksi} \qquad \text{(E24)}$

COMMENT

As in Method 1, there are three concepts that are combined. But the sequence in which the problem is solved is different. In Method 1 we used Mohr's circle (for strain) first and Hooke's law (in principal coordinates) second. In Method 2 we used Hooke's law (Cartesian coordinates) first and Mohr's circle (for stress) second. The number of calculations differs only with respect to ε_3, which is not calculated in Method 2.

EXAMPLE 9.11

The strain gage at point A recorded a value of $\varepsilon_A = -200\ \mu$. Determine the load P that caused the strain for the three cases shown in Figure 9.36. In each case the strain gage is $30°$ clockwise to the longitudinal axis (x axis). Use $E = 10,000$ ksi, $G = 4000$ ksi, and $v = 0.25$.

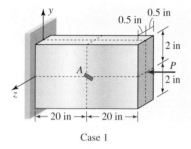
Case 1

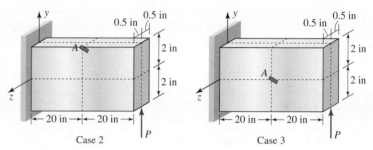
Case 2   Case 3

Figure 9.36 Three beams in Example 9.11.

PLAN

The axial stress σ_{xx} in case 1, the bending normal stress σ_{xx} in case 2, and the bending shear stress τ_{xy} in case 3 can be found in terms of P using Equations (4.9), (6.12), and (6.23), respectively. All other stress components are zero. Strains ε_{xx}, ε_{yy}, and γ_{xy} can be found in terms of P for each case, using the generalized Hooke's law. Substituting the strains and $\theta_A = -30°$ into Equation (9.10), the strain in the gage can be found in terms of P and equated to the given value of $-200\ \mu$ to obtain the value of P.

Solution

Stress calculations: The cross-sectional area A, the area moment of inertia I_{zz}, and the first moment Q_z of the area A_s shown in Figure 9.37a can be calculated as

$$A = 1 \times 4 = 4\ \text{in}^2 \tag{E1}$$

$$I_{zz} = \frac{1 \times 4^3}{12} = 5.33\ \text{in}^4 \tag{E2}$$

$$Q_z = 1 \times 2 \times 1 = 2\ \text{in}^3 \tag{E3}$$

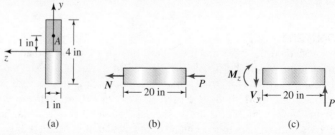

(a)   (b)   (c)

Figure 9.37 Calculation of geometric and internal quantities in Example 9.11.

Recollect that A_s is the area between the free surface and point A, where shear stress is to be found.

Figure 9.37b and c shows the free-body diagrams of the axial member and the beam after making the imaginary cut through point A. Using force and moment equilibrium equations, the internal forces and moment can be found,

$$N = -P \tag{E4}$$

$$V_y = P \tag{E5}$$

$$M_z = 20P \tag{E6}$$

Substituting Equations (E1) and (E4) into Equation (4.9), the axial stress in case 1 can be found,

$$\sigma_{xx} = \frac{N}{A} = \frac{-P}{4} = -0.25P \tag{E7}$$

Substituting Equations (E2), (E6), and $y = 2$ in into Equation (6.12), the bending normal stress in case 2 can be found,

$$\sigma_{xx} = -\frac{M_z y}{I_{zz}} = -\frac{20P \times 2}{5.33} = -7.5P \tag{E8}$$

Substituting Equations (E2), (E3), (E6), and $t = 1$ into Equation (6.23), the magnitude of τ_{xy} in case 3 can be found,

$$|\tau_{xy}| = \left| \frac{V_y Q_z}{I_{zz} t} \right| = \left| \frac{P \times 2}{5.33 \times 1} \right| = 0.375P$$

Noting that τ_{xy} must have the same sign as V_y, we obtain the sign of τ_{xy} (see Section 6.6.5),

$$\tau_{xy} = 0.375P \tag{E9}$$

Strain calculations: The only two nonzero stress components are given by Equations (E7), (E8), and (E9) for each case. Using the generalized Hooke's law [or Equations (4.12) and (6.25)], we obtain the strains as follows:

• Case 1:

$$\varepsilon_{xx} = \frac{\sigma_{xx}}{E} = \frac{-0.25P}{10000} = -25P\mu \tag{E10}$$

$$\varepsilon_{yy} = -\frac{\nu \sigma_{xx}}{E} = -\nu \varepsilon_{xx} = 6.25P\mu \tag{E11}$$

$$\gamma_{xy} = 0 \tag{E12}$$

• Case 2:

$$\varepsilon_{xx} = \frac{\sigma_{xx}}{E} = \frac{-7.5P}{10000} = -750P\,\mu \qquad (E13)$$

$$\varepsilon_{yy} = -\frac{\nu\sigma_{xx}}{E} = -\nu\varepsilon_{xx} = 187.5P\,\mu \qquad (E14)$$

$$\gamma_{xy} = 0 \qquad (E15)$$

• Case 3:

$$\varepsilon_{xx} = 0 \qquad (E16)$$

$$\varepsilon_{yy} = 0 \qquad (E17)$$

$$\gamma_{xy} = \frac{\tau_{xy}}{G} = \frac{0.375P}{4000} = 93.75P\,\mu \qquad (E18)$$

Case 1: Substituting $\theta_A = -30°$ and ε_{xx}, ε_{yy}, and γ_{xy} from Equations (E10), (E11), and (E12) into Equation (9.10) and equating the result to $-200\,\mu$, the value of load P can be found,

$$\varepsilon_A = -25P\,\cos^2(-30) + 6.25P\,\sin^2(-30) + 0\,\sin(-30)\cos(-30)$$

$$= -17.19P\,\mu = -200\,\mu$$

or

ANS. $P = 11.6$ kips $\qquad (E19)$

Case 2: Substituting $\theta_A = -30°$ and ε_{xx}, ε_{yy}, and γ_{xy} from Equations (E13), (E14), and (E15) into Equation (9.10) and equating the result to $-200\,\mu$, the value of load P can be found,

$$\varepsilon_A = -750P\,\cos^2(-30) + 187.5P\,\sin^2(-30) + 0\,\sin(-30)\cos(-30)$$

$$= -515.63P\,\mu = -200\,\mu$$

or

ANS. $P = 0.39$ kips $\qquad (E20)$

Case 3: Substituting $\theta_A = -30°$ and ε_{xx}, ε_{yy}, and γ_{xy} from Equations (E16), (E17), and (E18) into Equation (9.10) and equating the result to $-200\,\mu$, the value of load P can be found,

$$\varepsilon_A = 0\,\cos^2(-30) + 0\,\sin^2(-30) + 93.75P\,\sin(-30)\cos(-30)$$

$$= -40.59P\,\mu = -200\,\mu$$

or

ANS. $P = 4.93$ kips $\qquad (E21)$

COMMENTS

1. This example demonstrates one of the basic principles used in the design of load transducers, also called load cells. Load transducers are used for measuring, applying, and controlling forces and moments on a structure. This example showed how one may measure a force by using strain gage readings and mechanics of materials formulas. For applying and controlling forces, use is made of the fact that the signal from the strain gage is electrical in its nature, which can be processed further and correlated with the intensity of force or a moment.

2. In this example the strain in the gage was caused by a single force. When there are multiple forces or moments acting on a structure, then to correlate strain gage readings to the applied forces and moments we need to supplement the formulas of mechanics and materials with the formulas for the Wheatstone bridge. See Section 9.6 for additional details on the Wheatstone bridge.

3. In Examples 9.5, 9.10, and this example we saw the use of the generalized Hooke's law. The alternative to using the generalized Hooke's law is to use formulas that are derived from the generalized Hooke's law. This is one important reason for memorizing the generalized Hooke's law.

QUICK TEST 9.2   Time: 15 minutes/Total: 20 points

Grade yourself with the answers given in Appendix G. Each question is worth two points.

1. The strain gage recorded a strain of 800 μ. What is ε_{yy} for the two cases shown?

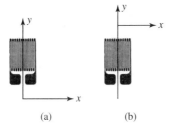

(a)   (b)

In Questions 2 through 4, report the smallest positive and the smallest negative angle θ that can be substituted in the strain transformation equation relating the strain gage reading to strains in Cartesian coordinates.

2.   **3.**   **4.**

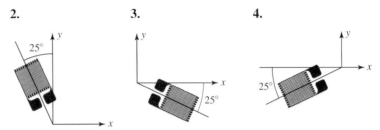

In Questions 5 through 7, Mohr's circles for strains for points in plane stress are as shown. The modulus of elasticity of the material is $E = 10,000$ ksi and Poisson's ratio is 0.25. What is the maximum shear strain in each question?

5.   **6.**   **7.**

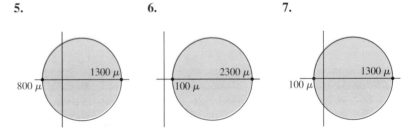

In Questions 8 through 10, answer true or false. If false, then give the correct explanation.

8. In plane strain there are two principal strains, but in plane stress there are three principal strains.

9. Since strain values change with the coordinate system, the principal strains at a point depend on the coordinate system used in finding the strains.

10. The principal coordinate axis for stresses and strains is always the same, irrespective of the stress–strain relationship.

PROBLEM SET 9.3

Strain gages

9.56 Show that by substituting $\theta \pm 180°$ in place of θ, the strain transformation equation, Equation (9.10), is unchanged.

9.57 At a point on a free surface the strain components in the x, y coordinates are calculated as $\varepsilon_{xx} = 400$ μin/in, $\varepsilon_{yy} = -200$ μin/in, and $\gamma_{xy} = 500$ μin/in. Predict the strains that the strain gages shown in Figure P9.57 would record.

9.58 At a point on a free surface the strains recorded by the three strain gages shown in Figure P9.57 are $\varepsilon_a = 200$ μin/in, $\varepsilon_b = 100$ μin/in, and $\varepsilon_c = -400$ μin/in. Determine strains ε_{xx}, ε_{yy}, and γ_{xy}.

9.59 At a point on a free surface of an aluminum machine component ($E = 10,000$ ksi

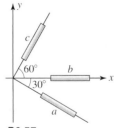

Figure P9.57

and $G = 4000$ ksi) the stress components in the x, y coordinates were calculated by the finite-element method as $\sigma_{xx} = 22$ ksi (T), $\sigma_{yy} = 15$ ksi (C), and $\tau_{xy} = -10$ ksi. Predict the strains that the strain gages shown in Figure P9.59 would show.

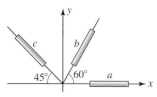

Figure P9.59

9.60 At a point on a free surface of aluminum ($E = 10{,}000$ ksi and $G = 4000$ ksi) the strains recorded by the three strain gages shown in Figure P9.59 are $\varepsilon_a = -600$ μin/in, $\varepsilon_b = 500$ μin/in, and $\varepsilon_c = 400$ μin/in. Determine stresses σ_{xx}, σ_{yy}, and τ_{xy}.

9.61 At a point on a free surface of a machine component ($E = 80$ GPa and $G = 32$ GPa) the stress components in the x, y coordinates were calculated by the finite-element method as $\sigma_{xx} = 50$ MPa (T), $\sigma_{yy} = 20$ MPa (C), and $\tau_{xy} = 96$ MPa. Predict the strains that the strain gages shown in Figure P9.61 would show.

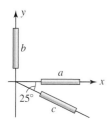

Figure P9.61

9.62 At a point on a free surface of a machine component ($E = 80$ GPa and $G = 32$ GPa) the strains recorded by the three strain gages shown in Figure P9.61 are $\varepsilon_a = 1000$ μm/m, $\varepsilon_b = 1500$ μm/m, and $\varepsilon_c = -450$ μm/m. Determine stresses σ_{xx}, σ_{yy}, and τ_{xy}.

9.63 On a free surface of steel ($E = 210$ GPa and $\nu = 0.28$) the strains recorded by the three strain gages shown in Figure P9.63 are $\varepsilon_a = -800$ μm/m, $\varepsilon_b = -300$ μm/m, and $\varepsilon_c = -700$ μm/m. Determine the principal

strains, principal angle 1, and the maximum shear *strain*.

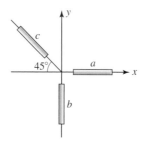

Figure P9.63

9.64 On a free surface of steel ($E = 210$ GPa and $\nu = 0.28$) the strains recorded by the three strain gages shown in Figure P9.63 are $\varepsilon_a = 200$ μm/m, $\varepsilon_b = 100$ μm/m, and $\varepsilon_c = 0$. Determine the principal *stresses*, principal angle 1, and the maximum shear *stress*.

9.65 On a free surface of an aluminum machine component ($E = 10{,}000$ ksi and $\nu = 0.25$) the strains recorded by the three strain gages shown in Figure P9.65 are $\varepsilon_a = -100$ μin/in, $\varepsilon_b = 200$ μin/in, and $\varepsilon_c = 300$ μin/in. Determine the principal *strains*, principal angle 1, and the maximum shear *strain*.

9.66 On a free surface of an aluminum machine component ($E = 10{,}000$ ksi and $\nu = 0.25$) the strains recorded by the three strain gages shown in Figure P9.65 are $\varepsilon_a = 500$ μin/in, $\varepsilon_b = 500$ μin/in, and $\varepsilon_c = 500$ μin/in. Determine the principal *stresses*, principal angle 1, and the maximum shear *stress*.

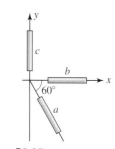

Figure P9.65

Strain gages on structural elements

9.67 An aluminum 50-mm × 50-mm square bar is axially loaded with a force $F = 100$ kN. The modulus of elasticity $E = 70$ GPa and the shear modulus $G = 28$ GPa. Determine the strain that will be recorded by the strain gage shown in Figure P9.67.

Figure P9.67

9.68 A circular steel bar is axially loaded. It has a diameter of 2 in, a modulus of elasticity $E = 30{,}000$ ksi, and a Poisson's ratio $\nu = 0.3$. Determine the applied axial force F when the strain gage shown in Figure P9.68 records a reading of 1000 μin/in.

Figure P9.68

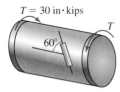

Figure P9.69

Figure P9.70

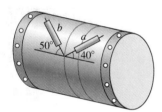

Figure P9.71

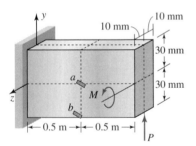

Figure P9.72

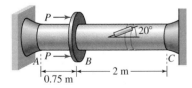

Figure P9.74

9.69 A circular shaft of 2-in diameter has a torque applied to it as shown in Figure P9.69. The shaft material has a modulus of elasticity of 30,000 ksi and a Poisson's ratio of 0.3. Determine the strain that will be recorded by a strain gage.

9.70 A circular shaft of 50-mm diameter has a torque applied to it as shown in Figure P9.70. The shaft material has a modulus of elasticity $E = 70$ GPa and a shear modulus $G = 28$ GPa. If the strain gage shows a reading of $-600~\mu$, determine the applied torque T.

9.71 The steel cylindrical pressure vessel ($E = 210$ GPa and $\nu = 0.28$) shown in Figure P9.71 has a mean diameter of 1000 mm. The wall of the cylinder is 10 mm thick and the gas pressure is 200 kPa. Determine the strain recorded by the two strain gages attached on the surface of the cylinder.

9.72 An aluminum beam ($E = 70$ GPa and $\nu = 0.25$) is loaded by a force $P = 10$ kN and moment $M = 5$ kN·m at the free end, as shown in Figure P9.72. If the two strain gages shown are at an angle of $25°$ to the longitudinal axis, determine the strains in the gages.

9.73 An aluminum beam ($E = 70$ GPa and $\nu = 0.25$) is loaded by a force P and a moment M at the free end, as shown in Figure P9.72. Two strain gages at $30°$ to the longitudinal axis recorded the following strains: $\varepsilon_a = -386~\mu$m/m and $\varepsilon_b = 4092~\mu$m/m. Determine the applied force P and applied moment M.

9.74 A steel rod ($E = 210$ GPa and $\nu = 0.28$) of 50-mm diameter is loaded by axial forces $P = 100$ kN, as shown in Figure P9.74. Determine the strain that will be recorded by the strain gage.

9.75 The strain gage mounted on the surface of the solid axial steel rod ($E = 210$ GPa and $\nu = 0.28$) illustrated in Figure P9.74 showed a strain of $-214~\mu$m/m. If the diameter of the shaft is 50 mm, determine the applied axial force P.

9.76 A steel shaft ($E = 210$ GPa and $\nu = 0.28$) of 50-mm diameter is loaded by a torque $T = 10$ kN·m, as shown in Figure P9.76.

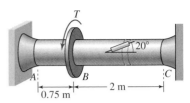

Figure P9.76

Determine the strain that will be recorded by the strain gage.

9.77 The strain gage mounted on the surface of the solid steel shaft ($E = 210$ GPa and $\nu = 0.28$) shown in Figure P9.76 recorded a strain of $1088~\mu$m/m. If the diameter of the shaft is 75 mm, determine the applied torque T.

Stretch yourself

In Problems 9.78 through 9.80, Equations (9.24a) and (9.24b) are transformation equations relating the x, y coordinates to the n, t coordinates of a point (Figure P9.78),

$$n = x\cos\theta + y\sin\theta \qquad (9.24a)$$

$$t = -x\sin\theta + y\cos\theta \qquad (9.24b)$$

Figure P9.78

Equations (9.24c) and (9.24d) are transformation equations relating displacements u and v in the x and y directions to the displacements u_n and u_t in the n and t directions, respectively,

$$u_n = u\cos\theta + v\sin\theta \qquad (9.24c)$$

$$u_t = -u\sin\theta + v\cos\theta \qquad (9.24d)$$

Solve each problem using Equations (9.24a) through (9.24d).

9.78 Starting with $\varepsilon_{nn} = \partial u_n/\partial n$, using Equations (9.24a) through (9.24d) and the chain rule for differentiation, derive Equation (9.10).

9.79 Starting with $\varepsilon_{tt} = \partial u_t/\partial t$, using Equations (9.24a) through (9.24d) and the chain rule for differentiation, derive Equation (9.11).

9.80 Starting with $\gamma_{nt} = \partial u_t/\partial n + \partial u_n/\partial t$, using Equations (9.24a) through (9.24d) and the chain rule for differentiation, derive Equation (9.12).

9.81 Starting from Equation (9.22), show that for isotropic materials $E_x = E_y$ and $G_{xy} = E_x/2(1 + v)$.

Computer problems

9.82 The displacements u and v in the x and y directions are given by the equations

$$u = [0.5(x^2 - y^2) + 0.5xy + 0.25x]10^{-3} \text{ mm}$$

$$v = [0.25(x^2 - y^2) - xy]10^{-3} \text{ mm}$$

Assuming plane strain, determine the principal strains, principal angle 1, and the maximum shear strain every 30° on a circle of radius 1 around the origin. Use a spreadsheet or write a computer program for the calculation.

9.83 On an aluminum beam ($E = 70$ GPa and $v = 0.25$) two strain gages were attached to monitor loads P and w, which vary slowly over time (Figure P9.83). The strain gage readings are given in Table P9.83. Determine the values of P and w at the times the strains were measured.

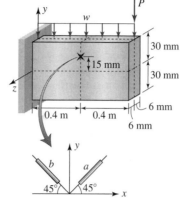

Figure P9.83

TABLE P9.83 Strain values

	ε_a (μ)	ε_b (μ)
1	1501	2368
2	1433	2276
3	1385	2193
4	1483	2336
5	1470	2331
6	1380	2191
7	1448	2282
8	1496	2366
9	1398	2223
10	1411	2228

*9.6 GENERAL INFORMATION

The history of strain gages is an interesting story that heralds modern universities' difficulties and pitfalls in maintaining the delicate balance between research for knowledge and commercial benefits that can accrue from research.

The strain gage was invented by two Americans independently but nearly simultaneously. In 1938 Professor Arthur C. Ruge of the Massachusetts Institute of Technology (MIT) needed to measure low-level strains in a elevated thin-walled water tank during an earthquake. He solved this problem by inventing the strain gage. When he sought to register his invention with the MIT patent committee in 1939, the committee felt that the commercial use of the invention was unlikely to be of major importance and released the invention to Professor Ruge. Around the same time, Edward E. Simmons was a graduate student at the California Institute of Technology (Caltech) who was studying the stress–strain characteristics of metals during impact. He invented the strain gage as part of a dynamometer for measuring the power of impact. A legal battle for the rights of the strain gage patent was fought between Caltech and Simmons. Simmons won because he was a student and not a Caltech salaried employee. The patent claims of Ruge and Simmons were subsequently resolved to each one's satisfaction.

Today strain gages are the most popular strain-measuring devices. Strain gages are also used in applications involving measurements or control of forces and moments. Pressure transducers, force transducers, torque transducers, load cells, and dynamometers are some examples of industrial applications of strain gages, whereas a bathroom scale is an example of a household product using strain gages. The popularity of strain gages comes from its cost-effectiveness in measuring strains as small as 1 μmm/mm to strains as large as 50,000 μmm/mm over a large range of temperatures.

In the early days strain gages were built by taking a very thin wire and going back and forth a number of times over a small area. This construction technique was based on the observation that the resistance R of a wire is related to its length L, its cross-sectional area A, and the material-specific resistance ρ by the expression $R = \rho L/A$. For a given value of strain, a longer wire would result in a larger change in L, hence a larger change in the resistance of a wire, which can be measured more easily. By using a small cross-sectional area of the transverse effect of Poisson's ratio could be reduced. Winding the long wire in a small region lead to a better average strain value. Though the idea of using a long thin conductor in a small region still dictates the design of modern strain gages, the manufacturing process has changed. In the production of strain gages photoetching, in which material is removed by a chemical process to produce a desired pattern, has replaced winding a wire.

The sensitivity of a strain gage is called *gage factor,* which is the ratio of percentage change in resistance to percentage change in length (strain). Metal foil gages have gage factors of between 2 and 4. Ideally we would like a linear relationship between changes in resistance to strain, that is, a constant gage factor value over the range of measurements. To be as close as possible to this ideal value of a constant gage factor, different types of materials are used for the construction of strain gages for different applications. The most commonly used material is constantan or Advance, which is an alloy of copper (55%) and nickel (45%). The thermal conductivity of the two metals in the alloy is such that the gage is a self-temperature-compensated gage, that is, the gage does not undergo thermal expansion over a large range of temperatures ($-75°C$ to $175°C$). Annealed constantan is useful in large strain measurements (as high as 20%). For high-temperature applications an alloy of iron (70%), chromium (20%), and aluminum (10%), called Armour D, is used. Strain gages using semiconductor materials (doped silicon wafers) have gage factors of between 50 to 200 and are used for small-strain measurements, but they require extreme care during installation because of the brittle nature of the silicon wafers.

By measuring the change in resistance and knowing the gage factor one can find the strain from a strain gage. The most common means of measuring changes in resistance is the Wheatstone bridge circuit, shown in Figure 9.38. The voltage V_0 can be related to V as follows:

$$V_0 = V \frac{R_1 R_3 - R_2 R_4}{(R_1 + R_2)(R_3 + R_4)}$$

Clearly, if $R_1 R_3 = R_2 R_4$ then the voltage V_0 will be zero and the bridge is said to be balanced. Suppose one of the resistances is a strain gage, say R_1. Before the material is loaded (strained) the bridge is balanced. When the load is applied, the resistance R_1 changes. By adjusting the values of the other resistances by a known amount, the bridge can be balanced again, and using $R_1 R_3 = R_2 R_4$ the new resistance of R_1 can be found. Hence the change in R_1 can be found, and from it the corresponding strain can be calculated. Very small changes in resistances can be measured by the Wheatstone bridge, and it is this sensitivity that makes the Wheatstone bridge circuit so important in strain measurements by strain gages. Since we need to use only one of the resistances for balancing the bridge, by using two or more gages in creative ways, strains due to different causes can be separated.

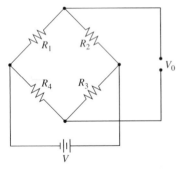

Figure 9.38 Wheatstone bridge circuit.

9.7 CLOSURE

In this chapter we studied the relationship of strains in different coordinate systems and methods to determine the maximum normal strains and maximum shear strains. We noted that the principal axes form an orthogonal coordinate system. Hence by using the generalized Hooke's law we can determine the principal stresses from the principal strains. These principal stresses can then be used in failure theories to determine whether or not a material would fail, as described in the closure of Chapter 4.

We also learned about strain gages as means of measuring strains at a point on a material. In Chapters 4 through 7 we studied one-dimensional structural elements and developed theories that let us compute strains in the x, y, z coordinate systems which an applied load produces. From these strains, which our theory predicts, we are able to determine the strain that a strain gage would record at any orientation. This idea of relating external loads to the reading of a strain gage can be used for monitoring and controlling forces and moments that are applied on a structure.

POINTS AND FORMULAS TO REMEMBER

- Strain transformation equations relate strains *at a point* in different coordinate systems:

$$\varepsilon_{nn} = \varepsilon_{xx} \cos^2\theta + \varepsilon_{yy} \sin^2\theta + \gamma_{xy} \sin\theta \cos\theta \quad (9.10)$$

$$\gamma_{nt} = -2\varepsilon_{xx} \sin\theta \cos\theta + 2\varepsilon_{yy} \sin\theta \cos\theta + \gamma_{xy}(\cos^2\theta - \sin^2\theta) \quad (9.11)$$

- Directions of the principal coordinates are the axes in which the shear strain is zero.
- Normal strains in principal directions are called *principal strains.*
- The greatest principal strain is called *principal strain 1.*
- The angles the principal axis makes with the global coordinate system are called *principal angles.*
- The angle of principal axis 1 from the x axis is only reported in describing the principal coordinate system in two-dimensional problems. Counterclockwise rotation from the x axis is defined as positive.
- Principal directions are orthogonal.
- Maximum and minimum normal strains at a point are the principal strains.
- The maximum shear strain in coordinate systems that can be obtained by rotating about one of the three axes (usually the z axis) is called *in-plane maximum shear strain.*
- The maximum shear strain at a point is the absolute maximum shear strain that can be obtained in a coordinate system by considering rotation about all three axes.
- Maximum shear strain exists in two coordinate systems that are $45°$ to the principal coordinate system.

$$\tan 2\theta_p = \frac{\gamma_{xy}}{\varepsilon_{xx} - \varepsilon_{yy}} \quad (9.13) \qquad \varepsilon_{1,2} = \frac{\varepsilon_{xx} + \varepsilon_{yy}}{2} \pm \sqrt{\left(\frac{\varepsilon_{xx} - \varepsilon_{yy}}{2}\right)^2 + \left(\frac{\gamma_{xy}}{2}\right)^2} \quad (9.14) \qquad \left|\frac{\gamma_p}{2}\right| = \left|\frac{\varepsilon_1 - \varepsilon_2}{2}\right| \quad (9.18)$$

where θ_p is the angle to either principal plane 1 or 2, ε_1 and ε_2 are the principal stresses, γ_p is the in-plane maximum shear stress.

$$\varepsilon_{nn} + \varepsilon_{tt} = \varepsilon_{xx} + \varepsilon_{yy} = \varepsilon_1 + \varepsilon_2 \quad (9.15)$$

- $$\varepsilon_3 = \begin{cases} 0, & \text{plane strain} \\ -\dfrac{\nu}{1-\nu}(\varepsilon_{xx} + \varepsilon_{yy}) & \text{plane stress} \end{cases} \quad (9.16) \qquad \frac{\gamma_{\max}}{2} = \left|\max\left(\frac{\varepsilon_1 - \varepsilon_2}{2}, \frac{\varepsilon_2 - \varepsilon_3}{2}, \frac{\varepsilon_3 - \varepsilon_1}{2}\right)\right| \quad (9.17)$$

- Each point on Mohr's circle represents a unique direction passing through the point at which the strains are specified. The coordinates of each point on the circle are the strains (ε_{nn}, $\gamma_{nt}/2$).
- The maximum shear strain at a point is the radius of the biggest of the three circles that can be drawn between the three principal strains.
- The principal directions for stresses and strains are the same for isotropic materials.
- Generalized Hooke's law in principal coordinates:

$$\varepsilon_1 = \frac{\sigma_1 - \nu(\sigma_2 + \sigma_3)}{E} \quad (9.21a) \qquad \varepsilon_2 = \frac{\sigma_2 - \nu(\sigma_3 + \sigma_1)}{E} \quad (9.21b) \qquad \varepsilon_3 = \frac{\sigma_3 - \nu(\sigma_1 + \sigma_2)}{E} \quad (9.21c)$$

- Strain gages measure only normal strains directly.
- Strain gages are bonded to a free surface, i.e., the strains are in a state of plane stress and not plane strain.
- Strain gages measure average strain at a point.
- The change in strain gage orientation by $\pm 180°$ makes no difference to the strain values.

CHAPTER TEN
DESIGN AND FAILURE

10.0 OVERVIEW

The propeller on a boat, shown in Figure 10.1*a*, subjects the shaft to an axial force as it pushes the water backward and a torsional load as it turns through the water. Gravity subjects the Washington Monument, shown in Figure 10.1*b*, to a distributed axial load and the wind pressure of a storm will subject the monument to large bending loads. Wind pressure on the highway sign shown in Figure 10.1*c* subjects the base to bending as well as to torsional loads. These examples are among the countless engineering applications in which structural members are subjected to combined axial, torsional, and/or bending loads simultaneously. This chapter synthesizes and applies the concepts developed in the previous nine chapters to the analysis and design of structures that are subjected to combined loading.

The two major learning objectives in this chapter are:

1. Learn the computation of stresses and strains in structural members subjected to combined axial, torsion, and bending loads.

2. Develop the analysis skills for the computation of internal forces and moments on individual members that compromise a structure.

(a)   (b)   (c)

Figure 10.1 Examples of combined loadings.

10.1 COMBINED LOADING

The theories for axial members in Section 4.1, for torsion of circular shafts in Section 5.2, and for symmetric bending about the z axis in Section 6.2 are all linear theories. Thus the stresses and/or strains at a point when axial, torsional, and bending loads are applied simultaneously to a structure can be obtained by the principle of superposition discussed in this section.

Equations (10.1), (10.2), (10.3a), and (10.3b), listed for convenience in Table 10.1, are the stress formulas that were derived earlier. The extension of the formulas that were developed for symmetric bending about the z axis [Equations (10.3a) and (10.3b)], to symmetric bending about the y axis, given by Equations (10.4a) and (10.4b), is discussed in Section 10.1.3.

A thin hollow cylinder, shown in Figure 10.2, will be subjected to axial, torsional, and bending loads in order to elaborate the important points about the application of superposition to determine stresses under combined loading. Stresses will be determined at four points A, B, C, and D. Figure 10.2 shows stress cubes at these points. The four points are on the surface of the shaft, which is a free surface, that is, stress free. Hence irrespective of the loading, there will be no stresses acting on this surface. The stresses from various loadings will be on the other surfaces of the stress cube, and will depend on the type of loading.

TABLE 10.1   Stresses and strains in one-dimensional structural members

	Stresses			Strains		
Axial	$\sigma_{xx} = \dfrac{N}{A}$		(10.1)	$\varepsilon_{xx} = \dfrac{\sigma_{xx}}{E}$	$\varepsilon_{yy} = -\dfrac{v\sigma_{xx}}{E}$	$\varepsilon_{zz} = -\dfrac{v\sigma_{xx}}{E}$
	$\sigma_{yy} = 0 \quad \sigma_{zz} = 0$			$\gamma_{xy} = 0$	$\gamma_{xz} = 0$	$\gamma_{yz} = 0$
	$\tau_{xy} = 0 \quad \tau_{xz} = 0 \quad \tau_{yz} = 0$					
Torsion	$\tau_{x\theta} = \dfrac{T\rho}{J}$		(10.2)	$\gamma_{x\theta} = \dfrac{\tau_{x\theta}}{G}$		
	$\sigma_{xx} = 0 \quad \sigma_{yy} = 0 \quad \sigma_{zz} = 0$			$\varepsilon_{xx} = 0$	$\varepsilon_{yy} = 0$	$\varepsilon_{zz} = 0$
	$\tau_{yz} = 0$			$\gamma_{yz} = 0$		
Symmetric bending about z axis	$\sigma_{xx} = -\dfrac{M_z y}{I_{zz}}$		(10.3a)	$\varepsilon_{xx} = \dfrac{\sigma_{xx}}{E}$	$\varepsilon_{yy} = -\dfrac{v\sigma_{xx}}{E}$	$\varepsilon_{zz} = -\dfrac{v\sigma_{xx}}{E}$
	$\tau_{xs} = -\dfrac{V_y Q_z}{I_{zz} t}$		(10.3b)	$\gamma_{xs} = \dfrac{\tau_{xs}}{G}$	$\gamma_{yz} = 0$	
	$\sigma_{yy} = 0 \quad \sigma_{zz} = 0 \quad \tau_{yz} = 0$					
Symmetric bending about y axis	$\sigma_{xx} = -\dfrac{M_y z}{I_{yy}}$		(10.4a)	$\varepsilon_{xx} = \dfrac{\sigma_{xx}}{E}$	$\varepsilon_{yy} = -\dfrac{v\sigma_{xx}}{E}$	$\varepsilon_{zz} = -\dfrac{v\sigma_{xx}}{E}$
	$\tau_{xs} = -\dfrac{V_z Q_y}{I_{yy} t}$		(10.4b)	$\gamma_{xs} = \dfrac{\tau_{xs}}{G}$	$\gamma_{yz} = 0$	
	$\sigma_{yy} = 0 \quad \sigma_{zz} = 0 \quad \tau_{yz} = 0$					

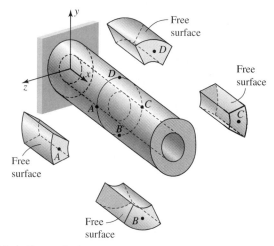

Figure 10.2 Thin hollow cylinder.

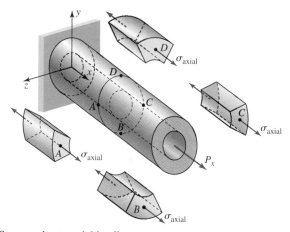

Figure 10.3 Stresses due to axial loading.

The free surfaces at points B and D have outward normals in the y direction. Thus τ_{yx},[1] which acts on this surface, has to be zero. As $\tau_{xy} = \tau_{yx}$, it follows that τ_{xy} at points B and D will be zero *irrespective* of the loading. Similarly we note that the free surfaces at points A and C have outward normals in the z direction, and hence $\tau_{zx} = 0$, which in turn implies that τ_{xz} is zero at these points, irrespective of the loading.

10.1.1 Combined Axial and Torsional Loading

We first consider the axial load and the torsional load individually and then consider the combined effect on the cylinder.

Axial load: Figure 10.3 shows the thin hollow cylinder subjected to an axial load P_x. The normal stress σ_{xx} will be uniform across the cross section and can be determined using

---

[1]Recollect, the first subscript on stress components is the direction of the outward normal to the surface on which the stress component acts.

Equation (10.1). Symbolically we represent the magnitude of this axial stress as σ_{axial} and show it as a tensile stress on the stress cubes associated with the four points A, B, C, and D.

Torsional load: Figure 10.4 shows the thin hollow cylinder subjected to a torque. The torsional shear stress $\tau_{x\theta}$ can be found from Equation (10.2). The direction of the shear stress $\tau_{x\theta}$ can be determined by either inspection or using subscripts, as was elaborated in Section 5.2.5. Symbolically we represent the magnitude of this torsional shear stress as τ_{tor} and show it on the stress cubes associated with the four points A, B, C, and D in Figure 10.4.

Combined axial and torsional loading: Figure 10.5 shows the thin hollow cylinder subjected to a combined axial and torsional load. Since the theories for axial members and torsion of circular shafts are linear theories, the state of stress at each point shown

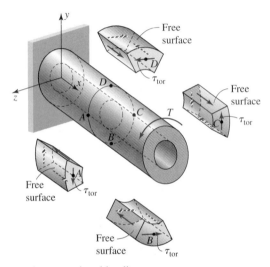

Figure 10.4 Stresses due to torsional loading.

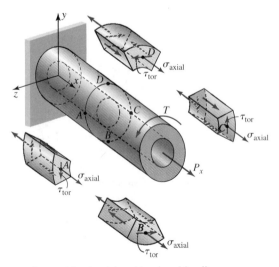

Figure 10.5 Stresses in combined axial and torsional loading.

Figures 10.3 and 10.4 by using the superposition principle, as elaborated in the next paragraph.

Notice that we did not add the stress component from torsion to that from axial load. Stress components in different directions cannot be added or subtracted from each other. So what did we add, that is, superpose? A simple answer is: we superposed the states of stress due to individual loads to obtain a state of stress in combined loading. To clarify this answer, we consider two stress components σ_{xx} and τ_{xy} at point C. In axial loading $\sigma_{xx} = \sigma_{axial}$ and $\tau_{xy} = 0$; in torsional loading $\sigma_{xx} = 0$ and $\tau_{xy} = \tau_{tor}$. When we add (or subtract), we add (or subtract) the same component in each loading, that is, the total state of stress at point C is $\sigma_{xx} = \sigma_{axial} + 0 = \sigma_{axial}$ and $\tau_{xy} = 0 + \tau_{tor} = \tau_{tor}$. The state of stress at point C in combined loading (Figure 10.5) is very different from the states of stress in individual loadings (Figures 10.3 and 10.4)—think of Mohr's circle associated with the state of stress at point C in Figure 10.5 and of Mohr's circle associated with the state of stress at C in Figures 10.3 and 10.4. This point is further elaborated in Example 10.1.

10.1.2   Combined Axial, Torsional, and Bending Loads about *z* Axis

We first consider the bending about the z axis alone. Then we superpose the results to those obtained previously for combined axial and torsional loads.

Bending about z axis: Figure 10.6 shows the thin hollow cylinder subjected to a load that bends the cylinder about the z axis. Points B and D are on the free surface. Hence bending shear stress is zero at these points. The bending normal stress at points B and D can be found from Equation (10.3a). The direction of the bending normal stress can be found either by inspection or by using subscripts, as elaborated in Section 6.2.6. Symbolically we represent the magnitude of this bending normal stress as $\sigma_{bend\text{-}z}$ and show it on the stress cubes associated with points B and D in Figure 10.6.

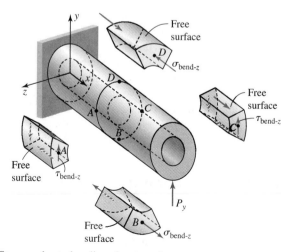

Figure 10.6   Stresses due to bending about z axis.

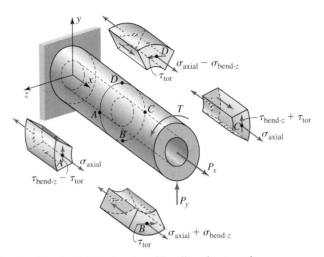

Figure 10.7   Combined axial, torsional, and bending about z axis.

Points A and C are on the neutral axis, hence the bending normal stress is zero at these points. The bending shear stress at points A and C can be found from Equation (10.3b). The direction of shear stress can be found by either inspection or using subscripts, as elaborated in Sections 6.6.1 and 6.6.2. Symbolically we represent the magnitude of this bending shear stress as $\tau_{\text{bend-}z}$ and show it on the stress cubes associated with points A and C in Figure 10.6.

Combined axial, torsional, and bending about z axis: We superpose the stress states for bending at the four points shown in Figure 10.6 to the stress states for the combined axial and torsional loads at the same points shown in Figure 10.5 to obtain the stress states shown in Figure 10.7.

In Figure 10.7 at point D the bending normal stress is compressive whereas the axial stress is tensile. Thus the resultant normal stress σ_{xx} is the difference between the two stress values. At point B both the bending normal stress and the axial stress are tensile, thus the resultant normal stress σ_{xx} is the sum of the two stress values. If the axial normal stress at point D is greater than the bending normal stress, then the total normal stress at point D will be in the direction (tensile) shown in Figure 10.7. If the bending normal stress is greater than the axial stress, then the direction will be opposite (compressive).

At point A the torsional shear stress is downward whereas the bending shear stress is upward. Thus the resultant shear stress τ_{xy} is the difference between the two stress values. At point C both the torsional shear stress and the bending shear stress are upward, thus the resultant shear stress τ_{xy} is the sum of the two stress values. If the bending shear stress at point A is greater than the torsional shear stress, then the total shear stress at point A will be in the direction (positive τ_{xy}) shown in Figure 10.7. If the torsional shear stress is greater than the bending shear stress, then the direction will be opposite (negative τ_{xy}).

10.1.3   Extension to Symmetric Bending about *y* Axis

The stress formulas given by Equations (10.4a) and (10.4b) are extensions of the formulas derived for symmetric bending about the z axis. If we assume that the xz plane is also a plane of symmetry and the loads lie in the plane of symmetry, then we can

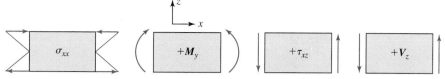

Figure 10.8 Sign convention for internal bending moments and shear force in bending about y axis.

obtain stress formulas, Equations (10.4*a*) and (10.4*b*), for bending about the y axis by interchanging the y and z subscripts in Equations (10.3*a*) and (10.3*b*). The sign conventions for the internal moment M_y and the shear force V_z in Equations (10.4*a*) and (10.4*b*) are simple extensions of M_z and V_y and are shown in Figure 10.8.

Definition 1 The direction of positive internal moment M_y on a free-body diagram must be such that it puts a point in the positive z direction into compression.

Definition 2 The direction of positive internal shear force V_z on a free-body diagram is in the direction of positive shear stress τ_{xz} on the surface.

The direction of shear stress in Equation (10.4*b*) can be determined either by using the subscripts or by inspection, as we did for symmetric bending about the z axis. To use the subscripts, recall that the s coordinate is defined from the free surface (see Section 6.6.1) used in the calculation of Q_y. The shear flow (shear stress), due to bending about the y axis only, is drawn along the centerline of the cross section in such a direction as to satisfy the following rules:

1. The resultant force in the z direction is in the same direction as V_z.
2. The resultant force in the y direction is zero.
3. It is symmetric about the z axis. This requires that shear flow change direction as one crosses the y axis on the centerline. Sometimes this will imply that shear stress is zero at the point(s) where the centerline intersects the z axis.

10.1.4 Combined Axial, Torsional, and Bending Loads about y and z Axes

Once more, we first consider bending about the y axis alone and then superpose the results to those obtained previously for combined axial, torsional, and bending about the z axis.

Bending about y axis: Figure 10.9 shows the thin hollow cylinder subjected to a load that bends the cylinder about the y axis. Points A and C are on the free surface, hence bending shear stress is zero at these points. The bending normal stress at points A and C can be found from Equation (10.4*a*). The direction of the bending normal stress can be found either by inspection or by using subscripts, as described briefly in Section 10.1.3 Symbolically we represent the magnitude of this bending normal stress as $\sigma_{\text{bend-}y}$ and show it on the stress cubes associated with points A and C in Figure 10.9.

Points B and D are on the neutral axis [$z_B = 0$ and $z_D = 0$ in Equation (10.4*a*)]. Hence the bending normal stress is zero at these points. The bending shear stress at points B and D can be found from Equation (10.4*b*). The direction of shear stress can

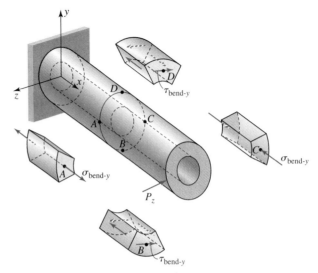

Figure 10.9 Stresses due to bending about y axis.

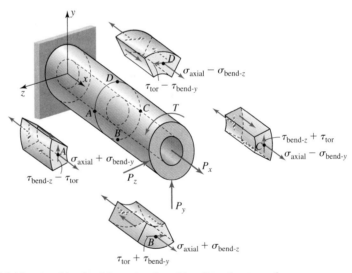

Figure 10.10 Combined axial, torsional, and bending about y and z axes.

be found either by inspection or by using subscripts, as described in Section 10.1.3. Symbolically we represent the magnitude of this bending shear stress as $\tau_{\text{bend-}y}$ and show it on the stress cubes associated with points A and C in Figure 10.9.

Combined axial, torsional, and bending about y and z axes: We superpose the states of stress in Figures 10.7 and 10.9 to obtain the states of stress for this combined loading, as shown in Figure 10.10. Thus the complex stress states shown in Figure 10.10 can be obtained in a simple manner by first calculating the stresses due to individual loadings and then superposing the stress states at each point.

To obtain strains in combined loading we can either superpose the strains given in Table 10.1 or superpose the stresses as discussed in the preceding sections and then use the generalized Hooke's law to convert these stresses to strains. The second approach, that is, superposing stresses and using the generalized Hooke's law, is preferable because often we will need to transform torsional shear stress $\tau_{x\theta}$ and bending shear stress τ_{xs} into the x, y, z coordinate system for reasons to be discussed.

The stress and strain transformation equations are written in reference to the stress and strain components in the x, y, z coordinate system. In order to use these transformation equations, the torsional shear stress $\tau_{x\theta}$ and the bending shear stress τ_{xs} have to be transformed to the x, y, z coordinate system, as was described in Sections 5.2.5 and 6.6.5. This transformation is achieved by drawing the stresses on a stress cube, either by inspection or by the use of subscripts, and then interpreting these stress components in the x, y, z coordinate system. Thus in Figure 10.10, at points A and C the shear stress shown is positive τ_{xy}, at point B the shear stress shown is negative τ_{xz}, and at point D the shear stress shown is positive τ_{xz}. The discussion in this paragraph emphasizes the importance of showing stresses on a stress element before proceeding to stress or strain transformation.

In studying stresses produced by individual loading, we often had prefixes to stresses such as maximum axial normal stress, maximum torsional shear stress, maximum bending normal stress, maximum bending shear stress, or maximum in-plane shear stress. In this chapter we are considering combined loading problems. Hence the maximum normal stress at a point will refer to the principal stress at the point and the maximum shear stress will refer to the absolute maximum shear stress. This implies that allowable normal stress refers to the principal stresses and allowable shear stress refers to the absolute maximum shear stress. The allowable tensile normal stress refers to principal stress 1, assuming it is tensile. The allowable compressive normal stress refers to principal stress 2, assuming it is compressive.

10.1.6   Important Points

The important points discussed in Sections 10.1.1 through 10.1.5, which should be kept in mind when solving problems involving combined loading, are listed as follows:

1. The complexity of finding the state of stress under combined loading can be simplified by first determining the states of stress due to individual loadings.

2. Superposition of stresses implies that a stress component at a specific point resulting from one loading can be added to or subtracted from a similar stress component from another loading. Stress components at different points cannot be added or subtracted nor can stress components which act on different planes or in different directions be added or subtracted.

3. The stress formulas in Table 10.1 give both the correct magnitude and the correct direction for each stress component if the internal forces and moments are drawn on the free-body diagrams according to the prescribed sign conventions. If the directions of internal forces and moments are not drawn according to the sign convention, but drawn to equilibrate external

forces and moments, then the directions of the stress components will have to be determined by inspection.

4. In a structure the structural members will have different orientations. In order to use subscripts to determine the directions (signs) of stress components, a local x, y, z coordinate system can be established for a structural member such that the x direction is normal to the cross section, that is, the x direction is along the axis of the structural member.

5. Table 10.1 shows that stresses σ_{yy} and σ_{zz} are zero for the four cases listed, emphasizing that the theories are for one-dimensional structural members. Additional stress components can be deduced to be zero by identifying the free surfaces.

6. The state of stress in combined loading should be shown on a stress cube before processing the stresses for the purpose of stress or strain transformation.

7. The strains at a point can be obtained from the superposed stress values using the generalized Hooke's law. As the normal stresses σ_{yy} and σ_{zz} are always zero in our structural members, the nonzero strains ε_{yy} and ε_{zz} are due to the Poisson effect, that is, $\varepsilon_{yy} = \varepsilon_{zz} = -\nu\varepsilon_{xx}$.

10.1.7 General Procedure for Combined Loading

A general procedure for calculating stresses in combined loading is outlined as follows:

Step 1: Identify the equations in Table 10.1 relevant for the problem and use the equations as a checklist for the quantities that must be calculated.

Step 2: Calculate the relevant geometric properties (A, I_{yy}, I_{zz}, J) of the cross section containing the points where stresses have to be found.

Step 3: At points where shear stress due to bending is to be found, draw a line perpendicular to the centerline through the point and calculate the first moments of the area (Q_y, Q_z) between the free surface and the drawn line. Record the s direction from the free surface toward the point where the stress is being calculated.

Step 4: Make an imaginary cut through the cross section and draw the free-body diagram. On the free-body diagram draw the internal forces and moments according to our sign conventions if subscripts are to be used in determining the directions of the stress components. Use equilibrium equations to calculate the internal forces and moments.

Step 5: Using the equations identified in Step 1, calculate the individual stress components due to each loading. Draw the torsional shear stress $\tau_{x\theta}$ and the bending shear stress τ_{xs} on a stress cube using subscripts or by inspection. By examining the shear stresses in the x, y, z coordinate system obtain τ_{xy} and τ_{xz} with proper signs.

Step 6: Superpose the stress components to obtain the total stress components at a point.

Step 7: Show the calculated stresses on a stress cube.

Step 8: Interpret the stresses shown on the stress cube in the x, y, z coordinate system before processing these stresses for the purpose of stress or strain transformation.

EXAMPLE 10.1

A hollow shaft that has an outside diameter of 100 mm and an inside diameter of 50 mm is loaded as shown in Figure 10.11. For the three cases shown, determine the principal stresses and the maximum shear stress at point A. Point A is on the surface of the shaft.

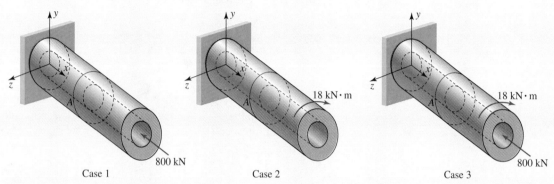

Figure 10.11   Hollow cylinder in Example 10.1.

PLAN

The axial normal stress in case 1 can be found from Equation (10.1). The torsional shear stress in case 2 can be found from Equation (10.2). The state of stress in case 3 is the superposition of the stress states in cases 1 and 2. The calculated stresses at point A can be drawn on a stress cube. Using Mohr's circle or the method of equations, we can find the principal stresses and the maximum shear stress in each case.

Solution

Step 1: Equations (10.1) and (10.2) are used for calculating the axial stress and the torsional shear stress.

Step 2: The cross-sectional area A and the polar area moment J of a cross section can be found as

$$A = \frac{\pi}{4}(100^2 - 50^2) = 5.89 \times 10^3 \text{ mm}^2$$

$$J = \frac{\pi}{32}(100^4 - 50^4) = 9.20 \times 10^6 \text{ mm}^4 \tag{E1}$$

Step 3:   This step is not needed as there is no bending.

Step 4:   We can make an imaginary cut and draw the free-body diagrams in Figure 10.12 using the right-hand part. The internal axial force and the internal torque are drawn according to our sign convention. Using equilibrium equations, the internal axial force and the

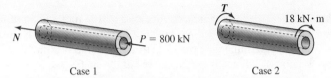

Case 1   Case 2

Figure 10.12 Free-body diagrams in Example 10.1.

internal torque can be determined,

$$N = -800 \text{ kN} \tag{E2}$$

$$T = -18 \text{ kN·m} \tag{E3}$$

Step 5:

    Case 1: The axial stress is uniform across the cross section and can be found from Equation (10.1),

$$\sigma_{xx} = \frac{N}{A} = \frac{-800 \times 10^3}{5.89 \times 10^{-3}} = -135.8 \times 10^6 \text{ N/m}^2 = -135.8 \text{ MPa} \tag{E4}$$

    Case 2: The torsional shear stress varies linearly and is maximum on the surface ($\rho = 0.05$ m) of the shaft. It can be found from Equation (10.2),

$$\tau_{x\theta} = \frac{T\rho}{J} = \frac{-18 \times 10^3 \times 0.05}{9.20 \times 10^{-6}}$$

$$= -97.83 \times 10^6 \text{ N/m}^2 = -97.83 \text{ MPa} \tag{E5}$$

Steps 6, 7: We draw the stress cube and show the stresses calculated in Equations (E4) and (E5).

    Case 1: The axial stress is compressive, as shown Figure 10.13*a*.

    Case 2: From Equation (E5) we note that $\tau_{x\theta}$ is negative. The θ direction in positive counterclockwise with respect to the x axis, as shown in Figure 10.13*b*. At point A the outward normal to the surface is in the positive x direction and the positive θ direction at A is downward. Hence a negative $\tau_{x\theta}$ will be upward at point A, as shown in Figure 10.13*b*.

Intuitive check: Figure 10.14 shows the hollow shaft with the applied torque on the right end and the reaction torque at the wall on the left end. The left part of the shaft would rotate counterclockwise with respect to the right part. Thus the surface of the cube at point A would be moving downward. The shear stress would oppose this impending

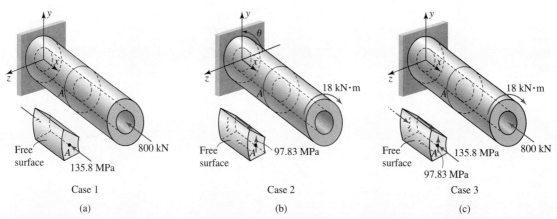

Figure 10.13 Stresses on stress cubes in Example 10.1.

motion by acting upward at point A, as shown in Figure 10.14, confirming the direction shown in Figure 10.13b.

Case 3: The state of stress is a superposition of the states of stress shown on the stress cubes for cases 1 and 2 and is illustrated in Figure 10.13c.

Step 8: We can redraw the stress cubes in two dimensions and follow the procedure for constructing Mohr's circle[2] for each case, as shown in Figure 10.15. The radius of Mohr's circle can be found for each case, and principal stresses 1 and 2 can be obtained as shown. Also note that point A being on the free surface, we deduce that the third principal stress is zero. The in-plane maximum shear stress and the absolute maximum shear stress are the same for all cases and the value of this maximum shear stress is equal to the radius of Mohr's circle as shown.

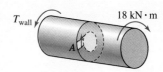

Figure 10.14 Direction of shear stress by inspection.

- *Case 1:*

$$R = 67.9$$

ANS. $\sigma_1 = 0$

ANS. $\sigma_2 = 135.8$ MPa (C)

ANS. $\sigma_3 = 0$

ANS. $\tau_{max} = 67.9$ MPa

---

[2]Alternatively, we can use the method of equations, as described in comment 2.

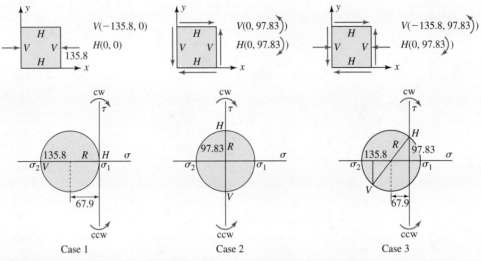

Figure 10.15 Mohr's circles in Example 10.1.

- *Case 2:*

$$R = 97.83$$

ANS. $\sigma_1 = 97.8$ MPa (T)

ANS. $\sigma_2 = 97.8$ MPa (C)

ANS. $\sigma_3 = 0$

ANS. $\tau_{max} = 97.8$ MPa

- *Case 3:*

$$R = \sqrt{67.9^2 + 97.83^2} = 119.1$$

$$\sigma_1 = -67.9 + 119.1$$

ANS. $\sigma_1 = 51.2$ MPa (T)

$$\sigma_2 = -67.9 - 119.1$$

ANS. $\sigma_2 = 187$ MPa (C)

ANS. $\sigma_3 = 0$

ANS. $\tau_{max} = 119.1$ MPa

COMMENTS

1. The results for the three cases show that the principal stresses and the maximum shear stress for case 3 cannot be obtained by superposition of the principal stresses and the maximum shear stress calculated for cases 1 and 2. This point is emphasized graphically in Figure 10.15. Mohr's circle of case 3 cannot be obtained by superposing Mohr's circle for cases 1 and 2. The reason why the superposition principle is not applicable to principal stresses is that the principal planes for the three cases are different. We cannot add (or subtract) stresses on different planes. If we calculated the stresses for the three cases on the same plane, then we could apply the superposition principle.

2. Substituting $\sigma_{xx} = -135.8$ MPa, $\tau_{xy} = +97.8$ MPa, and $\sigma_{yy} = 0$ into Equation (8.6), that is,

$$\sigma_{1,2} = \frac{\sigma_{xx} + \sigma_{yy}}{2} \pm \sqrt{\left(\frac{\sigma_{xx} - \sigma_{yy}}{2}\right)^2 + \tau_{xy}^2}$$

we can find σ_1 and σ_2 as before for case 3. Noting that $\sigma_3 = 0$, we can find τ_{max} from Equation (8.14),

$$\tau_{max} = \left| \max\left(\frac{\sigma_1 - \sigma_2}{2}, \frac{\sigma_2 - \sigma_3}{2}, \frac{\sigma_3 - \sigma_1}{2}\right) \right|$$

EXAMPLE 10.2

A hollow shaft that has an outside diameter of 100 mm and an inside diameter of 50 mm, is shown in Figure 10.16. Strain gages are mounted on the surface of the shaft at 30° to the axis. For each case determine the applied axial load P and the applied torque T_{ext} if the strain gage readings are $\varepsilon_a = -500 \, \mu$ and $\varepsilon_b = 400 \, \mu$. Use $E = 200$ GPa, $G = 80$ GPa, and $v = 0.25$.

PLAN

The stresses at point A in terms of P and T_{ext} can be found as in Example 10.1. Using the generalized Hooke's law, we can find the strains in terms of P and T_{ext}. Using the strain transformation equation, Equation (9.10), the normal strain in direction of the strain gage can be

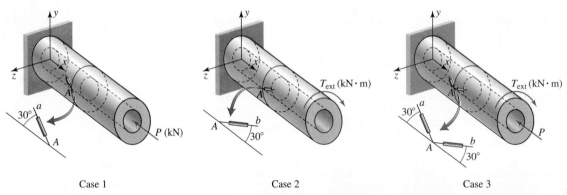

Case 1   Case 2   Case 3

Figure 10.16 Hollow cylinder in Example 10.2.

found in terms of P and T_{ext}. The values of P and T_{ext} can be determined from the given strain gage readings.

Solution

Step 1: Equations (10.1) and (10.2) will be used for calculating the axial stress and the torsional shear stress.

Step 2: From Example 10.1, the cross-sectional area A and the polar area moment J of a cross section are as given,

$$A = 5.89 \times 10^3 \text{ mm}^2 \qquad J = 9.20 \times 10^6 \text{ mm}^4 \qquad \text{(E1)}$$

Step 3: This step is not needed as there is no bending.

Step 4: We can make an imaginary cut and draw the free-body diagrams in Figure 10.17 using the right-hand part. The internal axial force and the internal torque are drawn according to our sign convention. Using equilibrium equations, the internal axial force and the internal torque can be determined,

$$N = -P \text{ kN} \qquad \text{(E2)}$$

$$T = -T_{\text{ext}} \text{ kN·m} \qquad \text{(E3)}$$

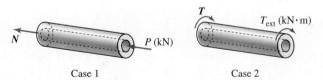

Case 1   Case 2

Figure 10.17 Free-body diagrams in Example 10.2.

Step 5:

Case 1: The axial stress is uniform across the cross section and can be found from Equation (10.1),

$$\sigma_{xx} = \frac{N}{A} = \frac{-P \times 10^3}{5.89 \times 10^{-3}} = -0.17P \times 10^6 \text{ N/m}^2 = -0.170P \text{ MPa}$$

$$(E4)$$

Case 2: The torsional shear stress on the surface ($\rho = 0.05$ m) of the shaft can be found from Equation (10.2),

$$\tau_{x\theta} = \frac{T\rho}{J} = \frac{-T_{ext} \times 10^3 \times 0.05}{9.20 \times 10^{-6}} = -5.435 T_{ext} \times 10^6 \text{ N/m}^2$$

$$= -5.435 T_{ext} \text{ MPa} \qquad (E5)$$

Steps 6, 7: Figure 10.18 shows the stresses on the stress elements calculated using Equations (E4) and (E5), as was done in Example 10.1.

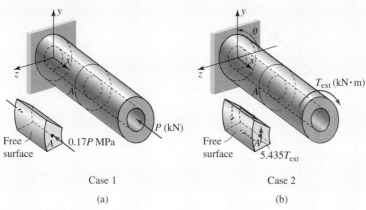
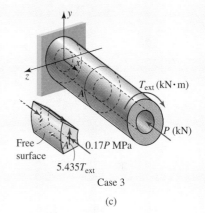

Case 1	Case 2	Case 3
(a)	(b)	(c)

Figure 10.18 Stresses on stress cubes in Example 10.2.

Step 8:

Case 1: We note that the only nonzero stress is the axial stress given in Equation (E4). From the generalized Hooke's law we obtain the strains as given by

$$\varepsilon_{xx} = \frac{\sigma_{xx}}{E} = \frac{-0.170P \times 10^6}{200 \times 10^9} = -0.85P \times 10^{-6} = -0.85P \ \mu \quad (E6)$$

$$\varepsilon_{yy} = -\nu\varepsilon_{xx} = -0.25(-0.85P) = 0.213P \ \mu \qquad (E7)$$

$$\gamma_{xy} = \frac{\tau_{xy}}{G} = 0 \qquad (E8)$$

Case 2: From Figure 10.18 we note that the shear stress $\tau_{xy} = +5.435 T_{ext}$. The normal stresses are all zero. From the generalized Hooke's law we obtain the strains as given by

$$\varepsilon_{xx} = 0 \tag{E9}$$

$$\varepsilon_{yy} = 0 \tag{E10}$$

$$\gamma_{xy} = \frac{\tau_{xy}}{G} = \frac{5.435 T_{ext} \times 10^6}{80 \times 10^9} = 67.94 T_{ext} \ \mu \tag{E11}$$

Case 3: The state of strain is the superposition of the state of strain for cases 1 and 2 as given by

$$\varepsilon_{xx} = -0.85 P \ \mu \tag{E12}$$

$$\varepsilon_{yy} = 0.213 P \ \mu \tag{E13}$$

$$\gamma_{xy} = 67.94 T_{ext} \ \mu \tag{E14}$$

Load calculations

Case 1: Substituting $\theta_a = 150°$ or $-30°$ and Equations (E6), (E7), and (E8) into the strain transformation equation, Equation (9.10), the normal strain in terms of P can be found and equated to the given value of $\varepsilon_a = -500 \ \mu$. The value of P can be found as

$$\varepsilon_a = -0.85 P \cos^2(-30) + 0.213 P \sin^2(-30) = -500$$

$$(-0.638 + 0.053) P = -500 \quad \text{or} \quad \text{ANS.} \quad P = 855 \text{ kN} \tag{E15}$$

Case 2: Substituting $\theta_b = 30°$ and Equations (E9), (E10), and (E11) into the strain transformation equation, Equation (9.10), the normal strain in terms of T_{ext} can be found and equated to the given value of $\varepsilon_b = 400 \ \mu$. The value of T_{ext} can be found as

$$\varepsilon_b = 0 + 0 + 67.94 T_{ext} \sin(30) \cos(30) = 400$$

$$29.42 T_{ext} = 400 \quad \text{or} \quad \text{ANS.} \quad T_{ext} = 13.6 \text{ kN·m} \tag{E16}$$

Case 3: Substituting $\theta_a = -30$, $\theta_b = 30$, and Equations (E12), (E13), and (E14) into the strain transformation equation, Equation (9.10), and using the given strain values, we obtain

$$\varepsilon_a = -0.85 P \cos^2(-30) + 0.213 P \sin^2(-30)$$
$$+ 67.94 T_{ext} \sin(-30) \cos(-30) = -500 \mu$$

or

$$-0.585P - 29.42T_{ext} = -500 \qquad \text{(E17)}$$

$$\varepsilon_b = -0.85P\cos^2(30) + 0.213P\sin^2(30)$$
$$+ 67.94T_{ext}\sin(30)\cos(30) = 400\,\mu$$

or

$$-0.585P + 29.42T_{ext} = 400 \qquad \text{(E18)}$$

Equations (E17) and (E18) can be solved simultaneously to obtain

$$\text{ANS.} \qquad P = 85.4 \text{ kN} \qquad T_{ext} = 15.3 \text{ kN·m} \qquad \text{(E19)}$$

COMMENTS

1. The values of P and T_{ext} for combined loading in Equation (E19) are different than the values obtained for individual loadings, given by Equations (E15) and (E16). The reason for this difference is discussed in the next comment.

2. If we had been given P and T_{ext} and were required to predict the strains in the gages, we could have calculated strains along the strain gage direction for individual loads and superposed to get the total strain in the gages for combined loading. But as the results in this example demonstrate, the strains in the gages (total strain) for combined loading cannot be separated into strain due to axial load and strain due to torsion. Loads P and T_{ext} affect both strain gages simultaneously, and these effects cannot be decoupled into effects of individual loadings.

3. In this example and the previous one we solved the problem by separating axial and torsion problems and calculated internal axial force and internal torque using separate free-body diagrams. We could have used a single free-body diagram, as shown in Figure 10.19, to calculate the internal quantities. In subsequent examples we shall construct a single free-body diagram for the calculation of the internal quantities,

Figure 10.19 Single free-body diagram for combined loading.

$$N = -P \text{ kN}$$
$$T = -T_{ext} \text{ kN·m}$$

This choice is not only less tedious but may be necessary. A single force may produce axial, torsion, and bending, which cannot be separated on a free-body diagram.

EXAMPLE 10.3

A box column is constructed from $\frac{1}{4}$-in-thick sheet metal and sub-jected to the loads shown in Figure 10.20.

(a) Determine the normal and shear stresses in the x, y, z coordinate system at points A and B and show the results on stress cubes.

(b) A surface crack at point B is oriented as shown. Determine the normal and shear stresses on the plane containing the crack.

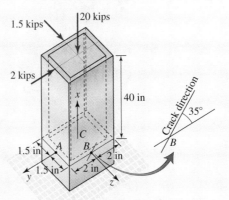

Figure 10.20 Beam and loading in Example 10.3.

PLAN

(a) We can follow the procedure in Section 10.1.7. The 20-kips force is an axial force, whereas the 2-kips and 1.5-kips forces produce bending about the z and y axes, respectively. Thus Equations (10.1), (10.3a), (10.3b), (10.4a), and (10.4b) will be used for calculating stresses. These formulas can be used as a checklist of the quantities that must be calculated in finding the individual stress components. By superposition the total stress at points A and B can be obtained.

(b) Using the method of equations or Mohr's circle, the normal and shear stresses on the plane containing the crack can be found from the stresses determined at point B.

Solution

Step 1: Equations (10.1), (10.3a), (10.3b), (10.4a), and (10.4b) will be used for calculating the stress components.

Step 2: The geometric properties of the cross section can be found as

$$A = 4 \times 3 - 3.5 \times 2.5 = 3.25 \text{ in}^2 \tag{E1}$$

$$I_{yy} = \frac{1}{12} 4 \times 3^3 - \frac{1}{12} 3.5 \times 2.5^3 = 4.443 \text{ in}^4 \tag{E2}$$

$$I_{zz} = \frac{1}{12}3 \times 4^3 - \frac{1}{12}2.5 \times 3.5^3 = 7.068 \text{ in}^4 \qquad \text{(E3)}$$

Step 3: At point *A* we draw a line perpendicular to the centerline of the cross section. The line cuts the thin column at two places, as shown in Figure 10.21. Thus the thickness of the material at *A* is twice the thickness of the column, as given by

$$t_A = 0.25 + 0.25 = 0.5 \text{ in} \qquad \text{(E4)}$$

The area between the free surface at the bottom and the line passing through *A* is used in the calculation of $(Q_y)_A$ and $(Q_z)_A$. The area shown in Figure 10.21 is viewed in parts. The area of each part is multiplied by the distance from the centroid to the *y* axis, and the products are added to obtain $(Q_y)_A$,

$$(Q_y)_A = 2 \times 1.5 \times 0.25 \times 0.75 + 3.5 \times 0.25(1.5 - 0.125)$$
$$= 1.766 \text{ in}^3 \qquad \text{(E5)}$$

By symmetry, $(Q_z)_A$ is zero, as given by

$$(Q_z)_A = 0 \qquad \text{(E6)}$$

At point *B* we once more draw a line perpendicular to the centerline of the cross section and repeat the calculations described for point *A* using the area shown in Figure 10.22,

$$t_B = 0.25 + 0.25 = 0.5 \text{ in} \qquad \text{(E7)}$$

$$(Q_y)_B = 0 \qquad \text{(E8)}$$

$$(Q_z)_B = 2 \times 2 \times 0.25 \times 1 + 2.5 \times 0.25(2 - 0.125)$$
$$= 2.172 \text{ in}^3 \qquad \text{(E9)}$$

Step 4: We can make an imaginary cut through the cross section containing points *A* and *B* and draw the free-body diagram shown in Figure 10.23. Internal forces and moments are drawn according to our sign convention. Using equilibrium equations, the internal forces and moments can be found,

$$N = -20 \text{ kips} \qquad \text{(E10)}$$
$$V_y = -2.0 \text{ kips} \qquad \text{(E11)}$$
$$V_z = 1.5 \text{ kips} \qquad \text{(E12)}$$
$$M_y = 60 \text{ in·kips} \qquad \text{(E13)}$$
$$M_z = -80 \text{ in·kips} \qquad \text{(E14)}$$

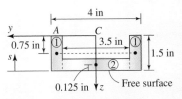

Figure 10.21 Calculation of Q_y and Q_z at point *A* in Example 10.3.

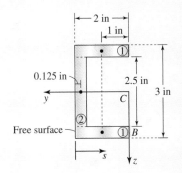

Figure 10.22 Calculation of Q_y and Q_z at point *B* in Example 10.3.

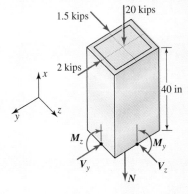

Figure 10.23 Free-body diagram in Example 10.3.

Step 5: The stress components due to each loading are calculated next.

Axial stress calculations: The axial stresses are uniform across the cross section. Substituting Equations (E2) and (E11) into Equation (10.1), the axial stresses at points A and B can be found,

$$(\sigma_{xx})_{A,B} = -\frac{20}{3.25} = -6.154 \text{ ksi} \qquad (E15)$$

Stresses due to bending about the y axis: We note that $z_A = 0$ and $z_B = 1.5$. Substituting these values of z and Equations (E2) and (E13) into Equation (10.4a), we obtain the normal stresses at points A and B due to bending about the y axis,

$$(\sigma_{xx})_A = 0 \qquad (E16)$$

$$(\sigma_{xx})_B = -\frac{60 \times 1.5}{4.443} = -20.258 \text{ ksi} \qquad (E17)$$

Substituting Equations (E2), (E4), (E5), and (E11) into Equation (10.3b), we obtain the shear stress τ_{xs}. From Figure 10.21 we note that the s direction is in the negative z direction at point A. Thus $(\tau_{xz})_A = -(\tau_{xs})_A$, and we obtain the value of shear stress at point A due to bending about the y axis. At point B, τ_{xs} is zero because $Q_y = 0$, as given by Equation (E6). Hence, τ_{xy} at B is zero. Then,

$$(\tau_{xs})_A = -\frac{1.5 \times 1.766}{4.443 \times 0.5} = -1.192 \text{ ksi}$$

$$(\tau_{xz})_A = -(\tau_{xs})_A = 1.19 \text{ ksi} \qquad (E18)$$

$$(\tau_{xs})_B = 0 \qquad (\tau_{xy})_B = 0 \qquad (E19)$$

Stresses due to bending about the z axis: We note that $y_A = 2$ and $y_B = 0$. Substituting these values of y and Equations (E3) and (E14) into Equation (10.3a), we obtain the normal stresses at points A and B due to bending about the z axis,

$$(\sigma_{xx})_A = -\frac{(-80)2}{7.068} = 22.638 \text{ ksi} \qquad (E20)$$

$$(\sigma_{xx})_B = 0 \qquad (E21)$$

Substituting Equations (E3), (E7), (E9), and (E12) into Equation (10.3b), we obtain the shear stress τ_{xs}. From Figure 10.22 we note that the s direction is in the negative y direction at point B. Thus $(\tau_{xy})_B = -(\tau_{xs})_B$, and we obtain the value of shear stress at point B due to bending about the y axis. At point A, τ_{xs} is zero because $Q_z = 0$, as given by Equation (E8). Hence,

τ_{xz} at A is zero. Then,

$$(\tau_{xs})_B = -\frac{(-2)2.172}{7.068 \times 0.5} = 1.229 \text{ ksi}$$

(E22)

$$(\tau_{xy})_B = -(\tau_{xs})_B = -1.23 \text{ ksi}$$

$$(\tau_{xs})_A = 0 \qquad (\tau_{xz})_A = 0$$

(E23)

Step 6: Superposition

Normal stress calculations: The normal stress at point A can be obtained by superposing the values in Equations (E15), (E16), and (E20),

$$(\sigma_{xx})_A = -6.154 + 0 + 22.638 = 16.484 \qquad \text{(E24)}$$

ANS. $\qquad (\sigma_{xx})_A = 16.5 \text{ ksi (T)}$

Similarly, the normal stress at point B can be obtained by superposition of Equations (E15), (E17), and (E21),

$$(\sigma_{xx})_B = -6.154 - 20.258 + 0 = -26.412 \qquad \text{(E25)}$$

ANS. $\qquad (\sigma_{xx})_B = 26.4 \text{ ksi (C)}$

Intuitive check on normal stress calculations: The axial stress σ_{axial} due to a 20-kips force will be compressive. Figure 10.24 shows the exaggerated deformed shapes due to bending about the y and z axes. These deformed shapes are drawn for the purpose of explanation, but can be visualized without drawing the figures. From Figure 10.24a it can be seen that the line passing through A will be in tension, that is, the

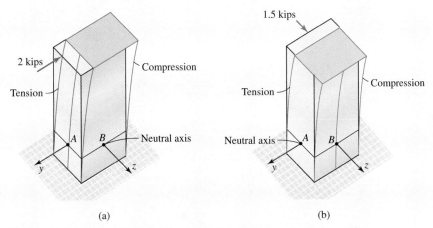

(a)   (b)

Figure 10.24 Determination of normal stress components by inspection. (*a*) Bending about z axis. (*b*) Bending about y axis.

normal stress due to bending about the z axis $\sigma_{bend\text{-}z}$ will be tensile. From 10.24b it can be seen that point A is on the neutral (bending) axis. Hence the normal stress due to bending about the y axis $\sigma_{bend\text{-}y} = 0$. Thus the total normal stress at point A is $(\sigma_{xx})_A = \sigma_{bend\text{-}z} - \sigma_{axial}$. Substituting the magnitude of $\sigma_{bend\text{-}z} = 22.638$ and $\sigma_{axial} = 6.154$, we obtain the result in Equation (E24).

From Figure 10.24b it can be seen that the line passing though B will be in compression, that is, the normal stress due to bending about the y axis $\sigma_{bend\text{-}y}$ will be compressive. From Figure 10.24a it can be seen that point B is on the neutral (bending) axis, hence $\sigma_{bens\text{-}z} = 0$. Thus the total normal stress at point B can be written as $(\sigma_{xx})_B = -\sigma_{axial} - \sigma_{bend\text{-}y}$. Substituting the magnitude of $\sigma_{bend\text{-}y} = 20.258$ and $\sigma_{axial} = 6.154$ we obtain the result in Equation (E25).

Shear stress calculations: The shear stresses at point A can be obtained from superposing the values in Equations and . The shear stress at point B can be obtained from superposing the values in Equations (E19) and (E23). The result for shear stress at points A and B is

$$(\tau_{xz})_A = 1.2 \text{ ksi} \qquad (\tau_{xy})_B = -1.2 \text{ ksi} \qquad \text{(E26)}$$

Intuitive check on shear stress calculations: Figure 10.25a shows two segments of the thin column after making an imaginary cut through the cross section containing points A and B. By inspection we deduce that the shear force on the bottom segment containing points A and B is in the negative y direction. As elaborated in Section 6.6.1, the shear stress distribution must be such that the resultant shear force is in the negative y direction; the resultant shear force in the z direction is zero; and it is symmetric about the y axis—we obtain the shear stress distribution

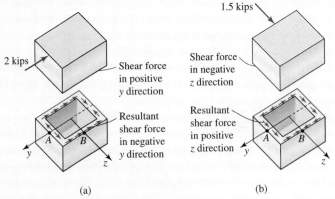

Figure 10.25 Direction of shear stress components by inspection. (*a*) Bending about z axis. (*b*) Bending about y axis.

shown in Figure 10.25b. Notice that at the point where the y axis intersects the centerline, the shear stress is zero because at this point it changes direction to ensure symmetry about the y axis. Repeating the foregoing arguments for the shear stress distribution due to bending about the y axis, we obtain the shear stress distribution in Figure 10.25b.

At point A Figure 10.25a shows that the shear stress due to bending about the z axis is zero. Figure 10.25b shows that the shear stress due to bending about the y axis is in the positive z direction. Thus the total shear stress at A is $(\tau_{xz})_A = \tau_{\text{bend-}y}$. At point B, Figure 10.25a shows that the shear stress due to bending about the z axis is in the negative y direction. Figure 10.25b shows that the shear stress due to bending about the z axis is zero. Thus the total shear stress at B is $(\tau_{xy})_B = -\tau_{\text{bend-}z}$. Substituting the magnitude of the shear stresses from Equations and $\tau_{\text{bend-}y} = 1.19$ and $\tau_{\text{bend-}z} = 1.23$ we obtain the result in Equation (E26).

Step 7: The stresses at points A and B can now be drawn on a stress cube, as shown in Figure 10.26.

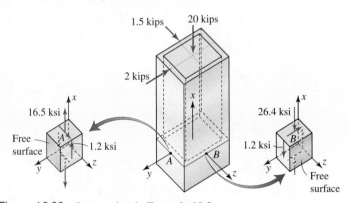

Figure 10.26   Stress cubes in Example 10.3.

Step 8: Figure 10.27 shows the plane containing the crack. From geometry we conclude that the angle that the outward normal makes with the x axis is 35°. Substituting $\theta = 35°$, $(\sigma_{xx})_B = -26.4$ ksi, $(\tau_{xy})_B = -1.2$ ksi, and $(\sigma_{yy})_B = 0$ into Equations (8.1) and (8.2), we obtain the normal and shear stresses on the plane containing the crack,

$$\sigma_{nn} = -26.4 \cos^2 35 + 2(-1.2) \sin 35 \cos 35$$

$$\text{ANS.} \qquad \sigma_{nn} = 18.84 \text{ ksi (C)}$$

$$\tau_{nt} = -(-26.4) \cos 35 \, \sin 35 + (-1.2)(\cos^2 35 - \sin^2 35)$$

$$\text{ANS.} \qquad \tau_{nt} = 11.99 \text{ ksi}$$

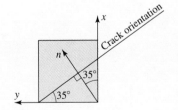

Figure 10.27   Angle of normal to plane containing crack.

COMMENTS

1. It may seem that the intuitive checks take as much effort as the calculation of the stresses by the procedural approach. But recognize that a great deal of the description and figures are given for the purposes of explanation. Most of the intuitive check is done by inspection. The real benefit of the intuitive check is the development of an intuitive sense about stresses under combined loading.

2. If drawing a three-dimensional free-body diagram with the internal forces and moments according to the sign convention proves difficult, then you may prefer drawing two perspectives of the free-body diagram, as shown in Figure 10.28. The free-body diagram in Figure 10.28a is constructed by looking down the y axis, whereas the free-body diagram in Figure 10.28b is the perspective looking down the z axis. Equations (E10) through (E14) can be obtained by writing equilibrium equations using the free-body diagrams shown in Figure 10.28.

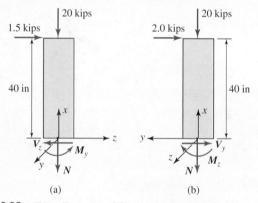

(a)   (b)

Figure 10.28   Two-dimensional free-body diagrams in Example 10.3.

3. In calculating bending stresses by inspection care must be taken to ensure that the correct area moment of inertia is used in the formula for rectangular cross sections as I_{yy} is not the same as I_{zz}. The subscripts in the formulas emphasize that the moment of inertia to be used is the value about the bending axis.

4. The stresses on the plane containing the crack are used for assessing whether or not a crack will grow and break the body. See Section 10.3 for more details.

EXAMPLE 10.4

A thin cylinder with an outer diameter of 100 mm and a thickness of 10 mm is loaded as shown in Figure 10.29. At point A, which is on the surface of the cylinder, determine the normal and shear stresses in the x, y, z coordinate system and show your results on a stress cube.

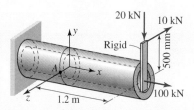

Figure 10.29 Geometry and loading in Example 10.4.

PLAN

We can follow the procedure outlined in Section 10.1.7. The 100 kN is an axial force. The 20-kN force will produce bending about the z axis. The 10-kN force will produce bending about the y axis and will also produce torque. Thus we need all the stress equations listed in Table 10.1. We can use these equations as a checklist of the quantities that must be calculated and determine the stress at point A by superposition.

Solution

Step 1: All the stress equations in Table 10.1 will be used in solving the problem.

Step 2: The geometric properties of the cross section can be found as

$$A = \pi(50^2 - 40^2) = 2.827 \times 10^3 \text{ mm}^2 \qquad (E1)$$

$$J = \frac{\pi}{2}(50^4 - 40^4) = 5.796 \times 10^6 \text{ mm}^4 \qquad (E2)$$

$$I_{yy} = I_{zz} = \frac{J}{2} = 2.898 \times 10^6 \text{ mm}^4 \qquad (E3)$$

Step 3: At point A we draw a line perpendicular to the centerline of the cross section. The line cuts the thin cylinder at two places, as shown in Figure 10.30. Thus the thickness of the material at A is twice the thickness of the cylinder, as given by

$$t_A = 20 \text{ mm} \qquad (E4)$$

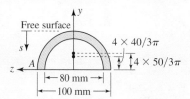

Figure 10.30 Calculation of Q_z in Example 10.4.

The area between the top surface and the line passing through A is used in the calculation of $(Q_y)_A$ and $(Q_z)_A$. By symmetry,

$$(Q_y)_A = 0 \qquad (E5)$$

To find $(Q_z)_A$, we use the formula $4r/3\pi$, given in Table A.2, for the location of the centroid for a half-disc of radius r. By subtracting the first moment of the area of the inner disc of radius 40 mm from the first moment of the outer disc of radius 50 mm, we obtain $(Q_z)_A$,

$$(Q_z)_A = \left(\frac{\pi \times 50^2}{2}\right)\left(\frac{4 \times 50}{3\pi}\right) - \left(\frac{\pi \times 40^2}{2}\right)\left(\frac{4 \times 40}{3\pi}\right)$$

$$= 40.667 \times 10^3 \text{ mm}^3 \qquad (E6)$$

Step 4: We can make an imaginary cut through the section at $x = 0$ and draw the free-body diagram shown in Figure 10.31. The internal forces and moments are drawn according to our sign convention for these quantities. Using equilibrium equations, the internal forces and moments can be found,

$$N = 100 \text{ kN} \tag{E7}$$

$$V_y = -20 \text{ kN} \tag{E8}$$

$$V_z = -10 \text{ kN} \tag{E9}$$

$$T = -5 \text{ kN·m} \tag{E10}$$

$$M_y = -12 \text{ kN·m} \tag{E11}$$

$$M_z = -24 \text{ kN·m} \tag{E12}$$

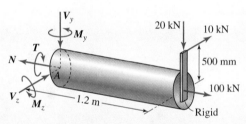

Figure 10.31 Free-body diagram in Example 10.4.

Step 5: The stress components due to each loading are calculated next.

Axial stress calculations: The axial stress at point A can be found by substituting Equations (E1) and (E7) into Equation (10.1),

$$(\sigma_{xx})_A = \frac{100 \times 10^3}{2.827 \times 10^{-3}} = 35.373 \times 10^6 \text{ N/m}^2 = 35.373 \text{ MPa} \tag{E13}$$

Torsional shear stress calculations: The torsional shear stress can be found by substituting $\rho_A = 50 \times 10^{-3}$ m and Equations (E2) and (E10) into Equation (10.2),

$$(\tau_{x\theta})_A = \frac{-5 \times 10^3 \times 50 \times 10^{-3}}{5.796 \times 10^{-6}} = -43.133 \times 10^6$$

Using the subscripts, the shear stress can be drawn on a stress cube, as shown in Figure 10.32. By examining the direction of shear stress in

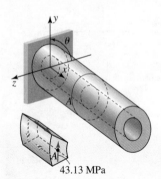

43.13 MPa

Figure 10.32 Direction of torsional shear stress in Example 10.4.

the x, y coordinate system, $(\tau_{xy})_A$ can be found,

$$(\tau_{xy})_A = 43.133 \text{ MPa} \qquad \text{(E14)}$$

Stresses due to bending about the y axis: Substituting $z_A = 50 \times 10^{-3}$ m and Equations (E3) and (E11) into Equation (10.4a), the bending normal stresses at point A can be found,

$$(\sigma_{xx})_A = -\frac{(-12)10^3 \times 50 \times 10^{-3}}{2.898 \times 10^{-6}} = 207.04 \times 10^6 \text{ N/m}^2$$

$$= 207.04 \text{ MPa} \qquad \text{(E15)}$$

As $(Q_y)_A$, the shear stress at A due to bending about the y axis, is zero, as per Equation (10.4b),

$$(\tau_{xs})_A = 0 \qquad \text{or} \qquad (\tau_{xy})_A = 0 \qquad \text{(E16)}$$

Stresses due to bending about the z axis: As $y_A = 0$, the normal stress at A due to bending about the z axis is zero,

$$(\sigma_{xx})_A = 0 \qquad \text{(E17)}$$

Substituting Equations (E3), (E4), (E6), and (E8) into Equation (10.3b), we obtain the shear stress,

$$(\tau_{xs})_A = -\frac{(-20)10^3 \times 40.667 \times 10^{-6}}{2.898 \times 10^{-6} \times 20 \times 10^{-3}} = 14.033 \times 10^6 \text{ N/m}^2$$

From Figure 10.30a we note that the s direction is in the negative y direction at point A. Thus,

$$(\tau_{xy})_A = -(\tau_{xs})_A = -14.033 \text{ MPa} \qquad \text{(E18)}$$

Step 6: Superposition

Normal stress calculations: The normal stress at point A can be obtained by superposing the values in Equations (E13), (E15), and (E17),

$$(\sigma_{xx})_A = 35.373 + 207.04 + 0 = 242.412 \qquad \text{(E19)}$$

$$\text{ANS.} \qquad (\sigma_{xx})_A = 242.4 \text{ MPa (T)}$$

Intuitive check on normal stress calculations: The axial stress σ_{axial} due to a 100-kN force will be tensile. Figure 10.33 shows the exaggerated deformed shapes due to bending about the y and z axes. These deformed shapes are drawn for the purpose of explanation, but can be visualized without drawing the figures. From Figure 10.33a, it can be seen that line AB will be in tension, that is, the normal stress due to bending about the y axis $\sigma_{\text{bend-}y}$ will be tensile at point A. From Figure 10.33b it can be seen that point A is on the neutral (bending)

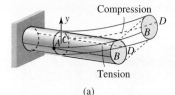

(a)

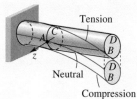

(b)

Figure 10.33 Determination of normal stress components by inspection. (*a*) Bending about y axis from 10-kN force. (*b*) Bending about z axis from 20-kN force.

axis. Hence the normal stress due to bending about the z axis $\sigma_{bend\text{-}z} = 0$. Thus the total normal stress at point A can be written as $(\sigma_{xx})_A = \sigma_{axial} + \sigma_{bend\text{-}y}$. Substituting the magnitude of $\sigma_{bend\text{-}y} = 207.04$ and $\sigma_{axial} = 35.373$, we obtain the result in Equation (E19).

Shear stress calculations: The shear stress at point A can be obtained by superposing the values in Equations (E14), (E16), and (E18),

$$(\tau_{xy})_A = 43.133 + 0 - 14.033 = 29.10 \qquad \text{(E20)}$$

$$\text{ANS.} \qquad (\tau_{xy})_A = 29.10 \text{ MPa}$$

Intuitive check on shear stress calculations: The direction of the torque acting on a small segment of the circular pipe shown in Figure 10.34a can be obtained by inspection of Figure 10.29. By visualizing the motion of the left part containing the stress cube we obtain the direction of shear stress, as was discussed in Section 5.2.5.

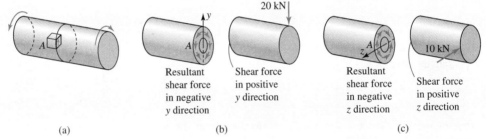

Resultant shear force in negative y direction   Shear force in positive y direction   Resultant shear force in negative z direction   Shear force in positive z direction

(a)   (b)   (c)

Figure 10.34 Direction of shear stress components by inspection. (*a*) Torsion from 10-kN force. (*b*) Shear stress from bending about z axis. (*c*) Shear stress from bending about y axis.

Figure 10.34b shows two segments of the pipe after making an imaginary cut through point A. By inspection we can deduce that the shear force on the left segment, which contains point A, is in the negative y direction. As elaborated in Section 6.6.1, the shear stress distribution must be such that the resultant shear force is in the negative y direction, the resultant shear force in the z direction is zero, and it is symmetric about the y axis—we obtain the shear stress distribution shown in Figure 10.34b. Repeating these arguments for the shear stress distribution due to bending about the y axis, we obtain the shear stress distribution in Figure 10.34c.

Figure 10.34 shows that at point A the torsional shear stress τ_{tor} is upward, the shear stress due to bending about the z axis $\tau_{bend\text{-}z}$ is downward, and the shear stress due to bending about y axis $\tau_{bend\text{-}y}$ is zero. Thus the total shear stress at A can be written as $(\tau_{xy})_A = \tau_{tor} - \tau_{bend\text{-}z}$. Substituting the magnitude of the shear stresses $\tau_{tor} = 43.133$ and $\tau_{bend\text{-}z} = 14.033$, we obtain the result in Equation (E20).

Step 7: Figure 10.35 shows the result of stresses on a stress cube.

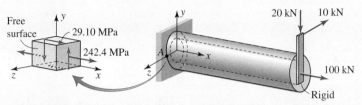

Figure 10.35 Results on stress cube in Example 10.4.

COMMENTS

1. The stresses shown on the stress cube in Figure 10.35 can be processed further if necessary. We could find principal stresses as in Example 10.1, or stresses on a plane as in Example 10.3, or strains along the direction of a gage as in Example 10.2.

2. The advantage of solving combined loading problems using the procedure outlined in Section 10.1.7 is that it is a methodical approach, which breaks the complexity of the problem into a sequence of simple steps, as shown in this and previous examples. The shortcoming of this procedural approach is that it does not exploit any simplification that may be intrinsic to the problem.

3. Solving a problem by inspection has two distinct advantages—it helps build an intuitive understanding of stress behavior and it can reduce the computational effort significantly by exploiting the intrinsic simplifications in the problem.

4. The possibility of making an error when solving the problem primarily by inspection is higher than with the procedural method. This is because a significant amount of thinking is done in the head rather than on paper. Internal forces and moments are equal and opposite on the two surfaces created by an imaginary cut and confusing one surface with another is possible if all visualization is done in the head, particularly for the calculation of shear stress. Therefore it is recommended that rough sketches be drawn for the calculation of shear stresses by inspection.

5. You may find it more effective for yourself to solve part of the problem in a procedural manner and part of the problem by inspection. For example, you could solve for normal stresses by inspection and for shear stresses in a procedural manner.

Consolidate your knowledge   Write a procedure you would use for solving combined loading problems.

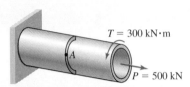

Figure 10.36 Cylinder and loading in Example 10.5*.

Figure 10.37 Free-body diagram in Example 10.5*.

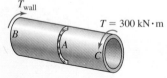

Figure 10.38 Direction of shear stress in Example 10.5*.

*EXAMPLE 10.5

The cylinder of 800-mm outer diameter shown in Figure 10.36 has a wall thickness of 15 mm. In addition to the axial and torsional loads the cylinder is pressurized to 150 kPa. Determine the normal and shear stresses at point A and show them on a stress element in a cylindrical coordinate system.

PLAN

The stress at any point is from three different sources. We use Equation (4.9) to find the axial stress, Equations (4.33) and (4.34) to find the axial and hoop stresses due to pressure on the thin cylinder, and Equation (5.26) to find the torsional shear stress in the thin tube.

Solution We make an imaginary cut, draw the free-body diagram in Figure 10.37, and determine the internal axial force and internal torque as

$$T_A = 300 \text{ kN·m} \tag{E1}$$

$$N_A = 500 \text{ kN} \tag{E2}$$

$$p = 150 \text{ kPa} \tag{E3}$$

Axial stress calculation: The outer radius $R_o = 400$ mm and the inner radius $R_i = 385$ m. Thus the cross-sectional area is $A = \pi(400^2 - 385^2) = 36.99 \times 10^3$ mm². From Equations (E2) and (4.9), we obtain

$$\sigma_{xx} = \frac{N_A}{A} = \frac{500 \times 10^3}{36.99 \times 10^{-3}} = 13.52 \times 10^6 \text{ N/m}^2 = 13.52 \text{ MPa} \tag{E4}$$

Stresses due to pressure of thin-walled cylinders: The mean radius is $R_m = (400 + 385)/2 = 392.5$ mm. From Equations (4.33) and (4.34), we obtain

$$\sigma_{xx} = \frac{pR_m}{2t} = \frac{150 \times 10^3 \times 392.5 \times 10^{-3}}{2 \times 15 \times 10^{-3}}$$

$$= 1.96 \times 10^6 \text{ N/m}^2 = 1.96 \text{ MPa} \tag{E5}$$

$$\sigma_{\theta\theta} = \frac{pR_m}{t} = 3.92 \text{ MPa} \tag{E6}$$

Torsion of thin-walled cylinders: The enclosed area is $A_E = \pi R_m^2 = 484.0 \times 10^3$ mm². From Equation (5.26) we can find the magnitude of the shear stress. The section BA will tend to rotate clockwise relative to section AC, as shown in Figure 10.38. Thus the shear stress will have to act in a counterclockwise direction, which at point A is downward,

$$\tau = \frac{T_A}{2tA_E} = \frac{300 \times 10^3}{2 \times 15 \times 10^{-3} \times 484.0 \times 10^{-3}} = 20.67 \text{ MPa} \quad \text{(E7)}$$

The total stress in the x direction is the sum of the axial stress and the stress due to the pressure of the thin-walled cylinder. Adding Equations (E4) and (E5) we obtain the total axial stress in the x direction,

$$\sigma_{xx} = 13.52 + 1.96 = 15.48 \text{ MPa} \quad \text{(E8)}$$

Figure 10.39 shows the stresses in Equations (E6), (E7), and (E8) on a stress element.

COMMENT

From the stress state in Figure 10.39, we could find principal stresses, or strains in any direction.

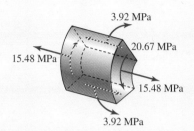

Figure 10.39 Stress element in Example 10.5*.

---

*10.2 ANALYSIS AND DESIGN OF STRUCTURES

Structures are composed of many members, which are joined together. Analyzing (or designing) these structures requires that we create a mathematical model that is an approximation of the actual structure. Many decisions go into the creation of a mathematical model, including those made with respect to the modeling of joints and supports. A conservative but immensely simplifying assumption is the approximation of joints by pins. Pin joints do not transmit moments. In other words, we are neglecting the joints' intrinsic moment-carrying ability. Thus the internal forces and moments our model will predict will be higher than those that will actually be present, making the pin joint approximation a conservative assumption.

Analyzing complex mathematical models requires numerical solutions. Here we shall consider simple structures made up of few members. Some members of the structure may be subjected to one type of combined loading whereas some other members may be subjected to another combination of loading. There are two major steps in the solution of problems related to the analysis and design of structures:

1. Analysis of forces and moments that act on individual members.

2. Computation of stresses on members under combined loading.

For statically determinate structures, the internal forces and moments can be found using the principles of statics, as shown in Example 10.6. For statically indeterminate structures, we will also need the deformation equations developed in this course to complement the analysis skills learned in statics. This is elaborated in Example 10.7.

10.2.1 Failure Envelope

An important design concept called *failure envelope* is elaborated in this section.

Consider a circular shaft that is subjected to axial loads and torsion, as shown in Figure 10.40a. Suppose the design limitation is that the maximum shear stress should not exceed 15 ksi. Further suppose that calculations show that the maximum shear stress at point A on the surface of the shaft is given by $\tau_{max} = 0.3183\sqrt{P^2 + 4T^2}$ ksi. Now τ_{max} should be less than or equal to 15 ksi, which gives us the following result: $P^2 + 4T^2 \leq 2220$. We can now make a plot of T versus P, as shown in Figure 10.40b.

The shaded area consists of all possible values of T and P for which the maximum shear stress will be less than 15 ksi and hence represents our acceptable design space. The region beyond the shaded area represents values of T and P for which the shear stress is greater than 15 ksi and hence represents the failure space. On the curve $P^2 + 4T^2 = 2220$ all values of P and T would result in a maximum shear stress of 15 ksi, and we are at incipient failure. This curve represents the failure envelope, which separates the design space from the failure space.

The preceding discussion can now be generalized to a design problem containing many, say n, variables, which could be geometric variables, material constants, or loads, as in Figure 10.40. We may need to find the values of the variables to meet several design constraints. If one took each design variable and plotted it on an axis, then one would obtain a n-dimensional space containing all possible combinations of the n variables. Some of these combinations of the variables would result in failure. A failure envelope separates the space of acceptable values of these variables from the unacceptable values. On the failure envelope the values of the design variables correspond to impending failure. The sum total of all the design constraints defines the failure envelope. We shall also use the concept of failure envelope in Section 10.4 to describe failure theories in which the variables are principal stresses that are plotted on an axis.

Definition 3 A failure envelope separates the acceptable design space from the unacceptable values of the variables affecting design.

Within the failure envelope we can compare different designs with respect to other criteria, such as cost, weight, and aesthetics. The search for a set of design variables to maximize or minimize an objective is called *design optimization*. In structures the objective of design optimization is usually to minimize the weight of a structure. Example 10.8 elaborates the construction and use of the failure envelope.

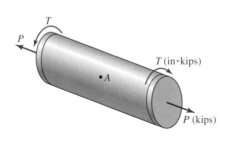

(a)

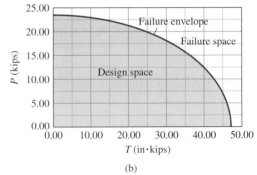

(b)

Figure 10.40 Failure envelope.

EXAMPLE 10.6

A hoist is to be designed for lifting a maximum weight $W = 300$ lb.
Space considerations have established the length dimensions shown
in Figure 10.41. The hoist will be constructed using lumber and
assembled using steel bolts. The dimensions of the lumber cross sections are listed in Table 10.2. The bolted joints will be modeled as
pins in single shear. Same-size bolts will be used in all joints. The
allowable normal stress in the wood is 1.2 ksi and the allowable shear
stress in the bolts is 6 ksi. Design the lightest hoist by choosing the
lumber from Table 10.2 and the bolt size to the nearest $\frac{1}{8}$-in diameter.

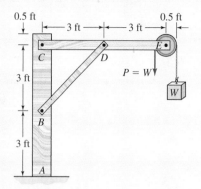

Figure 10.41 Hoist in Example
10.6.

TABLE 10.2   Dimensions of available lumber

Cross-Section Dimensions	Cross-Section Dimensions
2 in × 4 in	4 in × 8 in
2 in × 6 in	6 in × 6 in
2 in × 8 in	6 in × 8 in
4 in × 4 in	8 in × 8 in
4 in × 6 in	

PLAN

We analyze the problem in two steps. First we find the forces and
moments on individual members and then we find the stresses.

1. *BD* is a two-force axial member that will be in compression. Members *ABC* and *CDE* are multiforce members subjected to axial and
bending loads. Free-body diagrams of members *ABC* and *CDE* will
permit calculation of the forces at pin *C*, the axial force in *BD*—
forces on pins *B* and *D* are thus known, as well as the reaction
forces and the reaction moment at *A*.

2. We compute the maximum stresses from the forces calculated in
step 1 and, using the limiting values on the maximum stresses, compute the dimensions of the pin and the wooden members. From the
possible set of dimensions that satisfy the limiting criteria we
choose those that will result in the lightest structure.

Solution

Calculation of forces and moments on structural members: Figure
10.42 shows the free-body diagrams of members *CDE* and *ABC*. Using
the moment equilibrium at point *C* in Figure 10.42a, N_{BD} is found,

$$(N_{BD} \sin 45)3 - 300 \times 5.5 - 300 \times 6.5 = 0$$

or

$$N_{BD} = 1697 \text{ lb} \qquad (E1)$$

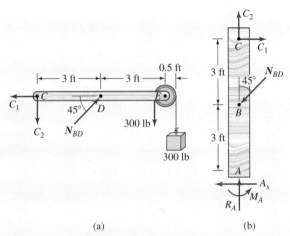

Figure 10.42 Free-body diagrams in Example 10.6.

In Figure 10.42a, using force equilibrium in the x and y directions, the forces at C can be found,

$$C_1 - N_{BD} \cos 45 = 0 \qquad \text{or} \qquad C_1 = 1200 \text{ lb} \quad (E2)$$

$$N_{BD} \sin 45 - C_2 - 600 = 0 \qquad \text{or} \qquad C_2 = 600 \text{ lb} \quad (E3)$$

In Figure 10.42b, using force equilibrium in the x and y direction, the reaction forces at A can be found,

$$C_1 - N_{BD} \sin 45 - A_x = 0 \qquad \text{or} \qquad A_x = 0$$
$$(E4)$$
$$R_A + C_2 - N_{BD} \cos 45 = 0 \qquad \text{or} \qquad R_A = 600 \text{ lb}$$

Moment equilibrium at point A yields the moment reaction,

$$M_A - C_1 \times 6 + (N_{BD} \cos 45)3 = 0 \qquad \text{or} \qquad M_A = 3600 \text{ ft·lb}$$
$$(E5)$$

Bolt size calculations: The shear force acting on each bolt can be found from the forces calculated,

$$V_B = V_D = N_{BD} = 1697 \text{ lb}$$

$$V_C = \sqrt{C_1^2 + C_2^2} = 1342 \text{ lb} \qquad V_E = 600 \text{ lb}$$

The maximum shear stress will be in bolts B and D. This maximum shear stress should be less than 6 ksi. The cross-sectional area can be found and the diameter of the bolt calculated,

$$\tau_{max} = \frac{1697}{\pi d^2/4} \leq 6000 \quad \text{or} \quad d \geq \sqrt{\frac{4 \times 1697}{6000\pi}} \geq 0.60 \text{ in} \quad \text{(E6)}$$

The nearest $\frac{1}{8}$-in size that is greater than the numerical value in Equation (E6) is $d = 0.625$ in.

ANS. $\frac{5}{8}$-in-size bolts should be used.

Lumber selection: The normal axial stress in member *BD* has to be less than 1200 psi. The cross-sectional area for member *BD* can be found as

$$\sigma_{BD} = \frac{N_{BD}}{A_{BD}} = \frac{1697}{A_{BD}} \leq 1200 \quad \text{or} \quad A_{BD} \geq 1.414 \text{ in}^2 \quad \text{(E7)}$$

The 2-in × 4-in lumber has a cross-sectional area of 8 in², which is the smallest cross section that meets the restriction of Equation (E7).

ANS. For member *BD* use lumber with the cross-section dimensions of 2 in × 4 in.

Shear force and bending moment diagrams for members *ABC* and *CDE* can be drawn after resolving the force N_{BD} into components parallel and perpendicular to the axis, as shown in Figure 10.43. A local *x, y, z* coordinate system for each member is established to facilitate drawing the shear and moment diagrams.

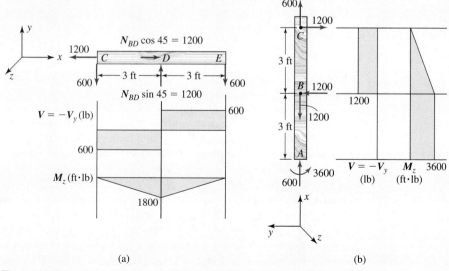

(a) (b)

Figure 10.43 Shear and moment diagrams in Example 10.6. (*a*) Member *CDE*. (*b*) Member *ABC*.

From Figure 10.43*a* it can be seen that the maximum axial force is 1200 lb tensile in segment *CD*. The bending moment is maximum on the cross section at *D* in member *CDE*. Due to bending, the top surface will be in tension and the bottom will be in compression. Thus the maximum normal stress in *CDE* will be at the top surface just before *D* and will be the sum of tensile stresses due to axial and bending loads. Using an axial force of 1200 lb and a bending moment of 1800 ft·lb = 21,600 in·lb, the maximum normal stress in *CDE* can be written as

$$\sigma_{CD} = \frac{1200}{A_{CDE}} + \frac{21,600}{S_{CDE}} \tag{E8}$$

where A_{CDE} and S_{CDE} are the cross-sectional area and the section modulus (with respect to the *z* axis) of member *CDE*.

From Figure 10.43*b* it can be seen that the axial force in *AB* is 600 lb compressive and the axial force in *BC* is 600 lb tensile. The bending moment is a maximum of 1800 ft·lb = 43,200 in·lb throughout segment *AB*. Due to bending, the right side of member *ABC* will be in compression and the left side will be in tension. Thus the maximum normal stress in member *ABC* will be on the right surface, just before *B* in segment *AB*, and will be the sum of compressive stresses due to axial and bending loads. Using an axial force of 600 lb and a bending moment of 3600 ft·lb = 43,200 in·lb, the maximum compressive normal stress in *ABC* can be written as

$$\sigma_{AB} = \frac{600}{A_{ABC}} + \frac{43,200}{S_{ABC}} \tag{E9}$$

where A_{ABC} and S_{ABC} are the cross-sectional area and the section modulus (with respect to the *z* axis) of member *ABC*.

For the list of available lumber given in Table 10.2, the cross-sectional area *A* and the section modulus *S* can be determined assuming that the smaller dimension *a* is parallel to the *z* axis (bending axis or dimension out of the plane of the paper) and the larger dimension *b* is in the plane of the paper. With this stipulation $I_{zz} = ab^3/12$ and $y_{max} = b/2$. Hence $S = I_{zz}/y_{max} = ab^2/6$ (see local coordinates in Figure 10.43. Substituting the values of *A* and *S* into Equations (E8) and (E9), the stress values σ_{CD} and σ_{AB} can be found. Using a spreadsheet, Table 10.3 can be created. Cross-section dimensions for which the normal stress is less than 1200 psi meet the strength limitation and are identified in bold in Table 10.3. But for the design of the lightest hoist we choose the cross section with the smallest area among the

TABLE 10.3   Cross-section properties and stresses in Example 10.6

a (in)	b (in)	$A = ab$ (in^2)	$S = ab^2/6$ (in^3)	σ_{CD} (psi)	σ_{AB} (psi)
2	4	8	5.3	4200.0	8175.0
2	6	12	12.0	1900.0	3650.0
2	8	16	21.3	**1087.5**	2062.5
4	4	16	10.7	2100.0	4087.5
4	6	24	24.0	**950.0**	1825.0
4	8	32	42.7	**543.8**	**1031.3**
6	6	36	36.0	**633.3**	1216.7
6	8	48	64.0	**362.5**	**687.5**
8	8	64	85.3	**271.9**	**515.6**

bold values. Thus the result are:

ANS.   For member *ABC*, use lumber with the cross-section dimensions of 4 in × 8 in.

ANS.   For member *CDE*, use lumber with the cross-section dimensions of 2 in × 8 in.

COMMENTS

1. Members in axial compression such as *BD* must be designed for strength as well as checked for buckling failure, as will be elaborated in the next chapter.

2. In actual design it may be preferable to use two pieces of 1-in × 8-in lumber for member *CDE* so that the pulley is in the middle of the two members. This will change pins at *C*, *D*, and *E* from single shear into double shear, thus also reducing the shear stresses in the pins.

3. The equality constraint in Equation (E8) defines the curve of the failure envelope for member *CDE*. Taking the equality sign, substituting for the area and the section modulus in terms of *a* and *b*, and solving for *a* in terms of *b*, we obtain the equation of the curve defining the failure envelope as $a = 0.5/b + 216/b^2$. Figure 10.44 shows the failure envelope. As can be seen, the three possible solutions in Table 10.3 for member *CDE* fall in the design space and the remaining cross sections fall in the failure space. If we were to choose any value for *a* and *b*, then the failure envelope would identify all possible solutions.

4. The bending shear stress was not considered in selecting the lumber cross sections for members *ABC* and *CDE*. This is based on the consideration that the maximum bending normal stress is significantly (~10 times) greater than the maximum bending shear stress.

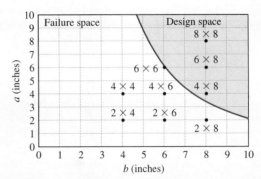

Figure 10.44 Failure envelope for member *CDE*.

Thus the principal stress at the top or bottom of the member, where the bending normal stress is maximum, will be greater than the principal shear stress at the neutral axis, where bending shear stress is maximum. We check these statements for the selected sizes of members *ABC* and *CDE* as follows.

For rectangular cross sections it can be shown (see comment 1 in Example 6.14) that the maximum shear stress in bending at a cross section is $\tau_{\max} = 1.5V/A$, where A is the cross-sectional area. From Figure 10.43 the maximum shear force is 600 lb and 1200 lb in members *CDE* and *ABC*, respectively. Substituting these shear force values and the values of 16 in^2 and 32 in^2 for the areas, we obtain the maximum shear stresses of $\tau_{CD} = 56.25$ psi and $\tau_{AB} = 56.25$ psi in members *CDE* and *ABC*, respectively. Comparing these maximum shear stress values to the maximum normal stress values of 1087.5 psi and 1031.3 psi for members *CDE* and *ABC*, which are given in Table 10.3, we conclude that the shear stresses can be ignored in the selection of lumber.

EXAMPLE 10.7

A rectangular wooden beam of 60 mm × 180 mm cross section is supported at the right end by an aluminum circular rod of 8-mm diameter, as shown in Figure 10.45. The allowable normal stress in the wood is 14 MPa and the allowable shear stress in aluminum is 60 MPa. The moduli of elasticity for wood and aluminum are $E_w = 12.6$ GPa and

$E_{al} = 70$ GPa. Determine the maximum intensity w of the distributed
load that the structure can support.

PLAN

We analyze the problem in two steps. First we find the forces and
moments on individual members and then we find the stresses.

1. To solve this statically indeterminate problem, the deflection of the
   beam at A can be equated to the axial deformation of the aluminum
   rod. This permits the calculation of the internal axial force in the
   aluminum rod in terms of w.

2. The axial stress in the aluminum rod in terms of w can be found
   from the internal axial force calculated in step 1. Using Mohr's circle, the maximum shear stress in the axial rod can be found in terms
   of w, and one limit on w can be obtained. The maximum bending
   moment at B can be found in terms of w and the maximum bending
   normal stress calculated in terms of w. Using the allowable value of
   14 MPa, another limit on w can be found and a decision made on
   the maximum value of w.

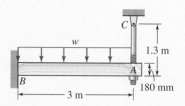

Figure 10.45 Beam in Example
10.7.

Solution

Calculation of forces on structural members: The area moment of
inertia of the wood and the cross-sectional area of the aluminum rod
can be calculated,

$$I_w = \frac{1}{12}60 \times 180^3 = 29.16 \times 10^6 \text{ mm}^4 = 29.16 \times 10^{-6} \text{ m}^4 \quad (E1)$$

$$A_{al} = \frac{\pi}{4}8^2 = 50.265 \text{ mm}^2 = 50.265 \times 10^{-6} \text{ m}^2 \quad (E2)$$

Making an imaginary cut through the aluminum rod, we obtain
the beam and loading shown in Figure 10.46a. The total loading on the
beam can be considered as the sum of the two loadings shown in Figure 10.46b and c.

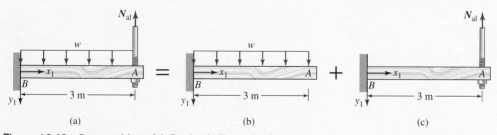

Figure 10.46 Superposition of deflection in Example 10.7.

Comparing the two beam loadings in Figure 10.46*b* and *c* to that shown for cases 1 and 3 in Table 7.1, we obtain $P = -N_{al}$ and $p = w$, $a = 3$, $b = 0$, $E = 12.6 \times 10^9$ N/m^2, and $I = I_w = 29.16 \times 10^{-6}$ m^4. Noting that v_{max} shown in Table 7.1 for the cantilever beam occurs at point A, we can substitute the load values and superpose to obtain deflection v_A,

$$v_A = \frac{w \times 3^4}{8 \times 12.6 \times 10^9 \times 29.16 \times 10^{-6}} + \frac{-N_{al} \times 3^3}{3 \times 12.6 \times 10^9 \times 29.16 \times 10^{-6}}$$

$$= 27.56 \times 10^{-6} w - 24.50 \times 10^{-6} N_{al} \qquad (E3)$$

The extension of the aluminum rod can be found using Equation (4.27),

$$\delta_{al} = \frac{N_{al}L_{al}}{E_{al}A_{al}} = \frac{N_{al} \times 1.3}{70 \times 10^9 \times 50.265 \times 10^{-6}} = 0.369 \times 10^{-6} N_{al} \qquad (E4)$$

The extension of the aluminum rod should equal the deflection of the beam at A. Equating Equations (E3) and (E4), the internal force N_{al} can be found in terms of w,

$$27.56 \times 10^{-6} w - 24.50 \times 10^{-6} N_{al} = 0.369 \times 10^{-6} N_{al}$$

or

$$N_{al} = 1.11 w \qquad (E5)$$

Figure 10.47 shows the free-body diagram of the beam with the distributed force replaced by an equivalent force. By force and moment equilibrium the reaction force and the reaction moment at the wall can be determined,

$$R_B - 3w + N_{al} = 0 \quad \text{or} \quad R_B = 1.89 w \qquad (E6)$$

$$M_B - 3w \times 1.5 + N_{al} \times 3 = 0 \quad \text{or} \quad M_B = 1.17 w \qquad (E7)$$

Figure 10.48 shows the shear and moment diagrams for the beam. From the moment diagram we see that the maximum moment is at the wall, and its value is

$$M_{max} = -1.17 w \qquad (E8)$$

Stress in aluminum: The axial stress in aluminum in terms of w can be found from Equations (E5) and (E1),

$$\sigma_{al} = \frac{N_{al}}{A_{al}} = \frac{1.11 w}{50.265 \times 10^{-6}} = 22.04 \times 10^3 w \text{ N/m}^2 \qquad (E9)$$

By constructing Mohr's circle in Figure 10.49, the maximum shear stress in aluminum can be found,

$$\tau_{max} = \frac{\sigma_{al}}{2} = 11.02 \times 10^3 w \text{ N/m}^2 \qquad (E10)$$

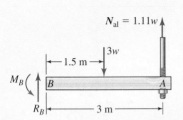

Figure 10.47 Free-body diagram in Example 10.7.

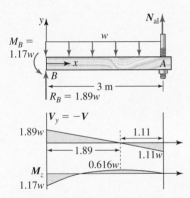

Figure 10.48 Shear force and bending moment diagrams in Example 10.7.

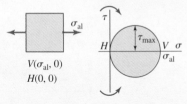

Figure 10.49 Mohr's circle in Example 10.7.

The maximum shear stress in Equation (E10) should be less than 60 MPa, yielding a limit on w,

$$\tau_{max} = 11.02 \times 10^3 w \le 60 \times 10^6 \quad \text{or} \quad w \le 5.44 \times 10^3 \text{ N/m}$$
(E11)

Stress in wood: The bending normal stress will be maximum at the top and bottom surfaces at the wall. Substituting $y_{max} = 0.09$ m and Equations (E8) and (E2) into Equation (10.3a), the magnitude of the maximum bending normal stress can be found and should be less than 14 MPa, yielding another limit on w,

$$\sigma_w = \left| \frac{M_{max} y_{max}}{I_w} \right| = \frac{1.17 w \times 0.09}{29.16 \times 10^{-6}}$$

$$= 3.61 \times 10^3 w < 14 \times 10^6 \quad \text{(E12)}$$

or

$$w \le 3.88 \times 10^3 \text{ N/m}$$

The value in Equation (E12) also satisfies the inequality in Equation (E10). Thus the maximum intensity of the distributed load is

ANS. $\quad w_{max} = 3.88$ kN/m

COMMENT

At joint A we ensured continuity of displacement by enforcing the condition that the deformation of the axial member be the same as the deflection of the beam. We also enforced equilibrium of forces by using the same force N_{al} in the axial member and acting on the beam. These two conditions, continuity of displacement and equilibrium of forces, must be satisfied by all joints in more complex structures.

EXAMPLE 10.8

A circular member was repaired by welding a crack at point A that was $30°$ to the axis of the shaft, as shown in Figure 10.50. The allowable shear stress at point A is 24 ksi and the maximum normal stress the weld material can support is 9 ksi (T). Calculations show that the stresses at point A are $\sigma_{xx} = 9.55 P_2$ ksi (T) and $\tau_{xy} = -6.79 P_1$ ksi.

(a) Draw the failure envelope for the applied loads P_1 and P_2.

(b) Determine the optimum values of loads P_1 and P_2.

(c) If $P_1 = 2$ kips and $P_2 = 1.5$ kips, determine the factor of safety.

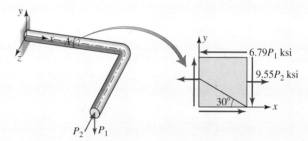

Figure 10.50 Problem geometry in Example 10.8.

PLAN

(a) By substituting the given stresses into Equation (8.13) we obtain the maximum shear stress in the material in terms of P_1 and P_2. Noting that the maximum shear stress is limited to 24 ksi, we obtain one equation relating P_1 and P_2. By substituting the given stresses into Equation (8.1) as well as the angle of the normal to the weld, we can obtain the normal stress on the weld, which gives us another equation relating P_1 and P_2. We can sketch both equations and obtain the failure envelope.

(b) We can find the values of P_1 and P_2 that satisfy the two equations in part (a) and obtain the optimum value of the loads.

(c) We can calculate the maximum in-plane shear stress in the material and the normal stress in the weld from the equations obtained in part (a) and compute two factors of safety. The lower value is the factor of safety for the system.

Solution

(a) Substituting the values of the given stresses into Equation (8.13), the maximum shear stress in the material can be obtained. Noting that it should be less than 24 ksi, we obtain one equation on P_1 and P_2,

$$\tau_{max} = \sqrt{\left(\frac{9.55P_2}{2}\right)^2 + (-6.79P_1)^2} = \sqrt{46.1P_1^2 + 22.8P_2^2} \quad \text{(E1)}$$

and

$$\sqrt{46.1P_1^2 + 22.8P_2^2} \le 24 \qquad \text{or} \qquad 46.1P_1^2 + 22.8P_2^2 \le 576 \quad \text{(E2)}$$

The normal to the weld makes an angle of 60° to the x axis. Substituting the given stresses and $\theta = 60°$ into Equation (8.1), the normal

stress on the weld must be less than 9 ksi (T), we obtain another
equation on P_1 and P_2,

$$\sigma_{weld} = 9.55 P_2 \cos^2 60 + 2(-6.79 P_1) \cos 60 \, \sin 60$$
$$= 2.387 P_2 - 5.881 P_1 \tag{E3}$$

and

$$2.387 P_2 - 5.881 P_1 \leq 9 \tag{E4}$$

The maximum value of P_1 that will satisfy Equation (E2) corresponds to $P_2 = 0$. This maximum value of P_1 is 3.534 kips. We
consider values of P_1 between zero and 3.534 in steps of 0.3 and
solve for P_2 from Equation (E2). For the same values of P_1 we can
also find values of P_2 from Equation (E4), as shown in Table 10.4,
which was produced on a spreadsheet. We can plot the values in
Table 10.4, as shown in Figure 10.51. The design space is the
shaded region and the failure envelope is the boundary $ABCD$.

TABLE 10.4   Values of loads in Example 10.8

P_1 (kips)	P_2 from Eq. (E2) (kips)	P_2 from Eq. (E4) (kips)
0.000	5.027	3.770
0.300	5.008	4.509
0.600	4.954	5.248
0.900	4.861	5.987
1.200	4.728	6.726
1.500	4.551	7.465
1.800	4.326	8.204
2.100	4.043	8.943
2.400	3.690	9.682
2.700	3.244	10.421
3.000	2.657	11.160
3.300	1.800	11.899

(b) The optimum values of P_1 and P_2 correspond to the maximum values of the loads that satisfy Equations (E2) and (E4). Using the
equality sign in Equation (E4), we can solve for P_2,

$$P_2 = 3.770 + 2.4634 P_1 \tag{E5}$$

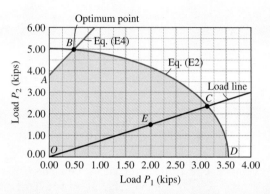

Figure 10.51   Failure envelope in Example 10.8.

We substitute Equation (E5) into Equation (E2) with the equality sign to obtain a quadratic equation in P_1,

$$46.11P_1^2 + 22.80(4.19 + 2.46P_1)^2 = 576$$

or   (E6)

$$184.45P_1^2 + 423.2P_1 - 252 = 0$$

Solving Equation (E6) we obtain two roots for P_1, 0.4905 and −2.784. Only the positive root is admissible. Substituting $P_1 = 0.4905$ into Equation (E5), we obtain $P_2 = 4.978$. Thus the optimum values of the loads are

ANS. $P_1 = 0.49$ kips $P_2 = 4.98$ kips

(c) Substituting $P_1 = 2$ kips and $P_2 = 1.5$ into Equations (E1) and (E3), we obtain

$$\tau_{max} = \sqrt{46.1 \times 2^2 + 22.8 \times 1.5^2} = 15.35 \text{ ksi}$$

$$\sigma_{weld} = 2.387 \times 1.5 - 5.881 \times 2 = -8.18 \text{ ksi}$$

For the given loads the normal stress in the weld is compressive. Hence it won't fail due to the specified failure in tension. The factor of safety is thus calculated from the maximum shear stress and can be found from Equation (3.23) as

$$k = \frac{24 \text{ ksi}}{15.35 \text{ ksi}} = 1.56$$

ANS. $k = 1.56$

COMMENTS

1. In this example we generated the failure envelope using analytical equations. For more complex structures, the failure envelope can be created using numerical methods, such as the finite-element method described in Section 4.9.

2. The optimum values of loads P_1 and P_2 in part (b) correspond to point B, the intersection of the two curves, as shown in Figure 10.51. If there are more than two constraints on a design, the optimum value usually lies at the intersection of different constraints. Each intersection point represents a set of acceptable values of design variables. Which set of values of design variables would represent an optimum solution? A typical answer in structural optimization would be: the set of values of design variables that minimizes the weight[3] of the structure.

3. In Figure 10.51 line AB, representing Equation (E2), would go downward if the direction of load P_1 were reversed [substitute $-P_1$ in place of P_1 in Equation (E2)]. If line AB went downward, it would cut the design space considerably. Thus not only is the magnitude of the loads important in design, but the direction of the load can be as critical. Failure envelopes can reveal such characteristics in a very visual manner.

4. The line joining the origin to point E is called *load line,* on which the loads vary proportionally. It's significance is that it can help give a graphical interpretation of the factor of safety. Along a load line, the distance of a point from the failure envelope is the margin of safety. In Figure 10.51 the factor of safety is the ratio of length OC to length OE. It can be verified that the coordinates of point C are $P_1 = 3.1263$ kips and $P_2 = 2.344$. Thus the length $OC = 3.9074$, whereas the length OE is 2.5. The factor of safety therefore is $k = 3.9074/2.5 = 1.56$, as before.

---

[3] In optimization methods the function that is being minimized is called *objective function.* Computer programs in design optimization search for the intersection points of the constraints in a methodical manner to minimize a given objective function.

PROBLEM SET 10.1

Combined axial and torsion forces

10.1 Determine the normal and shear stresses in the seam of the shaft passing through point A, as shown in Figure P10.1 The seam is at an angle of 60° to the axis of a solid shaft of 2-in diameter.

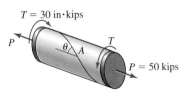

Figure P10.1

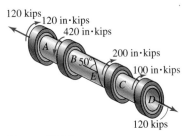

Figure P10.2

Figure P10.3

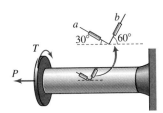

Figure P10.6

10.2 A 4-in-diameter solid circular steel shaft is loaded as shown in Figure P10.2. Determine the shear stress and the normal stress on a plane passing through point E. Point E is on the surface of the shaft.

In Problems 10.3 through 10.5, a solid shaft of 75-mm diameter is loaded as shown in Figure P10.3. The strain gage is 20° to the axis of the shaft and the shaft material has a modulus of elasticity $E = 250$ GPa and a Poisson ratio $v = 0.3$.

10.3 If $T = 20$ kN·m and $P = 50$ kN, what strain will the strain gage show?

10.4 If the strain gage shows a reading of -450 μm/m and $T = 10$ kN, determine the axial load P.

10.5 If the strain gage shows a reading of -300 μm/m and $P = 55$ kN, determine the applied torque T.

In Problems 10.6 and 10.7, a solid shaft of 2-in diameter is loaded as shown in Figure P10.6 The shaft material has a modulus of elasticity $E = 30,000$ ksi and a Poisson ratio $v = 0.3$.

10.6 Determine the strains the gages would show if $P = 70$ kips and $T = 50$ in·kips.

10.7 The strain gages mounted on the surface of the shaft recorded the strain values $\varepsilon_a = 2078$ μ and $\varepsilon_b = -1410$ μ. Determine the axial force P and the torque T.

10.8 Two solid circular steel shafts ($E_s = 200$ GPa, $G_s = 80$ GPa) and a solid circular bronze shaft ($E_{br} = 100$ GPa, $G_{br} = 40$ GPa) are securely connected and loaded as shown Figure P10.8 Determine the maximum normal and shear stresses in the shaft.

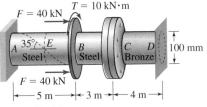

Figure P10.8

10.9 Determine the normal and shear stresses on a plane 35° to the axis of the shaft at point E in Figure P10.8. Point E is on the surface of the shaft.

Combined axial and bending forces

10.10 A 6-in × 4-in rectangular hollow member is constructed from a $\frac{1}{2}$-in-thick sheet metal and loaded as shown in Figure P10.10. Determine the normal and shear stresses at points A and B and show them on the stress cubes for $P_1 = 72$ kips, $P_2 = 0$, and $P_3 = 6$ kips.

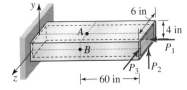

Figure P10.10

10.11 Determine the principal stresses and the maximum shear stress at points A and B in Figure P10.10 for $P_1 = 72$ kips, $P_2 = 3$ kips, and $P_3 = 0$.

10.12 Determine the strain shown by the strain gages in Figure P10.12 if $P_1 = 3$ kN, $P_2 = 40$ kN, the modulus of elasticity is 200 GPa, and Poisson's ratio is 0.3. The strain gages are parallel to the axis of the beam.

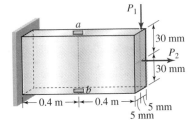

Figure P10.12

10.13 The strain gages shown in Figure P10.12 recorded the strain values $\varepsilon_a = 1000\ \mu$ and $\varepsilon_b = -750\ \mu$. Determine loads P_1 and P_2. The modulus of elasticity is 200 GPa and Poisson's ratio is 0.3.

10.14 Determine the strain shown by the strain gages in Figure P10.14 if $P_1 = 3$ kN, $P_2 = 40$ kN, the modulus of elasticity is 200 GPa, and Poisson's ratio is 0.3.

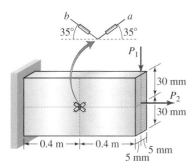

Figure P10.14

10.15 The strain gages shown in Figure P10.14 recorded the strain values $\varepsilon_a = 133\ \mu$ and $\varepsilon_b = 159\ \mu$. Determine loads P_1 and P_2. The modulus of elasticity is 200 GPa and Poisson's ratio is 0.3.

10.16 Determine the strain recorded by the gages at points A and B in Figure P10.16. Both gages are at 30° to the axis of the beam. The modulus of elasticity $E = 30,000$ ksi and $v = 0.3$.

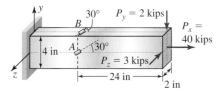

Figure P10.16

Combined axial, torsion, and bending forces

10.17 A thin cylinder with an outer diameter of 100 mm and a thickness of 10 mm is loaded as shown in Figure P10.17. Points A and B are on the surface of the shaft. Deter-mine the normal and shear stresses at points A and B in the x, y, z coordinate system and show your results on stress cubes.

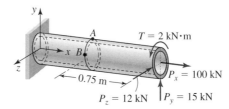

Figure P10.17

10.18 Determine the principal stresses and the maximum shear stress at point B on the shaft shown in Figure P10.17.

In Problems 10.19 through 10.30, a load is applied to bent pipes, as shown for each problem. By inspection determine and show the total stresses at points A and B on stress cubes using the following notation for the magnitude of the stress components:

- σ_{axial}—*axial normal stress*
- σ_{bend-y}—*normal stress due to bending about y axis*
- σ_{bend-z}—*normal stress due to bending about z axis*
- τ_{tor}—*torsional shear stress*
- τ_{bend-y}—*shear stress due to bending about y axis*
- τ_{bend-z}—*shear stress due to bending about z axis*

10.19 Using Figure P10.19 solve the problems as described in the text above.

Figure P10.19

10.20 Using Figure P10.20 solve the problems as described in the text above.

10.21 Using Figure P10.21 solve the problems as described in the text above.

Figure P10.20

Figure P10.21

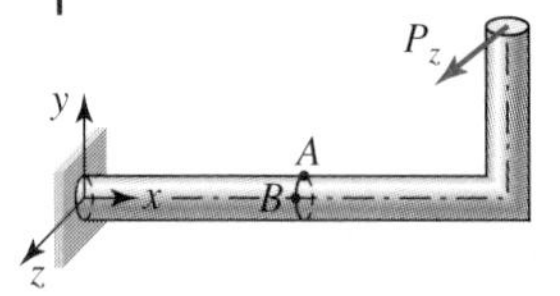

Figure P10.22

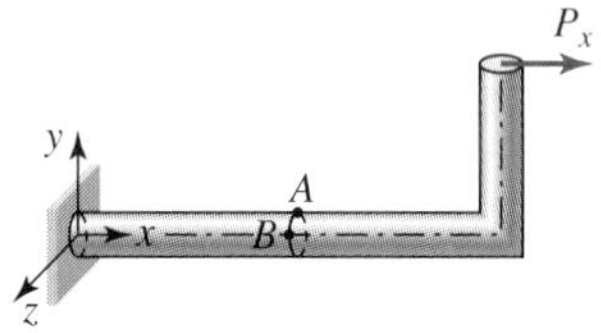

Figure P10.23

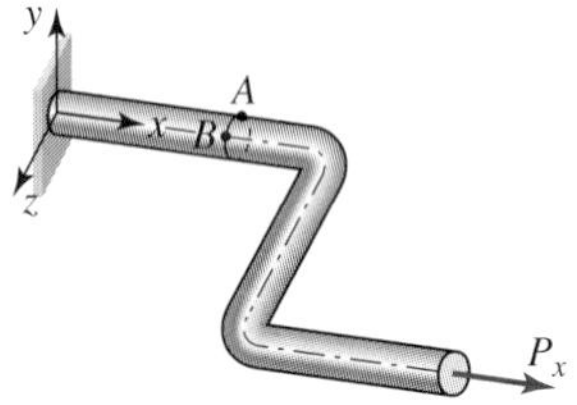

Figure P10.24

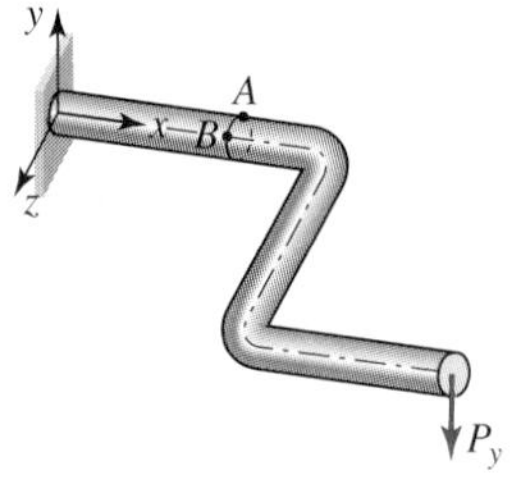

Figure P10.25

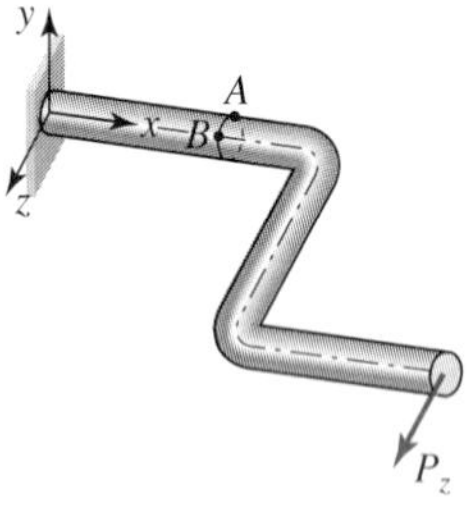

Figure P10.26

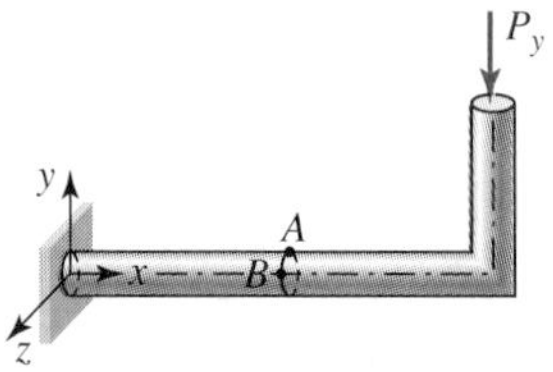

Figure P10.27

10.22 Using Figure P10.22 solve the problems as described in the text above.

10.23 Using Figure P10.23 solve the problems as described in the text above.

10.24 Using Figure P10.24 solve the problems as described in the text above.

10.25 Using Figure P10.25 solve the problems as described in the text above.

10.26 Using Figure P10.26 solve the problems as described in the text above.

10.27 Using Figure P10.27 solve the problems as described in the text above.

10.28 Using Figure P10.28 solve the problems as described in the text above.

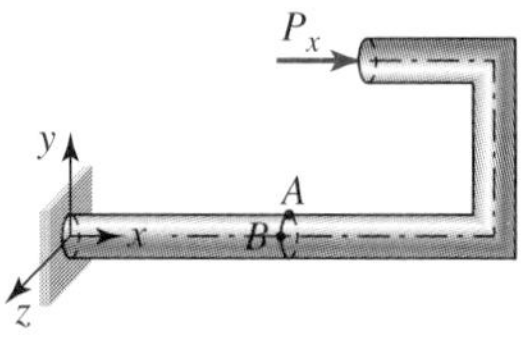

Figure P10.28

10.29 Using Figure P10.29 solve the problems as described in the text above.

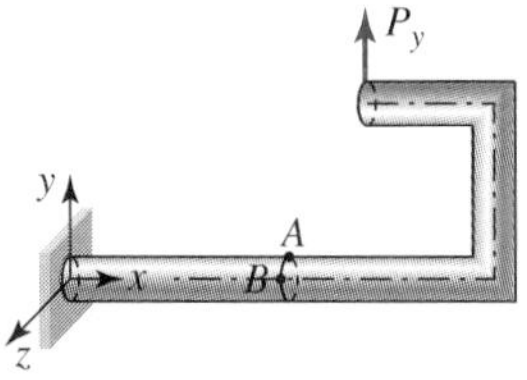

Figure P10.29

10.30 Using Figure P10.30 solve the problems as described in the text above.

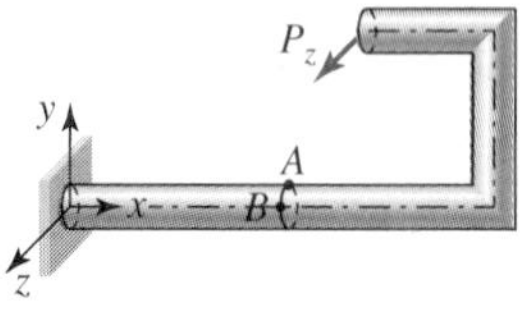

Figure P10.30

10.31 A pipe with an outside diameter of 2.0 in and a wall thickness of $\frac{1}{4}$ in is loaded as shown in Figure P10.31. Determine the normal and shear stresses at points A and B in the x, y, z coordinate system and show them on a stress cube. Points A and B are on the surface of the pipe.

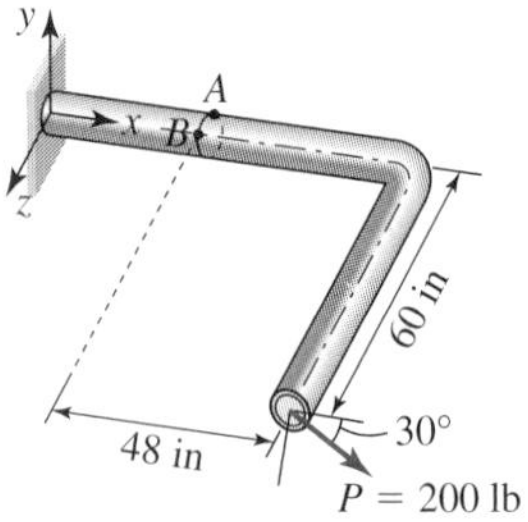

Figure P10.31

10.32 Determine the maximum normal stress and the maximum shear stress at point B on the pipe shown in Figure P10.31.

10.33 A pipe with an outside diameter of 40 mm and a wall thickness of 10 mm is loaded as shown in Figure P10.33. Determine the normal and shear stresses at points A and B in the x, y, z coordinate system and show them on a stress cube. Points A and B are on the surface of the pipe.

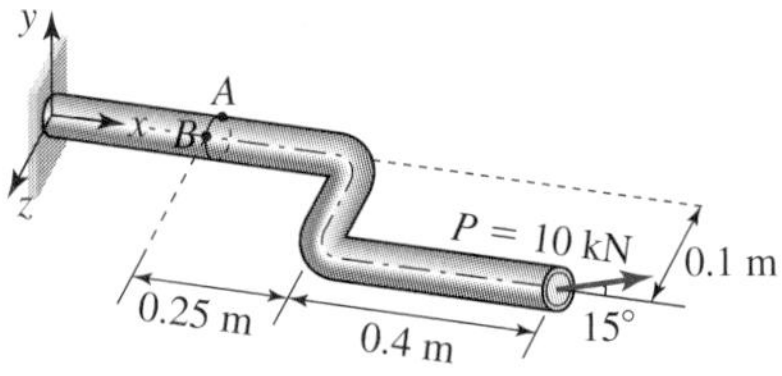

Figure P10.33

10.34 Determine the maximum normal stress and the maximum shear stress at point B on the pipe shown in Figure P10.33.

10.35 A bent pipe of 2-in outside diameter and a wall thickness of $\frac{1}{4}$ in is loaded as shown in Figure P10.35. Determine the stress components at point A, which is on the surface of the shaft. Show your answer on a stress cube.

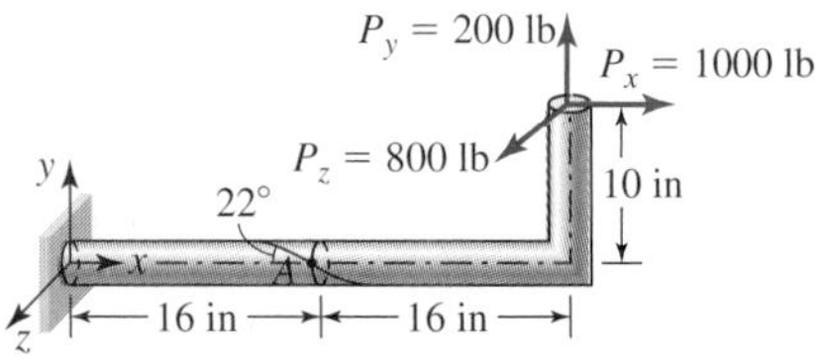

Figure P10.35

10.36 Determine the normal and shear stresses on a seam through point A that is $22°$ to the axis of the pipe shown in Figure P10.35.

10.37 The hollow steel shaft shown in Figure P10.37 has an outside diameter of 4 in and an inside diameter of 3 in. Two pulleys of 24-in diameter carry belts that have the given tensions. The shaft is supported at the walls using flexible bearings, permitting rotation in all directions. Determine the maximum normal and shear stresses in the shaft.

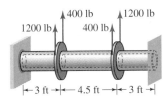

Figure P10.37

10.38 A thin cylinder is subjected to a uniform pressure of 300 psi and torques as shown in Figure P10.38. The cylinder has a outer radius of 10 in and a wall thickness of 0.25 in. Determine the normal and shear stresses at point A and show them on a stress element in cylindrical coordinates.

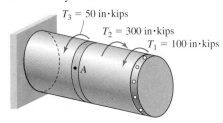

Figure P10.38

Structural analysis and design

10.39 A hollow shaft that has an outside diameter of 100 mm and an inside diameter of 50 mm is loaded as shown in Figure P10.39. The normal stress and the shear stress in the shaft must be limited to 200 MPa and 115 MPa, respectively. (a) Determine the maximum value of the torque T that can be applied to the shaft. (b) Using the result of part (a), determine the strain that will be shown by the strain gage that is mounted on the surface at an angle of $35°$ to the axis of the shaft. Use $E = 200$ GPa, $G = 80$ GPa, $\nu = 0.25$.

10.40 On the C clamp shown in Figure P10.40a determine the maximum clamping force P if the allowable normal stress is 160 MPa in tension and 120 MPa in compression.

(a)

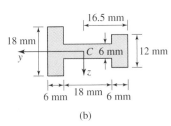

(b)

Figure P10.40

10.41 The T cross section of the beam was constructed by gluing two rectangular pieces together. A small crack was detected in the glue joint at section AA. Determine the maximum value of the applied load P if the normal stress in the glue at section AA is to be limited to 20 MPa in tension and 12 MPa in shear. The load P acts at the centroid of the cross section at C, as shown in Figure P10.41.

10.42 The bars in the pin connected structure shown in Figure P10.42 are circular bars of diameters that are available in increments of 5 mm. The allowable shear stress in the bars is 90 MPa. Determine the diameters of the bars for designing the lightest structure to support a force of $P = 40$ kN.

10.43 Member AB has a circular cross section with a diameter of 0.75 in as shown in Figure P10.43. Member BC has a square cross section of 2 in × 2 in. Determine the maximum normal stress in members AB and BC.

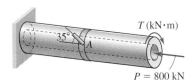

Figure P10.39

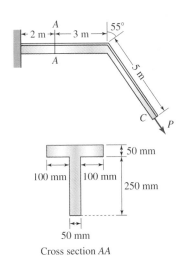

Cross section AA

Figure P10.41

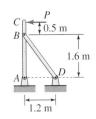

Figure P10.42

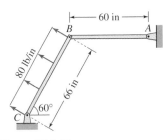

Figure P10.43

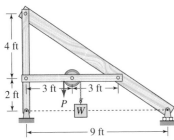

Figure P10.44

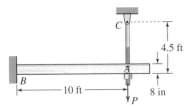

Figure P10.45

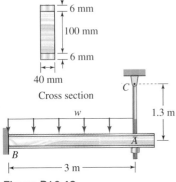

Figure P10.46

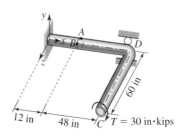

Figure P10.48

10.44 The members of the structure shown in Figure P10.44 have rectangular cross sections and are pin connected. Cross-section dimensions for members are 100 mm × 150 mm for *ABC*, 100 mm × 200 mm for *CDE*, and 100 mm × 50 mm for *BD*. The allowable normal stress in the members is 20 MPa. Determine the maximum intensity of the distributed load *w*.

10.45 A hoist is to be designed for lifting a maximum weight $W = 300$ lb, as shown in Figure P10.45 The hoist will be installed at a certain height above ground and will be constructed using lumber and assembled using steel bolts. The lumber rectangular cross-section dimensions are listed in Table 10.2. The bolt joints will be modeled as pins in single shear. Same-size bolts will be used in all joints. The allowable normal stress in the wood is 1.2 ksi and the allowable shear stress in the bolts is 6 ksi. Design the lightest hoist by choosing the lumber from Table 10.2 and the bolt size to the nearest $\frac{1}{8}$-in diameter.

10.46 A rectangular wooden beam of 4-in × 8-in cross section is supported at the right end by an aluminum circular rod of $\frac{1}{2}$-in diameter, as shown in Figure P10.46. The allowable normal stress in the wood is 1.5 ksi and the allowable shear stress in aluminum is 8 ksi. The moduli of elasticity for wood and aluminum are $E_w = 1800$ ksi and $E_{al} = 10,000$ ksi. Determine the maximum force *P* that the structure can support.

10.47 A steel pipe with an outside diameter of 1.5 in and a wall thickness of $\frac{1}{4}$ in is simply supported at *D*. A torque of 30 in·kips is applied as shown in Figure P10.47. Determine the normal and shear stresses at points *A* and *B* in the *x, y, z* coordinate system and show them on a stress cube. Points *A* and *B*

Figure P10.47

are on the surface of the pipe. The modulus of elasticity is $E = 30,000$ ksi and Poisson's ratio is $v = 0.28$.

10.48 A composite beam is constructed by attaching steel strips at the top and bottom of a wooden beam, as shown in Figure P10.48 The beam is supported at the right end by an aluminum circular rod of 8-mm diameter. The allowable normal stresses in the wood and steel are 14 MPa and 140 MPa, respectively. The allowable shear stress in aluminum is 60 MPa. The moduli of elasticity for wood, steel, and aluminum are $E_w = 12.6$ GPa, $E_s = 200$ GPa, and $E_{al} = 70$ GPa, respectively. Determine the maximum intensity *w* of the distributed load that the structure can support.

10.49 A park structure is modeled with pin joints at the points shown in Figure P10.49. Members *BD* and *CE* have cross-sectional dimensions of 6 in × 6 in, whereas members *AB*, *AC*, and *BC* have cross-sectional dimensions of 2 in × 8 in. Determine the maximum normal and shear stresses in each of the members due to the estimated snow load shown on the structure.

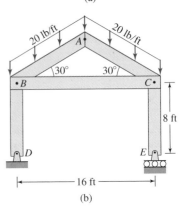

Figure P10.49

10.50 A highway sign uses a 16-in hollow pipe as a vertical post and 12-in hollow pipes for horizontal arms, as shown in Figure P10.50. The pipes are 1 in thick. Assume that a uniform wind pressure of 20 lb/ft² acts on the sign boards and the pipes. Note that the pressure on the pipes acts on the projected area Ld, where L is the length of pipe and d is the pipe diameter. Neglecting the weight of the pipe, determine the normal and shear stresses at points A and B and show these stresses on stress cubes.

10.51 A bicycle rack is made from thin aluminum tubes of $\frac{1}{16}$-in thickness and 1-in outer diameter. The weight of the bicycles is supported by the belts from C to D and the members between C and B. Member AC carries negligible force and is neglected in the stress analysis, as shown on the model in Figure P10.51b. If the allowable normal stress in the steel tubes is 12 ksi and the allowable shear stress is 8 ksi, determine the maximum weight W to the nearest lb of each bicycle that can be put on the rack.

10.52 The hoist shown in Figure P10.52 was used to lift heavy loads in a mining operation. Member EF supported load only if the load being lifted was asymmetric with respect to the pulley; otherwise it carried no load and can be neglected in the stress analysis. If the allowable normal stress in steel is 18 ksi, determine the maximum load W that could be lifted using the hoist.[4]

---

[4]Though the load on section BB is not passing through the plane of symmetry, the theory of symmetric bending can still be used because of the structure symmetry.

Figure P10.50

(a)

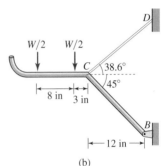

(b)

Figure P10.51

(a)

Figure P10.52

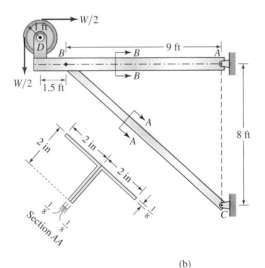

(b)

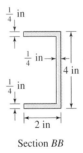

Section BB

Figure P10.53

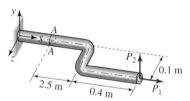

Figure P10.56

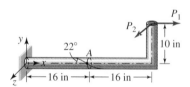

Figure P10.57

Failure envelopes

10.53 A solid shaft of 50-mm diameter is made from a brittle material that has an allowable tensile stress of 100 MPa, as shown in Figure P10.53. Draw a failure envelope representing the maximum permissible positive values of T and P.

10.54 The shaft shown in Figure P10.53 is made from a ductile material and has an allowable shear stress of 75 MPa. Draw a failure envelope representing the maximum permissible positive values of T and P.

10.55 The shaft in Problem 10.53 is 1.5 m long and has a modulus of elasticity $E = 200$ GPa and a modulus of rigidity $G = 80$ GPa. Modify the failure envelope of Problem 10.53 to incorporate the limitation that the elongation cannot exceed 0.5 mm and the relative rotation of the right end with respect to the left end cannot exceed $3°$.

10.56 A pipe with an outside diameter of 40 mm and a wall thickness of 10 mm is loaded as shown in Figure P10.56. At section AA the allowable shear stress is 60 MPa. Draw the failure envelope for the applied loads P_1 and P_2.

10.57 A bent pipe of 2-in outside diameter and a wall thickness of $\frac{1}{4}$-in is loaded as shown in Figure P10.57. The pipe has a weld at $22°$ to the axis as shown. The maximum shear stress the pipe material can support is 24 ksi. Draw the failure envelope for the applied loads P_1 and P_2.

Computer problems

10.58 A hollow aluminum shaft of 5-ft length is to carry a torque of 200 in·kips and an axial force of 100 kips. The inner radius of the shaft is 1 in. If the allowable shear stress in the shaft is 10 ksi, determine the outer radius of the lightest shaft.

10.59 The hollow cylinder shown in Figure P10.59 is fabricated from a sheet metal of 15-mm thickness. Determine the minimum outer radius to the nearest millimeter if the allowable normal stress is 150 MPa in tension or compression.

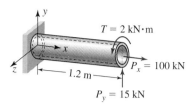

Figure P10.59

10.60 Table P10.60 shows the measured radii of the solid tapered member shown in Figure P10.60 at several points along the axis of the shaft. The member is subjected to a torque $T = 30$ kN·m and an axial force $P = 100$ kN. Plot the maximum normal and shear stresses as a function of x.

TABLE P10.60

x (m)	$R(x)$ (mm)
0.0	100.6
0.1	92.7
0.2	82.6
0.3	79.6
0.4	75.9
0.5	68.8
0.6	68.0
0.7	65.9
0.8	60.1
0.9	60.3
1.0	59.1
1.1	54.0
1.2	54.8
1.3	54.1
1.4	49.4
1.5	50.6

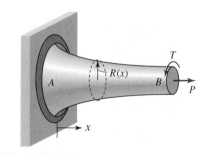

Figure P10.60

*10.3 STRESS INTENSITY FACTOR

In Section 3.7 we studied the concept of stress concentration factor, which made it possible to extend the results of elementary theories into regions with complex stress states for the purpose of design. Stress intensity factor is another concept that permits us to extend our elementary theories and formulas to materials containing flaws such as small cracks. The analysis of stresses and deformation of materials containing cracks is called *fracture mechanics,* which is beyond the scope of this book. But the concept of stress intensity factor and its application in design can be described using the conclusions derived mathematically in fracture mechanics.

Consider a small elliptical hole in an infinite plate, as shown in Figure 10.52. A solution obtained using the theory of elasticity[5] shows that the maximum stress at the tip of major axes (point A) can be found using the stress concentration factor $K_{conc} = 1 + 2a/b$, where a and b are half the diameter of the minor and major axes. If we let the minor axis diameter b go to zero, then the ellipse will become a crack, but our stress concentration factor would become infinite—the maximum stress at point A would become infinite. In other words, use of the stress concentration concept for cracks implies that the moment a tiny crack is formed, the entire body should break because no material can sustain an infinite stress. We know this cannot be correct, because we have often seen tiny cracks in materials and structures, which continue to function just fine. The explanation is that the tip of the crack gets slightly rounded,[6] but the blunting of the crack tip cannot be accurately predicted and accounted for in the theoretical elastic models. Furthermore we know that sometimes cracks grow and the machine component (or a structural member) breaks. Stress intensity factor is a concept that helps reconcile the infinite stress prediction of theoretical models to the reality of having stable or unstable cracks in materials. This is accomplished by using results from elasticity that show that the stress components in the immediate vicinity of the crack tip can be written as

$$\sigma_{ij} = K_{inten.} f_{ij}$$

where f_{ij} are some functions[7] that depend on the location of the point at which stress is being found relative to the location of the crack tip. The functions f_{ij} become infinite at the crack tip, which reflects the fact that the theoretical model ignores blunting of the crack tip. The factor K_{inten} is the stress intensity factor, which depends on the stress level and crack length for small cracks in very large bodies[8] (modeled as infinite bodies). When the value of K_{inten} reaches a critical value K_{crit}, the crack starts to grow rapidly and the body breaks. The critical value of the stress intensity factor K_{crit} is a material property. The higher the value of K_{crit}, the greater is the material's resistance to crack growth and the tougher is the material. Recollect that in Section 3.1.4 we discussed the modulus of toughness, which was equal to the area under the stress–strain curve. Critical stress intensity factors are related to the modulus of toughness. We had

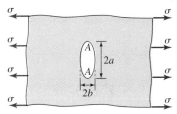

Figure 10.52 Elliptical hole in infinite plate.

---

[5]The solution was derived by Inglis in 1913.

[6]The rounding of crack tips is due to plasticity in ductile materials and due to microfracturing in brittle materials.

[7]The functions are proportional to $1/\sqrt{r}$, where r is the radial distance from the crack tip. See Problem 10.84.

[8]For finite-size bodies the stress intensity factor also depends on variables defining the geometry of the body.

also made the distinction between strong materials, which are characterized by high ultimate stress, and tough materials, which are characterized by large moduli of toughness or high values of the critical stress intensity factor. We record the following observations:

- The stress intensity factor depends on the stress level and the length of the crack.

- The critical stress intensity factor is a material property that is independent of the stress level or crack length.

- A crack becomes unstable (material breaks) when the stress intensity factor exceeds the *critical* stress intensity factor.

We need one more concept related to crack growth before we can use the stress intensity factor in design. Figure 10.53 shows three possible modes in which the two surfaces on either side of the crack can move relative to each other. Mode 1 is an opening mode, in which the crack grows due to tensile stresses. Mode 2 is a sliding mode due to in-plane shear stress. Mode 3 is a tearing mode such as when we tear a paper with two hands. Mode 1 predominates in most failures and hence is studied extensively. Mixed-mode failures are the subjects of ongoing research. We will consider a simple model of mixed modes 1 and 2 in this book.

Fracture mechanics shows that the stress intensity factors for modes 1 and 2 for a through crack in a thin infinite plate are as given by

$$K_{\mathrm{I}} = \sigma_{\mathrm{nom}}\sqrt{\pi a} \tag{10.5a}$$

$$K_{\mathrm{II}} = \tau_{\mathrm{nom}}\sqrt{\pi a} \tag{10.5b}$$

where σ_{nom} and τ_{nom} are the nominal normal and shear stresses obtained from elementary theories, as will be described. (See also Figure 10.54.) When both modes are

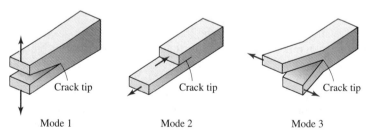

Figure 10.53 Three modes of relative crack surface movement.

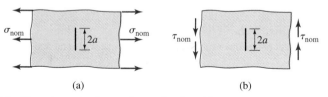

Figure 10.54 Stress intensity factors for modes 1 and 2.

present, then we will use an equivalent stress intensity factor given as

$$K_{equiv} = \sqrt{K_I^2 + K_{II}^2}$$ (10.6)

When the equivalent stress intensity factor exceeds the critical stress intensity factor, then the crack becomes unstable. Handbooks list formulas and graphs that can be used for obtaining stress intensity factors for a variety of situations.

Following are the steps of incorporating an existing crack into our analysis:

Step 1: Determine the state of stress at the point of crack using elementary theory that does not include a crack.

Step 2: Find the normal stress and the shear stress on the inclined crack surface using Mohr's circle or the method of equations. This normal stress and shear stress on the crack surface are the nominal stresses.

Step 3: Find stress intensity factors K_I and K_{II} using Equations (10.5a) and (10.5b).

Step 4: Find the equivalent stress intensity factor using Equation (10.6).

Step 5: Compare the equivalent stress intensity factor with the critical stress intensity factor for a material and decide whether or not the crack is stable.

If no existing crack orientation is specified, then we shall assume microcracks that will grow in mode 1 on a plane of maximum tensile normal stress, that is, on principal plane 1. We record this observation for future use.

Definition 4 Microcracks will be assumed to grow in mode 1 due to principal stress 1 if it is in tension, assuming there is no preexisting finite crack.

EXAMPLE 10.9

The propeller shaft of a submarine is subjected to a tensile axial stress and a torsional shear stress when the submarine reverses its direction. The propeller shaft of a submarine on display showed a crack at an angle of 27° to the axis of the shaft. At the point where the crack was seen, the stresses are estimated as shown in Figure 10.55. The shaft material has a critical stress intensity factor of 140 ksi $\sqrt{in}$. If the submarine was still in operation, then at what crack length would you recommend that the submarine be pulled out of the water for repairs, assuming:

(a) The detected crack could grow.

(b) There is no preexisting crack.

PLAN

The submarine has to be pulled out of the water before the crack reaches critical length.

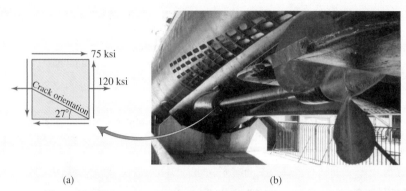

(a)   (b)

Figure 10.55 Crack in propeller in Example 10.9.

(a) Step 1 of the procedure in this section has already been done as the stresses are given to us. As per Step 2, we can find the normal stress and the shear stress on a plane containing the crack using Mohr's circle or the method of equations. As per Step 3, we can find the stress intensity factor for modes 1 and 2 in terms of the crack length, using Equations (10.5a) and (10.5b). As per Step 4, we can then find the equivalent stress intensity factor in terms of crack length using Equation (10.6). By equating the equivalent stress intensity factor to the critical value of the stress intensity factor, we can determine the critical crack length.

(b) We can determine principal stress 1 from Mohr's circle in part (a) or from Equation (8.6). Using Equation (10.5a) we can determine the stress intensity factor for mode 1, and by equating it to the given critical value, we determine the critical length.

Solution

(a) *Step 2:* We can find the normal and shear stresses on the plane containing the crack by either Mohr's circle or the method of equations.

Mohr's circle method: The coordinates of the vertical and horizontal planes are $V(120, 75 \text{ ccw})$ and $H(0, 75 \text{ cw})$, as shown in Figure 10.56. We can locate the points and find the center C as 60. Noting that $CA = 60$ and $AV = 75$, we can find the radius of the circle and the angle OCH,

$$R = \sqrt{60^2 + 75^2} = 96.0 \text{ ksi}$$

$$2\theta_p = \arctan(75/60) = 51.34°$$

We note that the plane containing the crack is at 27° from the horizontal plane in the clockwise direction. Hence from point H on Mohr's circle we rotate by twice the angle of 54° clockwise to

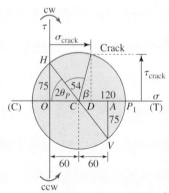

Figure 10.56 Stresses on plane containing crack using Mohr's circle.

obtain the plane containing the crack. We can then find the angle β and the normal and shear stresses,

$$\beta = 180° - 51.34° - 54° = 74.7°$$

$$\sigma_{\text{crack}} = OC + CD = 60 + 96.0 \cos 74.7 = 85.4 \text{ ksi} \quad \text{(E1)}$$

$$|\tau_{\text{crack}}| = 96.0 \sin 74.7 = 92.6 \text{ ksi} \quad \text{(E2)}$$

Method of equations: The angle of the normal to the plane containing the crack is $90° - 27° = 63°$. Substituting this angle and $\sigma_{xx} = 120$, $\sigma_{yy} = 0$, and $\tau_{xy} = 75$ into Equations (8.1) and (8.2), we can obtain the normal and shear stresses on the plane containing the crack,

$$\sigma_{\text{crack}} = 120 \cos^2 63 + 2 \times 75 \sin 63 \cos 63 = 85.4 \text{ ksi} \quad \text{(E3)}$$

$$\tau_{\text{crack}} = -120 \cos 63 \sin 63 + 75(\cos^2 63 - \sin^2 63) = -92.6 \text{ ksi} \quad \text{(E4)}$$

Step 3: Substituting the magnitudes of the stresses on the crack in Equations (10.5a) and (10.5b) we obtain the stress intensity factors for modes 1 and 2,

$$K_{\text{I}} = 85.4 \sqrt{\pi a} = 151.4 \sqrt{a} \text{ ksi} \sqrt{\text{in}} \quad \text{(E5)}$$

$$K_{\text{II}} = 92.6 \sqrt{\pi a} = 164.1 \sqrt{a} \text{ ksi} \sqrt{\text{in}} \quad \text{(E6)}$$

Steps 4, 5: Substituting Equations (E5) and (E6) into Equation (10.6) we obtain the equivalent stress intensity factor, which we can equate to the critical value of 140 ksi $\sqrt{\text{in}}$,

$$K_{\text{equiv}} = \sqrt{\left(151.4 \sqrt{a_{\text{cr}}}\right)^2 + \left(164.1 \sqrt{a_{\text{cr}}}\right)^2} = 223.3 \sqrt{a_{\text{cr}}} = 140$$

We thus obtain the critical crack length ($2a_{\text{cr}}$) as

$$\text{ANS.} \quad 2a_{\text{cr}} = 0.786 \text{ in}$$

(b) From Figure 10.56 principal stress $1(OP_1)$ can be found as $OC + CP_1$, that is, $\sigma_1 = 156$ ksi (T). We can also find it using Equation (8.6) as $\sigma_1 = 120/2 + \sqrt{(120/2)^2 + 75^2} = 156$ ksi. From Equation (10.5a) we can find the mode 1 stress intensity factor and equate it to the critical value of 140 ksi $\sqrt{\text{in}}$,

$$K_{\text{I}} = 156 \sqrt{\pi a_1} = 276.5 \sqrt{a_1} = 140$$

We thus obtain the critical crack length ($2a_1$) as

$$\text{ANS.} \quad 2a_1 = 0.513 \text{ in}$$

COMMENTS

1. If there are no visible cracks, then the crack growth occurs due to mode 1 fracture. But if a crack has been formed from other causes, such as the shaft hitting something, then the crack growth may be dictated by a mixed-mode fracture. But as the critical crack length for mode 1 is smaller than the critical crack length in mixed mode, it is possible that new cracks may start and be the cause of failure before the existing crack becomes critical.

2. It is not surprising that the critical crack length in part (b) is smaller than that in part (a). For this problem the magnitude of principal stress 1 is greater than the maximum shear stress on any plane. Hence mode 1 is the dominant fracture mode.

*10.4 FAILURE THEORIES

The maximum strength of a material is its atomic strength. In bulk materials, however, the distribution of flaws (such as impurities, microholes, or microcracks) create a local stress concentration. The effect of these local stress concentrations is that the bulk strength of the material is orders of magnitude lower than the atomic strength of the material. Failure theories assume a homogeneous material in which the effects of flaws have been averaged[9] in some manner. This assumption of homogeneity results in average material strength values, which are adequate for most engineering design and analysis.

For a homogeneous isotropic material the characteristic failure stress is either the yield stress or the ultimate stress that is usually obtained from the uniaxial tensile test. In the uniaxial tension test there is only one nonzero stress component. How do we relate the one stress component of uniaxial stress state to the stress components in two- and three-dimensional states of stress? There is no one answer that is applicable to all materials, and the theories that attempt to answer this questions are called failure theories.

Definition 5 Failure theory is a statement on the relationship of the stress components to the values of material failure characteristics.

---

[9]Micromechanics tries to account for the some of the flaws and nonhomogeneity in predicting the strength of a material, but to extrapolate from micro levels to macro levels requires some form of averaging or homogenization.

TABLE 10.5 Synopsis of failure theories   10.4 FAILURE THEORIES   **673**

	Ductile Material	**Brittle Material**
Characteristic failure stress	Yield stress	Ultimate stress
Theories	1. Maximum shear stress	1. Maximum normal stress
	2. Maximum energy distortion	2. Coulomb–Mohr

We shall consider four theories. The maximum shear stress theory and the maximum energy distortion theory are generally used for ductile materials. In ductile materials failure is characterized by yield stress. The maximum normal stress theory and Mohr's theory are generally used for brittle material. In brittle materials failure is characterized by ultimate stress. Table 10.5 gives a synopsis of the statements in this paragraph.

10.4.1   Maximum Shear Stress Theory

The theory predicts:

Definition 6 A material will fail when the maximum shear stress exceeds the shear stress at yield that is obtained from a uniaxial tensile test.

The theory gives reasonable results for ductile materials. Figure 10.57 shows that the maximum shear stress at yield in a tension test is half that of the normal yield stress. We obtain the following the failure criterion:

$$\boxed{\tau_{max} \le \frac{\sigma_{yield}}{2}} \tag{10.7}$$

Equation (10.7) is also referred to as *Tresca's yield criterion*. The maximum shear stress at a point is given by Equation (8.14). If we substitute Equation (8.14) into Equation (10.7), we obtain

$$\left|\max(\sigma_1 - \sigma_2, \sigma_2 - \sigma_3, \sigma_3 - \sigma_1)\right| \le \sigma_{yield} \tag{10.8}$$

If we plot each principal stress on an axis, then Equation (10.8) gives us the failure envelope. For plane stress problems the failure envelope is shown in Figure 10.59 seen later.

10.4.2   Maximum Distortion Energy Theory

The theory predicts:

Definition 7 A material will fail when the maximum distortion strain energy density is equal to the strain energy density at yield point in a tension test.

It has been observed that materials can withstand very large hydrostatic pressures without yielding. Hydrostatic pressure produces volume change[10] but no distortion of

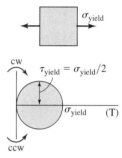

Figure 10.57   Shear stress at yield in tension test.

---

[10]See Problems 3.92 and 3.93.

the original shape of a body. We want to subtract the strain energy density due to hydrostatic stress from the total strain energy density to obtain the distortion strain energy density. To accomplish this we start from Equation (3.7), which gives us the linear strain energy density as $U_0 = \frac{1}{2}\sigma\varepsilon$ for the uniaxial state of stress. When all three principal stresses and strains are nonzero, then the total strain energy density is

$$U_0 = \frac{1}{2}(\sigma_1\varepsilon_1 + \sigma_2\varepsilon_2 + \sigma_3\varepsilon_3) \tag{10.9}$$

Substituting principal strains in terms of principal stresses using the generalized Hooke's law given by Equations (9.21a) through (9.21c), we obtain

$$U_0 = \frac{1}{2E}[\sigma_1^2 + \sigma_2^2 + \sigma_3^2 - 2\nu(\sigma_1\sigma_2 + \sigma_2\sigma_3 + \sigma_3\sigma_1)] \tag{10.10}$$

The hydrostatic[11] stress corresponds to $\sigma_{hydro} = (\sigma_1 + \sigma_2 + \sigma_3)/3$. Figure 10.58 shows that by subtracting and adding σ_{hydro} from each of the principal stresses we generate a stress cube which is subjected to a uniform stress corresponding to the hydrostatic stress. We can find the strain energy density for this state by substituting σ_{hydro} for all three principal stresses into Equation (10.10) to obtain

$$U_{hydro} = \frac{3(1-2\nu)}{2E}\sigma_{hydro}^2 = \frac{3(1-2\nu)}{2E}\left(\frac{\sigma_1 + \sigma_2 + \sigma_3}{3}\right)^2 \tag{10.11}$$

Subtracting Equation (10.11) from Equation (10.10) and after some algebraic manipulation we obtain the distortion strain energy,

$$U_{dist} = \frac{1+\nu}{6E}[(\sigma_1 - \sigma_2)^2 + (\sigma_2 - \sigma_3)^2 + (\sigma_3 - \sigma_1)^2]$$

We substitute $\sigma_1 = \sigma_{yield}$, $\sigma_2 = 0$, and $\sigma_3 = 0$ in the preceding equation to obtain the distortion strain energy at yield in a tension test,

$$U_{yield} = \frac{1+\nu}{3E}\sigma_{yield}^2$$

The failure criterion is $U_{dist} \leq U_{yield}$, which yields the following result:

$$\frac{1}{\sqrt{2}}\sqrt{(\sigma_1 - \sigma_2)^2 + (\sigma_2 - \sigma_3)^2 + (\sigma_3 - \sigma_1)^2} \leq \sigma_{yield} \tag{10.12}$$

Equation (10.12) is written in an alternative form using the following definition:

$$\boxed{\sigma_{von} = \frac{1}{\sqrt{2}}\sqrt{(\sigma_1 - \sigma_2)^2 + (\sigma_2 - \sigma_3)^2 + (\sigma_3 - \sigma_1)^2}} \tag{10.13}$$

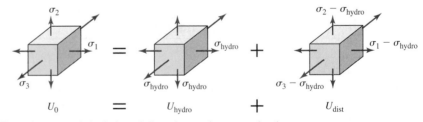

Figure 10.58 Calculation of distortion strain energy density.

---

[11]At a point in fluid the pressure, and hence the normal stress at a point, are the same in all directions. For this reason, when stress is uniform in all directions, the state of stress is referred to as hydrostatic state of stress.

where σ_{von} is referred to as equivalent von-Mises stress or, simply, as *von-Mises stress,* and the failure criterion represented by Equation (10.12) is sometimes referred to as von-Mises yield criterion[12] and is stated as

$$\boxed{\sigma_{\text{von}} \leq \sigma_{\text{yield}}} \tag{10.14}$$

Equations (10.7) and (10.12) are failure envelopes[13] in a space in which the axes are principal stresses. For a plane stress ($\sigma_3 = 0$) problem we can show these failure envelopes as in Figure 10.59. Notice that the maximum distortion energy envelope encompasses the maximum shear stress envelope. Experiments show that for most ductile materials the maximum distortion energy theory gives better results than the maximum shear stress theory, but the maximum shear stress theory is very simple to use.

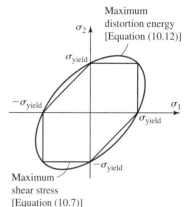

Figure 10.59 Failure envelopes for ductile materials in plane stress.

10.4.3 Maximum Normal Stress Theory

The theory predicts:

Definition 8 A material will fail when the maximum normal stress at a point exceeds the ultimate normal stress obtained from a uniaxial tension test.

The theory gives good results for brittle materials provided principal stress 1 is tensile, or if the tensile yield stress has the same magnitude as the yield stress in compression. Thus the failure criterion is given as

$$\boxed{\left| \max(\sigma_1, \sigma_2, \sigma_3) \right| \leq \sigma_{\text{ult}}} \tag{10.15}$$

For most materials the ultimate stress in tension is usually far less than the ultimate stress in compression because microcracks tend to grow in tension and tend to close in compression. But the simplicity of the failure criterion makes the theory attractive, and it can be used if principal stress 1 is tensile and is the dominant principal stress.

10.4.4 Mohr's Theory

The theory predicts:

Definition 9 A material will fail if a stress state is on the envelope that is tangent to the three Mohr's circles corresponding to uniaxial ultimate stress in tension, uniaxial ultimate stress in compression, and pure shear.

---

[12]It is also called maximum octahedral shear stress criterion. See Problem 8.68 for a description of octahedral planes and Equation (8.18) for the similarity of the von-Mises stress to the shear stress on octahedral planes.

[13]In drawing failure envelopes, the convention that $\sigma_1 > \sigma_2$ is ignored. If the convention were enforced, then there would be no envelope in the second quadrant and only the envelope below a 45° line would be admissible in the third quadrant, resulting in a very strange looking envelope rather than the symmetric envelope that is shown in Figure 10.59.

We can conduct three experiments and determine the ultimate stress in tension σ_T, the ultimate stress in compression σ_C, and the ultimate shear stress in pure shear τ_U. The three stress states are shown on the stress cubes in Figure 10.60a. Mohr's circles for each of the three stress states are drawn in Figure 10.60b. Finally an envelope that is tangent to the three circles is drawn, which represents the failure envelope. If Mohr's circle corresponding to a stress state just touches the envelope at any point, then the material is at incipient failure. If any part of Mohr's circle for a stress state is outside the envelope, then the material has failed at that point.

We can also plot the failure envelopes of Figure 10.60 using principal stresses as the coordinate axes. In plane stress this envelope is represented by the solid line in Figure 10.61. For most brittle materials the pure shear test is often ignored. In such a case the tangent line to the circles of uniaxial compression and tension would be a straight line in Figure 10.60. The resulting simplification for plane stress is shown as dotted lines in Figure 10.61 and is called *modified Mohr's theory.*

Figure 10.61 emphasizes the following:

1. If both principal stresses are tensile, then the maximum normal stress has to be less than the ultimate tensile strength.

2. If both principal stresses are negative, then the maximum normal stress must be less than the ultimate compressive strength.

3. If the principal stresses are of different signs, then for the modified Mohr's theory the failure is governed by

$$\left| \frac{\sigma_2}{\sigma_C} - \frac{\sigma_1}{\sigma_T} \right| \leq 1 \tag{10.16}$$

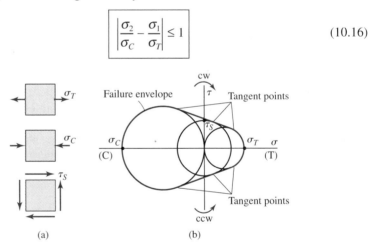

(a)   (b)

Figure 10.60   Mohr's failure envelope.

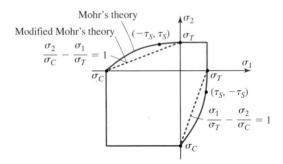

Figure 10.61 Mohr's failure envelope for plane stress.

EXAMPLE 10.10

At a critical point on a machine part made of steel, the stress components were found to be $\sigma_{xx} = 100$ MPa (T), $\sigma_{yy} = 50$ MPa (C), and $\tau_{xy} = 30$ MPa. Assuming that the point is in plane stress and the yield stress in tension is 220 MPa, determine the factor of safety using: (a) the maximum shear stress theory; (b) the maximum distortion energy theory.

PLAN

We can find the principal stresses and maximum shear stress by Mohr's circle or by the method of equations.

(a) From Equation (10.7) we know that failure stress for the maximum shear stress theory is half the yield stress in tension. Using Equation (3.23) we can find the factor of safety.

(b) We can find the von-Mises stress from Equation (10.13), which gives us the denominator in Equation (3.23), and noting that the numerator of Equation (3.23) is the yield stress in tension, we obtain the factor of safety.

Solution

Mohr's circle method: We draw the stress cube, record the coordinates of planes V and H, draw Mohr's circle as shown in Figure 10.62, and find the principal stresses and maximum shear stress,

$$R = \sqrt{75^2 + 30^2} = 80.8 \text{ MPa}$$

$$\sigma_1 = OC + OP_1 = 25 + 80.8 = 105.8 \text{ MPa} \qquad \text{(E1)}$$

$$\sigma_2 = OC - OP_1 = 25 - 80.8 = -55.8 \text{ MPa} \qquad \text{(E2)}$$

$$\tau_{\max} = R = 80.8 \text{ MPa} \qquad \text{(E3)}$$

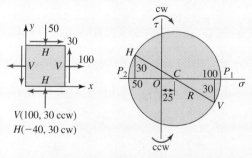

Figure 10.62   Calculation using Mohr's circle in Example 10.10.

We note that the third principal stress is zero as it is a plane stress problem.

Method of equations: From Equations (8.6) and (8.14) we can obtain the principal stresses and the maximum shear stress,

$$\sigma_{1,2} = \frac{100-50}{2} \pm \sqrt{\left(\frac{100+50}{2}\right)^2 + 30^2} = 25 \pm 80.8$$

$$\sigma_1 = 105.8 \text{ MPa} \tag{E4}$$

$$\sigma_2 = -55.8 \text{ MPa} \tag{E5}$$

$$\tau_{max} = \frac{\sigma_1 - \sigma_2}{2} = 80.8 \text{ MPa} \tag{E6}$$

(a) The failure shear stress is half the yield stress in tension, that is, 110 MPa. As per Equation (3.23), we divide this value by the maximum shear stress [Equation (E3) or (E6)] to obtain the factor of safety,

$$k_\tau = \frac{110}{80.8}$$

ANS. $k_\tau = 1.36$

(b) The von-Mises stress can be found from Equation (10.13),

$$\sigma_{von} = \frac{1}{\sqrt{2}} \sqrt{[105.8-(-55.8)]^2 + (-55.8^2) + 105.8^2} = 142.2 \text{ MPa} \tag{E7}$$

The failure stress is 220 MPa, which we can divide by the von-Mises stress to get the following factor of safety,

$$k_\sigma = \frac{220}{142.2}$$

ANS. $k_\sigma = 1.55$

COMMENTS

1. The failure envelopes corresponding to the yield stress of 250 MPa are shown in Figure 10.63. In comment 4 of Example 10.8 it was shown that graphically the factor of safety could be found by taking ratios of distances from the origin along the load line. If we plot the coordinates $\sigma_1 = 105.8$ and $\sigma_2 = -55.8$, we obtain point S. If we join the origin O to point S and draw the line, we get the load line for the given stress values. It may be verified by measuring (or calculating coordinates of T and V) that the following is true: $k_\tau = OS/OT = 1.36$ and $k_\sigma = OS/OV = 1.55$, as in Example 10.8.

2. Because the failure envelope for the maximum shear stress criterion is always inscribed inside the failure envelope of maximum distortion energy, the factor of safety based on the maximum distortion energy will always be greater than the factor of safety based on maximum shear stress.

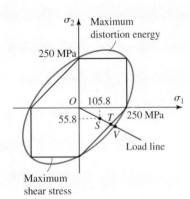

Figure 10.63 Failure envelopes in Example 10.10.

EXAMPLE 10.11

Due to a load P the stresses at a point on a free surface were found to be $\sigma_{xx} = 3P$ ksi (C), $\sigma_{yy} = 5P$ ksi (T), and $\tau_{xy} = -2P$ ksi, where P is measured in kips. The brittle material has a tensile strength of 18 ksi and a compressive strength of 36 ksi. Determine the maximum value of load P that can be applied on the structure using the modified Mohr's theory.

PLAN

We can determine the principal stresses in terms of P by Mohr's circle or by the method of equations. As the given normal stresses are of opposite signs, we can expect that the principal stresses will have opposite signs. Using Equation (10.16) we can determine the maximum value of P.

Solution

Mohr's circle method: We draw the stress cube, record the coordinates of planes V and H, draw Mohr's circle as shown in Figure 10.64, and find the principal stresses and the maximum shear stress,

$$R = \sqrt{(4P)^2 + (2P)^2} = 4.47P$$

$$\sigma_1 = OC + OP_1 = P + 4.47P = 5.57P \qquad \text{(E1)}$$

$$\sigma_2 = OC - OP_1 = P - 4.47P = -3.37P \qquad \text{(E2)}$$

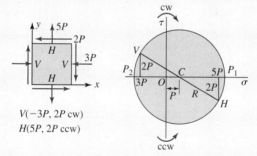

$V(-3P, 2P \text{ cw})$
$H(5P, 2P \text{ ccw})$

Figure 10.64 Calculation using Mohr's circle in Example 10.11.

We note that the third principal stress is zero as it is a plane stress problem.

Method of equations: From Equation (8.6) we can obtain the principal stresses as

$$\sigma_{1,2} = \frac{-3P + 5P}{2} \pm \sqrt{\left(\frac{-3P - 5P}{2}\right)^2 + (2P)^2} = P \pm 4.47P \quad \text{(E3)}$$

$$\sigma_1 = 5.57P \qquad \text{(E4)}$$

$$\sigma_2 = -3.37P \qquad \text{(E5)}$$

Substituting the principal stresses into Equation (10.16) and noting that $\sigma_T = 18$ ksi and $\sigma_C = 36$ ksi, we can obtain the maximum value of P,

$$\left| \frac{-3.37P}{-36} - \frac{5.57P}{18} \right| \leq 1 \qquad \text{or} \qquad 0.2158P \leq 1 \qquad \text{or}$$

$$P \leq 4.633 \qquad \text{or} \qquad \text{ANS.} \qquad P_{max} = 4.63 \text{ kips}$$

COMMENT

We could not have used the maximum normal stress theory for this material, since the tensile and compressive strengths are significantly different and it is the compressive strength that is the dominant strength and not the tensile-strength.

PROBLEM SET 10.2

Stress intensity factor

In Problems 10.61 through 10.63, an axial bar was welded together along the weld line shown in Figure P10.61. The bar material has a critical stress intensity factor of 20 $ksi\sqrt{in}$. Use these data to solve the problems that follow.

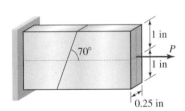

Figure P10.61

10.61 The applied load $P = 12$ kips. What is the critical length of a crack at which the axial member should be taken out of service?

10.62 Before the axial load $P = 10$ kips was applied, a crack was observed along the weld line shown in Figure P10.61. (a) What is the critical length of a crack along the weld line? (b) What is the critical length of a crack at which the axial member should be taken out of service?

10.63 The smallest crack length of 0.06 in represents the limit of the flaw-detection system. What is the maximum axial force that should be permitted to act on the member, assuming there is no preexisting crack?

In Problems 10.64 through 10.66, a thin tube with a seam as shown in Figure P10.64 has a centerline radius of 100 mm and a thickness of 10 mm. The tube material has a critical stress intensity factor of 33 $MPa\sqrt{m}$. Use these data to solve the problems that follow.

10.64 The applied torque on the tube is $T = 100$ kN·m. What is the critical length of a crack at which the tube should be taken out of service?

10.65 Before the torque $T = 150$ kN·m was applied, a crack was observed along the seam that is $40°$ to the axis of the shaft, as shown in Figure P10.64. (a) What is the critical length of a crack along the seam? (b) What is the critical length of a crack at which the tube should be taken out of service?

10.66 The smallest crack length of 5 mm represents the limit of the flaw-detection system. What is the maximum torque that should be permitted to act on the tube, assuming there is no preexisting crack?

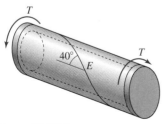

Figure P10.64

In Problems 10.67 and 10.68, the beam shown in Figure P10.67 is made from a material with a critical stress intensity factor of 33 MPa$\sqrt{m}$. Use these data to solve the problems that follow.

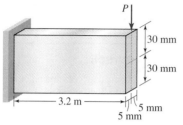

Figure P10.67

10.67   The applied load on the beam is $P = 150$ N. What is the critical length of a crack at which the beam should be taken out of service?

10.68   The smallest crack length of 2 mm represents the limit of the flaw-detection system. What is the maximum load P that should be permitted to act on the beam?

In Problems 10.69 through 10.71, the cylindrical tank shown in Figure P10.69 has a mean diameter of 4 m and a wall thickness of 5 mm. The critical stress intensity factor of the tank material is 45 MPa$\sqrt{m}$. Use these data to solve the problems that follow.

10.69   The pressure in the tank is 500 kPa. What is the critical length of a crack at which the cylinder should be taken out of service, assuming any leak can be ignored?

Figure P10.69

10.70   For a gas pressure of 750 kPa, what would be the critical crack length along the weld line in Figure P10.69?

10.71   A critical crack length of 4 mm is used for estimating the maximum pressure the tank (Figure P10.69) can sustain. What is the maximum permissible pressure in the tank?

In Problems 10.72 and 10.73, an aluminum alloy hollow shaft (G = 4000 ksi) has a critical stress intensity factor of 22 ksi$\sqrt{in}$. The shaft shown in Figure P10.72 has an outside diameter of 2 in and a wall thickness of $\frac{1}{4}$ in. Use these data to solve the problems that follow.

10.72   At what critical length of a crack should the shaft be taken out of service if $T = 60$ in · kips in Figure P10.72?

10.73   If the smallest crack length that can be detected is 0.04 in, determine the maximum torque that can be applied to the shaft in Figure P10.72.

10.74   A thin cylindrical tube with an outer diameter of 5 in is fabricated by butt welding a $\frac{1}{16}$ -in-thick plate along a spiral seam, as shown in Figure P10.74. If the critical stress intensity factor for the material is 22 ksi$\sqrt{in}$, determine the critical crack length of the crack along the seam.

Failure theories

10.75   Due to a force P measured in kN, the stress components at a critical point that is in plane stress were found to be $\sigma_{xx} = 10P$ MPa (T), $\sigma_{yy} = 20P$ MPa (C), and $\tau_{xy} = 5P$ MPa. The material has a yield stress of 160 MPa as determined in a tension test. If yielding must be avoided, predict the maximum value of force P using: (a) maximum shear stress theory; (b) maximum distortion energy theory.

10.76   Due to a force P the stress components at a critical point that is in plane stress were found to be $\sigma_{xx} = 4P$ ksi (C), $\sigma_{yy} = 3P$ ksi (T), and $\tau_{xy} = -5P$ ksi. The material has a tensile rupture strength of 18 ksi and a compressive rupture strength of 32 ksi. Determine the maximum value of force P using the modified Mohr's theory.

10.77   A material has a tensile rupture strength of 18 ksi and a compressive rupture strength of 32 ksi. During usage a component made from this plastic showed the following stresses on a free surface at a critical point: $\sigma_{xx} = 9$ ksi (T), $\sigma_{yy} = 6$ ksi (T), and

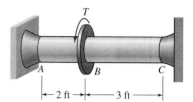

Figure P10.72

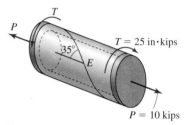

Figure P10.74

$\tau_{xy} = -4$ ksi. Determine the factor of safety using the modified Mohr's theory.

10.78 On a free surface of aluminum ($E = 10{,}000$ ksi, $v = 0.25$, $\sigma_{yield} = 24$ ksi) the strains recorded by the three strain gages shown in Figure P10.78 are $\varepsilon_a = -600$ μin/in, $\varepsilon_b = 500$ μin/in, and $\varepsilon_c = 400$ μin/in. By how much can the loads be scaled without exceeding the yield stress of aluminum at the point? Use maximum shear stress theory.

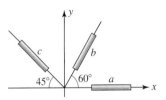

Figure P10.78

10.79 On a free surface of steel ($E = 200$ GPa, $v = 0.28$, $\sigma_{yield} = 210$ MPa) the strains recorded by the three strain gages shown in Figure P10.79 are $\varepsilon_a = -800$ μm/m, $\varepsilon_b = -300$ μm/m, and $\varepsilon_c = -700$ μm/m. By how much can the loads be scaled without exceeding the yield stress of steel at the point? Use the maximum distortion energy theory.

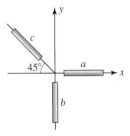

Figure P10.79

10.80 A thin-walled cylindrical gas vessel has a mean radius of 3 ft and a wall thickness of $\frac{1}{2}$ in. The yield stress of the material is 30 ksi. Using the von-Mises failure criterion, determine the maximum pressure of the gas inside the cylinder if yielding is to be avoided.

10.81 A thin cylindrical boiler can have a minimum mean radius of 18 in and a maximum mean radius of 36 in. The boiler will be subjected to a pressure of 750 psi. A sheet metal with a yield stress of 60 ksi is to be used with a factor of safety of 1.5. Construct a failure envelope with the mean radius R and the sheet metal thickness t as axes using the maximum distortion energy theory.

10.82 For plane stress show that the von-Mises stress of Equation (10.13) can be written as

$$\sigma_{von} = \sqrt{\sigma_{xx}^2 + \sigma_{yy}^2 - \sigma_{xx}\sigma_{yy} + 3\tau_{xy}^2} \quad (10.17)$$

Stretch yourself

10.83 In Cartesian coordinates the von-Mises stress in three dimensions is given by

$$\sigma_{von} = \sqrt{\begin{array}{c}\sigma_{xx}^2 + \sigma_{yy}^2 + \sigma_{zz}^2 - \sigma_{xx}\sigma_{yy} - \sigma_{yy}\sigma_{zz} \\ -\,\sigma_{zz}\sigma_{xx} + 3\tau_{xy}^2 + 3\tau_{yz}^2 + 3\tau_{zx}^2\end{array}}$$

$$(10.18)$$

Show that for plane strain Equation (10.18) reduces to

$$\sigma_{von} = \sqrt{\begin{array}{c}(\sigma_{xx}^2 + \sigma_{yy}^2)(1 + v^2 - v) \\ -\,\sigma_{xx}\sigma_{yy}(1 + 2v - 2v^2) + 3\tau_{xy}^2\end{array}}$$

$$(10.19)$$

where v is Poisson's ratio of the material.

10.84 Fracture mechanics shows that the stresses in model in the vicinity of the crack tip shown in Figure P10.84 are given by

$$\sigma_{xx} = \frac{K_I}{\sqrt{2\pi r}} \cos\frac{\theta}{2}\left(1 - \sin\frac{\theta}{2}\sin\frac{3\theta}{2}\right)$$

$$(10.20a)$$

$$\sigma_{yy} = \frac{K_I}{\sqrt{2\pi r}} \cos\frac{\theta}{2}\left(1 + \sin\frac{\theta}{2}\sin\frac{3\theta}{2}\right)$$

$$(10.20b)$$

$$\tau_{xy} = \frac{K_I}{\sqrt{2\pi r}} \sin\frac{\theta}{2}\cos\frac{\theta}{2}\cos\frac{3\theta}{2} \quad (10.20c)$$

Notice that at $\theta = \pi$, that is, at the crack surface, all stresses are zero. In terms of K_I and r, obtain the von-Mises stress at $\theta = 0$ and $\theta = \pi/2$, assuming plane stress.

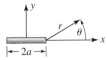

Figure P10.84

*10.5   GENERAL INFORMATION

Out of one hundred axial members in service, if two fail, then we can say our design of axial members is 98% reliable. But how do we design to obtain a desired reliability? Concepts related to reliability, and a design methodology that incorporates it, are described in this section.

10.5.1   Reliability

In Section 3.6 the concept of factor of safety was introduced as a measure of safety to account for the uncertainties regarding material properties, manufacturing processes, control and estimate of loads, and so on. But choosing a factor of safety was a compromise based on experience and influenced by several factors. So the question that arises is: how reliable is our design?

To understand the relationship between factor of safety and reliability, consider the following example. An axial member with a factor of safety of 1.3 is to be designed. The material strength of the axial member is 130 MPa. Thus we design the axial member using an allowable stress of 100 MPa.

The actual axial stress in the member may be quite different than the design stress of 100 MPa, because of such factors as manufacturing tolerances, variability of applied loads, and environmental effects such as temperature and humidity. If we measured the axial stress in different members and made a plot of the number of members (frequency) at a given stress level, we would get a distribution such as shown by the left curve in Figure 10.65. The material strength, that is, the failure stress for different batches of material, may vary due to impurities, material processing, and so on. Once more, suppose the distribution of the material strength is given by a curve that looks like the right curve in Figure 10.65.

The distributions in Figure 10.65 show that the mean axial stress in the members is 100 MPa and the mean strength of all materials is 130 MPa. However, there will be some axial members with stresses greater than 100 MPa but made from materials that have failure stresses less than 130 MPa and hence are likely to fail. These members

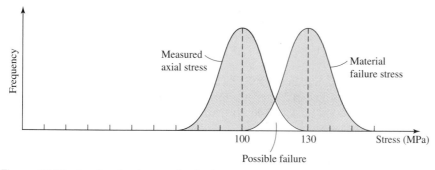

Figure 10.65   Load and resistance distribution curves.

are in the region common to both distributions and are labeled "possible failure" in Figure 10.65. Knowing the two distributions, the possible failure region can be found.

The parameters usually used to describe statistical distributions are *mean value* and *standard deviation*. Reliability formulas can be developed using these two parameters for different distributions, and used in design. If the reliability of a design as predicted from the formula is unacceptable, then a different factor of safety can be used for design to obtain the desired reliability.

10.5.2 Load and Resistance Factor Design (LRFD)

In civil engineering, for steel structures a design methodology has evolved called *load and resistance factor design* (LRFD), which can be used for designing structures to a specified reliability. Several ideas and the terminology used in LRFD are described first before describing LRFD.

The words "strength" and "resistance" (of material) are often used interchangeably in LRFD, reflecting the historical evolution of the concept of strength as was elaborated in Section 1.3. The loads on a structure are categorized as follows and are specified in building codes: dead load (D)—load due to the weight of the structural elements and permanent features in the structure; live load (L)—load from people, equipment, and other movable objects that comes from occupancy; snow load (S) and rain or ice loads (R)—loads that appear on the roof of a structure; L_r—roof load from movable objects during construction and maintenance (cranes, etc.) and occupancy (air conditioners, etc.); W—wind load; E—earthquake load.

LRFD incorporates ideas that were used in two other design methodologies, *allowable stress design* (ASD) and *plastic design* (PD). ASD is based on elastic analysis, using factors of safety that vary with the primary function of the structural member. For example, a factor of safety of 1.5 is used for beams, and 1.67 for tension members. These factors of safety are specified in the code and are usually based on statistical analysis. The sum total of the stresses from the various loads described must be less than or equal to the allowable stress.

In PD a load factor that varies with a combination of loads is used in determining the design load on the structure. Thus if only dead and live loads are considered, then the load factor is 1.7 [written as 1.7 ($D + L$)] but if the dead, live, and wind loads are considered, then the load factor is 1.3 [written as 1.3 ($D + L + W$)]. The factored loads are applied to the structure, and a nonlinear analysis is conducted to determine the strength of the member at structure collapse. By nonlinear analysis we mean that the stress values of many members in the structure are between yield stress and ultimate stress, that is, in the plastic region. The member strengths must equal or exceed the required strengths calculated using factored loads.

There are several features about ASD and PD that have not been elaborated here. From a viewpoint of obtaining a consistent reliability in design, neither of the two methods is very accurate. In ASD the factor of safety is used to account for all variability in loads and material strength. In PD the variability in material strength is ignored. Furthermore, all loads do not have the same degree of variability. These shortcomings are overcome in the LRFD method.

The nominal failure strength of a member is multiplied by the appropriate resistance factor (from Table 10.6 to obtain the design strength. This accounts for

TABLE 10.6 Some resistance factors

10.6 CLOSURE **685**

Tension members, failure due to yielding	0.9
Tension member, failure due to rupture	0.75
Axial compression	0.85
Beams	0.9
High-strength bolts, failure in tension	0.75

TABLE 10.7 Load factors and load combinations

$1.4D$

$1.2D + 1.6L + 0.5 (L_r \text{ or } S \text{ or } R)$

$1.2D + 1.6 (L_r \text{ or } S \text{ or } R) + (0.5L \text{ or } 0.8W)$

$1.2D + 1.3W + 0.5 + 0.5 (L_r \text{ or } S \text{ or } R)$

$12D \pm 1.0E + 0.5L + 0.2S$

$0.9D \pm (1.3W \text{ or } 1.0E)$

variability in material strength and inaccuracies in dimensions and modeling. In Table 10.7 the numerical values are the load factors, which account for the variability of the individual load components and the probability of combinations, such as live and snow loads, acting together. Using Table 10.7, factored loads are determined for a specific load combination, applied to the structure, and the member strength is calculated. This computed member strength must be less than or equal to the design strength computed using the resistance factor. Since variations of load and member strength are taken into account separately, LRFD gives a more consistent level of reliability.

10.6 CLOSURE

This chapter synthesized and applied the concepts of all previous chapters. The use of subscripts and formulas to determine the magnitude and direction of stress results in a systematic but slower approach to solving the problem. Determining the stress directions by inspection can reduce the algebra significantly, but it requires more care and depends on the problem being solved. It is important to find your individual mix of solving the problem by inspection, subscripts, and formulas. No matter what your individual preference is, the importance of a systematic approach to the problem cannot be overstated. The complexity of the design and analysis of structures is such that without a systematic approach the chances of error rise dramatically.

So far the designs were based on material strength or structure stiffness. Structure instability, however, can cause a structure to fail at stresses far lower than the material strength. What is structure instability, and how does it get incorporated into the design of columns? This will be described in the next and last chapter.

POINTS AND FORMULAS TO REMEMBER

- Superposition of stresses is addition or subtraction of stress components in the same direction acting on the same surface at a point.

$$\sigma_{xx} = \frac{N}{A} \quad (10.1) \qquad \tau_{x\theta} = \frac{T\rho}{J} \quad (10.2) \qquad \sigma_{xx} = -\frac{M_z y}{I_{zz}} \quad (10.3a) \qquad \tau_{xs} = -\frac{V_y Q_z}{I_{zz}t} \quad (10.3b)$$

$$\sigma_{xx} = -\frac{M_y z}{I_{yy}} \quad (10.4a) \qquad \tau_{xs} = -\frac{V_z Q_y}{I_{yy}t} \quad (10.4b)$$

Sign convention for internal forces and moments:

- Internal forces and moments on a free-body diagram must be drawn according to the sign conventions if subscripts are to be used for determining the direction of stress components.
- Direction of stress components must be determined by inspection if internal forces and moments are drawn on the free-body diagram to equilibrate the external forces and moments.
- Direction of shear flow (τ_{xs}) on a cross section due to bending about z axis (y axis) must be such that (i) the resultant force in the y direction (z direction) is in the same direction as V_y (V_z); (ii) the resultant force in the z direction (y direction) is zero; (iii) it is symmetric about the y axis (z axis).
- A local x, y, z coordinate system can be established such that the x direction is along the axis of the long structural member.
- Stress components should be drawn on a stress cube and interpreted in the x, y, z coordinate system for use in stress and strain transformation equations.
- Normal stresses perpendicular to the axis of the member are zero: $\sigma_{yy} = 0$, $\sigma_{zz} = 0$, $\tau_{yz} = 0$.
- Normal strains perpendicular to the axis of the member can be obtained by multiplying the normal strains in the axis direction by Poisson's ratio.
- Superpose stresses, then use the generalized Hooke's law to obtain strains in combined loading:

$$\varepsilon_{xx} = \frac{\sigma_{xx}}{E} \qquad \varepsilon_{yy} = -\frac{v\sigma_{xx}}{E} \qquad \varepsilon_{zz} = -\frac{v\sigma_{xx}}{E} \qquad \gamma_{xy} = \frac{\tau_{xy}}{G} \qquad \gamma_{xz} = \frac{\tau_{xz}}{G} \qquad \gamma_{yz} = 0$$

- Allowable normal and shear stresses refer to principal stresses and absolute maximum shear stress at a point, respectively.
- An individualized procedure that is a mix of subscripts, formulas, and inspection should be developed for analysis of stresses under combined loading.
- There are two major steps in the analysis and design of structures: (i) analysis of forces and moments that act on individual members; (ii) computation of stresses on members under combined loading.

CHAPTER ELEVEN
STABILITY OF COLUMNS

11.0 OVERVIEW

The weight of the building exerts a compressive axial force on the columns shown in Figure 11.1*a*. If a compressive axial force is applied to a long thin wooden strip, then it bends significantly, as shown in Figure 11.1*b*. If the columns were to bend the same way as the wooden strip, then the columns and the building would collapse. Under what conditions will a compressive axial force produce only axial contraction, and when does it produce bending? When is the bending caused by axial loads catastrophic? How do we design to prevent catastrophic failure from axial loads? We study answers to these questions in this chapter.

Bending due to a compressive axial load is called *buckling*. Structural members that support compressive axial loads are called *columns*. As shall be seen in this chapter, the phenomenon of buckling is the study of the *stability* of a structure's equilibrium. This stability problem is a critical design issue because when a structure collapses due to instability, the collapse is sudden and usually catastrophic. Fortunately it is possible to identify members in a structure that are likely to collapse. One critical requirement for instability is that the structural members be in compression.

(a)   (b)

Figure 11.1   Examples of columns.

Geometry, material, boundary conditions, and imperfections are some of the important factors that affect the stability of columns.

The two learning objectives in this chapter are:

1. Develop an appreciation of the phenomenon of buckling and the various types of structure instabilities.
2. Understand the development and use of buckling formulas in the analysis and design of structures.

11.1 BUCKLING PHENOMENON

There are many approaches, hence perspectives, for studying buckling. Though we cannot study all the approaches in detail in this book, the different perspectives help us understand the various aspects of buckling.

Figure 11.2 shows a marble that occupies an equilibrium position on different types of surfaces. Consider disturbing the marble to the shaded position. When the surface is concave, as in Figure 11.2a, then the marble will return to the original equilibrium position, and the marble is said to be in a stable equilibrium position. When the surface is flat, as in Figure 11.2b, then the marble will acquire a new equilibrium state at the disturbed position, and the marble is said to be in a neutral equilibrium position. When the surface is convex, as in Figure 11.2c, then the marble will roll off on disturbing the equilibrium state, and the marble is said to be in an unstable equilibrium state.

The marble analogy shown in Figure 11.2 is useful in understanding an *energy* method approach to the buckling problem. Every deformed structure has a potential energy associated with it. This potential energy depends on the strain energy stored in the deformed structure and the work done by the external load. If the potential energy function is concave at the equilibrium position, as in Figure 11.2a, then the structure is in a stable equilibrium. If the potential energy function is convex, as in Figure 11.2c, then the structure is in an unstable equilibrium position. In the energy method approach, the external load at which the potential energy function changes from concave to convex is called the critical load. This approach is beyond the scope of this book.

Another perspective on the stability of equilibrium is the *bifurcation problem*. Figure 11.3a shows a rigid bar with a torsional spring at one end and a compressive axial load at the other end. Figure 11.3b shows the free-body diagram of the bar. By equilibrium of the moment at O, we obtain $PL \sin \theta = K_\theta \theta$ or $PL/K_\theta = \theta/\sin \theta$. Figure 11.3c shows the plot of PL/K_θ versus θ. The equilibrium line separates the unstable region from the stable region. Below point A, as we increase the load P, the bar remains in the vertical position ($\theta = 0$). The bar will return to the vertical position if it is disturbed (rotated) slightly to the left or right. Once the load value takes us beyond point A, then *even if* the bar is in the equilibrium vertical position, any disturbance will send it either

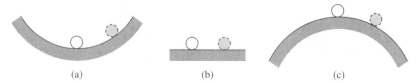

(a)   (b)   (c)

Figure 11.2   Stability of equilibrium using marble analogy. (*a*) Stable equilibrium. (*b*) Neutral equilibrium. (*c*) Unstable equilibrium.

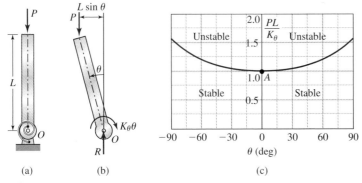

Figure 11.3 Bifurcation and eigenvalue problems.

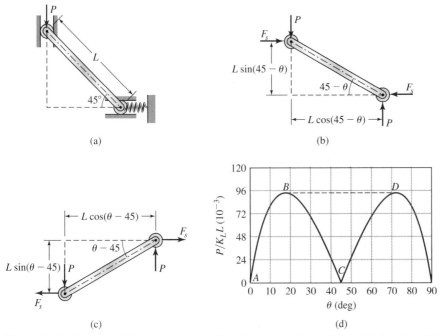

Figure 11.4 Snap buckling problem. (*a*) Undeformed position, $\theta = 0$. (*b*) $0 < \theta < 45°$. (*c*) $\theta > 45°$. (*d*) Load versus θ.

to the left branch or to the right branch of the curve, where the bar acquires a new equilibrium position. Point A is the *bifurcation point,* at which there are three possible solutions. The load P at the bifurcation point is called *critical load.*

Another way of looking at the problem described in Figure 11.3 is as follows. The moment equilibrium equation is satisfied if $\theta = 0$—we call it a trivial solution to the problem. We ask the following question: at what value of P does there exist a nontrivial solution to the problem? This is the classical statement of an *eigenvalue problem,* and the critical value of P for which the nontrivial solution exists is called *eigenvalue.*

In Section 11.2 the perspectives of the bifurcation problem and the eigenvalue problem are further elaborated.

In *snap buckling* a structure jumps from one equilibrium configuration to a dramatically different equilibrium configuration. To show and explain this phenomenon consider a bar that can slide in a smooth slot. It has a spring attached to it at the right end and a force P applied to it at the left end, as shown in Figure 11.4. As we increase

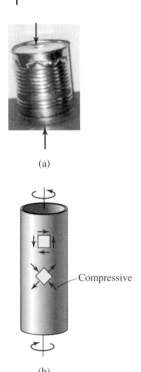

(a)

(b)

Figure 11.5 Local buckling. (*a*) Due to axial loads. (*b*) Due to torsional loads.

Compressive

the force P, the inclination of the bar moves closer to the horizontal position, where it reaches a new equilibrium position. But there is an inclination at which the bar suddenly jumps across the horizontal line to a position below the horizontal line.

To explain the phenomenon of snap buckling mathematically, we consider the equilibrium of the bar in the two configurations before and after the horizontal line. Suppose the spring is in the unstretched position, as shown in Figure 11.4a. We define the inclination of the bar by the angle θ measured from the undeformed position. The free-body diagrams of the bar before and after the horizontal position are shown in Figure 11.4b and c. The spring force must reverse direction as the bar crosses the horizontal position to ensure moment equilibrium. The deformation of the spring before the horizontal position is $L \cos(45 - \theta) - L \cos 45$. Thus the spring force is $F_s = K_L[L \cos(45 - \theta) - L \cos 45]$. By moment equilibrium we obtain

$$\frac{P}{K_L L} = [\cos(45 - \theta) - \cos 45] \tan(45 - \theta), \qquad 0 < \theta < 45°$$

In a similar manner, by considering the moment equilibrium in Figure 11.4c, we obtain

$$\frac{P}{K_L L} = [\cos(\theta - 45) - \cos 45] \tan(\theta - 45), \qquad \theta > 45°$$

Figure 11.4d shows a plot of $P/K_L L$ versus θ. As we increase P, we move along the curve until we reach point B. At B rather than following paths BC and CD, the bar jumps (snaps) from point B to point D. It should be emphasized that each point on paths BC and CD represents an equilibrium position, but it is not a stable equilibrium position that can be maintained.

These perspectives on the buckling problem were about structural stability. Besides the instability of a structure, we can have *local instabilities*. Figure 11.5a shows the crinkling of a aluminum can under compressive axial loads. This crinkling is the local buckling of the thin walls of the can. Figure 11.5b shows a thin cylindrical shaft under torsion. The stress cube at the top shows the torsional shear stresses. But if we consider a stress cube in principal coordinates, then we see that principal stress 2 is compressive. This compressive principal stress can also cause local buckling, though the orientation of the crinkles will be different than those from the crushing of the aluminum can. We will not be studying local buckling phenomena in this book.

The foregoing discussion highlights that failure from to compressive loads may occur due to buckling at loads significantly lower than those that produce material strength failure.

11.2 EULER BUCKLING

In this section we develop a theory for a straight column that is simply supported at either end. This theory was first developed by Leonard Euler[1] and is named after him.

Figure 11.6a shows a simply supported column that is axially loaded with a force P. We shall initially assume that bending is about the z axis, as our equations in Chapter 7 on beam deflection were developed with this assumption. We shall relax this assumption at the end to generate the buckling formula for a critical buckling load.

Let the bending deflection at any location x be given by $v(x)$, as shown in Figure 11.6b. An imaginary cut is made at some location x, and the internal bending moment is drawn according to our sign convention. The internal axial force N will

---

[1]See Section 11.4 for additional information on Euler.

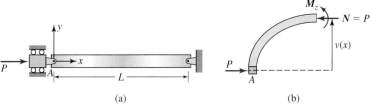

Figure 11.6 Simply supported column.

be equal to P. By balancing the moment at point A we obtain $M_z + Pv = 0$. Substituting the moment–curvature relationship of Equation (7.1), we obtain the following differential equation:

$$EI_{zz}\frac{d^2v}{dx^2} + Pv = 0 \tag{11.1}$$

or

$$\frac{d^2v}{dx^2} + \lambda^2 v = 0 \tag{11.2}$$

where

$$\lambda = \sqrt{\frac{P}{EI_{zz}}} \tag{11.3}$$

The boundary conditions are

$$v(0) = 0 \tag{11.4a}$$

$$v(L) = 0 \tag{11.4b}$$

Clearly $v = 0$ would satisfy the boundary-value problem represented by Equations (11.1), (11.4a), and (11.4b). This trivial solution represents purely axial deformation due to compressive axial forces. Our interest is to find the value of P that would cause bending; in other words, a nontrivial ($v \neq 0$) solution to the boundary-value problem. Alternatively stated: at what value of P does there exist a nontrivial solution to the boundary-value problem?, which, as was observed in Section 11.1, is the classical statement of an eigenvalue problem.

The solution to the differential equation, Equation (11.2), is

$$v(x) = A \cos \lambda x + B \sin \lambda x \tag{11.5}$$

From the boundary condition (11.4a) we obtain

$$v(0) = A \cos (0) + B \sin (0) = 0 \quad \text{or} \quad A = 0$$

From boundary condition (11.4b),[2] we obtain

$$v(L) = A \cos \lambda L + B \sin \lambda L = 0 \quad \text{or} \quad B \sin \lambda L = 0$$

---

[2] A matrix form may be a more familiar form as an eigenvalue problem. The boundary condition equations can be written in matrix form as

$$\begin{bmatrix} 1 & 0 \\ \cos\left(\sqrt{\frac{P}{EI_{zz}}}L\right) & \sin\left(\sqrt{\frac{P}{EI_{zz}}}L\right) \end{bmatrix} \begin{Bmatrix} A \\ B \end{Bmatrix} = \begin{Bmatrix} 0 \\ 0 \end{Bmatrix}$$

For a nontrivial solution, that is, when A and B are not both zero, the condition is that the determinant of the matrix be zero. This would yield $\sin\left(\sqrt{(P/EI_{zz})}L\right) = 0$, the same condition as described.

If $B = 0$, then we obtain a trivial solution. For a nontrivial solution the sine function must equal zero, that is,

$$\sin \lambda L = 0 \tag{11.6}$$

Equation (11.6) is called *characteristic equation,* also referred to as *buckling equation.*

Equation (11.6) is satisfied if $\lambda L = n\pi$. Substituting for λ and solving for P, we obtain

$$P_n = \frac{n^2 \pi^2 E I_{zz}}{L^2}, \qquad n = 1, 2, 3 \ldots \tag{11.7}$$

Equation (11.7) represents the value of load P (eigenvalue) at which buckling would occur. The question we ask is: what is the lowest value of P at which buckling will occur? Clearly, for the lowest value of P, n should be equal to 1 in Equation (11.7). But if buckling can occur about any axis and not just the z axis, as we initially assumed, then the area moment of inertia that should be used in the formula should be about that axis. We drop the subscripts zz in the area moment of inertia and write the critical buckling load as

$$\boxed{P_{cr} = \frac{\pi^2 E I}{L^2}} \tag{11.8}$$

where buckling occurs about an axis that has a minimum value of I. P_{cr}, the critical buckling load, is also called *Euler load.*[3]

The solution for v can be written as

$$v = B \sin\left(n\pi \frac{x}{L}\right) \tag{11.9}$$

Equation (11.9) represents the buckled mode (eigenvectors). Notice that the constant B in Equation (11.9) is undetermined. This is typical in eigenvalue problems. The importance of each buckled mode shape can be appreciated by examining Figure 11.7. If buckled mode 1 is prevented from occurring by installing a restraint (support), then the column would buckle at the next higher mode at critical load values that are higher than those for the lower modes. Point I on the deflection curves describing the mode shapes has two attributes: it is an inflection point and the magnitude of deflection at this point is zero. Recollect that the curvature d^2v/dx^2 at an inflection point is zero. Hence the internal moment M_z at this point is zero. If roller supports are put at any other points than the inflection points I, as predicted by Equation (11.9), then the boundary value problem must be resolved as it shall have different eigenvalues[4] (critical loads) and eigenvectors (mode shapes).

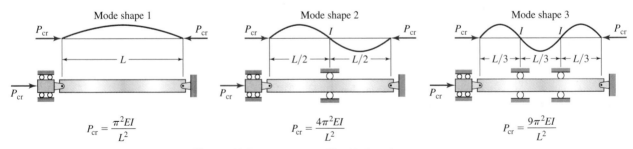

Figure 11.7 Importance of buckled modes.

---

[3]See Section 11.4 for additional details.
[4]See Problem 11.26.

In many situations it may not be possible to put roller supports in order to change a mode to a higher critical buckling load. But buckling modes and buckling loads can also be changed by using elastic supports. Figure 11.8 shows a water tank on columns. The two rings are the elastic supports. Elastic supports can be modeled as springs and formulas for buckling loads developed as shown in Example 11.3.

11.2.1 Effects of End Conditions

Equation (11.8) is only applicable to simply supported columns. However, the process used to obtain the formula can be used for other types of supports. Table 11.1 shows the critical elements in the derivation process and the results for three other supports. The

Figure 11.8 Elastic supports on columns of a water tank.

TABLE 11.1 Buckling of columns with different supports

	Case 1	Case 2	Case 3	Case 4*
	Pinned at both ends	One end fixed, other end free	One end fixed, other end pinned	Fixed at both ends
Differential equation	$EI\dfrac{d^2v}{dx^2} + Pv = 0$	$EI\dfrac{d^2v}{dx^2} + Pv = Pv(L)$	$EI\dfrac{d^2v}{dx^2} + Pv = R_B(L-x)$	$EI\dfrac{d^2v}{dx^2} + Pv = R_B(L-x) + M_B$
Boundary conditions	$v(0) = 0$ $v(L) = 0$	$v(0) = 0$ $\dfrac{dv}{dx}(0) = 0$	$v(0) = 0$ $\dfrac{dv}{dx}(0) = 0$ $v(L) = 0$	$v(0) = 0$ $\dfrac{dv}{dx}(0) = 0$ $v(L) = 0$ $\dfrac{dv}{dx}(L) = 0$
Characteristic equation $\lambda = \sqrt{\dfrac{P}{EI}}$	$\sin \lambda L = 0$	$\cos \lambda L = 0$	$\tan \lambda L = \lambda L$ [†]	$2(1 - \cos \lambda L) - \lambda L \sin \lambda L = 0$
Critical load P_{cr}	$\dfrac{\pi^2 EI}{L^2}$	$\dfrac{\pi^2 EI}{4L^2} = \dfrac{\pi^2 EI}{(2L)^2}$	$\dfrac{20.13 EI}{L^2} = \dfrac{\pi^2 EI}{(0.7L)^2}$	$\dfrac{4\pi^2 EI}{L^2} = \dfrac{\pi^2 EI}{(0.5L)^2}$
Effective length L_{eff}	L	$2L$	$0.7L$	$0.5L$

*R_B and M_B are the reaction force and moment of B.

[†]The roots of the equation have to be found iteratively. The two smallest roots of the equation are $\lambda L = 4.4934$ and $\lambda L = 7.7253$.

formula for critical loads for all cases shown in Table 11.1 can be written as

$$P_{cr} = \frac{\pi^2 EI}{L_{eff}^2} \tag{11.10}$$

where L_{eff} is the effective length of the column. The effective length for each case is given in the last row of Table 11.1. This definition of effective length will permit us to extend results that will be derived in Section 11.3 for simply supported imperfect columns to imperfect columns with supports shown in cases 2 through 4 in Table 11.1.

In Equation (11.8), I can be replaced by Ar^2, where A is the cross-sectional area and r is the radius of gyration [see Equation (A.10)]. We obtain the following equation:

$$\sigma_{cr} = \frac{P_{cr}}{A} = \frac{\pi^2 E}{(L_{eff}/r)^2} \tag{11.11}$$

where L_{eff}/r is the *slenderness ratio* and σ_{cr} is the compressive axial stress just before the column would buckle.

Equation (11.11) is valid only in the elastic region, that is, if $\sigma_{cr} < \sigma_{yield}$. If $\sigma_{cr} > \sigma_{yield}$, then elastic failure will be due to stress exceeding the material strength. Thus $\sigma_{cr} = \sigma_{yield}$ defines the failure envelope for a column. Figure 11.9 shows the failure envelopes for steel, aluminum, and wood using the material properties given in Table D.1. As nondimensional variables are used in the plots in Figure 11.9, the same plots will be obtained if metric units are used as given in Table D.2. Note that the slenderness ratio is defined using effective lengths; hence these plots are applicable to columns with different supports.

The failure envelopes in Figure 11.9 show that as the slenderness ratio increases, the failure due to buckling will occur at stress values significantly lower than the yield stress—hence the importance of buckling in the design of members under compression.

The failure envelopes, as shown in Figure 11.9, depend only on the material property and are applicable to columns of different lengths, shapes, and types of support. These failure envelopes are used for classifying columns as short or long.[5] Short column design is based on using yield stress as the failure stress. Long column design is based on using critical buckling stress as the failure stress. The slenderness ratio at point A for each material is used for separating short columns from long columns for that material. Point A is the intersection point of the straight line representing elastic material failure and the hyperbola curve representing buckling failure.

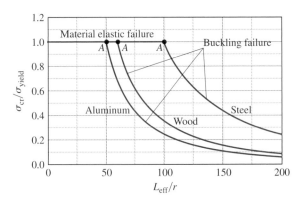

Figure 11.9 Failure envelopes for Euler columns.

---

[5]Intermediate column is a third classification, used if the critical stress is between yield stress and ultimate stress. See Equation (11.21) and Problems 11.49 and 11.50 for additional details.

EXAMPLE 11.1

A hollow circular steel column ($E = 30{,}000$ ksi) is simply supported over a length of 20 ft. The inner and outer diameters of the cross section are 3 in and 4 in, respectively. Determine:

(a) The slenderness ratio.

(b) The critical buckling load.

(c) The axial stress at the critical buckling load.

(d) If roller supports are added at the midpoint, what would be the new critical buckling load?

PLAN

(a) The area moment of inertia I for a hollow cylinder is same about all axes and can be found using the formula in Table A.2. From the value of I the radius of gyration can be found. The ratio of the given length to the radius of gyration gives the slenderness ratio.

(b) In Equation (11.8) the given values of E and L, as well as the calculated value of I in part (a), can be substituted to obtain the critical buckling load P_{cr}.

(c) Dividing P_{cr} by the cross-sectional area, the axial stress σ_{cr} can be found.

(d) The column will buckle at the next higher buckling load, which can be found by substituting $n = 2$ and E, I, and L into Equation (11.7).

Solution

(a) The outer diameter $d_o = 4$ in and the inner diameter $d_i = 3$ in. From Table A.2 the area moment of inertia for the hollow cylinder, the cross-sectional area A, and the radius of gyration r can be calculated using Equation (A.10),

$$I = \frac{\pi(d_o^4 - d_i^4)}{64} = \frac{\pi(4^4 - 3^4)}{64} = 8.590 \text{ in}^4 \qquad \text{(E1)}$$

$$A = \frac{\pi(d_o^2 - d_i^2)}{4} = \frac{\pi(4^2 - 3^2)}{4} = 5.498 \text{ in}^4 \qquad \text{(E2)}$$

$$r = \sqrt{\frac{I}{A}} = \sqrt{\frac{8.590}{5.498}} = 1.250 \qquad \text{(E3)}$$

The length $L = 20$ ft $= 240$ in. Thus the slenderness ratio is

$$\frac{L}{r} = \frac{240}{1.25} \qquad \text{or} \qquad \text{ANS.} \qquad \frac{L}{r} = 192$$

(b) Substituting $E = 30,000$ ksi, $L = 240$ in, and $I = 8.59$ in^4 into Equation (11.8), we obtain the critical buckling load,

$$P_{cr} = \frac{\pi^2 EI}{L^2} = \frac{\pi^2 \times 30,000 \times 8.590}{240^2}$$

ANS. $P_{cr} = 44.15$ kips

(c) The axial stress at the critical buckling load can be found as

$$\sigma_{cr} = \frac{P_{cr}}{A} = \frac{44.15}{5.498}$$

ANS. $\sigma_{cr} = 8.03$ ksi (C)

(d) With the support in the middle, the buckling would occur in mode 2. Substituting $n = 2$ and E, I, and L into Equation (11.7) we obtain the critical buckling load,

$$P_{cr} = \frac{n^2 \pi^2 EI}{L^2} = \frac{2^2 \pi^2 \times 30,000 \times 8.590}{240^2}$$

ANS. $P_{cr} = 176.6$ kips

COMMENTS

1. The example highlights the basic definitions of variables and equations used in buckling problems.

2. In part (d) we used the observation that the middle support forces the column into the mode 2 buckling mode. Another perspective is to look at the column as two simply supported columns, each with an effective length of half the column, that is, $L_{eff} = 120$ in. Substituting this into Equation (11.10), we obtain the same value as in part (d).

EXAMPLE 11.2

A hoist is constructed using two wooden bars ($E = 1800$ ksi), as shown in Figure 11.10. The allowable normal stress is 2 ksi. Determine the maximum permissible weight W that can be lifted using the hoist for the two cases: (a) $L = 4$ ft; (b) $L = 5$ ft.

PLAN

By inspection we see that member BC will be in compression. Therefore to determine the maximum value of W, we must consider buckling failure of member BC and strength failure of both members due to the axial stress exceeding the given allowable stress. By drawing the free-body diagram of the pulley, the internal axial forces in members BC and CD can be found in terms of W. The axial stresses in the member found are compared with the given allowable values to determine one set of limits on W. To determine the critical buckling load, the smaller moment of inertia about the y axis in Equation (11.8) should be used, and the upper limit on W to prevent buckling failure can be found. The maximum value of W that satisfies the strength and buckling criteria can now be determined.

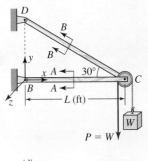

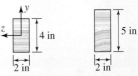

Cross section AA   Cross section BB

Figure 11.10 Hoist in Example 11.2.

Solution   The free-body diagram of the pulley is shown in Figure 11.11 with the force in BC drawn as compressive and the force in CD as tensile. The internal axial forces are found as

$$\sum F_y \quad N_{CD} \sin 30 = 2W \qquad N_{CD} = 4W \qquad (E1)$$

$$\sum F_x \quad N_{BC} = N_{CD} \cos 30 \qquad N_{BC} = 3.464W \qquad (E2)$$

The cross-sectional areas for the two members are $A_{BC} = 8$ in^2 and $A_{CD} = 10$ in^2. The axial stresses in terms of W can be found, and these should be less than 2 ksi, from which we get two limits on W,

$$\sigma_{CD} = \frac{N_{CD}}{A_{CD}} = \frac{4W}{10} \leq 2 \text{ ksi} \qquad \text{or} \qquad W \leq 5.0 \text{ kips} \qquad (E3)$$

$$\sigma_{BC} = \frac{N_{BC}}{A_{BC}} = \frac{3.463W}{8} \leq 2 \text{ ksi} \qquad \text{or} \qquad W \leq 4.62 \text{ kips} \qquad (E4)$$

For cross-section AA, we note that $I_{zz} = \frac{1}{12} \times 2 \times 4^3 = 10.67$ in^4 and $I_{yy} = \frac{1}{12} \times 4 \times 2^3 = 2.667$ in^4. Thus in the calculation of the critical buckling load from Equation (11.8) we use I_{yy}, as it is the smaller of the two area moments of inertia.

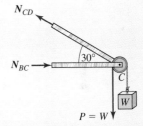

Figure 11.11 Free-body diagram in Example 11.2.

(a) Substituting $E = 1800$ ksi, $L = 4$ ft $= 48$ in, and $I = 2.667$ in^4, we obtain the critical buckling load for BC,

$$P_{cr} = \frac{\pi^2 EI}{L^2} = \frac{\pi^2 \times 1800 \times 2.667}{48^2} = 20.56 \text{ kips} \qquad (E5)$$

N_{BC} should be less than P_{cr},

$$N_{BC} \leq 20.56 \qquad \text{or} \qquad 3.464W \leq 20.56$$

From it we find the maximum that would not cause buckling,

$$W \le 5.94 \text{ kips} \qquad \text{(E6)}$$

The maximum value of W must satisfy Equations (E3), (E4), and (E6).

ANS. $\quad W_{max} = 4.6$ kips.

(b) Substituting $E = 1800$ ksi, $L = 5$ ft $= 60$ in, and $I = 2.667$ in^4, we obtain the critical buckling load for BC,

$$P_{cr} = \frac{\pi^2 EI}{L^2} = \frac{\pi^2 \times 1800 \times 2.667}{60^2} = 13.2 \text{ kips} \qquad \text{(E5)}$$

N_{BC} should be less than P_{cr},

$$N_{BC} \le 13.2 \quad \text{or} \quad 3.464W \le 13.2$$

From it we find the maximum that would not cause buckling,

$$W \le 3.81 \text{ kips} \qquad \text{(E6)}$$

The maximum value of W must satisfy Equations (E3), (E4), and (E6).

ANS. $\quad W_{max} = 3.8$ kips.

COMMENTS

1. This example highlights the importance of identifying compression members such as BC, so that buckling failure is properly accounted for in design.

2. The example also emphasizes that the minimum area moment of inertia that must be used is Euler buckling. Had we used I_{zz} instead of I_{yy}, we would have found $P_{cr} = 52.7$ kips and incorrectly concluded that the failure would be due to strength failure and not buckling.

3. In case (a) material strength governed the design, whereas in case (b) buckling governed the design. If we had several bars of different lengths and different cross-sectional dimensions (such as in Problems 11.12 and 11.13), then it would save a significant amount of work to calculate the slenderness ratio that would separate long columns from short columns. Substituting $\sigma_{cr} = \sigma_{allow} = 2$ ksi into Equation (11.11), we find that $L/r = 94.2$ is the ratio that separates long columns from short columns. It can be checked that the slenderness ratio in case (a) is 83.1, hence material strength governed W_{max}. In case (b) the slenderness ratio is 103.9, hence buckling governed W_{max}.

EXAMPLE 11.3

Linear springs are attached at the free end of a column, as shown in Figure 11.12. Assume that bending about the y axis is prevented. (a) Determine the characteristic equation for this buckling problem. Show that the critical load P_{cr} for (b) $k = 0$ and (c) $k = \infty$ is as given in Table 11.1 for cases 2 and 3, respectively.

PLAN

The spring exerts a spring force $kv(L)$ at the upper end that must be incorporated into the moment equation, and hence into the differential equation. The boundary conditions are that the deflection and slope at $x = 0$ are zero. (a) The characteristic equation will be generated while solving the boundary-value problem. (b), (c) The roots of the characteristic equation for the two cases will give P_{cr}.

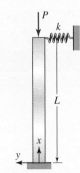

Figure 11.12 Column with elastic support in Example 11.3.

Solution We make an imaginary cut at some location x and take the top part for the free-body diagram, as shown in Figure 11.13. Moment M_z and shear force V_y are drawn according to our sign convention. By balancing the moment at point O, we obtain an expression for moment M_z,

$$M_z - P[v(L) - v(x)] + kv(L)(L - x) = 0$$

or

$$M_z + Pv(x) = Pv(L) - kv(L)(L - x) \tag{E1}$$

Substituting into Equation (7.1), we obtain the differential equation

$$EI_{zz}\frac{d^2v}{dx^2} + Pv(x) = Pv(L) - kv(L)(L - x) \tag{E2}$$

(a) Using Equation (11.3), Equation (E2) can be written as

$$\frac{d^2v}{dx^2} + \lambda^2 v = \lambda^2 v(L) - \frac{kv(L)}{EI}(L - x) \tag{E3}$$

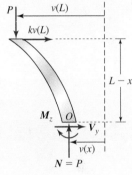

Figure 11.13 Free-body diagram in Example 11.3.

The zero deflection and slope boundary condition are also written to complete the statement of the boundary-value problem,

$$v(0) = 0 \tag{E4}$$

$$\frac{dv}{dx}(0) = 0 \tag{E5}$$

The homogeneous solution v_H to Equation (E3) is given by Equation (11.5). The particular solution to Equation (E3) is

$$v_P = v(L) - \frac{kv(L)}{\lambda^2 EI}(L - x)$$

Thus the total solution $v_H + v_P$ can be written as

$$v(x) = A \cos \lambda x + B \sin \lambda x + v(L) - \frac{kv(L)}{\lambda^2 EI}(L - x) \quad \text{(E6)}$$

Substituting $x = 0$ into Equation (E6) and using Equation (E4), we obtain

$$v(0) = A \cos(0) + B \sin(0) + v(L) - \frac{kv(L)}{\lambda^2 EI}(L - 0) = 0$$

or

$$A = \left(\frac{kL}{\lambda^2 EI} - 1\right)v(L) \quad \text{(E7)}$$

Differentiating Equation (E6), then substituting $x = 0$ and using Equation (E5), we obtain

$$\frac{dv}{dx}(0) = -\lambda A \sin(0) + B\lambda \cos(0) + \frac{kv(L)}{\lambda^2 EI} = 0$$

or

$$B = -\frac{k}{\lambda^3 EI}v(L) \quad \text{(E8)}$$

Substituting the values of A and B from Equations (E7) and (E8) into Equation (E6), we obtain

$$v(x) = \left[\left(\frac{kL}{\lambda^2 EI} - 1\right)\cos \lambda x - \frac{k}{\lambda^3 EI}\sin \lambda x + 1 - \frac{k}{\lambda^2 EI}(L - x)\right]v(L) \quad \text{(E9)}$$

Substituting $x = L$ into Equation (E9), we obtain

$$v(L) = \left[\left(\frac{kL}{\lambda^2 EI} - 1\right)\cos \lambda L - \frac{k}{\lambda^3 EI}\sin \lambda L + 1 - 0\right]v(L) \quad \text{(E10)}$$

Noting that $v(L)$ is a common factor, Equation (E10) can be simplified to the following *characteristic equation:*

$$\text{ANS.} \quad \left(\frac{kL}{\lambda^2 EI} - 1\right)\cos \lambda L - \frac{k}{\lambda^3 EI}\sin \lambda L = 0 \quad \text{(E11)}$$

(b) Substituting $k = 0$ into Equation (E11), we obtain $\cos \lambda L = 0$, which is the characteristic equation for case 2 in Table 11.1. Thus the P_{cr} value corresponding to the smallest root will be as given in Table 11.1 for case 2.

(c) We rewrite Equation (E11) as

$$\tan \lambda L = \lambda L - \frac{\lambda^3 EI}{k}$$

As k tends to infinity, the second term tends to zero and we obtain tan $\lambda L = \lambda L$, which is the characteristic equation for case 3 in Table 11.1. Thus the P_{cr} value corresponding to the smallest root will be as given in Table 11.1 for case 3.

COMMENTS

1. This example shows that a spring could simulate an imperfect support that provides some restraint to deflection. The restraining effect is more than zero (free end) but not as much as a roller support.

2. The spring could also represent other beams that are pin connected at the top end. These pin-connected beams provide elastic restraint to deflection but no restraint to the slope. If the beams were welded rather than pin connected, then we would have to include a torsional spring also at the end.

3. The example also demonstrates that the critical buckling loads can be changed by installing some elastic restraints, such as rings, to support the columns of the water tank in Figure 11.8.

EXAMPLE 11.4

Determine the maximum deflection of the column shown in Figure 11.14 in terms of the modulus of elasticity E, the length of the column L, the area moment of inertia I, the axial force P, and the intensity of the distributed force w.

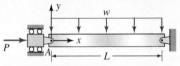

Figure 11.14 Buckling of beam with distributed load in Example 11.4.

PLAN

The moment from the distributed load can be added to the moment for case 1 in Table 11.1 and the differential equation written. The boundary conditions are that the deflection at $x = 0$ and $x = L$ is zero. The boundary-value problem can be solved, and the deflection at $x = L/2$ evaluated to obtain the maximum deflection.

Solution The reaction force in the y direction is half the total load wL acting on the beam. An imaginary cut at some location x can be made and the free-body diagram of the left part drawn as shown in Figure 11.15. By balancing the moment at point O, we obtain an expression for the moment M_z,

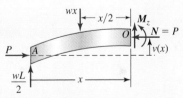

Figure 11.15 Free-body diagram in Example 11.4.

$$M_z + Pv(x) - \frac{wL}{2}x + \frac{wx^2}{2} = 0 \qquad \text{(E1)}$$

Substituting it into Equation (7.1), we obtain the differential equation

$$EI_{zz}\frac{d^2v}{dx^2} + Pv(x) = \frac{wL}{2}x - \frac{wx^2}{2} \qquad (E2)$$

Using Equation (11.3), Equation (E2) can be written as

$$\frac{d^2v}{dx^2} + \lambda^2 v = \frac{wLx}{2EI} - \frac{wx^2}{2EI} \qquad (E3)$$

The zero-deflection boundary conditions at either end are written as

$$v(0) = 0 \qquad (E4)$$

$$v(L) = 0 \qquad (E5)$$

to complete the statement of the boundary-value problem.

To find the particular solution, we substitute $v_P = a + bx + cx^2$ into Equation (E3) and simplify,

$$2c + \lambda^2(a + bx + cx^2) = \frac{wLx}{2EI} - \frac{wx^2}{2EI}$$

or

$$(2c + \lambda^2 a) + \left(\lambda^2 b - \frac{wL}{2EI}\right)x + \left(\lambda^2 c + \frac{w}{2EI}\right)x^2 = 0 \qquad (E6)$$

If Equation (E6) is to be valid for any value of x, then each of the terms in parentheses must be zero and we obtain the values of constants a, b, and c,

$$c = -\frac{w}{2\lambda^2 EI} \qquad b = \frac{wL}{2\lambda^2 EI} \qquad a = -\frac{2c}{\lambda^2} = \frac{w}{\lambda^4 EI}$$

Hence the particular solution is

$$v_P = \frac{w}{\lambda^4 EI} + \frac{wL}{2\lambda^2 EI}x - \frac{w}{2\lambda^2 EI}x^2 \qquad (E7)$$

The homogeneous solution v_H to Equation (E3) is given by Equation (11.5). Thus the total solution $v_H + v_P$ can be written as

$$v(x) = A\cos\lambda x + B\sin\lambda x + \frac{w}{\lambda^4 EI} + \frac{wL}{2\lambda^2 EI}x - \frac{w}{2\lambda^2 EI}x^2 \qquad (E8)$$

Substituting $x = 0$ into Equation (E8) and using Equation (E4), we obtain

$$v(0) = A\cos(0) + B\sin(0) + \frac{w}{\lambda^4 EI} + 0 - 0 = 0 \quad \text{or} \quad A = -\frac{w}{\lambda^4 EI}$$

$$(E9)$$

Substituting $x = L$ into Equation (E8) and using Equations (E5) and (E9), we obtain

$$v(L) = A \cos \lambda L + B \sin \lambda L + \frac{w}{\lambda^4 EI} + \frac{wL^2}{2\lambda^2 EI} - \frac{wL^2}{2\lambda^2 EI} = 0$$

$$-\frac{w}{\lambda^4 EI} \cos \lambda L + B \sin \lambda L + \frac{w}{\lambda^4 EI} = 0$$

Thus,

$$B = -\frac{w}{\lambda^4 EI} \frac{1 - \cos \lambda L}{\sin \lambda L} = -\frac{w}{\lambda^4 EI} \tan\left(\frac{\lambda L}{2}\right) \qquad \text{(E10)}$$

The simplification in Equation (E10) was done by noting that $\sin \lambda L = 2 \sin(\lambda L/2) \cos(\lambda L/2)$ and $1 - \cos \lambda L = 2 \sin^2(\lambda L/2)$. By symmetry the maximum deflection will occur at midpoint. Substituting $x = L/2$ into Equation (E8) and using Equations (E9) and (E10), we obtain

$$v_{max} = v\left(\frac{L}{2}\right) = A \cos\left(\frac{\lambda L}{2}\right) + B \sin\left(\frac{\lambda L}{2}\right) + \frac{w}{\lambda^4 EI} + \frac{wL^2}{4\lambda^2 EI} - \frac{wL^2}{8\lambda^2 EI}$$

$$= \frac{w}{\lambda^4 EI}\left[-\cos\left(\frac{\lambda L}{2}\right) - \tan\left(\frac{\lambda L}{2}\right) \sin\left(\frac{\lambda L}{2}\right)\right] + \frac{w}{\lambda^4 EI} + \frac{wL^2}{8\lambda^2 EI}$$

$$\text{(E11)}$$

Equation (E11) can be simplified by substituting the tangent function in terms of the sine and cosine functions to obtain

$$v_{max} = -\frac{w}{\lambda^4 EI}\left[\sec\left(\frac{\lambda L}{2}\right) - 1\right] + \frac{wL^2}{8\lambda^2 EI} \qquad \text{(E12)}$$

Substituting for λ, the maximum deflection can be written as

$$\text{ANS.} \qquad v_{max} = -\frac{wEI}{P^2}\left[\sec\left(\frac{L}{2}\sqrt{\frac{P}{EI}}\right) - 1\right] + \frac{wL^2}{8P}$$

COMMENTS

1. In Equation (E12) as $\lambda L \to \pi$, the secant function tends to infinity and the maximum displacement becomes unbounded, which means the column becomes unstable. $\lambda L = \pi$ corresponds to the Euler buckling load of Equation (11.8). Thus the distributed load does not change the critical load on the column.

2. Though the critical buckling load does not change with the distributed load, the failure mode can be significantly affected by the distributed load. The maximum normal stress will be the sum of axial stress and maximum bending normal stress, $\sigma_{max} = P/A + M_{max}y_{max}/I$. The maximum bending moment will be at $x = L/2$ and can be found from Equation (E1) as $M_{max} = wL^2/8 - Pv_{max}$. Substituting and simplifying, the maximum normal stress is given as

$$\sigma_{max} = \frac{P}{A} + \frac{wEy_{max}}{P}\left[\sec\left(\frac{L}{2}\sqrt{\frac{P}{EI}}\right) - 1\right]$$

By equating the maximum normal stress to the yield stress, we obtain a failure envelope, which clearly depends on the value of w.

*11.3 IMPERFECT COLUMNS

In the development of the theory for axial members and the symmetric bending of beams, it was shown that the condition for decoupling axial deformation from bending deformation for linear elastic homogeneous material was that the applied loads passed through the centroid of the cross sections, and the centroids of all cross sections were on a straight line. However, requirements for decoupling the axial from the bending problem may not be met for a number of reasons, some of which are given here:

- The column material may contain small holes, minute cracks, or other material inclusions. Hence the homogeneity requirement or the requirement that the centroids of all cross sections be on a straight line may not be met.
- The material processing may cause local strain hardening. Hence the condition of linear elastic material behavior across the entire cross section may not be met.
- The theoretical design centroid and the actual centroid are offset due manufacturing tolerances.
- Local conditions at the support result in the reaction force to be offset from the centroid.
- The transfer of loads from one member to another may not occur at the centroid.

This partial list can be considered as imperfections in the column, which cause the application of axial loads to be offset from the centroid of the cross section. This offset loading is termed *eccentric loading* on columns. In this section we study the impact of eccentricity in loading on buckling.

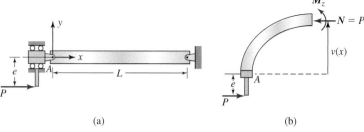

Figure 11.16 Eccentrically loaded column.

Figure 11.16*a* shows a simply supported column on which an eccentric compressive axial load is applied at a distance *e* from the centroid of the cross section. Figure 11.16*b* shows the free-body diagram of the column segment. By balancing the moment at point *A* we obtain $M_z + P(v + e) = 0$. Substituting the moment–curvature relationship of Equation (7.1), we obtain the differential equation

$$\frac{d^2v}{dx^2} + \lambda^2 v = -\frac{Pe}{EI} \tag{11.12}$$

where λ is given by Equation (11.3). The boundary conditions are that displacements at $x = 0$ and $x = L$ are zero, as given by Equation (11.4a) and (11.4b). The homogeneous solution to Equation (11.12) is given by Equation (11.5), that is, $v_H(x) = A \cos \lambda x + B \sin \lambda x$. The particular solution to Equation (11.12) is $v_P(x) = -e$. Thus the total solution $v_H + v_P$ is

$$v(x) = A \cos \lambda x + B \sin \lambda x - e \tag{11.13}$$

From boundary condition (11.4a) we obtain

$$v(0) = A \cos(0) + B \sin(0) - e = 0 \quad \text{or} \quad A = e$$

From boundary condition (11.4b) we obtain

$$v(L) = A \cos \lambda L + B \sin \lambda L - e = 0 \quad \text{or} \quad B = \frac{e(1 - \cos \lambda L)}{\sin \lambda L}$$

Using trigonometric identities,[6] *B* can be written as $B = e \tan(\lambda L/2)$. Substituting for *A* and *B* in Equation (11.13), we obtain the deflection as

$$v(x) = e\left[\cos \lambda x + \tan\left(\frac{\lambda L}{2}\right) \sin \lambda x - 1\right] \tag{11.14}$$

As $\lambda L/2 \to \pi/2$, the function $\tan(\lambda L/2) \to \infty$ and the displacement function $v(x)$ becomes unbounded. Thus the critical load value can be found by substituting for λ in the equation $\lambda L/2 = \pi/2$ to obtain the same critical value as given by Equation (11.8). In other words, the buckling load value does not change with the eccentricity of the loading. We will make use of this observation to extend our formulas to other types of support conditions.

In the eigenvalue approach discussed in Section 11.2 we were unable to determine the displacement function because we had an undetermined constant *B* in Equation (11.9). But here the displacement function is completely determined by

---

[6]The trigonometric identities used are $\sin \lambda L = 2 \sin(\lambda L/2) \cos(\lambda L/2)$ and $1 - \cos \lambda L = 2 \sin^2(\lambda L/2)$.

Equation (11.14). The maximum deflection (by symmetry) will be at midpoint. Substituting $x = L/2$ into Equation (11.14), we obtain

$$v_{max} = e\left[\cos\left(\frac{\lambda L}{2}\right) + \tan\left(\frac{\lambda L}{2}\right)\sin\left(\frac{\lambda L}{2}\right) - 1\right]$$

Using trigonometric identities, this equation can be simplified as $v_{max} = e[\sec(\lambda L/2) - 1]$. Substituting for λ from Equation (11.3), we obtain

$$v_{max} = e\left[\sec\left(\frac{L}{2}\sqrt{\frac{P}{EI}}\right) - 1\right] \tag{11.15}$$

We can write

$$\sqrt{\frac{P}{EI}} = \sqrt{\frac{PP_{cr}}{P_{cr}EI}} = \frac{\pi}{L}\sqrt{\frac{P}{P_{cr}}}$$

We obtain the maximum deflection equation as

$$\boxed{v_{max} = e\left[\sec\left(\frac{\pi}{2}\sqrt{\frac{P}{P_{cr}}}\right) - 1\right]} \tag{11.16}$$

The maximum normal stress is the sum of compressive axial stress and maximum compressive bending stress, that is,

$$\sigma_{max} = \frac{P}{A} + \frac{M_{max}y_{max}}{I}$$

The maximum bending moment will be at the midpoint of the column, and its value is $M_{max} = P(e + v_{max})$. Substituting for v_{max} we obtain the following equation for maximum normal stress:

$$\sigma_{max} = \frac{P}{A} + \frac{Py_{max}}{I}e\left[\sec\left(\frac{L}{2}\sqrt{\frac{P}{EI}}\right)\right] \tag{11.17}$$

Equation (11.17) was derived for simply supported columns. We can extend the results to other supports by changing the length of the column to the effective length L_{eff}, as given in Table 11.1. We also substitute $y_{max} = c$, where c represents the maximum distance from the buckling (bending) axis to a point on the cross section. Substituting $I = Ar^2$, where A is the cross-sectional area and r is the radius of gyration, we obtain

$$\boxed{\sigma_{max} = \frac{P}{A}\left[1 + \frac{ec}{r^2}\sec\left(\frac{L_{eff}}{2r}\sqrt{\frac{P}{EA}}\right)\right]} \tag{11.18}$$

Equation (11.18) is called the *secant formula*. The quantity ec/r^2 is called the *eccentricity ratio*.

By equating σ_{max} to failure stress σ_{fail} in Equation (11.18), we obtain the failure envelope for an imperfect column. The failure envelope equation can be written in nondimensional form as

$$\frac{P/A}{\sigma_{\text{fail}}}\left[1 + \frac{ec}{r^2}\sec\left(\frac{L_{\text{eff}}}{2r}\sqrt{\left(\frac{\sigma_{\text{fail}}}{E}\right)\frac{P/A}{\sigma_{\text{fail}}}}\right)\right] = 1 \qquad (11.19)$$

Equation (11.19) can be plotted for different materials, as shown in Figure 11.17. These curves can be used for metric as well for U.S. customary units[7] as the variables used in creating the plots are nondimensional. The failure stress in the cases of steel and aluminum would be the yield stress σ_{yield}, whereas for wood it would be the ultimate stress σ_{ult}. The curves can also be used for different end conditions by using the appropriate L_{eff} as given in Table 11.1.

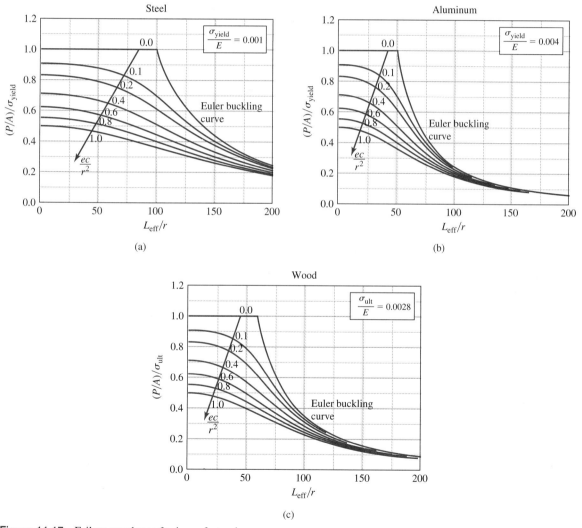

Figure 11.17   Failure envelopes for imperfect columns.

---

[7]The curves can be used for any material that has the same value for σ_{yield}/E.

EXAMPLE 11.5

A wooden box column ($E = 1800$ ksi) is constructed by joining four pieces of lumber together, as shown in Figure 11.18. The applied load is $P = 80$ kips. (a) If the length is $L = 10$ ft, what are the maximum stress and the maximum deflection? (b) If the allowable stress is 3 ksi, what is the maximum permissible length L to the nearest inch?

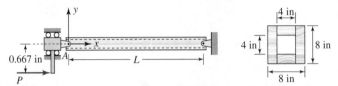

Figure 11.18 Eccentrically loaded box column.

PLAN

The cross-sectional area A, the area moment of inertia I, the radius of gyration r, and the maximum distance c from the bending (buckling) axis can be found from the cross-section dimensions. The effective length is the actual length L as the column is pin held at each end. (a) Substituting $L_{\text{eff}} = 120$ in and the values of the other variables into Equations (11.16) and (11.18), the maximum stress and the maximum deflection can be found. (b) Equating σ_{max} in Equation (11.18) to 3 ksi and substituting the remaining variables, the length L can be found.

Solution From the given cross section, the cross-sectional area A, the area moment of inertia I, and the radius of gyration r can be found:

$$A = 8 \times 8 - 4 \times 4 = 48 \text{ in}^2 \tag{E1}$$

$$I = \frac{1}{12}(8^4 - 4^4) = 320 \text{ in}^4 \tag{E2}$$

$$r = \sqrt{\frac{I}{A}} = 2.582 \text{ in} \tag{E3}$$

(a) As the column is pinned at both ends, $L_{\text{eff}} = L = 10 \text{ ft} = 120$ in. Substituting L_{eff}, I, and $E = 1800$ ksi into Equation (11.10), the critical buckling load can be found as

$$P_{\text{cr}} = \frac{\pi^2 \times 1800 \times 320}{120^2} = 394.8 \text{ kips} \tag{E4}$$

Substituting $e = 0.667$, $P = 80$ kips, and Equation (E4) into Equation (11.16), we obtain the maximum deflection,

$$v_{max} = 0.667 \left[\sec\left(\frac{\pi}{2} \sqrt{\frac{80}{394.8}} \right) - 1 \right] = 0.2103 \qquad \text{(E5)}$$

ANS. $v_{max} = 0.21$ in

Substituting $c = 4$ in, $e = 0.667$ in, $r = 2.582$ in, $P = 80$ kips, $E = 1800$ ksi, and $A = 48$ in^2 into Equation (11.18), we obtain the maximum normal stress,

$$\sigma_{max} = \frac{80}{48}\left[1 + \frac{0.667 \times 4}{2.582^2} \sec\left(\frac{120}{2 \times 2.582} \sqrt{\frac{80}{1800 \times 48}} \right) \right] = 2.544 \text{ ksi}$$

ANS. $\sigma_{max} = 2.5$ ksi (C)

(b) Substituting $\sigma_{max} = 3$ ksi, $c = 4$ in, $e = 0.667$ in, $r = 2.582$ in, $P = 80$ kips, $E = 1800$ ksi, and $A = 48$ in^2 into Equation (11.18), $L_{eff} = L$ can be found,

$$3 = \frac{80}{48}\left[1 + \frac{0.667 \times 4}{2.582^2} \sec\left(\frac{L}{2 \times 2.582} \sqrt{\frac{80}{1800 \times 48}} \right) \right]$$

or

$$\sec(5.892 \times 10^{-3} L) = 2 \qquad \text{or} \qquad \cos(5.892 \times 10^{-3} L) = 0.5$$

or

$$L = 177.7 \text{ in} \qquad \text{(E6)}$$

Rounding downward, the maximum permissible length is:

ANS. $L = 177$ in.

COMMENTS

1. The axial stress $P/A = 80/48 = 1.667$ ksi, but the normal stress due to bending from eccentricity causes the normal stress to be significantly higher, as seen by the value of σ_{max}.

2. If the right end of the column shown in Figure 11.18 were built in rather than held by a pin, then from case 3 in Table 11.1, $L_{eff} = 0.7L = 84$ in. Using this value, we can find $P_{cr} = 805.7$ kips, $v_{max} = 0.091$ in, and $\sigma_{max} = 2.42$ ksi.

3. In Equation (E6) we rounded downward, as shorter columns will result in a stress that is less than allowable.

EXAMPLE 11.6

A wooden box column ($E = 1800$ ksi) is constructed by joining four pieces of lumber together, as shown in Figure 11.18. The ultimate stress is 5 ksi. Determine the maximum load P that can be applied.

PLAN

The eccentricity ratio and the slenderness ratio can be found using the values of the geometric quantities calculated in Example 11.5. Noting that $\sigma_{ult}/E = 0.0028$, the failure envelopes for wood that are shown in Figure 11.17 can be used and $(P/A)/\sigma_{ult}$ can be found, from which the maximum load P can be determined.

Solution From Equation (E3) in Example 11.5, $r = 2.582$ in. Thus the slenderness ratio $L_{eff}/r = 120/2.582 = 46.48$. From Figure 11.18, $c = 4$ in and $e = 0.667$ in. Thus the eccentricity ratio $ec/r^2 = 0.400$.

From the failure envelope for wood in Figure 11.17, for a slenderness ratio of 46.48 and an eccentricity ratio of 0.4, we estimate the value of $(P/A)/\sigma_{ult} = 0.6$. Substituting $\sigma_{ult} = 5$ ksi and $A = 48$ in^2, we obtain the maximum load $P_{max} = 0.6 \times 5 \times 48$.

ANS. $P_{max} = 144$ kips.

COMMENT

If we let x represent $(P/A)/\sigma_{ult}$ and substitute the remaining variables in Equation (11.19), we obtain the following nonlinear equation: $x[1 + 0.4 \sec(1.2297\sqrt{x})] = 1$. The root of the equation can be found using a numerical method such as discussed in Appendix B. The value of the root to the third-place decimal is 0.593, which would yield a value of $P_{max} = 142.3$ kips, a difference of 1.18% from that reported in our example. The difference is small and an acceptable engineering approximation. Use of the plots in Figure 11.17 was a quick way of finding the load value with reasonable engineering approximation.

Consolidate your knowledge Write in one page all you understand about the buckling of columns.

QUICK TEST 11.1

Time: 15 minutes/Total: 20 points

Answer true or false. If false, give the correct explanation. Each question is worth two points. Use the solutions given in Appendix G to grade yourself.

1. Column buckling can be caused by tensile axial forces.
2. Buckling occurs about an axis with minimum area moment of inertia of the cross section.
3. If buckling is avoided at the Euler buckling load by the addition of supports in the middle, then the column will not buckle.
4. By changing the supports at the column end, the critical buckling load can be changed.
5. The addition of uniform transversely distributed forces decreases the critical buckling load on a column.
6. The addition of springs in the middle of the column decreases the critical buckling load.
7. Eccentricity in loading decreases the critical buckling load.
8. Increasing the slenderness ratio increases the critical buckling load.
9. Increasing the eccentricity ratio increases the normal stress in a column.
10. Material strength governs the failure of short columns and Euler buckling governs the failure of long columns.

PROBLEM SET 11.1

Stability of discrete systems

In Problems 11.1 through 11.3, in terms of the linear spring constant k and the length of the rigid bar L, determine the critical load value P_{cr}. The springs can be in tension or compression.

11.1 (Figure P11.1).

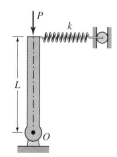

Figure P11.1

11.2 (Figure P11.2).

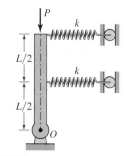

Figure P11.2

11.3 (Figure P11.3).

In Problems 11.4 through 11.6, linear deflection springs and torsional springs are attached to rigid bars as shown. The springs

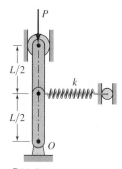

Figure P11.3

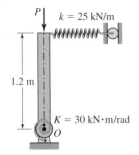

Figure P11.4

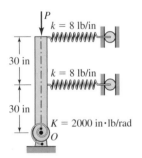

Figure P11.5

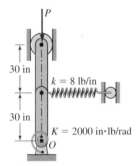

Figure P11.6

1. Square   2. Circle   3. Equilateral
triangle

Figure P11.14

can act in tension or in compression and resist rotation in either direction. Determine the critical load value P_{cr}.

11.4   (Figure P11.4).

11.5   (Figure P11.5).

11.6   (Figure P11.6).

Euler buckling

11.7   A hollow circular steel column ($E = 200$ GPa) is simply supported over a length of 5 m. The inner and outer diameters of the cross section are 75 mm and 100 mm. Determine: (a) the slenderness ratio; (b) the critical buckling load; (c) the axial stress at the critical buckling load. (d) If roller supports are added at the midpoint, what would be new critical buckling load?

11.8   A 30-ft-long hollow square steel column ($E = 30,000$ ksi) is built into the wall at either end. The column is constructed from $\frac{1}{2}$-in-thick sheet metal and has outer dimensions of 4 in × 4 in. Determine: (a) the slenderness ratio; (b) the critical buckling load; (c) the axial stress at the critical buckling load.

11.9   A 20-ft-long wooden column ($E = 1800$ ksi) has cross-section dimensions of 8 in × 8 in. The column is built in at one end and simply supported at the other end. Determine: (a) the slenderness ratio; (b) the critical buckling load; (c) the axial stress at the critical buckling load.

11.10   A W12 × 35 steel section (see Appendix E) is used for a 21-ft column that is simply supported at each end. Use $E = 30,000$ ksi and determine: (a) the slenderness ratio; (b) the critical buckling load; (c) the axial stress at the critical buckling load. (d) If roller supports are added at intervals of 7 ft, what would be the critical buckling load?

11.11   An S200 × 34 steel section (see Appendix E) is used as a 6-m column that is built in at each end. Use $E = 200$ GPa and determine: (a) the slenderness ratio; (b) the critical buckling load; (c) the axial stress at the critical buckling load.

11.12   Columns made from alloy will be used in the construction of a frame. The cross section of the columns is a hollow square of 0.125-in thickness and outer dimensions of a in. The modulus of elasticity $E = 9000$ ksi and the yield stress $\sigma_{yield} = 90$ ksi. Table 11.12 lists the lengths L and outer square dimensions a. Identify the long and short columns. Assume the ends will be simply supported.

TABLE P11.12   Column geometric properties

L (ft)	a (in)
1	1.125
1.5	1.5
2.0	1.75
2.5	2.75
3.0	3.0
3.5	3.0
4.0	3.0

11.13   Columns made from alloy will be used in the construction of a frame. The cross section of the columns is a hollow cylinder of 10-mm thickness and an outer diameter of d mm. The modulus of elasticity $E = 100$ GPa and the yield stress $\sigma_{yield} = 600$ MPa. Table P11.13 lists the lengths L and outer diameters d. Identify the long and short columns. Assume the ends of the column are built in.

TABLE P11.13   Column geometric properties

L (m)	d (mm)
1	60
2	80
3	100
4	150
5	200
6	225
7	250

11.14   Three column cross sections are shown in Figure P11.14. The area of each of the three cross sections is equal to A. Determine the ratios of critical loads $P_{cr1} : P_{cr2} : P_{cr3}$ assuming: (a) the ends are simply supported; (b) the ends are built in. (c) How do you expect the ratios to change if the end conditions were as in cases 2 and 3 of Table 11.1?

In Problems 11.15 through 11.17, a force
$F = 750$ lb is applied. Both bars have a diameter $d = \frac{1}{4}$ in, a modulus of elasticity
$E = 30,000$ ksi, and a yield stress
$\sigma_{yield} = 30$ ksi. Bars AP and BP have lengths
$L_{AP} = 8$ in and $L_{BP} = 10$ in. Determine the factor of safety for the two-bar structures.

11.15 (Figure P11.15).

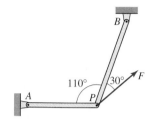

Figure P11.15

11.16 (Figure P11.16).

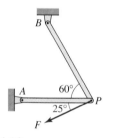

Figure P11.16

11.17 (Figure P11.17).

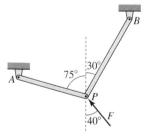

Figure P11.17

In Problems 11.18 through 11.20, a force
$F = 10$ kN is applied to the roller that slides
inside a slot. Both bars have a diameter
$d = 10$ mm, a modulus of elasticity $E =$
200 GPa, and a yield stress $\sigma_{yield} = 360$ MPa.
Bars AP and BP have lengths $L_{AP} = 200$ mm

and $L_{BP} = 300$ mm. Determine the factor of
safety for the two-bar structures.

11.18 (Figure P11.18).

11.19 (Figure P11.19).

11.20 (Figure P11.20).

Formulation and solutions

11.21 (a) Solve the boundary-value problem for case 2 in Table 11.1 and obtain the
critical load value P_{cr} that is given in the table.
(b) If buckling in mode 1 is prevented, then
what would be the P_{cr} value?

11.22 Solve the boundary-value problem
for case 3 in Table 11.1 and obtain the critical
load value P_{cr} that is given in the table. (b) If
buckling in mode 1 is prevented, then what
would be the P_{cr} value?

11.23 Solve the boundary-value problem
for case 4 in Table 11.1 and obtain the critical
load value P_{cr} that is given in the table. (b) If
buckling in mode 1 is prevented, then what
would be the P_{cr} value?

11.24 A torsional spring with a spring constant K is attached at one end of a column, as
shown in Figure P11.24. Assume that bending
about the y axis is prevented. (a) Determine
the characteristic equation for this buckling
problem. (b) Show that for $K = 0$ and $K = \infty$
the critical load P_{cr} is as given in Table 11.1
for cases 1 and 3, respectively.

Figure P11.24

11.25 A torsional spring with a spring constant K is attached at one end of a column, as
shown in Figure P11.25. Assume that bending
about the y axis is prevented. (a) Determine

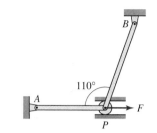

Figure P11.18

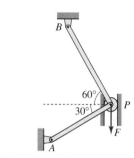

Figure P11.19

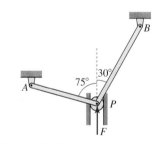

Figure P11.20

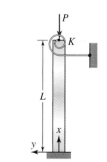

Figure P11.25

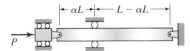

Figure P11.26

Figure P11.27

Figure P11.28

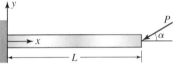

Figure P11.29

Figure P11.30

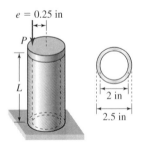

Figure P11.31

the characteristic equation for this buckling problem. (b) Show that for $K = 0$ the critical load P_{cr} is as given for case 2 in Table 11.1. (c) For $K = \infty$ obtain the critical load P_{cr}.

11.26   Consider the column shown in Figure P11.26. (a) Determine the critical buckling in terms of E, I, L, and α. (b) Show that when $\alpha = 0.5$, the critical load corresponds to mode 2, as shown in Figure 11.7.

In Problems 11.27 through 11.29, in terms of the modulus of elasticity E, the column length L, the area moment of inertia I, and the axial force P, determine (a) the deflection at x = L. (b) the critical load P_{cr}.

11.27   (Figure P11.27).

11.28   (Figure P11.28).

11.29   (Figure P11.29).

Imperfect columns

11.30   A column built in on one end and free at the other end has a load that is eccentrically applied at a distance e from the centroid, as shown in Figure P11.30. Show that the deflection curve is given by the equation

$$v(x) = \frac{e(1 - \cos \lambda x)}{\cos \lambda L}$$

where λ is as given by Equation (11.3).

11.31   On the cylinder shown in Figure P11.31 the applied load $P = 3$ kips, the length $L = 5$ ft, and the modulus of elasticity $E = 30,000$ ksi. What are the maximum stress and the maximum deflection?

11.32   On the cylinder shown in Figure P11.31 the applied load $P = 3$ kips and the modulus of elasticity $E = 30,000$ ksi. If the allowable normal stress is 8 ksi, what is the maximum permissible length L of the cylinder?

11.33   The length of the cylinder shown in Figure P11.31 is $L = 5$ ft. The yield stress of steel used in the cylinder is 30 ksi, and the modulus of elasticity $E = 30,000$ ksi. Determine the maximum load P that can be applied. Use the plot for steel in Figure 11.17.

11.34   On the column shown in Figure P11.34 the applied load $P = 100$ kN, the length $L = 2.0$ m, and the modulus of elasticity $E = 70$ GPa. What are the maximum stress and the maximum deflection?

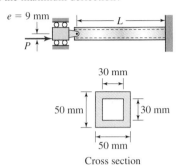

Cross section

Figure P11.34

11.35   On the column shown in Figure P11.34 the applied load $P = 100$ kN and the modulus of elasticity $E = 70$ GPa. If the allowable normal stress is 250 MPa, what is the maximum permissible length L of the column?

11.36   The length of the column shown in Figure P11.34 is $L = 2.0$. The yield stress of aluminum used in the column is 280 MPa, and the modulus of elasticity $E = 70$ GPa. Determine the maximum load P that can be applied. Use the plot for aluminum in Figure 11.17.

11.37   A wide-flange W8 × 18 member is used as a column, as shown in Figure P11.37. The applied load $P = 20$ kips, the length $L = 9$ ft, and the modulus of elasticity $E = 30,000$ ksi. What are the maximum stress and the maximum deflection?

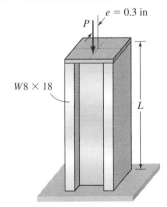

Figure P11.37

11.38 On the column shown in Figure P11.37 the applied load $P = 20$ kips and the modulus of elasticity $E = 30,000$ ksi. If the allowable normal stress is 24 ksi, what is the maximum permissible length L of the column?

11.39 The length of the column shown in Figure P11.37 is $L = 9$ ft. The yield stress of steel is 30 ksi, and the modulus of elasticity $E = 30,000$ ksi. Determine the maximum load P that can be applied. Use the plot for steel in Figure 11.17.

Design problems

11.40 A hoist is constructed using two wooden bars to lift a weight of 5 kips, as shown in Figure P11.40. The modulus of elasticity for wood $E = 1800$ ksi and the allowable normal stress is 3.0 ksi. Determine the maximum value of L to the nearest inch that can be used in constructing the hoist.

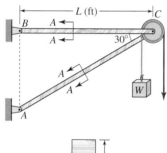

Cross section AA

Figure P11.40

11.41 Two steel cylinders ($E = 30,000$ ksi and $\sigma_{yield} = 30$ ksi) AB and CD are loaded as shown in Figure P11.41. Determine the maxi-

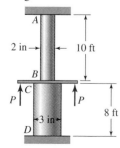

Figure P11.41

mum load P to the nearest lb, if a factor of safety of 2 is desired. Model the ends of column AB as built in.

11.42 A spreader is to be made from an aluminum pipe ($E = 10,000$ ksi) of $\frac{1}{8}$-in thickness and an outer diameter of 2 in, as shown in Figure P11.42. The pipe lengths available for design start from 4 ft in 6 in steps up to 8 ft. The allowable normal stress is 40 ksi. Develop a table for the lengths of pipe and the maximum force F the spreader can support.

11.43 Two 200-mm × 50-mm pieces of lumber ($E = 12.6$ GPa) form a part of a deck that is modeled as shown in Figure P11.43. The allowable stress for the lumber is 18 MPa. (a) Determine the maximum intensity of the distributed load w. (b) What is the factor of safety for column BD corresponding to the answer in part (a)?

11.44 Two 200-mm × 50-mm pieces of lumber ($E = 12.6$ GPa) form a part of a deck that is modeled as shown in Figure P11.44. The allowable stress for the lumber is 18 MPa. (a) Determine the maximum intensity of the distributed load w. (b) What is the factor of safety for column BC corresponding to the answer in part (a)?

11.45 A rigid bar hinged at point O has a force P applied to it, as shown in Figure P11.45. Bars A and B are made of steel with a modulus of elasticity $E = 30,000$ ksi and an allowable stress of 25 ksi. Bars A and B have circular cross sections with areas $A_A = 1$ in^2 and $A_B = 2$ in^2, respectively. Determine the maximum force P that can be applied.

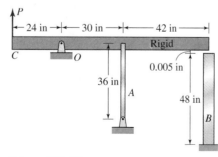

Figure P11.45

Stretch yourself

11.46 Show that for a beam with a constant bending rigidity EI, the fourth-order

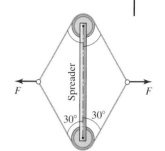

Figure P11.42

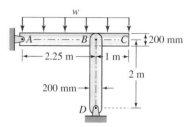

Figure P11.43

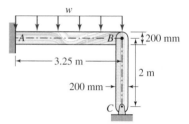

Figure P11.44

differential equation for solving buckling problems is given by

$$EI \frac{d^4v}{dx^4} + P \frac{d^2v}{dx^2} = p_y \qquad (11.20)$$

where P is a compressive axial force and p_y is the distributed force in the y direction.

11.47 Using Equation (11.20), solve Example 11.4.

11.48 Show that the critical change of temperature at which the beam shown in Figure P11.48 will buckle is given by

$$\Delta T_{\text{crit}} = \frac{\pi^2}{\alpha(L/r)^2}$$

where α is the thermal coefficient of expansion and r is the radius of gyration.

In Problems 11.49 and 11.50, the critical stress in intermediate columns is between yield stress and ultimate stress. The tangent modulus theory of buckling accounts for it by replacing the modulus of elasticity by the tangent modulus of elasticity (see Figure 3.7), that is,

$$P_{\text{cr}} = \frac{\pi^2 E_t I}{L_{\text{eff}}^2} \qquad (11.21)$$

where E_t is the tangent modulus, which depends on the stress level P_{cr}/A. Using an iterative trial and error procedure and Equation (11.21), the critical buckling load can be determined.

11.49 A simply supported 6-ft pipe has an outside diameter of 3 in and a thickness of $\frac{1}{8}$ in. The pipe material has the stress–strain curve shown in Figure P11.49. Using Equation (11.21), determine the critical buckling load.

11.50 A square box column is constructed from a sheet of 10-mm thickness. The outside dimensions of the square are 75 mm × 75 mm and the column has a length of 0.75 m. The material stress–strain curve is approximated as shown in Figure P11.50. Using Equation (11.21), determine the critical buckling load.

11.51 A column that is pin held at its ends has a small initial curvature, which is approximated by the sine function shown in Figure P11.51. Show that the elastic curve of

the column is given by

$$v(x) = \frac{v_0}{1 - P/P_{\text{cr}}} \sin \frac{\pi x}{L}$$

11.52 A column with a constant bending rigidity EI rests on an elastic foundation as shown in Figure P11.52. The foundation modulus is k, which exerts a spring force per unit length of kv. Show that the governing differential equation is

$$EI \frac{d^4v}{dx^4} + P \frac{d^2v}{dx^2} + kv = 0 \qquad (11.22)$$

(*Hint:* See Problems 7.31 and 11.46.)

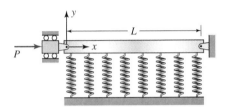

Figure P11.52

11.53 Show that the buckling load (eigenvalues) for the column on an elastic foundation described in Problem 11.52 is given by

$$P_n = \frac{\pi^2 EI}{L^2} \left[n^2 + \frac{1}{n^2} \left(\frac{kL^4}{\pi^4 EI} \right) \right],$$

$$n = 1, 2, 3, \dots \qquad (11.23)$$

Note: For $n = 1$ and $k = 0$ Equation (11.23) gives the Euler buckling load.

11.54 For a simply supported column with a symmetric composite cross section, show that the critical load P_{cr} is given by

$$P_{\text{cr}} = \frac{\pi^2 \sum_{i=1}^{n} E_i I_i}{L_{\text{eff}}^2} \qquad (11.24)$$

where L_{eff} = the effective length of the column, E_i is the modulus of elasticity for the ith material, I_i is the area moment of inertia about the buckling axis, and n is the number of materials in the cross section. [See Equations (6.30) and (11.1).]

11.55 In *double modulus theory,* also known as *reduced modulus theory* for intermediate columns, it was recognized that the bending action during buckling increases the compressive axial stress on the concave side of the beam but decreases the compressive stress on the convex

Figure P11.48

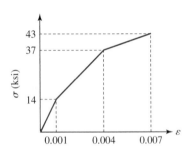

Figure P11.49

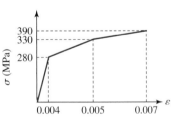

Figure P11.50

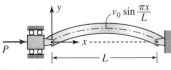

Figure P11.51

side of the beam. Thus the use of the tangent modulus of elasticity E_t is appropriate on the concave side, but on the convex side of the beam it may be better to use the original modulus of elasticity. Modeling the cross section material with the two moduli E_t and E and using Equation (11.24), show

$$P_{cr} = \frac{\pi^2 E_r I}{L_{eff}^2} \qquad E_r = E_t \frac{I_1}{I} + E \frac{I_2}{I} \quad (11.25)$$

where E_r is the *reduced modulus of elasticity,* I_1 and I_2 are the moments of inertia of the areas on the concave and convex sides of the axis passing through the centroid, and I is the moment of inertia of the entire cross section.

11.56 A composite column has the cross section shown in Figure P11.56. The modulus of elasticity of the outside material is twice that of the inside material. In terms of E, d, and L, determine the critical buckling load.

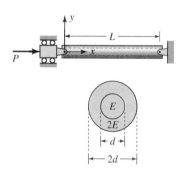

Figure P11.56

11.57 Two strips of material of a modulus of elasticity of $2E$ are attached to a material with a modulus of elasticity E to form a composite cross section of the column shown in Figure P11.57. In terms of E, a, and L, determine the critical buckling load. The column is free to buckle in any direction.

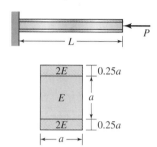

Figure P11.57

Computer problems

11.58 A circular marble column of 2-ft diameter and 20-ft length has a load P applied to it at a distance of 2 in from the center. The modulus of elasticity is 8000 ksi and the allowable stress is 20 ksi. Determine the maximum load P the column can support, assuming that both ends are (a) pinned; (b) built in.

11.59 Determine the maximum load P to the nearest Newton in Problem 11.36.

11.60 Determine the maximum load P to the nearest lb in Problem 11.39.

---

*11.4 GENERAL INFORMATION

Leonard Euler (1707–1782) is one of the most prolific mathematicians who ever lived (Figure 11.19). He has 866 entries in his bibliography. Born in Basel, he went to the University of Basel which at that time was known for its research in mathematics. After studying under John Bernoulli (see Section 7.6) he started work in 1727 at the Russian Academy at St. Petersburg, where he developed analytical methods for solving mechanics problems. At the invitation of King Frederick II of Prussia, he moved to Berlin in 1741, where he wrote three books: *Introduction to Calculus, Differential Calculus,* and *Integral Calculus.* In 1766 Catherine II, the empress of Russia, wooed him back to St. Petersburg. Even as he was going blind from cataract, he continued his prolific publications with the help of assistants.

As with the deflection of beams, the initial development with regard to the buckling of columns was primarily mathematical. Joseph-Louis Lagrange (1736–1813),

Figure 11.19 Leonard Euler.

Figure 11.20 Joseph-Louis Lagrange.

another pioneer in the establishment of analytical methods for problems in mechanics (Figure 11.20), took the next step after Euler and developed a complete set of buckling loads and the associated buckling modes given by Equations (11.7) and (11.9). Columns with eccentric loads (Problem 11.30) and columns with initial curvatures (Problem 11.51) were first formulated and studied by Thomas Young (1773–1829). He also initiated the consideration of columns with variable cross sections. Unfortunately Young was neither a good teacher nor a writer, and much of his work went unappreciated. Young's biographer Lord Rayleigh is quoted[8] as saying: "Young...from various causes did not succeed in gaining due attention from his contemporaries. Positions which he had already occupied were in more than one instance reconquered by his successors at great expense of intellectual energy." The importance of oral and writing skills in technical communications, which is emphasized today, has a historical lesson over two hundred years old.

There was another reason why in the early 1800s the mathematical developments in column buckling were unappreciated by the practicing engineer. At that time Euler buckling did not accurately predict compression failure in the structural members then in use. The impact of end conditions, imperfections, and the validity of the Euler buckling formula below proportional limits were later developments. The experiments of E. Hodgkinson in 1840 gave new life to the Euler buckling theory. In 1845 E. Lamarle, a French engineer, proposed correctly that the Euler formula should be used below the proportional limit, and after that experimentally determined formulas should be used for shorter columns. In 1889 F. Engesser, a German engineer, proposed the tangent modulus theory (see Problems 11.49 and 11.50), in which the elastic modulus was replaced by the tangent modulus of elasticity when proportional stress was exceeded. Also in 1889, the French engineer A. G. Considere, based on a series of tests, proposed that if buckling occurred above yield stress, then the elastic modulus in the Euler formula should be replaced by a reduced modulus of elasticity, which was between the elastic modulus and the tangent modulus. Engesser, on learning of Considere's work, incorporated Considere's suggestion and developed the *reduced modulus theory,* also known as *double modulus theory* (see Problem 11.51). In 1905 Johnson, Bryan, and Turneaure recommended that for steel columns the Euler formula modified by an experimentally determined constant for different supports should be used. This was the beginning of the concept of effective length to account for different end conditions. In 1946 F. R. Shanley, an American aeronautical engineering professor, refined the previous theories and resolved the paradoxes that separated the proponents of the reduced modulus theory and the tangent modulus theory.

For all the refinements and limitations, three centuries later the Euler buckling formula is still being used for column design and is still valid for long columns with pin-supported ends. Such is the power of logical thinking.

11.5 CLOSURE

Previous to this chapter, our analysis was based on the equilibrium of forces and moments. This chapter emphasized that not only is equilibrium important, but the stability of the equilibrium is as important a consideration in design. There are many

---

[8]Quotation is from S. P. Timoshenko, *History of Strength of Materials.*

types of instabilities. We studied one in which axial and bending deformation was coupled to produce the buckling of columns. The coupling emphasized the need for caution in decoupling phenomena for ease of understanding.

This chapter also concludes this text. The concepts of stress, strain, and the constitutive relationship between stress and strain are fundamental concepts in the mechanics of materials. These fundamental concepts are used in science to explain natural phenomena such as geological movements and the formation of hard and soft wood. The fundamental concepts of stress, strain, and constitutive relationships are used in engineering to develop formulas for the design of machine elements, buildings, and aircraft structures. The development of formulas follows a logic which is also used in more complex structural elements such as plates and shells. The modular nature of the logic provides means by which additional complexities can be incorporated into the theory. But nature being more complex than we can envisage, the formulas from our theory are usually modified by including experimentally determined factors prior to their use in design.

POINTS AND FORMULAS TO REMEMBER

- Buckling is the instability in equilibrium of a structure.
- Structural members that support compressive axial loads are called columns.
- Buckling is caused by compressive axial loads or stresses.
- Study of buckling as a *bifurcation* problem requires determining the critical buckling load at the point where two or more solutions exist for deformation.
- Study of buckling by the *energy method* requires determining the critical buckling load at the point the potential energy changes from a concave to a convex function.
- Study of buckling as an *eigenvalue* problem requires determining the critical buckling load at the point where a nontrivial solution exists for bending deformation due to axial loading.
- In *snap buckling* the structure snaps (jumps) from one equilibrium configuration to a very different equilibrium configuration at the critical buckling load.
- *Local buckling* of thin structural members occurs due to compressive stresses.
- Buckling of columns occurs about an axis that has a minimum value of area moment of inertia.

- Euler buckling load: $P_{cr} = \dfrac{\pi^2 EI}{L^2}$ (11.8)

  where P_{cr} is the critical buckling load, E is the modulus of elasticity, L is the length of the column, and I is the minimum area moment of inertia of the cross section.
- Equation (11.8) is only valid for elastic columns with pin-held ends.
- Effect of supports at the end can be incorporated by defining an effective length L_{eff} of a column and calculating the critical buckling load from

$$P_{cr} = \frac{\pi^2 EI}{L_{eff}^2} \qquad (11.10)$$

- *Slenderness ratio* is defined as L_{eff}/r, where r is the radius of gyration about the buckling axis.

- The slenderness ratio corresponding to the maximum normal stress being equal to the yield stress separates the short columns from the long columns in Euler buckling.

- Failure of short columns is governed by material strength.

- Failure of long columns is governed by Euler buckling loads.

- Eccentricity in loading does not affect the critical buckling load, but the maximum normal stress becomes significantly larger than the axial stress due to the addition of bending normal stress,

$$v_{max} = e\left[\sec\left(\frac{\pi}{2}\sqrt{\frac{P}{P_{cr}}}\right) - 1\right] \quad (11.16) \qquad \sigma_{max} = \frac{P}{A}\left[1 + \frac{ec}{r^2}\sec\left(\frac{L_{eff}}{2r}\sqrt{\frac{P}{EA}}\right)\right] \quad (11.18)$$

where v_{max} is the maximum deflection, e is the eccentricity in loading, P is the applied axial load, P_{cr} is the Euler buckling load for the column, σ_{max} is the maximum normal stress in the column, r is the radius of gyration about the buckling (bending) axis, c is the maximum distance perpendicular to the buckling (bending) axis, A is the cross-sectional area, L_{eff} is the effective length of the column.

- *Eccentricity ratio* is defined as ec/r^2.

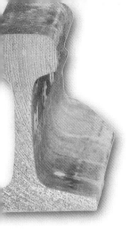

APPENDIX A
STATICS REVIEW

A.0 OVERVIEW

Statics is the foundation course for mechanics of materials. This appendix briefly reviews statics from a perspective of this course. The presentation presupposes that you are familiar with the concepts in statics. If you took a course in statics some time ago, then you may need to review your statics textbook before the brevity of this appendix serves you adequately. Review exams at the end of this appendix can be used for self-assessment.

A.1 TYPES OF FORCES AND MOMENTS

We can classify the forces and moments that we encounter in this book into three categories: external, internal, and reaction forces and moments.

A.1.1 External Forces and Moments

These are forces and moments that are applied to the body and are often referred to as the *load* on the body. They are assumed known in an analysis, though sometimes we carry external forces and moments as variables so that we may answer questions such as: how much load can a structure support? or what loads are needed to produce a given deformation?

Surface forces and moments are external forces (moments), which act on the surface and are transmitted to the body by contact. Surface forces (moments) applied at a point are called *concentrated* forces (moment or couple). Surface forces (moments) applied along a line or over a surface are called *distributed* forces (moments).

Body forces are external forces that act at every point on the body. Body forces are not transmitted by contact. Gravitational forces and electromagnetic forces are two examples of body forces. A body force has units of force per unit volume.

A.1.2 Reaction Forces and Moments

These are forces and moments that are developed at the supports of a body to resist movement due to the external forces (moments). Reaction forces (moments) are usually not known and must be calculated before further analysis can be conducted. The following three principles are used to decide whether there is a reaction force (reaction moment) at the support:

1. If a point cannot move in a given direction, then a reaction force opposite to the direction acts at that support point.
2. If a line cannot rotate about an axis in a given direction, then a reaction moment opposite to the direction acts at that support.
3. The support in isolation and not the entire body is considered in making decisions about the movement of a point or the rotation of a line at the support. Exceptions to the rule exist in three-dimensional problems, such as bodies supported by balanced hinges or balanced bearings (rollers). These types of three-dimensional problems will not be covered in this book.

Table A.1 shows several types of support that can be replaced by reaction forces and moments using the principles described in Section A.1.2.

A.1.3 Internal Forces and Moments

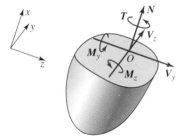

Figure A.1 Internal forces and moments.

A body is held together by internal forces. Internal forces exist irrespective of whether or not we apply external forces. The material resists changes due to applied forces and moments by increasing the internal forces. Our interest is in the resistance the material offers to the applied loads, that is, in the internal forces. Internal forces always exist in pairs that are equal and opposite on the two surfaces produced by an imaginary cut. The internal forces are shown in Figure A.1 and according to the convention in this book, all internal forces and moments are printed in bold italics: N = axial force; V_y, V_z = shear force; T = torque; M_y, M_z = bending moment. They are defined as follows:

Definition 1 Forces that are normal to the imaginary cut surface are called *normal forces*. The normal force that points away from the surface (pulls the surface) is called *tensile force*. The normal force that points into the surface (pushes the surface) is called *compressive force*.

Definition 2 The normal force acting in the direction of the axis of the body is called *axial force*.

Definition 3 Forces that are tangent to the imaginary cut surface are called *shear forces*.

Definition 4 Internal moments about an axis normal to the imaginary cut surface are called *torsional moments* or *torque*.

Definition 5 Internal moments about an axis tangent to the imaginary cut are called *bending moments*.

Type of Support	Reactions	Comments
 Roller on smooth surface	 R	Only downward translation is prevented. Hence the reaction force is upward.
 Smooth pin	 R_x R_y	Translation in the horizontal and vertical directions is prevented. Hence the reaction forces R_x and R_y can be in the directions shown, or opposite.
 Fixed support	 R_x M_z R_y	Beside translation in the horizontal and vertical directions, rotation about the z axis is prevented. Hence the reactions R_x and R_y and M_z can be in the directions shown, or opposite.
 Roller in smooth slot	 R	Translation perpendicular to slot is prevented. The reaction force R can be in the direction shown, or opposite.
 Ball and socket	 R_x R_z R_y	Translation in all directions is prevented. The reaction forces can be in the directions shown, or opposite.
 Hinge	 R_y M_y M_x R_x R_z	Except for rotation about the hinge axis, translation and rotation are prevented in all directions. Hence the reaction forces and moments can be in the directions shown, or opposite.
 Journal bearing	 M_z R_z M_y R_y	Translation and rotation are prevented in all directions, except in the direction of the shaft axis. Hence the reaction forces and moments can be in the directions shown, or opposite.
 Smooth slot	 M_x R_x M_z R_z M_y	Translation in the z direction and rotation about any axis are prevented. Hence the reaction force R_z and reaction moments can be in the directions shown, or opposite. Translation in the x direction into the slot is prevented but not out of it. Hence the reaction force R_x should be in the direction shown.

A.2 FREE-BODY DIAGRAMS

Newton's laws are only applicable to free bodies. By "free" we mean that if a body is not in equilibrium, it will move. If there are supports, then these supports must be replaced by appropriate reaction forces and moments using the principles described in Section A.1.2.

Definition 6 The diagram showing all the forces acting on a free body is called *free-body diagram* (FBD).

Additional free-body diagrams may be created by making imaginary cuts for the calculation of internal quantities. Each imaginary cut will produce two additional free-body diagrams. Either of the two free-body diagrams can be used for calculating internal forces and moments.

A body is in static equilibrium if the vector sum of all forces acting on a free body and the vector sum of all moments about any point in space are zero. Mathematically this is stated as

$$\sum \bar{F} = 0 \qquad \sum \bar{M} = 0 \tag{A.1}$$

where $\sum$ represents summation and the overbar represents a vector quantity. In a three-dimensional Cartesian coordinate system Equations (A.1) in scalar form are

$$\sum F_x = 0 \qquad \sum F_y = 0 \qquad \sum F_z = 0$$
$$\sum M_x = 0 \qquad \sum M_y = 0 \qquad \sum M_z = 0 \tag{A.2}$$

Equations (A.2) imply that there are six independent equations in three dimensions. In other words, we can at most solve for six unknowns from a free-body diagram in three dimensions.

In two dimensions the sum of the forces in the z direction and the sum of the moments about the x and y axes are automatically satisfied as all forces must lie in the x, y plane. The remaining equilibrium equations in two dimensions that have to be satisfied are

$$\sum F_x = 0 \qquad \sum F_y = 0 \qquad \sum M_z = 0 \tag{A.3}$$

Equations (A.3) imply that there are three independent equations per free-body diagram in two dimensions. In other words, we can at most solve for three unknowns from a free-body diagram in two dimensions.

The following observations can be used to reduce the computational effort:

- Balancing the moments at a point through which an unknown force (or forces) passes reduces the computational effort because such forces do not appear in the moment equation.

- Balancing the forces and/or moments perpendicular to the direction of an unknown force or moment reduces the computational effort because such forces do not appear in the equation.

Definition 7 A body on which the number of unknown reaction forces and moments is greater than the number of equilibrium equations (six in three dimensions and three in two dimensions) is called a *statically indeterminate body*.

Statically indeterminate problems arise when more supports than needed are used to support a structure. Extra supports may be used for safety considerations or for the purpose of increasing the stiffness of a structure. We define the following:

Definition 8 Degree of static redundancy = number of unknown reactions − number of equilibrium equations.

To solve a statically indeterminate problem we have to generate equations on the displacement or rotation at the support points. A mistake sometimes made is to take moments at many points in order to generate enough equations for the unknowns. A statically indeterminate problem cannot be solved from equilibrium equations alone. There are only three independent equations of static equilibrium in two dimensions and six independent equations of static equilibrium in three dimensions. Additional equations must come from displacements or rotation conditions at the support.

The number of equations on the displacement or rotation needed to solve a statically indeterminate problem is equal to the degree of static redundancy. There are two exceptions: (i) Symmetric structures with symmetric loadings. By using the arguments of symmetry one can reduce the total number of unknown reactions. (ii) Certain pin-connected structures. Pin connections do not transmit moments from one part of a structure to another. Thus it is possible that a seemingly indeterminate pin structure may be a determinate structure. We will not consider such pin-connected structures in this book.

Definition 9 A structural member on which there is no moment couple and forces act at two points only is called a *two-force member*.

Figure A.2 shows a two-force member. By balancing the moments at either point A or B we can conclude that the resultant forces at A and B must act along the line joining the two points. Notice that the shape of the member is immaterial. Identifying two-force members by inspection can save significant computation effort.

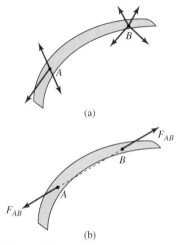

Figure A.2 Two-force member.

A.3 TRUSSES

Definition 10 A truss is a structure made up of two-force members.

The method of joints and the method of sections are two methods of calculating the internal forces in truss members.

In the method of joints a free-body diagram is created by making imaginary cuts on all members joined at the pin. If a force is directed away from the pin, then the two-force member is assumed to be in tension, and if it is directed into the pin, then the member is assumed to be in compression. By conducting force balance in two (three) dimensions two (three) equations per pin can be written.

In the method of sections an imaginary cut is made through the truss to produce a free-body diagram. The imaginary cut can be of any shape that will permit a quick calculation of the force in a member. Three equations in two dimensions or six equations in three dimensions can be written per free-body diagram produced from a single imaginary cut.

Definition 11 A zero-force member in a truss is a member that carries no internal force.

Identifying zero-force members can save significant computation time. Zero-force members can be identified by conducting the method of joints mentally. Usually if two members are collinear at a joint *and* if there is no external force, then the zero-force member is the member that is inclined to the collinear members.

A.4 CENTROIDS

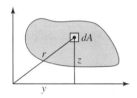

Figure A.3 Area moments.

The y and z coordinates of the centroid of the two-dimensional body shown in Figure A.3 are defined by

$$y_c = \frac{\int_A y \, dA}{\int_A dA} \qquad z_c = \frac{\int_A z \, dA}{\int_A dA} \tag{A.4}$$

The numerator in Equations (A.4) is referred to as the first moment of the area. If there is an axis of symmetry, then the area moment about the symmetric axis from one part of the body is canceled by the moment from the symmetric part, and hence we conclude that the centroid lies on the axis of symmetry.

Consider a coordinate system fixed to the centroid of the area. If we now consider the first moment of the area in this coordinate system and it turns out to be nonzero, then it would imply that the centroid is not located at the origin, thus contradicting our starting assumption. We therefore conclude:

- The first moment of the area calculated in a coordinate system fixed to the centroid of the area is zero.

The centroid for a composite body in which the centroids of the individual bodies are known can be calculated from the following equations:

$$y_c = \frac{\sum_{i=1}^{n} y_{c_i} A_i}{\sum_{i=1}^{n} A_i} \qquad z_c = \frac{\sum_{i=1}^{n} z_{c_i} A_i}{\sum_{i=1}^{n} A_i} \tag{A.5}$$

where y_{c_i} and z_{c_i} are the known coordinates of the centroids of the area A_i. Table A.2 shows the locations of the centroids of some common shapes that will be useful in solving problems in this book.

A.5 AREA MOMENTS OF INERTIA

The area moments of inertia, also referred to as second area moments, are defined as

$$I_{yy} = \int_A z^2 \, dA \qquad I_{zz} = \int_A y^2 \, dA \qquad I_{yz} = \int_A yz \, dA \tag{A.6}$$

TABLE A.2 Areas, centroids, and second area moments of inertia

Shapes*	Areas	Second Area Moments of Inertia
Rectangle	$A = ah$	$I_{zz} = \dfrac{1}{12}ah^3$
Circle	$A = \pi r^2$	$I_{zz} = \dfrac{1}{4}\pi r^4 \qquad J = \dfrac{1}{2}\pi r^4$
Triangle	$A = \dfrac{ah}{2}$	$I_{zz} = \dfrac{1}{36}ah^3$
Semicircle	$A = \dfrac{\pi r^2}{2}$	$I_{zz} = \dfrac{1}{8}\pi r^4$
Trapezoid	$A = \dfrac{h(a+b)}{2}$	
Quadratic curve	$A_1 = \dfrac{ah}{3}$ $A_2 = \dfrac{2ah}{3}$	$(I_{zz})_1 = \dfrac{1}{21}ah^3$ $(I_{zz})_2 = \dfrac{2}{7}ah^3$
Cubic curve	$A_1 = \dfrac{ah}{4}$ $A_2 = \dfrac{3ah}{4}$	$(I_{zz})_1 = \dfrac{1}{30}ah^3$ $(I_{zz})_2 = \dfrac{3}{10}ah^3$

*C—location of centroid.

The polar moment of inertia is defined as,

$$J = \int_A r^2 \, dA = I_{yy} + I_{zz} \tag{A.7}$$

with the relation to I_{yy} and I_{zz} deduced using Figure A.3.

If we know the area moment of inertia in a coordinate system fixed to the centroid, then we can compute the area moments about an axis parallel to the coordinate axis by the parallel-axis theorem illustrated in Figure A.4 and given by

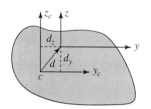

Figure A.4 Parallel-axis theorem.

$$
\begin{aligned}
I_{yy} &= I_{y_c y_c} + A d_y^2 \\
I_{zz} &= I_{z_c z_c} + A d_z^2 \\
I_{yz} &= I_{y_c z_c} + A d_y d_z \\
J &= J_c + A d^2
\end{aligned}
\tag{A.8}
$$

where the subscript c refers to the axis fixed to the centroid of the body. The quantities y^2, z^2, r^2, A, d_y^2, d_z^2, and d^2 are always positive. Therefore from Equations (A.6) through (A.8) we conclude that I_{yy}, I_{zz}, and J are always positive and minimum about the axis passing through the centroid of the body. However, I_{yz} can be positive or negative, as y, z, d_y, and d_z can be positive or negative in Equation (A.6). If either y or z is an axis of symmetry, then the integral in I_{yz} on the positive side will cancel the integral on the negative side in Equation (A.6), and hence I_{yz} will be zero. We record the observations as follows:

- I_{yy}, I_{zz}, and J are always positive and minimum about the axis passing through the centroid of the body.
- If either the y or the z axis is an axis of symmetry, then I_{yz} will be zero.

The moment of inertia of a composite body in which we know the moments of inertia of the individual bodies about its centroid can be calculated from

$$
\begin{aligned}
I_{yy} &= \sum_{i=1}^{n} (I_{y_{c_i} y_{c_i}} + A_i d_{y_i}^2) \\
I_{zz} &= \sum_{i=1}^{n} (I_{z_{c_i} z_{c_i}} + A_i d_{z_i}^2) \\
I_{yz} &= \sum_{i=1}^{n} (I_{y_{c_i} z_{c_i}} + A_i d_{y_i} d_{z_i}) \\
J &= \sum_{i=1}^{n} (J_{c_i} + A_i d_i^2)
\end{aligned}
\tag{A.9}
$$

where $I_{y_{c_i} y_{c_i}}$, $I_{z_{c_i} z_{c_i}}$, $I_{y_{c_i} z_{c_i}}$, and J_{c_i} are the area moments of inertia about the axes passing through the centroid of the ith body. Table A.2 shows the area moments of inertia about an axis passing through the centroid of some common shapes that will be useful in solving the problems in this book.

The radius of gyration $\hat{r}$ about an axis is defined by

$$\hat{r} = \sqrt{\frac{I}{A}} \qquad \text{or} \qquad I = A\hat{r}^2 \tag{A.10}$$

where I is the area moment of inertia about the same axis about which the radius of gyration $\hat{r}$ is being calculated.

A.6 STATICALLY EQUIVALENT LOAD SYSTEMS

Definition 12 Two systems of forces that generate the same resultant force and moment are called *statically equivalent* load systems.

If one system satisfies the equilibrium, then the statically equivalent system also satisfies the equilibrium as the resultant force and the resultant moment must be zero in both systems. The concept of a statically equivalent system can simplify an analysis significantly and is most often used in problems with distributed loads.

A.6.1 Distributed Force on a Line

Let $p(x)$ be a distributed force per unit length, which varies with x. We can replace this distributed force by a force and moment acting at any point or by a single force acting at point x_c, as shown in Figure A.5.

For two systems to be statically equivalent the resultant force and the resultant moment about any point (origin) must be the same. This implies

$$F = \int_L p(x)\, dx \qquad x_c = \frac{\int_L x\, p(x)\, dx}{F} \qquad (A.11)$$

The force F is equal to the area under the curve and x_c represents the location of the centroid of the distribution. This idea is used in replacing a uniform or a linearly varying distribution by a statically equivalent force, as shown in Figure A.6.

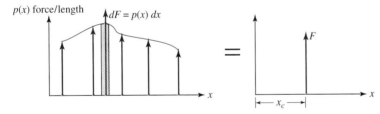

Figure A.5 Static equivalency for distributed force on a line.

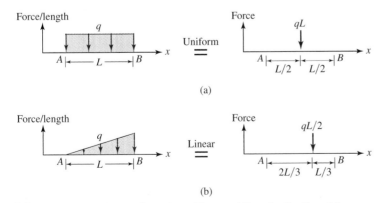

Figure A.6 Statically equivalent force for uniform and linearly distributed forces on a line.

Two statically equivalent systems are *not* identical systems. The deformation (change of shape of bodies) in two statically equivalent systems is different. The distribution of the internal forces and internal moments of two statically equivalent systems is different. The following rule must be remembered:

- The imaginary cut for the calculation of internal forces and moments must be made on the original body and not on the statically equivalent body.

A.6.2 Distributed Force on a Surface

Let $\sigma(y, z)$ be a distributed force per unit area that varies in intensity with y and z. We would like to replace it by a single force, as shown in Figure A.7.

For the two systems shown in Figure A.7 to be statically equivalent load systems, the resultant force and the resultant moment about the y axis and on the z axis must be the same. This implies

$$F = \iint_A \sigma(y, z) \, dy \, dz \qquad y_c = \frac{\iint_A y\sigma(y, z) \, dy \, dz}{F} \qquad z_c = \frac{\iint_A z\sigma(y, z) \, dy \, dz}{F} \tag{A.12}$$

The force F is equal to the volume under the curve. y_c and z_c represent the locations of the centroid of the distribution, which can be different from the centroid of the area on which the distributed force acts. The centroid of the area depends only on the geometry of that area. The centroid of the distribution depends on how the intensity of the distributed load $\sigma(y, z)$ varies over the area.

Figure A.8 shows a uniform and a linearly varying distributed force, which can be replaced by a single force at the centroid of the distribution. Notice that for the uniformly distributed force, the centroid of the distributed force is the same as the centroid of the rectangular area, but for the linearly varying distributed force, the centroid of the distributed force is different from the centroid of the area. If we were to place the equivalent force at the centroid of the area, then we would also need a moment at that point.

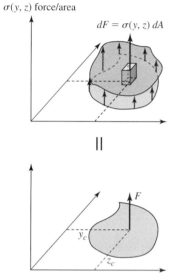

Figure A.7 Static equivalency for distributed force on a surface.

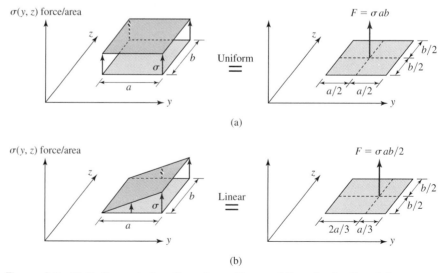

Figure A.8 Statically equivalent force for uniform and linearly distributed forces on a surface.

QUICK TEST A.1

Time: 15 minutes/Total: 20 points

Grade yourself using the answers and points given in Appendix G.

1. Three pin-connected structures are shown: (a) How many two-force members are there in each structure? (b) Which are the two-force member

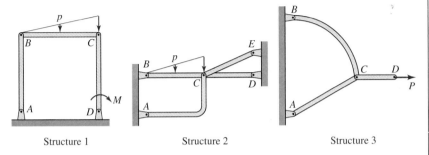

Structure 1   Structure 2   Structure 3

2. Identify all the zero-force members in the truss shown.

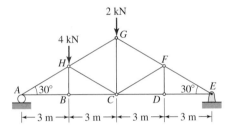

3. Determine the degree of static redundancy in each of the following structures and identify the statically determinate and indeterminate structures. Force P and torques T_1 and T_2 are known external loads.

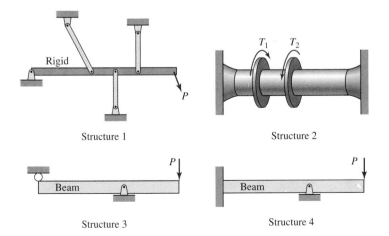

Structure 1   Structure 2

Structure 3   Structure 4

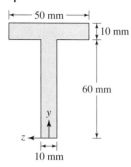

50 mm
10 mm
60 mm
y
z
10 mm

Figure R1.1

STATIC REVIEW EXAM 1

To get full credit, you must draw a free-body diagram anytime you use equilibrium equations to calculate forces or moments. Grade yourself using the solution and grading scheme given in Appendix F. Each question is worth 20points.

1. Determine: (a) the coordinates (y_c, z_c) of the centroid of the cross section shown in Figure R1.1; (b) the area moment of inertia about an axis passing through the centroid of the cross section and parallel to the z axis.

2. A linearly varying distributed load acts on a symmetric T section, as shown in Figure R1.2. Determine the force F and its location $(x_F, y_F$ coordinates) that is statically equivalent to the distributed load.

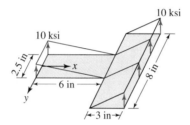

10 ksi
10 ksi
2.5 in
x
6 in
8 in
y
3 in

Figure R1.2

3. Find the internal axial force (indicate tension or compression) and the internal torque (magnitude and direction) acting on an imaginary cut through point E in Figure R1.3.

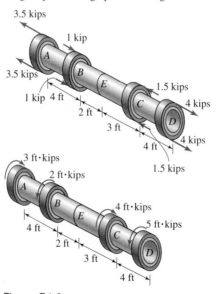

3.5 kips
1 kip
A
3.5 kips
B
E
1.5 kips
1 kip   4 ft
2 ft
C
4 kips
3 ft
D
4 ft
4 kips
1.5 kips

3 ft·kips
2 ft·kips
A
B
E
4 ft·kips
4 ft
5 ft·kips
2 ft
C
3 ft
D
4 ft

Figure R1.3

4. Determine the internal shear force and the internal bending moment acting at the section passing through A in Figure R1.4.

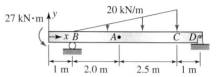

27 kN·m
y
20 kN/m
x B
A•
C   D
1 m   2.0 m   2.5 m   1 m

Figure R1.4

5. A system of pipes is subjected to a force P, as shown in Figure R1.5 By inspection (or by drawing a free-body diagram) identify the zero and nonzero internal forces and moments. Also indicate in the table the coordinate directions in which the internal shear forces and internal bending moments act.

Internal Force/ Moment	Section *AA* (zero/nonzero)	Section *BB* (zero/nonzero)
Axial force	_____	_____
Shear force	_____	_____
	in___ direction	in___ direction
Shear force	_____	_____
	in___ direction	in___ direction
Torque	_____	_____
Bending moment	in___ direction	in___ direction
Bending moment	in___ direction	in___ direction

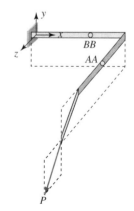

y
x
z
BB
AA
P

Figure R1.5

To get full credit, you must draw a free-body diagram anytime you use equilibrium equations to calculate forces or moments. Discuss the solution to this exam with your instructor.

1. Determine: (a) the coordinates (y_c, z_c) of the centroid of the cross section in Figure R2.1; (b) the area moment of inertia about an axis passing through the centroid of the cross section and parallel to the z axis.

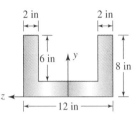

Figure R2.1

2. A distributed load acts on a symmetric C section, as shown in Figure R2.2. Determine the force F and its location $(x_F, y_F$ coordinates) that is statically equivalent to the distributed load.

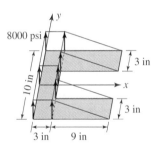

Figure R2.2

3. Find the internal axial force (indicate tension or compression) and the internal torque (magnitude and direction) acting on an imaginary cut through point E in Figure R2.3.

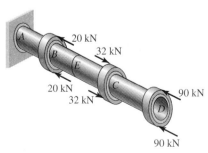

Figure R2.3

4. A simply supported beam is loaded by a uniformly distributed force of intensity 0.1 kip/in applied at 60°, as shown in Figure R2.4. Also applied is a force F at the centroid of the beam. Neglecting the effect of beam thickness, determine at section C the internal axial force, the internal shear force, and the internal bending moment.

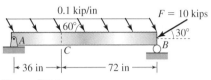

Figure R2.4

5. A system of pipes is subjected to a force P, as shown in Figure R2.5. By inspection (or by drawing a free-body diagram) identify the zero and nonzero internal forces and moments. Also indicate in the table the coordinate directions in which the internal shear forces and internal bending moments act.

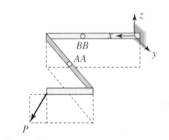

Figure R2.5

Internal Force/ Moment	Section *AA* (zero/nonzero)	Section *BB* (zero/nonzero)
Axial force	_____	_____
Shear force	_____	_____
	in___ direction	in___ direction
Shear force	_____	_____
	in___ direction	in___ direction
Torque	_____	_____
Bending moment	_____	_____
	in___ direction	in___ direction
Bending moment	_____	_____
	in___ direction	in___ direction

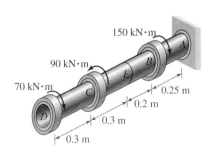

POINTS TO REMEMBER

- If a point cannot move in a given direction, then a reaction force opposite to the direction acts at that support point.
- If a line cannot rotate about an axis in a given direction, then a reaction moment opposite to the direction acts at that support.
- The support in isolation and not the entire body is considered in making decisions about the reaction at the support.
- Forces that are normal to the imaginary cut surface are called *normal forces*.
- The normal force that points away from the surface (pulls the surface) is called *tensile force*.
- The normal force that points into the surface (pushes the surface) is called *compressive force*.
- The normal force acting in the direction of the axis of the body is called *axial force*.
- Forces that are tangent to the imaginary cut surface are called *shear forces*.
- The internal moment about an axis normal to the imaginary cut surface is called torsional moment or *torque*.
- Internal moments about axes tangent to the imaginary cut are called *bending moments*.
- Calculation of internal forces and/or moments requires drawing a free-body diagram after making an imaginary cut.
- There are six independent equations in three dimensions and three independent equations in two dimensions per free-body diagram.
- A structure on which the number of unknown reaction forces and moments is greater than the number of equilibrium equations (6 in 3-D and 3 in 2-D) is called a statically *indeterminate* structure.
- Degree of static redundancy = number of unknown reactions − number of equilibrium equations.
- The number of equations on displacement and/or rotation we need to solve a statically indeterminate problem is equal to the degree of static redundancy.
- A structural member on which there is no moment couple and forces act at two points only is called a *two-force member*.
- The centroid lies on the axis of symmetry.
- The first moment of the area calculated in a coordinate system fixed to the centroid of the area is zero.
- I_{yy}, I_{zz}, and J are always positive and minimum about the axis passing through the centroid of the body.
- I_{yz} can be positive or negative.
- If either the y or the z axis is an axis of symmetry, then I_{yz} will be zero.
- Two systems that generate the same resultant force and moment are called *statically equivalent load systems*.
- The imaginary cut for the calculation of internal forces and moments must be made on the original body and not on the statically equivalent body.

APPENDIX B
ALGORITHMS FOR NUMERICAL METHODS

This appendix describes simple numerical techniques for evaluating the value of an integral, determining a root of a nonlinear equation, and finding constants of a polynomial by the least-squares method. Algorithms are given that can be programmed in any language. Also shown are methods of solving the same problems using a spreadsheet. If there are no other curriculum or pedagogical considerations, the author would recommend the use of spreadsheets based on the observation that spreadsheets are as ubiquitous as word processors and are easy to learn and use.

B.1   NUMERICAL INTEGRATION

We seek to numerically evaluate the integral

$$I = \int_a^b f(x)\, dx \tag{B.1}$$

where the function $f(x)$ and the limits a and b are assumed known.

This integral represents the area underneath the curve $f(x)$ in the interval defined by $x = a$ and $x = b$. The interval between a and b can be subdivided into N parts, as shown in Figure B.1. In each of the subintervals the function can be approximated by a straight-line segment. The area under the curve in each subinterval is the area of a trapezoid. Thus in the ith interval the area is $(\Delta x_i)[f(x_i) + f(x_{i-1})]/2$. By summing all the areas we obtain an approximate value of the total area represented by the integral in Equation (B.1),

$$I \cong \sum_{i=1}^{N} (\Delta x_i)\, \frac{f(x_i) + f(x_{i-1})}{2} \tag{B.2}$$

By increasing the value of N in Equation (B.2) we can improve the accuracy in our approximation of the integral. More sophisticated numerical integration schemes such as Gauss quadrature may be needed with increased complexity of the function $f(x)$.

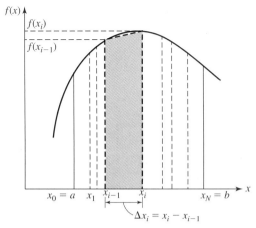

Figure B.1 Numerical integration by trapezoidal rule.

For the functions that will be seen in this book, integration by the trapezoidal rule given by Equation (B.2) will give adequate accuracy.

B.1.1 Algorithm for Numerical Integration

Following are the steps in the algorithm for computing numerically the value of an integral of a function, assuming that the function value $f(x_i)$ is known at $N + 1$ points x_i, where i varies from 0 to N.

  1. Read the value of N.
  2. Read the values of x_i and $f(x_i)$ for $i = 0$ to N.
  3. Initialize $I = 0$.
  4. For $i = 1$ to N, calculate $I = I + (x_i - x_{i-1})[f(x_i) + f(x_{i-1})]/2$.
  5. Print the value of I.

B.1.2 Use of Spreadsheet for Numerical Integration

Figure B.2 shows a sample spreadsheet that can be used to evaluate an integral numerically by the trapezoidal rule given by Equation (B.2). The data x_i and $f(x_i)$ can be either typed or imported into columns A and B of the spreadsheet, starting at row 2. In cells A2 and B2 are the values of x_0 and $f(x_0)$, and in cells A3 and B3 are the values of x_1 and $f(x_1)$. Using these values, the first term ($i = 1$) of the summation in Equation (B.2) can be found, as shown in cell C2. In a similar manner the second term of the summation in Equation (B.2) can be found and added to the result of the first term in cell C2. On copying the formula of cell C3, the spreadsheet automatically updates the column and row entries. Thus in all but the last entry we add one term of the summation at a time to the result of the previous row and obtain the final result.

	A	B	C	D	
1	x_i	$f(x_i)$	I		← Comment row
2	•	•	=(A3−A2)*(B3+B2)/2		
3	•	•	=C2+(A4−A3)*(B4+B3)/2		
4	•	•	Copy formula in cell C3		
5	•	•			
6	•	•			
7	•	•			

Figure B.2   Numerical integration algorithm on a spreadsheet.

B.2   ROOT OF A FUNCTION

We seek the value of x in a function that satisfies the equation

$$f(x) = 0 \tag{B.3}$$

We are trying to find that value of x at which $f(x)$ crosses the x axis. Suppose we can find two values of x for which the function $f(x)$ has different signs. Then we know that the root of Equation (B.3) will be bracketed by these values. Let the two values of x that bracket the root from the left and the right be represented by x_L and x_R. Let the corresponding function values be $f_L = f(x_L)$ and $f_R = f(x_R)$, as shown in Figure B.3a. We can find the mean value $x_N = (x_L + x_R)/2$ and calculate the function value $f_N = f(x_N)$. We compare the sign of f_N to those of f_L and f_R and replace the one with the same sign, as elaborated below.

   In iteration 1, f_N has the same sign as f_L; hence in iteration 2 we make the x_L value as x_N and the f_L value as f_N. In so doing we ensure that the root of the equation is still

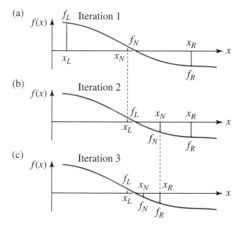

Figure B.3   Roots of an equation by halving the interval.

bracketed by x_L and x_R, but the size of the interval bracketing the root has been halved. On repeating the process in iteration 2, we find the mean value x_N, and the corresponding value f_N has the same sign as f_R. Thus for iteration 3, x_R and f_R are replaced by x_N and f_N found in iteration 2. In each iteration the root is bracketed by an interval that is half the interval in the previous iteration. When f_N reaches a small enough value, the iteration is stopped and x_N is the approximate root of Equation (B.3). This iterative technique for finding the root is called *halfinterval method* or *bisection method*.

B.2.1 Algorithm for Finding the Root of an Equation

The steps in the algorithm for computing the root of Equation (B.3) numerically are listed here. The computation of $f(x)$ should be done in a subprogram, which is not shown in the algorithm. It is assumed that the x_L and x_R values that bracket the root are known, but the algorithm checks to ensure that the root is bracketed by x_L and x_R. Note that if two functions have the same sign, then the product will yield a positive value.

1. Read the values of x_L and x_R.
2. Calculate $f_L = f(x_L)$ and $f_R = f(x_R)$.
3. If the product $f_L f_R > 0$, print "root of equation not bracketed" and stop.
4. Calculate $x_N = (x_L + x_R)/2$ and $f_N = f(x_N)$.
5. If the absolute value of f_N is less than 0.0001 (or a user-specified small number), then go to step 8.
6. If the product $f_L f_N > 0$, then replace x_L by x_N, and f_L by f_N. Go to step 4.
7. If the product $f_R f_N > 0$, then replace x_R by x_N, and f_R by f_N. Go to step 4.
8. Print the value of x_N as the root of the equation and stop.

B.2.2 Use of Spreadsheet for Finding the Root of a Function

Finding the roots of a function on a spreadsheet can be done without the algorithm described. The method is in essence a digital equivalent to making a plot to find the value of x where the function $f(x)$ crosses the x axis.

To demonstrate the use of a spreadsheet for finding the root of a function, consider the function $f(x) = x^2 - 28.54x + 88.5$. We guess that the root is likely to be a value of x between 0 and 10.

Trial 1: In cell A2 of Figure B.4*a* we enter our starting guess as $x = 0$. In cell A3 we increment the value of cell A2 by 1, then copy the formula in the next nine cells (copying into more cells will not be incorrect or cause problems). In cell B2 we write our formula for finding $f(x)$ and then copy it into the cells below. The results of this trial are shown in Figure B.4*b*. We note that the function value changes sign between $x = 3$ and $x = 4$ in trial 1.

Trial 2: Based on our results of trial 1, we set $x = 3$ as our starting guess in cell D2. In cell D3 we increment the value of cell D2 by 0.1 and then copy the formula into the cells below. We copy the formula for $f(x)$ from cell B2 into the column starting at cell E2. The results of this trial are given in Figure B.4*b*. The function changes sign between $x = 3.5$ and $x = 3.6$.

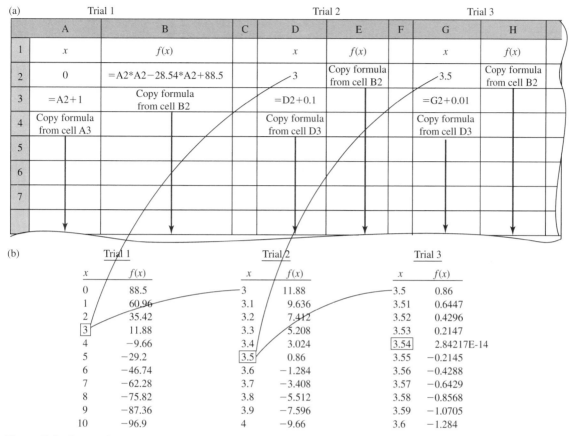

Figure B.4 Roots of an equation using spreadsheet.

Trial 3: Based on our results of trial 2, we set $x = 3.5$ as our starting guess in cell G2. In cell G3 we increment the value of cell G2 by 0.01 and then copy the formula into the cells below. We copy the formula for $f(x)$ from cell B2 into the column starting at cell H2. The results of this trial are given in Figure B.4b. The function value is nearly zero at $x = 3.54$, which gives us our root of the function.

The starting value and the increments in x are all educated guesses that will not be difficult to make for the problems in this book. If there are multiple roots, these too can be determined and, based on the problem, the correct root chosen.

B.3   DETERMINING COEFFICIENTS OF A POLYNOMIAL

We assume that at N points x_i we know the values of a function f_i. Often the values of x_i and f_i are known from an experiment. We would like to approximate the function by the quadratic function

$$f(x) = a_0 + a_1 x + a_2 x^2 \tag{B.4}$$

If $N = 3$, then there is a unique solution to the values of a_0, a_1, and a_2. However, if $N > 3$, then we are trying find the coefficients a_0, a_1, and a_2 such that the error of approximation is minimized. One such method of defining and minimizing the error in approximation is the least-squares method elaborated next.

If we substitute $x = x_i$ in Equation (B.4), the value of the function $f(x_i)$ may be different than the value f_i. This difference is the error e_i, which can be written as

$$e_i = f_i - f(x_i) = f_i - (a_0 + a_1 x_i + a_2 x_i^2) \tag{B.5}$$

In the least-squares method an error E is defined as $E = \sum_{i=1}^{N} e_i^2$. This error E is then minimized with respect to the coefficients a_0, a_1, and a_2 and to generate a set of linear algebraic equations. These equations are then solved to obtain the coefficients.

Minimizing E implies setting the first derivative of E with respect to the coefficients equal to zero, as follows. In these equations all summations are performed for $i = 1$ to N.

$$\frac{\partial E}{\partial a_0} = 0 \quad \text{or} \quad \sum 2 e_i \frac{\partial e_i}{\partial a_0} = 0 \quad \text{or} \quad \sum 2 [f_i - (a_0 + a_1 x_i + a_2 x_i^2)][-1] = 0$$

$$\frac{\partial E}{\partial a_1} = 0 \quad \text{or} \quad \sum 2 e_i \frac{\partial e_i}{\partial a_1} = 0 \quad \text{or} \quad \sum 2 [f_i - (a_0 + a_1 x_i + a_2 x_i^2)][-x_i] = 0$$

$$\frac{\partial E}{\partial a_2} = 0 \quad \text{or} \quad \sum 2 e_i \frac{\partial e_i}{\partial a_2} = 0 \quad \text{or} \quad \sum 2 [f_i - (a_0 + a_1 x_i + a_2 x_i^2)][-x_i^2] = 0$$

The equations on the right can be rearranged and written in matrix form,

$$\begin{bmatrix} N & \sum x_i & \sum x_i^2 \\ \sum x_i & \sum x_i^2 & \sum x_i^3 \\ \sum x_i^2 & \sum x_i^3 & \sum x_i^4 \end{bmatrix} \begin{Bmatrix} a_0 \\ a_1 \\ a_2 \end{Bmatrix} = \begin{Bmatrix} \sum f_i \\ \sum x_i f_i \\ \sum x_i^2 f_i \end{Bmatrix} \quad \text{or} \quad \begin{bmatrix} b_{11} & b_{12} & b_{13} \\ b_{21} & b_{22} & b_{23} \\ b_{31} & b_{32} & b_{33} \end{bmatrix} \begin{Bmatrix} a_0 \\ a_1 \\ a_2 \end{Bmatrix} = \begin{Bmatrix} r_1 \\ r_2 \\ r_3 \end{Bmatrix} \tag{B.6}$$

The coefficients of the b matrix and the r vector can be determined by comparison to the matrix form of the equations on the left. The coefficients a_0, a_1, and a_2 can be determined by Cramer's rule. Let D represent the determinant of the b matrix. By Cramer's rule, we replace the first column in the matrix of b's by the right-hand side, find the determinant of the so constructed matrix, and divide by D. Thus, the coefficients a_0, a_1, and a_2 can be written as

$$a_0 = \frac{\begin{vmatrix} r_1 & b_{12} & b_{13} \\ r_2 & b_{22} & b_{23} \\ r_3 & b_{32} & b_{33} \end{vmatrix}}{D} \qquad a_1 = \frac{\begin{vmatrix} b_{11} & r_1 & b_{13} \\ b_{21} & r_2 & b_{23} \\ b_{31} & r_3 & b_{33} \end{vmatrix}}{D} \qquad a_2 = \frac{\begin{vmatrix} b_{11} & b_{12} & r_1 \\ b_{21} & b_{22} & r_2 \\ b_{31} & b_{32} & r_3 \end{vmatrix}}{D}$$

Evaluating the determinant by expanding about the r elements, we obtain the values of a_0, a_1, and a_2

$$D = b_{11}C_{11} + b_{12}C_{12} + b_{13}C_{13} \tag{B.7}$$

$$a_0 = [C_{11}r_1 + C_{12}r_2 + C_{13}r_3]/D \tag{B.8}$$

$$a_1 = [C_{21}r_1 + C_{22}r_2 + C_{23}r_3]/D \tag{B.9}$$

$$a_2 = [C_{31}r_1 + C_{32}r_2 + C_{33}r_3]/D \tag{B.10}$$

where

$$C_{11} = b_{22}b_{33} - b_{23}b_{32} \qquad\qquad C_{12} = C_{21} = -[b_{21}b_{33} - b_{23}b_{31}]$$

$$C_{13} = C_{31} = b_{21}b_{32} - b_{22}b_{31} \qquad C_{22} = b_{11}b_{33} - b_{13}b_{31} \tag{B.11}$$

$$C_{23} = C_{32} = -[b_{11}b_{23} - b_{13}b_{21}] \qquad C_{33} = b_{11}b_{22} - b_{12}b_{21}$$

It is not difficult to extend these equations to higher order polynomials. However, a numerical method for solving the algebraic equations will be needed as the size of the b matrix grows. For problems in this book a quadratic representation of the function is adequate.

B.3.1   Algorithm for Finding Polynomial Coefficients

The steps in the algorithm for computing the coefficients of a quadratic function numerically by the least-squares method are listed here. It is assumed that x_i and f_i are known values at N points.

1. Read the value of N.
2. Read the values of x_i and f_i for $i = 1$ to N.
3. Initialize the matrix coefficients b and r to zero.
4. Set $b_{11} = N$.
5. For $i = 1$ to N, execute the following computations:

$$b_{12} = b_{12} + x_i \qquad b_{13} = b_{13} + x_i^2 \qquad b_{23} = b_{23} + x_i^3 \qquad b_{33} = b_{33} + x_i^4$$

$$r_1 = r_1 + f_i \qquad r_2 = r_2 + x_i f_i \qquad r_3 = r_3 + x_i^2 f_i$$

6. Set $b_{21} = b_{12}$, $b_{22} = b_{13}$, $b_{31} = b_{13}$, $b_{32} = b_{23}$.
7. Determine D using Equation (B.7).
8. Determine the coefficients a_0, a_1, and a_2 using Equations (B.8), (B.9), and (B.10).

B.3.2   Use of Spreadsheet for Finding Polynomial Coefficients

Figure B.5 shows a sample spreadsheet that can be used to evaluate the coefficients in a quadratic polynomial numerically. The data x_i and $f(x_i)$ can be either typed or imported into columns A and B of the spreadsheet, starting at row 2. In cells C2 through G2, the various quantities shown in row 1 can be found and the formulas copied to the rows below. We assume that the data fill up to row 50, that is, $N = 49$. In cell A51 the sum of the cells between cells A2 and A50 can be found using the summation command in the spreadsheet. By copying the formula to cells B51 through G51, the remaining sums in Equation (B.6) can be found. The coefficients in the b matrix and the right-hand-side r vector in Equation (B.6) can be identified as shown in comment row 52. The formulas in Equations (B.7) through (B.10) in terms of cell numbers can be entered in row 53, and D, a_0, a_1, and a_2 can be found.

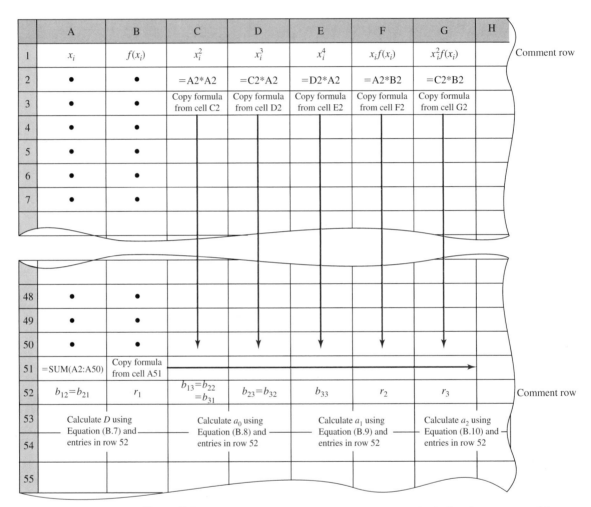

Figure B.5 Numerical evaluation of coefficients in a quadratic function on a spreadsheet.

APPENDIX C
CHARTS OF STRESS CONCENTRATION FACTORS

The stress concentration factor charts given in this appendix are approximate. For more accurate values the reader should consult a handbook. As per Equation (3.24) the stress concentration factor is defined as

$$K = \frac{\sigma_{max}}{\sigma_{nom}} \qquad\qquad \text{(C.1)}$$

where σ_{max} is the maximum stress and σ_{nom} the nominal stress obtained from elementary theories.

C.1 FINITE PLATE WITH A CENTRAL HOLE

Figure C.1 shows two stress concentration factors that differ because of the cross-sectional area used in the calculation of the nominal stress. If the gross cross-sectional area Ht of the plate is used, then we obtain the nominal stress $(\sigma_{nom})_{gross}$,

$$(\sigma_{nom})_{gross} = \frac{P}{Ht} \qquad\qquad \text{(C.2a)}$$

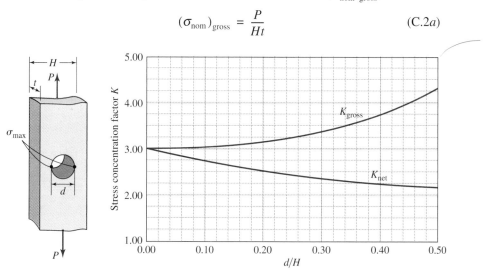

Figure C.1 Stress concentration factor for plate with a central hole.

and the top line in Figure C.1 should be used for the stress concentration factor. If the net area at the hole $(H - d)t$ is used, then we obtain the nominal stress $(\sigma_{nom})_{net}$

$$(\sigma_{nom})_{net} = \frac{P}{(H - d)t} \tag{C.2b}$$

and the bottom line in Figure C.1 should be used for the stress concentration factor. The two stress concentration factors are related as shown in the equation

$$K_{net} = \left(1 - \frac{d}{H}\right)K_{gross} \tag{C.2c}$$

C.2 STEPPED AXIAL CIRCULAR BARS WITH SHOULDER FILLET

The maximum axial stress in a stepped circular bar with shoulder fillet will depend on the values of the diameters D and d of the two circular bars and the radius of the fillet r. From these three variables we can create two nondimensional variables D/d and r/d for showing the variation of the stress concentration factor, as illustrated in Figure C.2. The maximum nominal axial stress will be in the smaller diameter bar and, as per Equation (4.9), is given by

$$\sigma_{nom} = \frac{4P}{\pi d^2} \tag{C.3}$$

C.3 STEPPED CIRCULAR SHAFTS WITH SHOULDER FILLET IN TORSION

The maximum shear stress in a stepped circular shaft with shoulder fillet will depend on the values of the diameters D and d of the two circular shafts and the radius of the fillet r. From these three variables we can create two nondimensional variables D/d and r/d for showing the variation of the stress concentration factor, as illustrated in Figure C.3. The maximum nominal shear stress will be on the outer surface of the smaller diameter bar and, as per Equation (5.8), is given by

$$\tau_{nom} = \frac{16T}{\pi d^3} \tag{C.4}$$

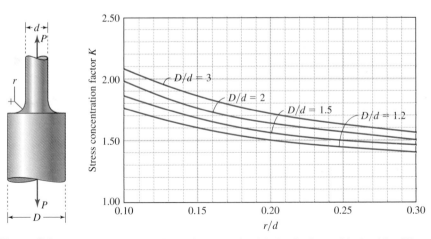

Figure C.2 Stress concentration factor for stepped axial circular bars with shoulder fillet.

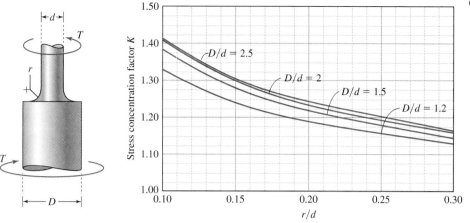

Figure C.3   Stress concentration factor for stepped circular shaft with shoulder fillet.

C.4 STEPPED CIRCULAR BEAM WITH SHOULDER FILLET IN BENDING

The maximum bending normal stress in a stepped circular beam with shoulder fillet will depend on the values of the diameters D and d of the two circular shafts and the radius of the fillet r. From these three variables we can create two nondimensional variables D/d and r/d for showing the variation of the stress concentration factor, as illustrated in Figure C.4. The maximum nominal bending normal stress will be on the outer surface in the smaller diameter bar and, as per Equation (6.12), is given by

$$\sigma_{\text{nom}} = \frac{32M}{\pi d^3} \qquad\qquad (C.5)$$

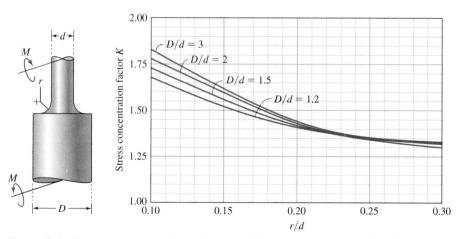

Figure C.4   Stress concentration factor for stepped circular beam with shoulder fillet.

APPENDIX D
PROPERTIES OF SELECTED MATERIALS

Material properties depend on many variables and vary widely. The properties given here are approximate mean values. Elastic strength may be represented by yield stress, proportional limit, or offset yield stress. Both elastic strength and ultimate strength refer to tensile strength unless stated otherwise.

TABLE D.1 Material properties in U.S. customary units

Material	Specific Weight (lb/in³)	Modulus of Elasticity E (ksi)	Poisson's Ratio ν	Coefficient of Thermal Expansion α (μ/°F)	Elastic Strength (ksi)	Ultimate Strength (ksi)	Ductility (% elongation)
Aluminum	0.100	10,000	0.25	12.5	40	45	17
Bronze	0.320	15,000	0.34	9.4	20	50	20
Concrete	0.087	4000	0.15	6.0		2*	
Copper	0.316	15,000	0.35	9.8	12	35	35
Cast iron	0.266	25,000	0.25	6.0	25*	50*	
Glass	0.095	7500	0.20	4.5		10	
Plastic	0.035	400	0.4	50		9	50
Rock	0.098	8000	0.25	4	12*	78*	
Rubber	0.041	0.3	0.5	90	0.5	2	300
Steel	0.284	30,000	0.28	6.6	30	90	30
Titanium	0.162	14,000	0.33	5.3	135	155	13
Wood	0.02	1800	0.30			5*	

*Compressive strength.

TABLE D.2   Material properties in metric units

Material	Density (mg/m³)	Modulus of Elasticity E (GPa)	Poisson's Ratio ν	Coefficient of Thermal Expansion α ($\mu/°C$)	Elastic Strength (MPa)	Ultimate Strength (MPa)	Ductility (% elongation)
Aluminum	2.77	70	0.25	12.5	280	315	17
Bronze	8.86	105	0.34	9.4	140	350	20
Concrete	2.41	28	0.15	6.0		14*	
Copper	8.75	105	0.35	9.8	84	245	35
Cast iron	7.37	175	0.25	6.0	175*	350*	
Glass	2.63	52.5	0.20	4.5		70	0
Plastic	0.97	2.8	0.4	50		63	50
Rock	2.72	56	0.25	4	84*	546*	
Rubber	1.14	2.1	0.5	90	3.5	14	300
Steel	7.87	210	0.28	6.6	210	630	30
Titanium	4.49	98	0.33	5.3	945	1185	13
Wood	0.55	12.6	0.30			35*	

*Compressive strength.

APPENDIX E
GEOMETRIC PROPERTIES OF STRUCTURAL STEEL MEMBERS

TABLE E.1 Wide-flange sections (FPS units)

Designation (in × lb/ft)	Depth d (in)	Area A (in²)	Web Thickness t_W (in)	Flange Width b_F (in)	Flange Thickness t_F (in)	I_{zz} (in⁴)	S_z (in³)	r_z (in)	I_{yy} (in⁴)	S_y (in³)	r_y (in)
W12 × 35	12.50	10.3	0.300	6.560	0.520	285.0	45.6	5.25	24.5	7.47	1.54
W12 × 30	12.34	8.79	0.260	6.520	0.440	238	38.6	5.21	20.3	6.24	1.52
W10 × 30	10.47	8.84	0.300	5.81	0.510	170	32.4	4.38	16.7	5.75	1.37
W10 × 22	10.17	6.49	0.240	5.75	0.360	118	23.2	4.27	11.4	3.97	1.33
W8 × 18	8.14	5.26	0.230	5.250	0.330	61.9	15.2	3.43	7.97	3.04	1.23
W8 × 15	8.11	4.44	0.245	4.015	0.315	48	11.8	3.29	3.41	1.70	0.876
W6 × 20	6.20	5.87	0.260	6.020	0.365	41.4	13.4	2.66	13.3	4.41	1.50
W6 × 16	6.28	4.74	0.260	4.03	0.405	32.1	10.2	2.60	4.43	2.20	0.967

TABLE E.2 Wide-flange sections (metric units)

Designation (mm × kg/m)	Depth d (mm)	Area A (mm²)	Web Thickness t_W (mm)	Flange Width b_F (mm)	Flange Thickness t_F (mm)	I_{zz} (10⁶ mm⁴)	S_z (10³ mm³)	r_z (mm)	I_{yy} (10⁶ mm⁴)	S_y (10³ mm³)	r_y (mm)
W310 × 52	317	6650	7.6	167	13.2	118.6	748	133.4	10.20	122.2	39.1
W310 × 44.5	313	5670	6.6	166	11.2	99.1	633	132.3	8.45	101.8	38.6
W250 × 44.8	266	5700	7.6	148	13.0	70.8	532	111.3	6.95	93.9	34.8
W250 × 32.7	258	4190	6.1	146	9.1	49.1	381	108.5	4.75	65.1	33.8
W200 × 26.6	207	3390	5.8	133	8.4	25.8	249	87.1	3.32	49.9	31.2
W200 × 22.5	206	2860	6.2	102	8.0	20.0	194.2	83.6	1.419	27.8	22.3
W150 × 29.8	157	3790	6.6	153	9.3	17.23	219	67.6	5.54	72.4	28.1
W150 × 24	160	3060	6.6	102	10.3	13.36	167	66	1.844	36.2	24.6

TABLE E.3 S shapes (FPS units)

Designation (in × lb/ft)	Depth d (in)	Area A (in^2)	Web Thickness t_W (in)	Flange Width b_F (in)	Flange Thickness t_F (in)	z Axis I_{zz} (in^4)	z Axis S_z (in^3)	z Axis r_z (in)	y Axis I_{yy} (in^4)	y Axis S_y (in^3)	y Axis r_y (in)
S12 × 35	12	10.3	0.428	5.078	0.544	229	38.4	4.72	9.87	3.89	0.98
S12 × 31.8	12	9.35	0.350	5.000	0.544	218	36.4	4.83	9.36	3.74	1.0
S10 × 35	10	10.3	0.594	4.944	0.491	147	29.4	3.78	8.36	3.38	0.901
S10 × 25.4	10	7.46	0.311	4.661	0.491	124	24.7	4.07	6.79	2.91	0.954
S8 × 23	8	6.77	0.411	4.171	0.426	64.9	16.2	3.10	4.31	2.07	.798
S8 × 18.4	8	5.41	0.271	4.001	0.426	57.6	14.4	3.26	3.73	1.86	0.831
S7 × 20	7	5.88	0.450	3.860	0.392	42.4	12.1	2.69	3.17	1.64	0.734
S7 × 15.3	7	4.50	0.252	3.662	0.392	36.9	10.5	2.86	2.64	1.44	0.766

TABLE E.4 S shapes (metric units)

Designation (mm × kg/m)	Depth d (mm)	Area A (mm^2)	Web Thickness t_W (mm)	Flange Width b_F (mm)	Flange Thickness t_F (mm)	z Axis I_{zz} (10^{6}mm^4)	z Axis S_z (10^{3}mm^3)	z Axis r_z (mm)	y Axis I_{yy} (10^{6}mm^4)	y Axis S_y (10^{3}mm^3)	y Axis r_y (mm)
S310 × 52	305	6640	10.9	129	13.8	95.3	625	119.9	4.11	63.7	24.9
S310 × 47.3	305	6032	8.9	127	13.8	90.7	595	122.7	3.90	61.4	25.4
S250 × 52	254	6640	15.1	126	12.5	61.2	482	96.0	3.48	55.2	22.9
S250 × 37.8	254	4806	7.9	118	12.5	51.6	406	103.4	2.83	48.0	24.2
S200 × 34	203	4368	11.2	106	10.8	27.0	266	78.7	1.794	33.8	20.3
S200 × 27.4	203	3484	6.9	102	10.8	24	236	82.8	1.553	30.4	21.1
S180 × 30	178	3794	11.4	97	10.0	17.65	198.3	68.3	1.319	27.2	18.64
S180 × 22.8	178	2890	6.4	92	10.0	15.28	171.7	72.6	1.099	23.9	19.45

APPENDIX F

SOLUTIONS TO STATIC REVIEW EXAM

REVIEW EXAM 1

1 point

1. As the y axis is the axis of symmetry, the centroid will lie on the y axis. Thus $z_c = 0$.

Equations (A.5) and (A.9) can be used to find the y coordinate of the centroid and the area moment of inertia.

Calculation of centroid (Figure F.1).

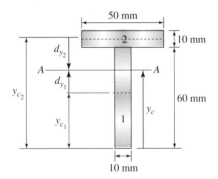

Figure F.1

Section	y_{c_i} (mm)	A_i (mm²)	$y_{c_i} A_i$ (mm³)
1	30	60 × 10 = 600	18,000
2	65	50 × 10 = 500	32,500
Total		1100	50,500

1 point for each correct entry
Total 8 points

From Equation (A.5) we obtain

1 point for correct answer and units

$$y_c = \frac{50,500}{1100} = 45.9 \text{ mm}$$

750

Section	$d_{z_i} = y_c - y_{c_i}$ (mm)	$I_{z_i z_i} = \frac{1}{12} a_i b_i^3$ (mm^4)	$I_{z_i z_i} + A_i d_{z_i}^2$ (mm^4)
1	15.9	$10 \times 60^3/12 = 180 \times 10^3$	331.7×10^3
2	19.1	$50 \times 10^3/12 = 4.2 \times 10^3$	186.6×10^3
	1 point for each correct entry	2 points for each correct entry	1 point for each correct entry

From Equation (A.9) we obtain

$$I_{AA} = (331.7 + 186.6)10^3 = 518.\,3 \times 10^3 \text{ mm}^4$$

1 point for correct answer
1 point for correct units

2. We can replace each linear loading by an equivalent force, as shown in Figure A.8, then replace it by a single force. Using Figure F.2 we obtain

$$F_1 = \frac{10 \times 2.5 \times 6}{2} = 75 \text{ kips}$$

$$F_2 = \frac{10 \times 8 \times 3}{2} = 120 \text{ kips}$$

For correct locations of forces F_1 and F_2, 3 points/force; for correct calculations of F_1 and F_2, 3 points/force—total 12 points

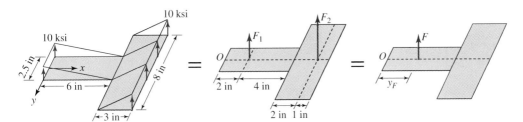

Figure F.2

The resultant forces for the two systems on the right in Figure F.2 must be the same. We thus obtain

2 points for correct answer
1 point for correct units

$$F = F_1 + F_2 = 75 + 120 = 195 \text{ kips}$$

The resultant moments about any point (point O) for the two systems on the right of Figure F.2 must also be the same. We obtain

$$2F_1 + 8F_2 = y_F F \qquad \text{or} \qquad y_F = \frac{150 + 960}{195} = 5.69 \text{ in}$$

1 point for each correct entry in this equation

1 point for correct answer
1 point for correct units

3. We make an imaginary cut at E and draw the free-body diagrams shown in Figures F.3 and F.4.

Internal axial force calculations (Figure F.3)

Either FBD is acceptable.
For drawing: 7-kip force at *A* or 8 kips
at *D*—2 points

2 kips at *B* or 3 kips at *C*—
2 points

Normal force at *E*—
2 points

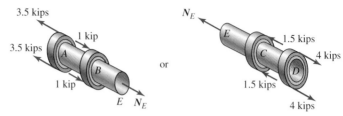

Figure F.3

1 point for correct equation

1 point for correct answer
1 point for correct units
1 point for reporting tension

$$N_E - 7 + 2 = 0 \quad \text{or} \quad N_E - 8 + 3 = 0$$

$$N_E = 5 \text{ kips (T)}$$

Internal torque calculations (Figure F.4)

Either FBD is acceptable.
For drawing: torque at *A* or *D*—
2 points

torque at *B* or *C*—
2 points

torque at *E* (either direction)—2 points

Figure F.4

2 points for correct equation

1 point for correct answer
1 point for correct units

$$T_E - 3 + 2 = 0 \quad \text{or} \quad T_E - 5 + 4 = 0$$

$$T_E = 1 \text{ ft·kips}$$

4. We can draw the free-body diagram of the entire beam as shown in Figure F.5*a*. By
balancing the moment at *D* we find the reaction at *B*,

$$5.5R_B - 27 - 45 \times 2.5 = 0 \quad \text{or} \quad R_B = 25.36 \text{ kN}$$

We can then make an imaginary cut at *A* on the original beam and draw the free-body
diagram in Figure F.5*b*.

1 point for each force or moment and
1 point for correct location of *F*—total
5 points

1 point for each term in equation—
total 3 points

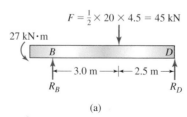

$F = \frac{1}{2} \times 20 \times 4.5 = 45$ kN

(a)

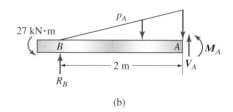

(b)

Figure F.5

We can find the intensity of the distributed force at A by similar triangles (Figure 6a),

$$\frac{p_A}{2} = \frac{20}{4.5} \quad \text{or} \quad p_A = 8.89 \text{ kN/m}$$

We can then replace the distributed force on the beam that is cut at A and draw the free-body diagram shown in Figure F.6b.

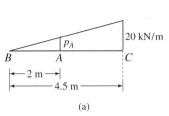

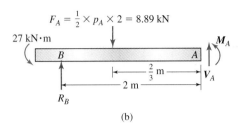

$F_A = \frac{1}{2} \times p_A \times 2 = 8.89$ kN

(a)   (b)

Figure F.6

1 point for correct value of F_A
1 point for correct location of F_A

3 points for correct calculation of p_A 2 points for showing V_A and M_A irrespective of direction

Balancing the forces in the y direction we obtain

$$V_A + 25.36 - 8.89 = 0 \quad \text{or} \quad V_A = -16.5 \text{ kN}$$

1 point for correct equation corresponding to your direction of V_A 1 point for correct answer

Balancing moments about point A we obtain

$$M_A + 27 - 25.36 \times 2 + 8.89 \times 2/3 = 0 \quad \text{or} \quad M_A = 17.8 \text{ kN·m}$$

2 points for correct equation corresponding to your direction of M_A 1 point for correct answer

5. By inspection we can write the following answers.

Internal Force/Moment	Section *AA* (zero/nonzero)	Section *BB* (zero/nonzero)
Axial force	Nonzero	Zero
Shear force	Nonzero in y direction	Nonzero in y direction
Shear force	Zero in x direction	Nonzero in z direction
Torque	Zero	Nonzero
Bending moment	Nonzero in x direction	Nonzero in y direction
Bending moment	Zero in y direction	Nonzero in z direction

1 point for each correct zero/nonzero entry
1 point for each correct direction
Total 20 points

ANSWERS TO QUICK TESTS

1 point for each correct true/false
1 point for each correct explanation

QUICK TEST 1.1

1. False. Stress is an internal quantity that can only be inferred but cannot be measured directly.

2. True. A surface has a unique normal, and normal stress is the internally distributed force in the direction of the normal.

3. True. Shear stress is an internally distributed force, and internal forces are equal and opposite in direction on the two surfaces produced by an imaginary cut.

4. False. Tension implies that the normal stress pulls the imaginary surface outward, which will result in opposite directions for the stresses on the two surfaces produced by the imaginary cut.

5. False. kips are units of force not stress.

6. False. The normal stress should be reported as tension.

7. False. 1 GPa equals 10^9 Pa.

8. False. 1 psi nearly equals 7 kPa not 7 Pa.

9. False. Failure stress values are in millions of pascals for metals.

10. False. Pressure on a surface is always normal to the surface and compressive. Stress on a surface can be normal or tangential to it, and the normal component can be tensile or compressive.

QUICK TEST 1.2

1. False. Stress at a point is a second-order tensor.

2. True. Each of the two subscripts can have three values, resulting in nine possible combinations.

3. True. The remaining three components can be found from the symmetry of shear stresses.

4. True. The fourth component can be found from the symmetry of shear stresses.

5. False. A point in plane stress has four nonzero components; thus only five components are zero in general.

6. False. The sign of stress incorporates both the direction of the force and the direction of the imaginary surface.

7. True. A stress element is an imaginary object representing a point.

8. False. The normals of the surface of a stress element have to be in the direction of the coordinate system in which the stress at a point is defined.

9. True. Stress is an internally distributed force system that is equal and in opposite directions on the two surfaces of an imaginary cut.

10. False. The sign of stress incorporates the direction of the force and the direction of the imaginary surface. Alternately, the sign of stress at a point is independent of the orientation of the imaginary cut surface.

QUICK TEST 2.1

1. Displacement is the movement of a point with respect to a fixed coordinate system, whereas deformation is the relative movement of a point with respect to another point on the body.

2. The reference geometry is the original undeformed geometry in Lagrangian strain and the deformed geometry in Eulerian strain.

3. The value of normal strain is $0.3/100 = 0.003$.

4. The value of normal strain is $2000 \times 10^{-6} = 0.002$.

5. Positive shear strain corresponds to a decrease in the angle from right angle.

6. The strain will be positive as it corresponds to extension and is independent of the orientation of the rod.

7. No. We have defined small strain to correspond to strains less than 1%.

8. There are nine strain nonzero components in three dimensions.

9. There are four nonzero strain components in plane strain.

10. There are only three independent strain components in plane strain as the fourth strain component can be determined from the symmetry of shear strains.

Each correct answer is worth 2 points

QUICK TEST 3.1

1. The modulus of elasticity has units of pascals or newtons per square meter. For metals it is usually gigapascals (GPa). Poisson's ratio has no units as it is dimensionless.

2. Offset yield stress is the stress value corresponding to a plastic strain equal to a specified offset strain.

3. Strain hardening is the increase in yield stress that occurs whenever yield stress is exceeded.

4. Necking is the sudden decrease in cross-sectional area after the ultimate stress.

5. Proportional limit defines the end of the *linear* region, whereas yield point defines the end of the *elastic* region.

6. A brittle material exhibits little plastic deformation before rupture, whereas a ductile material can undergo large plastic deformation before rupture.

7. A linear material behavior implies that stress and strain be linearly related. An elastic material behavior implies that when the loads are removed, the material returns to the undeformed state but the stress–strain relationship can be nonlinear, such as in rubber.

8. Strain energy is the energy due to deformation in a volume of material, whereas strain energy density is the strain energy per unit volume.

9. The modulus of resilience is a measure of recoverable energy and represents the strain energy density at yield point. The modulus of toughness is a measure of total energy that a material can absorb through elastic as well as plastic deformation and represents the strain energy density at ultimate stress.

10. A strong material has a high ultimate stress, whereas a tough material may not have high ultimate stress but has a large strain energy density at ultimate stress.

Each correct answer is worth 2 points

QUICK TEST 3.2

1. In an isotropic material the stress–strain relationship is the same in all directions but can differ at different points. In a homogeneous material the stress–strain relationship is the same at all points provided the directions are the same.

or

In an isotropic material the material constants are independent of the orientation of the coordinate system but can change with the coordinate locations. In a homogeneous material the material constants are independent of the locations of the coordinates but can change with the orientation of the coordinate system.

2. There are only two independent material constants in an isotropic linear elastic material.

3. 21 material constants are needed to specify the most general linear elastic anisotropic material.

4. There are three independent stress components in plane stress problems.

5. There are three independent strain components in plane stress problems.

6. There are four nonzero strain components in plane stress problems.

7. There are three independent strain components in plane strain problems.

8. There are three independent stress components in plane strain problems.

9. There are four nonzero stress components in plane strain problems.

10. For most materials E is greater than G as Poisson's ratio is greater than zero and $G = E/2(1 + v)$. In composites, however, Poisson's ratio can be negative; in such a case E will be less than G.

1 point for each correct true/false
1 point for each correct explanation

QUICK TEST 4.1

1. True. Material models do not affect the kinematic equation of a uniform strain.

2. False. Stress is uniform over each material but changes as the modulus of elasticity changes with the material in a nonhomogeneous cross section.

3. True. In the formulas A is the value of a cross-sectional area at a given value of x.

4. False. The formula is only valid if N, E, and A do not change between x_1 and x_2. For a tapered bar A is changing with x.

5. True. The formula does not depend on external load. External loads affect the value of N but not the relationship of N to σ_{xx}.

6. False. The formula is only valid if N, E, and A do not change between x_1 and x_2. For a segment with distributed load N changes with x.

7. False. The equation represents static equivalency of N and σ_{xx}, which is independent of material models.

8. False. The equation represents static equivalency of N and σ_{xx} over the entire cross section and is independent of material models.

9. True. The uniform axial stress distribution for a homogeneous cross section is represented by an equivalent internal force acting at the centroid which will be also collinear with external forces. Thus no moment will be necessary for equilibrium.

10. True. The equilibrium of a segment created by making an imaginary cut just to the left and just to the right of the section where an external load is applied shows the jump in internal forces.

QUICK TEST 5.1

1. True. Torsional shear strain for circular shafts varies linearly.

2. True. The shear strain variation is independent of material behavior across the cross section.

3. False. If the shear modulus of a material on the inside is significantly greater than that of the material on the outside, then it is possible for the shear stress on the outer edge of the inside material to be higher than that at the outermost surface.

4. True. The shear stress value depends on the J at the section containing the point and not on the taper.

5. False. The formula is obtained assuming J is constant between x_1 and x_2.

6. True. The shear stress value depends on the T at the section. The equilibrium equation relating T to external torque is a separate equation.

7. False. The formula is obtained assuming T is constant between x_1 and x_2, but in the presence of distributed torque, T is a function of x.

8. False. The equation represents static equivalency and is independent of material models.

9. True. Same reason as in question 8.

10. True. Equilibrium equations require that the difference between internal torques on either side of the applied torque equal the value of the applied torque.

QUICK TEST 6.1

1 point for each correct true/false
1 point for each correct explanation

1. True. Bending normal strain varies linearly and is zero at the centroid of the cross section. By knowing the strain at another point the equation of a straight line can be found.

2. True. Bending normal stress varies linearly and is zero at the centroid and maximum at the point farthest from the centroid. Knowing the stress at two points on a cross section, the equation of a straight line can be found.

3. False. The larger moment of inertia is about the axis parallel to the 2-in side, which requires that the bending forces be parallel to the 4-in side.

4. True. The stresses are smallest near the centroid. Alternatively, the loss in moment of inertia is minimum when the hole is at the centroid.

5. False. y is measured from the centroid of the beam cross section.

6. True. The formula is valid at any cross section of the beam. I_{zz} has to be found at the section where the stress is being evaluated.

7. False. The equations are independent of the material model and are obtained from static equivalency principles, and the bending normal stress distribution is such that the net axial force on a cross section is zero.

8. True. The equation is independent of the material model and is obtained from the static equivalency principle.

9. True. The equilibrium of forces requires that the internal shear force jump by the value of the applied transverse force as one crosses applied force from left to right.

10. True. The equilibrium of moments requires that the internal moment jump by the value of the applied moment as one crosses applied moment from left to right.

QUICK TEST 8.1

2 points for each correct answer

1. $\theta = 115°$ or $295°$ or $-65°$
2. $\theta = 245°$ or $65°$ or $-115°$
3. $\theta = 155°$ or $-25°$ or $335°$
4. $\sigma_1 = 5$ ksi (T)
5. $\sigma_1 = 5$ ksi (C)
6. $\tau_{max} = 10$ ksi
7. $\tau_{max} = 12.5$ ksi
8. $\tau_{max} = 10$ ksi
9. $\theta_1 = 55°$ or $-125°$
10. $\theta_1 = -35°$

2 points for each correct answer

QUICK TEST 8.2

1. D
2. A
3. E
4. 12° ccw or 168° cw
5. 102° ccw or 78° cw
6. 78° ccw or 102° cw
7. D
8. A
9. B
10. $\sigma = 30$ MPa (T), $\tau = -40$ MPa

2 points for each correct true
1 point for each correct false
1 point for each correct explanation

QUICK TEST 8.3

1. False. There are always three principal stresses. In two-dimensional problems the third principal stress is not independent and can be found from the other two.
2. True.
3. False. Material may affect the state of stress, but the principal stresses are unique for a given state of stress at a point.
4. False. The unique value of principal stress depends only on the state of stress at the point and not on how these stresses are measured or described.
5. False. Planes of maximum shear stress are always at 45° to the principal planes, and not 90°.
6. True.
7. True.
8. False. Depends on the value of the third principal stress.
9. False. Each plane is represented by a single point on Mohr's circle.
10. False. Each point on Mohr's circle represents a single plane

QUICK TEST 9.1

1.	D	2 points for correct answer
2.	C	2 points for correct answer
3.	B	2 points for correct answer
4.	C	2 points for correct answer
5.	D	2 points for correct answer
6.	108° ccw or 72° cw	1 point for correct answer
7.	18° ccw or 162° cw	1 point for correct answer
8.	$\varepsilon_1 = 1300\ \mu$, $\gamma_{max} = 2000\ \mu$	1 point for each correct answer
9.	$\varepsilon_1 = 2300\ \mu$, $\gamma_{max} = 2300\ \mu$	1 point for each correct answer
10.	$\varepsilon_1 = -300\ \mu$, $\gamma_{max} = 2300\ \mu$	1 point for each correct answer

QUICK TEST 9.2

1. (a) $\varepsilon_{yy} = 800\ \mu$; (b) $\varepsilon_{yy} = 800\ \mu$ — 1 point for each correct answer

2. $\theta = +115°$ or $-65°$ — 1 point for correct answer

3. $\theta = +155°$ or $-25°$ — 1 point for correct answer

4. $\theta = +25°$ or $-155°$ — 1 point for correct answer

5. $\gamma_{max} = 2100\ \mu$ — 2 points for correct answer

6. $\gamma_{max} = 3100\ \mu$ — 2 points for correct answer

7. $\gamma_{max} = 1700\ \mu$ — 2 points for correct answer

8. False. There are always three principal strains. In two-dimensional problems the third principal strain is not independent and can be found from the other two. — 1 point for reporting false / 1 point for correct explanation

9. False. The unique value of principal strains depends only on the state of strain at the point and not on how these strains are measured or described. — 1 point for reporting false / 1 point for correct explanation

10. False. Only for isotropic materials the principal coordinates for stresses and strains are the same, but for any anisotropic materials the principal coordinates for stresses and strains are different. — 1 point for reporting false / 1 point for correct explanation

QUICK TEST 11.1

1. False. Only compressive axial forces can cause column buckling.

2. True.

3. False. There are infinite buckling loads. The addition of supports changes the buckling mode to the next higher critical buckling load.

4. True.

5. False. The critical buckling load does not change with the addition of uniform transverse distributed forces, but the increase in normal stress may cause the column to fail at lower loads.

6. False. Springs and elastic supports in the middle increase the critical buckling load.

7. False. The critical buckling load does not change with eccentricity, but an increase in normal stress causes the column to fail at lower loads with increasing eccentricity.

8. False. The critical buckling load decreases with increasing slenderness ratio.

9. True.

10. True.

QUICK TEST A.1

1. Structure 1: One; *AB*
   Structure 2: Three; *AC, CD, CE*
   Structure 3: Three; *AC, BC, CD*

2. *DF, CF, HB*

3. Structure 1: Two; indeterminate
   Structure 2: One; indeterminate
   Structure 3: Zero; determinate
   Structure 4: One; indeterminate

APPENDIX H
ANSWERS TO SELECTED PROBLEMS

CHAPTER 1

1.1 $\sigma = 1019$ psi (T)   1.3 $W_{max} = 125.6$ lb   1.6 $d_{min} = 1.5$ mm   1.8 $\sigma = 145.7$ psi (T)   1.12 (a) $\sigma_{col} = 50.9$ ksi (C);
(b) $\sigma_b = 2.3$ ksi (C)   1.15 (a) $\sigma_{col} = 156$ MPa (C); (b) $\sigma_b = 8.33$ MPa (C)   1.19 $\sigma_b = 3$ MPa (C)   1.21 $\tau = 9947$ Pa
1.22 $P = 3\ aL\tau$ 1.25 $P_{max} = 13.2$ kips   1.28 $P = \tau\pi(d_o + d_i)t$ 1.29 $\tau = 3.18$ MPa   1.31 $\tau = 226.3$ MPa
1.34 $W_{max} = 125.6$ lb   1.37 $\sigma_{AA} = 3.286$ ksi (T); $\tau_{AA} = 1.53$ ksi   1.41 $\sigma = 11.9$ psi (T); $V = 19$ lbs
1.42 $\sigma_{HA} = 38$ MPa (C); $\sigma_{HB} = 16$ MPa (T); $\sigma_{HG} = 22$ MPa (C); $\sigma_{HC} = 16$ MPa (C)   1.43 $(\tau_H)_{max} = 53.76$ MPa
1.49 $\sigma_{BD} = 100$ MPa (T); $\tau_{max} = 259$ MPa   1.50 $N = 200$ kips; $y_N = 4.4$ in   1.51 $N = 15.5$ kips; $y_N = 3.03$ in
1.52 $F_2 = 50$ kips; $F_3 = 80$ kips   1.53 $F_2 = 234.5$ kips; $F_3 = 264.5$ kips   1.56 (a) $P = 150$ kN;
(b) $M_{ext} = 5$ kN-m; (c) $\tau = 2.5$ MPa   1.59 (a) $N = 1250$ kN; $M_z = 104.2$ kN-m; (b) $P = 2179$ kN;
$w = 1213$ kN/m; $\tau = 74.2$ MPa   1.62 $P_{max} = 70.6$ kN   1.67 $P_{max} = 5684$ lb   1.69 $L = 5.2$ in   1.71 (a) $\tau = 8.5$ psi;
(b) $T = 6.7$ in-lb   1.74 $d_{CG} = 30$ mm; $d_{CD} = 27$ mm; $d_{CB} = 23$ mm   1.75 $d_C = 22$ mm; sequence: CB, CG, CD

1.77   1.80   1.83   1.85

1.88   1.92   1.94

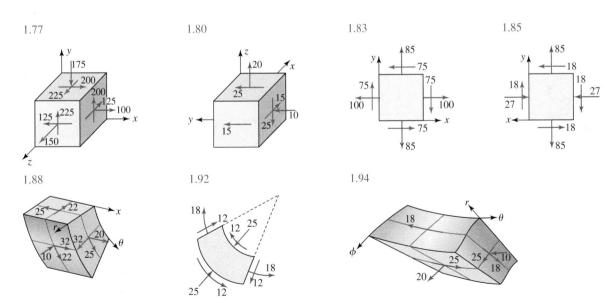

CHAPTER 2

2.1 $\varepsilon = 0.9294$ cm/cm   2.4 $\varepsilon = 0.321$ in/in   2.6 $u_D - u_A = 2.5$ mm   2.7 $\varepsilon_A = 393.3$ μin/in; $\varepsilon_B = -1500$ μin/in
2.9 $\varepsilon_A = -0.0125$ in/in   2.11 $\varepsilon_A = -0.0108$ in/in   2.13 $\varepsilon_A = -0.0108$ in/in; $\varepsilon_F = -0.003$ in/in
2.16 $\delta_B = 2$ mm to the left   2.18 $\delta_B = 2.5$ mm to the left   2.19 $\varepsilon_A = -416.7$ μmm/mm; $\varepsilon_F = 533.3$ μmm/mm
2.22 $\gamma_A = -3000$ μrad   2.25 $\gamma_A = 5400$ μrad   2.27 $\gamma_A = 1296$ μrad   2.31 $\gamma_A = -928$ μrad   2.33 $\gamma_A = -1332$ μrad
2.36 (a) $\varepsilon_{AP} = 1174.7$ μmm/mm; (b) $\varepsilon_{AP} = 1174.6$ μmm/mm; (c) $\varepsilon_{AP} = 1174.6$ μmm/mm
2.39 $\delta_{AP} = 0.0647$ mm extension; $\delta_{BP} = 0.2165$ mm extension   2.42 $\delta_{AP} = 0.0035$ in contraction;
$\delta_{BP} = 0.0188$ in contraction   2.45 $\varepsilon_{BC} = 4200$ μmm/mm; $\varepsilon_{CF} = -2973$ μmm/mm; $\varepsilon_{FE} = -2100$ μmm/mm
2.48 $\varepsilon_{BC} = 500$ μmm/mm; $\varepsilon_{CG} = -833$ μmm/mm; $\varepsilon_{GB} = 0$; $\varepsilon_{CD} = 667.5$ μmm/mm   2.52 $\varepsilon_{xx} = -128$ μmm/mm;
$\varepsilon_{yy} = -666.7$ μmm/mm; $\gamma_{xy} = 3600$ μrad   2.55 $\varepsilon_{xx} = 1750$ μmm/mm; $\varepsilon_{yy} = -1625$ μmm/mm; $\gamma_{xy} = -1125$ μrad
2.58 $\varepsilon_{xx}(24) = 555$ μin/in   2.61 $u(20) = 0.005$ in   2.64 $u(1250) = 1.516$ mm   2.69 $\varepsilon = 42.2$ μmm/mm
2.71 $\varepsilon = 47\%$

CHAPTER 3

3.1 (a) $P = 23.56$ kN; (b) $P = 35.34$ kN   3.2 $\delta = 3.25$ mm   3.3 (a) $\sigma_{alt} = 510$ MPa; (b) $\sigma_{frac} = 480$ MPa
3.4 (a) $E = 7.5$ GPa; (b) $\sigma_{prop} = 300$ MPa   3.5 $\varepsilon_{total} = 0.065$; $\varepsilon_{elas} = 0.056$; $\varepsilon_{plas} = 0.009$ 3.6 $P = 36.9$ kN
3.7 $\sigma_{yield} = 300$ MPa   3.8 (a) $E_t = 2.5$ GPa; (b) $E_s = 6.5$ GPa   3.17 (a) $E = 300$ GPa; (b) $\sigma_{prop} = 1022$ MPa;
(c) $\sigma_{yield} = 1060$ MPa; (d) $E_t = 1.72$ GPa; (e) $E_s = 11.2$ GPa; (f) $\varepsilon_{plas} = 0.1203$ 3.19 $E = 25,000$ ksi; $v = 0.2$
3.21 $G = 4000$ ksi   3.24 $P = 33$ kips; $\Delta d = -0.00064$ in   3.26 0.0327%   3.30 $U = 125$ in-lbs   3.35 (a) 300 kN-m/m^3;
(b) 21,960 kN-m/m^3; (c) 5,340 kN-m/m^3; (d) 57,623 kN-m/m^3 3.37 (a) 1734 kN-m/m^3; (b) 157 MN-m/m^3; (c) 18 MN-m/m^3;
(d) 264 MN-m/m^3 3.40 $F = 22.1$ kN   3.43 $F = 16.7$ kN   3.44 $F = 0.795$ lb; $\theta = 65.96°$ 3.47 $P_1 = 0$;
$P_2 = 2$ kN   3.52 $N = 60$ kips; $M_z = 30$ in-kips   3.55 (a) $a = 1062.1$ MPa; $b = 4493.3$ MPa; $c = -12993.1$ MPa;
(b) $U_o = 236.8$ MJ/m^3; $E_T = 2.87$ GPa   3.59 $P = 70.1$ lbs   3.65 (a) $\sigma_{zz} = 0$; $\varepsilon_{xx} = -3661$ μ; $\varepsilon_{yy} = 2589$ μ;
$\gamma_{xy} = 5357$ μrad; $\varepsilon_{zz} = 357$ μ; (b) $\varepsilon_{zz} = 0$; $\sigma_{zz} = 25$ MPa (C); $\varepsilon_{xx} = -3571$ μ; $\varepsilon_{yy} = 2679$ μ; $\gamma_{xy} = 5357$ μrad
3.69 (a) $\sigma_{zz} = 0$; $\varepsilon_{xx} = 0.0344$; $\varepsilon_{yy} = 0.05$; $\gamma_{xy} = -0.0625$; $\varepsilon_{zz} = -0.0281$; (b) $\varepsilon_{zz} = 0$; $\sigma_{zz} = 56.25$ psi (T);
$\varepsilon_{xx} = 0.0273$; $\varepsilon_{yy} = 0.043$; $\gamma_{xy} = -0.0625$ 3.72 $\sigma_{zz} = 0$; $\sigma_{yy} = 40.9$ ksi (C); $\sigma_{xx} = 36.26$ ksi (C);
$\varepsilon_{zz} = 771$ μin/in; $\tau_{xy} = -5.77$ ksi   3.74 $\sigma_{zz} = 0$; $\sigma_{yy} = 60$ MPa (T); $\sigma_{xx} = 60$ MPa (C); $\varepsilon_{zz} = 0$; $\tau_{xy} = 18$ MPa
3.77 $\sigma_{xx} = 16$ ksi (C); $\sigma_{yy} = 4$ ksi (C)   3.80 $a = 50.06$ mm; $b = 50.1725$ mm   3.95 $\varepsilon_{xx} = -936$ μ; $\varepsilon_{yy} = -2180$ μ;
$\gamma_{xy} = -5333$ μ 3.99 $\sigma_{xx} = 19.07$ ksi (C); $\sigma_{yy} = 0.99$ ksi (C); $\tau_{xy} = 0.6$ ksi   3.103 $h = 4\frac{3}{8}$ in; $d = 1\frac{1}{8}$ in
3.107 $d_{min} = 23$ mm   3.109 $K = 2.4$ 3.111 $\sigma_{max} = 45.3$ ksi   3.117 $\theta = 0.34°$ 3.120 $\sigma_{xx} = 47.4$ ksi (C);
$\sigma_{yy} = 52.02$ ksi (C); $\varepsilon_{zz} = 1254$ μ; $\tau_{xy} = -5.77$ ksi   3.127 (a) $T = 33.33$ hours; (b) $T = 133.33$ hours; (c) $T = \infty$
3.129 $n = 400,000$ cycles   3.132 (a) $E_1 = 15,000$ ksi; (b) $E_2 = 64.15$ ksi; (c) $n = 0.1694$; $E = 56.2$ ksi

CHAPTER 4

4.2 $F_1 = 108.5$ kN; $F_2 = 45.2$ kN; $F_3 = 94.3$ kN   4.4 $F = 11.25$ kips   4.11 $u_D - u_A = -0.175$ in
4.15 (a) $u_D - u_A = -0.0234$ in; (b) $\sigma_{max} = 3.75$ ksi (C)   4.17 $u_B - u_A = 0.126$ mm   4.18 $u = 0.4621$ P/EK
4.20 (a) $u_C - u_A = 0.034$ in; (b) $\sigma_{max} = 33.95$ ksi (T)   4.21 $u_B = -\gamma L^2/2E$ 4.26 $\delta = 0.045$ in
4.30 $F_{max} = 4886$ lb   4.32 $d_p = 0.5$ in; $a_b = 1\frac{1}{8}$ in; $b_s = 1\frac{5}{16}$ in   4.39 (a) $\Delta u = 0.60$ mm; (b) $\sigma_{max} = 62.2$ MPa (T)
4.42 $a = 224.40$; $b = -23.60$; $c = -0.40$; $u_A = 0.017$ in to the left   4.43 (a) $N = 64$ kN (T); (b) $y_N = 46.25$ mm
4.45 (a) $h = 3.3$ in; (b) $(\sigma_{max})_s = 33.9$ ksi (T); $(\sigma_{max})_w = 2.3$ ksi (T)   4.48 (a) $(\sigma_{BC})_{max} = 7.4$ MPa (T);
(b) $u_D - u_A = 0.11$ mm   4.50 $F_{max} = 4.8$ kips   4.54 $F = 46.9$ kips   4.58 $\delta_P = 0.23$ mm   4.59 $\sigma_A = 8.0$ ksi (C);
$\delta_B = 0.0021$ in   4.62 $\delta_P = 0.24$ mm; $\sigma_A = 118$ MPa (C)   4.64 (a) $\delta_p = 0.0265$ in; (b) $\Delta d_s = 0.00074$ in;

$\Delta d_{\mathrm{al}} = -0.00066$ in 4.66 $\sigma_A = 22.5$ ksi (C); $\sigma_B = 17.2$ ksi (T) 4.69 $F_{\max} = 555$ kN 4.73 $F_{\max} = 17.2$ kN
4.76 $w_{\max} = 9.4$ MPa 4.79 $(\sigma_i)_{\max} = 11.6$ ksi (C); $(\sigma_C)_{\max} = 2.1$ ksi (C) 4.80 $P_{\max} = 106.7$ kips
4.82 $A_{BC} = 1.1$ in^2; $d = 1.71$ in 4.84 $F_{\max} = 148.6$ kN 4.86 $F_{\max} = 181.9$ kN 4.87 $\sigma_A = 5.2$ ksi (T);
$\sigma_B = 3.5$ ksi (T) 4.91 $\sigma_{xx} = 0$; $u(L/2) = \alpha T_L L/24$ 4.92 $\sigma_{xx} = E\alpha T_L/3$ (C); $u(L/2) = -\alpha T_L L/8$
4.96 $\sigma_A = 25.70$ ksi (T) 4.99 $F_c = 28.12$ kN; $\delta_c = 0.966$ mm 4.100 $F_c = 61.65$ kN; $\delta_c = 10$ mm
4.102 $F_c = 24.3$ kN; $\delta_c = 6.310$ mm 4.104 $\sigma_{\theta\theta} = 10$ MPa (T); $\tau_r = 40$ MPa 4.108 $t_{\min} = 0.05$ in;
$d_{\mathrm{noz}} = 0.206$ in 4.110 $p_{\max} = 500$ psi; $d_{\mathrm{riv}} = 0.85$ in

CHAPTER 5

5.1 $\gamma_D = 2400$ μrad 5.2 $T = 64.8$ in-kips 5.7 $\phi_1 = 0.0400$ rad; $\phi_2 = 0.0243$ rad; $\phi_3 = 0.0957$ rad
5.9 $T = -495.2$ in-kips 5.12 $T = 10.9$ kN-m 5.14 (a) $(\tau_{xy})_A > 0$; (b) $(\tau_{xy})_B < 0$ 5.17 (a) $(\tau_{xy})_A > 0$; (b) $(\tau_{xy})_B < 0$
5.24 $\phi_D - \phi_A = 0.00711$ rads CW 5.27 $\phi_D = 0.0163$ rads CW; $(\tau_{x\theta})_E = -4.4$ ksi 5.29 $\phi_A = 1676$ μrads CW;
$(\tau_{x\theta})_E = 15.1$ MPa 5.31 (a) $\phi_B = 0.0523(T_{\mathrm{ext}}L/Gr^4)$ CW; (b) $\tau_{\max} = 0.1188T_{\mathrm{ext}}/r^3$ 5.33 $\phi_A = (aL^2/GJ)$ CW
5.34 $T = 69.2$ in-kips 5.36 $(r_i)_{\max} = 24$ mm 5.40 $d_{\min} = 21$ mm; $\tau_{AB} = 52.5$ MPa 5.47 $R_o = 2\frac{3}{8}$ in
5.49 $\Delta\phi = 0.085$ rad; $\tau_{\max} = 172$ MPa 5.50 $\Delta\phi = 0.088$ rad 5.53 $\phi_B = 0.0516$ rads ccw; $\tau_{\max} = 25.8$ ksi
5.56 $\phi_C = 0.006$ rads CCW; $T = 205.9$ in-kips 5.57 $\phi_B = 0.0438$ rads CW 5.63 $\phi_B = 5TL/9GJ$ CCW;
$\tau_{\max} = 80/9\pi\, T/d^3$ 5.65 $T_{\max} = 32$ kN-m; $\phi_B = 0.048$ rads CCW; $\tau_{\max} = 130.4$ MPa 5.66 $d_{\min} = 89$ mm;
$\phi_B = 0.0487$ rads CCW; $\tau_{\max} = 116$ MPa 5.67 $d_{\min} = 108$ mm; $\phi_B = 0.025$ rads CCW; $\tau_{\max} = 58.62$ MPa
5.72 $\phi_D - \phi_A = 0.0358$ rads CW; $(\tau_{ir})_{\max} = 30.1$ MPa; $(\tau_{c4})_{\max} = 70.2$ MPa 5.74 $\tau_{\max} = 39.54$ MPa;
$\phi_1 = 0.0062$ rads CCW 5.78 $r = 0.135$ in 5.80 $\phi_B = T_{\mathrm{ext}}^2 L/1507.4\ G^2 R^7$ 5.81 $T = 123.3$ in-kips
5.83 $T = 148.4$ in-kips 5.90 $T_{\mathrm{ext}} = 111.9$ kN-m; $\phi_A = 0.1715$ rad ccw 5.92 $\phi_B = 0.15$ rad; $T_{\mathrm{ext}} = 90.0$ kN-m
5.96 $|\tau_{\max}| = 10.8$ MPa 5.102 $|\tau_{\max}| = 10.8$ MPa

CHAPTER 6

6.1 $\psi = 2.41°$ 6.3 $\varepsilon_1 = 182$ μm/m; $\varepsilon_3 = -109.1$ μm/m; $\varepsilon_4 = -654$ μm/m; $\varepsilon_6 = 393$ μm/m 6.6 $P = 1454N$;
$M_z = 123.6$ N-m 6.7 $P_1 = 14.58$ kN; $M_1 = 130.3$ N-m; $P_2 = 9.88$ kN; $M_2 = 64.0$ N-m 6.9 $M_z = 9.13$ in-kips
6.13 $M_z = -2134$ kN-m 6.21 $\sigma_{\max} = 56$ MPa (C) or (T) $M_z = 1620$ N-m 6.23 $\sigma_A = 1224$ psi (C);
$\sigma_B = 735$ psi (C); $\sigma_D = 1714$ ksi (T) 6.29 σ_A is (C); σ_B is (T) 6.32 σ_A is (T); σ_B is (C) 6.36 (a) $\sigma_{3.0} = 2.96$ ksi (C);
(b) $\sigma_{\max} = 6.93$ ksi (C) or (T) 6.39 $\sigma_A = 4.17$ ksi (C); $\sigma_{\max} = 12.5$ ksi (C) or (T) 6.43 $\sigma_A = 15.6$ ksi (C);
$\sigma_{\max} = 28.9$ ksi (T) 6.45 $\varepsilon_A = -1500$ μ 6.47 $\varepsilon_A = -327$ μ 6.50 (a) $V_y = 3(72 - x)$ kips;
(b) $M_z = -1.5(72 - x)^2$ in-kips 6.52 (a) $V_y = [108 - \frac{1}{48}x^2]$ kips; (b) $M_z = [5184 - 108x + \frac{1}{144}x^3]$ in-kips
6.54 $V_y = -wL$ kips $0 \le x < L$; $M_z = wLx$ in-kips $0 \le x < L$; $V_y = [w(x - L) - wL]$ kips $L < x \le 2L$;
$M_z = [wLx - \frac{w}{2}(x - L)^2]$ in-kips $L < x \le 2L$ 6.58 $V_y = (76 - 12x)$ kN 5 m $< x <$ 9 m;
$M_z = (6x^2 - 76x + 154)$ kN-m 5 m $< x <$ 9 m; $V_y = -20$ kN 9 m $< x <$ 12 m;
$M_z = (20x - 240)$ kN-m 9 m $< x <$ 12 m 6.59 $V_y = -6x$ kN $0 \le x < 3$ m; $M_z = 3x^2$ kN-m $0 \le x < 3$ m;
$V_y = -8$ kN 3 m $< x <$ 5 m; $M_z = (8x - 7)$ kN-m 3 m $< x <$ 5 m 6.64 $(V_y)_{\max} = \pm7.5$ kN;
$(M_z)_{\max} = 5.625$ kN-m 6.68 $(V_y)_{\max} = 36$ kN; $(M_z)_{\max} = -86.67$ kN-m 6.72 $(V_y)_{\max} = 9$ kN;
$(M_z)_{\max} = -23.625$ kN-m 6.75 $(V_y)_{\max} = \pm6$ kips; $(M_z)_{\max} = -16$ in-kips 6.76 $w_{\max} = 154.3$ lb/in
6.83 $a_i = 11\frac{7}{8}$ in 6.84 $r = 3.75$ mm 6.86 $P = 165.7$ N 6.89 Point A: negative τ_{xz}; Point B: positive τ_{xy};
Point C: negative τ_{xz}; Point D: positive τ_{xz} 6.94 Point A: positive τ_{xy}; Point B: negative τ_{xz}; Point C: zero; Point D: positive τ_{xy}
6.96 $|\sigma_{\max}| = 348.4$ MPa; $|\tau_{\max}| = 6.84$ MPa; $(\sigma_{xx})_A = 48$ MPa (C); $(\tau_{xy})_A = -3.2$ MPa 6.98 $V_{AB} = 614.4$ lbs;
$V_{BC} = 921.6$ lbs 6.102 $M_{\mathrm{ext}} = 8333.33$ in-lbs 6.103 $P_{\max} = 202$ N; $\Delta s = 16$ cm 6.115 $R_O = 2\frac{3}{16}$ in
6.116 $|\sigma_{\max}| = 9185$ psi; $|\tau_{\max}| = 295$ psi 6.118 $(\sigma_1)_{\max} = 400$ psi (T); $(\sigma_2)_{\max} = 400$ psi (C); $M_z = -2667$ in-lb
6.122 $w = 21.1$ lb/in; $|\tau_{\max}| = 190.1$ psi 6.124 $|(\sigma_s)_{\max}| = 21.5$ MPa (T) or (C) $|(\sigma_{\mathrm{al}})_{\max}| = 8.7$ MPa (T) or (C)

$|\tau_{max}| = 0.60$ MPa   6.127 $w_{max} = 11.0$ kN/m   6.129 $M_z = 864.5$ kN-m   6.131 $M_z = 1061$ kN-m
6.139 $a = 2.392$ in; $M_z = 997.6$ in-kip   6.141 $w = 426.7$ lb/in

CHAPTER 7

7.2 $v(x) = -(wx/24EI)(x^3 - 2Lx^2 + L^3)$; $v(L/2) = -(5wL^4/384EI)$ 7.4 $v(x) = -(wx^2/24EI)(x^2 - 4Lx + 6L^2)$;

$v(L) = -(wL^4/8EI)$ 7.8 $v_A = PL^3/3EI$ 7.9 $v(x) = \begin{cases} (wLx/9EI)(x^2 - 5L^2) & 0 \le x \le L \\ (w/9EI)(Lx^3 - 5L^3x) + w/6EI(x-L)^3 & L \le x \le 3L \end{cases}$;

$v(L) = -(4wL^4/9EI)$ 7.11 $v(x) = \begin{cases} (wLx/48EI)(2x^2 - 3L^2) & 0 \le x \le L \\ (w/EI)(2Lx^3 - 7\ L^3x)/48 + (w/24EI)(x-L)^4 & L \le x \le 2L \end{cases}$;

$v(L) = -5wL^4/48EI$ 7.15 $h(x) = \sqrt{6Px/b\sigma}$; $v_{max} = -\sqrt{8b\sigma^3 L^3/27PE^2}$ 7.18 $\sigma_{max} = 128PL/27\pi d_0^3$;
$v_{max} = -8PL^3/3E\pi d_0^4$ 7.20 $R_A = 16.2$ kips up; $M_A = 10.8$ in-kips CCW   7.22 $R_A = 5P/2$;

$v(x) = \begin{cases} (P/12EI)[2(x-2L)^3 - 5(x-L)^3 - 9L^2x + 11L^3] & 0 \le x \le L \\ (P/12EI)[2(x-2L)^3 - 9L^2x + 11L^3] & L \le x \le 2L \end{cases}$ 7.24 $\dfrac{dv}{dx}(L) = \dfrac{wL^3}{80EI}$; $R_A = \dfrac{61wL}{120}$ up;

$M_A = 11wL^2/120$ CW   7.40 $v_A = -41wL^4/24EI$ 7.44 $v_A = -PL^3/96EI$ 7.47 $v_A = -wL^4/136EI$;
$R_C = 11wL/17$; $M_C = 5wL^2/34$ 7.49 $v(x) = (w/18EI)[2Lx^3 - 3L\langle x-L\rangle^3 - 10L^3x]$; $v(L) = -4wL^4/9EI$
7.53 $v(x) = (w/24EI)[x^4 - \langle x-L\rangle^4 - 4L\langle x-2L\rangle^3 - 12L^2\langle x-2L\rangle^2 - 40L^3x + 71L^4]$; $v(2L) = 23wL^4/12EI$;
$(dv/dx)(2L) = -wL^3/2EI$ 7.54 $v(x) = (P/12EI)[3Lx^2 - 3x^3 + 5\langle x-L\rangle^3]$; $R_A = 5P/2$
7.57 $v'(x_A) = -(wL^3/6EI)$; $v(x_A) = -(wL^4/8EI)$

CHAPTER 8

8.1 σ_{nn} is (C); τ_{nt} is positive   8.6 σ_{nn} is (C); τ_{nt} can't say   8.12 $\sigma_{nn} = 8.66$ ksi (C) $\tau_{nt} = 5.0$ ksi   8.15 Compression
8.19 $\sigma_{nn} = 50$ MPa (C); $\tau_{nt} = 40$ MPa   8.24 $\sigma_{nn} = 45.36$ ksi (C); $\tau_{nt} = 1.84$ ksi   8.26 $P_{max} = 84.9$ lb
8.33 (a) $\sigma_{nn} = \sigma$ (T); $\tau_{nt} = 0$; (b) $\sigma_{nn} = 0$; $\tau_{nt} = -\sigma$ 8.42 $\sigma_{nn} = 7$ MPa (T); $\tau_{nt} = -59.7$ MPa;
$\sigma_1 = 75.2$ MPa (T); $\sigma_2 = 45.2$ MPa (C); $\sigma_3 = 0$; $\theta_1 = 69.2°$; $\tau_{max} = 60.2$ MPa   8.44 $\sigma_{nn} = 45.4$ ksi (C);
$\tau_{nt} = 1.84$ ksi; $\sigma_1 = 15.4$ ksi (T); $\sigma_2 = 45.4$ ksi(C); $\sigma_3 = 0$; $\theta_1 = 40.3°$; $\tau_{max} = 30.4$ ksi
8.47 $\sigma_{nn} = 0.63$ ksi(C); $\tau_{nt} = -7.06$ ksi; $\sigma_1 = 0.62$ ksi (T); $\sigma_2 = 40.62$ ksi (C); $\sigma_3 = 12$ ksi (C) $\theta_1 = 128°$;
$\tau_{max} = 20.62$ ksi   8.49 $\sigma_1 = 67.9$ MPa (T); $\sigma_2 = 207.9$ MPa(C); $\sigma_3 = 0$; $\theta_1 = 78°$; $\tau_{max} = 137.9$ MPa
8.52 $\sigma_{xx} = 7.54$ ksi(C); $\sigma_{yy} = 9.46$ ksi (C); $\tau_{xy} = 1.15$ ksi   8.55 $\sigma_{nn} = 16.5$ ksi (C); $\tau_{nt} = -9.55$ ksi
8.60 $P_{max} = 40$ kN

CHAPTER 9

9.2 $\varepsilon_{nn} = -234.7\ \mu$; $\phi = 196.96\ \mu$rad CW   9.6 $\varepsilon_{nn} = 150\ \mu$; $\varepsilon_{tt} = 450\ \mu$; $\gamma_{nt} = -519.6\ \mu$ 9.9 $\varepsilon_{nn} = -70.2\ \mu$;
$\varepsilon_{tt} = -529.8\ \mu$; $\gamma_{nt} = -385.67\ \mu$ 9.15 $\varepsilon_{nn} = -192.8\ \mu$; $\varepsilon_{tt} = 192.8\ \mu$; $\gamma_{nt} = -459.6\ \mu$ 9.19 Sectors 8 and 2 or
Sectors 4 and 6   9.34 $\varepsilon_1 = 659\ \mu$; $\varepsilon_2 = -459\ \mu$; $\varepsilon_3 = 0$; $\gamma_{max} = 1118\ \mu$; $\theta_1 = 103.3°$; $\varepsilon_{nn} = 643.7\ \mu$;
$\varepsilon_{tt} = -443.7\ \mu$; $\gamma_{nt} = -259.8\ \mu$ 9.40 $\varepsilon_1 = 1246.5\ \mu$; $\varepsilon_2 = -196.5\ \mu$; $\varepsilon_3 = 0$; $\gamma_{max} = 1443\ \mu$; $\theta_1 = 52°$;
$\varepsilon_{nn} = 326.6\ \mu$; $\varepsilon_{tt} = 723.4\ \mu$; $\gamma_{nt} = -1387.4\ \mu$ 9.45 $\varepsilon_{xx} = -466\ \mu$; $\varepsilon_{yy} = 1266\ \mu$; $\gamma_{xy} = -1000\ \mu$

9.47 $\varepsilon_1 = 767.9\ \mu$; $\varepsilon_2 = -312.5\ \mu$; $\varepsilon_3 = -151.8\ \mu$; $\gamma_{max} = 1080.4\ \mu$; $\theta_1 = -26.57°$　9.49 $\varepsilon_1 = 982.1\ \mu$; $\varepsilon_2 = -2008.9\ \mu$; $\varepsilon_3 = 0$; $\gamma_{max} = 2991\ \mu$; $\theta_1 = 116.6°$　9.52 $\varepsilon_1 = 659\ \mu$; $\varepsilon_2 = -459\ \mu$; $\theta_1 = 103.3°$; $\sigma_1 = 1.37$ ksi (T); $\sigma_2 = 2.78$ ksi (C); $\theta_1 = 98.8°$　9.57 $\varepsilon_a = 33.49\ \mu$; $\varepsilon_b = 400\ \mu$; $\varepsilon_c = 166.5\ \mu$　9.61 $\varepsilon_a = 687.5\ \mu$; $\varepsilon_b = -406.3\ \mu$; $\varepsilon_c = -656.9\ \mu$　9.65 $\varepsilon_1 = 685.9\ \mu$; $\varepsilon_2 = -185.9\ \mu$; $\varepsilon_3 = -166.7$; $\gamma_{max} = 871.8\ \mu$; $\theta_1 = 48.3°$　9.67 $\varepsilon = 392.9\ \mu$　9.69 $\varepsilon = 716.7\ \mu$　9.74 $\varepsilon = -112.5\ \mu$

CHAPTER 10

10.1 $\sigma_{nn} = 5.15$ ksi (C); $\tau_{nt} = -16.13$ ksi　10.4 $P = 60.76$ kN　10.6 $\varepsilon_a = 1696\ \mu$; $\varepsilon_b = -1176\ \mu$　10.12 $\varepsilon_a = 1333\ \mu$; $\varepsilon_b = -666.66\ \mu$　10.22 $(\sigma_{xx})_A = 0$; $(\sigma_{xx})_B = -\sigma_{bend-y}$; $(\tau_{yz})_A = \tau_{tor} + \tau_{bend-y}$; $(\tau_{xy})_B = -\tau_{tor}$　10.26 $(\sigma_{xx})_A = 0$; $(\sigma_{xx})_B = -\sigma_{bend-y}$; $(\tau_{xz})_A = \tau_{bend-y}$; $(\tau_{xy})_B = 0$　10.30 $(\sigma_{xx})_A = 0$; $(\sigma_{xx})_B = -\sigma_{bend-y}$; $(\tau_{xz})_A = \tau_{tor} + \tau_{bend-y}$; $(\tau_{xy})_B = -\tau_{tor}$　10.35 $(\sigma_{xx})_A = 23.1$ ksi (C); $(\tau_{xy})_A = -7.2$ ksi　10.36 $\sigma_{nn} = 8219$ psi (C); $\tau_{nt} = 3180$ psi　10.40 $P_{max} = 4.3$ kN　10.44 $w = 791.2$ N/m　10.49 $\sigma_{BD} = \sigma_{CE} = 5.13$ psi (C); $\sigma_{BC} = 10$ psi (T); $\sigma_{AB} = 167.4$ psi (C)　10.51 $W_{max} = 67$ lb　10.58 $R_o = 2.529$ in　10.61 $2a_{crit} = 0.442$　10.66 $T_{max} = 223$ kN-m　10.67 $2a_{crit} = 108.3$ mm　10.74 $2a_{crit} = 1.236$ in　10.75 (a) $P_{max} = 5$ kN; (b) $P_{max} = 6.4$ kN　10.76 $P_{max} = 9.5$ kips　10.79 $K = 1.22$

CHAPTER 11

11.2 $P_{cr} = 5/4$ kL　11.6 $P_{cr} = 153.3$ lb　11.9 $L/r = 72.7$; $P_{cr} = 214.7$ kip; $\sigma_{cr} = 3.36$ ksi (C)　11.15 $K = 1.106$　11.19 $K = 3.633$　11.31 $\sigma_{max} = 2.68$ ksi (C); $v_{max} = 0.0458$ in　11.35 $L_{max} = 2.09$ m　11.39 $P_{max} = 39.45$ kip　11.40 $L_{max} = 42$ in　11.43 $w_{max} = 12$ kN/m; $K_{BD} = 2.3$　11.49 $P_{cr} = 17.0$ kip

INDEX

The Lives of Bats

Wilfried Schober

The Lives of Bats

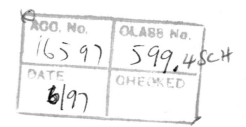

CROOM HELM
London & Canberra

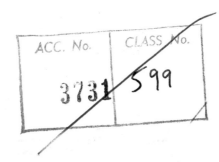
Translated from the German by Sylvia Furness
Revised by A. M. Hutson

© 1984 Edition Leipzig
Croom Helm Ltd, Provident House, Burrell Row,
Beckenham, Kent BR3 1AT

British Library Cataloguing in Publication Data
Schober, Wilfried
 The lives of bats.
 1. Bats
 I. Title   II. Fledertiere, *English*
 599 4   QL737 C5
 ISBN 0-7099-2389-9

Design: Traudl Schneehagen
Drawings: Michael Lissmann
Printed and bound in the German Democratic Republic

Contents

Foreword

Bats—strange beings? They fly with their "hands", they "see" with their ears and hang themselves up to sleep by the toes of their hind feet.

Of all species of mammals living on the earth today, more than 950, that is, approximately one in every five, belong to the order of Chiroptera—"insectivorous" and "fruit" bats. A substantial number! For many years, these animals have been a rewarding subject of scientific research, yielding a wealth of interesting information. Although much fascinating material on the biology of bats has been published in recent decades, the findings, on the whole, have drawn the attention only of a small group of specialists.

Bats alone among mammals have mastered the art of active flight. This has enabled them to conquer a new environment. Hand in hand with the development of the organ of flight, bats have evolved a perfect system of echolocation, by means of which they can perceive obstacles and prey in the darkest night. Because they have acquired the capacity for flight, they are able to undertake migrations of the kind we are familiar with in migratory birds. In autumn, many species of the temperate latitudes leave the summer roosts in which they have given birth to their young, and migrate to frost-free winter quarters sometimes more than a hundred kilometres distant, where they survive the cold season while maintaining their metabolism at a reduced rate.

Most people, even in our highly developed industrial countries, know very little about the way of life briefly touched upon here, or about the diversity of these animals.

Indeed, the very word bat is associated with the idea of mystical creatures that are menacing, that may become entangled in a woman's hair and that bode nothing but ill. Their nocturnal habit and silent flight were undoubtedly responsible for the belief that grew up among many people that bats are in league with the Powers of Evil. For this reason, they have played a major role in the superstitious beliefs of men. Still today, in many countries, bats are quite unjustifiably persecuted and destroyed. Yet only very few species are able to cause any real harm to man and his possessions. The majority of bats are harmless or even positively beneficial.

However, it must be noted with concern that in the last twenty years, bat numbers in the industrial countries of the Old and the New World have shown a steady decline. Fear and superstition are not the prime causes of the reduction in numbers, but to a much greater extent, the increasingly rapid changes in the environment. An animal group that has developed and become specialized over a period of millions of years, cannot adapt itself overnight to new conditions. Everyone should be aware of the danger that exists. Legal measures alone are by no means enough to safeguard bat numbers. Only when people are prepared to take active measures to conserve the creatures and their habitats and to avert dangers that threaten them, is there any chance of preventing these curious mammals from being eliminated from our list of fauna.

The purpose of this book is on the one hand to enlighten and to bring about an understanding of the life of bats, and on the other, to foster a positive attitude to bats among people who are prepared to help in the conservation of these admirable creatures.

Leipzig, 1983   *Wilfried Schober*

Bird or mammal, deity or demon

Two characteristic features of bats, namely their ability to fly and their predominantly nocturnal habit caused them on the one hand to be regarded for centuries not as mammals but as birds, and on the other hand to be looked upon in folkloristic tradition as a symbol of evil and mystery—as indeed they still are to some extent today. Even in the age of the atom and the era of space flight, there are still people who run off shrieking at the sight of a bat, or who ascribe supernatural powers to the creatures.

The old superstition that bats will become inextricably entangled in a woman's hair or will make their way into smoke larders in order to eat the bacon there, has persisted to the present day in many countries in Europe. It is said that among the rural population of Finland, the idea still exists that during sleep, the soul leaves the body and flies about in the form of a bat.

The origin of many prejudices against these harmless animals is difficult to establish. Nevertheless, the blind belief that bats are endowed with magic powers is a deeply rooted one. Ancient man, living close to elemental nature was, after all, enormously impressed by the striking appearance or the physical superiority of animals. Since animals were able to perform feats of which man was incapable, supernatural powers were attributed to them. Among certain tribes living close to nature there still exist today interrelationships between man and animals that are difficult for us to comprehend.

For three thousand years, Christians and Jews have been able to read in Leviticus 11 about clean and unclean meats: ". . . And these are they which ye shall have in abomination among the fowls; they shall not be eaten, they are an abomination: the eagle, and the ossifrage, and the osprey, . . . And the stork, the heron after her kind, and the lapwing, and the bat." In a similar list in Deuteronomy, the bat is again denoted as a bird.

In ancient Babylon, Persia and Arabia, bats were similarly numbered among the birds. A Persian manuscript from the ninth century reads: "Among birds, two were created which have a character different from the others. These are the griffon and the bat, which have teeth within the mouth and which feed their young with milk from the teats." Perhaps the griffon mentioned here is not a mythological creature but a representative of the Old World Fruit bats, the Flying Foxes. An ancient Oriental folk-tale tells how bats were at one time birds which were not satisfied with their outward appearance. Instead, they wanted to be changed into men. Their desire was granted only in part. They grew hair and teeth and developed other human features. But their body continued to resemble that of a bird. The bats were dismayed at this, and ashamed to be seen by the birds. From that time on, in order to avoid meeting them, they ventured out only at night.

There are many stories which tell how bats have attempted by guile to profit from their duality of form—as bird and bat. In the sixth century B.C. the Greek slave Aesop, whose fables brought him fame and freedom, related the following story. A weasel that had caught a bat informed its captive that all birds were its enemies. Quick-wittedly, the victim claimed that it was not a bird but a bat, and so regained its freedom. Some time later, the same bat was captured by another weasel, who this time declared that mice were its favourite food. Once again the bat was able to save its life, this time by maintaining that it was a bird. In another fable, Aesop tells how the birds once waged a war with the beasts of the forest, in which victory went to either side in turn. But at each change of fortune, the cowardly bat aligned itself with the victorious side. When the war was over, neither of the contenders wanted to have anything to do with the bat, and it ended as a solitary creature of the night.

A Roman fable tells the following story: When the birds in council passed an edict to exile bats from the kingdom, the bats claimed that they were mice. Then the birds ordained that all mice were to be held in contempt. Now the bats protested that they were birds. Finally, all the beasts in the land became angry, and the bats feared for their life and thereafter dared to fly out only at night.

Among certain peoples, the bat plays the role of a weather prophet. There are many examples of old peasant lore which links the behaviour of animals and the weather. These dicta undoubtedly have a valid scientific basis. So universal was their acceptance that about a hundred years ago, guidelines for forecasting the weather were still being published by the Intelligence Department of the American War Office which were based on precepts of this kind. Some of those concerning bats were of the following nature: "If bats fly late at night, the weather will be fair", or "When bats flutter excitedly and beetles buzz about, the next day will be a fine one." The curious thing about these "service regulations" was that the officer responsible for the compilation of the rules of this kind dealt with bats under the heading "Birds", although zoologists had long since classified them as mammals.

The fact that bats are not birds but rank among the mammals was expressed by John Swan as early as 1635 in the words:

"(A bat) . . . is no bird but a winged mouse;
for she creeps
with her wings, is without feathers and
flyeth with a kinde of skin, as bees and flies do;
excepting that the Bats wings hath
a farre thicker and stronger skin.
And this creature thus mungrell-like, cannot
look very lovely."

This somewhat unflattering description did little to remove the stigma of evil associations borne by bats.

When man, accustomed to light and sun, finds himself in the darkness of night, and his sense of vision is able to tell him little about his immediate environment, he is inclined to give free rein to his imagination. It is easy to attach an aura of mystery and ghostliness to activities of the night. For instance, the croaking of toads becomes a portent of bad tidings or the cry of the little owl an omen of death. Was it not natural then to associate mysterious powers of darkness with bats which dwell in gloomy caverns, only emerging under the cloak of night to flit silently round trees and dwellings, and to compare their fluttering with the phantasmal dance of evil spirits? It is scarcely surprising that not only owls but also bats were nailed to barn doors to ward off evil, disease and witches.

The devil is sometimes depicted with bats' wings, as is that sinister creature of fable, the dragon. The Roman writer Divus Basilius said: "The nature of the bat is related by blood to that of the devil." Not only evil and mystery but also magic and demonic powers are ascribed to bats. The witch doctors of many primitive tribes wore bat amulets and used parts of the bodies of bats to concoct their mixtures and vile-tasting potions. Their purpose was to ward off evil, to heal diseases or to compel the affection of a loved one. It is said that in Anatolia today there are still those who carry secretly upon their person the bone of a bat as a love charm.

In early Egyptian manuscripts dating from the first century A.D. and in the writings of Arabian scholars and physicians there are prescriptions in which whole bats or parts of the bodies are used. They are recommended as cures for asthma and rheumatism as well as for sore throats and baldness. In India, live flying foxes are sold in the bazaars. The skin of these large animals is removed and placed upon the diseased part of the body.

The gods of the Mayas in Central America were often depicted as figures with the head of a bat. They were usually modelled on Spear-nosed bats (Phyllostomidae). The illustration shows a bat god and a human sacrifice with the heart torn out (from Brentjes, 1971).

This is said to cure lumbago and rheumatism. It seems almost incredible that in the sixties, a news item in the Daily Mail reported that in New York, the authorities had banned the sale of bat's blood in the shops there.

Since nothing was known about the bat's capacity for ultrasonic navigation until a few decades ago, it was generally believed that all species of bats had nocturnal vision. Some well-meaning recommendations were, of course, derived from this. For example, in the thirteenth century, in his book *De mirabilibus mundi* (Of the marvels of the world), Albertus Magnus wrote:

If thou wilt see a thing drowned,
or se depe in the water in the
nyghte, and that it shall not bee
more hyd to the than in the daye
and readde bookes in a darcke night.
Anoynte thy face with the
bloude of the Reremouse or backe
and it shal be done that I saye.

(From the *Boke of the mervels of the worlde* translated from Latin and published in London, 1560.)

In the Middle Ages, these "witches' birds" often brought ruin to the people in whose houses they lodged. In the popular imagination, witches and bats had become so closely linked that many an unfortunate soul in whose home bats had found shelter was accused of being a witch and even punished by death at the stake.

In the ancient civilizations of Central America, bats played an important part in the history of religion. One of the deities of the Mayas was Zotzilaha Chamalcan. This god was represented in the form of a man with extended bat wings and the head of a bat. He is depicted on altars, stone columns and on a large number of earthenware vessels which have been excavated in the vicinity of the temples that were built some two thousand years ago. In the picture writing of the Mayas, the hieroglyph for "bat" occurs frequently. It consists of very clearly recognizable bats' heads, the bats usually featured being representatives of the Spear-nosed bats and Vampires. For the faithful who looked upon blood as food of the gods to be obtained by human sacrifice, the blood-sucking Vampire bat must have appeared as a god. Yet the bat chosen to serve as model in the original graphic representations and to be revered as a "blood sucker" was the Spear-nosed bat, probably because its facial appearance was so much more bizarre. To the ancient Zapotecs, the bat was the god of death. For this reason, many burial urns and grave reliefs bear the form of bats. The Underworld Kingdom of Darkness and Death was ruled by the Death Bat Cama Zotz, and all who ventured to descend into this kingdom were slain by him.

Although these gods and their legends have lost their significance in the Central America of today, the old Mayan word for bat has not yet been forgotten. In the uplands of Guatemala, there is still a tribe with the name of "Zotzil" (belonging to the bat). Their god is the bat and their capital city Zimacantlan (the Place of the Bats).

The bat as a heraldic animal. Coat of arms of the town of Valencia in Spain. It is the emblem of the royal house of James I of Aragon (thirteenth century).

Among various peoples still very close to nature, the bat serves as a totem figure. It is believed to possess supernatural powers and is revered accordingly. In the same way, the bat holds a special place in the culture of the North American Indians. One of their traditional stories tells how a powerful and proud warrior, who had been cast ashore on a craggy coast, was saved by a bat in the guise of an old man or old woman.

The Californian Indians believed that they could locate the source of a fire with the help of bats. They also claimed that a bat which has eaten volcanic rock can spew forth particularly fine arrows. From this, one can con-clude, as the American bat specialist Allen writes, that the Red Indians were very familiar with the Californian Long-eared bat which bears on its nose a quite distinct, arrow-shaped cutaneous outgrowth.

Not only did bats make their way into the religious and intellectual history of various peoples, but also into the works of certain leading writers. In his epic poem describing the wanderings of Ulysses, Homer himself made use of the simile of the bat. On his travels which were to lead him home from the Trojan Wars, Ulysses descended into the Underworld. So greatly did he disturb the "shades" that they flew after him fluttering like bats startled from a tree.

Many of the writers of antiquity in whose works bats appear also link them with the forces of evil. Characters who deserve punishment may well be turned into bats. Portrayals of this kind kept alive the antipathy to these animals felt in wide circles of the population.

Shakespeare mentions bats quite frequently in his works. In *A Midsummer Night's Dream*, Queen Titania lets her fairy attendants fashion cloaks from the "leathern wings" of bats. The three witches in *Macbeth* brew a poisonous draught of ingredients which include

"Eye of newt, and toe of frog
Wool of bat, and tongue of dog."

In children's rhymes found in the Anglo-American linguistic area, the bat is often mentioned, frequently with an allusion to its fondness for bacon. This notion has even provided the vernacular name of "bacon mouse", although it appears to be quite unsubstantiated.

Feelings of aversion to bats are all too frequently exploited in film and television. In horror films, bat-like creatures fly across the screen and their ghastly appearance and gigantic size increase the loathing felt for these animals. The bats torture their victims in gloomy cellars or serve as accomplice and tool of the powers of darkness. We need only recall the figure of Count Dracula who, with his vampires, has struck terror into the hearts of thousands.

Yet there are exceptions to this general rejection. In their wisdom, the ancient Chinese placed the bat in a position of high esteem. In China it became a symbol of happiness, and the word "fu" (bat) is also the term for happiness. In their paintings and carvings, Chinese artists made wide use of the bat as a decorative element. Bat medallions are found embroidered on Oriental robes. A talisman commonly worn in China is in the shape of a coin bearing the symbol of the Tree of Life (a tree with roots and branches) around which five bats with wings outspread are arranged in a circle. This talisman of the five bats, called in Chinese a "wu fu", symbolizes the greatest joys of man, contentment, happiness, prosperity, health and longevity.

Bats are also revered on Bali, where the flying foxes that live in thousands in the temple grottoes are protected from all disturbance.

Animal worship and animal cults played a particularly important role in Ancient Egypt. Animals considered sacred included the bull, dog, jackal, snake, ibis, scarab and many others. But remarkably enough, neither the Pharaohs and priests nor the peasantry represented bats as godlike beings. Yet the creatures were not at all rare,

The Chinese look upon bats as bearers of good fortune. Detail from an eighteenth-century robe shows ornamentation in which five bats encircle a Tree of Life. They incorporate the concepts of Health, Prosperity, Long Life, Happiness and Contentment.

for they inhabited the temples and burial chambers in their thousands. Realistic portrayals of bats in wall paintings some 4000 years old also show that the creatures were by no means unfamiliar to the Egyptians.

The examples given here of the centuries old interrelationship between man and bat belong mainly to the sphere of fable or religion, yet for us today they can prove just as interesting as the most impressive results of recent scientific research on bats.

Are bats flying mice?

It has taken a long time for bats to find their place within the animal kingdom scientifically confirmed. In the sixteenth century, the naturalist Konrad Gesner of Zurich wrote in his *Historia Animalium*: "The bat is an intermediate animal between a bird and a mouse, so that it can reasonably be called a flying mouse, although it can be numbered neither among birds nor among mice, since it has something of the form of both." Today this statement seems a curious one, although there are still people who are not too sure what kind of animals bats really are. Indeed, superficial examination reveals characteristics of great similarity between bats and mice. In the case of our indigenous species, we need think only of the size of body, the colour of the fur, the shape of the ears. Over and again we are struck by this similarity to mice and it is sometimes reflected in the names given to bats, as for example, the Large Mouse-eared bat *(Myotis myotis)*. Indeed a whole genus of bats bears the Greek name *Myotis* which simply means "mouse ear". In other countries as well, the popular names for the Chiroptera show that these animals were likened to mice or indeed regarded as such. In Germany a bat is a *Fledermaus* (flutter mouse), in France a *chauve-souris* (naked mouse), in Mexico bats are called *ratones voladores* or flying rats, and old vernacular English names include flittermouse and reremouse.

The scientific name for this order of mammals is Chiroptera, in English, "hand wings". This refers to the most important feature common to all species that are included in the order: the development of the front limbs into an organ of flight. In the process, the forearm, the metacarpal bones and the second to fifth fingers have been greatly extended.

The English zoologist Edward Wotton, a contemporary of Gesner, was the first to recognize the true nature of bats and to classify them as mammals. However, the exact position of the bat within this class, represented a considerable problem to zoologists.

Some 250 years ago, the Swedish physician and naturalist Carl von Linné (Linnaeus) began to make the first comprehensive systematic survey of minerals, plants and animals. Starting out from the concept of "species" which in each case comprises all those animals which are mutually fertile, he classified living beings into a system based on degrees of relationship. His great work *Systema naturae* was first published in 1735. At that time, Linnaeus recognized six species of bat; two each in Europe, Asia and America. At first he classified them with the carnivores, basing his grouping upon characteristics of dentition—many species possess extremely sharp teeth. Later, Linnaeus met with a seventh species, the Fisherman bat from South America, which he placed among the rodents. Since new criteria were constantly being established that were of significance for the classification of animals, Linnaeus' book underwent frequent revision. It is not surprising that at a later stage he replaced the class Quadrupeda (quadrupeds) by that of Mammalia (mammals), recognizing that whales also belonged to this group. Thirty years after bats had first been described, he placed them among the primates, which were further subdivided into four categories: Homo (man), Simia (ape), Lemur (lemuroid) and Vespertilio (bat). Linnaeus based this classification upon the location of the lacteal glands.

The material available to Linnaeus at that time was much too meagre for him to recognize that these animals constitute an order of their own among mammals. But it was only a question of time before the correct zoological classification of bats could be achieved.

After Linnaeus there followed an epoch of discovery and description of new species of animals, in which fresh impetus was given to faunistic and systematic research.

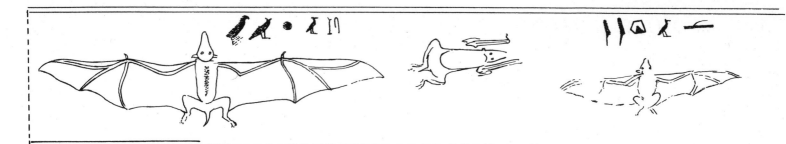

Wall painting with bats (Emballonuridae). These paintings were found in a tomb near Beni Hasan (Egypt) and date from 1800 B. C. (from Brentjes).

Thus began the age of voyages of discovery and expeditions to foreign lands. One result was that over the next 100 years, valuable zoological material, including of course that concerning bats, was gathered, evaluated scientifically and, as a consequence, knowledge of species was enormously enriched.

When the zoologist Koch published his book *Das Wesentliche der Chiropteren* (A Fundamental Study of the Chiroptera) in 1863, he already knew of more than 300 species. Of the Megachiroptera, the Old World Fruit bats, some 50 species were known at this time.

The discovery and description of new species has continued throughout the last hundred years. In all the great museums of the world, specialists are working on bats and reporting on new species from areas of the world that until now have been virtually inaccessible. Specialized literature shows that at the present time there are some 950 species of bat in the world.

The evolution of this group of mammals over millions of years has produced a vast diversity of form. In respect of number of species, bats are surpassed only by rodents —"mice" in popular terms—with some 1800 species.

The development of such a wealth of species within the order of bats can be attributed primarily to the acquisition of the ability to fly. The same phenomenon can be observed in the other members of the animal kingdom that are capable of flight, namely birds and insects. These classes also have an enormous number of species, and have established themselves in the most varied of biotopes. But whereas most people are aware of the great diversity within the bird and insect world, only few know anything about the variety that exists among bats.

Another factor that has furthered this development is that because of their mode of locomotion, it was not necessary for bats to compete for food with the great army of rodents. No bat eats grass or leaves. And competition with birds could not arise since the latter are active for the most part during the day, whereas at night, air space belongs almost exclusively to the bats.

The numerous species that make up the order Chiroptera are classified by the systematist into 16 to 18 families. It is not possible to specify an exact number here since those features which suffice in the opinion of one taxonomist to justify the establishment of a new family may not seem adequate to another. On one point though they are all in agreement, namely that Chiroptera fall naturally into two suborders. The first is the Megachiroptera, the generally large forms known as fruit bats or "flying foxes" (Eisentraut's *Flughunde*, flying dogs). In these cases, the names again show a comparison drawn with an animal that was already familiar. And indeed it is possible to see similarities between the head or face of many of the Megachiroptera and that of a fox or dog. The Flying Foxes inhabit the tropics and subtropics of the Old World; the number of species living today is given as 175.

The second suborder is that of the Microchiroptera, the mainly small forms (Eisentraut's *Fledermäuse*), those known simply as bats or sometimes insectivorous bats. Their feeding habits are varied, their distribution world-wide.

15

Bats evolved with the dinosaurs

1 *The Temptation of Saint Anthony in the Grotto* by David Teniers the Younger (1610–1690)

The capacity for flight has been developed by bats as a secondary acquisition. Their ancestors were almost certainly quadrupedal mammals. So the interesting question arises "When did these ancestors live, and what did they look like?"

It is necessary to go far back in the history of the earth and in the history of mammalian evolution to find an answer. If man had lived 50 million years ago, he would —as fossil finds prove—even then have seen bats which looked virtually indistinguishable from the species we see today. Bats, then, are a very ancient though highly specialized group of mammals.

In order to reconstruct the evolution of these experts in flight, it is interesting to ask whether there are some other species among mammals living today which have also attempted to conquer the air and could be considered as possible predecessors. And we can indeed find representatives in various orders which show signs of incipient development of flight membranes. Among marsupials, whose best-known representatives are the kangaroos, there is the group of gliding possums or phalangers. They vary greatly in size, but all of them possess a broad fold of skin stretching between the fore and hind legs, forming a patagium, which, when the legs are spread, acts as a parachute, carrying the creature on a gliding flight down through the air. It is not inconceivable that this passive, parachuting flight could represent a preliminary stage in the development of active flight, but the marsupials are such an ancient and isolated branch in the mammalian genealogical tree that there can be no question of their having been ancestors of the bat.

What about the "gliders" among the rodents? In both the Old World and the New, the rodents have produced several species that can be grouped under the name of "flying squirrels". Compared with the phalangers, their adaptation to flight shows some further advances. Firstly, the flight membranes at the sides of the body are broader, and secondly, additional skin flaps have developed between forelegs and neck as well as between hind legs and tail. Anatomical refinements in the structure of this "flight organ" show that it is not invariable in form in all flying squirrels. But the most important point —and this is true also of the phalangers—is that the fore limbs show none of the alterations in the region of the metacarpal bones and fingers that are typical of the Chiroptera. Since adaptation to gliding flight is a secondary acquisition in flying squirrels, and in any case, there is no possibility of rodents being the ancestors of bats, they too can be ruled out of our considerations.

Among mammals, a third group remains which shows clearly the development of a flight membrane or patagium. These are the "flying lemurs", colugos or cobegos of Southeast Asia. The two species so far described are such oddities among mammals that systematists place them in an order of their own. They are, however, more closely related to bats than are marsupials or rodents.

The patagium extends from the neck along the fore and hind limbs and then to the tip of the tail. But these species are not capable of powered flight. Still more significant is the fact that even the colugos show no sign of extension of the finger joints. The patagium does not extend between the fingers. It is precisely this characteristic which, in the case of bats, is a typical feature of the development of the fore limbs into real wings.

Of all the "gliders" living today, none fits in with the line of evolution which has been followed by our bats on their path to active flight. The bonds existing between bats and other mammalian orders in which "gliders" have evolved, are very tenuous. Examination of their "flight organs" is merely in the nature of setting up a prototype to provide suggestions as to how the flight organ of bats might have developed. In all, it can be said

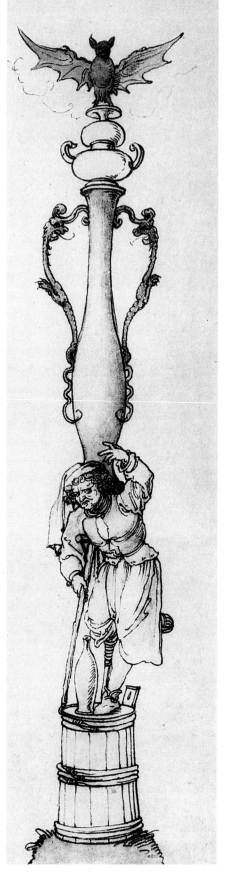

2 A devil with the wings of a bat presents a child as a sacrificial offering to an elephant-headed demon. (French illustration to *The Arabian Nights*. Stuttgart, 1838)

3 *Saturnian Column*. Among the drawings of Dürer are some in which bats are depicted to suggest magic and mystical associations.

18

Following page :

6 Wall hanging with representation of a bat·god.
Folk art of the Cuna Indians of the San Blas Islands (Panama). Murcielago is the Spanish word for bat.

7 Raffia basket with bat motif. Woven by Maidu Indians on the north-west coast of the U.S.A. (about 1880)

8 Representations of fabulous animals also show hybrid beings with the wings of bats.

4 *The Siren of Ravenna.*
A hybrid being with the wings of a bat

5 In the visual arts, Evil is frequently depicted in the form of witches and devils with the wings of bats. This drawing is Number 48 of *Los Caprichos* by Goya, entitled "The Blowers".

19

that among the species of mammals living today, none can be considered as the probable ancestor of the bat. In order to obtain information about the evolution of bats and about their forebears, it is necessary to consider results obtained from palaeontological research.

What is the situation as far as bats are concerned? Among the numerous mammalian fossil finds, there are some fossils of both major divisions of bats. But in comparison with the fossils of some other groups of mammals, they are not very numerous. The reason for this paucity of fossil evidence could well be that many Chiroptera were inhabitants of tropical forests where, after death, they would be devoured in their entirety by the vast army of ground-dwelling organisms. A more favourable environment for fossilization is provided by caves, although here again, finds dating back several million years are rare. Some of the fossil remains of bats excavated up to now have been in a very good state of preservation. The earliest date from the middle Eocene epoch; that is, they are about 50 million years old.

Because of the excellent condition of some of these fossils, it has been possible to establish that the creatures living at that time had very much the appearance of the species of today. Body form and proportions were basically identical. Forearm, metacarpal bones and fingers are lengthened, and dentition resembles that of the insectivorous bats living today.

A skeleton, completely preserved, of the earliest known bat (Icaronycteris) dates from the early Eocene in Wyoming, U.S.A. It combines features of the two present-day suborders of bats. Accordingly, it is clear that as a type, the bat went through a single process of evolution and this process was already completed some 50 million years ago. The great antiquity of the order of bats is underlined by the fact that it was another 42 million years before the first ape-men evolved on the earth.

Fossil bats from the Eocene have also been found in Europe. In most cases, strata that have been found to hold a wealth of fossil material are those containing gypsum, lignite or mineral oil, laid down millions of years ago in lagoons or great inland lakes.

One of the famous sites of fossil finds of vertebrates in the Federal Republic of Germany is the former oil-shale mine at Messel near Darmstadt. More than a hundred years ago, miners excavating here discovered the first remains of fossil fauna from a tropical-subtropical primeval forest. Careful excavations recently carried out show that all the animals are in an astonishingly good state of preservation, probably because the deeper layers of water in a lake contained no oxygen, as a result of which decomposition of any animal which lay on the bottom was impeded. In addition to fish, reptiles and birds, complete skeletons of insectivores, rodents and primitive horses were found in Messel. It is the fossil bats in which we are particularly interested. The exceptional state of preservation of the skeletons, indeed even of remnants of soft parts with wing membrane and fur, confirm that the general structure of these creatures corresponded closely to that of present-day genera. In some of the bats found, it was even possible to analyze the contents of the stomach. The presence of tegulae (wing scales) is evidence that bats fed on moths even at this time. From this, it can be concluded that the creatures had already evolved an efficient system of echolocation.

A rich source of vertebrate finds from the Eocene was the lignite deposits in the Geiseltal region near Halle (German Democratic Republic). Since humic acids of plant origin liberated in the process of carbonization will normally dissolve bone, so that no remnants are preserved, conditions must have existed here which made it possible for the dead organisms to undergo rapid preservation. In the Geiseltal region, it is very probable that the calcareous ground waters which penetrated from the surroundings into the bog, counteracted the destructive action of the humic acids. A further factor was that the animals died quickly in floods and were embedded in mud; since bacteria and oxygen were excluded, decomposition could not take place. As a result, fragments of skin, feathers and hair can still be distinguished in addition to the bones. The thousands of exhibits in the Geiseltal Museum in Halle provide a glimpse of a tropical fauna and flora inconceivable today, which existed in this region some 40 to 50 million years ago. In addition to plants and invertebrates, the Geiseltal also held fish, amphibians, reptiles, birds and numerous mammals concealed within it. There are fragments of 25 specimens of bats, most of which were excavated during the thirties. Some are complete skeletons, others skull and bone fragments. The entire material is assignable to the species *Cecilionycteris prisca*. Geiseltal bats show certain primitive characteristics of dentition, although the development of the flight apparatus scarcely differs from that of modern bats.

The earliest known fruit bat *(Archaeopteropus)* was found in Italy and is assigned to the Oligocene; that is, it is about 40 million years old. But the separation of the fruit bats from the archetypal bats happened earlier than this. Various characteristics such as the longer skull and the invariable retention of a claw on the second digit have given rise to the conjecture that the Megachiroptera may be more primitive than the Microchiroptera. This assumption has, however, recently been questioned on the basis of extensive examinations carried out on the brains of fruit bats and insectivorous bats. The brain of the fruit bat is undoubtedly more highly developed. In addition, the fruit bats exhibit so many features of specialization (reduced dentition, large eyes, regression of the tail and tail membrane), that they cannot possibly provide a model for the primitive form of the bats of today.

Even though bats evolved to become a distinct group more than 50 million years ago, developments have continued to take place since then. There has been constant refinement of apparently primitive features, adaptation to new biotopes, specialization in respect of feeding, and—of vital significance for the bats—extension of their ultrasonic system of orientation.

Within the system of mammals, the order of insectivores includes the most primitive of mammals which possess a placenta. The ancestors of today's hedgehogs and shrews can also be considered as the forebears of all other placental mammals that inhabit the earth today. It is not easy to imagine that the ancestors of, for instance, great beasts of prey or elephants, of whales or ungulates might well have been shrew-like animals. Yet a wealth of fossil finds dating from various geological epochs has led palaeozoologists to this conclusion. Bats hold a position very close to insectivores. Apart from the capacity for flight and adaptations to particular ways of life that were acquired later, many species still exhibit primitive characteristics. These include skull structure, dentition, the degree of brain development and the uncomplicated structure of the intestinal tract. Therefore it is beyond doubt that bats are derived from a primitive insectivore-like stock. Existing fossil finds show that such forms lived as long ago as in the Upper Cretaceous period, that is, at the threshold of the Caenozoic era some 70 million years ago. Since the earliest finds of fully developed bats date back 50 million years, the process of branching off from the original insectivores must have taken place before this. The ground was prepared in the Upper Cretaceous for the development which was to make the Caenozoic era into the Age of Mammals. In addition to insectivores, new arrivals on the scene included lemurs and the ancestors of the ungulates. They were mostly creatures of small stature, leading a secretive existence, for

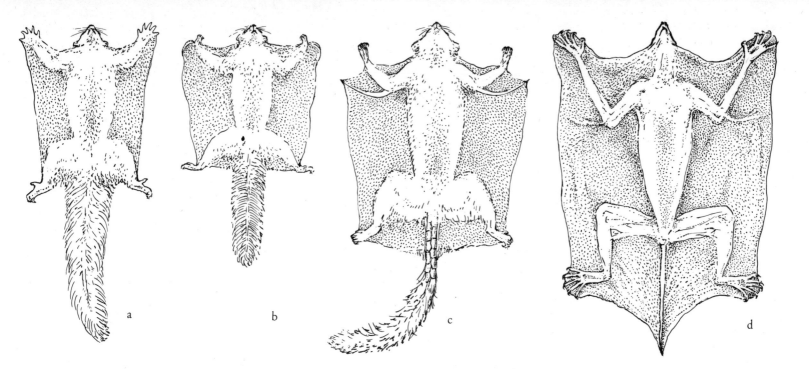

Various kinds of gliding mammals
a flying phalanger or gliding possum *(Petaurus breviceps)*;
b flying squirrel *(Glaucomys volans)*, the flight membrane is extended by a gristly rod at the wrist;
c African flying squirrel *(Anomalurus sp.)*, the patagium is given tension by a rod of cartilage extending outwards from the elbow which acts as a strut;
d colugo or cobego, misnamed "flying lemur" *(Cynocephalus sp.)*

they still lived under the shadow of the mighty dinosaurs. Not until these great saurians became extinct, was the way clear for an explosive advance in the development of mammals, which, as we can see today, led to an enormously vast abundance of forms.

In members of various orders of mammals, there are indications of incipient evolutionary development of flight capacity, but only bats achieved true flight. At what stage this occurred, we do not know. From what is already known about the genealogical history and embryological development of mammals, it is possible to gain some simplified picture of the kind of varied adaptations that might have produced flying mammals. Primitive arboreal insectivores undoubtedly were the starting point. The claws on their feet made it possible for them to climb easily and safely up tree trunks and along branches with the sort of skill shown by the squirrels of

today. Their life was spent almost entirely in trees and they became experts in leaping from branch to branch. Since they fed on insects, they would doubtlessly have attempted to catch flying species as they leaped. The gliding mammals mentioned above are also tree dwellers. Although there can be no question of a phylogenetic link with the bat, these examples may perhaps provide some clue to the kind of biological developments that could have led to the evolution of active flight. It is conceivable that the extensive parachute-like flight membranes which make gliding flight possible predated the evolution of the flight organ in bats. A glance at the embryogeny of bats supports this assumption. It is an old biological precept that embryonic development (ontogenesis) passes through the same stages in much abbreviated form as evolutionary development (phylogenesis). What can we learn in this respect from the embryology of bats? In the early stages of growth within the womb there is an initial development of folds of skin along the sides of the body, and only after this do the bones of the fingers begin to lengthen. The wing of the new-born bat is still undeveloped. It is only after a period of growth outside the mother's body that the hand of the young bat reaches its final proportions.

During the course of evolution, when those elements that proved an advantage in coping with the challenges of

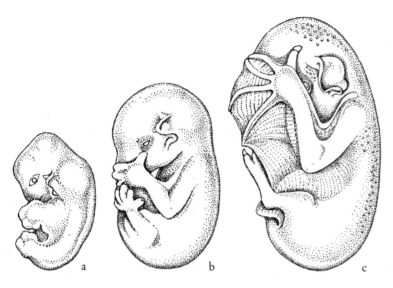

Embryos of Large Mouse-eared bats *(Myotis myotis)* of different ages.
Crown of head to rump length: a 9 mm, b 12 mm, c 17 mm.
Stage c shows the beginning of the modification of the fore extremities into
an organ of flight. The first elastic membranes between the bones of the arm
and fingers are developing (after Schumacher, from Grassé, 1955).

life prevailed and were transmitted, a process began in which the metacarpal bones and fingers lengthened. Undoubtedly this evolutionary feature proved useful. When the fore legs were straddled out stiffly, the wing membrane which stretched between the fingers and extended along the back legs to the tail, functioned as a lifting surface. The flight membrane proved advantageous to the creature when it was catching insects in gliding flight, since it increased manoeuvrability and enhanced the bat's efficiency as a hunter. Once the lifting surface had reached a certain size, the next stage followed in which it could be raised and lowered by the specially adapted front legs. Active flight had become possible. The bats, as a new kind of mammal, had taken possession of the earth, and, like birds, could lift themselves up into the sky.

Flight

Among vertebrates, bats as well as birds have been able to conquer the skies in active flight as a result of modification of arm and hand into a flight organ. But comparison of the bat's wing with the flight organ of a bird or with that of the extinct pterosaurs, shows greater similarities in structure to the pterosaurs. Birds have achieved the capacity for flight by different modifications in the structure of their fore limbs.

In bats, tension and support of the wing membrane is achieved primarily by the arm and hand. In the pterosaurs, only one finger (the fourth) is involved in the system of support and tension, whereas in bats, both the extended fore arm and the entire vastly elongated hand play a vital part. The thumb is something of an exception, for it has not been lengthened along with the other digits. In the Microchiroptera, it is the only finger still to have a claw; the Megachiroptera still have a claw of this kind on the end of the second finger.

In contrast to those of birds, the fore limbs of bats have not lost their significance for locomotion on the ground, in spite of their modification into organs of flight. The powerful claw of the thumb enables bats to climb with great agility along roof timbers or among rocks in caves, and the fruit-eating species to seize fruit and carry it to their mouth. But not only are bats skilful climbers, they are also able to walk on the ground, supporting themselves on the joints of the wrists. It is curious to see how they draw their folded wings along the ground beside them. This locomotion on all fours shows little of the skill usual among mammals.

In spite of being winged, bats are also able to swim to a certain extent. So it is not too serious for them if they accidently land in water. In addition to their function in walking and climbing, the hind legs play another important role. With their sharply pointed, curved claws they serve the bats as hooks by which the body is suspended.

In America and Madagascar, there are certain species which possess sucker-like discs on the thumbs and back feet, by means of which they are able to maintain a hold on smooth surfaces. The bats' ability to suspend themselves in a head-down roosting position, often with the belly against the support, is possible only because the legs of bats, unlike those of other mammals, are directed backwards from the hips.

If a man tried to hang by the toes on wall bars in such a way that his belly lay against the wall, he would not be able to do so. It would be possible only if the knee or foot joints could be turned outwards. In bats, this is precisely what has happened to the hind limbs. Since the leg can be flexed at the knee joint both upwards and outwards, it can take up a position which makes such suspension possible. This rotation of the hind limbs is also an advantage in flight, since they can participate in the movement of the wings.

But to return to the fore legs. The X-ray photograph in Ill. 13 clearly shows the modification of the skeleton of the hand into an organ of flight, and the drawing on page 28 illustrates the essential functional modifications in the anatomy of the fore limbs of bats.

The wing membrane, the patagium, which extends from the side of the body both between the fingers and between the limbs as far as the tail, can be divided into different parts. That part which extends in front of the wings proper along the front edge of the upper and lower arm, is the forward or antebrachial membrane (propatagium). Between the arm and hind margin of the wing is the arm membrane (plagiopatagium). The part which extends between digits 2 and 5 is called the finger membrane (dactylopatagium or chiropatagium). The tail membrane (uropatagium) stretches between the hind legs and the tail. The latter is often supported by a spur extending from the foot joint, the calcar. In certain species of bats, the uropatagium is reduced, and in some groups the tail projects freely as in a mouse. This particularly striking feature has earned them vernacular names such as Free-tailed bats or Mouse-tailed bats. In fruit bats, the tail vertebrae are absent and the tail membrane is only rudimentary.

Embryonic development shows that the wing membrane develops from a fold of skin on the body. Although this double membrane is very delicate, being no more than 0.03 mm thick in small species, it encloses a large number of blood vessels, nerves and elastic fibres as well as small muscle bundles. The muscles of the wing membrane do not contribute to locomotion, but have a supplementary bracing effect and prevent wing flutter caused by air flow. They also facilitate folding of the wing membrane when the bat is at rest. Usually this is achieved by drawing the wing membranes close to the body, while some species wrap themselves in their wings as in a sleeping bag. Since the extensive surfaces of the patagium play an important part in heat regulation, a great variety of different wing positions can be observed among roosting bats, particularly fruit bats, depending upon the external temperature.

In comparison with the feathered wings of birds, the fragile hairless patagia seem much more susceptible to injury. Yet the elastic fibres lend them enormous strength. In addition, the regenerative capacity of an injured patagium is very high. Holes 2 cm in diameter in the wing membranes of captive fruit bats have knitted together again within 28 days.

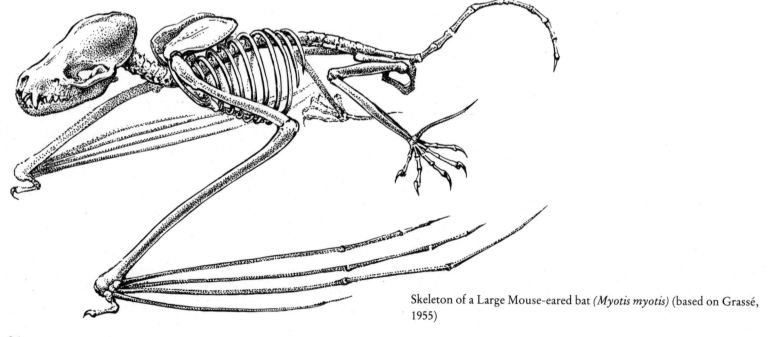

Skeleton of a Large Mouse-eared bat *(Myotis myotis)* (based on Grassé, 1955)

The development of sucker discs on the wrists and ankles of American Disc-winged bats (Thyropteridae). The discs produce a sticky secretion and have muscles that allow them to function as suckers. Adhesion is so effective that a single disc can support the entire weight of the bat (from Yalden and Morris, 1976).

The flight capacity of bats dominates every aspect of their life. It is the essential pre-requisite to obtaining food, and enables them to cover great distances between roost and feeding area. It is remarkable how particular kinds of food specializations are reflected in the adaptations of the organ of flight. Each species has developed the style of flight most appropriate to the requirements of its own way of life.

In the Megachiroptera, the wings are primarily a means of transportation from the roost to trees in feeding grounds many kilometres away. Therefore the demands made on them are quite different from those in species which require, for instance, a high degree of manoeuvrability in order to catch insects in flight. In avoiding obstacles or pursuing their prey, many species of bats show an agility which can be matched by few species of birds.

Body size and wing area are closely linked. A large, heavy body—and this is true of all flying animals—requires a larger lifting surface to support it in the air. Since, with an increase in the size of the animal by a given amount, the wing area increases by the square of this amount, while the volume, and therefore the weight, increases by the power of three, wing loading becomes increasingly heavy. But at a certain point, the animals reach a limit in body size beyond which flight is not possible. Among birds there are species incapable of flight —these are the cursorial birds such as the ostrich— whose bodies have attained a size which precludes the possibility of flight. No parallels exist among bats. There is no species with a body size which prevents it from flying.

The large Flying Foxes represent the limit of potential expansion of the wing area. This size of wing permits a slow but steady flight which is quite adequate for the bats as they move from place to place. More rapid and agile hunting flight such as is shown by many species of insectivorous bats is not possible with these large wings.

Flight in bats can be described as rowing flight. Many species have also mastered hovering flight such as is seen in raptors and humming-birds. Keeping the body erect, they are able to remain stationary in mid-air, without forward motion, like a helicopter. In this way, they can extract nectar from blossoms or pick insects off leaves.

Rowing flight can occasionally merge into a brief gliding flight. Extended periods of gliding, or soaring flight, is unknown among bats. The individual phases of move-

27

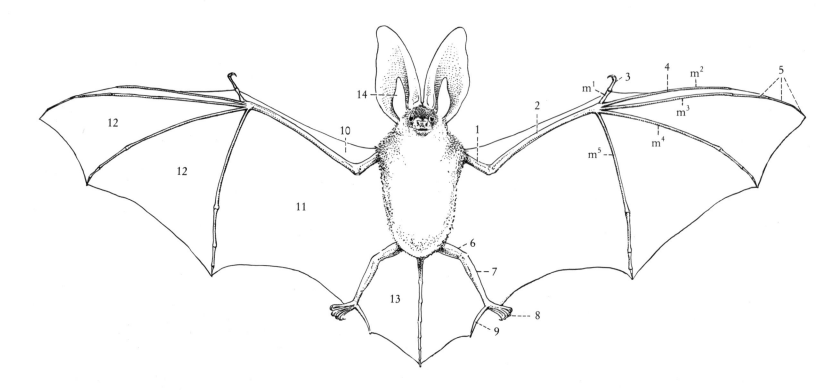

Diagram of the flight organ of a bat; supporting structures and the components of the membrane:
1 upper arm, 2 forearm, 3 thumb, 4 metacarpal bones (m^2, m^3, m^4, m^5), 5 digits, 6 upper leg, 7 lower leg, 8 hind foot, 9 calcar, 10 forward membrane (propatagium), 11 arm membrane (plagiopatagium), 12 finger membrane (chiropatagium or dactylopatagium), 13 tail or interfemoral membrane (uropatagium), 14 tragus or ear shield

ment in the course of a wing beat in rowing flight can be seen in the sketches on page 61. At the start of the downstroke, the wing is directed backwards and upwards, with the tips of the wings behind the body's centre of gravity. Then the wings are swung outwards and downwards and after this, drawn forwards. At the lowest position of the wing stroke, the tips of the wings are a long way in front of the head. In this phase, the edges of the wings are drawn downwards, so that the wings have the shape of an open umbrella. The upstroke is not exactly the reverse movement, but now the wings are moved first upwards and then backwards. If one were able to

mark out the track of the wing tip, it would—in relation to the body—describe an ellipse. In order to reduce air resistance during the upward stroke, many insectivorous bats fold the wings somewhat; the fruit bats raise them tightly folded past the sides of the body. The downbeat of the wing is a very important element of flight; it achieves both thrust and lift.

The number of wing beats per unit of time, that is the frequency of stroke, is much higher for insectivorous than for fruit bats. Large Mouse-eared bats raise and lower the wings an average of 11 to 12 times per second, and Lesser Horseshoe bats as many as 16 to 18 times. Only seven wing beats per second are reported for the fruit bat *Eidolon* sp. With this small number of strokes, the large fruit bats reach the same speed on an open stretch as most small insectivorous bats.

It has already been pointed out that the total extent of wing area has a significant effect on flight capacity. If the wing area is small in proportion to body size, loading is high and the stalling speed is increased. At first, the

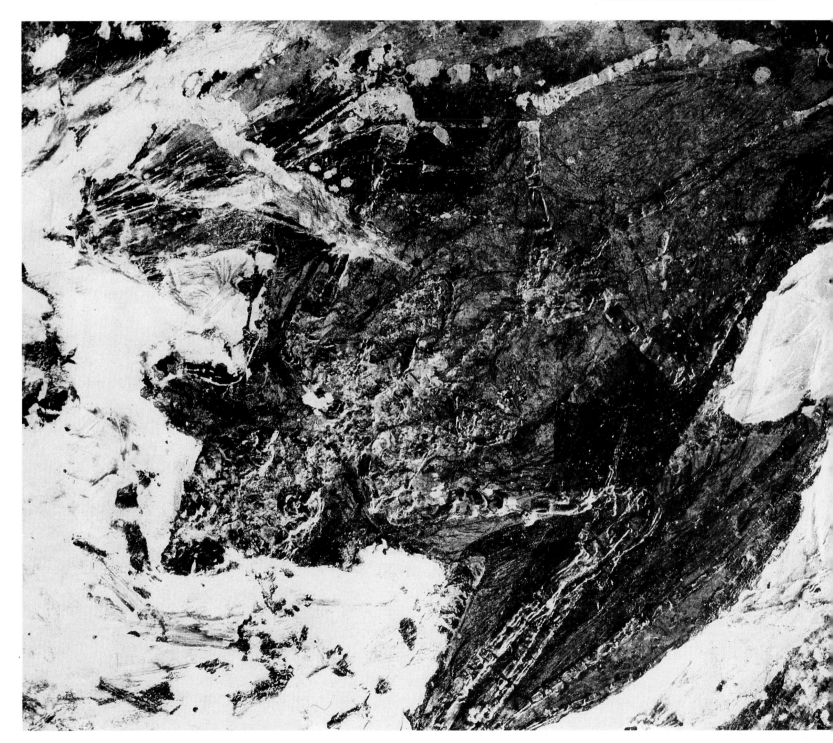

10   Greater Horseshoe bat
(*Rhinolophus ferrumequinum*) in
flight. These bats have a wing
span of up to 40 cm. Note the
band attached to the left forearm
to mark the bat.

11 Grey Long-eared bat *(Plecotus austriacus)* in flight. Wing spread about 25 cm. The tail membrane is well developed. The oversized ears are the outstanding recognition character of this genus.

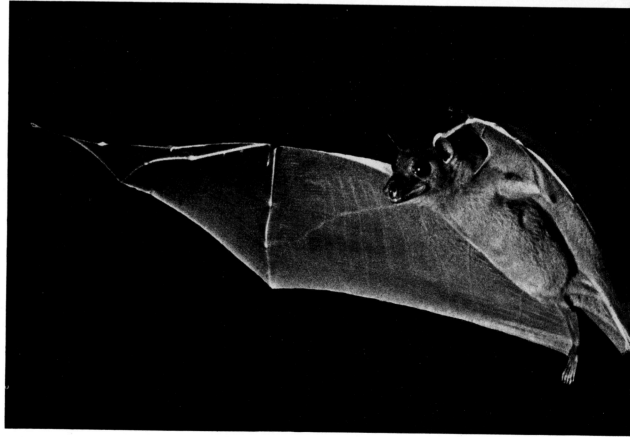

12 Egyptian Fruit bat *(Rousettus aegyptiacus)* in flight. Wing span reaches about 80 cm. In these bats, no tail membrane has developed.

13 X-ray photograph of a Large Mouse-eared bat *(Myotis myotis)*. It shows clearly the changes that have taken place in the skeleton of the fore limbs as they were modified into an organ of flight: the lower arm, particularly the metacarpal bones and the finger bones are enormously elongated. They are vital elements in supporting the wing membrane. The first digit (thumb) was not involved in these changes. The powerful development of the chest, as the point of attachment of the flight muscles, is striking.

14 Three-day-old Pipistrelle *(Pipistrellus pipistrellus)*. It is clear that at birth, wing growth has already begun but the wings are only slightly developed. The great extension in length, particularly of the bones of the fingers, begins only after birth.

15 Egyptian Fruit bat *(Rousettus aegyptiacus)*. Left wing, spread wide, seen from below. The various parts that make up the patagium extend between the lengthened bones of the arm and hand. The arm membrane is given additional tension by sets of powerful muscles.

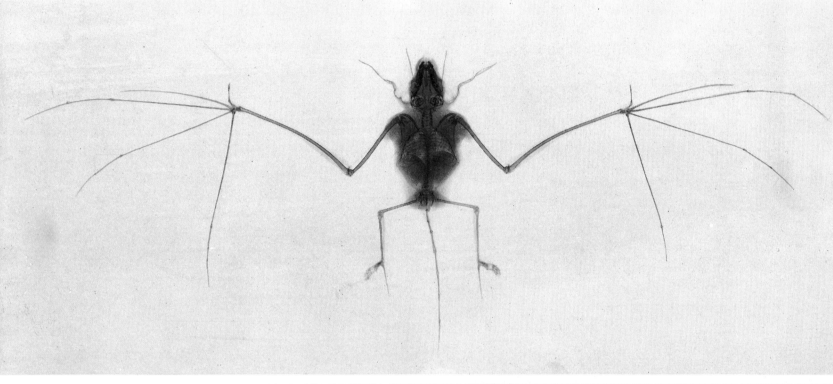

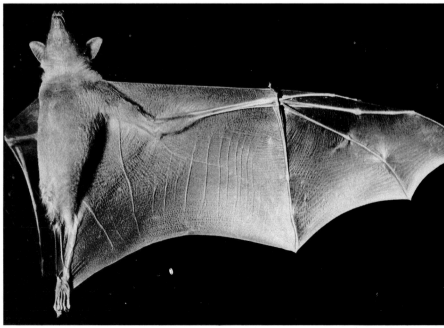

animals are able to compensate for this by increasing the stroke frequency. But this is possible only to a limited extent, as it in turn depends upon the functional capacity of the flight muscles. The only solution is an enlargement of the wing area. It is interesting that among the suborder Microchiroptera, increase of the wing area always lags a little behind that of body size, while in the Megachiroptera, the size of the wing area runs ahead of that of the body. This is associated with the fact that among fruit bats, it has been possible for certain species to evolve which are larger than any of the insectivorous bats. This in turn has meant that the large species of fruit bats have a much lower stalling speed and can manage with a slower wing beat. Of course, such large wings, reaching spans of up to 2.00 m, are much more cumbersome since they meet with considerably greater air resistance than do wings of small area. Flight appears slow and ponderous. From a biological point of view, these wings fulfil all the demands made upon them by the fruit bat. It is not unusual for them to carry the bat a distance of 50 to 80 kilometres in one night from sleeping quarters to feeding area.

Among insectivorous bats there are few species that cover very long distances at night in search of food. The hunting territory of many Molossid bats, such as the Mexican Free-tailed bat (the Guano bat, *Tadarida brasiliensis*) which lives in millions in caves in the southern states of the U.S.A., may have a radius of about 75 kilometres. To make it possible for these vast numbers of bats to obtain sufficient food on their nocturnal flights, they are forced to cover distances of this order. But they are exceptional.

For most of the insectivorous bats, the capture of prey and the nature of the habitat within which the hunting is carried out calls for great flying skill. Within woodlands, the bat must alter direction with lightning speed and

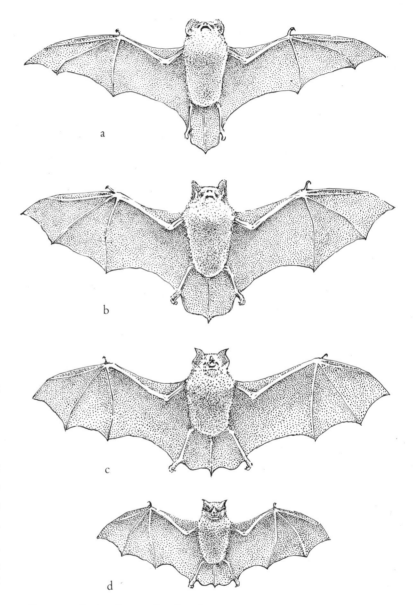

Silhouette of wing shapes of various bats
a Common Noctule *(Nyctalus noctula)*,
b Large Mouse-eared bat *(Myotis myotis)*,
c Greater Horseshoe bat *(Rhinolophus ferrumequinum)*,
d Lesser Horseshoe bat *(Rhinolophus hipposideros)*.
Characteristic of rapid hunting flight are long, narrow wings such as those of the Noctule. Broad wings such as those of the Lesser Horseshoe bat allow only slow but dexterous flight (after Gaisler, from Natuschke, 1960).

33

Diagram showing wing flight patterns for the Long-eared bat (*Plecotus* sp.) in hovering flight. The first 7 pictures show the downward movement of the wing (from Norberg, 1976).

great agility to avoid the many obstacles in its path. These requirements are best met by a short broad wing which also allows for hovering flight. On open ground, however, a greater wing area and a longer wing are advantageous. The bat can fly rapidly, catch food over a large area and escape from enemies more readily. A long, narrow wing is also found in those species of bats which cover long distances between their summer and winter roosts. Modifications in wing form and area occur primarily in the region of the hand membrane. There is also great variation in uropatagium associated with variation in leg, calcar and tail length. Few statistics exist for the speed of flight in bats, since it is a difficult measurement to make. For the Big Brown bat (*Eptesicus fuscus*) which is widely distributed in the U.S.A., speeds of up to 75 kilometres an hour are claimed. The Long-winged bat (*Miniopterus schreibersi*), with 70 kilometres an hour, is said to be the fastest flying species in Europe. A speed of 50 kilometres per hour is given for the Common Noctule (*Nyctalus noctula*). The Large Mouse-eared bat (*Myotis myotis*), on the other hand, maintains a more leisurely stroke and covers only 15 kilometres in an hour. It is reported that fruit bats reach speeds of 15 to 30 kilometres an hour.

The tail membrane has little influence on the capability for flight. This can be deduced from the fact that it shows such great variation in the Microchiroptera, and is completely absent in a number of species. When present, it is

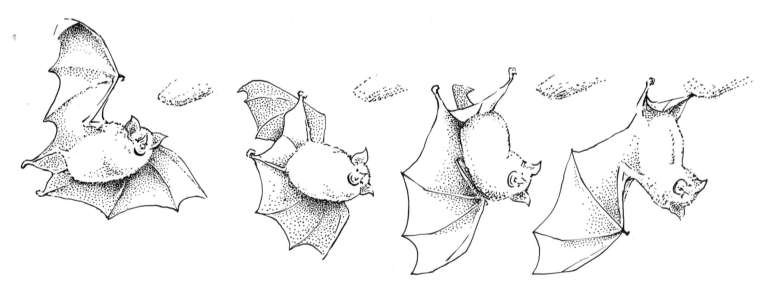

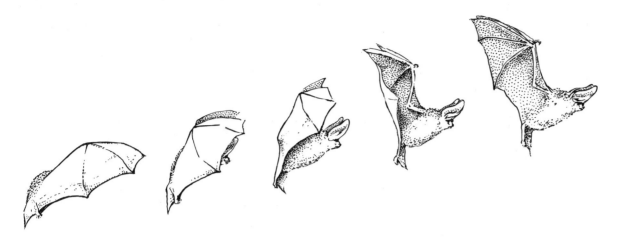

curved in flight and is used less for steering than for reducing the speed of flight when the bat is landing. The landing manoeuvre of bats at small projections on a wall or on rough wooden beams is quite an acrobatic feat. The finer points of landing technique differ from species to species. But all of them twist with a stroke of the wing away from the resting place on which they want to hang, in such a way that the feet can swing upwards and the talons immediately gain a firm hold. Although the widely spread wings slow down the descent like a parachute, the impetus of flight sometimes sends the bat hurtling downwards, particularly if the surface is very smooth, and then it must make another attempt.

Many fruit bats, on the other hand, land with the belly on the branches of their tree roosts. Then the claws of the hind feet and of the thumb seize branches and twigs and the bat takes up its roosting position.

In many species of the Microchiroptera, the tail membrane has taken on an additional function: it serves as a pouch when prey is being caught.

Flight demands a great expenditure of energy. The muscles of flight alone need an enormous supply of oxygen. In order to meet fuel requirements which in flight are four times greater than at rest, considerable demands are made on the organs of respiration and circulation. The high loading brought about during flight causes a massive increase in the cardiac and respiration rates. For

Diagram showing the landing manoeuvre of a Greater Horseshoe bat *(Rhinolophus ferrumequinum)*. In the last phase of approach to the wall or branch, the bat rolls to the right, round the axis of its own body, and lowering the body, it seizes hold with one foot. The extended wings check the momentum and the second foot seizes hold. The manoeuvre lasts about 110 milliseconds (from photographs by Kulzer and Weigold, 1978).

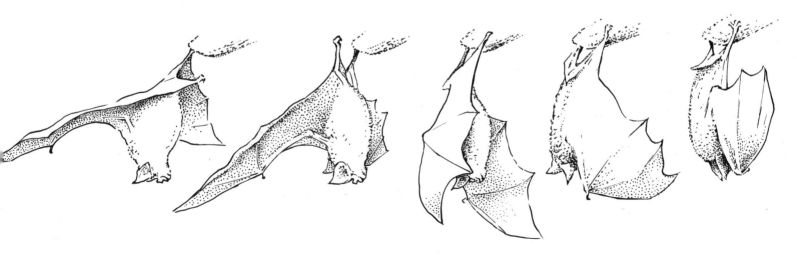

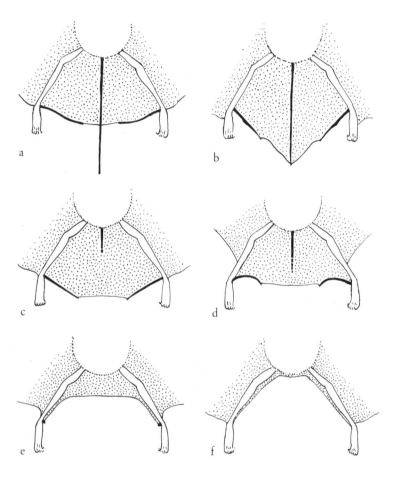

Differences in development of the tail and tail membrane (from Husson, 1962)
a *Eumops geijskesi* (Molossidae)
b *Lasiurus borealis* (Vespertilionidae)
c *Pteropteryx kappleri* (Emballonuridae)
d *Chilonycteris rubiginosa* (Chilonycteridae)
e *Desmodus rotundus* (Desmodontidae)
f *Sturnira lilium* (Phyllostomidae)

example, in the Spear-nosed bat *(Phyllostomus hastatus)*, the cardiac rate increases from 522 beats per minute at rest to 822 per minute in flight. Respiration rate increases from 180 to 560 per minute. The heart is considerably larger than in other mammals of the same size, and therefore is able to pump a much greater volume of blood through the body in the same length of time. In the blood, oxygen is carried to the muscles. Here, in comparison with other mammals, there is another difference. Whereas in the majority, the capacity of the blood to absorb oxygen is only 18 per cent by volume, in bats it reaches 27.

Physical work creates heat, and humans sweat. Bats are not able to maintain as constant a body temperature as, for example, carnivores or apes, so their temperature rises in flight. Unlimited, this could prove dangerous. After a maximum temperature is reached, bats benefit from a useful cooling system which, to some extent, helps to keep temperatures at a constant level. When body temperature rises, the small blood vessels within the membranes dilate and the increased quantities of blood flowing through are cooled by the cold stream of air passing the wings.

The structure of the bat's body shows various adaptations to flight. The neck is short, the chest is massive on account of the powerfully developed thoracic muscles. The adjoining abdomen is narrow and tapering. There are many other special biological features that are the consequence of the bat's power of flight, and which will be mentioned in other chapters of this book.

Distribution of bats throughout the world

For land mammals, the shoreline of the oceans and quite often great lakes and rivers represent significant obstacles which normally they are unable to cross. Bats, on the other hand, are able to cross at least lakes and rivers with no difficulty. Crossing oceans, however, is rarely within their capability. But since many species are capable of covering fifty kilometres and more in continuous flight, they are able to negotiate broad arms of the sea and to reach distant islands.

One example which shows that bats are often the first mammals to colonize an island is that of the island of Krakatau in the straits between Sumatra and Java. It lies 40 kilometres from the mainland and 18 from the closest neighbouring island. In a volcanic eruption in 1883, parts of the island were submerged and the rest buried beneath a shower of ash. All animal and plant life was destroyed. But the seemingly dead island did not keep its desolate appearance for long. Within three years it was green again as a new carpet of vegetation spread over it.

Man deliberately refrained from interfering and it was possible for detailed studies to be carried out on the natural process of recolonization by plants and animals. After 25 years, it could be shown that the animals which had gained a foothold there included 240 species of insects, 16 of birds and 2 of reptiles. After 37 years, there were more than 500 different species of animals including two species of bats as the first mammals.

The bat's capacity for flight is undoubtedly the reason why, in its 50 million years of existence on earth, the great army of bats has succeeded in extending its territory across every continent and to many islands.

Today they show a variety of form which is much greater in wealth of species in the tropics and subtropics than elsewhere. In both the northern and southern latitudes, as the distance from the equator increases, the number of species decreases very rapidly. It is logical therefore to look for centres of differentiation and speciation in the tropics.

The total number of species of bat found in Europe is 30; north of the Alps there are only 20 species, that is, not even 5 per cent of all species living in the world today. Apart from the temperature, it is food supply which more than any other factor presents the bats of the northern latitudes—they are all insectivores—with difficulties unknown to the tropical forms. They have adapted to the cold winters when food is scarce by hibernating during this critical period.

It is surprising to find that although all bats are capable of flight, there are certain families that occur on every continent, while others live only on one particular island. One inevitably asks: why do we find members of the suborder Megachiroptera—the fruit bats—in Asia and Africa, whereas the numerous family of Spear-nosed bats (Phyllostomidae) is encountered only in America?

In addition to temperature and food supply, another important factor in determining the distribution of bats was the displacement of continental land masses that took place long ago in geological history. Another point perhaps of considerable significance in answering these questions is that in spite of the ability to fly, not all species show equally strong migratory tendencies. Newly evolved species did not begin immediately to disperse over a wider area. Often the specific environment of a region or island and perhaps a certain food that was available suited them well, and they remained in that particular locality. In other cases, if the situation was less favourable, the creatures moved across country and settled in new areas. Millions of years ago, land masses drifted apart and land bridges were severed, so that individual populations were separated from their place of origin, and over thousands of years, they underwent changes. New species evolved. This is why islands such as Borneo, Java and Madagascar possess such an independent fauna. Of the species living today, there are some which occur in several continents. Interestingly enough, they are always found in similar biotopes; even at the opposite ends of their area of distribution, these species inhabit the same type of living quarters or the same landscape.

This constancy in respect of environment shown by individual species can be used to detect changes in our environment. The arrival of new species, or—as is much more common in Europe—their disappearance, are certain signs of such change.

If we consider the families of bats living in the world today from the point of view of their distribution, we can distinguish three major groups:
1) those living exclusively on the continents and islands of the Old World (nine families);
2) those occurring only in the New World (six families);
3) those which are found both in the Old and the New World (three families).
Among the nine Old World families, there are three which are represented by only a single species. These are the Short-tailed bats (Mystacinidae) which exist in New Zealand, the Sucker-footed bats (Myzopodidae) found only in Madagascar and the Hog-faced bats (Craseonycteridae) found only in Thailand. The typical feature distinguishing members of the second family is the presence of suction discs on the feet and the balls of the thumb. Quite independent of the Madagascan Sucker-footed bats, suction pads of this kind have also been developed by the two species of the family of American Disc-winged bats (Thyropteridae) living in tropical South America. This is a good example of convergent evolution within the animal kingdom, which in no way implies close relationships.

The six remaining families which inhabit only the eastern hemisphere are concentrated mainly in the tropics and subtropics. The many species of Megachiroptera make up a single family, the fruit bats or Flying Foxes (Pteropodidae). They live in India, Southeast Asia, Africa and the northern part of Australia. The distribution of the true Flying Foxes (*Pteropus* spp.) is of particular zoo-geographical interest, for although they are absent from the African mainland, they are found on the small island of Pemba which lies only 60 kilometres off the east coast of Africa, and Mafia, 16 kilometres from the same coast. It is not clear how the creatures got here, particularly since they are closer to the species living in the Malayan regions than to the Flying Foxes of India and Sri Lanka. Could it be that a storm cast them up like flotsam off the coast of Africa, or even that their ancestors lived on the African mainland? In any case, other fruit-eating Megachiroptera, usually smaller species, are found today only in Equatorial and South Africa.

The distribution of the Slit-faced bats (Nycteridae) of the suborder Microchiroptera suggests that continuous forests once extended from the Congo to Indonesia. Members of the family are found particularly in the primeval forests of Africa. They are entirely absent from the Indian region, yet they are found again in Southeast Asia. Early authors described the members of this family as wholly forest-dwelling. More recent observations have shown that they also extend into steppe regions, and certain species have become indigenous there.

The many species of the family of Horseshoe bats (Rhinolophidae) are found primarily in tropical regions. Their area of distribution extends from Africa across India, southern China and as far as the East Indies. But they are absent from both Madagascar and various islands of the Indo-Australian archipelago. Certain species of this family are found in temperate latitudes. As relics of warmer days, the fauna of central Europe still includes the Greater Horseshoe bat (*Rhinolophus ferrum-*

equinum) and the Lesser Horseshoe bat *(Rhinolophus hipposideros)*. Unfortunately, the Greater Horseshoe bat has become quite a rarity north of the Alps, and in recent years, an alarming decline in the numbers of Lesser Horseshoe bats has been reported. In Europe today they are numbered among the endangered species. The Old World Leaf-nosed bats of the family Hipposideridae, which also contains many species, are closely related to the Rhinolophidae. Their area of distribution coincides almost exactly with that of the latter, except that since they are unable to hibernate, members of this family are not found in temperate latitudes.

The small number of species in the families of Mouse-tailed bats (Rhinopomatidae) and False Vampires or "cannibal bats" (Megadermatidae) are today exclusively tropical forms. Fossil finds of Megadermatids from the Oligocene and the Miocene show that this family—and doubtless others as well—was also widespread in Europe millions of years ago.

Six families of bats occur exclusively in the New World. The majority of them live in tropical and subtropical regions. The greatest variety of forms, which is reflected in a large number of species, has been developed by the Spear-nosed bats (Phyllostomidae). It has been assumed that this family developed and achieved its high degree of differentiation in South America. Only very few species, in the northernmost extent of their range, reach the south-west of the U.S.A.

Purely tropical forms are the blood-feeding Vampire bats (Desmodontidae), which are probably descended from Spear-nosed bats. They live in an area which extends from southern Mexico to Brazil.

Another group of specialized feeders among the Chiroptera of the New World which has approximately the same area of distribution is the family of Fisherman bats (Noctilionidae).

Area of distribution of Vampire bats (Desmodontidae) in Central and South America (after Greenhall, 1975)

A parallel to the distribution of various species of Old World fruit-eating bats that occur only on particular islands in the East Indies is provided in the New World by the Funnel-eared bats (Natalidae). Many species of this family are found only on a few islands in the West Indies.

Of the three families with species occurring both in the Old and the New World, the Sheath-tailed bats (Emballonuridae) and the Free-tailed bats (Molossidae) are restricted in their distribution to the tropics and subtropics. Representatives of these families can be found in sites at opposite ends of the earth, thousands of miles apart. A comparison in terms of morphological and ecological specialization of bats of different genera that are indigenous to either Africa or South America, shows interesting similarities. It is tempting to conclude from these that the species of the New World derive from those of the Old. But this is not justified, since the similarities result rather from parallel developments that have taken place in like

39

but widely separated biotopes. Nevertheless, the genera have indeed a common ancestral stock. But that dates far back in the history of the world.

Among the Molossidae, only a few species live in the warmer regions of the temperate zone. The Free-tailed bat *Tadarida teniotis* is found today in Spain, Italy and Greece. Other species of this genus extend into the southern states of the U.S.A. Best known of all is the Guano bat, *Tadarida brasiliensis*, which spends the summer in caves, roosting in its millions.

About a quarter of the 320 species of Vespertilionid bats (sometimes called Vesper bats; Vespertilionidae) are not restricted in their distribution to the warm regions of the earth, but also inhabit temperate zones. Most of the bats of Europe, North America and Japan belong to this family. They are also found on many islands in the Pacific and the Atlantic (Bermudas, Azores, Galápagos, Hawaii).

Numerous morphological features together with the fact that they are insectivorous suggest that many species of this family are very primitive forms. Since they have not developed a high degree of specialization, it has been possible for members of this family to adapt easily, compared to others, and to occupy niches in which specialized species would be incapable of survival. For example, we find members of the genus *Myotis* (Mouse-eared bats) in the most northerly regions of the eastern and western hemispheres. They have extended the scope of their distribution to the northern limit of tree growth between Norway and Kamchatka, and are also found on the Alaskan and Labrador peninsulas. The Little Brown bat *(Myotis lucifugus)* is distributed across the whole of North America. Closely related species inhabit Central and South America as far as Chile and Argentina. In the Old World, the Daubenton's or Water bat *(Myotis daubentoni)* extends its range from Western Europe across Northern Asia to the Pacific coast. Some members of the large genera of Pipistrelles *(Pipistrellus)* and Big Brown bats *(Eptesicus)* are similarly very widely distributed, although the northern limits of their range do not extend as far as those of the Mouse-eared bats. It is known from the evidence of fossil plants and animals that some 50 million years ago, average temperatures in Europe were much higher than they are today; consequently the spectrum of Vespertilionids was much wider. In addition, many more species from other families—even Flying Foxes—were found here. The decline in temperature which reached its peak during the glacial period forced a great many of the warmth-loving species to retreat into areas where conditions suited them better. There was a great reduction in species in temperate latitudes.

It is interesting to take a look at Australia, an island continent with a highly specialized mammalian fauna. Anyone who considers only the egg-laying mammals or the many marsupials may be led to assume that this is an oasis for the earliest and most primitive mammals. But a glance at the bat fauna gives a very different picture. There are only a few very old "endemic" species, such as the Tube-nosed Fruit bats (Nyctimeninae). Much more common are quite "modern" representatives of Chiroptera that occur on the neighbouring islands of Southeast Asia. The cosmopolitan genera *Myotis*, *Pipistrellus* and *Eptesicus* are also represented in Australia.

Many species of bats only established themselves in Australia considerably later than the primitive population of mammals such as the spiny anteater, duck-billed platypus and the marsupials. The sea channels between the islands proved no barrier to the expansion of the Indo-Malayan bat fauna into the Australian region. The result is that, in contrast to other groups of mammals, fewer differences exist between the bat fauna of Australia and the rest of the Old World than between the bat fauna of the Old World and the Americas.

17   Epauletted Fruit bats
(*Epomophorus anurus*; Pteropodidae). These bats rest with
their wing membranes folded
round them, as if in a protective
blanket.

42

18 Zenker's Fruit bats (*Scoto-nycteris zenkeri*; Pteropodidae). These small representatives of the Epauletted Fruit bats lack the usual hair tufts on the shoulder. They reach a body length of 8 cm and a wing span of 30 cm.

19 Tube-nosed Fruit bat (*Nycti-mene* sp.; Pteropodidae). So called because they have extended tubular nostrils.

20 Wahlberg's Epauletted Fruit bat (*Epomophorus wahlbergi*; Pteropodidae). The shoulder pouches of these bats are surrounded by tufts of light-coloured hair (epaulettes). The white hairs on the front and back edge of the ear are typical of this species.

21 Dwarf Long-tongued Fruit bat (*Macroglossus minimus*; Pteropodidae). These bats are among the smallest of the fruit bats, up to 7 cm in length of body, with a wing span of 25 cm. The elongated head indicates specialization in nectar feeding.

22 *Rousettus stresemanni* (Pteropodidae); one of the cave-dwelling fruit bats

23   Indian Flying Fox *(Pteropus giganteus)* engaged in grooming

24   The Hammer-headed Fruit bat *(Hypsignathus monstrosus*; Pteropodidae). The largest species of fruit bat in Africa. The hammer-shaped muzzle of this bat carries pendulous mouth lappets. The males of the species are larger than the females (body length up to 20 cm); the wing span is 90 cm.

25 Indian Flying Foxes (*Pteropus giganteus*; Pteropodidae). With their large eyes, Flying Foxes are able to find their way easily at night.

26 Greater Mouse-tailed bat (*Rhinopoma microphyllum*; Rhinopomatidae). A representative of the "primitive" family of Mouse-tailed bats. The tail extends freely from the body with only its basal part incorporated into the narrow tail membrane.

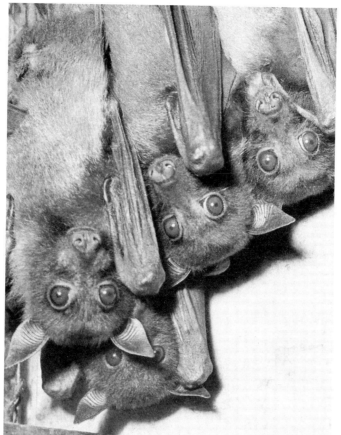

27 Mauritius Tomb bat (*Taphozous mauritianus*; Emballonuridae). A representative of the Old World Sheath-tailed bats. These bats are not found exclusively in underground haunts, but sometimes roost during the day on tree trunks and walls.

28 *Taphozous saccolaimus* (Emballonuridae); one of the group of Tomb bats

29 Fisherman bat (*Noctilio leporinus*; Noctilionidae). This is the larger of the two species in the family of Fisherman bats (body size about 13 cm). It feeds primarily on fish.

30 Javanese Slit-faced bat (*Nycteris javanica*; Nycteridae). In contrast to the majority of Slit-faced bats that live in Africa, this species inhabits Southeast Asia.

31 *Hipposideros larvatus* (Hipposideridae) of the family of Old World Leaf-nosed bats

32 *Hipposideros ruber* (Hipposideridae). Red Leaf-nosed bat from Africa

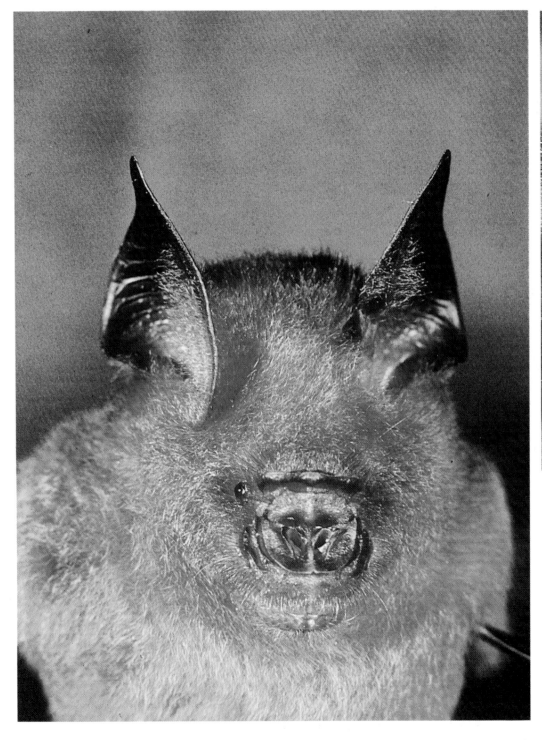

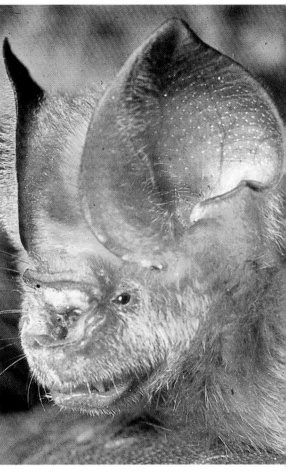

33   False Vampire (*Megaderma lyra*; Megadermatidae). Large ears and a triple-lobed nose leaf are distinguishing features of this species.

34   Heart-nosed False Vampire (*Megaderma cor*; Megadermatidae). This small African bat takes its name from its heart-shaped nose leaf. The relatively large eyes are a striking feature.

35   Long-tongued bat (*Glossophaga soricina*; Phyllostomidae). This small Long-tongued bat feeds on nectar and pollen. The extended snout carries a pointed nose leaf at its tip.

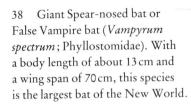

38 Giant Spear-nosed bat or False Vampire bat (*Vampyrum spectrum*; Phyllostomidae). With a body length of about 13 cm and a wing span of 70 cm, this species is the largest bat of the New World. Apart from insects and fruit, it also eats small vertebrates. It was incorrectly believed to feed on blood and named accordingly.

36 *Sturnira lilium* (Phyllostomidae). One of the Yellow-shouldered Spear-nosed bats. In this species, the tail has regressed completely.

37 *Uroderma bilobatum* (Phyllostomidae). This species belongs to the group of frugivorous Spear-nosed bats. Typical are the four light-coloured longitudinal lines on the head.

39 *Sphaeronycteris toxophyllum* (Phyllostomidae). This species, with its curious cutaneous outgrowths above the nose and its chin lappets, is found only in the New World, and belongs to the family of Spear-nosed bats. The significance of this facial ornamentation is not clear.

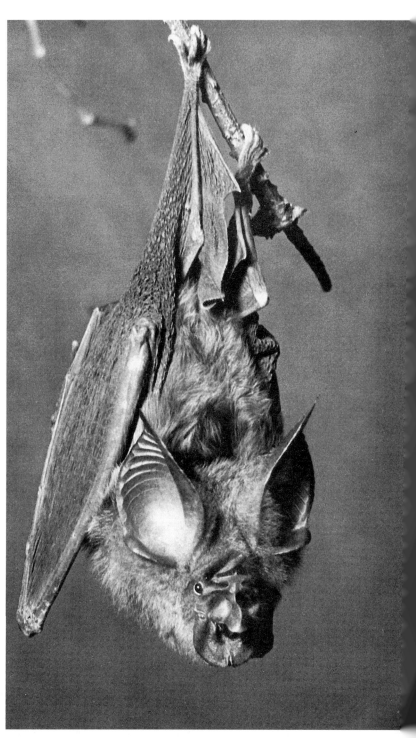

42 Greater Horseshoe bat (*Rhinolophus ferrumequinum*; Rhinolophidae). This species, now rare in Europe, has a wing span of 35 to 40 cm. The horseshoe-shaped nose leaf surrounding the nostrils is a clear recognition character. Above it are two more cutaneous lobes in the shape of a pointed lancet. The ears have no tragus.

40 Lesser Horseshoe bat (*Rhinolophus hipposideros*; Rhinolophidae). Like its large relation, the smaller species of the two European representatives of the Horseshoe bats is becoming a rarity in our bat fauna. The bats weigh about 5 to 9 g and have a wing spread of 20 to 25 cm. On its right forearm, this one carries a band (wing clip).

41 Hildebrandt's Horseshoe bat (*Rhinolophus hildebrandti*; Rhinolophidae). Habitat: Africa

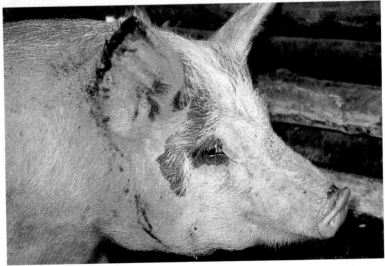

43   Common Vampire bat (*Desmodus rotundus*; Desmodontidae)

44   Domestic pig with numerous vampire bites on ear and neck

Fruit bats (Megachiroptera)—an Old World family of bats with many species

In what way do they differ from the Microchiroptera?

The Megachiroptera (great bats) is the name used to denote the suborder of fruit bats ("flying foxes"), consisting of about 175 species. The name was chosen when only the large species of Flying Foxes were known. A look at the members of the suborder today shows that the name is a misleading one. The largest of the bats are indeed found among the Flying Foxes, but there are also certain small species in the suborder which are exceeded in size by some of the Microchiroptera.

If it is not size, is it perhaps choice of diet which distinguishes the Megachiroptera from the Microchiroptera? In older literature, they were described as the fruit eaters in contrast to the Microchiroptera that were assumed to be purely insectivorous. But neither is this distinction a valid one. It has long since been established that some of the Microchiroptera are also fruit-eating species.

What then are the differences? There are certain anatomical features that justify the division of bats into two major suborders. Of these, an important one is the presence, exclusive to the Megachiroptera, of a small claw on the second digit. This feature alone is not sufficient to make the distinction in every case, since members of the Bare-backed fruit bats (*Dobsonia* sp.) and a few other groups possess no claw on the second digit.

Another feature typical of although not exclusive to fruit bats is the reduction of the caudal vertebrae and, as a result, the absence of an externally visible tail. In only very few species is there a fully developed tail similar to that of the fossil bat *Archaeopteropus*. On the other hand, reduction of the tail is also found among the Microchiroptera, particularly in certain of the fruit-eating Spear-nosed bats of the New World. As a result of this feature, most fruit-eating bats have no tail membrane or only a rudimentary one that can be seen as a narrow fold of skin on the inside of the lower leg. Thus, no single feature but a series of features makes it possible to decide whether a species is assignable to the Megachiroptera or not.

In spite of their nocturnal habit, fruit bats on the whole navigate by means of visual perception. For this purpose, like the nocturnally active owl, they have comparatively large eyes. Orientation by means of echolocation which the Microchiroptera master to perfection has not been evolved by the Megachiroptera. There is one exception: fruit bats of the genera *Rousettus* and *Lissonycteris* are capable of both optical and acoustical orientation by means of echolocation.

Examination of the eyes of fruit bats has shown that they are extremely well adapted for nocturnal vision. Two features possessed by the eyes of other mammals are not found in those of fruit bats, since they are unimportant for nocturnal vision; firstly, fruit bats lack the ability to distinguish colour and secondly, they are unable to alternate between near and distant vision. The lens of the eye is particularly thick and thus achieves a high refractive power. In this way, only small but very intense images are produced on the retina. In addition, the number of light-sensitive receptors in the retina of the fruit bat is very great, exceeding those in the eye of the owl. Experiments have shown that the enhanced efficiency of the fruit bat's eye takes the form of a greatly improved sharpness of vision even in semi-darkness. On their nocturnal flights in search of food, the creatures show great skill in avoiding obstacles and reaching their feeding places safely.

The sense of smell plays an important part in the location of food. A sensitive nose is an undoubted asset in detecting the aromatic scent of ripe fruit in darkness. Experiments have shown Rousette Fruit bats to be capable of locating small pieces of banana only 10 mg in weight that were hidden in their cage. The creatures could also distinguish easily between the smell of real bananas and a chemically produced aroma. Examination of the brain of the fruit bat has also shown that the sense of vision and the sense of smell are of vital importance to fruit bats in establishing contact with the environment and particularly in finding food. The brain centres associated with particular sense organs clearly reflect in their degree of development and differentiation, differences in the functional efficiency of those sense organs. Quantitative analyses of the appropriate brain centres have shown that in the case of fruit bats, not only the olfactory centre of the brain but also and to an even greater extent the optical centres are much more highly developed than they are in insectivorous bats. Conversely, the acoustic centres in the brain of fruit bats show only slight or, more accurately, normal development.

Extensive analyses of the size of the brain or separate parts of the brain of numerous species of bats carried out in recent years have called into question the widely accepted assumption that the Megachiroptera are more primitive than the Microchiroptera. In assessing the level of development in bats, emphasis was always laid on the degree of refinement achieved in the development of flight, that is, on a feature of specialization. Up to now, little consideration has been given to the extent of brain development, that is, to centralization. Research carried out by the Frankfurt brain specialist Stephan showed that there is considerable variation in the development of the brains of individual species of bats. There is a close correlation between diet and the degree of brain specialization, quite independent of the family to which the species belongs. The fruit bats are something of an exception. On the basis of measurements made of the neocortex, they are found to have achieved the highest degree of brain development among bats. The neocortex is four to five times larger than that of primitive insectivorous species. In this respect, the fruit bats are surpassed only by the highly specialized Vampires and the fish-eating bats.

Giants and dwarfs among the Megachiroptera

The great variation in body size that can occur among different species of fruit bats is illustrated in the contrast between certain bats with a body weight of 1000 g and others weighing only 5 g. Wing spread varies accordingly, ranging from 200 cm to 25 cm.

Apart from differences in size, most species of fruit bats are fairly similar in appearance. Striking facial appendages are usually absent. Only the Australasian Tube-nosed bat (*Nyctimene* sp.) and the Hammer-headed or Horse-faced Fruit bat (*Hypsignathus monstrosus*) of Africa, with its grotesque muzzle, deviate from the general pattern. Adaptation to the nectar-feeding habit has produced in some species a long narrow snout rather different in appearance from the usual fox-like or dog-like head of the Megachiroptera.

Since bats in zoos do not have the same appeal to the public as, for instance, monkeys, nor are they easy to look after, there is very little opportunity to get to know them in captivity. Those few bats that are kept, however, are usually large species of flying foxes belonging to the true Fruit bats (Pteropodinae).

The largest African representative of this group is the grey to reddish-brown Straw-coloured bat (*Eidolon helvum*), which measures 20 cm from head to rump. These creatures are very gregarious and, during the day, assemble in large communal roosts on high trees. Roosts of this kind are sometimes found in the centre of large towns, where even heavy traffic fails to disturb these otherwise timid creatures. In a large gathering of this kind, it is striking to see how the individual animals carry out certain activities such as grooming almost simultaneously.

After sunset, the *Eidolon* roosting communities become active and prepare to fly off towards the trees where they feed. The general departure is accompanied by a good deal of clamour as wing membranes are once again thoroughly cleaned and groomed. Then the bats fly off quietly in small groups. On the way, they visit watering places at which they quench their thirst. To drink, the bats fly low over the surface of the water, and in flight, scoop up water with the mouth. They eat noisily throughout the night. In doing so, they often maintain a hold on a branch with only one foot, while with the other they grip the fruit, pressing it against the chest and biting pieces from it. They spit out hard parts and fibrous matter, which fall to the ground. Feeding is by no means a peaceful process. Fruit bats are voracious feeders, and fighting for the best places with the ripest fruits is accompanied by a constant screeching and fluttering. At intervals, they rest and fly to other trees in the vicinity to feed. Only with the break of day do the bats return to their roosts on trees that are often many kilometres away. Even here they do not immediately fall quiet. Grooming is carried out and there are hours of noisy quarrelling over the best roosting places until the community eventually settles to sleep.

Although most fruit bats roost on trees, members of the genera *Rousettus, Eonycteris, Notopteris* and *Lissonycteris* roost in caves. They are medium-sized species which are found in the tropical regions of the Old World. They do not occur in Australia. Little is known of the habits of most of the species since they are active at night and it is difficult to observe them in their natural habitats.

Probably more is known about the habits of the Egyptian Fruit bat (*Rousettus aegyptiacus)* than of any other. In recent years, this species has been the subject of de-tailed investigation, and for this purpose has frequently been kept in laboratories.

The Egyptian Fruit bats, the distribution of which extends into the Mediterranean area as far as Cyprus, also live in large assemblies, spending the day in tombs, caves and the cellars of old buildings. Their sleep is not a deep one, and the slightest noise wakens them. At any disturbance, the creatures retreat into any chinks and crevices they can find. After dark, they leave their gloomy quarters and fly to the trees on which they feed. On light nights, they can be seen in towns among avenues of trees, if these provide the food they require. At first, they hover over the fruit to examine its scent before landing on the tree. Here they clamber about, using their keen sense of smell to lead them to the ripest fruits. Feeding is followed by an extensive period of grooming, in which fur and wing membranes are licked clean and fragments of food removed from the claws of the thumbs and feet using the teeth. Since these species—and the cave-dwelling Angola Fruit bat (*Lissonycteris angolensis)* to an equal extent —are known to use ultrasonic orientation in addition to having eyes with highly efficient nocturnal vision, they are able to inhabit dark recesses of caves where no light penetrates. In this darkness, they are able to find their way by means of their "radar system". The sounds emitted by *Rousettus* are produced not by the larynx as in the Microchiroptera, but by the tongue. As soon as the bats approach the lighter entrance to the cave, they change over from the acoustical to the optical system of orientation. This optimal development of two sense organs as important for orientation as ear and eye, has enabled this group of fruits bats to colonize habitats which remain unavailable to most other species.

The daily alternation of light-dark, or day-night, provides the stimulus for the phases of rest and activity. When the bats are hanging from trees, they have no difficulty in determining the onset of dusk. But how do cave dwellers know when it is getting dark and the time is near to set out after food? Experiments have shown that in addition to the direct effect of light, a kind of "internal clock" helps to determine the daily routine. Towards evening it stimulates the cave-dwelling bats. They fly to the cave entrance to check external conditions of darkness. If it is still too light, departure is delayed. Experiments in which the dark phase of the day was reduced to two hours, had the interesting effect of reducing the activity phase to this length as well. For the Egyptian Fruit bat in particular, the periodic unit of the internal clock is approximately 24 hours. In this way, the bat is so well adapted to the passage of the natural day that it always wakens at the correct time. Combined with ultrasonic orientation, this periodicity is a supplementary and very useful adaptation to nocturnal activity and life in caves.

In their day-time roost, the mass assemblies of fruit bats usually include males, females and, in the appropriate season, young animals. Purely female colonies, such as are known to exist in many species of Microchiroptera in the form of maternity colonies have scarcely been observed among fruit bats. However, Eisentraut reports that on his journeys through Cameroon, he came across assemblies of Angola Fruit bats in various caves, which consisted entirely of females. The animals were hanging close together from the roofs of caves and were in late stages of pregnancy or had already borne young. So the possibility cannot be ruled out that some species of fruit bats also set up maternity groups.

Among fruit bats, the genus *Pteropus* which is found from islands of extreme western Indian Ocean through Southeast Asia to Australia and in many of the islands of the South Pacific, contains the largest number of species including those known as Flying Foxes. Many of them have a very strictly limited area of distribution. In many cases, each island has its own particular species. One of the largest species is the Flying Fox of India *Pteropus giganteus*, which is light to dark brown in colouring and has a wing span of 120 cm. While its home is principally India and Sri Lanka as well as the slopes of the Himalayas, its larger relation, the Kalong (*Pteropus vampyrus*) inhabits the Malayan Peninsula and the islands of Indonesia and the Philippines. A giant among the Flying Foxes is the Javanese Kalong with a body length of 40 cm and a wing spread of 170 cm. In comparison, the Grey-headed Flying Fox (*Pteropus poliocephalus*) from the eastern coastal regions of Australia, with a wing span of only 100 cm, and the Rufous Flying Fox (*Pteropus rufus*) found on Madagascar, with about the same dimensions, seem moderately small creatures.

The large species of Flying Foxes of Asia and Australia also spend the day sleeping in roosting colonies or "camps". Often they seek out very tall trees, which can even be in the middle of towns or villages. Like the Microchiroptera, they remain faithful to their roosting sites, sometimes for several generations. As a result of constant colonization, the trees are partly or entirely denuded of foliage. It is an impressive sight for the traveller who comes upon a "classic" colony of this kind with thousands of bats hanging like pendulous fruit from the branches of trees. Each animal can be distinguished separately, for in spite of their gregarious nature, they avoid any physical contact and are careful to maintain a certain distance between themselves and their neighbours, in order to sleep undisturbed.

In these roosts, the hundreds or even thousands of bats are exposed to all weathers. Since the branches from which they hang are often bare, the bats have to endure the full heat of the sun as well as storm and rain. In addition, they must compensate for seasonal fluctuations in temperature in the different biotopes.

Diagram showing wing flight patterns for the fruit bat *Eidolon helvum*
a from in front, b from the side (from Kulzer, 1968)

The large wing membranes serve as a valuable means of protection. In the full heat of the midday sun, the bats are usually wide awake, and with extended wings, they waft cool air towards their body in a fanning action. They can also lose a lot of heat from their large, well-vasculated wing membranes. When it is cold and wet, the patagium is folded tightly round the body, providing good protection from the inclemencies of the weather.

The Tübingen physiologist Kulzer examined heat regulation in Flying Foxes under experimental conditions, and was able to confirm the pattern of behaviour shown by the bats in roosting communities. At temperatures of between 18°C and 30°C, the creatures hold their wings loosely folded against the body. If the temperature is raised, the bats spread out their wings and begin to fan themselves with cool air. It is dangerous and may easily be fatal if the normal body temperature is exceeded by only a few degrees. In the conditions described above, Flying Foxes are particularly susceptible to heatstroke. It is impressive to observe the way in which they protect themselves, even under experimental conditions. When the ambient temperature reaches 37°C and over, the animals make use of cooling by evaporation to dissipate excess heat. Since they are unable to sweat, they begin to lick their body and wing membranes extensively. They almost look as if they had been submerged in water. At the same time, they intensify the fanning action of the wings so that more cool air is brought into contact with the body. This increases evaporation, and the Flying Foxes achieve the same result as, for example, man and many animals do when they sweat. With a normal body temperature of about 37°C, Flying Foxes can tolerate an increase to 40°C or more for only a short time. As soon as these refrigerative measures cease to be effective, the creatures fly off. It can be assumed, that in natural conditions, Flying Foxes avoid dangerous overheating by

leaving the roosting sites. Since bats are usually active only at night, additional heat generated by muscle activity, particularly in flight, is very important. The considerable drop in temperature at night in the tropics causes cooling of the body. This can be counterbalanced by the generation of additional heat in flying.

When the temperature of the environment was reduced in the experiments, it became essential for the animals to avoid any unnecessary loss of body warmth and to reduce to a minimum heat loss by radiation from the surface of the body. They wrapped themselves firmly in their wing membranes as in a cloak; head and nose and even one of the feet were all hidden beneath the patagium. In their natural habitat, the temperatures in the open never sink to freezing point, so the Flying Foxes are able in this way to maintain a constant body temperature even during cool periods. These findings are in direct contrast to those concerning the regulation of body temperature in insectivorous bats (Microchiroptera) in temperate regions. When temperatures are low, they enter a state of torpor which causes a considerable lowering of the body temperature. The experiments with Flying Foxes, however, suggest that most species of the Megachiroptera are true warm-blooded animals (homoiotherms).

In many mammals it is possible to distinguish the sexes by externally visible features. Secondary sexual characteristics of this kind are absent in many of the Chiroptera. Without close examination, it is not usually possible to tell male and female apart. Certain of the Flying Foxes, however, show sexual dimorphism. They include the group of Epauletted Fruit bats (*Epomops, Epomophorus*) living in the forests of Africa. The males of these species are usually rather larger than the females and in addition possess glandular pouches in the skin of the shoulder which are surrounded by light-coloured tufts of hair. These produce the effect of epaulettes. Little is known about the habits of these fruit bats. During the day, they remain concealed in small groups among bushes along the banks of streams in primitive forests, and are difficult to observe.

Included in this group of Epauletted bats, although lacking the tufts of hair on the shoulders, is the Hammer-headed or Horse-faced Fruit bat *(Hypsignathus monstrosus)*, which, with its massive head, has a monstrous and bizarre appearance. In this species, the males are much larger than the females. With a wing span of some 90 cm, they are giants among the African Flying Foxes. A striking feature of the ponderous head is the greatly swollen hammer-shaped muzzle. The pendulous lips are particularly suitable for enclosing whole fruits and squeezing the juice from them.

Eisentraut observed that at mating time, the males will spend hours at a time producing curious deep booming sounds. A much enlarged larynx is the structural basis for this vocal effect. Calls of this kind are otherwise virtually unknown among Chiroptera. It is very probable that, as in the case of many other animals, they serve a sexual purpose, helping to attract the female.

All the species mentioned so far belong to the Pteropodinae or Long-nosed Flying Foxes. They contrast with the subfamily of Short-nosed Fruit bats (Cynopterinae), the species of which live only in the Indo-Malayan region. Most of them are quite small forms with a wing spread of 30 to 45 cm. They are scarcely larger than our own indigenous bats. Most species live communally, roosting during the day on trees. In many areas they hang on branches of palm trees where they are well protected from sun and rain by the large leaves. Short-nosed Fruit bats have also been found in caves and hollow trees. In many places they live in human habitations.

The majority of the Short-nosed Fruit bats are fruit eaters. They have also been observed to visit blossoms.

But no special adaptation to this type of feeding is discernible. On the other hand, the members of the subfamily of Long-tongued Fruit bats (Macroglossinae) are exclusively nectar eating. These bats live in the Asiatic-Australian region and one species in Africa. They are small species, which with their narrow extended skull and long slender tongue are extremely well adapted to visiting blossoms. They are able to insert their long snout deep into a blossom and extract nectar or pollen with the tongue. The Macroglossinae include the smallest of the fruit bats. The Asiatic Long-tongued Fruit bat *(Macroglossus minimus)* has a body length of only 7 cm and a wing span of 25 cm. With these dimensions, it is considerably smaller than many of the Microchiroptera.

There is one group which stands out from the majority of fruit bats on account of curious external features. These are the Tube-nosed Fruit bats (*Nyctimene* sp.). As their name reflects, the nostrils of these species are prolonged as scroll-like tubes some 6 to 7 mm in length. These tubular nostrils lend the creatures a strange, somewhat eerie appearance. It is not known whether they have any functional significance. If the bats make use of any form of supersonic orientation—and it is not known that they do—the nasal tubes might possibly be linked with a particular kind of sound emission. In the case of many insectivorous bats, curious elaborations of the tissues of the face have proved to fulfil a vitally important function. In addition to this extraordinary nasal modification, the markings of the body are interesting, being of a kind rarely met with in bats. The Tube-nosed Fruit bats have a random distribution of yellow spots over the entire body and wing membranes. They stand out distinctly against the brown background of the creature's fur. This marking provides useful camouflage protection to the small animals as they hang suspended among foliage on trees during the day.

Nose and ears— distinguishing features of bats (Microchiroptera)

It is astonishing what an abundance of forms mammals have evolved in the course of millions of years, since their prototypes first appeared in the early Tertiary. The Microchiroptera in particular, that is, those known as "insectivorous" bats, illustrate this diversity with nearly 800 species. There is no doubt that the capacity for flight has been a significant factor in contributing to the development of variety within this group of mammals. Since they are able to fly and are active during the hours of night, bats have scarcely any enemies that pursue them. In the tropics of the Old World, it is the Bat-hawk (*Machaerhamphus alcinus*) and in those of the New World, the Bat-falcon (*Falco rufigularis*) that specialize in hunting bats. Certain other kinds of vertebrates will lie in wait at the entrance to a cave to catch a bat as it flies out in the evening. But on the whole, the ancestors of bats were able to evolve and specialize undisturbed, and over millions of years, to produce a great wealth of forms.

And yet a glance at the European species of bats may well produce the impression that they hardly vary at all from one another. Apart from the Horseshoe bats with their complex nose leaf, all other forms appear inconspicuous, with a "normal" shape of head and with only slight differences in body size, ear structure and colour of pelage. Even the expert sometimes finds it no easy task to distinguish between such similar pairs as the Common Pipistrelle and Nathusius' Pipistrelle, or the Grey Long-eared bat and the Brown Long-eared bat. So it is not surprising if, for most people, just as one mouse is like any other mouse, so one bat is like any other bat. It is rare to have the opportunity to see several species together at one time, since, to a large extent, they elude observation by their nocturnal, secretive habits.

Until recently, all that was known of many species, even to specialists, was the description and place of origin. A comparison of the numerous species of bat, however, shows that one bat is certainly not very much like every other. In the tropics in particular, a long process of evolution has produced some really exciting forms.

Of great diversity, and therefore an essential feature in distinguishing species, is the form of the head. In some species, the face has an appearance little short of grotesque. The fleshy nasal appendages found in many species show enormous variation. But their form is so characteristic that they have been responsible for the common names of some families, such as the Horseshoe, Spear-nosed, Leaf-nosed or Slit-faced bats. For most species, the significance of these appendages is not known. As will be shown in connection with echolocation, they certainly do not represent mere whims of nature, but may have a vitally important function. Like the nose leaf, the ears of many species are of exceptional dimensions and have a distinctly individual form. Even the tragus, a membranous lobe which projects at the front of the ear orifice, may be smooth or notched, broad or long, round or pointed. A third feature which can give the face a characteristic appearance is the form of the lips. They are not always smooth but may be indented or wrinkled and covered with wart-like growths. All these elements produce faces that can often seem grotesque or hideous in appearance.

The Microchiroptera are small forms with a body size in general smaller than that of most fruit bats. The smallest of them, *Craseonycteris thonglongyai*, was discovered in Thailand only some 10 years ago. The small size of this species (length of head and body 29 to 33 mm; weight 1.7 to 2.0 g) makes it the smallest living mammal. The largest species can reach 18 cm (the Giant Spear-nosed or Linné's False Vampire bat, *Vampyrum spectrum* of tropical America; the Australian False Vampire bat, *Macroderma gigas*). Among Microchiroptera, the wing spread ranges from 15 to 90 cm.

A feature common to them all is the absence of a claw on the second digit. Most species have a well-developed tail with an interfemoral membrane developed accordingly. Among Spear-nosed bats there is some variation here; the tail and interfemoral membrane show various degrees of regression. As an adaptation to a very wide range of feeding habits, the teeth of bats are often reduced in number. In addition, the entire skull may be modified, varying from elongate to short with "pug-like" proportions. The fur covering usually does not extend to any part of the patagium, the ears and facial embellishments. A fact of interest to the bat systematist is that the individual hair shows typical structural features in its outer cortex. It may appear scaly like a fir cone or show transverse protruberances or a spiral thread. This surface conformation is often so specific that it can help to determine genera.

The fur of most bats is grey or brownish in colour, with varying gradations of lightness. Bright colours such as orange-red or rufous-brown are rare. The same is true of markings on the fur; only a few species show spots or stripes. Striking differences in colour in the skin of the body and of the patagium are found rarely. The markings of the East African species *Myotis welwitschi* are outstanding. The yellow colour of the fur, which is extended across the tail membrane, contrasts strongly with the mat black of the wing membranes. In the Microchiroptera, as in most Megachiroptera, colouration varies hardly at all between the sexes. Sometimes the females are rather more intensely coloured than the males, as for example, in the American Red bat *(Lasiurus borealis)*.

Although the Microchiroptera are distributed all over the world, they are nevertheless heat-loving animals, essentially requiring warmth, and are therefore encountered in very large numbers in the tropics and subtropics. But some species have extended their territo-

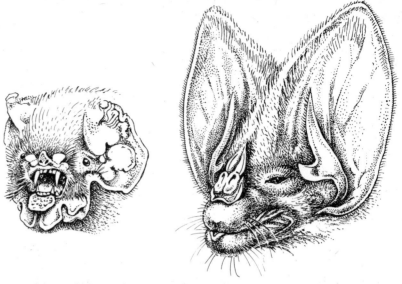

Face of the Leaf-chinned bat *Mormoops megalophylla*. The numerous lobes and folds near the ears and on the lower lip have earned it the name of the Wrinkle-faced bat.

This False Vampire (Megadermatidae) owes its popular name of Ghost bat to the shape of its head (from Felten, 1958).

ry to the edge of the Arctic zone, stopping only where the insect food supply ceases. Those bats indigenous to Europe are representatives of the families of Vespertilionid bats (Vespertilionidae), Horseshoe bats (Rhinolophidae) and one species of Free-tailed bats (Molossidae). So vast is the number of species which comprise the seventeen families of the Microchiroptera that here we are able to consider only a few in detail, but they will probably suffice to illustrate the great diversity in form and habit.

Mouse-tailed bats (Rhinopomatidae)

This primitive family contains only 3 species and they differ somewhat from the generally accepted "typical" bat form. The long, thin tail has only a narrow edging of skin tissue at its root. Is the interfemoral membrane rudimentary, or are these species so primitive that they do not yet possess a tail membrane? The ears are large and are connected at the anterior base. The snout is naked and has an extended appearance; at the centre of the upper lip there is a rudimentary nose leaf.

65

Bats of this family are found round the Sahara region, and also in the Near East, India and Southeast Asia. They live communally in caves and old buildings, and are commonly found in ancient tombs in Egypt. Thick layers of droppings that have piled up beneath roosting places show that the colonies have inhabited the same quarters continuously for hundreds of years. Mouse-tailed bats are insectivorous and are found in very hot, arid regions. During cool periods and when food is scarce, the bats become torpid and live on subcutaneous reserves of fat that they have accumulated at the base of the tail.

Slit-faced bats (Nycteridae)

This family is also restricted to the Old World. Its representatives are distinguished by a particular structural feature of the skull; the nostrils lie within a deep hollow in the front of the skull that extends to the brow, and which is bordered by mobile cutaneous outgrowths. The very large ears are fused in front at the base. A very typical feature of this family is the T-shape of the last joint of the tail which supports the hind edge of the interfemoral patagium.

Far from its African relations, the Javanese Slit-faced bat (*Nycteris javanica*) is found in Southeast Asia.

False Vampires (Megadermatidae)

The species of this family are closely related to the Slit-faced bats. Many systematists still combine them into a single family. The family of Old World False Vampires contains only a few species. They are distinguished by a large, conspicuous leaf-like expansion of skin around the nostrils which can take many forms. It lends the bat a very bizarre appearance. For example, the curled edges of the nose leaf of the African Yellow-winged bat (*Lavia frons*) give it the shape of a three-pronged harpoon blade. That of another species is lyre-shaped and has earned it its name of *Megaderma lyra*. The ears of these bats are large and the tragus is bifid. In spite of the absence of caudal vertebrae, the interfemoral patagium is well developed.

The Australian False Vampire bat (*Macroderma gigas*) is among the largest of the Microchiroptera. Although these bats are not blood feeders, but live on a diet of insects and small vertebrates, their grotesque appearance has made them the prototype of the vampire in illustrations and in horror films. For this reason, the family was given the vernacular name of False Vampires.

And what is so fearful about these bats? Undoubtedly the total impression they produce. The light-coloured creatures with a face almost pure white, achieve a wing span of more than 70 cm. A massive skull with large eyes, a huge nose leaf and great ears which also show only very slight pigmentation all intensify the extraordinary impression.

Since the False Vampire bats are solitary, it is not easy to discover them when they have withdrawn into their retreats during the day. So far, little is known about the habits of this exciting species.

Horseshoe bats (Rhinolophidae) and Old World Leaf-nosed bats (Hipposideridae)

These two Old World families are so closely linked that in the literature they are quite frequently dealt with as one. Both possess curious fleshy appendages surrounding the nostrils, from which the vernacular names are derived.

In the Rhinolophids, a "horseshoe-shaped" flap of skin covers the upper lip and surrounds the nostrils. Further membranous structures over the nose include a narrow, lengthwise ridge and an upright, pointed blade, the lancet. In each species, these features are species-characteristic and provide a reliable guide to the systematist. We shall consider later their function in focussing the beams of sound during the emission of ultrasonic pulses. The ears of Horseshoe bats are relatively large. They lack the tragus present in the members of other families of bats.

There are about 70 species, most of which are cave dwellers. They live communally, but when roosting, are careful to ensure that a certain distance separates each bat from its neighbour. The normal position at rest is a typical one in which the delicate wing membranes are folded round the body like a cloak. Found in their winter roost, the Horseshoe bats resemble dried fruits hanging freely suspended from the roof of the cave. In most species, including the European, the colour of the fur is an inconspicuous dark grey or dark brown. But some are of a striking orange-red. The hairs are long and fine, producing a soft, furry impression. All Horseshoe bats are insectivorous. Those which live in the temperate zone, hibernate during cold winter weather.

North of the Alps, as far as the Hercynean mountains and in the south of England, only two species can be found, and unfortunately their numbers are declining all the time; they are the Greater and Lesser Horseshoe bats (*Rhinolophus ferrumequinum* and *Rhinolophus hipposideros*). Whereas in the former, the length of head and body together reaches 7cm, in the latter it is no more than 4cm, making it the dwarf in this family.

The Hipposiderids, which are closely related to the Rhinolophids, consist of about 60 species. They inhabit tropical and subtropical Africa, Asia Minor, India and Southeast Asia as far as Australia. No species has extended its range northwards as far as Europe. In most of the members of the family, a nose leaf of rather simpler construction surrounds the nostrils. In some species, these skin flaps are grouped like the petals of a flower round the nostrils (*Anthops ornatus*). A structure corresponding to the lancet of the Rhinolophids, only this time shallow oval in form, is also found in *Hipposideros commersoni*. Or it can be subdivided into several leaflets, as in the Trefoil Leaf-nosed bat (*Triaenops persicus*) or the Trident Leaf-nosed bat (*Asellia tridens*).

Most members of this family are insectivorous, but some large species are also carnivorous. During the day, they roost in buildings or caves, either alone or in colonies which may consist of more than a thousand bats. In comparison with other families, the range of colouring among the Hipposiderids is strikingly wide with bright orange shades predominating. In the small South African Lesser Leaf-nosed bat (*Hipposideros caffer*) that is widespread in Africa, there are two colour variants, one being a dark grey-brown and one a glowing reddish-brown. Similar examples of dimorphism have also been described among certain Rhinolophids. The large species Commerson's Leaf-nosed bat (*Hipposideros commersoni*) of Africa reaches a head and body length of 11 cm, making it one of the larger forms among the Microchiroptera.

Spear-nosed bats (Phyllostomidae)

The New World family of Spear-nosed bats provides interesting parallels to the families of the Old World, which contain those species with the most complex and highly elaborated nasal attachments. As the name implies, members of this family also possess a nose leaf.

There are some 140 species in this family, and they live mainly in the tropical regions of Central and South America. In addition to the nose leaf, there are other features which provide evidence of evolutionary developments parallel with those found in the distant families of the Old World. Certain species without nose leaves have developed fleshy foliaceous outgrowths in the form of chin leaves. In others, the snout region is greatly extended, an indication, as in the case of the Long-tongued Fruit bats, that these species are nectar feeders. On the basis of these varied and sometimes extreme developments within the single family, the Spear-nosed bats are divided into several major subfamilies. Some systematists consider that certain of them represent distinct families. The family group of Spear-nosed bats includes both small forms with a head and body length of 4.5 cm, as well as some that are 15 cm long. Apart from body size, other features are extremely diverse. There are great differences in ear size; in some species the interfemoral membrane is fully developed, while in others it is reduced or completely absent. In the latter case, the caudal vertebrae are correspondingly regressed.

Among the typical Spear-nosed bats in the narrowest sense (Phyllostominae) there are species with very singular cuticular structures. These in many cases have earned the creatures names which are a reference to the shape of the nasal appendage. For example, the nose leaf of the Sword-nosed bat *(Lonchorhina aurita)* is in the shape of a spearhead and reaches the considerable length of 2 cm, while the small nose leaf of the Spear-nosed bat *(Phyllostomus hastatus)* is much like the tip of a lance. The Spear-nosed bats are large creatures with a wing span of up to 55 cm. They live on insects and fruit, but are also prepared to take small vertebrates. With their large canine teeth they are able to seize small birds and bats—in captivity, even mice—kill them by crushing and devour

them. Observations of these bats in captivity have shown that they always eat the prey head first, supporting it skilfully by the claw of the thumb.

The Spear-nosed bats roost during the day in underground chambers and caves as well as in buildings, churches and hollow trees. They are sometimes found in thousands.

This subfamily contains the largest species of "insectivorous" bat in the New World, the Giant Spear-nosed bat *(Vampyrum spectrum)*. It achieves a wing span of 70 to 90 cm. The large ears and dagger-like nasal appendage give the bat a grotesque appearance. It is not surprising that this species also accords well with the popular conception of a vampire, and for this reason was believed by early writers to feed on blood. Although it has long been recognized that this is not so, the name "False Vampire" has continued to be used in many places. Even the scientific name reflects this error.

The predominantly fruit-eating Flying Foxes of the Old World have their counterpart in the fruit-eating species of certain subfamilies of Spear-nosed bats. It is interesting to observe how this parallel development in feeding habits has led to the same morphological changes. The fruit eaters of the eastern hemisphere as of the western show considerable broadening and flattening of the molars.

Of the Short-tailed Spear-nosed bats (Carolliinae), the Seba's Short-tailed bat *(Carollia perspicillata)* is the most widespread. It inhabits an area extending from Mexico to Southern Brazil. These bats live mainly on wild figs, bananas and guavas. The subfamily of fruit-eating Spear-nosed bats with the largest number of species is the Stenoderminae. These medium-sized bats are stoutly built and in most genera are distinguished by a small nose leaf with a spear, with an additional horseshoe-shaped membranous structure at the base. In some cases, the dark brownish-grey fur typical of these species is marked with four light-coloured longitudinal stripes on the head (*Artibeus* spp.). In addition, some species, such as *Uroderma bilobatum* have a light stripe along the middle of the back.

All but a few species of Stenoderminae spend the day in small groups, roosting among the foliage of trees. Many observers believe that some small *Artibeus* species and *Uroderma* are able to form the leaves of palm trees into a tent-like roof which protects them from sun, rain and natural enemies. This feature has earned them the name of Tent-building bats or Tent bats.

Like Old World fruit bats, the "fruit vampires" also prefer to eat soft juicy fruits from which they frequently extract only the juice. The remains are allowed to drop to the ground where they accumulate in large quantities beneath feeding trees or regular resting sites.

The extremes which facial conformation can achieve in bats is shown among the Stenodermids by the Wrinkle-faced bat *(Centurio senex)*. This species, which occurs primarily in Central America, has no nose leaf, but the skin of the face which is hairless, is arranged in an intricate pattern of wrinkles, folds and fleshy lobes producing a grotesque appearance of senility.

A curious modification in the attachment of the wing membranes can be seen in the Naked-backed bats belonging to the subfamily of Leaf-chinned bats (Chilonycterinae). The name Naked-backed bat is, in fact, misleading, for the bat's back is furred quite normally. Since, however, both in the Greater *(Pteronotus suapurensis)* and in the Lesser Naked-backed bat *(Pteronotus davyi)* the wing membranes are attached at the centre line of the back, the impression is given that the back is bare. These bats roost during the day in caves, in colonies of various sizes. The Frankfurt zoologist Felten reports that they prefer day-time quarters with very high relative humidity and a temperature of 38°C. In this suffocatingly hot and humid atmosphere, where the nauseous vapours emanating from vast quantities of droppings make it impossible for humans to breathe, the creatures obviously feel perfectly at home, and even bring up their young here.

One of the small number of Spear-nosed bats found in the southern states of the U.S.A. is the Leaf-chinned bat *Mormoops megalophylla*. This reddish-brown, medium-sized species has been reported in Arizona and Texas. Colonies of up to 4000 bats spend summer and winter in caves, mine galleries and tunnels. Like Horseshoe bats, they do not cluster closely together, but carefully avoid direct bodily contact with their neighbour. The skull of this insectivorous species is extremely compressed. The neurocranium curves upwards in a dome almost at right angles to the facial cranium, in a manner reminiscent of the skull shape of the pug breed of dog.

Another inhabitant of Arizona is the Californian Spear-nosed bat *(Macrotus waterhousei)*. It has very large ears and can easily be distinguished from other species of similar appearance by the presence of a triangular nose leaf. This species is also insectivorous and roosts underground during the day.

Among those species of Phyllostomids in the subfamily of Nectar-feeding Spear-nosed bats (Glossophaginae), three representatives are found in the southern states of North America. These are the Mexican Long-tongued bat *(Choeronycteris mexicana)*, the Nectar bat *Leptonycteris nivalis* and Sanborn's Nectar bat *(Leptonycteris sanborni)*. All species possess an elongate skull and a small nose leaf at the end of the snout. The tongue is long and highly extensible with a concentration of bristle-like papillae at the tip. These features show that the bats are extremely well adapted to obtaining food from flowers.

The Long-nosed bats roost communally during the day in caves and mine galleries. At night, they seek their food from the flowers of agaves, yucca trees and various cactus plants. They will also eat insects. Since these species do not hibernate, they leave their habitats in the cold season and migrate south from the United States into neighbouring Mexico.

Of the remaining New World families of bats, only a few species are known and little research has been carried out into their way of life. True Vampire bats (Desmodontidae) will be considered in a separate chapter.

Fisherman bats (Noctilionidae)

These bats have no nose leaf, but a fleshy upper lip, which is divided by a vertical fold. This feature has given the bats the vernacular name of "hare-lipped bats". The family is particularly noteworthy, since one of its two species *(Noctilio leporinus)* is specialized as a fish eater. It also eats insects, as does the smaller species in this family, *Noctilio labialis*. As an adaptation to the fishing habit, the hind feet of the Fish-eating or Fisherman bats are much enlarged and are furnished with strong curved claws. The brilliant rufous colouring of the males distinguishes them from the greyish-brown of the females. These species are distributed across a wide range from Mexico to Argentina.

Funnel-eared bats (Natalidae)

Bats of this small family inhabit tropical Central and South America. Forms so far found have no nose leaf. They have a long tail and particularly long hind legs. Indeed, the length of the hind legs can exceed that of the head and body together. The large, funnel-shaped ears have earned the name of Funnel-eared bats for members of the genus *Natalus*. They are insect feeders and roost during the day in caves.

Thumbless or Smoky bats (Furipteridae)

This family is closely related to the Funnel-eared bats. In the two species, the thumb is so rudimentary that only a tiny non-functioning vestigial claw is visible externally. Representatives of this family are among the smallest species of bats. Head and body measured together scarcely reach 5 cm. The most striking feature of these small bats is their large, funnel-shaped ears.

Among those families that are distributed throughout the world are the Sheath-tailed bats (Emballonuridae), the Free-tailed bats (Molossidae) and the Vespertilionid bats (Vespertilionidae). Their members, like those of certain families already mentioned, have no nose leaf, that is, they are "simple-nosed". One could almost regret the lack of such an ideal identification feature, particularly since these three families comprise more than half of all species of bats. It is not always simple to detect the minute differences which distinguish the many species one from another.

Sheath-tailed bats (Emballonuridae)

The members of this family, some 50 species, are small to medium sized bats that are found in the tropical and subtropical regions of the Old World, and also in large numbers in the warmer regions of America. Their name refers to the particular structure of the tail, part of which extends freely beyond the interfemoral membrane. In flight, many of the species are able to retract this part into a sheath of skin in the membrane. Some species have unusual markings on the fur, which provide a useful, protective camouflage when they are roosting during the day on the branches of forest trees. Marking is sometimes in the form of two light-coloured stripes running the length of the back as far as the tail membrane, as, for example, in the Greater White-lined bat *(Saccopteryx bilineata)*. In the front part of the wing membrane between shoulder and elbow, this species has conspicuous wing sacs which produce a strong-smelling glandular secretion. They are more highly developed in the male and clearly have a sexual function. A very striking group within this family is the Ghost bats *(Diclidurus* spp.), that is entirely white. It is not clear whether this unusual colouring is of any biological advantage to the species.

The best-known and most widely distributed species of Emballonurids in Africa are the Tomb bats *(Taphozous* spp.). They are also found in Southern Asia, Australia and on various islands in the South Pacific. They are medium-sized bats weighing between 10 and 30 g. They were given the name of Tomb bats by scientists who accompanied Napoleon on his campaigns to Egypt and who first discovered them in the burial chambers of the Pyramids. But it has since been established that not all species roost in gloomy underground chambers. Many of them prefer open terrain and roost on tree trunks or under roofs.

Free-tailed bats (Molossidae)

This family contains some 90 species which are distributed throughout the warmer regions of the world. The Molossids are well adapted to various climatic conditions. In Africa, Kulzer found them in almost every climatic zone. In these bats, the end of the tail projects well beyond the edge of the interfemoral membrane; this feature has earned them their name. In flight, however, the interfemoral membrane can be slid down over the tail, producing an active enlargement of the bearing surface.

A striking feature of many Molossids is the broad, bulky head. The thick, angular ears lend the head a massive aspect, which in many species is given a grotesque quality by large numbers of folds and warts on the upper lip. The long, narrow wings indicate that these bats are excellent, long-distance flyers, but are not particularly manoeuvrable. On the ground, they can run with considerable agility on powerful legs, with wings tightly folded. The Molossids are insectivorous. They live communally, often in large colonies that can number up to a million individuals. With so many different species, it is not surprising to find them in a wide variety of habitats, in caves, on trees and in buildings. They have even been observed to roost under galvanized iron roofs where the temperatures at midday reach extraordinarily high levels. In Egypt, these species are often found in ancient burial chambers, in India they live in temples.

In some roosts that have been used continually for centuries, vast quantities of droppings have accumulated underneath the roosts as guano, which can be collected and marketed as a fertilizer. The most spectacular example of mass congregations of bats can be seen in the summer roosts of the Guano bats *(Tadarida brasiliensis)* in the Carlsbad Caverns in New Mexico, U.S.A. In the early sixties, the number of bats living here in quite a small space was estimated to be some four to five million. Among mammals, only bats have been found in vast concentrations of this kind. Unfortunately, their numbers have declined sharply in recent years. Later chapters will examine various aspects of life in vast colonies such as these.

Among Molossids, the large Naked bats (*Cheiromeles* spp.) are most bizarre in appearance. These species, which live in Southeast Asia, are almost completely hairless. Black, bristly hairs grow only on the large throat sack and the "collar" of loose folds of skin round the neck. The wings start at the mid point of the back, and folds of skin form a pouch along the sides of the body in which the wing membrane is stowed when the bat is resting.

Vespertilionid bats (Vespertilionidae)

This family comprises some 320 species, that is, a third of all known bats. Vespertilionids represent the largest part of the bat fauna of Europe and North America, both in number of species and number of individuals. 24 species of Vespertilionid bats live in Europe, while 30 species of this family are found in North America. It should be emphasized that some species are very common and widely distributed, whereas others are quite narrowly restricted in range and are considerable rarities among our fauna.

The wide variety of species within the family is reflected on the one hand in differences in body size and other morphological features; on the other hand, in a broad range of variations in habit and habitat. In addition to pigmy forms weighing no more than 3 g, there are quite large species which can weigh 80 g. With such an abundance of forms, it is difficult to specify features that are common to all members of this large family. As the British zoologists Yalden and Morris pertinently remark, "perhaps the best generalisation is that Vespertilionids are those bats which, externally at least, lack the special characters of other families".

In all species, the tail is well developed and usually integrated into the interfemoral membrane. Apart from a few exceptions in Australia, the bats have simple noses. The majority of them are insectivorous and are capable of nocturnal orientation by means of ultrasonic echolocation.

In a family so rich in species, it is impossible to attempt a comprehensive survey of all Vespertilionids. Certain European species have therefore been selected. In many cases, these bats or close relations of theirs, also inhabit other continents, where they exhibit numerous similarities in appearance and habit to their European cousins.

There are the Mouse-eared bats of the genus *Myotis*, which, as cave dwellers, spend at least the winter season in underground quarters. In summer, they are often found in buildings, where the Large Mouse-eared bat (*Myotis myotis*) (body size 8 cm, wing span up to 40 cm) shows a preference for the spacious roof vaults of old buildings, particularly churches. There are some 60 species of Mouse-eared bats; systematic differentiation is based particularly on the size and shape of ear, tragus and teeth. In the case of other groups of Vespertilionids, such as the Pipistrelle, Noctule, Barbastelle and Long-eared bat, the ear and teeth once again play an important part as distinguishing features in identifying species.

Of the 24 European species of Vespertilionid bats alone, ten belong to the genus of Mouse-eared bats, *Myotis*. Older accounts invariably state that the Large Mouse-eared bat is by far the most common species in Europe, where it is usually found in summer roosting in large colonies. Unfortunately, this can no longer be claimed today. The decline in the numbers of this admirable and beneficial species gives cause for concern.

The other *Myotis* species occurring in Europe, and closely related to the Large Mouse-eared bats, are medium-sized animals measuring 4 to 5 cm in length from snout to tail base, usually coloured greyish-brown on the back and lighter yellowish-grey or whitish-grey on

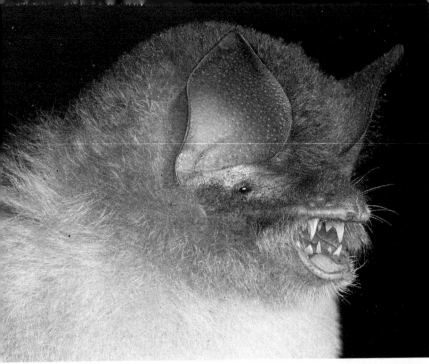

50 Daubenton's bat or Water bat (*Myotis daubentoni*; Vespertilionidae). Medium-sized species of the Vespertilionid bats (body weight 7–10 g). Found throughout Europe; usually in the vicinity of water.

51 Northern bat (*Eptesicus nils-soni;* Vespertilionidae)

52 Geoffroy's bat (*Myotis emarginatus;* Vespertilionidae)

53 Brandt's bat (*Myotis brandti;* Vespertilionidae)

76

54 Natterer's bat (*Myotis nattereri;* Vespertilionidae) taking flight. In this medium-sized bat, the hind margin of the interfemoral membrane is furnished with numerous stiff hairs.

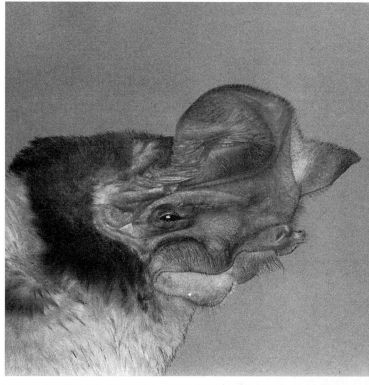

55   Brown Long-eared bat (*Plecotus auritus*; Vespertilionidae). The appearance of the head is dominated by the very long ears. The large number of ear pleats indicate that the ear can be furled and folded. During hibernation, the bats fold the ears down along the side of the body and tuck them under the wings. This species is frequently found in Europe. In summer, they often inhabit nesting boxes.

56   *Otomops martiensseni* (Molossidae). A Free-tailed bat from Africa. The remarkable structure of the snout region lends the bats an extremely odd appearance.

57   *Tadarida hindei* (Molossidae). A representative of the African Free-tailed bats.

58 European Free-tailed bat (*Tadarida teniotis*; Molossidae). The only species of the family of Free-tailed bats to occur in the Mediterranean countries.

About half of its tail extends freely beyond the flight membrane. These bats are a little larger than Large Mouse-eared bats.

the underside. They include the rare Bechstein's bat (*Myotis bechsteini*) which spends the summer in hollow trees and the nesting boxes of birds. Natterer's bat (*Myotis nattereri*) is distinguished by a fringe of fine hairs on the back edge of the interfemoral patagium. It inhabits well-wooded regions. A rather rarer species is the Notch-eared bat or Geoffroy's bat (*Myotis emarginatus*), found principally in southern Europe. A more common species is the Daubenton's or Water bat (*Myotis daubentoni*), frequently found near water. In the summer, they roost during the day in crevices in the walls of old bridges, in buildings and in hollow trees. At night, they skim low over the surface of ponds and streams to catch insects. Other relations of the Mouse-eared bat include the Whiskered bats. It was established only in the last ten years that there are two species in Europe, the Whiskered bat (*Myotis mystacinus*) and the Brandt's bat (*Myotis brandti*). They are rare creatures and little is known of their habits and their distribution.

The smallest species are found among the Pipistrelles that are also distributed throughout the world, apart from South America. The European species can easily be distinguished from other indigenous species. But the four found here are extremely similar in appearance, and the exact identification of a Pipistrelle (*Pipistrellus pipistrellus*), Nathusius' Pipistrelle (*Pipistrellus nathusii*), Kuhl's Pipistrelle (*Pipistrellus kuhli*) and Savi's Pipistrelle (*Pipistrellus savii*) can cause difficulty even to the specialist. Whereas the European species set up their summer roosts in and on buildings, in hollow trees and nesting boxes, the Banana bats (*Pipistrellus nanus*) living in Africa, often roost in young banana leaves that have not yet unrolled.

Pipistrelles have proved to be more common and more widely distributed in Europe than had hitherto been realized. Because of their small size and their habit of

concealing themselves in narrow crevices, they have to a great extent successfully eluded observation. Increasing numbers of new roosts have been reported in some rural areas, but in the U. K. monitored colonies have decreased markedly in recent years. The bats are obviously not excessively sensitive to cold and wet weather, and winter roosts with more than 1000 bats have been found, even in churches.

In the autumn, as they make their way to their winter sites, these small creatures can cause a good deal of disturbance. At this time, whole flocks of Pipistrelles sometimes invade some particular corner of a house which they have never entered previously. One can well imagine the reaction of the householders when, next morning, they discover these harmless creatures in, say, the folds of their bedroom curtains. After autumn intrusions of this kind, Pipistrelles have also been found inside vases, behind picture frames and door ledges, under floorboards and between double windows. An "attack" of this kind is no cause for panic. It is much more important to help the creatures to escape.

Many species of Vespertilionid bats find their ideal habitat on and in the dwellings of human beings, and for this reason are sometimes called house bats. Others seem never to roost in buildings; they live in trees and are designated "tree bats". A typical tree bat is the Common Noctule (*Nyctalus noctula*). It is one of the larger European species; the ears and roundish tragus are relatively small. Noctules often leave their roosts early, even before sunset, and it is sometimes possible to observe them in rapid flight high above the treetops. In contrast to the grey fur of the Mouse-eared bats, that of the Noctule is reddish-brown in colour. The short, dense fur has a silky sheen.

Another of the larger species is the Serotine or Broadwinged bat (*Eptesicus serotinus*). These bats enjoy the proximity of human settlements. They can often be observed on warm summer evenings flying low in search of insects near trees in town, in gardens and close to houses. Since the wings are much broader and less pointed than those of Noctules, flight is correspondingly slower and more leisurely. In their winter roosts, the Serotines rarely hang freely suspended from a roof but more often hide away in crevices in rocks. So it is very difficult to find them.

Closely related to the European species is the North American Big Brown bat *(Eptesicus fuscus)*. It too prefers buildings for its day-time roosting site. The genus *Eptesicus* comprises many very small Vespertilionids. In an African species of *Eptesicus*, the newly-born young are described as no larger than bees.

The European Northern bat *(Eptesicus nilssoni)* is characterized by a particularly high resistance to cold. The range of this species extends northwards as far as the Arctic Circle and it is found on mountains up to an altitude of 2000 m. The Red bat *(Lasiurus borealis)* and the larger Hoary bat *(Lasiurus cinereus)* are two American species that penetrate to the far north. In summer, these bats roost in trees and bushes where they suspend themselves from trunks and branches. Usually they are so well concealed among the foliage that they are protected both from inclement weather and from enemies. Since they often hang by one foot only, they can, at a cursory glance, easily be mistaken for a withered leaf. In these species, too, the "tree bats" in contrast to the "house bats" have richer and more varied colouring. In the male of the Red bat in particular, the orange-red colour is especially intense. Another striking feature is that in the *Lasiurus* species, the upper surface of the interfemoral membrane is completely or partially covered with dense fur. In autumn, these species cover long distances to the southern states of the U.S.A., where they spend the winter. But they have considerable resistance to cold, and the Red bat has been observed flying and feeding in the evenings of warmer winter days.

A feature unusual in *Lasiurus* is the number of young that are born. In contrast to most bat species, in which the female gives birth to only one young, multiple births appear to be fairly common among the *Lasiurus* species. In the case of Red bats, a total of up to four young is not unusual, while twins occur frequently among Hoary bats and also the European Common Noctule *(Nyctalus noctula)*.

Because of their exceptional powers of flight, these bats have been able to cross from the American mainland to the Hawaiian Islands almost 4000 km away. It is inconceivable that they could cover this distance in continuous flight. It must have been possible for them to "pause for breath" by resting on the rigging of ships on the way. But the fact remains that *Lasiurus cinereus* has been living there for more than a century as the only species of bat, and in that time has already developed the specific features of an island race. It has also been recorded several times in Iceland and once in Orkney.

Among those Vespertilionids that are widespread in Europe, mention should also be made of the Barbastelle *(Barbastella barbastellus)* and the Long-eared bat *(Plecotus* spp.). These are medium-sized species weighing between 5 and 8 g. They roost in summer on or in buildings. The former may be found behind window shutters and wall panelling, while the latter prefer to roost behind beams and rafters in roof spaces. They are also found in hollow trees and nesting boxes.

Barbastelles are dark, almost black, in colour, while the tips of the hairs on the back gleam white, giving the bat a frosted appearance. The broad, stunted ears are joined at the base by a ridge of skin. The nose is flattened, so that the face is not unlike that of a small pug dog.

A distinctive feature of the Grey and Brown Long-eared bats *(Plecotus austriacus* and *P. auritus)* is the exceptional length of the ears, which sometimes are three times the length of the head. These sound-sensing devices are so large that they quite overshadow the small face. But during hibernation and in periods of day-time sleep, nothing is seen of these great "sonar receivers". At these times, only the narrow tragus is left pointing forwards, while the ears themselves are furled, folded backwards and concealed under the wings.

Among the Long-eared species that inhabit the southwest of the U.S.A., mention should be made of the Pallid bat *(Antrozous pallidus)*. In this species, the fur on the upper surface of the body is a pale yellowish colour with brownish-grey tips. The underside is also light-coloured. The loss of dark pigmentation from the creature's coat has been associated with the nature of its habitat. It prefers to live in dry areas of steppe, where in summer, it roosts during the day in caves, ruins and buildings. It feeds largely on insects which it catches as it flies close to the ground, or it may also take them from the ground or from leaves. Remnants of food found in the roosts show that these bats also eat scorpions and small reptiles.

Where bats live

The life of bats, with its phases of rest and activity, is strictly adapted to the alternating rhythm of night and day. Like all mammals, bats have two territorial regions, one of which offers protection during periods of reduced activity when the bats are able to rest and satisfy social needs, and the other through which they move in search of food.

The day roost is just as important an element in the life of bats as is the hunting territory. Only if there is a day roost available to them, is it possible for bats to colonize a particular region. It is not the availability of food alone which affects the distribution of bats and their numbers in a particular region, but also the existence of suitable roosting quarters.

These are especially vital in the case of bats since these animals are unable to prepare refuges for themselves in the form of burrows, hollows or nests.

The simplest solution to the accommodation problem, although perhaps not the most ideal, is that adopted by the large fruit bats. During the day, they roost on the branches of large trees. We have already seen how they brave all weathers and how skilfully they manage to achieve a measure of protection. Most of the smaller forms of bats are not able to spend the day hanging from a branch. Sun, wind, rain, fluctuations in temperature and not least the attention of enemies would affect them and their young so adversely that the survival of the species would be endangered. A roosting site which provides more protection is essential.

Since they cannot construct one themselves, they must depend on finding and using shelters that already exist. Bats have settled in some quite surprising sites, and their roosts show a wide variety of forms. In addition to those offered by the natural environment, they also choose the homes of other animals as well as buildings inhabited by man. The fact that bats make use of existing caves and hollows in trees in which to live, or that they inhabit the underground burrows, nesting hollows or nests of other animals is usually termed ecological parasitism. For lack of natural retreats, bats in many areas have settled in or on buildings erected by man, where they have found refuges with a microclimate in keeping with their biological requirements. In addition, mine workings, quarries, bridges and tunnels, even the hollow concrete columns of street lights have all been considered as a welcome extension of the range of available accommodation and have been colonized accordingly.

Tree-dwelling bats

Today, many bats, like their ancestors, are arboreal. Most of the species found on or in trees, live in the tropics. The variety of potential dwellings which a tree can offer is surprisingly wide, and bats have discovered and exploited virtually every possibility for colonization. As many as twenty different dwelling sites on sound and decaying trees have been listed.

The roots alone are greatly in demand as a living area. The Indian Short-nosed Fruit bat (*Cynopterus sphinx*), for example, has been found living among the aerial roots of fig trees. In the evening, it climbs first up the roots and then up the trunk to a branch, before setting off on its nocturnal flight. In Africa, species of Epauletted Fruit bats (*Epomophorus anurus*) and of Old World Leafnosed bats (*Hipposideros beatus*) have been found among tree roots projecting from the sloping sides of river banks.

In the next stage of the tree, the region of the trunk, a few species are found which roost in the open, pressed up against the tree bark. This primitive form of roost is found in particular among representatives of the Sheath-tailed bats such as the African Tomb bat *(Taphozous mauritianus)* or the South American Greater White-lined bat *(Saccopteryx bilineata)*. Since it affords scarcely any protection, the bats get little rest during the day. As soon as danger approaches, they scuttle rapidly to the other side of the trunk or fly to a different tree. In Southeast Asia, two small species of Vespertilionid bats with flattened skulls *(Tylonycteris* spp.) have elected to roost in the hollows in bamboo stems excavated by insect larvae. They creep into the bamboo canes through narrow vertical slits made by beetles. Hollows of this kind have been found to contain individual bats, usually males, but occasionally, small colonies of ten to twelve animals of both sexes.

Large species of African Epauletted Fruit bats roost in the tangled branches and thick foliage of old trees. They prefer branches that extend far out across rivers or lakes. From here, they can easily take flight without becoming entangled in the dense thickets of the forest. On the other hand, the dark roof of leaves provides adequate protection from sun and rain. In contrast, the degree of protection afforded by the tree is insignificant in the case of the Asiatic species of *Pteropus*. Over the years, their traditional roosting trees have been stripped of bark and foliage by thousands of claws so that the bats hang in full view on the dry branches.

Although the tree roosts are often close to human habitations, the bats quickly learned that no danger threatens them from that quarter. The religion of the country forbids its adherents to kill bats. As a result, it has been possible for many colonies of fruit bats to become established in such populous cities as Bombay. The constant stream of people thronging the streets of the town seems only to enhance the bats' sense of security. Only an attempt to climb the tree will cause the entire colony to take flight.

Previously, large numbers of fruit bats in Australia also established themselves close to human settlements "for reasons of safety". But they were forced to move away, for here neither religion nor law protected them. They were increasingly persecuted and their numbers drastically reduced; they had no alternative but to withdraw into sparsely populated areas.

The small species of fruit bats and various members of the American Spear-nosed bats, which are also arboreal, are very difficult to discern when roosting. They hang quietly among dense foliage and many species gain additional protection from the striped markings on their fur, as the outline of the body is completely obliterated in the play of light and shade. The *Lasiurus* species which in summer roost in the deciduous forests of North America are forced to leave their summer quarters in autumn when the leaves begin to fall. Those species which live all the year round in the southern states are reported to move at this time from apricot trees on to evergreen orange trees.

Among cultivated trees, palm and banana trees offer particularly favourable living conditions to bats, and are regularly colonized by a large number of species. In addition to Short-nosed Fruit bats *(Cynopterus sphinx)*, representatives of other bat families are also found in the crowns of palm trees in India. In Cuba and other Central American countries, large colonies of Free-tailed bats *(Tadarida laticaudata yucatanica* and *Tadarida minuta)* live in the withered foliage of tall palms. One species of tree is such a typical domicile for bats that botanists gave it the scientific name of *Copernicia vespertilionum*, the Bat Palm. Those trees that house bats are easily recognized by a thick layer of droppings on the ground round the trunk. Since the lower part of the tufty foliage of these trees is a thick curtain of dried-out fronds, it offers excellent protection to its tenants. The number of bats on

Possible roosts for bats on and in trees (after Greenhall, 1968)

one tree has been estimated at between two and three thousand.

On banana trees, several species of bats may be found together beneath the central ribs of the large hanging leaves. Even in regions where these trees have been introduced commercially, bats have not been slow to recognize that the large leaves offer an ideal protection from sun and rain. In Africa and South America, small species of fruit-eating bats live in banana trees. Pipistrelles often roost on clusters of banana fruits. In Africa, small species of Mouse-eared bats and Pipistrelles and in America, Sucker-footed bats (*Thyroptera* spp.) have been found inside young leaves that are still rolled up into a cone. The latter sit in small groups among the leaves, and in contrast to the habitual position of bats, roost with the head upwards. The French zoologist Brosset reported finding Vespertilionid bats in that part of the tree where buds were forming on all banana plantations in Gabon. So specialized are the bats in seeking out this particular resting place that they follow in the wake of the sequence of planting. Each of the selected leaves thus provides accommodation for only twenty-four hours. After that time, it has unrolled so far that the bats are obliged to seek out a new banana leaf, still tightly rolled, for the following day.

Dead trees that have remained standing for decades, particularly in primeval forests, also offer ideal roosting sites. Many tropical species of forest-dwelling Rhinolophids (Horseshoe bats), Hipposiderids (Old World Leaf-nosed bats) and Molossids (Free-tailed bats) live in natural depressions and holes made by birds, and in spaces behind loose bark. In Africa, the Great Slit-faced bat

86

(Nycteris grandis) and the Old World Leaf-nosed bat *Hipposideros cyclops* rely on hollows in trees. In South America, the Giant Spear-nosed bat *(Vampyrum spectrum)* and the Common Vampire bat *(Desmodus rotundus)* are often found in hollow trees.

Certain European species are also found behind loose bark, in nesting holes made by woodpeckers and in knot-holes in branches. A typical arboreal bat is the Common Noctule *(Nyctalus noctula)* which is found in deciduous and coniferous forests. Hollow spaces inside trees situated close to river banks are favoured by the Daubenton's or Water bat *(Myotis daubentoni)*, while Long-eared bats *(Plecotus* spp.) can very often be found in hollows in fruit trees and in nesting boxes. If the bats make use of these roosts for a number of years, marks made by urine stains at the entry hole often betray their presence. It appears that decomposition of wood inside the tree brought about by the action of urine and faecal matter causes an increase in heat and humidity which is difficient to enable the Noctules to remain all winter in these quarters.

Inhabitants of natural and man-made caves

In addition to the large group of bats that live in trees and bushes, or even in grass and moss, there are probably just as many that have chosen to roost in natural or artificial habitats underground. In both tropical and temperate latitudes, many species in various families are found that at least occasionally are cave dwellers.

Whereas in Europe most species spend at least the cold season in underground habitats, in the tropic and subtropical regions of the world, underground chambers are the homes of bats throughout the year. Here they roost, here social activities are performed and the young are born and tended. Almost every continent has certain caves that have become legendary as the roosts of hundreds of thousands, even millions of bats. These places are in the nature of family property, going back to ancient times.

On the island of Borneo there is the Niah Cave (Sarawak), where some 300,000 bats have been found to live. In the "Cave of Thieves" near Bombay (India), there are about 100,000 Serotines *(Eptesicus serotinus)*; no less than 500,000 Old World Leaf-nosed bats of the species *Hipposideros caffer* roost in the Falcon Cave of Belinga (Gabon). Unimaginably large numbers make up the vast colonies of Guano bats *(Tadarida brasiliensis)* in certain caves in Mexico and the southern states of the U.S.A. It has been estimated that in some caves, the annual total of bats has reached ten to twenty million. It will be shown later how numbers have declined alarmingly in the last twenty years. As darkness falls and these vast numbers of bats set out on their nocturnal flight, pouring forth in thousands from the cave entrances, it looks to the distant observer like the column of smoke from a volcano rising skywards. This unique natural phenomenon is used as a tourist attraction by the authorities in charge of the Carlsbad Caverns in New Mexico, U.S.A. Every evening during the summer months, hundreds of sightseers wait to watch as some half a million bats take flight. The exact time of the spectacle cannot be predicted. It depends upon the arrival of dusk, and between May and October, has been recorded to take place between 4.30 in the afternoon and 8 o'clock in the evening. When the young bats born in June become independent, they join in the evening flight. So, from mid July, there is a spectacular increase in the number of animals issuing from the caves.

Before the sensational event begins for the tourists at the Carlsbad Caverns, growing excitement among the bats inside the caves signals the impending departure. Awakened by their "internal clock", the creatures yawn and stretch, rousing any of their neighbours that may still be dozing. This is unavoidable, since up to 3000 bats may roost in a square metre of roof area. The first bats take wing and stimulate others. More and more of them make for the cave entrance where, at first, they circle irresolutely, until some unknown signal indicates the start of the flight. The exodus from the Carlsbad Caverns does not always take the same form. Sometimes the bats emerge like clouds of smoke; that is, for 1 to 2 minutes, a great swarm flies out, and then there is an interval, frequently no more than a few seconds, before the next swarm appears. This continues for some minutes, but finally, so many bats press forward that the emerging stream becomes unbroken. They have also been observed to emerge straight away as a continuous swarm that streams out steadily for some 10 to 15 minutes and ends just as abruptly. No one knows why. After 30 or 40 minutes it begins again, and then the rest of the bats come flying out.

Generally the bats leave their day-time roost in continuous flight. Some 5000 per minute emerge from the vast cave entrance, and at the end of an hour, the cave is empty, the spectacle is concluded. At first the bats spiral upwards to a height of 50 to 60 m above the cave entrance and then fly off in the direction of their principal feeding grounds. These are wooded hillsides which provide insects in great numbers. Yet if such masses of hungry animals are to be satisfied, the individual may have to remain on the wing for eight to ten hours to obtain all the food it needs. Within this period of one night, these small mammals, weighing 15 g and with a wing span of 25 cm, cover a distance of 70 km or more. It is unusual for a bat to return in less than three or four hours. The majority do not get back to the cave until dawn is breaking.

The return flight to the roosts, and to the young bats waiting there, takes on a different form. Some eye-witnesses have claimed that it is an even more impressive sight. The bats come in from a great height, and streaming downwards in headlong flight, they draw in their wings and disappear into the Stygian gloom of the labyrinth of caves. Because of the high speed of flight, the air begins to vibrate as it streams past the wings, and when simultaneously hundreds of bats skim towards the cave mouth, curious, eerie sounds are produced. Just as suddenly as the exit began, the return flight of the bats suddenly comes to an end; an impressive natural phenomenon which still holds many mysteries, and which has been taking place day after day each summer for many thousands of years.

Vast colonies on this scale are, of course, exceptional. Depending upon the size of caves, and—a critical consideration—on the conditions of temperature, humidity and light existing within them, the number and composition of the species found in different regions of the world varies greatly. Even today we are not able to explain which factors cause bats to consider one roosting site particularly suitable and to colonize it in thousands, and yet to avoid another, which as far as we can see is entirely similar. These interesting questions are still waiting to be answered. It is especially important to bats that there should be no draughts, since they cannot tolerate draughty conditions. Therefore they select enclosed systems of caves as their summer roosts. If the cave has a blind end, warm air can accumulate there and the temperature does not fall to below the critical level of 15°C. Summer roosts of the Greater Horseshoe bat (*Rhinolophus ferrumequinum*) and of the Long-winged bat (*Miniopterus schreibersi*) have, however, been found in which the average ambient temperature was only 10 to 13°C.

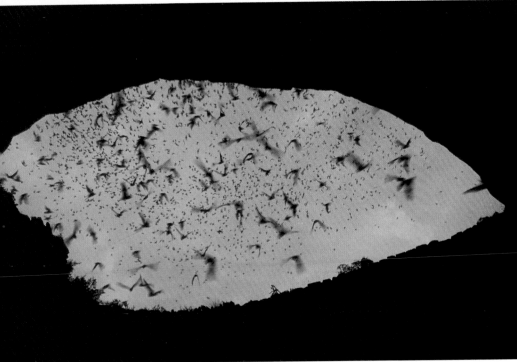

60   Swarms of Guano bats fly
across the evening sky on the way
to their feeding grounds.

61   Thousands of Guano bats
(*Tadarida brasiliensis*) set out in
the evening from their day-time
roost in the Carlsbad Caverns in
Texas, U.S.A.

62   Kalongs (*Pteropus vampyrus*)
leaving their tree roost

63 Fruit bats circling their tree roost. Large numbers of them sleep during the day hanging close together on the dry branches.

64　Dried-out river bed in the Namib Desert in South Africa. The roosting places of Robert's Flat-headed bats (*Sauromys petrophilus*; Molossidae) are to be found beneath small projecting ledges on the granite blocks in the foreground.

65　Roost of a solitary male Whiskered bat *(Myotis mystacinus)* among cut timber

66 Maternity colony of Large Mouse-eared bats *(Myotis myotis)* and Long-fingered bats *(Myotis capaccini)* against the roof of a cave in Bulgaria. In southern Europe, caves and cellars are frequently used as summer roosts. This colony consists of about 1500 animals.

67 Crack in a tree trunk. It is used by Daubenton's or Water bats *(Myotis daubentoni)* as the entrance to their roosting place inside the tree trunk. Bat droppings can be seen round the fissure.

68 Attic of a house with indications of a bat maternity colony. Large quantities of fresh doppings on the floor as well as remnants on the chair and the walls above the door, are clear evidence of the presence of a bat roost under the roof.

69　Holes in bamboo canes.
Originally made by insects, the
holes now serve the small Vespertilionid bat *Tylonycteris* sp. as the
entrance to the roost.

70　The day roost of the Red bat
(*Lasiurus borealis*) is among the
leaves of trees and bushes.

94

71   A Lesser Horseshoe bat
(*Rhinolophus hipposideros*) hibernating. The bat has enfolded itself
completely within its wing
membranes.

72   Three Natterer's bats (*Myotis
nattereri*) in hibernation

73   A group of Grey Mouse-
eared bats (*Myotis grisescens*) in
their winter roost.

74　A group of Large Mouse-eared bats *(Myotis myotis)* in their winter roost

75　Serotine *(Eptesicus serotinus)*. This bat, in a state of lethargy, has fallen from the wall of the roost, and at first is completely helpless. As a reflex action, it extends its wings and moves its extremities about in search of a support.

It is astonishing how well bats are able to tolerate the high concentrations of ammonia liberated from piles of droppings, and are still able to breathe in such an atmosphere. Experiments have shown that the strong smell of ammonia keeps enemies at bay, so that in effect it provides protection.

Just as in the European winter roosts, the individual species show distinct preferences for particular roosting places, so too within the caves of the tropics, there is a quite specific distribution of the species which is largely determined by conditions of light. Egyptian Fruit bats (*Rousettus aegyptiacus*) and members of the Sheath-tailed bats (Emballonurids) usually roost in the semi-darkness of the outer chambers. In the innermost recesses of the caves where no light penetrates, representatives of the Old World Leaf-nosed bats, Leaf-chinned bats and true Vampire bats can be found.

Bats have even succeeded in colonizing the desert sands of the Sahara, where they live in the subterranean channels of the irrigation systems. These are completely dark and pleasantly warm. Horseshoe bats and Vespertilionid bats are found there.

The great depth beneath the ground at which bats can be found is illustrated in reports from the U.S.A. Here, the galleries of a zinc mine north of New York serve as winter quarters to about 1000 Little Brown bats (*Myotis lucifugus*). The majority of the bats roost in the upper passages at a depth of 200 m and with an ambient temperature of about 5°C. But the deeper galleries are also visited, and contrary to all expectations, a bat has even been observed in the deepest gallery at a depth of 1160 m. Such depths are, of course, out of the question as winter roosts, since the prevailing temperature here is at least 25°C.

Many of the species that live underground avoid spacious and high-roofed cave systems. They prefer narrow

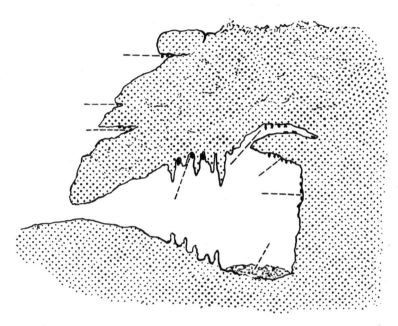

Roosts of bats in caves, tunnels and cellars (after Greenhall, 1968)

crevices in rocks and interstices between stones for their living quarters. In these cases, the bats are usually solitary or live in small groups. They conceal themselves very effectively and are difficult to observe.

In Egypt, Mouse-eared bats are found in the gravel of river beds, and on the Pacific coast of Mexico, the fish-eating bat *Pizonyx vivesi*, which is a Vespertilionid bat, has been found among loose stones and fragments of rock, where it often lives in company with the storm petrels that breed there. Rock crevices are a popular abode for various Free-tailed bats and Sheath-tailed bats. Any attempt to extract them from their roosts is unlikely to succeed since the slightest disturbance causes them to withdraw into the innermost recesses.

Inhabitants of buildings

Bats frequently find a modified form of the natural roosts that are available to them in and on buildings constructed by man. So it is not surprising that a wide range of species has adapted rapidly and taken over congenial houses, churches, temples, pyramids, towers and fortresses, occupying them "from cellar to attic". This explains why, in many regions, bat numbers are greater in towns and villages than in woodland and field. Quite often, the residents of a house do not realize that bats are sharing the same roof, unless the colonies assume such proportions that they become an annoyance to the occupants. Bats are frequently found in churches and places of worship, undoubtedly because here they are free from harassment. The lofty roof spaces of churches and castles are rarely disturbed, so that in addition to bats, other animals such as the barn owl *(Tyto alba)* will often roost and bear their young there. In Europe, Large Mouse-eared bats *(Myotis myotis)*, Serotines *(Eptesicus serotinus)* and Long-eared bats *(Plecotus* spp.) are found during the summer in the roof spaces and belfries of churches and castles. In the winter, Common Noctules *(Nyctalus noctula)* and Pipistrelles *(Pipistrellus pipistrellus)* may be discovered in concealed hiding places in churches. The Church of Our Lady in Dresden housed one such winter roost until it was destroyed in February 1945. For decades, bats had lived here, high above the town without their presence being suspected. Not until 1926, when repairs were being carried out on the cupola, did workmen discover a large number of Noctules in a small space above the antependium of the altar. They hung up against the sandstone walls in rows one above the other like tiles on a roof. Their numbers were assessed at 800 to 1000 bats, grouped into several clusters. From that time on, counts carried out annually showed that every win-

Possible roosts for bats on and in buildings (after Greenhall, 1968)

ter, large numbers of males and females assembled in the warmest parts of the small attic room. The temperature there suited the requirements of the Noctules perfectly, so that virtually no losses were reported, except in the hardest of winters, when numbers could be reduced by half.

Other highly esteemed roosting sites are the ruins of ancient cult burial places, fortresses and defensive structures. In the Orient, Tomb bats *(Taphozous nudiventris)* are found in thousands in the ruins of former mausoleums. The Horseshoe bats of Europe, Asia and Africa like to live in the brickwork of old castles. Depending upon their geographical distribution, they roost in the company of Tomb bats, Mouse-tailed bats or Fruit bats like *Rousettus* or *Dobsonia*. In America, Sac-winged bats *(Saccopteryx)* have colonized the arcades of ancient Mayan temples. Like their relations in the Old World, such as various of the Tomb bats, they attach themselves to the outer walls of houses with no protection whatsoever. When the heat of the sun becomes excessive, they move to a cooler wall of the house.

Houses with cellars are, on the whole, restricted to the temperate and cool regions of the world. So it is only here that bats have the opportunity to occupy cellars; and they do so willingly when conditions are favourable. In Europe, bats frequently use cellars as an overwintering site. They roost in quiet, damp corners or dark wall crevices.

It is rare for humans and bats to share the same living quarters. Usually the bats are driven out or even killed by man. There are exceptions. In certain Indian villages, families of Hindus, to whom all animals are sacred and therefore protected, live in the same room as bats. The small bat lodgers roost on the smoke-blackened walls, knowing that no one will harm them. If anyone gets too close, they do not fly away but start to bite!

In many countries, these so-called "house bats" inhabit not only roof spaces and the joints and crevices of roof beams, but also window shutters, fascias, wainscotting, the frames of blinds and many other structural embellishments. Tiny Pipistrelles even lodge in joints between roof tiles and in thatched roofs. Such roosting places are met with equally in Europe, Algeria and India. In tropical America, small Free-tailed bats *(Molossus)* are to be found creeping out from under the roofs and out of cracks in walls as darkness falls and they prepare to set out on their nocturnal flights.

Bats that live in the nests and burrows of other animals

Ecological parasitism among the Chiroptera is also shown in their habit of taking over or sharing the living quarters and nests of other animals. For example, in England, Whiskered bats *(Myotis mystacinus)* have been observed in the nesting holes of sand martins.

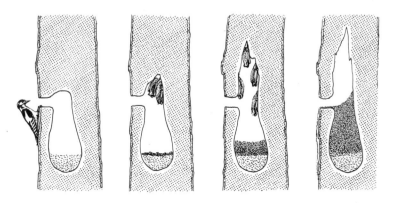

The development of a woodpecker's nesting hole into a bat roost (after Stratmann, 1978).

In various countries, colonies of bats have even been found in the sets, burrows and underground nests of the fox, badger, rabbit and porcupine. These holes in the ground are not always the most favourable of residences, particularly when the "owner" himself shares the occupancy. But if the environment offers nothing better, bats are prepared to live even there.

In Africa, Vespertilionids of the genus *Kerivoula* are often found roosting in the abandoned nests of birds, particularly weaver birds, and the American Mouse-eared bat *(Myotis velifer)* has been observed in the nesting holes of swallows under bridges. Even the nests of termites do not escape the attention of bats. In Trinidad and other places certain Spear-nosed bats *(Tonatia minuta)* have been known to establish themselves there, sharing these quarters with a small species of parrot.

Even more curious is the report of bats in spider's nests. Although there can be room there for only very few animals, it would seem from observations made in Gabon by the French zoologist Brosset that the Vespertilionid bat *Kerivoula harrisoni* regularly roosts under the webs of the *Agalena* spider. In doing so, the bat clings to the thin branches that support the filaments of the web.

99

The biological and ecological significance of the habitat

With such a large number and variety of dwellings and roosts as has been described here, one cannot but gain the impression that finding a suitable hiding place in which to rest during the day must present no problem to bats. It seems that nature and man provide roosting sites in abundance. But the impression is a false one.

Most bats are unable to maintain a constant body temperature; it rises and falls in accordance with the ambient temperature. Therefore the temperature provided by the selected roost is very important for the daily activities of the bats. Temperature affects metabolism, and this in turn determines how "active" the bats are. Most bats function most effectively at temperatures of about 30 °C. When the colder season of the year begins in our latitudes—usually the time when food also becomes scarce—the bats withdraw into frost-free winter quarters, reduce their metabolic rate considerably, and await the spring. More will be written of this interesting phenomenon in a later chapter.

In warmer regions it can also happen that the supply of food diminishes. In this case, the bats once again survive the critical period by finding a cooler place. The saving of energy by reduction of the metabolic rate and the resulting torpor can be observed even within the daily cycle. Many of the Free-tailed bats that live in rock crevices along coasts retreat to the furthest depths of these crevices in the morning. Since it is very cool there, they enter a state of lethargy which considerably retards metabolism. In the evening, they make their way to the opening of the crevices, warm themselves in the heat of the setting sun, and in this way, activate the metabolic process to such an extent that, within a short time, they are ready to set off on their flight in search of food.

Many bats exploit the different climatic conditions of the various localities within a house, so that they are able to organize the course of their life here throughout the entire year. The zoologist Gaisler of Brno describes how in summer the females in a colony of Lesser Horseshoe bats *(Rhinolophus hipposideros)* show a preference for the attics of houses. Here, the temperatures are usually higher and this is favourable for the development and rearing of young. Cool rainy days during the summer endanger the successful rearing of young bats. The situation is quite different for the males. They can afford to lower the energy balance and spend the day in a state of semi-torpor. So they are found in cooler parts of the house, for instance, under floorboards or even in the cellar. In autumn, depending on the temperature outdoors, females and young bats are found in various parts of the house. When winter comes, all the bats are re-united in the cellar where they roost in a dark, cool but frost-free situation. Similar observations have also been made of the Greater Horseshoe bat *(Rhinolophus ferrumequinum)* which, according to the season, is found in the various "storeys" of old castles.

In tropical regions, many species have principal and secondary sleeping quarters, although the day-time roosts show very constant temperatures. The secondary roosts are visited in particular when the bats are disturbed in their habitual quarters. Considerations of safety play a vital part in the selection of a roost. Branching and tortuous cave systems are much more advantageous and more sought-after than simple galleries.

It has been found that very large populations have several roosts at their disposal. For reasons that are not yet clear, the majority of the bats in the colony are found now in roost A, and then again in roost B or C. Thus, periodical assembling and dispersal takes place within such a colony of between 1000 and 100,000 individuals.

In this chapter, we have seen that bats live as solitary individuals or in large assemblies of various sizes, and that bats of the same sex sometimes group together in a particular roost. The latter phenomenon is exemplified in the maternity colonies. These are roosts in which female bats come together in order to give birth and rear their young.

As a generalization, it is clear that the majority of the many species of bats live a communal life. The Chiroptera are social creatures; solitary individuals are the exceptions. They are found at least in pairs, as is commonly the case among smaller species of fruit bats. But usually the pair consists not of male and female, but of mother and young. In almost all families, there are species that are less sociable and therefore found only in small groups. This may well be because the roost selected precludes the formation of a large congregation.

In temperate regions, where the females of most species form maternity colonies, the males adopt separate roosts, frequently close to the females. Since they live in small groups or as individuals and in addition are very secretive, they are difficult to find in the summer. There are also examples of totally male colonies.

In the tropics, maternity colonies are less usual. Here the entire social group, males, females and young, roost in the same place. Among Free-tailed bats (*Molossus*) and Tomb bats (*Taphozous*), males and females are sometimes observed to form separate groups.

The fact that in the northern latitudes, the sexes live in separate quarters in the summer, while in the tropical zones, this habit is rare, can be observed even within a single species. For instance, the Horseshoe bat *Rhinolophus rouxi* of Northern India shows strict segregation of the sexes, whereas in Sri Lanka, males and females are always found together. In the case of some fruit bats of Northern India, the sexes live apart for a short period during the summer months. This phenomenon is not found in the same species living in tropical regions.

Little research has been carried out into the extent of competitive rivalry for territory and roosting sites. In the winter quarters in temperate zones, several species are often found together. For example, an annual total of more than 3000 bats overwintering in a chalk mine near Berlin includes Large Mouse-eared bats (*Myotis myotis*), Daubenton's or Water bats (*Myotis daubentoni*), Whiskered bats (*Myotis mystacinus*), Natterer's bats (*Myotis nattereri*), Long-eared bats (*Plecotus auritus*) and Pipistrelles (*Pipistrellus pipistrellus*). In mine workings in Bohemia, ten different species have been observed in the winter. Usually, each individual species has its favourite roost, selected according to conditions of temperature and humidity. In the tropics, certain species share a common roosting site. This shows that the basic requirements determining the selection of living quarters are the same for many species, and that within these quarters, each species is very selective in its demands.

What bats eat

Diversity, such as we have met among bats in various fields so far, also characterizes the feeding habits of the Chiroptera. Although two major foods predominate —insects in the case of insectivorous bats (Microchiroptera) and fruits in the case of fruit bats (Megachiroptera)—various species have nevertheless become specialized in quite unexpected foods. This extension of the food spectrum shows that in the field of feeding, bats have once again been able to open up and occupy interesting new niches. In some cases, specialization has gone so far that certain species feed exclusively on blood or on pollen and nectar. On the other hand, no bat has adapted to an exclusive diet of grass or leaves. Clearly the energy gain from such a source is not sufficient for these flying mammals. The quantity of leaves or grass required to satisfy the requirements of bats would be too great, and its digestion would demand a considerably more voluminous intestinal tract, such as that of rodents or ungulates. The morphological adaptations involved would conflict with the bat's capacity for flight. But feeding on fruit presents no problems. The juicy flesh of fruit is rapidly digested and passes quickly through the intestines. Very often only the juice of the fruit is extracted and the fibrous matter rejected.

Insectivores

The majority of bats eat flying insects. Food of animal origin always has high nutritional value in comparison with the same volume of vegetable food. It can be consumed more rapidly and satisfies for a longer time. This last consideration is important for bats which must often go without food during the day for 14 to 18 hours.

The habit of feeding on insects led to the earlier subdivision of the order of Chiroptera into "insectivorous

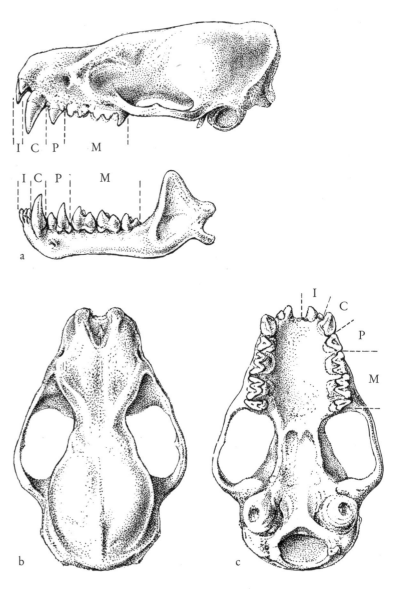

Skull of a Serotine *(Eptesicus serotinus)* seen from
a the side, b above and c below.
Dentition in the upper and lower jaw show absence or reduction of incisors and of premolars. The remaining teeth are well developed, pointed and in part knobby. They are reminiscent of the teeth of a beast of prey. The pointed canines seize the prey and the sharp molars grind down the chitinous shells of insects.
I incisors, C canines, P premolars, M molars.

bats" (Microchiroptera) and "fruit-eating bats" (Megachiroptera). In fact, every family of Microchiroptera—apart from Vampire bats—has representatives which feed on insects. But there are also fruit-eating members in some families of Microchiroptera. Insectivorous bats are found on every continent, and indeed in the temperate latitudes, the species are exclusively insectivorous. This form of feeding is undoubtedly the original, primitive one, since the Chiroptera are derived from insectivorelike stock. It was important for them to develop a system of location that would assist them in finding and seizing prey in the air, in the dark of night. With the evolution of ultrasonic echolocation, this problem was solved. Not only does it enable bats to find their way in the dark but it also directs them to a source of food. When bats came upon the scene, a new threat entered the life of the great host of nocturnal insects. From birds they had little to fear, for few birds hunt them at night, and during the day, when they are resting, their excellent protective coloration makes them virtually invisible. But for bats, they are major sources of food.

The precise hour of the evening at which bats set out on their nocturnal flight cannot be predicted. Differences exist between species, but even within a single species or even a single colony, variations can be observed from one day to the next. Clearly, the familiar "internal clock" controls the departure. Other influences in addition to meteorological factors may also be decisive, for example, the stage that has been reached in the rearing of young. It has been observed repeatedly that pregnant or lactating females start their flight rather earlier.

When colonies are small, the bats usually hunt in the vicinity of the day-time roost. Barbastelles *(Barbastella barbastellus)* are usually found within a radius of 500 m of the roost. Horseshoe bats *(Rhinolophus* spp.) also have a restricted range. However, if a colony consists of

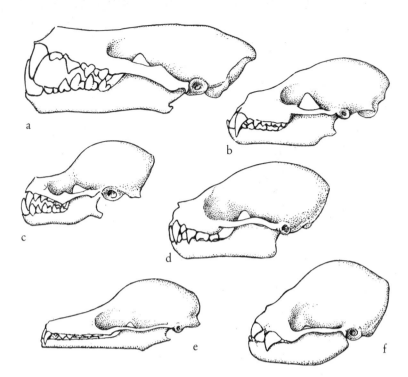

The adaptation of skull shape and dentition to different feeding habits in Spear-nosed bats and closely related Vampire bats (from Yalden and Morris, 1976)

a *Vampyrum*: small mammals, insects, fruits?,
b *Phyllostomus*: small mammals, insects, fruits,
c *Tonatia*: insects,
d *Artibeus*: fruits,
e *Anoura*: nectar and pollen,
f *Desmodus*: blood.

The insectivorous bats of the genus *Tonatia* show the least degree of specialization in dentition. The species *Vampyrum spectrum*, that lives mainly on small mammals, has large canines. As an adaptation to the blossom-visiting habit, the facial skull in *Anoura* is elongate and the teeth only weakly developed. The blood-feeding habit of the Common Vampire *(Desmodus)* makes chewing superfluous; the molars are greatly reduced. It is the function of the pointed, sharp incisors and canines to remove the hair covering at a suitable point on the body of the host animal and to make a small gash in the muscle from which blood will trickle.

many thousands of bats, the hunting territory extends over ten kilometres or more. Species that do not find their food in the immediate vicinity of the roost make use of regular "flight paths" or "flyways" to reach their hunting area. As dawn breaks, they fly back along the same route to their roost.

Bats are not active continuously throughout the entire night. How long they are on the wing and how persistently they hunt depends to a great extent on the amount

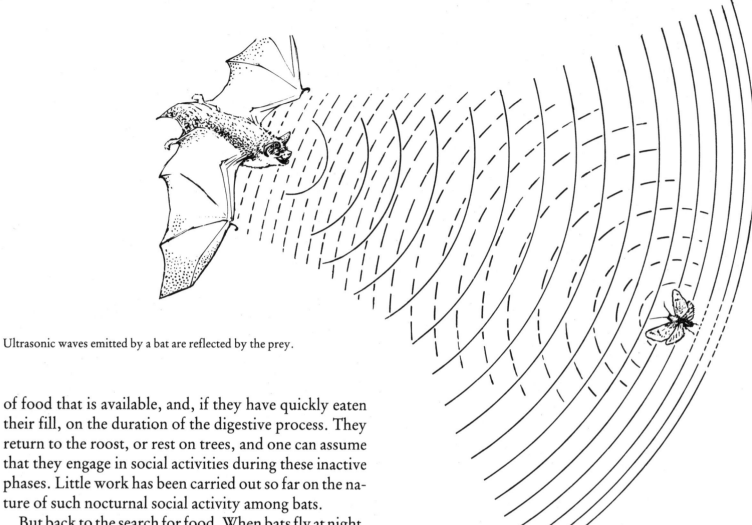

Ultrasonic waves emitted by a bat are reflected by the prey.

of food that is available, and, if they have quickly eaten their fill, on the duration of the digestive process. They return to the roost, or rest on trees, and one can assume that they engage in social activities during these inactive phases. Little work has been carried out so far on the nature of such nocturnal social activity among bats.

But back to the search for food. When bats fly at night, they emit short, regular sound pulses. The echoes that are bounced back provide the bats with information about obstacles in their flight path and about insects that are flying nearby. As soon as an insect is heard or enters the sound-sensing beam of a bat, the frequency of impulses is increased instantly to allow its exact location and pursuit. The noise produced, although inaudible to the human ear, registers an intensity greater than that of a pneumatic hammer.

Although the hunt for prey takes a different form in individual species, it always demands high manoeuvrability and almost acrobatic skill.

Common Noctules *(Nyctalus noctula)* and many tropical species of Free-tailed bats (Molossidae) ascend high into the air, in the manner of birds such as the swallow and the swift, and swoop down in plunging flight upon their prey. Pipistrelles *(Pipistrellus pipistrellus)* and

Serotines *(Eptesicus serotinus)*, on the other hand, fly in rapid twisting flight, low down close to the ground. They like tree-covered or bushy terrain and so are frequently to be found in the parks and gardens of towns and villages. Other species such as the Daubenton's or Water bat *(Myotis daubentoni)* skim in low flight over the surface of fairly large bodies of water to catch the insects flying there. They have been observed also to catch those insects that run on the surface of the water.

The insects, of course, try to evade their pursuer. The bat, swooping at great speed, does not always succeed in catching the insect in its widely open mouth. So it will often try to draw the insect towards it with its wing tip, or to catch it in the interfemoral membrane and then take it into the mouth. The capture of an insect is made in the

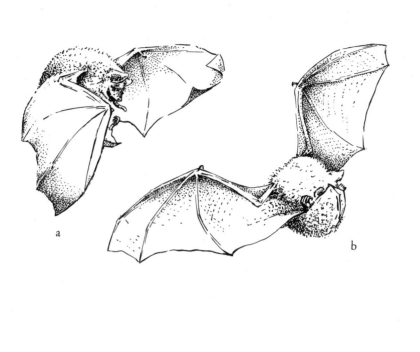

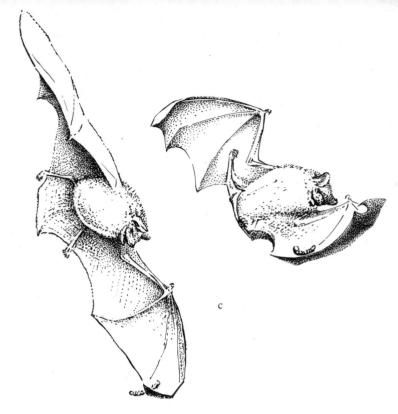

Variations on the capture of insect prey. Generally, small insects are seized directly with the mouth (a). Frequently bats catch larger insects initially in a "pouch" formed by the tail membrane then transfer it from there to the mouth (b). If the prey tries to escape, it is "seized" by the wing membrane and directed by a stroke towards the bat's head (c) (after Webster and Griffin, 1962).

fraction of a second. When the food supply is plentiful, the manoeuvre is repeated several times in a minute. If the insect that is caught is so large that it cannot be eaten on the wing, the bat repairs to a habitual feeding site where it roosts and consumes the "succulent" body of a moth or beetle, while the legs and wings are bitten off and allowed to fall to the ground.

Once an insect is held by the teeth of a bat, it has little prospect of escape. The dentition of this insectivore is reminiscent of that of a beast of prey. The large, needle-sharp, dagger-like canines pierce the prey and hold it fast, and the broad, intricately crenulated molars have no difficulty in chopping up the hard, chitinous body of an insect.

The size of the insect and the frequency of its wing beat are features which enable bats to distinguish between different species of insects when they are catching food. For the smaller species of bats, there is the added consideration that they are able to kill and eat only insects below a certain size. Since the different species of bats have specific hunting territories, and within these they pursue insects of a particular size, there is rarely any competition for food among the various bats in any one habitat.

The chitinous remains of the ectoskeletons of insects are only partially digested, and from an examination of the bat's excreta, it is possible to reconstruct its diet. In this way it was found that the Big Brown bat (*Eptesicus fuscus*) has a diet consisting of beetles (36%), hymenopterons such as bees, wasps and flying ants (26%), flies (13%), together with a small number of mayflies, caddis flies, stone flies and a few crickets and grasshoppers. A list of this kind is, of course, valid only for a particular region and a particular season of the year. The insect fauna in a different habitat during a different month could be quite dissimilar.

Moreover, food analysis of this kind is inevitably incomplete, since small insects are digested without leaving any identifiable remains. It is not merely by chance that beetles appear at the top of the list. Their hard wing covers are more readily preserved. They do not necessarily represent the major type of insect prey.

105

Phases of the flight manoeuvre in catching a mealworm tossed into the air (after Webster, 1967)

In examining the faeces of the Large Mouse-eared bat (*Myotis myotis*), the Bamberg zoologist Kolb found that during the entire summer half-year, these bats also include carabids (ground beetles) in their diet. If there is enough food of other kinds available, the proportion of carabids fluctuates. When cockchafers appear, they make up the greater part of the total food intake. In June, Kolb found large numbers of mole crickets, and when the mass flights of green oak-leaf rollers began, it was this moth which provided up to 90% of the food eaten by the Mouse-eared bats. In July, carabids came to the fore again and, for the sake of variety, grasshoppers were added to the menu. In autumn, it was found that a high proportion of dung beetles supplemented the diet of carabids. Clearly, demand is determined by supply.

There are many choice insects that can be caught by bats only with the greatest difficulty since they possess counterweapons. Some have developed silent flight which results from the existence of extremely fine hairs on every part of the body where air vortices could arise. In others, the body is enclosed in a soft, dense "fur" covering, which absorbs the sounds emitted by the bat, re-

turning only a weak echo. Many moths such as owlet moths, tiger moths and geometers are even capable of perceiving the bats from a considerable distance by means of their sense of hearing. In this case, the approaching bats betray themselves by their calls and cause the prey to take evasive action. Although the insects fly more slowly than their pursuers, they are usually able to escape with their life by a variety of strategic moves. Some drop to the ground like a stone, others set off in erratic zigzag flight which the bats cannot follow.

There are some moths which themselves emit a rapid succession of ultrasonic sounds when danger is most acute. But the noise is produced in quite a different way from that of bats. Rapid movements of the leg muscles cause vibrations in a chitinous plate situated where the hind legs join the body, which produces the sounds. The bats are irritated or frightened and give up the hunt.

Not all bats take their prey exclusively on the wing. They can also pick up non-flying insects from the ground. There have been reports that the African Slit-faced bat *(Nycteris thebaica)* picks up scorpions from the ground and eats them. According to investigations made by Kolb, Large Mouse-eared bats *(Myotis myotis)* can locate insects crawling on the ground. They fly towards them with unerring precision and track them down

among leaves or grass using their sense of smell. Experiments have shown that not all insects are acceptable as food. Mouse-eared bats could never be persuaded to accept potato (Colorado) beetles, either from the ground or in the air. Kolb was able to show that in the search for food, other species of bats also make use of their sense of smell in addition to their sense of hearing. From a distance of 10 to 20 cm, the bat's nose can tell it whether the creature being pursued is edible or not.

European Long-eared bats (Plecotus) are able to hover in the air like helicopters. This hovering flight, of which many raptors are also master, makes it possible for them to hover over foliage, tree trunks and walls of buildings and pick off insects, caterpillars and other creatures. All species of the Old and New World which pursue their food in this way show similar morphological modifications and adaptations.

Since the metabolism of small mammals is particularly high, and since in addition, flying requires a great deal of energy, bats are always voracious eaters. The great heaps of droppings under the roosts of their day-time quarters give some indication of the quantities of insects consumed by bats each night, namely, in small species something like half their body weight. Anyone who has ever kept bats in captivity knows that large species such as

the Large Mouse-eared bat and Common Noctule eat 30 to 40 mealworms a day or finish off 30 cockchafers in a very short time. Small Pipistrelles eat 40 to 50 flies in an hour. It is often difficult to provide them with a varied diet in such quantities. Therefore one cannot but advise in the strongest terms against keeping these useful animals simply as "pets", quite apart from the fact that it is illegal to keep protected animals in captivity. In addition to feeding, there are other necessities such as sleeping quarters and flight territory that can never be provided adequately for bats, so that sooner or later, captivity becomes a torment to them.

The following may serve to indicate the importance of the voracious appetite of these small mammals in the wild as an instrument of biological pest control. If a colony of 250,000 bats in which each individual weighs 15 to 20 g eats its fill in one night, it will have disposed of three to four tons of insects. Kolb estimated that during the mass flight of green oak-leaf rollers (Tortrix viridana), the 800 Mouse-eared bats in a maternity colony in Bamberg destroyed a daily total of 55,000 of these insect pests. Clearly they are an important regulatory factor in maintaining the balance of nature, yet one which has almost completely lost its significance in industrial countries.

107

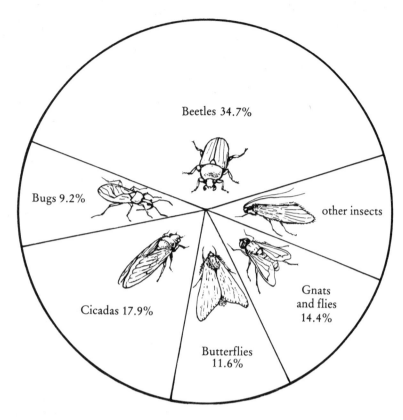

Beetles 34.7%

Bugs 9.2%

other insects

Cicadas 17.9%

Gnats
and flies
14.4%

Butterflies
11.6%

Dietary range of a cave-dwelling bat (*Myotis velifer*) from Kansas, U.S.A.
Average quantities of insects consumed for the months of June to September

Fishing bats

It is difficult to imagine how bats that were originally exclusively insectivorous managed in a few rare cases to specialize in quite unusual foods. The opportunity to avoid competing with other bats for food by occupying the niche that offers fish as food, is one that has been taken by three species.

These species are not related to one another and live in quite different parts of the world. It may well be that their ancestors frequently caught insects from the surface of the water and from time to time scooped up a fish as well. The bat found this new food congenial and, from then on, gave it selective preference, until finally it became the sole source of food.

The best-known of the fish-eating bats is the Fisherman bat (*Noctilio leporinus*) of tropical America. In addition, the Vespertilionid *Pizonyx vivesi*, found in the Gulf of California and one species of *Myotis* in the Far East have also specialized in catching fish. Common to all of them are certain adaptations that enable them to secure their unusual food. In particular, the hind feet are exceptionally large and are furnished with strong claws. To catch a fish, the bats fly slowly, close to the surface of lakes or calm coastal waters. When they locate a fish, they snatch it up with their back feet and lift it from the water.

Bats are not natural swimmers and hardly ever enter water voluntarily. They have never been seen to submerge the entire body in water in order to seize a fish, as seabirds will do. Since fish often swim in large swarms close to the surface of water, there is usually no shortage of food. Difficulties in obtaining food arise only when strong wind and heavy rain whip up the water and make flying difficult. In captivity, a fish-eating bat has been found to consume 30 to 40 small fish in one night, a number that would be difficult to match in the wild. When the fish is caught, it is transferred by foot to the mouth, and if its size allows, is immediately eaten on the wing. Larger prey is held in the bat's mouth and carried to a roost, where it is eaten.

It is not yet known for certain how the fish are discovered and seized successfully. Ultrasonic echolocation undoubtedly enters into it, but clearly does not locate the fish itself, but rather the slight ripples produced by the fish on the surface of the water.

Bats that specialize in small vertebrates

In addition to fish, there are other vertebrates that serve as food for various bats. Certain Vespertilionids of tropical America have developed into regular carnivores. It is known that the Spear-nosed bat (*Phyllostomus hastatus*) likes to eat small rodents, other bats and small birds, in addition to fruit. The Giant Spear-nosed bat *(Vampyrum spectrum)* kills mice with ease. It is able to locate its prey in the rustling undergrowth and to seize the small rodents.

In the Old World, various representatives of the False Vampires or "cannibal bats" (Megadermatidae) are reported to pursue small rodents and eat them. Their diet also includes scorpions, frogs, birds and other bats. These specialist feeders among the Megadermatids land skilfully on the ground, and even with a comparatively heavy-bodied prey, are able to take off with great facility. They carry prey to a roost, where remnants of the meal can often be found in the form of skull and bone fragments.

In contrast to other groups of animals that are primarily carnivorous, no species of bat feeds on carrion. If bats are found close to carrion, it is undoubtedly because they are hunting for insects which are likely to be there.

It is estimated that about 70 per cent of all bats feed on insects and only 2 per cent feed exclusively on or supplement their diet by the addition of small vertebrates, which category, of course, includes fish.

Fruit-eating bats

The change from a diet of insects to mixed feeding or even to a diet consisting exclusively of vertebrates required less modification of the digestive system than did the adoption of a purely vegetable diet. This was a slow process of readjustment, resulting in many morphological changes. Probably at first these species ate a mixed diet of animal and vegetable food. Since many insects eat ripe fruit and their larvae live inside fruits, insectivorous bats looking for food were also attracted there. As they picked off the insects, no doubt they occasionally also bit off fragments of the sweet, juicy flesh of the fruit. It is palatable and the nutritional value is high, so increasing numbers of species in the tropics specialized in this new source of food and eventually became pure vegetarians. Of the Chiroptera living today, about a quarter are frugivorous.

The development of this kind of food specialization took place on two occasions in the bat kingdom, each quite independently of the other. It is interesting to see how the vegetarian feeders of the Old and of the New World went through similar (convergent) stages of development in their adaptation to a purely fruit diet. Special mention should be made here of the teeth, which are not pointed but have wide flat masticatory surfaces, and of the palate which is crossed by a series of ridges.

In the Old World, it is the numerous species within the family of the Megachiroptera which—equipped with a good memory for places and a sensitive nose—make their way with unerring precision to the trees of their feeding grounds, and have become specialists with an exclusively fruit diet. In the New World, there are many species of Spear-nosed bats (Phyllostomidae) that exploit this potential food source. But this family also includes many insectivorous species. The fruit eaters among the Spear-nosed bats have an equally good sense of smell, but they find their way about, as do all the Microchiroptera, by means of ultrasonic echolocation.

In both the Old and the New World, certain species of fruit bats have specialized in frequenting flowers, and feed on pollen and nectar. They will be examined in more detail in the next chapter.

All frugivorous bats inhabit the tropics. In densely forested regions, the vegetation is so luxuriant and varied that throughout the entire year there are trees and shrubs bearing ripe fruit. In the darkness, the bats are guided to the aromatic, sweet and succulent fruits by their sensitive organ of smell. Originally, fruit bats fed only on wild fruits. With the increasing cultivation of bananas, figs, guavas, mangoes, oranges and grapefruit, the bats turned their attention with enthusiasm to these fruits also. In the course of a night, they easily consume an amount equal to their own body weight. Consequently, fruit bats can cause considerable damage to fruit crops in certain areas. To protect their plantations, the farmers tried to exterminate the bats. Their meagre success did not justify the expense and effort. It proved much more satisfactory to harvest the fruit before it had ripened.

The fruit bats and fruit-eating Spear-nosed bats consume some kinds of fruit whole, others they crush in the mouth and swallow the juice together with a little of the fruit pulp. The broad-surfaced teeth and the tongue, which compress the fruit against the ridged palate, are equipment well suited for the purpose of extracting nourishing juices. The bats spit out the fibrous matter that cannot be crushed. Their food, then, consists entirely of water and sugars. The great length of intestine usual and indeed essential in herbivores is not necessary. They do not require a host of bacteria to break down copious cellulose components. Observations of bats in captivity have shown digestion in fruit-eating bats to be such a rapid process that urine is excreted within a very short time of feeding.

The ecological significance of feeding habits

A particular biotope can provide food for only a certain number of individuals. Therefore it is utilized to the full. Even where an adequate supply of food is available, some degree of "division of labour" ensures that all the reserves within the biotope are exploited. An example of this is the way in which certain species specialize in hunting for food close to the surface of water, others along the margins of woods or avenues of trees, others again high above the tree tops. Even the insects on leaves and in the ground cover are included. In the tropics, particularly on islands, it has been found that in terms of diet, the composition of species shows quite typical distribution groups. For example, a particular area was inhabited by two fruit-eating species, two nectar-feeding species, two insectivorous species and one fish-eating species. Each species has its special food categories and hunting zones, and scarcely impinges upon the territory of others.

In many places, the feeding habits of bats are an important factor in maintaining the balance of nature. In the sphere of biological pest control, the destruction of insect pests by insectivorous bats, particularly where the bats occur in large colonies, has an importance which should not be underestimated.

The frugivorous species, on the other hand, play an important part in the distribution of tropical plants. Since they frequently carry off fruits and eat them on trees in the surrounding area, they also effect the dispersal of the seeds. It can be seen that many plants have adapted to the distribution of their seeds by bats. Their fruits are long-stemmed and when ripe, remain hanging on the branch for a considerable time. In addition, they give off an aroma that attracts the bats at night. Moreover, not all tropical plants bloom and bear fruit at the same time. If they did so, the fruit-eating Chiroptera would have no food at certain seasons of the year, and moreover, the excess of food at other seasons would diminish the chances for seed dispersal by bats for many species of trees and bushes.

The relationship between bat and plant is one of mutual advantage and profit. A true symbiosis exists between the two.

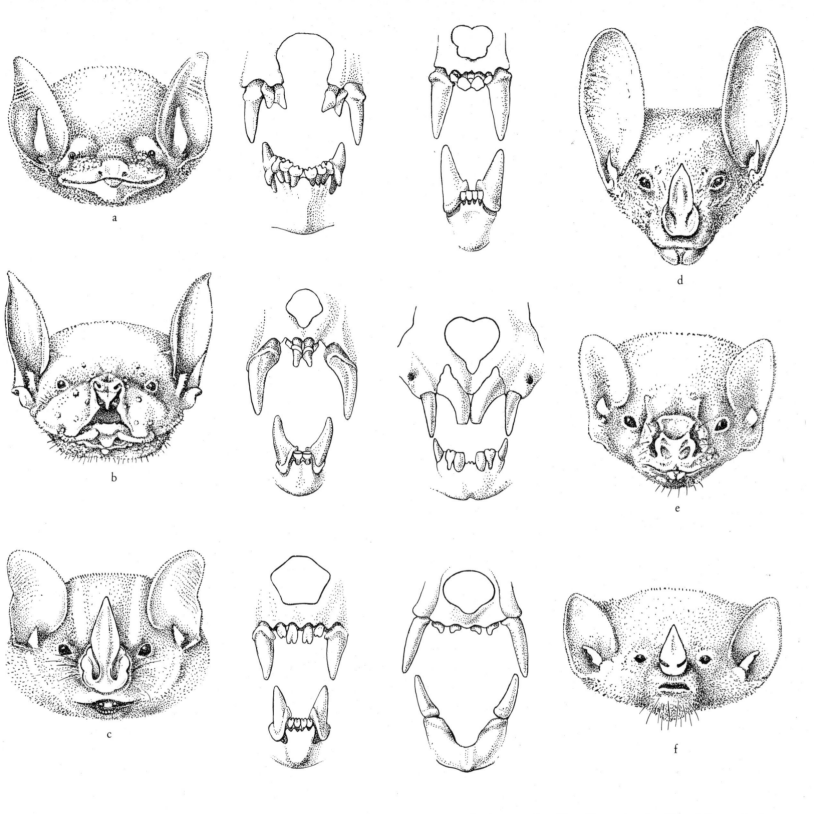

Sketches of various species of bats illustrating the effect of diet on the development of incisors and canines (based on Husson, 1962)
a *Myotis nigricans*: insects   b *Noctilio leporinus*: fish
c *Uroderma bilobatum*: fruits   d *Vampyrum spectrum*: small mammals
e *Diaemus youngi*: blood   f *Lichonycteris obscura*: nectar and pollen

Bat flowers and flower bats

Many flowering plants, in the extremely critical phase of pollination, depend entirely upon animals and moreover, animals capable of flight. Over the course of millions of years diverse and sometimes highly specialized relationships have developed between plant and animal. Not only have plants continually evolved new forms, colours, scents and even behaviour patterns, but in addition, many animals have adapted in various ways to the habit of frequenting blossoms. It is well known that bees, flies and butterflies as well as African sunbirds and South American humming-birds feed on the nectar of plants, and in doing so, pollinate those plants. Not many people are aware that there are also certain bats that feed on blossoms.

In flowering plants, the male generative cells, in the form of pollen, cannot make their way actively to the ova to fertilize them. Another means of transfer had to be found. Nature has solved the problem in three ways, making use of the mobility of three transfer media to carry the pollen. These are wind, water and animals that can fly.

So we distinguish between plants that are pollinated by wind action (anemophilous), those pollinated by water action (hydrophilous) and those by the action of animals (zoidiophilous). By comparing different blossoms, it soon becomes clear, even to the non-expert, to which type a flower belongs. The wind-pollinated blossoms have no need to attract attention; their blossoms are small and their colours restrained. Usually they are not at all conspicuous. Scarcely anyone notices the blossom of poplars, birches or alder trees. The blossoms of those plants pollinated by the action of animals are of quite a different kind, being large and striking in colour and scent. They are the ones that are really thought of as flowers. These flowers can consist of a single blossom, as tulips do, or can be an entire inflorescence, as is, for example, the sunflower.

The flower has the function of drawing attention and attracting and it takes on different forms according to which animals visit that particular plant. Insects and birds are wooed by visual stimuli. In addition, the fragrance of many flowers has an effect on insects, and in the case of certain tropical plants, on bats as well. Since bats are active only at night, this is when the blossoms open. Brilliance of colouring would be of no account at night, so the blossoms of chiropterophile flowers are inconspicuous in colouring.

Animals that visit blossoms do not fly there for the sake of the attractive colours or scents, but in search of food. If they found none, even the most beautiful flowers would lose their interest and would no longer be frequented. Usually the animals do not come in vain, but find sweet-tasting nectar, rich in carbohydrates, and in addition, pollen that is a valuable source of protein. Nectar alone, with a sugar content of 17 to 20 per cent—the rest is water—could not maintain the strength of the bats. They must also eat pollen in order to obtain the protein they need. It has been found that the pollen of chiropterophile flowers contains proportionally much more protein than the pollen of other flowers (agave, for example, 43 per cent).

Flowers visited by birds and bats provide much more nectar than those frequented only by insects. The American zoologist Howell was able to collect up to half a cup of nectar from a single inflorescence of many bat flowers. Even this amount of food is not sufficient to satisfy a bat. It is necessary for the creatures to visit many flowers every day to assuage their hunger.

It is an advantage to the plant that the individual blossoms contain only small quantities of nectar, since this necessitates visits to further blossoms, by which means the vital transfer of pollen is effected. The cunningly shaped and arranged anthers and stigmas inside the blossoms ensure that the pollen will adhere to parts of the body of the visiting animal that will come into contact with the reproductive organs at the next blossoms, and deposit pollen there. Pollination thus achieved is an important recompense to the plant for food supplied to its guests.

In order to obtain nectar from the tubular blossoms, many insects have evolved a long proboscis. Flower birds such as the humming-birds or Hawaiian honeycreepers possess long, thin beaks and a highly protrusible tongue with a brush-like tip. Adaptations of a similar kind are found in the bats that visit flowers. Extension of the facial skull and a very long, protrusible tongue are typical characteristics of a flower bat. On the other hand, the molar teeth are greatly reduced.

For much fascinating information on bat-pollinated plants and the reciprocal adaptations in flower and bat, we are indebted to the botanist Vogel. On his travels in South America, he discovered a large number of species of plants that are visited by bats. These were not only woody plants but also epiphytes and cactus plants.

Vogel was able to show that in Central and South America, the geographical distribution of plants that are pollinated by bats coincides to a large extent with the distribution of bat species that frequent blossoms. Their range extends northwards to the southern states of the U.S.A. (Texas, Arizona). Here, various species of cactus and agave are visited by bats. Most of the bats in question are glossophagine Nectar or Long-nosed bats (Leptonycteris nivalis and L. sanborni) and Mexican Long-tongued bats (Choeronycteris mexicana). These species occur here only as summer visitors, at the time when the bat flowers blossom. In South America, bat plants occur south of northern Argentina. Here, the Long-tongued bat (Glossophaga soricina) and the Pale Spear-nosed bat (Phyllostomus discolor) are the species that can be seen most frequently visiting blossoms. A similar correlation in the geographical distribution of bat flowers with that of flower-visiting bats was also established in Southeast Asia.

The first indications of bats visiting flowers are of comparatively recent date. Apart from sporadic statements made at about the turn of the century, it was not until the thirties that reports drew attention to the flower-visiting habits of bats in Southeast Asia, Africa and Central America. There were repeated reports of small bats having been found feeding in the flowers of the African baobab tree (Adansonia digitata). And in botanical gardens in Central America, small bats were again observed flying round baobab trees. The vital importance of bats in the propagation of these trees was illustrated in Hawaii. Numbers of baobab trees that were introduced here remained infertile because there were no bats to pollinate them.

One of the first trees to be recognized as a tree visited by bats, because of the shape of its blossom, was the calabash trees (Crescentia spp.)

Blossom-visiting by bats was at first presented as a curiosity. Up to that time, few people were prepared to believe that bats could feed on nectar and pollen. It seemed much more likely that the bats were visiting the flowers to catch the many insects flying round them.

The details of a bat's visit to a flower are difficult to capture, even using modern techniques. The event takes place only under the cloak of darkness, and moreover, usually in the tops of trees in tropical forests. Therefore in the case of many plants that are assumed to be pollinated by the action of bats (chiropterophile), ultimate proof of this—namely the observed visit of a bat to the blossom of these plants—still eludes us today. Frequently, the only indication of the visitor is the impression left on the petals by the claw of the thumb.

Many bat-pollinated plants facilitate their guest's visit in various ways. They have long-stemmed, large flowers that extend beyond the canopy of leaves; in others, the flowers lie close up against the trunk. Some plants shed their leaves when they bloom, and as a result, the bats can reach the blossom more easily. The inconspicuous colours of the blossoms vary between greenish-white, dull red, dingy brown and inky blue. Usually they open in the evening, shortly before or after darkness has fallen and often blossom only for a single night.

Many of the blossoms exude a "fragrance" that is difficult to describe; it has been called unpleasantly stale-smelling, cabbage-like and also musky. Others give off an odour of fermenting, over-ripe fruit; some of the species that visit them have evolved from fruit-eating species. The number of species of plants in the tropics that show specialization in pollination by bats contrasts with the considerably smaller number of species of bats that visit blossoms.

It is interesting to note that specialization in nectar and pollen feeding has taken place both in the Old and in the New World, quite independently of one another. In Africa and Asia, there are some 15 species of fruit bats belonging to the group of Long-tongued Fruit bats (Macroglossinae) and in America, about 40 members of the Spear-nosed bats, namely the Nectar-feeding Small Spear-nosed bats (Phyllonycterinae) and the Nectar-feeding Phyllostomids (Glossophaginae) that successfully occupy this niche. Also a number of Stenodermine Spear-nosed bats feed partly at flowers and partly at fruit.

Since not all bat plants blossom throughout the year, and certain differences in form exist among blossoms, the bats use various techniques to obtain nectar. They hover in the air in front of many of the flowers and insert only the tongue and tip of the snout; on other blossoms,

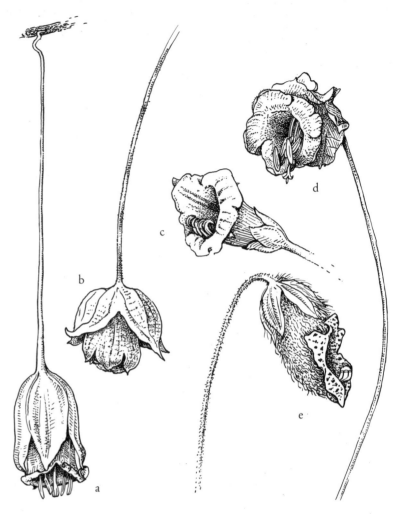

Long-stemmed flowers of plants pollinated by bats (chiropterophile) in South America (after Vogel, 1958)
a *Trianacea speciosa* (Solanaceae)
b *Cayaponia* sp. (Cucurbitaceae)
c *Symbolanthus latifolius* (Gentianaceae)
d *Cobaea scandens* (Polemoniaceae)
e *Campanea grandiflora* (Gesneriaceae).

they take a firm hold first of all with the claws of the thumbs or back legs, and then extract nectar and pollen. At the moment of drinking, the extended wings are motionless. After no more than one or two seconds, the bats allow themselves to fall, and fly away. They circle the tree briefly, and quick as a flash, fly to another blossom. Apart from short intervals, the flower bats are active in this way the entire night.

The visit of a Pale Spear-nosed bat *(Phyllostomus discolor)* to blossoms of the tree *Parkia auriculata* that is distributed throughout the world, is described by Vogel in

114

the following way. The inflorescences, that consist of three blossom kinds (generative, nectar and scent blossoms), are approached diagonally from below. Suddenly, the bats throw themselves round so that they can grasp hold with the back legs. The thumb plays no part in this. The wings are held extended and almost motionless. Head and neck curve round the lower, fertile flower head, so that the mouth approaches the nectar ring from below. It is licked clean in less than a second with the brush-like tip of the tongue. Then the bat draws back its head sharply, relinquishes its hold and flies diagonally downwards and away. Inadvertently it has brushed against the generative blossoms with its cheek, neck and belly, and pollinated them. The Short-nosed bats, which in India regularly visit the blossoms of the baobab tree, also hold on to the blossoms with their back feet and in addition, the claw of the thumb grips the upper part of the blossom. The weight of the bat bends the stem and blossom over; this causes the nectar to flow on to the lower petal from which it can easily be licked. In this way, even those species that are primarily fruit eaters and show no special adaptations in the region of the head to nectar feeding, can reach the precious food deep inside the blossom.

Many blossoms possess a deep calyx into which the bat must insert not only the entire head but also the shoulders, in order to obtain the sweet nectar.

The true flower bats are very small members of the Chiroptera. Both the Macroglossinae (Long-tongued Fruit bats) and the Glossophaginae (Nectar-feeding Phyllostomids) are among the smallest species of their families.

Since most of the fruit bats are of a quite impressive size, the nectar-feeding species, with a body only 6 to 7 cm in length, appear like dwarfs among giants. They have the advantage that they can adapt more easily to nectar feeding.

The adaptations described above, which animal and plant have undergone in order to develop this mutually beneficial cooperation, are the result of millions of years of evolution. It may well be that some of the nectar-feeding bats proceeded directly from feeding on insects, with which many species still supplement their diet, to feeding on nectar, whereas other species specialized in feeding on fruit first. Some fruit bats have been observed to consume petals as well. In such cases, they destroy the blossom and damage the plant.

The Indian Short-nosed or Dog-faced Fruit bat (*Cynopterus sphinx*) is said to be partial to the honey-sweet blossoms of the banana tree, and to cause widespread damage in many plantations. The Banana bat (*Musonycteris harrisoni*), discovered in Mexico only 20 years ago, on the other hand, lives exclusively on the nectar of the banana blossoms.

In the case of bat flowers and flower bats, a high degree of specialization brings mutual advantage to the partners involved. The situation is quite different in the adaptation of certain bats to blood feeding. The adoption of this source of food is true parasitism, unique among mammals. But of this, more in the next chapter.

Vampires—fable and fact

The picture of a vampire that most people hold in their imagination is only partially realized by the true Vampire bats. Their specialization in a diet of blood, that is, their habit of "sucking" blood or, rather more accurately, of drinking blood, is undoubtedly somewhat repugnant. But in their external appearance, they do not show such grotesque features as many other species of bats. And in the Old World and the New, it was other bats that came to be represented as vampires and still today reflect this fact in their names, the False Vampire (*Megaderma spasma*) and *Vampyrum spectrum* (the Giant Spear-nosed bat). Both were reputed to suck blood. And so they combined all the characteristics necessary to enable them to play the leading role in any tale of horror. Few other mammals have inspired in the human imagination such a combination of terror, respect and fascination as have "vampires".

What then are the plain facts about the true Vampires (Desmodontidae), about that family of bats whose members feed exclusively on the blood of vertebrates? It is certainly the most extreme example of food specialization among the Chiroptera. As a consequence, Vampire bats have become the only parasites among mammals. Not only this particular adaptation, unique among vertebrates, but also economic problems that have resulted from it in recent decades have made the Vampire bat the focus of particular interest.

Apart from the fruit-eating species of bats that from time to time do some damage to fruit plantations, true Vampires are the only members of the Chiroptera that cause extensive and serious harm to man. The three known species live exclusively in the New World. They have evolved from the large family of Spear-nosed bats. The best-known and most widely researched representative of the Vampire bats is the Common Vampire (*Desmodus rotundus*). Like the other two species, it inhabits tropical and subtropical regions of America, from Mexico to northern Chile and Argentina. Vampire bats are relatively small, with a body length of about 7 cm, and they weigh about 30 g. The fur is reddish-brown above with lighter, yellowish-brown underparts. They have no tail and the interfemoral membrane is correspondingly narrow. The head is short and rounded with relatively large eyes and a broad nose, with a fleshy nose leaf. The snout is compact, with a rather pig-like appearance. An indication of their feeding habits, in addition to the deeply grooved lower lip, is the abnormal dentition, in which the number of teeth is greatly reduced. In contrast to insectivorous species of bats that have up to 36 teeth, *Desmodus* has only 22. A diet of blood has rendered teeth superfluous, with the vital exception of the upper incisors and canines. These are very large, razor-sharp and set well forward. With them, the bat inflicts a shallow wound in the skin of its victim, through which the blood can trickle. The notched lower lip is pressed against the wound and blood is lapped up, using the tongue. For their attack, Vampire bats select those areas of their victim's body that are well supplied with blood, and where the skin is thin. *Desmodus* will usually bite large hoofed animals on the neck, the ears or the legs. When the victim is a human, the bat will, if possible, slash the big toe. The White-winged Vampire (*Diaemus youngi*) prefers feeding on the blood of birds. The victims are bitten on the neck or more commonly on the joints of the leg or the anal region.

Little is known of the habits of the third species, the Hairy-legged Vampire (*Diphylla ecaudata*). It is said to feed on the blood of birds rather than that of mammals.

An anticoagulant in the bat's saliva introduced into the wound prevents clotting of the blood. As a result, bleeding can persist for hours. Since the victims of the Vampire bat's attacks are usually large mammals that spend the night at rest, a system of echolocation is of less importance than it is to insectivorous species. In seeking out their source of food, the bats depend much more upon their relatively large eyes and acute sense of smell.

Although the first reports of "blood-sucking" bats reached Europe in the sixteenth century, it was only in the present century that these species gained such importance. In comments on the subject of Vampire bats written in 1857 by Kolenati in his *Naturgeschichte der Europäischen Chiropteren* (Natural History of the Chiroptera of Europe), the author was still able to say that "many of the evil reports of this notorious bloodsucker are without foundation". It was, of course, well known that they bite horses and pack animals, and take blood from them, but it was thought that this was probably owing to lack of insect food. Serious illness or death due to Vampires had never been established.

The situation changed when cattle breeding became increasingly extensive in many countries of Central and South America. Growing numbers of cattle herds provided these specialist feeders with a welcome addition to their list of potential hosts. Conditions were favourable for an increase of their own population.

In the early thirties, an epidemic of rabies broke out in Trinidad, which, within a short time, claimed 89 human victims and carried off thousands of cattle. It was recognized that Vampire bats had spread the fatal virus. Suddenly they were the centre of universal interest.

It is not clear how these specialized feeders evolved. Perhaps certain species of bats originally lived mainly on blood-feeding insects, and frequently would find them sitting on various mammals and birds, until finally they themselves began to lick blood directly from the wounds of the host.

Detailed investigations have shown that Vampire bats frequent all known domestic animals, that is, cattle, horses, goats, pigs as well as domestic poultry. They approach dogs more rarely, probably because these animals are more alert to attack. There are many reports of sleeping humans, particularly children, having been bitten. If bats are given the opportunity to select from a wide range of prey, they are found to prefer certain types of animals, even particular breeds. But it is not the taste of the blood that determines the preference, but the behaviour pattern characteristic of that species or breed. Colour of hide or details of the anatomical structure of the skin may also serve as criteria of choice. According to the zoologist Turner, calves are bitten more often than cows, because their periods of sleep are longer.

Vampire bats are very wary and leave their roost only under the darkness of night. Their period of activity reaches a peak before midnight and again soon after midnight, provided there is no moon. On bright, moonlit nights no Vampire bats are to be seen. They do not leave their roosts, in order, many workers believe, to avoid certain nocturnal raptors that are their natural enemies. However, detailed research carried out by Turner into the habits and behaviour of Vampires, shows that they have other reasons for not appearing on moonlit nights. On light nights, many animals at pasture move about as they graze, and it is not possible for the bats to draw blood from their victims. But when it is quite dark, they fly low across the ground in small groups, searching for their prey. Usually they circle above them several times before landing. Earlier writers also observed this circling action and wove phantastic tales from it. It was said that Vampires would hover on the wing in front of the cow's face, until the victim fell into a hypnotic sleep and so did not notice the blood-letting. Undoubtedly, Vampires set to work with insidious care once they have selected a victim, but they do not use hypnosis. They approach without the slightest sound, select a suitable area for attack and slash the skin so rapidly that the sleeping victim, man or animal, is not even wakened. Travellers who have spent the night in the open without adequate protection have reported that they did not even realize until next morning that they had been attacked by Vampire bats. Blood stains showed where the bats had struck. Since the wounds were made painlessly, the victims had not even been roused from sleep.

This manner of obtaining food makes it essential for Vampires to be agile, not only in the air but also on the ground. They can walk and run more nimbly than other bats, and hop forwards and backwards like small hobgoblins. These characteristics are extremely useful as they approach their prey, and equally so if they are disturbed and must make off rapidly. As they alight on the host, small soft pads of skin on the joints of the hand and soles of the feet cushion the impact. Once the landing has been effected, the body is held well above the surface with only the soft soles of the feet touching the host.

Often Vampires land at some distance from their victim and creep or hop cautiously towards it. Up to fourteen bats have been observed simultaneously on one horse, and a host animal may be bitten more than thirty times in one night. Usually each bat taps its own source. If the blood flows strongly, several bats will lick at one wound. Frequently a number of different bats will visit the same bite in turn. The length of time spent on the host varies greatly, and can be up to 40 minutes.

If the Vampire is allowed to feed undisturbed, it will gorge until its body seems ready to burst and it has difficulty in launching into flight. Sometimes it must wait for a while after feeding until it is able to fly again. Until then, it will creep into some nearby shelter, taking advantage of its agility on the ground.

Laboratory experiments have shown that a Common Vampire consumes 15 to 20 millilitres of blood in a day. This represents almost 40 per cent of its body weight. In the wild, the bat will probably consume more nearly its

own weight in blood. Because blood has a high water content, quantities of this magnitude are needed to satiate the bat. From these figures, it can be concluded that a Vampire takes about 7 litres of blood in a year, and a colony of 100 bats consumes a quantity of blood equal to that of 25 cows or 14,000 hens.

The structure of the bat's stomach is adapted to this manner of feeding. Not only is it long and coiled, but also extraordinarily expansible. Similar adaptations are found in other blood-feeding animals. When Vampires are satiated, they fly back to their day-time roosts, from which they may not emerge until several days later.

Vampires roost either as solitary individuals or in colonies of thousands in old buildings, dark caves and hollow trees. They often share a roosting site with other species of bats. When they return from feeding they devote themselves to a lengthy thorough grooming. The feet are used to clean the fur; flight membranes and thumbs are licked clean by the tongue.

The bite of the Vampire is in itself harmless, and any healthy mammal can withstand a single blood-letting with no difficulty whatsoever. But the general health of an animal can be affected if it is bitten repeatedly. In addition, flies and other insects often lay their eggs in the open wounds caused by Vampires. Serious infection can result. Grazing animals are debilitated and can become worthless. But by far the greatest danger that threatens man and animal is that the Vampires can spread dangerous diseases as they feed. They are greatly feared in the countries of tropical America as transmitters of disease. In addition to tetanus in cattle and an infectious disease fatal to horses, the major disease they spread is rabies.

The leaping capacity of Vampire bats is unique among the Chiroptera. They hop forwards, sideways or backwards like small hobgoblins. In this way, they are able to approach their victims stealthily or to evade them skilfully (sketches after photographs from Leen, 1976).

The incubation period for rabies is about two weeks in Vampires. For some of them, it is fatal, but others survive apparent exposure.

Rabies, transmitted by Vampire bats, threatens whole herds of cattle in Latin America, where it has become a serious veterinary and medical problem. Not the "false vampires" of the horror films but the true Vampires are the ones which, as it has turned out, have given many people sleepless nights. Recent estimates put the annual loss of cattle at a million head, representing a sum of more than 100 million dollars.

In 1966, the World Health Organization decided to send in a team of experts to deal with the problem. The initial measures included the vaccination of large numbers of livestock, but this brought little success. More important were the attempts to combat the transmitters of the disease. The authorities in the U.S.A. began to work out a long-term programme with the aim of recording vampire numbers and distribution, in order to analyze their habits and from the findings, to work out measures of control. Only when more was known about the preferred roosts, the numbers in the colonies and their methods of finding and taking food, reproductive patterns and social behaviour, and the influence of the weather on flight activity, would it be possible to introduce effective measures of control.

Meanwhile, the farmers took action of their own. They tried to destroy the Vampires in their roosts, using dynamite, gas and poison sprays. The widespread but uncoordinated vigilante activity of the cattle farmers did little to reduce the numbers of Vampires; they did, however, succeed in endangering and even destroying many other species of bat.

An American zoologist Greenhall, who had studied the vampire problem for decades, developed a method in which strychnine syrup was placed on the wound. If the Vampire returned to that particular wound, it would ingest the poison along with blood. It proved possible to exterminate a fairly large number of Vampire bats in this way, but success was only partial. Similarly, attempts at shooting or netting Vampire bats brought only temporary respite.

After years of practical research, American scientists developed a substance that causes the death of Vampires by internal bleeding. The anticoagulant, called Diphenadion, is injected into the bloodstream of grazing cattle in small quantities that are harmless to them. But when the substance is swallowed by Vampires as they drink the blood, it causes weakening of the vascular walls and internal bleeding. Coagulation of the blood is prevented and the bats die within two or three days.

On experimental farm stations, where the substance was first tested, there was a distinct decline in the number of fresh bites within a very short time. One great advantage of this material is its long-term effectiveness, since re-injection of the grazing cattle need be carried out only after three to five years.

A second method of dealing with Vampire bats exploits their strongly marked grooming instincts. An ointment again containing an anticoagulant is spread on the fur and wing membranes of Vampire bats that have been caught in nets. The bats are set free and return to their roosts, where they start the extensive process of grooming. This involves licking the fur of other bats, and ensures that many other Vampires also come in contact with the substance and are destroyed. In field tests, 6 bats from a colony of sixty living in a cave were caught, ointment was applied and the bats liberated. Checks made after one week showed that only one bat was still alive in this cave.

The measures of control are organized in such a way that they do not cause the indiscriminate destruction of all Vampires, but merely regulate the size of the bat population. This is essential in areas of intensive cattle farming. Even though all the methods so far developed still have certain drawbacks, chemical control is still a means of providing valuable help to the cattle farmers of tropical America.

The specialization of Vampire bats in blood feeding has shown how this otherwise insignificant creature can, as a result of particular circumstances, become an unexpectedly important economic factor within its area of distribution.

These efforts made to control Vampire bats emphasize how vitally important it is to study in detail the way of life and habits of every animal. Only then is it possible to regulate a population effectively on occasions when the balance of nature is disturbed.

"Seeing" with sound

There is no doubt that the development of a system of ultrasonic echolocation has been of great advantage to bats. Strictly speaking, it is only the Microchiroptera, the "insectivorous" bats, that have this perfect sensory capability at their disposal. The fruit bats (Megachiroptera) use visual orientation. A single exception is that of the cave-dwelling bats of the genus *Rousettus*. In addition to visual orientation, they also possess a well-developed ultrasonic direction-finding system, which they have evolved quite independently of the ultrasonic orientation of the Microchiroptera. In these bats, visual and acoustic systems support and supplement each other. When they return to their day roosts after their feeding flights, they switch from visual orientation to echolocation.

Bats are able to produce a wide range of vocal sounds that are also perceptible to the human ear. If bats are startled from sleep or if they disturb one another, high shrill sounds can often be heard, which can be described as twittering, whining or shrieking. These vocal sounds can also be heard in the reproductive season, as the sexes attract or pursue one another.

But this chapter is concerned with those sounds that human ears do not perceive, since they are in the ultrasonic range.

Making use of ultrasonic echolocation, bats conquered nocturnal air space and a whole new feeding area, to an extent that would have been impossible with visual orientation alone. Now they were in a position to track down the great army of nocturnal insects, even very small species, by making use of their ears. They were virtually in sole command of this abundant food supply.

The use of a system of echolocation is not exclusive to bats. Other animals make use of the same principle, although not to such a great extent, nor with such a degree of perfection. Shrews, for instance, find their way by means of ultrasonic emissions, and various species of toothed whales make use of echolocation in the ocean, in order to detect animal prey and avoid obstacles. It became known about thirty years ago that certain birds, such as the guacharo or oil-bird *(Steatornis)*, which is related to the nightjars, found in the north of South America, and the cave swiftlets (*Aerodromus* spp., formerly included in the genus *Collocalia*) of the order of swifts (Apodiformes) in Southeast Asia and Australia, are able to find their way by a system of echolocation. In contrast to our bats, the pitch of the sounds used by these birds, at about 10 kHz, is within the limits audible to the human ear. Accuracy of the auditory image is not very great. For guacharos and cave swiftlets, as for *Rousettus* fruit bats, echolocation plays little part in the acquisition of food. However, it allows the creatures to colonize underground quarters where no light penetrates. Here they nest in colonies of several thousand individuals, and use their "click" sounds to locate their roosting places. Even though they have large eyes that function in very dim light (scotopic vision), these alone would not enable them to make their way into roosts of this kind. Moreover, the birds use their sense of sight in seeking food, whereas the fruit bats also rely on their sense of smell.

It was only very recently that man was able to solve the mystery of how bats can move about with such assurance and to seek out and catch their prey in the darkest of nights. For 50 million years, they have had their own system of ultrasonic transmitters, direction finders and "radar" equipment, while man has been able to develop these technological aids only in the last few decades and with a vast expenditure of time and energy.

There has been no lack of attempts to discover the secret of the bat's capacity for nocturnal flight. Some of the earliest theories put forward were on the right lines, but it required the sophisticated apparatus of modern scientific practice to provide an explanation of this amazing achievement of the animal senses.

For many people, an explanation was easily found: these sinister "birds" that fly so confidently in the dark, catching even quite tiny insects, must be in possession of magic powers—undoubtedly they are the devil's own creatures.

At the end of the eighteenth century, an Italian naturalist and priest, Lazzaro Spallanzani, in defiance of his superstitious contemporaries, made the first attempt to lay this particular ghost. In his experiments on nocturnal animals, he found that an owl could fly across his room in half-light with no difficulty. But if his room was in total darkness, the owl blundered helplessly into every obstacle. When he repeated these experiments with bats, he discovered to his astonishment, that even in total darkness, they were able to avoid all obstacles in his study, as surely as if they could see them. In addition, he hung threads across the length and breadth of his room, but the result was the same: at no time did the bats strike them. However, when Spallanzani covered the head of each bat with a small hood of opaque material, they suddenly blundered against walls and fell to the ground. From this, he drew the conclusion that bats find their way visually, and obviously can do so even with exceedingly small amounts of light.

It might well have seemed that the problem was solved. However, Spallanzani carried out further experiments and discovered that covering or even removing the bat's eyes had no effect on the powers of orientation.

When the Swiss zoologist Jurine learned about this work, he added one more decisive experiment to the series. He plugged the ears of bats and found that their sense of direction failed. Spallanzani repeated these experiments many times and obtained the same result: bats require their sense of hearing in order to find their way.

But who at that time would believe such an assertion? Nobody was able to detect any sound, so proof of orientation by means of sound could not be furnished. In addition, the renowned and influential French naturalist Cuvier declared himself against the possibility of a nonvisual navigation system, maintaining that the bats in Spallanzani's experiments must have been damaged or disturbed. In Cuvier's opinion, a sense of touch in the body surface or wing membrane was the explanation of the bat's ability to avoid obstacles. At the time, this seemed a much more plausible explanation, and for another hundred years, the problem seemed to be solved.

Even when, in about 1900, two scientists were once again forced to the conclusion that bats are guided in flight by their sense of hearing, the idea that the phenomenon of nocturnal flight depended on sound was not able to gain general acceptance.

In 1920, Hartridge, a British physiologist, put forward the hypothesis that bats emit ultrasonic signals and receive back the echoes of these sounds. Just 18 years later, this fact was confirmed by the American zoologist Griffin working with the physicist Pierce. Pierce had constructed special apparatus that would detect ultrasonics. Insect interference to his studies prompted him to turn his attention to bioacoustics as a hobby. He had found that the shrill "singing" of insects is perceived only partially by the human ear, since it contains many frequencies that lie in the ultrasonic range. Griffin, who was working on bat migration, was interested in recording the shrill calls of bats as well. He obtained a few bats in 1938 and took them to Pierce to test Hartridge's theory. When the two workers placed the laboratory animals in front of the recording equipment, to their great delight, it registered sounds of great intensity, although to the human ear, the room seemed silent. They observed that when the noises were produced, the bat's mouth was always slightly open. So intrigued was Griffin by this result, that he began immediately to devote himself to a study of sound emission, echo detection and direction finding.

He refined the experiments and later, working with Galambos, established that closing the mouth of the bat led to disorientation. So it was clear: the transmission of sound is indeed by way of the mouth, while the larynx is responsible for creating the sound. Since then, it has become known that many species of bats fly with the mouth wide open, since it is in this way that the sound waves are projected. There are, however, a number of species that keep the mouth closed in flight, because in their case, the sound waves are sent out through the nose.

At almost the same time as Griffin, Pierce and Galambos, the zoologist Dijkgraaf working in Holland observed that even without the aid of electronic apparatus, but using only his extremely acute hearing, he was able to perceive the sound transmissions in bats as a quiet ticking like that of a wrist watch. He made the additional observation that the bat's sense of direction fails to function if the transmitter (mouth) or receiver (ear) is rendered non-functional.

So ultrasonic echolocation in bats was proved. Now it was known that obstacles are located by means of probe signals (ultrasonic sounds), the returning echo is received by the ears and thus perception of objects is made possible. The bats build up a sound picture of their environment. In the same way in which a landscape at night becomes visible to us as light thrown out by the headlights of a car is reflected from objects it strikes, so do bats recognize their surroundings hidden in darkness in the echo they receive from sounds they have emitted. They "see" by means of sound waves whether there is an obstacle in their flight path, requiring an alteration in the direction of flight. They "hear" the entrance to their sleeping quarters and the projecting ledge of rock on which they can roost.

After the Second World War, these findings were the prelude in a number of countries to a comprehensive programme of biophysical research into the further intricacies of this extraordinary sensory function in various species of bats.

Not every species transmits on the same frequency

Whereas the human ear perceives only sounds in the frequency range up to 20 kHz, that is, 20,000 cycles (oscillations) per second, the frequency range of the signals emitted by bats extends to 215 kHz. With an increasing number of oscillations per second, the length of the individual waves becomes shorter, and can be as little as 1.6 millimetres. And it is precisely the short waves that are best suited to produce usable echoes from small obstacles. They have the added advantage that they can be concentrated into a beam of sound. If the sound strikes a solid object, the energy density of the echoes bounced back can be quite high.

In order to analyze these sound waves, the scientist can display them on a cathode-ray oscillograph—equipment that depicts the oscillations in visual form on a screen. This will show that the signals made by the Vespertilionid bats that were tested are neither restricted to a single wave length, nor do they show a haphazard wave pattern. Rather are they in the form of a distinctive sequence of frequencies that in many species decreases sharply towards the end. For example, the frequency of the ultrasonic sound produced by the Little Brown bat (*Myotis lucifugus*), common in America, sweeps down from 100 kHz at the start of transmission to only 40 kHz at the end. Within one to two milliseconds, the call covers a frequency range three times as great as the total auditory range of the human ear. It may contain in all only about 50 sound waves, no two of which have the same wave length. If such a sound could be heard by man, it would not be a pure tone, but a chirp.

Many insectivorous bats of temperate and tropical regions belonging to the family of Vespertilionids that have since been investigated, produce this type of sound. This includes most of the European species of Vespertilionidae but not the Rhinolophidae. The emission of salvos of such chirps has led to the bats of this family sometimes being designated as "chirping bats". The frequency range in which the sounds lie varies between individual species, but always shows a drop of at least several kilohertz within the total duration of the sound. Minimum range of a sweep may be 15 or 10 kHz, but no Vespertilionid is known to call below 22 kHz, although some tropical bats are known much lower.

This characteristic modulation of the frequency range has earned another name for the bats that use this type of sound. They comprise the group of "frequency-modulating" or "FM" bats. But the "sound picture" mechanism is not the same in all bats. It soon became clear that diversity in the sphere of ultrasonic echolocation is greater than had at first been suspected. Other families were found to have quite different systems of echolocation, and many species use more than one system.

The Tübingen zoologist Möhres was the first to investigate the calls made by Horseshoe bats (Rhinolophidae), and found that these bats did not produce chirps but long-drawn, pure tones of up to 50 milliseconds duration with a short-frequency sweep at the end. As the bat takes wing or in cruising flight, the constant-frequency part of the pulses of sounds can reach a duration of 50 or even 100 milliseconds. In Möhres' opinion, these sounds are the purest of any produced by animals. They lie at a frequency which is characteristic of the particular species, although slight individual variations may occur. The Greater Horseshoe bat *(Rhinolophus ferrumequinum)* transmits at a frequency of 83 kHz, while the Lesser Horseshoe bat *(Rhinolophus hipposideros)* uses a frequency of 119 kHz.

Möhres made another very interesting discovery. He found that Horseshoe bats emit sounds through the nose, and as they do so, the complex nose leaf that earned the bats their vernacular name, serves to concentrate the high-frequency tones into a beam of sound that can be swept from side to side like a searchlight. The Rhinolophids can take bearings on their environment with great accuracy. What had seemed at first to be a useless ornament, a whim of nature, was suddenly shown to be an extremely practical, functional element in a sophisticated system of "auditory viewing".

The nasal excrescence carried by the family of Old World Leaf-nosed bats (Hipposideridae) has been found to fulfil the same function. Because of the different form of the nose leaf it does not focus the sound waves as sharply as that of the Rhinolophids. Analysis of the sounds showed a further variation in the Hipposiderids: the sound is in two parts. The first part is a very high, pure tone of 120 kHz. The final part corresponds to a chirp with a frequency range of 40 to 50 kHz. With this double system of acoustical location, the Rhinolophids and Hipposiderids are able to obtain information about nearby objects by means of the chirping part of the signal, and about objects considerably further away by means of the constant-frequency component.

In recent years, the zoologist Pye in Britain has examined the acoustical behaviour of representatives of many bat families. He established that both the frequency-modulating and the constant-frequency principles are met with in various families. Whether the sounds are emitted through the mouth or the nose does not affect the type of frequency pattern found. According to findings, certain representatives within individual families use the mouth as a transmitter, while other species in the same family emit sounds through the nose. Some groups have been found that use both mouth and nose as transmitter; for example the Long-eared bats *(Plecotus)*.

Measurement of distance by sound waves

In addition to the frequency and the ability to modulate it, another remarkable feature of many bats is the short duration of the ultrasonic sounds they emit. Within one millisecond, the sound waves travel 34 cm through the air. The echo from an obstacle at this distance returns to the bat's ear within two milliseconds. Since it is essential for the bat to hear the echo that returns from an obstacle or a quarry after its own emission of sound has ceased, these sounds must be of extremely short duration. It is quite possible for a chirp pulse to last for only 0.25 or at most 1 millisecond when an obstacle lies directly in front of the bat. The short, sound-free intervals are sufficient to prevent the sounds that are emitted and the echo that is returned from overlapping. Those species capable of frequency modulation judge the distance of an object by measuring the time delay between the outgoing sound and the returning echo. Information about the direction in which the reflecting object lies or in which an insect is flying is probably obtained from a comparison of the relative intensities of the echo reaching the right and the left ear. Of course, analysis of echoes received is not as simple as outlined here. In this sphere, there are many questions still unanswered today. One need consider only the

factor of time delay in echoes returning from objects at varying distances that would cause overlap.

The method of assessing range described here is problematical in the case of Horseshoe bats. Since in comparison with Vespertilionids, the tones they emit are of long duration, the echo returns while the sound is still being emitted. They have developed a different principle of echo reception. Their highly mobile ears move independently as they are turned towards the approaching echo and used as directional receivers. Depending upon the distance at which the obstacle lies, the echoes return with varying intensities, and from these differences in intensity, the Horseshoe bats are able to assess the distance to an object. In addition, an echo that an object reflects as it lies at an angle of 20° in front and to the left of the animal will have its maximum effect on the left ear, while its effect on the right ear will be less. These differences in intensity enable Horseshoe bats as well as Vespertilionids to assess the direction in which the obstacle lies. Anyone who is fortunate enough to observe a bat, particularly a Horseshoe bat, at rest before it takes flight, will be able to see from movements of the body, head and ears that for some considerable time beforehand, it emits ultrasonic sounds and thereby "examines" its environment.

Potentialities of ultrasonic echolocation

As bats fly into a cave, they emit 10 to 30 sound pulses per second. They have built up such an accurate sound picture of their day roost that they are familiar with every obstacle. Using their efficient place memory, they remember every detail, and so are able to move through their quarters without the aid of vision. It is estimated that, with their refined system of location, Horseshoe bats can perceive obstacles from a distance of 8 to 10 me-

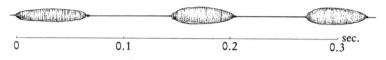

Diagram showing oscillographs of ultrasonic echolocation pulses. Three sounds emitted by a Horseshoe bat (*Rhinolophus* sp.). These are pure tones of comparative long duration that can last up to 0.1 seconds. Frequency is remarkably constant diminishing only in the final 1.5 milliseconds (from Kulzer, 1957).

tres. Vespertilionids, with their bursts of sound, cannot rival this. They must come closer to an obstacle before reacting to it. If the bats are approaching a place at which they want to land, or if unexpected obstacles present themselves in their path, the emission of sound increases immediately to between 50 and 100 pulses per second. In this way, bats are able to continually gather up-to-date information about the unknown object.

Laboratory experiments have shown the amazing feats of which they are capable. A number of threads —decreasing in thickness in the course of these experiments—were hung close together across a space, to see when the bats would be unable to avoid contact with them. The first experiments carried out with Vespertilionids brought striking results. Not until they reached threads with a diameter of only 0.10 mm—the thickness of a human hair—did the bats touch them. Later experiments with Greater Horseshoe bats showed that this species even outshines Vespertilionids. They were still able to detect threads with a diameter of 0.05 mm.

This also explained why it had not been possible, until a few years ago, to catch free-flying bats in the nets used

to catch birds. The threads were much too thick. Experiments show that bats respond to threads of 3 mm thickness from a distance of 2 m, and threads of 0.18 mm from a distance of 1 m. So they still have sufficient time to turn aside. Only since the finest of nylon nets have been used, has it been possible to catch bats on the wing.

Analysis of the calls made by bats has shown, in addition to the findings mentioned above, that volume of sound is not uniform in all species. Particularly in those that catch flying insects or other animal prey, the ultrasonic sound is of high intensity. These bats screech through the night. Other species, for example fruit and nectar-feeders, produce much softer pulses, so that it is often difficult to record them. Since fruit-eating bats track down their food with the nose, these "whispering" tones are sufficient as a means of detecting obstacles.

The group of "whispering bats" consists of a wide variety of bats—including our Long-eared bats (*Plecotus* spp.) and the False Vampires (Megadermatidae)—producing a similar quality of sound for a wide variety of reasons. Vampires also belong to the group of "whispering bats". Because of their particular feeding habits, it is adequate if they can locate obstacles from a distance. So the sounds they produce are of low intensity. For Vampires, the senses of sight and smell are more important.

A series of sounds emitted by the Large Mouse-eared bat *(Myotis myotis)*. The sounds begin with a sudden increase in pulse frequency and gradually die down. In 0.1 seconds, some 10 such sounds (chirps) are emitted. Duration of the individual sound is at most 0.005 seconds.

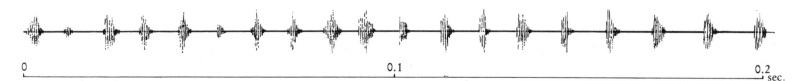

How do fish-eating bats locate their prey?

For a long time, the question of how or even whether fish-eating bats use their system of ultrasonic echolocation to catch their food was much discussed. It was known that the Fisherman bat *(Noctilio leporinus)* emits short sounds of high intensity through the mouth as it flies low above the surface of water. Observations of bats while they were fishing showed that this large bat does not rely on chance alone when it dips its hind feet with the sharp claws into water in order to catch a fish. Could it locate a fish swimming under water so accurately by means of sound waves? It is a fact of physics that only a small proportion of sound waves passing through the air will continue their path when they strike water. The surface of the water reflects almost the entire sound, and only about 0.1 per cent passes into the water. Similarly only 0.1 per cent of any echo would break out through the water/air interface. It is highly improbable that an echo from this faint sound bounced back from a fish would be perceived even by the sensitive organ of hearing of a bat.

Experiments carried out on bats in captivity finally solved the problem. Fisherman bats *(Noctilio leporinus)* were taught to take pieces of fish attached to the end of wires. When the wires were placed in a tank of water in such a way that the pieces of fish were entirely submerged, the bats were no longer able to locate their food beneath the surface. But when the surface of the water immediately above the bait was caused to move even slightly, the bat plunged its feet into the water and seized the prey. Clearly, *Noctilio* is not able to locate a completely submerged fish by means of ultrasonic echolocation, but is able to perceive the slightest of ripples in the water. So when a fish swimming beneath the surface causes waves on the surface, the fishing bat has no difficulty in locating the fish and seizing it.

Recognizing food by sound waves

The location of obstacles by echoes and the location and capture of insects are undoubtedly two skills of a different order. For years, scientists have been occupied with the problem of how bats are able to track down their prey at night. Spallanzani was one of the first to try to find an answer. His notebooks from the year 1794 tell how he blinded a number of bats and then released them. They returned to their roost in a bell tower in Pavia. Some days later, when he caught three of the test animals and examined the stomachs, he found that those of the blinded bats contained as many insects as those of normal bats. He was convinced that the bat's visual sense plays no part in the catching of insect food. But how could the insects be detected?

Later on, various scientists assumed that bats, with their acute hearing, are able to perceive the sounds made by insects in flight. But Griffin showed that this was unlikely. When he started to study the orientation sounds emitted by bats in the open country as well, he observed that bats catching insects intensified their succession of location calls, just as they did in manoeuvring to avoid obstacles. In experiments, the American Big Brown bat *(Eptesicus fuscus)*, a species closely related to the Serotine *(Eptesicus serotinus)*, was found to increase the rate of succession of its sounds as it flew towards prey from 10 to 150, even to 200 calls per second.

In the early sixties, after a good deal of preliminary work, Griffin and his colleagues were able to produce a laboratory analysis of the typical hunting behaviour of the Little Brown bat *(Myotis lucifugus)*. They released thousands of fruit flies *(Drosophila)* or mosquitoes *(Culex)*, and immediately the circling bats began to pursue them. From a distance of as much as one metre, the small insects were located and hunted with precision. The

Pipistrelles *(Pipistrellus pipis-trellus)* eating mealworms. Even this small species of European bat can learn to eat its food from a dish.

78 A Little Brown bat *(Myotis lucifugus)* pursues a Tiger moth *(Apantesis virgo)*. With acrobatic flight manoeuvres, it uses its wings to try to force the prey into the tail membrane.

79 Two Fruit bats of the genus *Rousettus* in captivity, eating a banana.

80   Nectar bat (*Leptonycteris sanborni*; Phyllostomidae) visiting a cactus flower. To reach the nectar at the base of the flower, the bat must insert its head deep into the blossom.

81   The long tongue with bristle-like papillae at the tip is well adapted for feeding in this way.

82   When the bats leave the flower, the head is thickly coated with pollen. Part of it is licked off and eaten, with the rest the next flower is fertilized.

83   A Little Brown bat (*Myotis lucifugus*) catching a moth (*Arctia caja*)

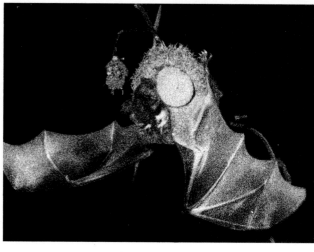

84 Pale Spear-nosed bat *(Phyl-lostomus discolor)* on a flower of *Hymenaea courbaril* (locust tree) in the Amazon region. The bat inclines its head over and down into the blossom, and licks up nectar in a second. The umbellated flowers are separated from the leaf shoots and so are easily accessible to bats.

85–87   The Pale Spear-nosed bat *(Phyllostomus discolor)* visiting a spadix of mimosa *Parkia auriculata* (Brazil), which consists of a large number of individual flowers. The blossoms extend beyond the foliage on long stems and so can be reached easily by the bats in flight.

The animals are scarcely able to gain a foothold on the flowers, so each visit lasts less than a second. Within this time, the bat licks nectar from the base of the fertile spadix while pollen from the blossom is transferred to its cheek, neck and chest.

88   Serotine *(Eptesicus serotinus)*. It is typical of the Vespertilionids that they emit ultrasonic waves through the open mouth in flight.

89   Brown Long-eared bat *(Plecotus auritus)* in flight. Among Vespertilionids, Long-eared bats are exceptions in that they can also emit sound waves through the nose. This bat is flying with the mouth closed.

135

90 Large Mouse-eared bat *(Myotis myotis)* in flight. Through its open mouth, it sends out sound waves into the night in order to locate obstacles and prey.

nearer the bat came to its prey, the greater was the rate at which the succession of sounds accelerated. At intervals of one or two seconds, and at the very moment when an insect was seized in an intricate flight manoeuvre, the cries reached a crescendo of sound which we hear as a "buzz" on a bat detector. In these experiments, it was also found that the bats are well able to distinguish between objects that are edible and those that are not. Apart from the insects mentioned above, mealworms, the larvae of meal beetles, were also thrown into the air. After about a week, the bats had learned to eat the mealworms. Now a number of plastic discs 3 mm in thickness and 16 mm in diameter were mixed with the mealworms which are about 2 cm long and 3 mm thick; and this "food mixture" was then tossed into the air. At first, the bats located both the mealworms and the plastic pellets, and rushed at the latter as if they were food. But within no more than a week, the first bats had learned to distinguish live food from simulated. 80 to 100 per cent of the mealworms were caught, while the plastic discs were rejected with a high degree of reliability (80-90 per cent). It is still not clear how the bats were able to distinguish one component from the other in the food supply.

The intricate and varied series of experiments illustrated definitively that bats locate and catch their insect prey by using echolocation.

This, of course, does not preclude the possibility that bats also perceive sounds made by insects. The zoologist Kolb in Bamberg produced some evidence for this. He was able to show that the Large Mouse-eared bat *(Myotis myotis)* could track down beetles that were not flying, but rustling among the floor litter of the test chamber.

The bat flew towards them accurately and then tracked them by nose.

Today, we are still a long way from being able to explain the entire phenomenon of ultrasonic echolocation. As each question is answered, the small creatures present new problems to the scientist. Bats are capable of performing feats that seem inexplicable to us.

The precision and accuracy with which the system of echolocation functions is of vital importance to them. Not only have they solved the problem of creating and emitting ultrasound, but also—and this is perhaps even more important—the problem of receiving back their own echo and distinguishing it from a multiplicity of competitive sounds. One can only marvel at the way in which even large numbers of bats flying about within an enclosed space seem not to disturb or confuse their fellows with their ultrasonic calls.

It is a remarkable achievement that the sense of hearing and the responsible brain centres are able to select a feeble echo out of the complex background of competitive noise made by other bats. Using two input channels, the ears, the brain can analyze accurately the direction of incoming noise and distinguish between the echo of its own location sounds and background noise that reaches the ear from other directions. It is not surprising that the acoustic centres of the bat's brain, which itself often measures no more than half a cubic centimetre, are highly developed and specialized. Research into the fine structure of these centres and into the functions of the brain that are the basis of the bat's amazing performances will continue to occupy scientists for many years to come.

Breeding habits

In previous chapters, so many peculiarities in the life of bats have already been described that it comes as no surprise to learn that the process of reproduction in bats also deviates from the general mammalian norm.

At about the turn of the century, scientists knew little about reproduction in bats. However, one of the first descriptions of parturition was given by the French naturalist Pierre Belon as early as 1555. Many of his observations accord well with accepted scientific findings of today. Until about twenty years ago, all that was known for certain about the reproductive biology of bats concerned various members of the family of Vespertilionid bats (Vespertilionidae) indigenous to Europe and North America and European Horseshoe bats (Rhinolophidae). But it is precisely these forms that show particularly great variation in their reproductive patterns. It was at one time assumed that the oestrus cycle in species of temperate latitudes did not begin until the spring, when they waken from hibernation. Only later was it discovered that most copulation occurs in autumn, although maturation of the egg does not take place at this time.

Since the habits of the many species are so varied, these findings could not be assumed to cover the reproductive biology of all Chiroptera. Recent research into tropical species in Africa, Asia and South America shows that there are many species-characteristic features of reproduction determined by geographical, primarily that is climatic, factors. For fruit bats in tropical regions, no separation in time has been recorded between insemination and fertilization. The deferment of fertilization until the spring in species living in temperate latitudes is a secondary adaptation to climatic conditions.

Findings on sexual behaviour, fertilization, period of gestation, birth and care of the young cannot be given general application. In many cases, information is available only on individual phases of the reproductive process. A good deal is known about the periods in which the bats assemble to give birth and rear young. In contrast, little is known of those times in which the bats disperse and live in isolation. Observation of tropical species of bats is rarely carried out continuously throughout an entire year, so that our knowledge of the reproductive biology of tropical bats is often fragmentary. Consequently, details described here apply only to selected representatives of individual families.

In temperate latitudes, the Microchiroptera do not possess secondary characteristics by means of which it is possible to distinguish the sexes. But in the tropics, in addition to those species that show no clear sexual dimorphism, there are some in which the males differ from the females by being larger and more intensely coloured. The nose leaf of the male of the Old World Leaf-nosed bat *Hipposideros larvatus* is said to differ in form from that of the female. In certain species, a localized, dense hair growth distinguishes male from female. The shoulder hair of the Epauletted Fruit bat *(Epomophorus)* has already been mentioned. In the Collared Fruit bat *(Myonycteris)*, the throat region has a dense covering of coarse hairs, and in various of the Tomb bats *(Taphozous)*, the males have such a distinctive "beard" that they can be recognized easily even at a distance.

Another sexual characteristic is especially important to bats in the breeding period, although it is barely detectable to the human senses. The bats have special skin glands that produce a strong-smelling, sexually stimulating secretion. In the mating season, these odorous substances are applied to the fur, wing membranes and even objects in the environment, in order to attract and stimulate sexual partners.

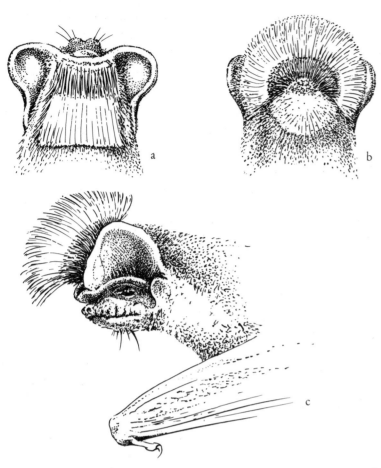

Adult male of the species *Chaerophon chapini* (Molossidae). An example of sexual dimorphism, a rare phenomenon in bats. Here, the male has a tuft of long hairs on the head (a) that stand erect during the phase of breeding (b, c) (after Allen, Lang and Chapin, from Brosset, 1966).

Factors affecting the reproductive cycle

In temperate latitudes, the reproductive process in bats, as in many other animals, is strongly affected by the annual seasonal cycle. In Europe, bats have only a few months during the summer in which to give birth to and rear their young. The food available between April and June must be used by the female to ensure the successful development of the embryo. An abundant supply of insects is vital in June and July to provide the mothers with enough nourishment to be able to feed the infant bats, while in August and September, the young bats that have become independent, as well as their parents, must consume a sufficient quantity of reserves to see them through six months of hibernation. The short summer of the Palaearctic and Nearctic regions provide just sufficient time to raise one litter.

The situation is more favourable for tropical species. They are not under such pressure in tending their young. Nevertheless, an annual rhythm of reproduction can also be observed in species living in regions with a very constant climate. The female again produces young only once each year. The birth occurs in the spring, or in the season that corresponds to our spring. This has produced an interesting finding. The same species that in the northern hemisphere gives birth in May/June, does so in the southern hemisphere in September/October. As the French zoologist Brosset discovered, the geographical equator does not coincide with the biological equator of bats. In Africa, a few degrees north of the equator, he observed species of *Rhinolophus* and *Hipposideros* with a typical seasonally-controlled rhythm of reproduction. But, like the populations south of the equator, they produced their offspring in September. Tomb bats *(Taphozous)* which in Bombay give birth to young in April, show a distinct delay in their reproductive cycle in Sri Lanka. Here, they do not bring their offspring into the world until September.

Particularly in the equatorial region, there are various species that are polyoestrus, they reproduce several times in a year. Among Microchiroptera, certain members of the Slit-faced bats *(Nycteris)*, Spear-nosed bats *(Artibeus, Glossophaga)*, Free-tailed bats *(Molossus)* and Sheath-tailed bats *(Taphozous)* have two mating seasons in the year. Most of these genera have two birth peaks: the first time in spring; a second birth follows towards the end of the summer.

Many species, Vampires for instance, show no recognizable annual rhythm. All year round, the colonies include pregnant and lactating females. In various species of fruit bats among the Rousette Fruit bats *(Rousettus)*

139

and Short-nosed Fruit bats *(Cynopterus)*, the process of reproduction in many areas is not seasonal.

The important factor determining the frequency of parturition is undoubtedly the supply of food. Insectivorous or frugivorous species must produce their young when an optimal supply of food is available. For the sanguinivorous species, this is no longer important, they can find host animals all year round.

Since the majority of tropical species have a one-year reproductive cycle, the question arises of what are the factors that control the rhythm of reproduction in these species. It can scarcely be length of day, since close to the equator, day and night are of almost equal length. Nor do the seasons show such fluctuations in temperature and food supply as they do in northern latitudes. The rainy season might be a factor, but there is no evidence that this is the case. Clearly, this is a question still to be answered.

Sexual maturity and reproductive behaviour in bats

As far as is known today, European bats reach sexual maturity in their second year of life. Mating occurs first at about 14 months and the females are about two years of age when they produce their first young. But there are exceptions in this respect; *Rhinolophus ferrumequinum* reaches maturity mainly in its fourth year, whereas a few species, such as Pipistrelles *(Pipistrellus pipistrellus)*, have been found to give birth within their first year. Apparently only the female matures at such an early age; the male is not ready for mating until its second year.

Similarly, most fruit bats do not reproduce until their second year. Only the Rousette Fruit bat has been observed to copulate before it is mature.

Information on sexual behaviour and reproductive biology in bats in temperate regions has been obtained

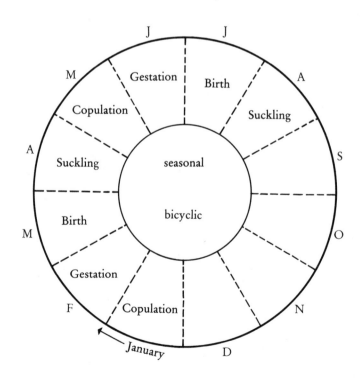

both by observing them in the field and studying them in captivity.

In Europe, when the young bats become independent in July or August, the maternity colonies gradually disintegrate. Old and young bats seek out different, temporary quarters before they make their way to the winter roosts. In these intermediate quarters, small groups consisting of males and females can be found. The sexual organs of the mature bats show that they are in a state in which fertile mating is possible. It is assumed that in many cases, pairing occurs there. Other indications suggest that the majority of females reach the winter roosts unmated, and that copulation occurs during the period of rest in the winter.

Mating in bats, in so far as it has been observed, is initiated by the male. In courting the female, the male makes use of acoustic as well as of olfactory and tactile stimuli. Since acoustic stimuli would have no effect on a sleeping female in winter quarters, it is probable that the sense of smell plays an important sexual role at this time.

Observation has shown that the selected female is awakened by bites on the neck, and when she has become active, copulation takes place. If the two sexes meet in their day-time roost, the male approaches the female and rubs his head against the side of the female's body. Usually this approach is accepted passively by the females, but those of certain species resist violently. Before copulation, the male of the Egyptian Fruit bat (*Rousettus aegyptiacus*) encloses the female in its wing membranes. The female objects to this strenuously and, shrieking, tries to escape. But usually, the male manages to seize hold of the female by an adroit bite on the back of the neck, and copulation can take place.

Tropical species mate at the day roost and only rarely outside it. The American Red bat (*Lasiurus borealis*) is alleged to copulate in flight. This seems unlikely.

Since bats are essentially promiscuous and apparently have no order of rank, it often happens that a single individual in breeding condition, male or female, may participate in acts of copulation with different partners several times in succession. Selection is entirely random and the partners probably never come into contact again.

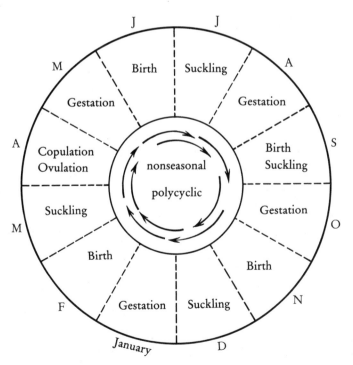

Reproductive cycles in various bats (in part after Wilson, 1973).
Seasonal, monocyclic reproduction in European bats. It is adapted to a food supply restricted to the summer half-year.
Seasonal, bicyclical reproduction in neotropical fruit-eating bats. No young are born in periods of reduced food supply.
Nonseasonal, polycyclical reproduction in Vampire bats. The individual stages of reproduction overlap and are not tied to particular months, since food is available to the bats all year round.

Problems of embryonic development in temperate latitudes

In those species that inhabit tropical regions, ripening of the sperm and ripening of the ovum are synchronized in such a way that fertilization takes place at insemination, and development of the embryo can begin at once. For bats of temperate regions, most of which mate in autumn, it would be unfavourable if embryonic development were to start immediately. The young would be born in the winter and would perish. So either fertilization does not occur at the time of copulation in autumn, or if it does, embryonic development is delayed or prolonged.

Either way, the result is that the offspring is not born until the early summer. Observation of indigenous species in captivity has shown that this involuntary delay in the process of reproduction is induced solely by the cold period of the winter months. If females that have been impregnated in autumn are kept in a warm room in winter and fed adequately, they remain awake and give birth to young much earlier than other members of their group living in the wild. That is, ripening of the ovum has started immediately and fertilization and normal embryonic development have followed.

These observations show that the sexual cycle in male and female does not normally progress synchronously. Whereas the male possesses fertile sperm from autumn until spring, it is only at the end of hibernation that ripening of the ovum in the female is initiated by means of a hormonal control mechanism. However, in most cases, insemination has already taken place months before this. The sperm must remain alive for five to seven months within the genital tract of the female, so that they can fertilize the egg in the spring. This is a phenomenon unique among mammals.

In spring, when the bats of temperate latitudes leave their winter roosts (hibernacula), the females group together and seek out their summer quarters, where they live in maternity colonies and give birth to their young. These quarters are familiar to the females; they return there year after year because the microclimate is a particularly favourable one. So strong are the bonds that attach the species to their summer quarters that they will even undertake long migrations from their winter quarters to their summer dwelling.

Maternity colonies—the "nests" of bats

Female Guano bats (*Tadarida brasiliensis*) cover distances of more than 1000 km in order to fly from Mexico to the southern states of the U.S.A., where they give birth to their young. Whereas in Europe, a maternity colony will consist of between a half-dozen and several hundred roosting females, the largest assembly of female Guano bats in the Eagle Creek Cave in Arizona at the beginning of the sixties was reported by American zoologists to comprise more than 20 million bats. The largest sea-bird colonies do not achieve concentrations on such a scale, and among terrestrial animals, they are found only in certain insects that live in social groups. The largest known "maternity colonies" of fruit bats are the mass assemblies found on trees at Cape York in Australia. It is estimated that four million bats rear their young there.

Observation of our indigenous species has shown that patterns of behaviour within the maternity colonies are very varied. On hot days, the colonies are loosely distributed, roosting under roofs or behind window shutters. When it is cool, the animals move closer together. The bats in a roost form a close community. But the nature of the bond that unites them is not clear. If they are constantly disturbed, the whole company makes off and settles in a different hanging area. In all, little is known of the behaviour in a maternity colony. It is, of course, very difficult to observe such large colonies in half-light and over a long period of time, particularly as they are in constant movement. In summer, the males are rarely to be found. They go their own way and take no interest in the fate of their progeny. The males of some species, such as the Parti-coloured bat *(Vespertilio murinus)* are also reported to live together in small colonies in the summer. In the Lesser Horseshoe bats, a number of males have also been observed living in the female maternity colonies.

In general, the number of young is limited to one. Twins occur rarely and multiple births have so far been observed only in a few species. The *Lasiurus* species of North America are recognized as particularly fecund. In their case, a litter may comprise two, three or four young.

This high number is not exceptional. The fact that the females each possess two pairs of nipples indicates how well they are adapted to a larger litter. Even in those species that bear only a single young, several egg cells (ova) are produced—in *Pipistrellus*, sometimes as many as seven. But it is not clear whether the remaining eggs are discharged unfertilized, or resorption of fertilized eggs takes place in the uterus. Only the embryo that is most favourably positioned within the uterus has the chance of further development.

The small number of offspring in bats represents an adaptation to flying. A gravid female carrying several embryos would be seriously impeded in its nocturnal feeding flight, which requires endurance and agility.

Various factors contribute to maintain bat numbers, in spite of the small number of young produced; for example, the considerable age to which bats can live, the long duration of reproductive capacity, and the small number of natural enemies. As to longevity, results obtained from bat banding have shown that the Greater Horseshoe bat *(Rhinolophus ferrumequinum)* can attain an age of 23 years, the Large Mouse-eared bat *(Myotis myotis)* 18 years and the Pipistrelle *(Pipistrellus pipistrellus)* 11 years. Large species of fruit bats have lived for more than 20 years in zoos. These maxima are only rarely attained. Average life expectancy is lower, standing perhaps at half these figures. Rats, mice and other rodents are known to live at most to only three to four years. Their numbers are threatened by many more enemies than are those of bats. It is not surprising that in order to maintain the species, rodents produce a higher number of young per litter and also more litters per year.

Environmental influences affect the period of gestation

For many species of bats of the Palaearctic and Nearctic regions, it is difficult to determine the duration of pregnancy, if impregnation occurred in autumn and it is not known at exactly what time in the spring fertilization took place. It is easier to determine the period of gestation for species of warmer regions, although here again, climatic factors can affect embryonic development. In comparison with other mammals that also give birth to young that are still considerably underdeveloped (nidicolous or altricial young), the long period of gestation in tropical fruit bats is surprising. It is five to six months for the large Flying Foxes, while smaller species give birth after four months.

Among the Microchiroptera, a gestation time of about three months has been estimated for small species of Old World Leaf-nosed bats (Hipposideridae) in India, but the Vampires of Central and South America do not give birth until after six to eight months. In cases where it has been possible to discover the period of gestation in species of temperate latitudes, it is found to be significantly shorter than for related species in hotter regions of the world. In temperate zones, a much extended pregnancy would be a disadvantage in biological terms, since the young must not only be born but also reared within the short summer months. Once independent, the young must make use of the late summer and autumn in order to achieve a peak of physical condition that will enable them to undertake the migratory flight to winter quarters and to overwinter there.

Small species *(Pipistrellus)* are pregnant for about 45 days and larger species *(Myotis)* for 50 to 65. But in temperate latitudes, climatic factors can cause great fluctuations in the period of gestation. If cold, wet weather in spring forces the female to reduce her activity and lower her metabolic rate, there is a delay in the development of the embryo. As a result, the period of pregnancy is extended. This environmentally induced retardation has been demonstrated experimentally. A moderate length of time spent in a warm room initiated the development of the embryo. After some time, a proportion of the animals were returned to cooler quarters (4 to 8 °C). They entered a state of hibernation and were examined three weeks later. It was found that in comparison with the animals that had continued to live in warm conditions, development of the embryos was distinctly retarded.

The phenomenon of delayed implantation, that is also found in other species of mammals, can be observed in the wild in those species of bats that lived originally only in warm areas, but which spread later into temperate latitudes. A typical example is the Long-winged bat *(Miniopterus schreibersi)*. In this species, the sexes have retained a synchronized reproductive cycle. When the female is mated in autumn, the ovum has already developed and fertilization is possible. But development of the embryo stops in the winter at the blastocyst stage, and the birth does not take place until early summer, that is, only when favourable conditions exist for rearing the young bat.

It is somewhat surprising to learn that this delayed embryonic development is also found in one tropical species. In the African Straw-coloured bat *(Eidolon helvum)*, fertilization takes place in early summer. Cleavage of the ovum, however, begins only in the rainy season in autumn. Thus, the embryo grows at the same time as the vegetation, and the young bat is born at a time of abundant food supply. Since embryonic development stagnates for about four months, and another four months pass before the birth, this species has a very long period of gestation of eight to nine months.

Links between mother and young

The process of birth and the position taken up by the female in labour differ among individual species. Most mothers do not give birth in the normal roosting position of hanging head downwards. In the maternity colonies, they have been observed to clamber up and away from the mass of bats, and to hang by the thumbs with head and body directed diagonally upwards. The young bat can emerge either head or hind quarters first, although breech deliveries are believed to predominate. During the birth, the female repeatedly attempts to assist the expulsion of the young, using the mouth or a hind leg. When the small, naked bat emerges, it does not fall into a warm nest, but in many species is caught in the tail membrane of the mother. The membrane is spread and curved into a pouch, and prevents the young bat from falling to the ground. As a result of their adaptation to life in the

92 Egyptian Fruit bat *(Rousettus aegyptiacus).* These bats are about to copulate. The male has enclosed the female within its wings.

93 A pair of Egyptian Fruit bats with young, engaged in grooming.

94 Egyptian Fruit bat twins at the mother's breast

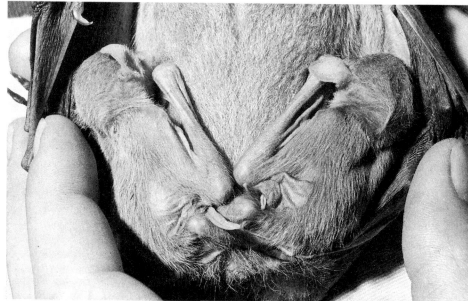

96 Pipistrelles *(Pipistrellus pipis-trellus)*. Two seven-day-old young. Pelage has not yet developed; the wings have grown somewhat, but are not yet functional.

From the first day of life, the pointed claws of the large back feet allow the bats to gain a firm hold as they cling to the fur of the mother or when they are left hanging alone at night in the roost.

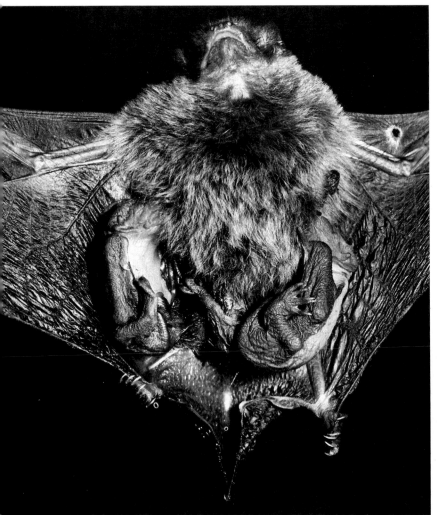

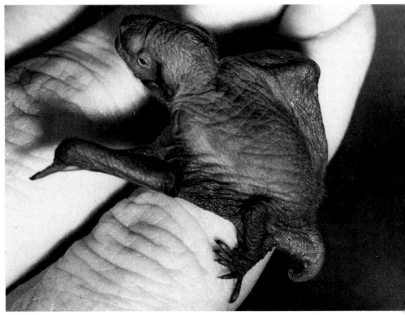

97 Pipistrelles *(Pipistrellus pipis-trellus)*. Female with new-born twins that have fastened themselves firmly to the breast.

98 Pipistrelle, three days old. The eyes are open, hair is scant, the wings still undeveloped.

100 Large Mouse-eared bats *(Myotis myotis)*. The young, about one week old, are still blind and almost hairless. They have been left at the roost while the mothers make their nocturnal flight in search of food.

101 Maternity colony of the Greater Horseshoe bat *(Rhinolophus ferrumequinum)*. The young are still incapable of flight. They have been left behind alone in the roof by the mother bats that are out catching insects.

102 Maternity colony of Large Mouse-eared bats (*Myotis myotis*) beneath a roof ridge. This large assembly consists mainly of young bats almost old enough to fly.

151

103　Angola Free-tailed bat *(Tadarida angolensis)*. Female with young on its back.

104　Angola Free-tailed bat. Young bat, a few days old, still naked. The clinging apparatus in the form of the large back-feet and the thumb are already well formed; the flight organ still incompletely developed.

air, bats have lost the ability to prepare a soft nest for their young. The mother herself provides a warm and protective refuge. The new-born bat crawls straight away into the fur of the belly and is covered by the wing membranes. Soon after the birth, it is hungry and seeks out the source of milk. It attaches itself to one of the two teats that are situated on the chest, and immediately begins to suck. Meanwhile, the mother licks the young clean and bites off the umbilical cord.

Although the period of gestation is long in comparison with that of many rodents, young bats at birth show a state of development that places them, together with mice and rabbits, in the category of nidicolous animals. In most species, the young are still hairless and blind, but already fairly large. Birth weight can be one fifth to one third of the weight of the mother. They are not yet able to fly, but crawl about in the parent's fur and maintain a firm hold with their relatively large feet and their thumb claws. Usually the mouth is attached firmly to a teat, providing a secure support.

Brood care is well developed in many fruit bats. For months, the instinct to hold fast to the mother is dominant in the young. In contrast to insectivorous bats, fruit bats take their young with them in the first few weeks on nocturnal flights in search of food. If the young bat moves too far away from the mother in the day roost, the latter calls it back again. Conversely, the young bat is able to communicate with the mother by characteristic sounds indicating that it is lost ("isolation calls"), whereupon the latter will begin to search for it. Since parent and young recognize one another by very specific scents, the bats in the large tree roosts always know their own.

The small bats attach themselves to the mother in a variety of ways, depending upon whether the mother's day roost is in the open or in a protected position inside a cave or building. Among fruit bats, they are often con- cealed beneath the protective wing membranes. In those species that do not hang freely from the branch of a tree or a roof, the baby bat finds a place of safety between the side of the belly and the surface against which the mother is hanging. In many of the Free-tailed bats, the mother carries its young on its back. Infant Horseshoe bats (Rhinolophidae) and Mouse-tailed bats (Rhinopomatidae) find special clinging aids on the body of the mother. She has two false teats in the pubic region which the young grip, attaching themselves by suction, during periods of rest.

Most species of bats show highly developed links between mother and young. Long-eared bats (Plecotus) or Pipistrelles (Pipistrellus) that have lost their young have frequently been observed to search for it intensively, and like fruit bats, they are able to locate it at quite considerable distances. Both acoustic and olfactory signals ensure recognition between parent and offspring. A new-born bat can already emit high-pitched sounds that lie just within the limit audible to the human ear. They are known as contact vocalizations or isolation calls.

For a long time, it was generally believed that in their first few days of life, young insectivorous bats were carried along with the mother at night. There are even photographs showing female bats in flight with their baby. But these pictures do not reflect the true situation. If, during the day, as the young bat clings to the mother, there is some sudden alarm, the female flies off with the young. But under normal conditions, the offspring is not taken on the nocturnal flights in search of food. From the very first evening, the new-born bats must remain behind alone. At this time, they cluster closely together and await the parent's return. Although they have, as yet, no system of heat regulation, they are not especially sensitive to low temperatures. When they are six days old, they are able to raise their metabolic rate spontaneously for the first time, but only when they have developed a full covering of fur are they capable of complete temperature control and able to adapt their body temperature to existing environmental conditions.

When the females return from their feeding flight, the piping calls of the young direct them back to their abandoned offspring. The mothers reply to these cries, land close to the young, crawl up to them and distinguish their own individual offspring by its scent and sound. The young immediately creeps to the mother and is provided with food, warmth and security.

In very large colonies of Vespertilionid and Free-tailed bats, the young bats hang together in flocks that are separated from the roosts of the mother bats. In the case of the Long-winged bat (*Miniopterus schreibersi*), these flocks can comprise tens of thousands of young. In the American Guano bat (*Tadarida brasiliensis*), there can even be millions of these small, naked, rosy bodies hanging close together in several large groups for mutual warmth. In face of such masses of young, the mothers are no longer able to seek out their own offspring. All accept shared responsibility for rearing this vast progeny.

When the female Guano bats return at dawn from their hunting flight, they land among the masses of hungry young. Those closest to them rush to this milk source to drink their fill. It hardly ever happens that a young bat fails to be fed, since two are able to be suckled by one female at one time, and the quantity of milk produced daily by the female can be as much as 16 per cent of her body weight. This is without precedent among mammals, as also is the manner in which the females care so selflessly for the well-being of the entire new generation, and behave with complete indifference to their own offspring.

This form of caring for the young can be compared only with the impersonal social organization in large insect communities, where care of the rising generation is also a collective task.

Apparently this is the only feasible way in which such vast nurseries can be organized and supplied. Of course, it has its disadvantages. For example, if a young bat that is still incapable of flight falls from the roof of the cave, it is doomed. There is no mother to respond to its cries, since none of them regards it as her own or is ready to show concern for it.

The development of the new-born and young bat

If the supply of food is abundant and regular, the small bats develop rapidly. Apparent disproportions in body form at birth, caused by the excessive size of the back legs and the undeveloped state of the wings, alter from day to

Deciduous teeth of a young Large Mouse-eared bat *(Myotis myotis)* at eight days old. The small, pointed milk teeth are hooked, enabling the young bat to attach itself firmly to the teats of the mother (after Eisentraut, from Natuschke, 1960).

154

day. The fore limbs grow rapidly and within six to eight weeks, the forearm has reached its adult length. Development is slower in the large Flying Foxes. They require almost a year to become fully grown.

Like humans, many mammals come into the world toothless. Nor do they have any need of teeth during the suckling period. But most species of bats deviate from this norm. The new-born bat already has well-developed dentition consisting of up to 22 milk teeth. These are known as "clutching teeth", since the hooked ends enable the baby bat to cling more readily to the mother. The milk teeth are shed in the first few weeks of life. There are also certain species such as Horseshoe bats *(Rhinolophus)* in which the milk teeth regress before birth, so no deciduous teeth are present. It is no coincidence that in these species, the mothers have "false teats" that can be gripped by the young with the mouth as an additional means of attachment.

In insectivorous bats, the various kinds of permanent teeth appear almost simultaneously, while in fruit bats, eruption of the molars may extend over several months.

Many of the details of development in the young, such as the opening of the eyes or the development of the pelage, depend on various ecological factors and differ slightly in almost every species.

Young Large Mouse-eared bats *(Myotis myotis)* have a clearly developed coat within four to five days of birth, while young Long-winged bats *(Miniopterus schreibersi)* still have only a slight hair covering after two weeks. The young of the Egyptian Fruit bat *(Rousettus aegyptiacus)* already have a fine covering of hair on the back at birth.

Many of the New World Fruit bats *(Artibeus)* have the eyes open from birth, young Mouse-eared bats can see when they are four to five days old, and young Pteropodid fruit bats do not open their eyes until seven to ten days after birth.

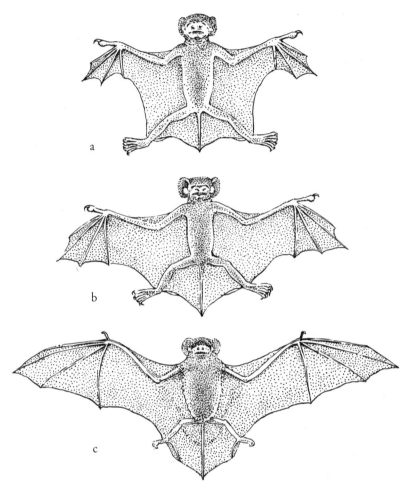

Alteration in body size-wing area ratio in the Common Noctule *(Nyctalus noctula)* in the course of development:
a 1 day old, b 28 days old, c adult (approx. 60 days old) (after Mohr, 1932).

With the development of the hair covering, links with the mother loosen, since temperature regulation is improving steadily. After two or three weeks, the young of many species roost on their own near to the mother, and go to her only to be suckled. The first attempts at flight are made at this time. It is important that the young do this in order to strengthen the muscles of flight.

In many species of the temperate latitudes, extremely high demands are made on the flight apparatus of young bats within only a few weeks of their becoming independent. Together with the older animals, they must undertake migratory flights to winter quarters, sometimes several hundreds of kilometres distant.

It is not altogether clear whether the mothers are able to satisfy the appetite of the young bats with milk alone until they are able to feed independently. Some reports suggest that the older bats frequently return to the young during the night to bring them supplementary insect food. This kind of mixed feeding is said to last for several weeks. But this seems very unlikely.

At the age of two months, the young are fully independent and must look for food themselves. The flight apparatus has completed its morphogenesis and is fully functional; it carries the creatures safely through the night. At first, flight is not as elegant and agile as it is in the adults, but the pattern of wing beats shows the character of typical bat flight. The growing bats do not have to learn from their parents how to fly. At a certain time they are simply able to fly.

The young of Vampire bats have the longest period of development among the Microchiroptera. They take nine months to become fully grown. For two months, they are fed only on milk, and then they must make the complicated readjustment to a diet of blood. At first, the mothers feed small quantities of blood by mouth, until at about six months, the young bats have developed sufficiently to be able to fly to the prey on their own. There is, however, no question of their being taught how to attack a victim and obtain blood. All the necessary behavioural patterns are innate in the animal.

Flight and self-sufficiency in feeding are possible only if the system of ultrasonic directional orientation is fully functional. Möhres was able to show that the isolation calls of the growing Vespertilionid bats gradually change over into typical orientation sounds. They are progressively diversified into an increasing number of pulses that become shorter and higher in frequency. During the first few nocturnal flights, orientation is not yet sufficiently perfected for the young to find their way alone. Möhres observed that they follow close "on the heels" of the mother, carried along in her sonic slipstream.

Comparative examinations of various species of bats have shown that the ability to emit sounds and the degree of differentiation in the first few days depends very much upon the stage of development of the new-born bat. If at birth the young are still very immature, without hair, with eyes and ears closed, their sound repertoire is restricted to faint cries of distress, the isolation calls, that are of somewhat extended duration. In those species in which the eyes and ears of the young are already open at birth, it is possible to distinguish several types of cry. But the frequencies are still lower than those of adult bats. Even quite soon after birth, the young of the Short-tailed Spear-nosed bats (*Carollia*) are able to emit the same frequency-modulated impulses as the parents.

Among large fruit bats, the growth and maternal care of the young extend over several months. Observation of the Indian Flying Fox (*Pteropus giganteus*) has shown the young still being suckled at five months and remaining with the mother until they are eight months old. Kulzer reports that the young of the Egyptian Fruit bat (*Rousettus aegyptiacus*) live in close association with the mother for four months. Long before that time, they are already capable of flight, and accompany the parents on nocturnal flights in search of food.

Development, which is rapid at first, later slows down and the young, although already self-sufficient, do not reach full body size until they are nine to twelve months old. In spite of the maternal care enjoyed by young bats of almost every species, not all grow to adulthood. Information on mortality in young bats is scarce and conflicting. Factors that can prove hazardous are very various. In many species, birth is a critical period, in others, the moment of separation from the mother.

In our latitudes, climatic factors more than any others threaten the life of the young bat. Cold, wet summer months, particularly at the time when the young are being suckled, can be disastrous for the new generation. If prolonged low temperatures prevent the mothers from flying for several days, or if they hunt for food in vain, the supply of milk decreases. As a result, the young are weakened and may even starve to death or be deserted.

Observations have shown that in summers with unfavourable weather conditions, mortality among the young in large maternity colonies of Mouse-eared bats is more than 40 per cent. The most important effect of bad weather is delayed births and weaning, making it impossible for the young bats to fatten up properly for the winter. Although there are no reliable data available, it is probable that excessively high temperatures also cause losses among young bats incapable of flight. And not least, man himself represents a threat to bats, particularly when he carries out measures of chemical conservation and insect pest control at the time when the young are being tended in the roosts.

Why some bats hibernate

The life of an animal depends very much upon the temperatures prevailing in its habitat. Only with optimal temperatures to which the animal is adapted will all bodily functions and behaviour patterns take their normal course.

In the earth's temperate zones, which include, for instance, Central and Northern Europe, the winter months with temperatures falling to below freezing point make special demands on the animal world. As a result, many species have adapted to these demands in such a way that they are able to survive the cold season of the year and avoid the danger of being frozen to death.

According to the manner in which the organisms adjust to the ambient temperature and alterations in it, they can be divided into cold-blooded animals or poikilotherms, warm-blooded animals or homoiotherms and as a subdivision of these, hibernating or aestivating animals or heterotherms.

The poikilothermic animals have a body temperature almost identical to that of the surroundings. They have limited ability to control the temperature of the body by metabolic processes, and therefore can tolerate extremes of temperature only with difficulty. In winter, unless they are able to find a place that is protected from frost, in which they can enter a state of torpor, they are unable to survive. These poikilotherms include not only the invertebrates but also certain vertebrates such as fish, amphibia and reptiles.

Only birds and mammals are able to maintain a constant body temperature. They constitute the group of homoiotherms. The ability to produce or dispose of heat as required makes it possible for them to hold the body temperature at the same level irrespective of changes in the ambient temperature.

If the temperature of the surroundings decreases, the animal must produce heat to maintain a constant body temperature. The production of heat presupposes a good supply of food. If food is scarce and there are no bodily reserves, the animals weaken, become cold and freeze to death.

Adaptations to the cold season

Most warm-blooded animals (homoiotherms) show a range of adaptations that help to prevent the winter months from becoming too great a burden for them. A change of pelage achieves better heat insulation, and reserves of fat laid down beneath the skin serve as energy stores. Many species of birds prefer to avoid the rigours of the cold weather and so leave these latitudes in autumn. They follow migratory routes to the south, where they survive the winter well in congenial temperatures and with a good supply of food.

For our bats, as well as for various rodents and insectivores, winter is a time in which not only the cold but also a shortage of food must be overcome. Certain species of bats, such as the Guano bat *(Tadarida brasiliensis)* are also known to undertake journeys in autumn, like migratory birds, in order to spend the winter in warmer regions.

But many of the species of temperate latitudes do not migrate in this way. They live through the cold season and the shortage of food by lowering the body temperature drastically, almost to freezing point. In doing so, metabolic processes are reduced so severely that the bats can survive the cold period without any intake of food, merely by using the energy reserves built up in the summer. In contrast to poikilothermic (cold-blooded) animals, they are capable at any time of active thermogenesis, that is, of waking from their sleeping state without an artificial input of heat. Mammals that employ this method are known as hibernating animals. Those of our indigenous European fauna include the hedgehog, the fat dormouse and the common dormouse, the souslik, marmot and hamster, as well as the bats. It is not long since bats were recognized as true hibernators. Some thirty years ago, they were still considered to be similar to cold-blooded animals. Nothing was known of the processes of temperature regulation and, contingent on them, the kind of safety measures bats have at their disposal to protect themselves from dying of cold.

Many hibernating animals bury themselves deep in the ground or build warm nests of grass and leaves. Not so our bats. They seek out caves or systems of galleries and cellars, and spend the winter there. Certain species live during the winter months in churches (e. g. the Pipistrelle, *Pipistrellus pipistrellus*) or hollow trees (e. g. the Common Noctule, *Nyctalus noctula*). Such quarters are suitable only if they fulfil two important conditions: they must be frost-free and humid. There are some species that can tolerate a few degrees of frost and are frequently found at the entrances to caves. Others prefer roosts with an air temperature of 5 or 10°C. High humidity is important since otherwise there is a danger of dehydration. For this reason, bats in their winter quarters are often found in places where the atmospheric humidity is so high that water vapour condenses on the bat's fur like pearls of dew when the surface temperature of the bat is less than that of the air. The method of temperature regulation employed during hibernation depends upon various metabolic processes and is slightly different in every hibernating species of mammal. With a considerable lowering of body temperature, bats enter a state of hibernatory torpor or lethargy. In summer months, a sudden fall in air temperature can induce a state of lethargy (torpor) during the day-time sleeping period. This phenomenon serves the same purpose as

hibernation. Lowering the body temperature, even by only 10°C, reduces metabolism and conserves energy. The closer an animal brings its temperature to that of its surroundings, the less heat it loses externally. So it is not surprising to find that in the winter, body temperature can fall to 10°C, 5°C or less, depending upon the ambient temperature.

In a state of lethargy, the bats hang with no sign of life on the walls and roof of their roosts. Externally, this state resembles the torpor induced by cold in poikilotherms. However, the two quiescent states cannot be equated. Since poikilotherms possess no means of thermoregulation, they must necessarily go through all the fluctuations in temperature of the environment. Bumble-bees, frogs and lizards warm up and cool down passively. In the first few days of life, young bats still react like poikilotherms or cold-blooded animals. Their regulatory system must first develop before they can actively raise their temperature by increasing metabolism. The existence of a regulatory device of this kind permits hibernating bats to survive at low temperatures, and to emerge from this state actively, by developing the higher body temperature, even though the ambient temperature remains low. Bats then, in contrast to cold-blooded or poikilothermic animals, are not entirely dependent on the environment for the level of their body temperature.

The period of hibernation starts in October/November and ends in March/April. The species most sensitive to cold appear in the roosts first and usually sleep for the longest period. Some cold-resistant species, such as the Pipistrelle, can still be seen flying about in the open in December. The last bats disappear with the arrival of prolonged frost, when the regular supply of insect food is insufficient.

Year after year, the bats return to the same roosting quarters, where they enter a state of lethargy, hanging in large clusters (e. g. Large Mouse-eared bats), in small groups (e. g. Daubenton's bat and the Pond bat) or individually (e. g. Long-eared bats, Lesser Horseshoe bats), with the hind feet hooked to the roof, walls or into crevices. Those species that insinuate themselves into cracks and niches in the deepest recesses of caves live in a microclimate that is virtually unaffected by fluctuations of temperature in the surroundings.

Most species show distinct preferences in the temperature of their roosts. Since they are capable of registering temperature fluctuations of less than 1°C, they will sometimes move to a different hanging place within the roosting area, if there is a change in temperature. On the other hand, a single species has a different preferred temperature for different months of the winter and young bats may select different temperatures from adults. The reasons for this are not clear, but are probably related to the decreasing amount of stored energy. In the winter roosts, several species are frequently found together.

A visit to such a roost is always an impressive experience. Seeing the creatures hanging there, it is difficult to believe that there is any life at all in the small bodies. The eyes are closed, the body cold to the touch and incapable of coordinated movement. In their hibernatory sleep, the bats are utterly vulnerable to any enemy. Fortunately, apart from man, they have few enemies.

Hibernation in mammals is not a prolonged sleep lasting for six or seven months, for the animals waken spontaneously at intervals of several days or weeks. In many of the hibernating animals, waking phases last hours or even days. The pattern of sleep varies greatly between species. For bats living in simulated winter conditions in the laboratory, sleep periods of up to 30 to 80 days have been recorded. This is considerably longer than the period of continuous sleep recorded for other hibernatory

mammals. Many winter sleepers use these interruptions in sleep for the excretion of urine and faeces. Many bats use warm winter spells to feed up and drink. There are good numbers of flying insects available from time to time during the winter. Bats of northern latitudes, often use the hours of waking to alter their hanging place or even to move to different quarters if the old ones have become cold or draughty or too warm. Mating may take place during these phases.

What signals the bats' departure for the winter roost?

What is it that tells bats that the time has come to make for winter quarters? Just as the alternation of light and dark establishes the diurnal rhythm of animals, so it is clear that light also affects their seasonal rhythm. In addition, other "signals" from the environment announce the approach of winter, such as the sudden onset of cold weather in the autumn, or a perceptible decrease in the supply of food. A characteristic adaptation to severe fluctuations in temperature that can be observed in bats even in summer, is that during a prolonged cold spell in autumn, the bats no longer waken every night, but extend the day-time period of lethargy over several nights.

All these "key stimuli", as they are called by the behaviourist, together with a certain instinctive hibernatory urge, prompt the bats to set off on one particular day on their flight towards winter quarters.

Once they have reached the accustomed roosts, a few days of adaptation to constantly low ambient temperatures of between 3°C and 8°C induces a state of lethargy in the bats. The speed at which the body cools, in spite of being under the control of the brain, depends primarily upon the temperature of the surroundings. The greater the difference between skin temperature and external temperature, the more rapidly does cooling proceed. Since there is a lower limit of temperature for developing the state of lethargy, the bats first "test" whether they are able to achieve this state or not in the place that they have selected as a roost. If it is too cold at the chosen site, there is a danger that the bats may not succeed in raising the level of heat production, and may die of cold.

Lowering of the body temperature and, with it, a gradual deceleration of all physical functions, is brought to a halt at an intermediate point. By increasing the cardiac rate and respiration, the bat manages to reawaken. Once again, there is a reduction of temperature, this time to somewhat lower levels, and the process of waking is again "practised". Finally the bats reach the level of hibernatory temperature at which they hang for days or weeks at a time.

For many tropical species that are able to enter a state of lethargy during their day-time sleep, the lower threshold temperature is still relatively high. They are able to waken from a state of lethargy only if the ambient temperature remains above a critical level. If it falls below it, active re-warming is no longer possible. In Vampires (Desmodus), the minimal level is 20°C and in Lesser Tube-nosed Fruit bats (Nyctimene), as high as 25 °C. Members of the family of Old World Leaf-nosed bats (Hipposideridae) living in the tropics also show only slight tolerance to persistent cold.

For the hibernating bats of our own latitudes, the ambient temperature can fall a good deal further. It has been found that it can be dangerously low without the animals waking. They are assured of active re-warming because the "thermo-regulator" within the animal can maintain the body temperature at a particular minimum—the lethargy level. Bats in hollow trees have been observed to continue sleeping even at air temperatures of −5°C to −8°C, and to adjust their minimum temperature. One very frost-resistant species is the American Red bat (Lasiurus borealis). It spends the winter outdoors on the branches of trees. Sometimes it has even been snowed under at temperatures below freezing point. To keep the body temperature above 0°C and to avoid heat loss, these bats bring the large hair-covered tail membrane up to completely cover the lower surface of the body.

The lowest recorded body temperatures during hibernation vary between 1°C and 5°C according to species. If the temperature falls below this, some supercooling occurs, but if it falls below freezing point, the bats freeze to death. But this stage is rarely reached. Very low temperatures have the effect of an alarm signal. The cardiac rate rises, the bats wake up and look for a more favourable roost.

Vital functions proceed "at the economy rate"

It is interesting to consider the changes that take place in the animal as a result of intense cooling, when all vital processes are reduced to a minimum. A drastically restricted provision of energy of this kind heats the body only just sufficiently to prevent death. The chemical processes in the cells of a homoiothermic (warm-blooded) organism are very well adapted to low temperatures. Yet very little is known in detail about them, even though this particular phenomenon could have extremely important human applications.

Retardation of circulation during hibernation can best be detected in the engine responsible for circulation, the heart. Whereas the heart of a hibernating animal continues to beat at temperatures of around freezing point, any cooling of a fully homoiothermic animal to temperatures of between 10 and 20°C has a critical effect on circulation and proves fatal. In hibernating bats that are sleeping deeply, the heart-beat is scarcely perceptible. Cardiac rate is drastically reduced. The heart of a Large Mouse-eared bat *(Myotis myotis)* that beats 400 times a minute when the bat is awake, and 800 times when in a state of excitement, now falls to only 15 to 20 beats a minute. It is not really possible to specify an exact cardiac rate since this feature is influenced by many factors and shows frequent spontaneous increases. Since with such a low pulse rate, blood flow is only slight and moreover, the blood vessels are greatly constricted, the "superfluous" blood is stored in the major veins of the body.

Not only does the heart beat more slowly than when the bat is awake, but respiration is also much reduced. Several seconds may elapse before an observer sees a hibernating bat draw breath at all. Sometimes there is a periodic grouping of breaths, in which series of breaths are drawn in and expelled, with intervals of several minutes between the groups.

If a bat is disturbed, regular breathing begins immediately. This shows that in deep hibernation, respiration, like circulation, is still under the control of the brain. The greater part of the stored reserves of fat are required for the activities of arousal, and so only a little is available for the actual process of metabolism. If bats had to remain awake throughout the winter in an ambient temperature of about 5°C, they would have to produce one hundred times the amount of heat. But there is not sufficient food available for this, so restriction of the energy expenditure is the only way to survive the cold season.

Because of the extremely low metabolism, waste products are few. The quantity of urine in particular is only 1 per cent of its level during the active season. This is a considerable advantage to bats, since otherwise important mineral salts would be lost from the body.

It has frequently been suggested that the state of lethargy is a primitive characteristic and indicates a certain inadequacy in the heat balance of hibernators. While it is quite true that their heat balance is organized differently from that of non-hibernating animals, they cannot, for this reason alone, be considered as primitive or underdeveloped. On the contrary, this specialization in heat regulation which allows bats to reduce the expenditure of energy during every period of sleep and enables all hibernators to survive the cold season of the year, is a functionally expedient adaptation to extreme living conditions.

By actively increasing the rate of heart-beat and respiration and at the same time, dilating the blood vessels, they are capable of re-warming the body at any time. The warming process is initiated in two ways: on the one hand, violent shivering produces heat, and on the other, the body temperature is raised in a special, "non-shivering" manner. The first form, known as shivering thermogenesis, is a familiar process. It is one we can observe in our own bodies, when, after remaining too long in cold water, rapid contractions of the muscles cause shivering and our teeth "chatter with cold". This action of the muscles increases metabolism enormously and is an important source of heat.

"Non-shivering" heat is also produced partly from the musculature but to a much greater extent from the tissue known as "brown fat". By eating extra food in autumn, bats, like other hibernators, build up a reserve of this fat, storing it between the scapulae and along the spine. The significance of the brown fat tissue was unclear for a long time, although it was given the name of the "hibernating gland". Today, it is known that it represents the bat's energy store, allowing for intensive chemical production of heat. The supply of fat is used slowly both to maintain vital functions at the much reduced "economy rate", and to allow spontaneous arousal from hibernation at fairly frequent intervals. From the loss of weight during the winter months, it is clear how large is the proportion of fat reserves that bats accumulate in autumn as winter provisions. Examination of various species shows a weight loss in spring, compared with the creature's weight in autumn, of 25 to 35 per cent. It is particularly important for the females to enter hibernation in autumn with energy stocks as large as possible, since in spring, these reserves must also ensure the start of embryonic development.

Brown fat occurs in non-hibernating mammals, including man, only for a short period after birth. As soon as the ability to raise body temperature by shivering has developed, this tissue loses its thermo-regulatory significance and regresses. It is clearly a primitive characteristic of mammals, which in its further development in hibernating animals became a factor of vital importance to them.

No oversleeping in the spring

It is surprising how bats waken from their winter sleep and emerge from their dark winter quarters at exactly the time when the cold season is past and new food supplies are available. What timing mechanism arouses them at the right time? The so-called "biological clock" that controls the innate circadian rhythm of activity in animals undoubtedly also has a certain effect on hibernation, but cannot bring it to an end.

Since the internal rhythm frequently deviates from a strict 24-hour periodicity, and during the winter months the bats do not waken for days or weeks at a time, the synchronization between the day's course and the biological clock must be lost in hibernation.

In those bats that roost near the entrance to the cave, the rise in temperatures outside in the spring undoubtedly plays a part as an arousal stimulus. But this cannot affect the many species that roost in the farthest depths of caves where temperature is constant. There must exist some other arousal mechanism within the animal itself.

One can assume that during the winter months, internal histological changes occur within the organism as a result of increases in the products of metabolism, so that the bat would die, if it did not waken at a particular time and restore balance in its inner processes. These changes apparently affect the relevant centres of the brain which bring about the restoration of normal body temperature, and thus cause arousal of the animal. It is also possible that these brain centres are stimulated by way of certain sense organs and then begin to function. In either case, arousal would be triggered off as a reflex.

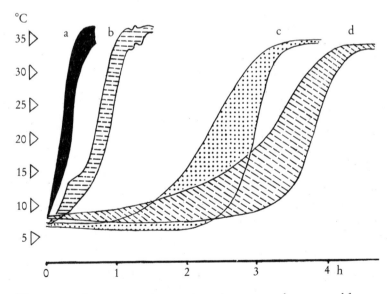

The pattern of temperature increase in various mammals on arousal from hibernation
a bat   b garden dormouse   c golden hamster   d common hamster (after Raths, modified, 1977)

163

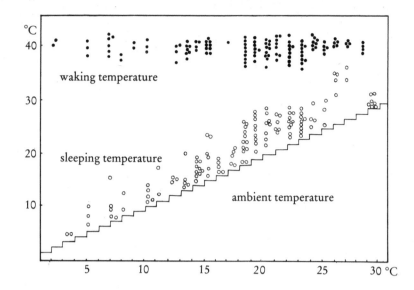

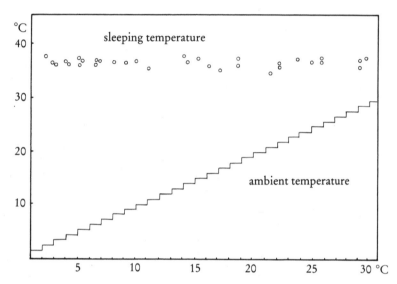

Sleeping and waking temperature correlated to ambient temperature in:
a Large Mouse-eared bat *(Myotis myotis)*. Sleeping temperature is only a few degrees above ambient temperature.
b Indian Flying Fox *(Pteropus giganteus)*. In fruit bats, body temperature is kept constant during sleep. It is entirely independent of the ambient temperature (after Kulzer, 1965).

But here one should not overlook the fact that at such low body temperatures, various of the animal's sense organs respond to stimuli only minimally or not at all. At temperatures of below 12°C, hearing in bats is reduced, and at less than 5°C, bats are effectively deaf. In hibernating hamsters, the retina of the eye is insensitive to stimuli. But the nerve tissue has not entirely lost its ability to function, since even in hibernation, bats are capable of

many reflex actions. If these apparently lifeless creatures are lifted from their hanging place, the legs perform searching movements and when the toes have found a support, they hook on to it firmly by reflex action. The sensory cells of the skin which react to pressure, temperature and pain are not entirely blocked by cold. So it is not impossible that the cutaneous receptors become so highly sensitive as a result of the altered internal condition of the animal that they give off spontaneous arousal stimuli.

When bats waken, it is essential that they warm up rapidly. The lower the ambient temperature and therefore also the body temperature, the slower is the process of arousal. Their relatively small body size is an advantage, since smaller animals warm themselves more rapidly than large ones. It has been found that the body temperature rises by 0.5°C to 1°C per minute, that is, with an ambient temperature of 5°C, the bats have attained their normal waking temperature of 38°C after approximately an hour.

It is interesting that the front part of the body warms much more quickly than the hind quarters. This is not mere chance. It means that the body mass which is heated initially is small, and that in particular, the vital centres of the head and chest region are enabled to function normally as quickly as possible.

Arousal requires a high expenditure of energy, yet the stored reserves of fat must be sufficient for the long period of hibernation. As a result of repeated spontaneous awakenings throughout the winter, it is estimated that under normal conditions, two thirds of the stored energy is consumed.

However, if the animals are disturbed frequently in winter, their energy reserves drop to critical levels. Since bats in hibernation react to every change in temperature and to every stimulus of touch and light by waking, it is essential that disturbance of winter roosts should be avoided.

Once the temperature of the head region has been raised to 15°C by heat produced from the brown fat, the muscles begin to shiver and help to raise the body temperature further. In addition, other stored energy reserves such as sugar, glycogen and even the body's own proteins are mobilized in order to advance the increase in temperature.

Since all processes of metabolic combustion are associated with a high consumption of oxygen, the bats can be seen to breathe heavily on arousal. Once the waking temperature is reached, normal body movements begin. The environment is examined by the bat's system of echolocation, a great quantity of urine is given off and extensive grooming begins. But the bats do not remain for long in the place that has sheltered them through the winter months. It is time to seek food and replenish the much depleted energy stocks. The bats fly out from caves and crevices in walls, rest initially in intermediate roosts before setting off on longer or shorter migration flights, to arrive in April, May or June in their maternity colonies.

It is still not clear how bats evolved this complicated thermo-regulatory system which makes it possible for them to hibernate and to enter a day-time state of torpor.

In the course of evolution, there was an initial segregation of bats into those species which always show a constant body temperature and those in which the body temperature adapts itself to that of the surroundings. All fruit bats belong to the first group; they are good homoiotherms. On the other hand, certain species in the families of Vespertilionid and Horseshoe bats can lower their temperature to almost 0°C. This has enabled those bats to adapt to the cold season in our latitudes.

Between these two extremes, there are various species that represent intermediate stages in thermoregulation. For example, Free-tailed bats can reduce their temperature to only about 20°C. If it falls below this, they are not able to warm up again. Although these species can enter a reversible state of heterothermia daily, they are not capable of true hibernation. So they are unable to spread further northwards. Individual representatives of this family have extended their territory as far as the south of Europe and of North America.

Species belonging to the families mentioned above (Vespertilionid and Horseshoe bats) that live in the tropics can lower their body temperature to that of their surroundings when they sleep. This decrease is much slighter than it is in our indigenous species, since the air temperature in the day roosts of many tropical Microchiroptera is often only a few degrees below the external temperature. But if the ambient temperature happens to drop too far, the tropical species react differently from those of the temperate regions. They do not become torpid but enter a state of alarm instead. All physical functions are intensified, heat is produced and the bats remain awake.

Since the supply of food in the tropics does not fall to such a low level as it does here in the winter months, it is unlikely that heat regulation will be threatened. However, in experiments, the Tübingen physiologist Kulzer was able to induce hibernation for a period of three weeks in Mouse-eared bats from tropical Australia. In that time, the creatures showed the same metabolic changes with which we are familiar in our own hibernating species.

These results indicate that the capacity for heterothermia is a character of bats which evolved even before they began to spread out across all the continents of the world.

1000 kilometres without a compass

It is well known that many mammals undertake regular seasonal migrations along routes which vary greatly in length. Among these migratory mammals there are various species of bats. Their capacity for flight is a considerable advantage, allowing them to cover long distances in a very short time.

Observers long ago noticed that some species of bats are to be seen during winter in areas in which they occur only in small numbers or not at all during the summer. From this they concluded that the species living in mountainous regions move to the lowlands in winter and species that extend almost as far as the polar circle make long migrations in autumn to regions further south. Because bats lead a very secretive life and restrict their activities largely to the hours of darkness, much still remains to be discovered about bat migrations.

Lack of food and cold weather compel a change of abode

There are various factors that can induce bats to move a longer or shorter distance to a new home. In the tropics, migration has been observed principally among large fruit-eating bats, the Megachiroptera. They leave their roosts at particular times of the year when food becomes scarce and move to areas in which it is plentiful. The Grey-headed Flying Fox (*Pteropus poliocephalus*) of Australia is reported to migrate in spring (October) from Queensland southwards along the east coast to New South Wales to reach the Sydney area in the middle of November. Huge flocks of the bats cover more than a thousand kilometres in the course of a few weeks. They arrive in the south at exactly the time when the wild figs and the many succulent fruits ripen in the large fruit plantations. Here they are assured an abundant supply of food. Even so, it is not certain that all the bats of the northern regions take part in these long migrations to the south.

Migratory flights that are closely linked to the time at which fruit ripens are also known to take place among African species of fruit bats. Wahlberg's Epauletted Fruit bat (*Epomophorus wahlbergi*) appears in South Africa at just the right time to help itself to the fruits ready waiting there. In the Congo, migrating Straw-coloured bats (*Eidolon helvum*) have been observed on their way to new feeding grounds.

Another reason given for movements on a large scale in tropical and subtropical species, in addition to a shortage of food, is the excessive heat of the summer months. Since the temperatures in certain zones reach intolerable levels, the only course of action open to bats is to fly into regions of more agreeable climate.

No individual studies exist on migratory behaviour, routes, direction of travel and distances covered by tropical bats, particularly Flying Foxes. So far, detailed research on these subjects has scarcely been possible, since the practice of marking bats by means of wing clips, much used in studying the bats of temperate regions, has rarely been applied to tropical populations. So far, information about the migrations of tropical bats has come largely from chance observations.

Reliable information on the migrations of bats exists in particular for the species of the Palaearctic and Nearctic region. In these areas, it is the cold season and the inadequacy of food which cause the bats to leave their summer quarters.

Among the species of the temperate latitudes that are unable to hibernate are the Guano bats (*Tadarida brasiliensis*), of which the females leave the summer roosts in the caves of Texas and New Mexico, U.S.A., in autumn after the young have become self-sufficient, to migrate southwards over distances of up to 1000 kilometres to Mexico, where they spend the winter. Observations have shown that they cover this migratory route very rapidly, for in no more than three weeks, the first Guano bats have reached their winter roost. Here, it is warm and food is plentiful. With the arrival of spring, millions of these bats set off once again on their migratory flight northwards to the maternity colonies. It has been found that most of the males of this species remain in Mexico and await the return of the females in the autumn. But it is clear that a good many females also elect not to join in the great migration, but give birth to their young in the winter roost. Why should this be? So far, no one has discovered what factors determine which of the bats fly off and which remain.

Most species of the temperate regions hibernate, and move from their summer quarters into frost-free caves, tunnels or buildings. The distance covered may be only a few kilometres, but occasionally can be more than a thousand.

Three North American species, the Red bat (*Lasiurus borealis*), the Hoary bat (*Lasiurus cinereus*) and the Silver-haired bat (*Lasionycteris noctivagans*) are known to migrate over long distances. They spend the summer in the north-east of the United States and in Canada. Here they show themselves to be particularly resistant to the inclemencies of the weather, since they hang during the day almost entirely without protection among foliage or against the trunks of trees. If the weather becomes too disagreeable, they leave their summer home and begin to migrate southwards in small troupes. From the end of August until November, representatives of these species are found in regions along the Atlantic seaboard of America where they are scarcely ever observed in summer. The females in particular travel as far as the southern states of Georgia and Florida. Here the winter months are so mild that they hibernate for only a short period. On their migratory flight along the Atlantic coast, these species of bats perform amazing feats of flying. There is no obstacle that can halt them as they cover this great distance. They cross ocean inlets in continuous flight. Seafarers have reported seeing these species far out at sea, where they sometimes land on the rigging of ships in order to rest, or else fall there from exhaustion. They have been found in autumn and spring on barren islands out in the ocean from which they are absent at other seasons of the year.

These observations indicate how it is possible for members of the *Lasiurus* species to reach even the Bermuda Islands that lie some 1200 km from the mainland.

The seasonal migrations of the three North American species of bats described here are probably those which approximate most closely to the migrations of birds. In most cases, the biological significance of migration is different for bats of the temperate regions than for migratory birds. At the beginning of the cold season, our bats leave their summer quarters to seek out frost-free winter sleeping places. Birds, on the other hand, move to the warmer south where they escape from the cold of winter, and where an adequate food supply makes it possible for them to continue their life in its normal way.

It is not possible for bats to inhabit the frost-free retreats throughout the whole of the year. In the summer, it is too cold in the caves and cellars: the bats would be in a continual state of torpor. So at the start of the warm season, they are obliged to move into summer quarters. Only the males of various species are still to be found in summer in rock crevices, cracks in walls and beneath bridges. They require less heat, indeed cooler temperatures are more favourable for spermatogenesis. Consequently they sometimes use the same roost throughout the winter and summer.

Bat banding with wing clips aids scientific research

In order to obtain more information on the migration of bats, it was necessary to mark the bats. The practice of marking birds by means of leg rings in order to discover their migration routes had already been introduced by ornithologists and had obtained remarkable results.

So in 1932, Eisentraut, the great authority on bat research, began bat banding in Germany. The method had first been employed in 1916 by the American zoologist Allen. He used bird rings, attaching them to the hind legs of the bats. Bat banding in the U.S.A. took on a much increased impetus in the thirties. From 1939 onwards, the rings were attached to the forearm, since the small hind legs are less suitable for the attachment of rings.

Right from the start of his planned programme of banding, Eisentraut used specially designed flanged aluminium "rings", or they might better be described as wing clips, which were fitted around the forearm. With a band of this kind, the bat carries an identification tag, usually for the rest of its life; whenever it is caught or wherever it flies, it can be identified at any time. In Germany, bat banding groups expanded rapidly and bat researchers in other European countries also adopted the method with enthusiasm. In 1932 Ryberg in Sweden, in 1936 Bels in the Netherlands, and in 1937 members of the Soviet Academy of Sciences in the U.S.S.R. took up the banding of bats. Many European countries as well as Australia, Canada and Mexico followed their lead, and a well-organized system of banding has been established. In contrast, it must be said that the use of bat banding for scientific research has scarcely been developed at all so far in the tropical regions of Africa, Asia and South America which, of course, are the principle territories of the bats.

In order to coordinate the increasing activities of marking, most countries have set up a central office of banding. These central offices organize the manufacture of the bands and their distribution to interested colleagues. Each band is stamped with an individual number and an abbreviated address so that it can be returned by the finder to the correct office.

In addition to this method of banding which is widely used today, other methods of marking bats have been tried out by various workers. For example, marked tags have been attached to the ears, or a coded pattern of perforations have been impressed or tattooed on the wing membrane. The ear tags have the advantage that they cannot be damaged by the teeth of the bat, but it is possible that they hinder to a considerable extent the reception of ultrasonic sound waves. Tattoo marks and perforations fade or knit together and so are good, but only for short-term studies.

It must be said that even the bands have certain disadvantages. To the bat, the band is a disagreeable foreign body, and so many of the creatures, particularly in the summer roosts, try to remove them with their sharp teeth. They bite at the metal so vigorously that the numbers and letters engraved upon it become illegible. The band can prove harmful if the bat compresses it in such a way that it injures the wing membrane or causes swelling of the forearm. Opinions differ on the extent of losses brought about in this way. Undoubtedly, banding represents a disturbing factor in the life of a bat, and therefore, in some countries, marking is no longer carried out on those species in which numbers are declining seriously. In the remaining species, every attempt is made to ensure that banding is carried out carefully, with no unnecessary disturbance in the roost. And certainly while the females are pregnant or suckling young, no programme of banding should be initiated.

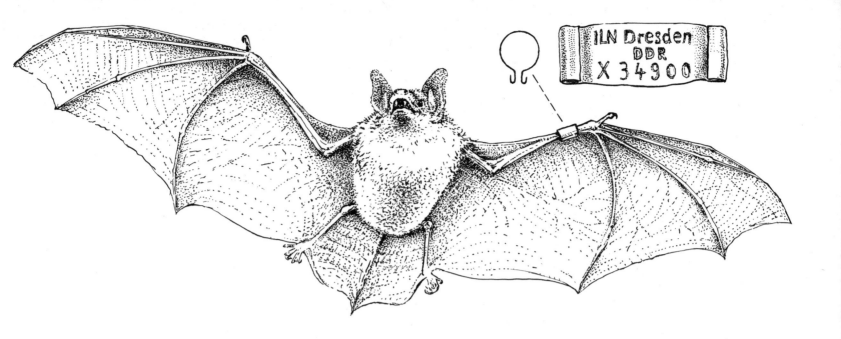

Bat band (arm clip). The band is applied to the forearm and can provide valuable scientific information.

Much of the information on individual species of bats that can be provided by banding has already been obtained for a large number of populations, and in these cases, banding should be restricted.

What information has banding provided?

Bat banding has provided valuable information, particularly when it has been possible to mark very large numbers of one species; because of the secretive habits of bats, only a small proportion of those banded are recovered.

When it is calculated that over the whole world, some millions of bats of a wide variety of species have been banded (in the U.S.A. alone, more than 162,000 Guano bats between 1952 and 1967 and in Europe well over 60,000 Mouse-eared bats), it seems an enormous total. But banding on such a scale is necessary when one considers that the number of long-distance recoveries is very small. The first recoveries of banded bats that had met with an accident or had been found in the course of checks on summer and winter roosts, provided interest-

ing information on the quarters, what direction they take in the flight to the summer roost and how far it is from the summer to the winter home.

Recoveries of Large Mouse-eared bats banded by Eisentraut showed that they end their winter rest and set off on the flight to the maternity colonies in early spring (March/April). Not infrequently, sudden cold spells at this time catch them unawares and can prove fatal.

Although there are both species that are fond of travelling and others that are sedentary, it is not possible to say that the former always cover long distances to get to their summer grounds. Whether they migrate, and how far they travel, depends very much upon the habitat and the available living quarters. In places where ideal summer and winter quarters lie close together, the move is often a local one of no more than a few kilometres. It can even happen that in summer, the cellar is simply exchanged for the attic within the same house. But this is rare. Usually the site of the maternity colony lies further from the winter roost than this.

How far the bat migrates, depends upon its flight capacity. It is much easier for species that are powerful fliers, such as the Common Noctule (*Nyctalus noctula*) to cover long distances than it is for species with a weaker flight behaviour. Included here are, for example, Long-eared bats (*Plecotus*) and Horseshoe (*Rhinolophus*). In summer, the latter travel only a few kilometres from

169

their winter roosts. Observations carried out over a number of years in England by the Hoopers showed that the Greater Horseshoe bat (*Rhinolophus ferrumequinum*), which is one of the larger species of European bats, rarely covers more than 30 kilometres in its seasonal migrations. Its summer roosts lie in the neighbourhood of the caves which in winter are vitally important to this bat. The zoologist Roer of Bonn divides European cave-dwelling and arboreal bats according to their migratory habits into three groups: 1. species committed to their habitat, 2. species with moderately well-developed migratory instincts, and 3. species with strong migratory tendencies.

He includes Horseshoe bats in the first group. It has been found that, like its larger relations in England, the Lesser Horseshoe bat (*Rhinolophus hipposideros*), which has been observed on the northern edge of the Hercynean Mountains and the Alps, also travels no more than 30 kilometres at most from its winter roost.

Most European species show a fairly marked inclination to migrate. The principal members of this group are the Large Mouse-eared bats (*Myotis myotis*). In summer, they are able to travel up to 100 kilometres, or even more, from their underground winter quarters. The limit of their distribution can extend further northwards if suitable caves are available there as winter quarters. Mine galleries or quarries built by man are often selected as the winter roost.

Barbastelles (*Barbastella barbastellus*) behave in a similar way to Large Mouse-eared bats. They also cover distances of up to 100 kilometres. Individuals have been found up to 150 kilometres north of the winter roosts. Geoffroy's bat (*Myotis emarginatus*) spends the winter in large numbers inside systems of caves in Southern Limburg (Netherlands). Recoveries of bats banded here have shown that they spread out in summer in a northwesterly to northeasterly direction, covering distances of up to 100 kilometres.

Pond bats (*Myotis dasycneme*) which also spend the winter in these caves, belong to the group of species with strong migratory tendencies. Since they are found very frequently in northern Europe in summer, while the caves that are their winter quarters lie on the fringes of the Hercynian Mountains, it is clear that they must cover 200 kilometres or more twice every year. The Long-winged bat (*Miniopterus schreibersi*) is also known to undertake extensive migrations. Some that were banded in the winter roost near Barcelona (Spain), were recovered in the south of France. They had flown northwards a distance of about 350 kilometres.

While banding Large Mouse-eared bats in the Berlin area in the thirties, Eisentraut found that in spring they fly in a northerly direction up to 200 kilometres from their winter roosts. Of the Common Noctules (*Nyctalus noctula*) that spent the winters in the Church of Our Lady in Dresden until 1945, the most distant recovery was reported from Lithuania. This bat had covered 750 kilometres in a north-northeasterly direction. Since then, large numbers of bats have been banded in many European countries, providing much information on the migratory behaviour of various species. Observations made by the Soviet zoologist Strelkov are particularly interesting. He has reported autumn migrations of Common Noctules in which distances of more than 500 kilometres have been covered. A record performance was established by a female Noctule that was banded in August 1957 near Voronezh, and was recovered at the beginning of January 1961 in southern Bulgaria, at a distance of 2347 kilometres from the banding site. Because of the interval of several years between banding and recovery, it is not certain whether the bat covered this distance in the course of a single migration.

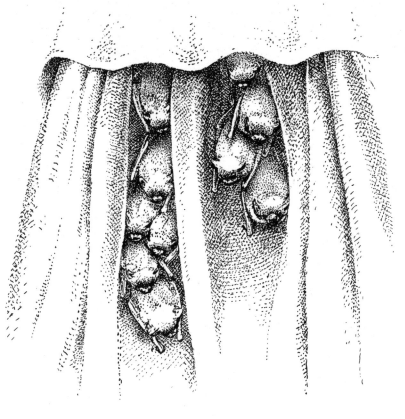

On their autumn migrations, Pipistrelles *(Pipistrellus pipistrellus)* sometimes take up temporary quarters in rooms in houses. But the discovery of a group of these small creatures one morning among the folds of the curtains is no cause for panic.

Some remarkable recoveries have also been made in Bulgaria and Greece of Pipistrelles *(Pipistrellus pipistrellus)* that had been banded in the Ukraine. They had flown in autumn more than 1000 kilometres in a south-southwesterly direction. It is possible that the extreme cold of a continental winter compels the bats to undertake such long journeys. Long-range migrations have also been reported for Pipistrelles that had been banded in the German Democratic Republic. A female, banded at the end of July 1970 near Neubrandenburg was recovered as early as November of the same year in Saint-Dizier (France). In just over three months, it had covered about 770 kilometres in a southwesterly direction.

The migratory behaviour of the small Pipistrelles remains something of an enigma, for it is known that many members of this genus living in Central Europe make no long-distance flights and can be classified as decidedly committed to their local habitat.

The correct interpretation of recoveries made at such distances as those quoted is difficult. Since they are usually single finds, they should initially be considered as exceptions. It is hardly possible that such vast distances need to be covered to find suitable winter quarters. But the possibility cannot be ruled out that in certain regions, the pressure of population becomes so great that the bats are forced, not to migrate, but to emigrate.

The large number of recoveries of banded bats shows that it is not possible to determine a fixed direction of migration. Depending upon the location of the winter roost, it can take the bats in various directions. The northern European species usually migrate south in autumn, since it is not until they reach the Central Mountains that they find ideal winter quarters.

As a result of bat banding and the regular check on the roost which it entails, it has also been found that bats are very faithful to one locality. Whether they undertake major migrations or remain throughout the year in the same district, they are still found again every summer in the same maternity colonies, and in winter, banded bats are discovered once again in the same cave in which they were banded. For this reason, banding has been able to provide information on the longevity of bats.

Long-distance orientation— an unsolved riddle

When bats have to cover very long distances to reach new quarters, they apparently make use of familiar migratory routes. This does not prevent them from interrupting their migratory flights for short periods of rest in intermediate roosts. The details of migration remain something of a mystery. Little is known about how the bats migrate, whether individually, in small groups or in mass

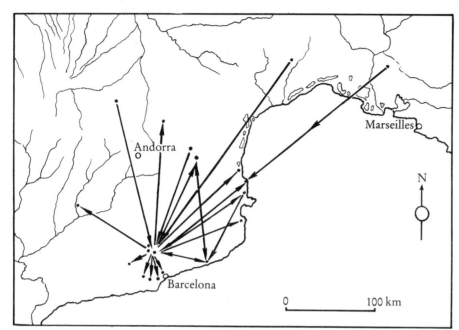

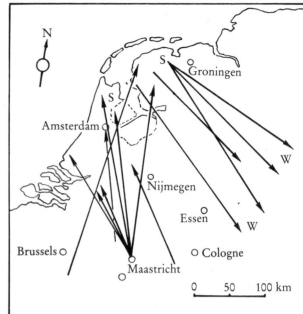

Migratory routes of European bats (after Roer, 1967),
a Migration routes of Long-winged bats *(Miniopterus schreibersi)* that were marked at the Banding Centre of *Avenc del Davi* near Barcelona. Arrows indicate the direction of flight within the area of northern Spain and southern France, from the place of banding to the point of recovery.
b Many Pond bats *(Myotis dasycneme)* in the Netherlands have their winter roost (W) in the caves of Southern Limburg (near Maastricht) and migrate northwards in spring to their summer roosts (S).

flights. It is possible that the distinctly gregarious nature of most species causes them to remain together in fairly large groups. But how is it possible for the young of a population to find the accustomed winter roost? Do they rely upon the "local knowledge" of the older bats until they themselves have imprinted the migratory route upon their memory, or can they manage without any adult guidance?

Eisentraut was the first to demonstrate the extent to which bats remain true to the original roost, in a series of large-scale "homing" experiments. If Large Mouse-eared bats were removed from one winter roost to another, they usually remained there for that particular winter, but by the following one, they had already returned to roost with the members of the same species in the original home. Even when they were transported for distances of up to 150 kilometres, they were found the next winter in the original roost once more. Homing experi-

ments carried out since then in other countries have confirmed Eisentraut's findings. The highest record in such an experiment is held by certain Big Brown bats *(Eptesicus fuscus)* in the U.S.A., which found their way home over a distance of almost 700 kilometres. A Common Noctule that was taken from its summer roost in Lund (Sweden) to Gothenburg 237 kilometres away, was found once again in Lund the following year. Another Noctule, released 125 kilometres from Amsterdam, appeared back at its home roost only two days later. On the other hand, homing experiments with Lesser Horseshoe bats, carried out both in Poland and in the Federal Republic of Germany, showed that at distances of more than 20 kilometres, the bats were unable to find their way back to their roosts. It is important not to generalize too widely from the results of these homing experiments, since in all cases, only a certain percentage of the released animals were found again in the original roosts. The greater the distances, the fewer bats could be shown to have returned.

Nevertheless, the results show that it is rarely possible to rid a house of bats simply by catching them and releasing them at another place. It should surprise no one if, a few days later, the small lodgers are found once again hanging in their accustomed roost in the attic.

This roost loyalty does not prevent bats from moving to new quarters. They are sensitive to every detail of their habitat, and either in summer or winter, may well remove quite suddenly, often for no obvious reason, to a new home.

The attachment to the chosen roost on the one hand, and on the other the fact that most species travel quite long distances between summer and winter quarters, raise the question of how bats are able to find their way when they migrate. Just as ornithologists try to solve the problem of long-distance orientation in migratory birds by means of observation and ringing, so, too, the question arises for the bat specialist of what are the sensory systems and criteria that help bats to find their way back to the place from which they set off.

Birds migrate by day and by night. They appear to orientate themselves particularly by the position of the sun, but there is increasing evidence of the importance of the earth's magnetic field. How can a bat manage, since it is active only at night? Can it make use of its system of echolocation in finding its way over great distances? It is known that bats are able to build up excellent sound-memory pictures of the immediate area of their biotope, using their system of echolocation. But it is impossible to say whether they are capable of storing memory pictures of this kind of the thousands of square kilometres they cover in the course of migration. It may be that bats possess some kind of sensory receptors for long-distance orientation that we are not yet able to recognize and analyze.

To begin with, the possibility cannot be ruled out that, even over wide areas, bats may learn to recognize salient features of the countryside, and use them as a means of orientation on their seasonal migratory flights. Yet the bats that were transported inside closed boxes in the homing experiments had no opportunity to familiarize themselves with the surroundings during the journey, and nevertheless, at least some of the bats found their way home.

It may be that they possess a sense of direction that acts like the needle of a compass, showing them which way to fly to reach the summer or winter roost. Do they perhaps take their bearings from the moon or stars, or even from variations in the brightness of the sky at dusk and at dawn? We do not know yet.

There is no doubt that migration and the system of orientation involved are instinctive. But certain learning processes obviously play a part in comprehending the environment. This would explain why older bats reach their destination with a higher success rate and in a shorter length of time.

In this sphere, a number of important questions still remain unanswered. Further observation and new methods of research are needed to explain this intriguing phenomenon in the life of bats.

Scientific methods employed today include that of attaching miniature transmitters to the backs of bats in order to record with a high degree of accuracy a continuous account of daily routine, migratory routes and speed of travel, in short, the entire activity of a population. Such techniques are already producing interesting information and promise exciting results in the future.

Bats need friends

By no means all aspects of the biology of bats have yet been explained. A number of interesting questions still remain, but will they ever be answered? With today's rapid developments in science and technology, there would appear to be cause for optimism, were it not for the sobering realization that the very existence of the bat is threatened.

Many species of the temperate latitudes, which to a large extent inhabit the highly developed industrial countries of the Old and the New World, have been showing a steady decline in numbers over the last fifteen to twenty years. The reasons for the drastic reduction in the stock of these harmless and indeed useful creatures are not only those depicted in the opening chapter of this book, namely, fear and superstition, but the increasingly rapid alterations in environmental conditions which provide the bats with scarcely any opportunity to come to terms with the new situation.

For a group of animals which has evolved so many different forms of specialization over the last 50 million years, it is not easy to adapt, overnight as it were, to new living conditions. Therefore it is urgently necessary to help the cause of bats, not only because, like every living creature, they are entitled to their place in nature, but because in many ways, directly and indirectly, they contribute to the affairs of men, both in the field of medicine and of economics.

Their value to medicine is no longer that of providing the basis of "cures" in the form of essences and salves; rather is it a question of turning certain abilities and skills that bats have evolved to the benefit of mankind as well. The discovery of ultrasonic orientation in bats opened up a field of research into the possibility of equipping blind people with a system of echolocation to facilitate spatial orientation and the perception of obstacles. Future experiments with bats will show how the creatures are able not only to locate objects, but also to differentiate the forms and structures of those objects. The solution of these problems may allow blind people one day to perceive the details of their surroundings by the use of technology akin to the system evolved by bats millions of years ago.

Research into all the metabolic processes on which hibernation is based is of paramount importance to medicine. Surgery carried out at a lowered body temperature with reduced consumption of oxygen, presupposes a knowledge of the control mechanisms and possible attendant symptoms of hypothermia. In addition, it has been found that in hibernation, animals age less rapidly and show increased resistance to infection. There is also a reduced susceptibility to X rays. These findings may also help to extend the biological potential of man. As a final example drawn from the biology of Chiroptera, one might well mention sperm conservation and delayed foetal development during hibernation. In this sphere, bats have been employing mechanisms for thousands of years which man has only just started to use in animal breeding, and which could even have valuable implications for human reproduction.

The role of bats in the economic sphere is probably of even greater importance. This is associated with their feeding habits. For example, in their area of distribution, the nectar- and pollen-eating species make a valuable contribution to the pollination of chiropterophile flowers and shrubs. But it is the insectivorous species that have particular economic value. Analysis of the food spectrum and of the amount of food consumed shows that the activities of these nocturnal hunters are of great benefit to forestry and agriculture. Depending upon their numbers, they can devour many hundredweights of insects annually, most of which are exclusively nocturnal and include many that are plant pests.

Since hardly any insectivorous birds are active at night, bird and bat complement one another as agents of biological pest control. The energy consumption of bats is high and their appetite is accordingly great. Large species, such as Large Mouse-eared bats and Common Noctules, consume more than 30 cockchafers each in the course of one night, so it is easy to see what vast quantities of insects large colonies demolish in the summer months. It is said that the American Little Brown bat (*Myotis lucifugus*) can eat 65 moths or 500 gnats in an hour. Noctules in captivity have devoured 115 mealworms in half an hour. This represents about a third of the animal's own body weight. Since digestion is rapid, the stomach is soon empty again and the search for food is renewed. A colony of 100 Noctules consumes an estimated 15 kg of insects during a single summer.

The numbers of bats in most regions of the temperate latitudes are, however, now too small to play a decisive part in pest control and thereby to contribute to maintaining a balanced ecological community. For a long time now, the use of chemicals has been widespread in agriculture and forestry as the principal means of protecting crops and stored foods from pests. It has been estimated that today, about 20 per cent of the world's crops are destroyed by insects. Although measures of chemical pest control are undoubtedly necessary, it is at the same time important to seize any opportunity of reducing their use. Chemical pest control spreads poison not only among specific pests but also throughout the whole environment, for it is impossible to avoid destroying useful creatures together with the pests. So every effort should be made to find an alternative to the use of chemicals against plant pests and, particularly in tropical areas, against insects of medical importance, and increasingly to develop and employ various biological methods of controlling harmful organisms. This includes measures to conserve and increase the numbers of natural enemies of pests.

Since bats consume such vast numbers of insects, they also produce considerable quantities of droppings, and these have proved very valuable. They accumulate under the roosts of large colonies as dunes of guano several metres high. The value of bat dung to agriculture, on account of its high nitrogen content, was recognized in America at about the turn of the century, and the commercial mining of bat guano began. In the first forty years of this century, a fertilizer company in California removed from caves during the winter months more than 100,000 tons of this organic fertilizer. Bats enabled the owners to amass a considerable fortune. In Europe and Africa (with the exception of Kenya), guano mining has not achieved commercial importance. But huge deposits of it are known to exist in Australia and Southeast Asia.

In the Carlsbad Caverns of New Mexico, U.S.A., layers of guano some 15 m in depth cover several hundred square metres of the floor of the caves beneath the roost of Guano bats (*Tadarida brasiliensis*). Examinations carried out by palaeontologists and archaeologists show that Guano bats have spent the summer in these caves for about 17,000 years, in which time they have produced these vast mountains of droppings.

The accumulations of guano had economic value during the American Civil War. In this case, they were mined not for the purpose of obtaining fertilizer, but for the extraction of nitrate from the guano for the manufacture of gunpowder.

This example of the indirect exploitation of bats is part of history. More topical are the culinary habits of certain people in Africa and Asia, who eat bats. Members of expeditions have reported that in various parts of these continents, large species of Flying Foxes in particular are hunted, because their flesh is considered a delicacy.

Many dangers threaten bat numbers

Bats need friends because they are endangered. Bat specialists in various countries of Europe, U. S. A. and Canada report with concern that in their countries, the existence of the creatures is threatened by a variety of direct and indirect dangers. In spite of this, not all European countries have placed the Chiroptera under protection. And in the U. S. A., not all the states have enacted legislation for the protection of bats.

But legal measures alone cannot ensure the preservation of bats. Only if man himself is prepared to take positive action to safeguard the existence of bats and their roosts and to ward off dangers that threaten them, is there any chance that these curious mammals will be preserved.

However, alterations that are made to the environment of many populations of bats are often so radical that the bats are no longer able to adapt to the new conditions. Structural changes to the roosting areas and the intensive use of land for agriculture have far-reaching consequences on the life of bats. Since bats make very specific demands on climatic conditions, on the nature of the roost and on the food that is available, they react with particular sensitivity to alterations made to their habitat. In the role of biological indicators, they can provide important information on the extent to which a balance exists between the ecological and economic aspects of the increasing industrialization that also prevails in agriculture and forestry.

The decline in the numbers of bats in many European countries indicates that this balance has been disturbed. The ecological quality of our environment has deteriorated.

As early as 1972, a report from England showed that the numbers of Greater Horseshoe bats (*Rhinolophus ferrumequinum*) had declined in the last 15 years by 80 to 90 per cent. At the end of the seventies, word came that the Lesser Horseshoe bat (*Rhinolophus hipposideros*) had become extinct in the Federal Republic of Germany. In spite of legal measures of protection, information provided to the public and special care of the roosts, it had not been possible to prevent the disappearance of this species.

In 1977 the Bonn zoologist Roer assessed the population trends in bats in the Federal Republic of Germany and reached the alarming conclusion "that in addition to Horseshoe bats, many other species also show a considerable decline, and that bats as a whole are endangered". The decline in the number of individuals living in the roosts of the Large Mouse-eared bat (*Myotis myotis*) is particularly striking. In many places in the Federal Republic of Germany, there are now only 10 per cent of the bats that lived there 15 years ago. A similar decline in numbers is seen among Large Mouse-eared bats throughout Western Europe and to a lesser extent Eastern Europe.

What are the particular reasons for this reduction? People still continue to persecute bats, since they do not feel comfortable about living under the same roof as the creatures. "Weird" noises and the soiling of the roosts are the excuses they make for driving out bats or moving them to different roosts. The latter practice is usually unsuccessful, since bats always return to the roosts they have used for decades or even centuries, and remain true to them. If it is absolutely necessary, for reasons of hygiene, to remove bats from a roost, first consult an expert. Perhaps the best way is to estimate the number of bats flying out on one or two evenings and then bar all means of access to the roost on a subsequent evening after counting the bats that fly out. Since not all the bats will necessarily have emerged on the first night, the main

106 Hibernatory community of the Lesser Horseshoe bat *(Rhinolophus hipposideros)* in a cave in Czechoslovakia. In their roost, Horseshoe bats do not bunch together but space themselves with intervals between individuals. All the bats have enveloped themselves in their flight membranes.

107 Hibernatory community of Long-winged bats *(Miniopterus schreibersi)* in a cave in Czechoslovakia

108 Mediterranean Horseshoe bats *(Rhinolophus euryale)* hibernating on the roof of a cave in Czechoslovakia.

109 Greater Horseshoe bats *(Rhinolophus ferrumequinum)* hibernating in a cave in England

110 Hibernating Daubenton's or Water bats *(Myotis daubentoni)*. Note the identification bands on the forearm of many of the bats.

111 Brown Long-eared bat *(Plecotus auritus)* in hibernation. Frequently the bats roost individually in rock crevices. The large ears are concealed beneath the wings; what appears like an ear is the tragus.

112 Bechstein's bat *(Myotis bechsteini)*. This large-eared species does not fold its ears away when it sleeps. Bechstein's bat is found rather rarely in Europe.

113   Greater Horseshoe bat *(Rhi-nolophus ferrumequinum)*. Typical hibernatory position. The wing membranes are wrapped round most of the body but the face is left free.

114   Large Mouse-eared bat *(Myotis myotis)* hibernating. Wing and tail membrane are drawn in close against the body. The small thumbs can be seen clearly.

115　Large Mouse-eared bat *(Myotis myotis)* hibernating. These bats often roost in places with very high humidity so that they become covered over with droplets of water.

116　Albinism is not unknown among bats. Here is a white Daubenton's or Water bat *(Myotis daubentoni).*

117　A Long-eared bat is banded. The wing clip is placed round the right forearm and carefully pressed together.

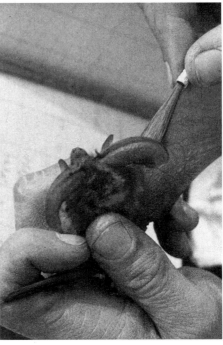

Following page:
118　Bats found in winter roosts are identified, weighed, measured and banded. Finally they are hung on the walls of the caves from which position they can return to their preferred roosts.

119　A check is made on a winter roost. Many bats crawl into deep rock crevices and can be extracted only by means of special pincers.

120 A Northern bat *(Eptesicus nilssoni)* has become entangled in a mist net. Only since extremely fine synthetic fibres have been used in the manufacture of nets has it been possible to catch bats by netting them. The threads are so fine that they cannot be located by the bat's "radar" system.

121 Bat roosting and breeding boxes. These are intended to encourage colonization by woodland bats, particularly in coniferous monocultures.

122 Inspecting a bat box

123 If a box is occupied, a gauze sack is attached beneath it to catch the bats. In this way they are collected, identified and banded.

accesses should be opened again the next evening. This should not be done between mid-June and late August when dependent young may be left in the roost. Nor should it be done during cold wet weather when few bats will emerge and those that do may have difficulty in finding alternative accommodation. Constant illumination of the roost can also cause the bats to move away. Certainly these methods are less unpleasant for the animals and householders than the ill-considered use of poison or gas. In any measures taken to remove bats from their roosts, it is worth considering whether the damage done to them does not outweigh the minor inconvenience to a few people of the bats' presence.

It is quite certain that every roost that is lost reduces the numbers of bats even further. This is equally true of tree-dwelling and of house bats. The modern forester in his efficiently managed forests tolerates no ancient and diseased trees, which often contain natural hollows and the nesting holes of woodpeckers. Yet it is in precisely these holes that certain species of bats like to set up their summer roosts; frequently they also spend the winter there.

Although no statistics are available on the population trends for bats in tropical forests, it is quite certain that recent extensive programmes of forest clearing in South America, Africa and Asia must increasingly restrict the habitats of the indigenous bat populations, and therefore reduce bat numbers in the tropical regions as well.

The loss of roosts in old buildings also leads to constant reduction in the number of maternity colonies to be found there. Nowadays, old buildings are demolished or often repaired in such a way that bats are denied access. In tropical countries, where bats in their thousands inhabit ancient sacred buildings, their existence is also threatened, because of the widespread reconstruction of these ancient architectural monuments as tourist attrac-

tions. If the bats manage to find another roost there, the smell of their urine is found to be unacceptable, and the daily work of removing droppings excessive. How much simpler it is to drive the creatures away or to destroy them!

Modern methods of building, both in towns and in rural areas, frequently do not suit the living requirements of bats. High, spacious roof frames made of timber are replaced by severe concrete structures. Window shutters and ornamental detail on houses, that are favourite hiding places for many species, are found only rarely today.

Programmes of timber preservation and insect control carried out in old buildings hardly ever take into consideration the bat populations of hundreds or thousands of individuals that may well be destroyed in the process. The danger to the bats does not exist only while such treatments are being carried out, but even months later, bats can absorb the poisons through contact with roof beams or rafters.

The loss of suitable winter roosts in the form of caves, mine galleries and other underground structures also has a detrimental effect on bat numbers. Added to this are disturbances of different kinds occurring during hibernation, either because systems of caves are visited all too frequently by speleologists, tourists, children and young people, or scientists studying bats, or as a result of the commercial use of old systems of mine workings and cellar vaults. The French zoologist Brosset reported that some 9000 bats of 11 species were living in the Rancogne cave (Western France) in 1950. Ten years later, they had almost all disappeared since the cave was visited by increasing numbers of biologists, speleologists and tourists. In many places, entrances to caves and systems of galleries are sealed off tightly, for reasons of safety, without any thought being given to bats that are imprisoned there or permanently excluded from their roosts.

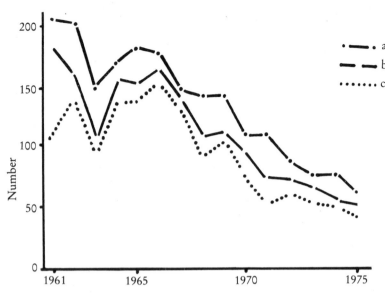

Decline in number of Large Mouse-eared bats *(Myotis myotis)* over a period of 15 years in a maternity colony consisting of three separate roosts, in the Eifel (Federal Republic of Germany) (after Roer, 1977)
a number of adult females,   b total number of births,   c number of surviving young

Modern agricultural practices effect both alterations to the biotope of bats and increasing deterioration in the supply of insect food. Extensive systems of monoculture, the removal of thickets from fields, measures of soil improvement, the drainage of small bodies of water and pools, and the implementation of unrestricted programmes of weed control reduce drastically the total numbers and the number of species of insects. In certain test areas in England, a decline of 50 per cent in the number of flying insects due to environmental factors was ascertained for the years 1947 to 1953.

In certain places within the countries of Central and South America, measures for the control of Vampire bats are not concentrated exclusively on this family, and here a great many harmless species of bats have also been destroyed. It is reported from Venezuela that in the sixties, 2.7 million bats of all species living there were killed in this way.In Brazil, as part of an anti-rabies campaign, all the bats living in 8240 caves were gassed or buried alive within a period of 5 years. The same has occurred in the Middle East in attempts to control fruit-eating Rousette bats.

Insecticides are not exclusively a benefit to man

Probably the greatest danger that threatens bats results from the use of pesticides. These are chemical poisons used in many countries to control animal and plant pests in agriculture and forestry. One group of these poisons comprises the insecticides, of which DDT has become known world-wide. In many countries, particularly in the tropics, it has primarily been used for some time, and still is today, to destroy pests of medical importance, such as Mosquitoes, Sand-flies, Black-flies and Tsetse

flies. But unfortunately it also kills all other insects, useful and insignificant.

It soon became clear that the excessive and uncontrolled use of measures of insect control is harmful both to higher animals and to man. Since even slight quantities in the body can be fatal, there were losses among domestic animals, fish died and innumerable birds were destroyed in the U.S.A., where the product was used very intensively in the years around 1960. The decline in numbers in the large bat colonies became increasingly apparent. It was only now that the danger of the introduction of the poison into the food chains was recognized.

Many of our bats, large numbers of birds and other vertebrates live on insects that show traces of these poisons in their body. Many insects are no longer affected by the poison; they have long since established resistance to it. But in the digestive process, the insecticides enter the body of the vertebrates where they are stored in fatty organs. Since it is virtually impossible for DDT, for example, to be broken down within the body, its concentration increases over a period of time. As long as these materials remain in the fatty tissue, they are harmless.

But if the body's fat reserves are reduced rapidly, as happens regularly in bats in the course of arousal from hibernation and during spring and autumn migrations, high concentrations of poison enter the circulation and reach the brain, where they have a damaging or even fatal effect. For our indigenous bats, the danger is particularly great in spring, since at the end of hibernation, they are already weakened and are scarcely able to cope with additional physical stresses.

Examinations carried out by American workers of the Big Brown bat *(Eptesicus fuscus)* have shown that these animals are significantly more sensitive to DDT than other mammals that were tested. At the end of hibernation, the lethal dose *(dosis letalis)* was between 25 and 40 mg/kg, whereas for rats, the lethal dose was between 200 and 800 mg/kg and for mice, between 175 and 450 mg/kg. Experiments carried out on Pipistrelles showed that in summer, sensitivity to chlorinated hydrocarbons is also greater than in other mammals.

On the basis of these findings, the accumulation of insecticide residues in the bodies of bats must be considered a substantial cause of the drastic decline in numbers or even the extermination of these creatures in many countries.

In her book *Silent Spring* (1962), the American biologist Rachel Carson first drew attention to the damage that had been caused to birds by the use of insecticides. Only the specialists realized that a similar startling balance sheet could be drawn up for bats. The absence of bird song in the spring might well cause general alarm, but who notices the presence or the numbers of bats flitting through parks and gardens in the evening.

American workers have reported that the Eagle Creek Cave in Arizona houses what is probably the world's largest colony of Guano bats *(Tadarida brasiliensis)*. In summer 1964, it was estimated that there were 25 million individuals there; to satisfy numbers of this kind would require 40 tons of insects a night. In June 1970, this maternity colony comprised only 600,000 bats. This means that for every 40 animals in 1964, only one was there six years later. And it is not only among Guano bats that this situation exists.

In the early seventies, the list of endangered and increasingly rare species already included 22 of the 78 species or subspecies living in the U.S.A. In addition to insectivorous bats, the blossom-visiting species are affected equally, since in many places they come into direct contact with the insecticides.

The annihilation of bat populations, as of other animals, does not begin when they die as a result of poisoning which has reached them through the food chain, but starts with the fact that even slight amounts of poison impair reproductive capacity and can lead to sterility. Since bats in any case have a low reproduction rate, this factor is vitally important for the preservation of the species. The new-born young are particularly threatened, since with their mother's milk, they absorb poisons which build up within their bodies, at a time when they are not able to go out and find their own food.

Since that time, the use of certain insecticides, particularly DDT, has been prohibited or restricted in many countries. In order to eliminate the dangers inherent in the chemicals, it would be necessary to develop rapidly degradable materials that are quite specific in their action. Another way, which in the interests of maintaining a healthy environment it is essential for us to pursue, is the extension of the practice of biological pest control.

The protection of bats—
an urgent task of Nature Conservation

The multiplicity of dangers to which bats are exposed and the resulting decline in numbers that has been reported in the last twenty years from many different countries, raise the question of whether there still remains any possibility at all of saving these useful animals in the Palaearctic and Nearctic regions from destruction.

Even if economic and ecological demands cannot always be reconciled, our awareness of the causes of the danger that threatens should make it possible for us to alleviate that danger, even if only partially.

Protection of the winter roost by fixing a grating over the entrances and exits to systems of caves and galleries.

In addition to legal measures of protection for bats, it is important to enlighten people and bring about an understanding of these harmless and useful mammals. Just as vital are concrete measures of protection and conservation of the bats and of their habitats.

Reports of local measures of protection in many different countries give grounds for hope that the populations living there may yet be saved. For example, the caves in Southern Limburg in the Netherlands, which for a long time have been the winter roosts of many species of bats living in north western Europe, were officially designated a Bat Protection Area by the Dutch authorities in 1970. This system of galleries comprises an area of about 40 km², in which every winter, more than 1000 bats of ten different species are found. In the thirties, Dutch workers began a large-scale programme of bat banding in these caves, and since then, have marked more than 20,000 individuals.

The Carlsbad Caverns in New Mexico, U.S.A., which in summer house such vast numbers of Guano bats, were designated a United States National Park in 1930. In this case, the reason why protection of the summer roosts was not sufficient to halt the drastic decline in numbers, was that the Guano bats here, like those in other caves, were also exposed to poisoning by DDT. In the thirties, the colony in the Carlsbad Caverns was estimated at 8 million animals. In the fifties, about 4 million bats assembled in the caves each summer. But today, numbers are estimated at 200,000 individuals.

Although the use of DDT has been prohibited in the U.S.A. for some years, the Guano bats continue to be seriously endangered during the winters spent in Mexico, where this insecticide was still being used in the seventies.

It has been pointed out in the U.S.A. that constant visits from scientists and student groups, with up to 40

188

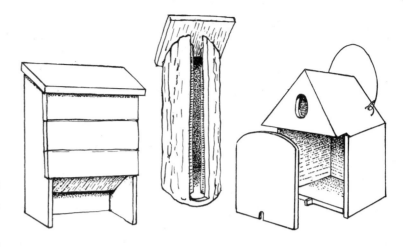

Sleeping and breeding boxes made of timber and wood concrete to encourage colonization by bats.

excursions a year, are also detrimental to large colonies of bats. Continuous disturbances of the maternity roosts upset the bats to such an extent that pregnant females frequently miscarry. If harassed, they remove to substitute roosts in which conditions are usually less favourable than those of the familiar roosts. In so doing, they may desert their young.

Protection of the winter roosts, where they occur in systems of caves or disused mine galleries, can often be achieved only by making it impossible for any unauthorized person to visit the caves. For this reason, the practice of closing off the entrance to caves has been started, while the use of firmly fixed lattice-work gates makes it possible for the bats to fly in and out freely, and for the roost attendants to visit the caves to carry out checks.

In various countries in which large numbers of roosts have been safeguarded in this way, it has been found that bat numbers have remained constant or in some cases have even increased. In carrying out such measures, it is vital that the circulation of air and the conditions of temperature inside the systems of tunnels should not be altered substantially, since bats react with great sensitivity to any change of the microclimate, and may even desert the carefully protected roost.

The necessity of providing protection for maternity colonies that are still occupied as well as for winter roosts has been taken into consideration in the nature conservation legislation of the GDR and of the U.K.

In the last 15 years, a start has been made on the provision of substitute roosts for bats. For the winter period, this is not an easy task. But the first attempts, in which hollow concrete blocks were fixed against the ceilings of unused cellars so that bats could crawl into the spaces, had a considerable success.

A simpler expedient is the construction of sleeping and breeding boxes which can provide woodland bats with an alternative roost if hollows in trees are no longer available. Bat boxes were first used in the 1930's in Eastern Europe. Species typically found in hollow trees, in the nesting holes of woodpeckers and also in the nesting boxes of birds are Common Noctules (*Nyctalus noctula*), Long-eared bats (*Plecotus* spp.), Bechstein's bat (*Myotis bechsteini*), Nathusius' Pipistrelle (*Pipistrellus nathusii*) and Pipistrelle (*Pipistrellus pipistrellus*).

In the Federal Republic of Germany in the fifties, as part of the project "Aid for the Forest", bats were encouraged to settle in artificial nesting hollows. As a substitute for old, hollow trees, suitable summer roosts were prepared in forests. Many variations on these bat boxes were developed after this in various European countries, and tried out with varying degrees of success. It is essential that they should be made of coarsely sawn timber so that the bats can climb and hang more easily on the rough surfaces of the boards. Treatment of the wood with preservatives is inadvisable, since the smell keeps the bats out of the boxes and may kill bats. In contrast to nesting boxes for birds, the most suitable entrance is in the form of a narrow slit near the floor of the box.

Experience of the use of such boxes has shown that success depends very much upon the selection of the site and the nature of the tree growth. Boxes hung in low-lying and deciduous woodlands are very rarely accepted. Here, the bats still manage to find a natural retreat. On the other hand, in forests that consist exclusively of conifers and where undergrowth is sparse, natural roosts

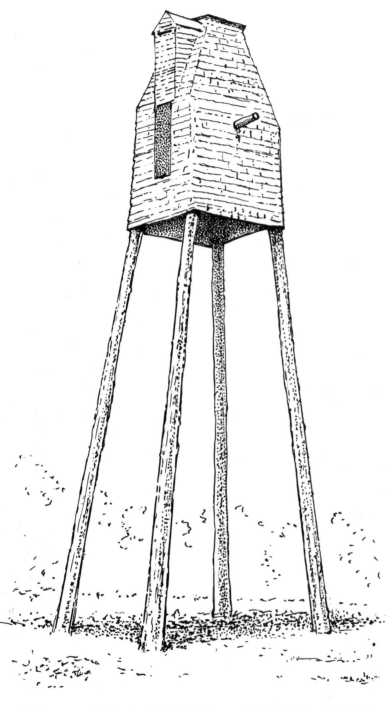

The bat tower built in the U.S.A. in the seventies by A.R. Rashig should help to provide additional roosting accommodation.

are few and the boxes are much used. Because bats are heat-loving animals, in summer they prefer those boxes that are warmed intensively and for long periods by the sun. Free access should not be impeded by twigs. The boxes are best hung in forest clearings protected from wind, on the edges of woods, forest lanes and paths, at a height of 4 to 6 m, facing in a south-east to south direction. Boxes facing different ways will provide roosts at different times of the year, e.g. North (East) for hibernation. Since bats are inclined to change roosts, it is advisable to put up about five boxes in one location.

Checks carried out over a number of years on bat sleeping and breeding boxes have shown that all common European species can be found using artificial roosts of this kind. Horseshoe bats are the single exception.

The provision of bat boxes has often been supported by the Forestry Authorities, since it has been recognized that the artificial colonization of bats makes a positive contribution to the control of many nocturnal insect pests. The strong attachment that bats show to a particular site is an undoubted advantage. Once they have accepted a box, it is likely that they will be found in that particular territory for a number of years.

There has been no lack of attempts to create suitable roosts for house bats. Since many species prefer spacious attics and church towers with a moderate temperature as summer roosts, a number of special bat towers have been constructed in the U.S.A. In 1911, Campbell built the first "malaria eradication guano-producing bat roost". Some very well known ones are the wooden towers built by R.C. Perky in the late twenties at Sugarloaf Key, Florida. Outwardly, they resembled huge windmills without sails. So solidly were they built that one of the towers has survived the intervening years, and can still be visited today as a tourist attraction. The towers were built to provide roosting space for half a million bats. They included a resting area, a "maternity station", a container to collect the guano and even a "cemetery".

In their day, these towers were not thought of as substitute roosts, but were to be an additional means of concentrating bats in that particular area in order to use them to control vast numbers of mosquitoes living there which transmitted malaria. At the same time, they were to provide an easy source of guano. As far as is known, bats did not settle there for any length of time. Nevertheless, it was on the model of this original tower that the zoologist R. A. Rashig constructed a similar building forty years later, but this time its purpose was to provide a roost for endangered bats. His wooden tower on the Eagle River (Wisconsin) which he built at his own expense in the early seventies, is a smaller edition of Parky's tower, designed to house only 150,000 bats. Up to 1978, Rashig had been able to observe only species of bats resting on their migratory flight, and so far it has not become a permanent summer roost. But he remains optimistic, in the knowledge that he has at least made an effort towards ensuring the preservation of bats.

Today, animal conservation is a task for which everyone should feel responsible. Yet often the measures taken are very one-sided and concentrate solely upon those animals that have a particular appeal to the wider public. But we have no right to select which animals living in the wild are to be given the chance to continue to live in today's altered environmental conditions, simply on the grounds that they are attractive or of economic importance.

Apart from the fact that bats have an undoubted economic value in eradicating insect pests, they, like every other creature, represent a link in the totality of nature with its own right to existence. If man fails to realize that he does not exist for himself alone, it is he who will one day pay the price. Man, animal, plant and landscape form a unity. They are inseparably linked and mutually dependent. It is the moral duty of all of us, for our own sake and that of our children and grandchildren, to preserve the diversity and harmony of our environment. Like the bright butterflies and the majestic birds of prey, bats, that have existed for millions of years, are equally worthy of being befriended and protected by man.

British bats and the law

In common with much of Europe, serious population declines have been observed in bats in Britain. Although this may be due partly to natural factors, these declines can also be attributed to increased disturbance of breeding roosts and hibernation sites, to their susceptibility to poisoning by pesticides and to changes in land use resulting in declines in diversity and numbers of their insect food. By the 1970s it was evident that the Greater Horseshoe bat had suffered a dramatic decline in the last 100 years, and that the Mouse-eared bat, which has never been common, was on the point of disappearing. Both species were given full protection under the Wild Creatures and Wild Plants Act 1975. However, disappearance of colonies and declines in population of monitored colonies of even our most common species, such as the Pipistrelle, were also being widely recorded. While this may partly reflect a natural fluctuation in numbers, industrialized countries often offer limited opportunities for species to recover from any population slump. Frequently the disturbance to bat sites could be avoided or minimized with a little forethought or advice. It was with such points in mind that all bat species were given

191

very wide protection under the. Wildlife and Countryside Act 1981.

This Act has far-reaching powers. It is an offence for anyone without a licence intentionally to kill, injure or handle any wild bat in Britain, to possess a bat (dead or alive), or to disturb a bat while roosting. Licences are required for ringing, marking, photographing in the roost, selling (or offering for sale), hiring, bartering or exchanging any wild bat (dead or alive). It is permitted to tend a disabled bat in order to release it upon its recovery, or to kill humanely a seriously disabled bat which has no reasonable chance of recovery.

The law on disturbance and handling applies even in houses and other buildings. Here, as elsewhere, it is an offence to damage, destroy or obstruct access to any place that a bat uses for shelter or protection or to disturb a bat while it is occupying such a place. The only exception is that this does not apply to bats appearing in the living area of a house.

The announcement of this law and its relevance to householders led to newspaper headlines such as "An Englishman's home is no longer his castle". On the face of it this does seem a particularly restrictive law and one that will be difficult to enforce. However, all the law requires is that if work on a building is planned and might affect bats or their roost, the Nature Conservancy Council (NCC) must be informed and given the opportunity to advise and, if necessary, investigate within a reasonable period of time. It is hoped that in most cases where it is appropriate a visit can be made to assess the problem, ascertain to what degree the work is likely to disturb the bats and decide if it can be done at such a time and in such a way as to minimize disturbance to the bats. Most people prove to be sympathetic, some remain uncertain about these strange secretive creatures that they are host to, while others become very enthusiastic about "their" bats.

To help make the law effective the NCC is using the resources available in existing local interested individuals and "bat-groups" and the growing number of "bat-groups" being formed by many country Naturalists' Trusts. Not only does this give the opportunity to advise and educate, but also to get a much better idea of the numbers, distribution, population changes and roost requirements of our bat species. But most of all, "bats need friends" and this offers the opportunity to assure them as many as possible. With increased knowledge it is becoming possible to improve sites for bats and it may become more practical to encourage bats away from sites where they are unwelcome and into sites where they would be welcome.

Much remains to be learnt of these extraordinary, fascinating little animals that are a major part of our limited mammal fauna and an important part of our animal community. This law should help to ensure that they remain a significant and interesting part of our fauna.

Further information on the reasons for the law, on its rulings and advice on conservation measures can be found in a booklet available from the Nature Conservancy Council: *Focus on Bats, their Conservation and the Law,* by R. E. Stebbings and D. J. Jefferies (published by the Nature Conservancy Council, 1982). Details of the law as it affects bats can be found in section 9–11 and 16–27 of the Wildlife and Countryside Act, 1981.

Appendix

Systematic survey of bat families

Classification		Number of known living species[1]
Suborder Megachiroptera		
family Pteropodidae	Fruit bats, Flying Foxes	175
Suborder Microchiroptera		
Superfamily Emballonuroidae		
family Rhinopomatidae	Mouse-tailed bats	3
family Craseonycteridae	Hog-faced bats	1
family Emballonuridae	Sheath-tailed bats	50
family Noctilionidae	Fisherman bats	2
Superfamily Rhinolophoidea		
family Nycteridae	Slit-faced bats	11
family Megadermatidae	False Vampires	5
family Rhinolophidae	Horseshoe bats	70
family Hipposideridae	Old World Leaf-nosed bats	60
Superfamily Phyllostomoidea		
family Phyllostomidae	Spear-nosed bats	148
family Desmodontidae	Vampire bats	3
Superfamily Vespertilionoidea		
family Vespertilionidae	Vespertilionid bats	320
family Natalidae	Funnel-eared bats	8
family Furipteridae	Smoky bats	2
family Thyropteridae	American Disc-winged bats	2
family Myzopodidae	Madagascan Sucker-footed bats	1
family Mystacinidae	Short-tailed bats	1
family Molossidae	Free-tailed bats	90

[1] Data on the number of known living species differ among individual authors, particularly in families with many species (after: Pye, 1969, Koopman and Jones, 1970, Yalden and Morris, 1975, Corbet and Hill, 1980).

Distribution and principal foods of bats

Family	Area of distribution	Food
Pteropodidae	only Old World; tropics and sub-tropics from Africa to Australia	fruit, flowers, pollen
subfamily Macroglossinae	Africa, Eastern Asia	fruit juices, nectar
Rhinopomatidae	North Africa, Asia to Sumatra	insects
Emballonuridae	world-wide (pan-tropical)	insects
Noctilionidae	Central and South America (tropics)	insects, fish
Nycteridae	Africa, Eastern Asia	insects
Megadermatidae	Africa, Asia, Australia (tropics)	insects, small vertebrates
Rhinolophidae	Old World	insects
Hipposideridae	Africa, Asia, Australia (tropics)	insects
Phyllostomidae	Central and South America	insects, small vertebrates, fruit, fruit juice, nectar
Desmodontidae	Central and South America (tropics)	blood of vertebrates
Vespertilionidae	world-wide (Nearctic, Palaearctic, Africa, Australia, South America; Hawaii, Iceland, New Zealand)	insects, individual species: fish and small vertebrates
Natalidae	Central America, Caribbean	insects
Furipteridae	Central America, tropical South America	insects
Thyropteridae	Central America, tropical South America	insects
Myzopodidae	Madagascar	insects
Mystacinidae	New Zealand	insects
Molossidae	world-wide (especially pan-tropical, also Southern Europe and the south of North America)	insects

Index

Bibliography

Allen, G. M.: *Bats*. Cambridge, 1939.

Baker, J. K.: *What about bats?* Carlsbad, New Mexico, 1961.

Barbour, R. W., and W. H. Davis: *Bats of America*. Kentucky, 1969.

Brentjes, B.: "Fledertiere in den Kulturen Altamerikas und des Alten Orients." In: *Milu* (1971), pp. 175–183.

Brosset, A.: *La Biologie des Chiroptères*. Paris, 1966.

Constantine, C. G.: "Bats in relation to health, welfare and economy of man." In: Wimsatt (Editor): *Biology of Bats*, 1970.

Corbet, G. B., and J. E. Hill: *A world list of mammalian species*. British Museum (Nat. Hist.), London, 1980.

Davis, W.-H.: "Hibernation: Ecology and physiological ecology." In: Wimsatt (Editor): *Biology of Bats*, 1970.

Eisentraut, M.: *Aus dem Leben der Fledermäuse und Flughunde*. Jena, 1957.

Eisentraut, M. (Editor): "Berichte und Ergebnisse von Markierungsversuchen an Fledermäusen in Deutschland und Österreich." In: *Bonner Zoolog. Beiträge*, Bonn, 1960.

Felten, H.: "Gespensterfledermaus und Röhrennasenflughund." In: *Natur und Volk*, 88 (1958), p. 361.

Geluso, K. N., J. S. Altenbach, and D. E. Wilson: "Bat mortality: Pesticide, Poisoning and Migratory Stress." In: *Science*, 194 (1976), pp. 184–186.

Grassé, P.-P.: *Traité de Zoologie*. Vol. XVII, *Mammifères*. Paris, 1955.

Greenhall, A. M., and J. L. Paradiso: *Bats and Bat Banding*. Washington, 1968.

Greenhall, A. M.: "La rage chez les chauves-souris." In: *La Revue de Médecine* (1975), pp. 751–754.

Griffin, D. R.: *Vom Echo zum Radar*. Munich, 1959.

Griffin, D. R.: *Echoes of bats and men*. 1960.

Husson, A. M.: *The bats of Surinam*. Thesis. Liège, 1962.

Koopman, K. F., and J. K. Jones: "Classification of bats." In: Slaughter and Walton (Editors): *About Bats*. Dallas, 1970, pp. 22–28.

Kulzer, E.: "Über die Orientierung der Fledermäuse." In: *Aus der Heimat*, 65 (1957), pp. 132–139.

Kulzer, E.: "Der Thermostat der Fledermäuse." In: *Natur und Museum*, 95 (1965), pp. 331–345.

Kulzer, E.: "Der Winterschlaf der Fledermäuse." In: *Umschau* (1969), pp. 195–201.

Kulzer, E.: "Der Flug des afrikanischen Flughundes *Eidolon helvum*." In: *Natur und Museum*, 98 (1968), pp. 181–194.

Kulzer, E., and A. Weigold: "Das Verhalten der Grossen Hufeisennase *(Rhinolophus ferrum-equinum)* bei einer Flugdressur." In: *Z. Tierpsychol.*, 47 (1978), pp. 268–280.

Leen, N.: *The Bat*. New York, 1976.

Lera, T. M., and S. Fortune: "Bat management in the United States." *NSS Bulletin*, 41 (1979), pp. 3–9.

Meise, W.: *Der Abendsegler*. Neue Brehm-Bücherei, No. 42, Leipzig, 1951.

Milne, L. J., and M. Milne: *Die Sinneswelt der Tiere und des Menschen*. Hamburg, Berlin, 1963.

Möhres, P.: "Bildhören – eine neuentdeckte Sinnesleistung der Tiere." In: *Umschau*, 60 (1960), pp. 673–678.

Natuschke, G.: *Heimische Fledermäuse*. Neue Brehm-Bücherei, No. 269, Wittenberg, 1960.

Norberg, U.: "Aerodynamics of hovering flight in the long-eared bat *Plecotus auritus*." In: *J. exp. Biol.*, 65 (1976), pp. 459–470.

Novick, A., and N. Leen: *The world of bats*. New York, 1970.

Pye, J. D.: "Echolocation by bats." In: *Endeavour*, 20 (1961), pp. 101–111.

Pye, J. D.: "The diversity of bats." In: *Science Journal* (1969), pp. 47–52.

Sources
of illustrations

Raths, P.: *Tiere im Winterschlaf*. Leipzig, Jena, Berlin, 1977.

Roer, H.: "Wanderungen der Fledermäuse." In: Hediger (Editor): *Die Strassen der Tiere*. 1967, pp. 102–118.

Roer, H. (Editor): "Berichte und Ergebnisse von Markierungsversuchen an Fledermäusen in Europa." In: *Decheniana*, No. 18, Bonn, 1971.

Roer, H.: "Zur Populationsentwicklung der Fledermäuse (Mammalia, Chiroptera) in der Bundesrepublik Deutschland unter besonderer Berücksichtigung der Situation im Rheinland." In: *Z. f. Säugetierkunde*, 42 (1977), pp. 265–278.

Schmidt, U.: *Vampirfledermäuse*. Neue Brehm-Bücherei, 515, Wittenberg, 1978.

Stratmann, B.: "Faunistisch-ökologische Beobachtungen an einer Population von *Nyctalus noctula*." In: *Nyctalus* (N.F.), 1 (1978), pp. 2–22.

Strelkov, P. S.: "Migratory and stationary bats of the European part of the Soviet Union." In: *Acta zool. Cracoviensia*, 14 (1969), pp. 393–439.

Turner, D. C.: *The Vampire Bat*. Baltimore, London, 1975.

Vogel, St.: "Fledermausblumen in Südamerika." In: *Österr. Botan. Zschr.*, 104 (1958), pp. 492–530.

Webster, F. A., and D. R. Griffin: "The role of the flight membrane in insect capture by bats." In: *Animal Behav.*, 10 (1962), pp. 332–340.

Webster, F. A.: *Some acoustical differences between bats and man*. Conference on sensory devices for the blind. 1967.

Wilson, D. E.: "Reproduction in neotropical bats." In: *Period. biol.*, 75 (1973), pp. 215–217.

Wimsatt, W. A.: *Biology of Bats*. 3 vols., Academic Press, London, New York, 1970.

Yalden, B. W., and P. A. Morris: *The Lives of Bats*. David and Charles, Newton Abbot, London, Vancouver, 1975.

Andera, M., Prague 114

Bilke, P., Naumburg 48, 71

Birnbaum, O., Halle (Saale) 8

Carlsbad Caverns Authorities, New Mexico 60, 61

Cerveny, J., Prague 65, 67, 112, 115, 116, 120

Devez, A. R., Gabon 18, 24

Franzen, L., Frankfort on the Main 9

Gaisler, J., Brno 66, 106, 108

Grimmberger, E., Eberswalde 14, 45, 47, 50, 51, 52, 53, 54, 72, 74, 77, 96, 97, 98, 99, 105, 113

Hooper, J., Staines 10, 101, 109

Hosking, E., London 58

Howell, D. J., West Lafayette 6, 7, 80, 81, 82

Hrabé, J., Brno 107

Klawitter, J., Berlin (West) 88

Kulzer, E., Tübingen 17, 23, 26, 33, 34, 41, 57, 75, 79, 92, 93, 94, 95, 103, 104

Kunz, T. H., Boston 70

Loskarn, P., Bülstringen 11

Mönch/OKAPIA, Frankfort on the Main 90

Morris, P., Ascot 20, 22, 29, 32, 38, 44

Museum der bildenden Künste, Graphische Sammlungen, Leipzig 3

Norberg, U., Göteborg 12, 15, 89

Ortlieb, R., Helbra 91

Publisher's archives 2, 4, 5

Pye, J. D., London 27, 35, 46, 56

Roedl, F., Valtiče 42

Roer, H., Bonn 64, 100, 102

Root/OKAPIA, Frankfort on the Main 63

Rudloff, K., Erfurt 68, 110, 111, 118, 119, 121, 122, 123

Schmidecker/OKAPIA, Frankfort on the Main 16

Schober, W., Leipzig 13, 25, 40, 49

Staatliche Kunstsammlungen Dresden, Gemäldegalerie "Alte Meister" 1

Stephan, H., Frankfort on the Main 19, 21, 28, 30, 31, 36, 37, 39, 62, 69

Tuttle, M. D., Milwaukee 43, 73

Vogel, J., Vienna 84, 85, 86, 87

Webster, F., Vermont 78, 83

Woloszyn, E. W., Cracow 55, 117

ZEFA GmbH, Düsseldorf 59, 76

Special thanks are due to all colleagues and friends who provided me with interesting and rare photographs.